FORMULAS FROM GEOMETRY

area A circumference (or perimeter) C volume V curved surface area

RIGHT TRIANGLE

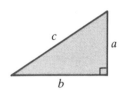

Pythagorean Theorem: $c^2 = a^2 + b^2$

TRIANGLE

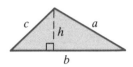

$A = \frac{1}{2}bh$ $C = a + b + c$

EQUILATERAL TRIANGLE

$h = \frac{\sqrt{3}}{2}s$ $A = \frac{\sqrt{3}}{4}s^2$

RECTANGLE

$A = lw$ $C = 2l + 2w$

PARALLELOGRAM

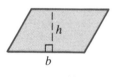

$A = bh$

TRAPEZOID

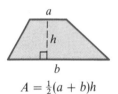

$A = \frac{1}{2}(a + b)h$

CIRCLE

$A = \pi r^2$ $C = 2\pi r$

CIRCULAR SECTOR

$A = \frac{1}{2}r^2\theta$ $s = r\theta$

CIRCULAR RING

$A = \pi(R^2 - r^2)$

RECTANGULAR BOX

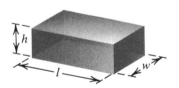

$V = lwh$ $S = 2(hl + lw + hw)$

SPHERE

$V = \frac{4}{3}\pi r^3$ $S = 4\pi r^2$

RIGHT CIRCULAR CYLINDER

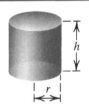

$V = \pi r^2 h$ $S = 2\pi rh$

RIGHT CIRCULAR CONE

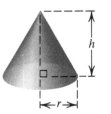

$V = \frac{1}{3}\pi r^2 h$ $S = \pi r\sqrt{r^2 + h^2}$

FRUSTUM OF A CONE

$V = \frac{1}{3}\pi h(r^2 + rR + R^2)$

PRISM

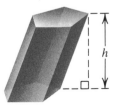

$V = Bh$ with B the area of the base

ANALYTIC GEOMETRY

DISTANCE FORMULA

$$d(P_1, P_2) = \sqrt{(x_2 - x_1)^2 + (y_2 - y_1)^2}$$

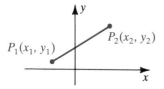

SLOPE m OF A LINE

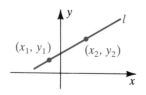

$$m = \frac{y_2 - y_1}{x_2 - x_1}$$

POINT-SLOPE FORM OF A LINE

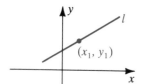

$$y - y_1 = m(x - x_1)$$

SLOPE-INTERCEPT FORM OF A LINE

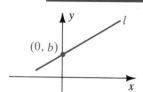

$$y = mx + b$$

INTERCEPT FORM OF A LINE

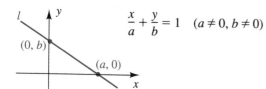

$$\frac{x}{a} + \frac{y}{b} = 1 \quad (a \neq 0, b \neq 0)$$

EQUATION OF A CIRCLE

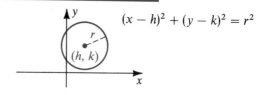

$$(x - h)^2 + (y - k)^2 = r^2$$

GRAPH OF A QUADRATIC FUNCTION

$$y = ax^2, a > 0 \qquad y = ax^2 + bx + c, a > 0$$

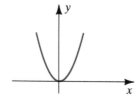 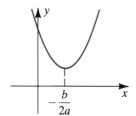

CONSTANTS

$\pi \approx 3.14159$

$e \approx 2.71828$

CONVERSIONS

1 centimeter ≈ 0.3937 inch

1 meter ≈ 3.2808 feet

1 kilometer ≈ 0.6214 mile

1 gram ≈ 0.0353 ounce

1 kilogram ≈ 2.2046 pounds

1 liter ≈ 0.2642 gallon

1 milliliter ≈ 0.0381 fluid ounce

1 joule ≈ 0.7376 foot-pound

1 newton ≈ 0.2248 pound

1 lumen ≈ 0.0015 watt

1 acre $= 43,560$ square feet

1 radian ≈ 57.296 degrees

1 degree ≈ 0.0175 radian

FUNDAMENTALS OF TRIGONOMETRY

TO THE MEMORY OF EARL W. SWOKOWSKI

NINTH EDITION

FUNDAMENTALS OF TRIGONOMETRY

EARL W. SWOKOWSKI

JEFFERY A. COLE
Anoka-Ramsey Community College

Brooks/Cole Publishing Company

ITP® **An International Thomson Publishing Company**

*Pacific Grove ■ Albany ■ Belmont ■ Bonn ■ Boston ■ Cincinnati ■ Detroit
Johannesburg ■ London ■ Madrid ■ Melbourne ■ Mexico City
New York ■ Paris ■ Singapore ■ Tokyo ■ Toronto ■ Washington*

Sponsoring Editor: *Margot Hanis*
Marketing Team: *Caroline Croley, Margaret Parks*
Editorial Assistant: *Kimberly Raburn*
Production Service: *Lifland et al., Bookmakers*
Interior Design: *Julia Gecha*

Technical Artwork: *Scientific Illustrators*
Cover Design: *Cassandra Chu*
Typesetting: *The Beacon Group, Inc.*
Cover Printing: *Phoenix Color Corporation*
Printing and Binding: *World Color Corporation*

For more information, contact:

BROOKS/COLE PUBLISHING COMPANY
511 Forest Lodge Road
Pacific Grove, CA   93950
USA

International Thomson Publishing Europe
Berkshire House 168–173
High Holborn
London WC1V 7AA
England

Thomas Nelson Australia
102 Dodds Street
South Melbourne, 3205
Victoria, Australia

Nelson Canada
1120 Birchmount Road
Scarborough, Ontario
Canada M1K 5G4

International Thomson Editores
Seneca 53
Col. Polanco
11560 México, D. F., México

International Thomson Publishing GmbH
Königswinterer Strasse 418
53227 Bonn
Germany

International Thomson Publishing Asia
60 Albert Street
#15-01 Albert Complex
Singapore 189969

International Thomson Publishing Japan
Hirakawacho Kyowa Building, 3F
2-2-1 Hirakawacho
Chiyoda-ku, Tokyo 102
Japan

Printed in the United States of America.

10   9   8   7   6   5

Library of Congress Cataloging-in-Publication Data

Swokowski, Earl William
    Fundamentals of trigonometry / Earl Swokowski and Jeffery Cole.—
9th ed.
        p.   cm.
    Includes indexes.
    ISBN 0-534-36128-5
    1. Trigonometry, Plane.   I. Cole, Jeffery A. (Jeffery Alan).
    II. Title.
QA533.S95   1998   98-38678
516.24′2—dc21   CIP

CONTENTS

CHAPTER 6   TOPICS FROM ANALYTIC GEOMETRY   373

APPENDIXES

PREFACE

The ninth edition of *Fundamentals of Trigonometry* improves upon the eighth edition in three important ways. First, discussions have been updated to enable the student to more easily understand the mathematical concepts presented. Second, exercises have been added that require the student to estimate, approximate, interpret a result, write a summary, create a model, explore, or find a generalization. Third, graphing calculators have been incorporated to a greater extent through the addition of examples and exercises as well as a revised version of the cross-referenced appendix on use of the TI-82/83. These changes have been incorporated without sacrificing the mathematical soundness that has been paramount to the success of this text.

The appendix on graphing calculators offers an excellent opportunity for both students and professors to familiarize themselves with the use of a graphing calculator by working through examples and exercises (some from the main text, some designed specifically for the appendix). The graphing calculator material in this text is versatile enough to satisfy the needs of every professor, from those who make use of the graphing calculator completely optional to those who require its use in every assignment.

The principal changes made to each chapter are highlighted below. For the reader who is unfamiliar with the previous editions of this text, a list of the general features of the text follows this list of changes.

CHANGES FOR THIS EDITION

CHAPTER 1   New to this edition is an emphasis on exercises in which the student is asked to produce and examine a table of values as an aid in solving a problem and exercises in which the student is asked to interpret some aspect of a given table of values. More exercises involving modeling, graphing calculators, and applications are included, making this a much stronger chapter than in the eighth edition. A section on inverse functions (Section 1.5) has been added to help students understand the general relationships between a function and its inverse.

CHAPTER 2 The topic of harmonic motion (formerly Section 2.9) is included in Section 2.8, Applied Problems. Additional sine wave modeling problems will help students relate realistic data to one of the most important mathematical functions. This chapter provides many routine exercises involving identities, applications, and the concept of the inverse trigonometric functions, before treating the more difficult exercises in these areas.

CHAPTER 3 The relationship between the inverse trigonometric functions and the definition of the trigonometric functions in terms of a unit circle is highlighted in Section 3.6. New alternative solutions of examples give students valuable insight into solving exercises using a variety of methods.

CHAPTER 4 Additional geometric interpretations of results from complex number operations have been included. New applied exercises highlight the role that complex numbers play in the study of electrical circuits. The review section on complex numbers now appears as Appendix V.

CHAPTER 5 Modeling problems and exercises involving tables have been added.

CHAPTER 6 Application of the translation of axes formulas has been replaced with more logical references to the general translations from Chapter 1. The section on parametric equations has been moved up to Section 6.4. New charts and their corresponding figures in Section 6.5 will help students better understand the relationship between r and θ. The section on lines, typically review material, has been moved to Appendix VI.

APPENDIXES The appendixes provide technological help and serve as a valuable reference.

Appendix I, "Using a TI-82/83 Graphing Calculator," has specific keystrokes for the TI-82/83 graphing calculator. Many of the examples in the appendix are examples or exercises from the text. The reader can easily work through this appendix in two different ways: (1) before studying any of the material in the text or (2) as each appendix cross-reference is encountered in the text. If students have no previous experience with a graphing calculator, both options are recommended.

Appendix II, "Common Graphs and Their Equations," is a pictorial summary of graphs and equations that students commonly encounter in precalculus mathematics.

Appendix III, "A Summary of Graph Transformations," is an illustrative synopsis of the basic graph transformations discussed in the text: shifting, stretching, compressing, and reflecting.

Appendix IV, "Graphs of Trigonometric Functions and Their Inverses," contains graphs, domains, and ranges of the six trigonometric functions and their inverses. Note to Professors: The range of the inverse secant and inverse cosecant functions has been selected so that the derivatives of these functions can take on both positive and negative values.

Appendix V, "Complex Numbers," provides review material on basic operations on complex numbers. You may wish to cover this appendix before Section 4.3 if your students need a refresher.

Appendix VI, "Lines," contains material on slopes as well as equations and graphs of lines. This appendix may be covered at any point in the course.

FEATURES

Illustrations   Brief demonstrations of the use of definitions, laws, and theorems are provided in the form of illustrations.

Charts   Charts give students easy access to summaries of properties, laws, graphs, relationships, and definitions. These charts often contain simple illustrations of the concepts that are being introduced.

Examples   Titled for easy reference, all examples provide detailed solutions of problems similar to those that appear in exercise sets. Many examples include graphs, charts, or tables to help the student understand procedures and solutions.

Step-by-Step Explanations   In order to help students follow them more easily, many of the solutions in examples contain step-by-step explanations.

Discussion Exercises   Each chapter now ends with several exercises that are suitable for small-group discussions. These exercises range from easy to difficult and from theoretical to application-oriented.

Checks   The solutions to some examples are explicitly checked, to remind students to verify that their solutions satisfy the conditions of the problems.

Graphing Calculator Examples   Wherever appropriate, examples requiring the use of a graphing utility have been added to the text. These are designated by a calculator icon (shown to the left) and illustrated with a figure reproduced from a graphing calculator screen.

Graphing Calculator Exercises   Exercises specifically designed to be solved with a graphing utility are included in appropriate sections. These exercises are also designated by an icon (**C**).

Applications   To arouse student interest and to help students relate the exercises to current real-life situations, applied exercises have been titled. New applications relate to such diverse topics as the freezing level in a cloud, magnetic pole drift, radio signal intensity, designing a solar collector, the flow of ground water, the study of an epidemic, government spending,

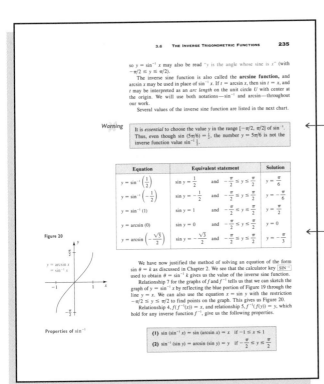

Warnings are interspersed throughout the text to alert students to common mistakes.

Many discussions contain graphs, charts, or tables to help the student understand procedures and solutions.

Exercise sets begin with drill problems and then progress to more challenging problems, including applications designed to show students how the mathematical procedures can be applied to current real-life situations.

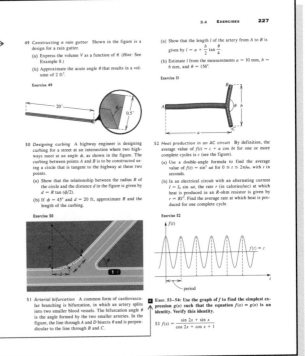

Graphing calculator exercises, designated by a **C**, are included in many sections.

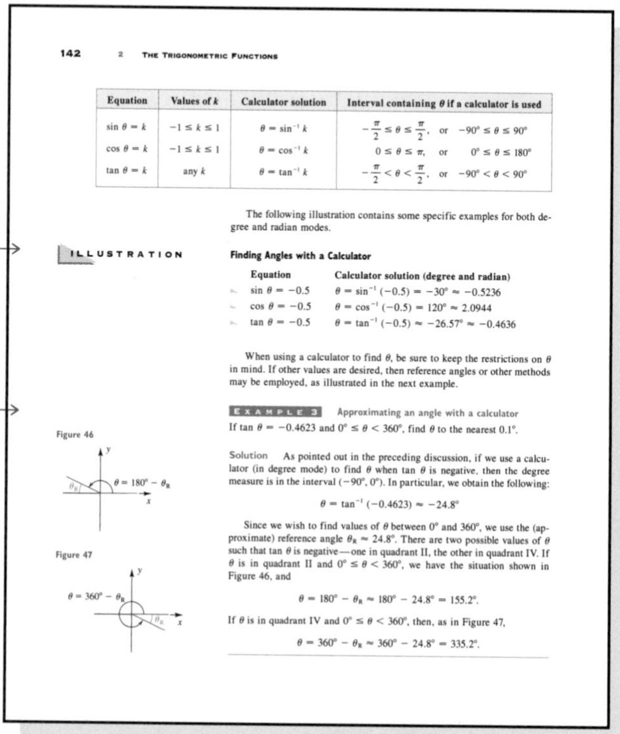

Illustrations provide brief demonstrations of the use of definitions, laws, and theorems.

Examples, well structured and graded by difficulty, are titled for easy reference.

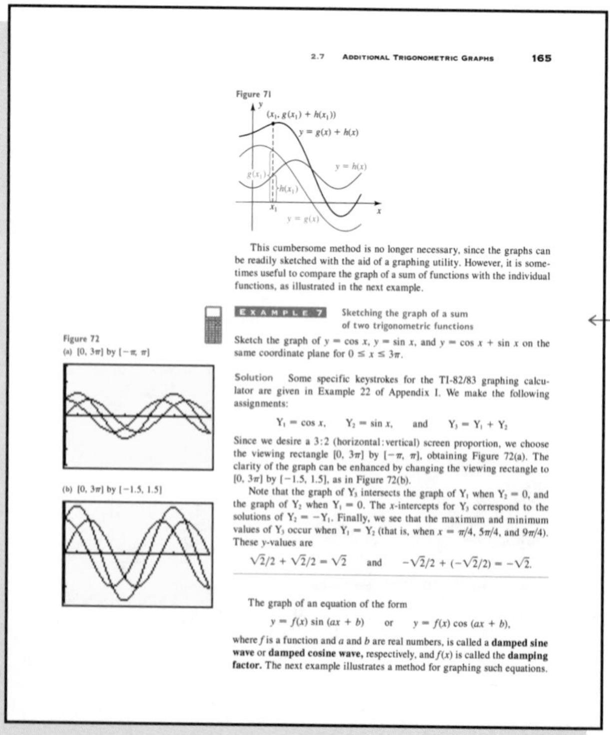

Examples requiring the use of a graphing calculator are included in the text wherever appropriate. These are designated by an icon and illustrated with a figure reproduced from a graphing calculator screen.

---

142   2   THE TRIGONOMETRIC FUNCTIONS

Equation	Values of k	Calculator solution	Interval containing θ if a calculator is used
$\sin\theta = k$	$-1 \le k \le 1$	$\theta = \sin^{-1} k$	$-\dfrac{\pi}{2} \le \theta \le \dfrac{\pi}{2}$, or $-90° \le \theta \le 90°$
$\cos\theta = k$	$-1 \le k \le 1$	$\theta = \cos^{-1} k$	$0 \le \theta \le \pi$, or $0° \le \theta \le 180°$
$\tan\theta = k$	any k	$\theta = \tan^{-1} k$	$-\dfrac{\pi}{2} < \theta < \dfrac{\pi}{2}$, or $-90° < \theta < 90°$

The following illustration contains some specific examples for both degree and radian modes.

ILLUSTRATION   **Finding Angles with a Calculator**

Equation	Calculator solution (degree and radian)
$\sin\theta = -0.5$	$\theta = \sin^{-1}(-0.5) = -30° \approx -0.5236$
$\cos\theta = -0.5$	$\theta = \cos^{-1}(-0.5) = 120° \approx 2.0944$
$\tan\theta = -0.5$	$\theta = \tan^{-1}(-0.5) \approx -26.57° \approx -0.4636$

When using a calculator to find θ, be sure to keep the restrictions on θ in mind. If other values are desired, then reference angles or other methods may be employed, as illustrated in the next example.

EXAMPLE 3   Approximating an angle with a calculator

If $\tan\theta = -0.4623$ and $0° \le \theta \le 360°$, find θ to the nearest 0.1°.

Figure 46

Solution   As pointed out in the preceding discussion, if we use a calculator (in degree mode) to find θ when $\tan\theta$ is negative, then the degree measure is in the interval $(-90°, 0°)$. In particular, we obtain the following:

$$\theta = \tan^{-1}(-0.4623) \approx -24.8°$$

Since we wish to find values of θ between 0° and 360°, we use the (approximate) reference angle $\theta_R \approx 24.8°$. There are two possible values of θ such that $\tan\theta$ is negative—one in quadrant II, the other in quadrant IV. If θ is in quadrant II and $0° \le \theta < 360°$, we have the situation shown in Figure 46, and

Figure 47

$$\theta = 180° - \theta_R \approx 180° - 24.8° = 155.2°.$$

If θ is in quadrant IV and $0° \le \theta < 360°$, then, as in Figure 47,

$$\theta = 360° - \theta_R \approx 360° - 24.8° = 335.2°.$$

2.7   ADDITIONAL TRIGONOMETRIC GRAPHS   165

Figure 71

This cumbersome method is no longer necessary, since the graphs can be readily sketched with the aid of a graphing utility. However, it is sometimes useful to compare the graph of a sum of functions with the individual functions, as illustrated in the next example.

EXAMPLE 7   Sketching the graph of a sum of two trigonometric functions

Figure 72
(a) [0, 3π] by [−π, π]

Sketch the graph of $y = \cos x$, $y = \sin x$, and $y = \cos x + \sin x$ on the same coordinate plane for $0 \le x \le 3\pi$.

Solution   Some specific keystrokes for the TI-82/83 graphing calculator are given in Example 22 of Appendix I. We make the following assignments:

$$Y_1 = \cos x, \quad Y_2 = \sin x, \quad \text{and} \quad Y_3 = Y_1 + Y_2$$

Since we desire a 3:2 (horizontal:vertical) screen proportion, we choose the viewing rectangle [0, 3π] by [−π, π], obtaining Figure 72(a). The clarity of the graph can be enhanced by changing the viewing rectangle to [0, 3π] by [−1.5, 1.5], as in Figure 72(b).

(b) [0, 3π] by [−1.5, 1.5]

Note that the graph of Y_3 intersects the graph of Y_1 when $Y_2 = 0$, and the graph of Y_2 when $Y_1 = 0$. The x-intercepts for Y_3 correspond to the solutions of $Y_2 = -Y_1$. Finally, we see that the maximum and minimum values of Y_3 occur when $Y_1 = Y_2$ (that is, when $x = \pi/4$, $5\pi/4$, and $9\pi/4$). These y-values are

$$\sqrt{2}/2 + \sqrt{2}/2 = \sqrt{2} \quad \text{and} \quad -\sqrt{2}/2 + (-\sqrt{2}/2) = -\sqrt{2}.$$

The graph of an equation of the form

$$y = f(x)\sin(ax + b) \quad \text{or} \quad y = f(x)\cos(ax + b),$$

where f is a function and a and b are real numbers, is called a **damped sine wave** or **damped cosine wave**, respectively, and $f(x)$ is called the **damping factor**. The next example illustrates a method for graphing such equations.

the effect of the ozone layer on skin cancer, and the relationship of cholesterol levels to heart disease. One look at the Index of Applications in the back of the book reveals the wide array of topics. Many professors have indicated that the applications constitute one of the strongest features of the text.

Exercises   Exercise sets begin with routine drill problems and gradually progress to more difficult problems. Many exercises containing graphs and tabular data are new to this edition. There are also many new problems requiring the student to find a mathematical model for the given data. Applied problems generally appear near the end of an exercise set, to allow students to gain confidence in working with the new ideas that have been presented before they attempt problems that require greater analysis and synthesis of these ideas. Review exercises at the end of each chapter may be used to prepare for examinations.

Guidelines   Boxed guidelines enumerate the steps in a procedure or technique, to help students solve problems in a systematic fashion.

Warnings   Interspersed throughout the text are warnings to alert students to common mistakes.

Text Art   Forming a total art package that is second to none, figures and graphs have been computer-generated for accuracy, using the latest technology. Colors are employed to distinguish between different parts of figures. For example, the graph of one function may be shown in blue and that of a second function in red. Labels are the same color as the parts of the figure they identify.

Text Design   The text has been designed to ensure that discussions are easy to follow and important concepts are highlighted. Color is used pedagogically to clarify complex graphs and to help students visualize applied problems. Previous adopters of the text have confirmed that the text strikes a very appealing balance in terms of color use.

Endpapers   The endpapers in the front and back of the text provide useful summaries from algebra, geometry, and trigonometry.

Answer Section   The answer section at the end of the text provides answers for most of the odd-numbered exercises, as well as answers for all chapter review exercises. Considerable thought and effort were devoted to making this section a learning device for the student instead of merely a place to check answers. For instance, verifications are given for trigonometric identities. Numerical answers for many exercises are stated in both an exact and an approximate form. Graphs, proofs, and hints are included whenever appropriate. Author-prepared solutions and answers ensure a high degree of consistency among the text, the solutions manuals, and the answers.

TEACHING TOOLS FOR THE INSTRUCTOR

Instructor's Solutions Manual by Jeff Cole (0-534-36129-3)   This author-prepared manual contains answers to all exercises and detailed solutions to many. The manual has been thoroughly reviewed for accuracy.

Test Items with Chapter Tests by Laurel Technical Services (0-534-36130-7) This printed manual contains all the test items found in the computerized test bank, plus two sample tests for each chapter.

Thomson World Class Learning™ Testing Tools (Windows: 0-534-36080-7; Mac: 0-534-36127-7)   This fully integrated set of testing tools includes tools for test creation, test delivery, and classroom management. The test programs provide text-specific testing and tutorial capabilities, which allow the instructor to create dynamic, algorithm-based questions and regenerate the values of variables and calculations to produce multiple versions of the same test.

Video Tutorial Series (0-534-36081-5)   A set of book-specific videotapes is available without charge to adopters. The videotapes feature instruction by Jeff Cole, who takes students step by step through the key concepts from the text, and computer-generated art.

LEARNING TOOLS FOR THE STUDENT

Student's Solutions Manual by Jeff Cole (0-534-36079-3)   This author-prepared manual contains detailed solutions for nearly all of the odd-numbered exercises, as well as strategies for solving additional exercises. Solutions of the more difficult applied problems are emphasized, and many helpful hints and warnings are included.

Quick Reference Card   Packaged with this edition of the text is a formula card for solving exercises. This perforated card, found in the front of the book, will aid students in mastering key formulas and minimize the need for page turning. Because the card reduces time spent on tedious tasks, the student can focus on the central concepts and principles of the course.

ACKNOWLEDGMENTS

Special thanks go to Gary Rockswold, of Mankato State University, for supplying most of the fine assortment of new applied problems and calculator exercises. Gary has been an invaluable source of knowledge and support. I also wish to thank my wife, Joan, for proofreading the galleys and George Morris, of Scientific Illustrators, for creating the mathematically precise art package.

Many changes for this edition are due to the following individuals, who reviewed the manuscript and/or made suggestions to increase the usefulness of the text for the student:

Daniel D. Anderson
University of Iowa

Joann Bossenbrock
Columbus State Community College

Louis T. Congelio
Westmoreland County Community College

Melissa Cordle
Columbus State Community College

Mary Marsha Cupitt
Durham Technical Community College

Gregory J. Davis
University of Wisconsin–Green Bay

Robert S. Doran
Texas Christian University

George Drake
Lake Tahoe Community College

Arthur Foss
Broward Community College

Florence Gossett
North Hennepin Community College

Karen Hinz
Anoka-Ramsey Community College

Cindy Jones
Indiana University Purdue University Indianapolis

Darrell Minor
Columbus State Community College

Michal Misiurewicz
Indiana University Purdue University Indianapolis

Henry Nixt
Shawnee State University

Jean E. Rubin
Purdue University

Katherine Struve
Columbus State Community College

Linda C. Vereen
Coastal Carolina University

S. K. Wong
The Ohio State University

I am also thankful for the excellent cooperation of the staff of Brooks/ Cole Publishing Company. Editor Margot Hanis supervised the project and supplied a wealth of information and advice. Beth Wilbur assembled the excellent ancillary package that accompanies the text. Sally Lifland, Denise Throckmorton, Quica Ostrander, and Gail Magin, all of Lifland et al., Bookmakers, took exceptional care in seeing that no inconsistencies occurred and offered many helpful suggestions.

In addition to all the persons named here, I would like to express my sincere gratitude to the many students and teachers who have helped shape my views on mathematics education. Please feel free to write to me about any aspect of this text—I value your opinion.

Jeffery A. Cole

TOPICS FROM ALGEBRA

This chapter contains topics that are prerequisites for the study of trigonometry. After a review of real numbers, coordinate systems, and graphs of equations in two variables, we turn our attention to one of the most important concepts in mathematics—the notion of function. We then examine some fundamental concepts of functions, including graphs and inverses. The last section contains a brief survey of exponential and logarithmic functions.

1.1 REAL NUMBERS

Real numbers are used throughout mathematics, and you should be acquainted with symbols that represent them, such as

$$1, \quad 73, \quad -5, \quad \tfrac{49}{12}, \quad \sqrt{2}, \quad 0, \quad \sqrt[3]{-85}, \quad 0.33333\ldots, \quad 596.25,$$

and so on. The **positive integers,** or **natural numbers,** are

$$1, \quad 2, \quad 3, \quad 4, \quad \ldots.$$

The **whole numbers** (or *nonnegative integers*) are the natural numbers combined with the number 0. The **integers** are often listed as follows:

$$\ldots, \quad -4, \quad -3, \quad -2, \quad -1, \quad 0, \quad 1, \quad 2, \quad 3, \quad 4, \quad \ldots$$

Throughout this text lowercase letters $a, b, c, x, y, \ldots$ represent arbitrary real numbers. If a and b denote the same real number, we write $a = b$, which is read "a **is equal to** b" and is called an **equality.** The notation $a \neq b$ is read "a **is not equal to** b."

A **rational number** is a real number that can be expressed in the form a/b, where a and b are integers and $b \neq 0$. Note that every integer a is a rational number, since it can be expressed in the form $a/1$. Every real number can be expressed as a decimal, and the decimal representations for rational numbers are either *terminating* or *nonterminating and repeating.* For example, we can show by using the arithmetic process of division that

$$\tfrac{5}{4} = 1.25 \quad \text{and} \quad \tfrac{177}{55} = 3.2181818\ldots,$$

where the digits 1 and 8 in the representation of $\tfrac{177}{55}$ repeat indefinitely (sometimes written $3.2\overline{18}$).

Real numbers that are not rational are **irrational numbers.** Decimal representations for irrational numbers are always *nonterminating and nonrepeating.* One common irrational number, denoted by π, is the ratio of the circumference of a circle to its diameter. We sometimes use the notation $\pi \approx 3.1416$ to indicate that π **is approximately equal to** 3.1416.

There is no rational number b such that $b^2 = 2$, where b^2 denotes $b \cdot b$. However, there is an irrational number, denoted by $\sqrt{2}$ (the **square root** of 2), such that $(\sqrt{2})^2 = 2$.

The system of **real numbers** consists of all rational and irrational numbers. Relationships among the types of numbers used in algebra are illustrated in the diagram in Figure 1, where a line connecting two rectangles means that the numbers named in the higher rectangle include those in the lower rectangle. The complex numbers, discussed in Appendix V, contain all real numbers.

Figure I

Types of numbers used in algebra

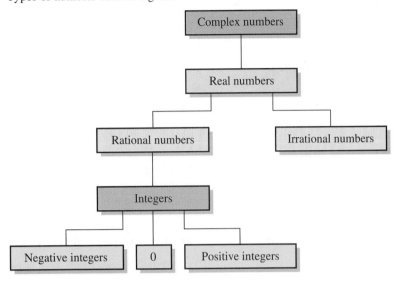

Real numbers may be represented by points on a line *l* such that to each real number *a* there corresponds exactly one point on *l* and to each point *P* on *l* there corresponds one real number. This is called a **one-to-one correspondence.**

In Figure 2, the number *a* that is associated with a point *A* on *l* is the **coordinate** of *A*. We refer to these coordinates as a **coordinate system** and call *l* a **coordinate line** or a **real line.** A direction can be assigned to *l* by taking the **positive direction** to the right and the **negative direction** to the left. The positive direction is noted by placing an arrowhead on *l*, as shown in Figure 2.

Figure 2

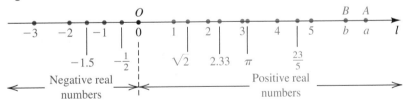

The numbers that correspond to points to the right of *O* (called the **origin**) in Figure 2 are **positive real numbers.** Numbers that correspond to points to the left of *O* are **negative real numbers.** *The real number* 0 *is neither positive nor negative.*

In the following chart we define the notions of **greater than** and **less than** for real numbers a and b. The symbols $>$ and $<$ are **inequality signs,** and the expressions $a > b$ and $a < b$ are called **strict inequalities** or simply **inequalities**.

Greater Than or Less Than

Notation	Definition	Terminology
$a > b$	$a - b$ is positive	a is greater than b
$a < b$	$a - b$ is negative	a is less than b

If points A and B on a coordinate line have coordinates a and b, respectively, then $a > b$ is equivalent to the statement "A is to the *right* of B," whereas $a < b$ is equivalent to "A is to the *left* of B."

I L L U S T R A T I O N

Greater Than ($>$) and Less Than ($<$)

- $5 > 3$, since $5 - 3 = 2$ is positive.
- $-6 < -2$, since $-6 - (-2) = -6 + 2 = -4$ is negative.
- $\frac{1}{3} > 0.33$, since $\frac{1}{3} - 0.33 = \frac{1}{3} - \frac{33}{100} = \frac{1}{300}$ is positive.
- $7 > 0$, since $7 - 0 = 7$ is positive.
- $-4 < 0$, since $-4 - 0 = -4$ is negative.

The next law enables us to compare, or *order*, any two real numbers.

Trichotomy Law

If a and b are real numbers, then exactly one of the following is true:
$$a = b, \qquad a > b, \qquad \text{or} \qquad a < b$$

We refer to the **sign** of a real number as positive if the number is positive, or negative if the number is negative. Two real numbers have *the same sign* if both are positive or both are negative. The numbers have *opposite signs* if one is positive and the other is negative. The following results about the signs of products and quotients of two real numbers a and b can be proved using properties of negatives and quotients.

Laws of Signs

(1) If a and b have the same sign, then ab and $\dfrac{a}{b}$ are positive.

(2) If a and b have opposite signs, then ab and $\dfrac{a}{b}$ are negative.

*If a theorem is written in the form "if *P*, then *Q*," where *P* and *Q* are mathematical statements called the *hypothesis* and *conclusion,* respectively, then the *converse* of the theorem has the form "if *Q*, then *P*." If both the theorem and its converse are true, we often write "*P* if and only if *Q*" (denoted *P* iff *Q*).

The **converses*** of the laws of signs are also true. For example, if a quotient is negative, then the numerator and denominator have opposite signs.

The notation $a \geq b$, read "*a* **is greater than or equal to** *b*," means that either $a > b$ or $a = b$ (but not both). For example, $a^2 \geq 0$ for every real number *a*. The symbol $a \leq b$, which is read "*a* **is less than or equal to** *b*," means that either $a < b$ or $a = b$. Expressions of the form $a \geq b$ and $a \leq b$ are called **nonstrict inequalities**, since *a* may be equal to *b*. As with the equality symbol, we may negate any inequality symbol by putting a slash through it—that is, $\not>$ means not greater than.

An expression of the form $a < b < c$ is called a **continued inequality** and means that both $a < b$ *and* $b < c$; we say "*b* **is between** *a* and *c*." Similarly, the expression $c > b > a$ means that both $c > b$ and $b > a$.

ILLUSTRATION

Ordering Three Real Numbers

- $1 < 5 < \frac{11}{2}$ - $-4 < \frac{2}{3} < \sqrt{2}$ - $3 > -6 > -10$

There are other types of inequalities. For example, $a < b \leq c$ means both $a < b$ and $b \leq c$. Similarly, $a \leq b < c$ means both $a \leq b$ and $b < c$. Finally, $a \leq b \leq c$ means both $a \leq b$ and $b \leq c$.

If *a* is an integer, then it is the coordinate of some point *A* on a coordinate line, and the symbol $|a|$ denotes the number of units between *A* and the origin, without regard to direction. The nonnegative number $|a|$ is called the *absolute value of a.* Referring to Figure 3, we see that for the point with coordinate -4 we have $|-4| = 4$. Similarly, $|4| = 4$. In general, *if a is negative, we change its sign to find* $|a|$; *if a is nonnegative, then* $|a| = a$. The next definition extends this concept to every real number.

Figure 3

DEFINITION OF ABSOLUTE VALUE

> The **absolute value** of a real number *a*, denoted by $|a|$, is defined as follows.
> **(1)** If $a \geq 0$, then $|a| = a$.
> **(2)** If $a < 0$, then $|a| = -a$.

Since *a* is negative in part (2) of the definition, $-a$ represents a *positive* real number. Some special cases of this definition are given in the following illustration.

ILLUSTRATION

The Absolute Value Notation $|a|$

- $|3| = 3$, since $3 > 0$.
- $|-3| = -(-3)$, since $-3 < 0$. Thus, $|-3| = 3$.
- $|2 - \sqrt{2}| = 2 - \sqrt{2}$, since $2 - \sqrt{2} > 0$.
- $|\sqrt{2} - 2| = -(\sqrt{2} - 2)$, since $\sqrt{2} - 2 < 0$. Thus, $|\sqrt{2} - 2| = 2 - \sqrt{2}$.

In the preceding illustration, $|3| = |-3|$ and $|2 - \sqrt{2}| = |\sqrt{2} - 2|$. In general, we have the following:

$$|a| = |-a| \quad \text{for every real number } a$$

E X A M P L E 1   Removing an absolute value symbol

If $x < 1$, rewrite $|x - 1|$ without using the absolute value symbol.

Solution   If $x < 1$, then $x - 1 < 0$; that is, $x - 1$ is negative. Hence, by part (2) of the definition of absolute value,

$$|x - 1| = -(x - 1) = -x + 1 = 1 - x.$$

Figure 4

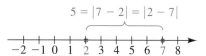

We shall use the concept of absolute value to define the distance between any two points on a coordinate line. First note that the distance between the points with coordinates 2 and 7, shown in Figure 4, equals 5 units. This distance is the difference obtained by subtracting the smaller (leftmost) coordinate from the larger (rightmost) coordinate ($7 - 2 = 5$). If we use absolute values, then, since $|7 - 2| = |2 - 7|$, it is unnecessary to be concerned about the order of subtraction. This fact motivates the next definition.

DEFINITION OF THE DISTANCE BETWEEN POINTS ON A COORDINATE LINE

Let a and b be the coordinates of two points A and B, respectively, on a coordinate line. The **distance between A and B,** denoted by $d(A, B)$, is defined by

$$d(A, B) = |b - a|.$$

The number $d(A, B)$ is the length of the line segment AB.

Since $d(B, A) = |a - b|$ and $|b - a| = |a - b|$, we see that

$$d(A, B) = d(B, A).$$

Note that the distance between the origin O and the point A is

$$d(O, A) = |a - 0| = |a|,$$

which agrees with the geometric interpretation of absolute value illustrated in Figure 4. The formula $d(A, B) = |b - a|$ is true regardless of the signs of a and b, as illustrated in the next example.

Figure 5

> **EXAMPLE 2**   Finding distances between points

Let A, B, C, and D have coordinates -5, -3, 1, and 6, respectively, on a coordinate line, as shown in Figure 5. Find $d(A, B)$, $d(C, B)$, $d(O, A)$, and $d(C, D)$.

Solution   Using the definition of the distance between points on a coordinate line, we obtain the distances:

$$d(A, B) = |-3 - (-5)| = |-3 + 5| = |2| = 2$$
$$d(C, B) = |-3 - 1| = |-4| = 4$$
$$d(O, A) = |-5 - 0| = |-5| = 5$$
$$d(C, D) = |6 - 1| = |5| = 5$$

The concept of absolute value has uses other than finding distances between points; it is employed whenever we are interested in the magnitude or numerical value of a real number without regard to its sign. We now turn our attention to the role that real numbers play in solving equations and inequalities.

Let x be a **variable**—that is, a letter that may be assigned different values from a given set of real numbers. These values are sometimes called the **permissible values** for the variable. Expressions such as

$$x + 3 = 0, \qquad x^2 - 5 = 4x, \qquad \text{and} \qquad (x^2 - 9)\sqrt[3]{x + 1} = 0$$

are **equations** in x. A number a is a **solution,** or **root,** of an equation if a true statement is obtained when a is substituted for x. We say that a **satisfies** the equation. The equation $x^2 - 5 = 4x$ has 5 as a solution, since substitution gives us $(5)^2 - 5 = 4(5)$, or $20 = 20$, which is a true statement. To **solve an equation** means to find all the solutions.

An equation in x is called an **identity** if every permissible value of x is a solution of the equation.

ILLUSTRATION

Identity

$$\frac{x}{x^2 - 4} = \frac{x}{(x + 2)(x - 2)}$$

An equation in x is called a **conditional equation** if there are permissible values of x that are *not* solutions. Two equations are **equivalent** if they have exactly the same solutions.

We shall assume some experience in finding solutions of equations in one variable. To solve a **linear equation** $ax + b = 0$, where a and b are real numbers and $a \neq 0$, we subtract b from both sides and divide by a as follows:

$$ax + b = 0, \quad ax = -b, \quad x = -\frac{b}{a}$$

Thus, the linear equation $ax + b = 0$ has exactly one solution, $-b/a$.

A **quadratic equation** in x is an equation that can be written in the form $ax^2 + bx + c = 0$, where $a \neq 0$. This form is called the **standard form** of a quadratic equation in x.

To enable us to solve quadratic and other types of equations, we will make use of the next theorem.

ZERO FACTOR THEOREM

> If p and q are algebraic expressions, then
>
> $$pq = 0 \quad \text{if and only if} \quad p = 0 \quad \text{or} \quad q = 0.$$

The zero factor theorem can be extended to any number of algebraic expressions—that is,

$$pqr = 0 \quad \text{if and only if} \quad p = 0 \quad \text{or} \quad q = 0 \quad \text{or} \quad r = 0,$$

and so on. It follows that if $ax^2 + bx + c$ can be written as a product of two first-degree polynomials, then solutions can be found by setting each factor equal to 0, as illustrated below. This technique is called the **method of factoring.** To use the method of factoring, *it is essential that only the number* 0 *appear on one side of the equation.*

ILLUSTRATION

Solving Quadratic Equations by Factoring

$$3x^2 = 10 - x \qquad\qquad x^2 + 16 = 8x$$
$$3x^2 + x - 10 = 0 \qquad\qquad x^2 - 8x + 16 = 0$$
$$(3x - 5)(x + 2) = 0 \qquad\qquad (x - 4)(x - 4) = 0$$
$$x = \tfrac{5}{3} \quad \text{or} \quad x = -2 \qquad\qquad x = 4$$

Since $x - 4$ appears as a factor twice in the second part of the previous illustration, we call 4 a **double root** or **root of multiplicity 2** of the equation $x^2 + 16 = 8x$.

If a quadratic equation has the form $x^2 = d$ for some $d > 0$, then $x^2 - d = 0$ or, equivalently,

$$(x + \sqrt{d})(x - \sqrt{d}) = 0.$$

Setting each factor equal to zero gives us the solutions $-\sqrt{d}$ and $\sqrt{d}$. We frequently use the symbol $\pm\sqrt{d}$ (*plus or minus* $\sqrt{d}$) to represent both $\sqrt{d}$ and $-\sqrt{d}$. Thus, for $d > 0$, we have proved the following result. (The case $d < 0$ requires the system of complex numbers discussed in Appendix V.)

A Special Quadratic Equation

> If $x^2 = d$, then $x = \pm\sqrt{d}$.

The process of solving $x^2 = d$ as indicated in the preceding box is referred to as *taking the square root of both sides of the equation.* Note that if $d > 0$ we obtain both a positive square root and a negative square root.

ILLUSTRATION

Solving Equations of the Form $x^2 = d$

$$x^2 = 169$$
$$x = \pm\sqrt{169}$$
$$= \pm 13$$

$$(x + 3)^2 = 5$$
$$x + 3 = \pm\sqrt{5}$$
$$x = -3 \pm\sqrt{5}$$

The solutions of a **quadratic equation** $ax^2 + bx + c = 0$, for $a \neq 0$, may be obtained by means of the following formula.

Quadratic Formula

If $a \neq 0$, the roots of $ax^2 + bx + c = 0$ are given by

$$x = \frac{-b \pm \sqrt{b^2 - 4ac}}{2a}.$$

Note that if the quadratic formula is executed properly, it is unnecessary to check the solutions because no *extraneous solutions* are introduced. The number $b^2 - 4ac$ under the radical sign in the quadratic formula is called the **discriminant** of the quadratic equation. If the discriminant is positive, there are two real and unequal roots of the quadratic equation. If it is zero, there is one root of multiplicity 2; and if it is negative, there are no real roots of the quadratic equation.

EXAMPLE 3   **Using the quadratic formula**

Solve the equation $2x(3 - x) = 3$.

Solution   To use the quadratic formula, we must write the equation in the standard form $ax^2 + bx + c = 0$. Doing so gives us the equation $-2x^2 + 6x = 3$ or, equivalently, $2x^2 - 6x + 3 = 0$. We now let $a = 2$, $b = -6$, and $c = 3$ in the quadratic formula, obtaining

$$x = \frac{-(-6) \pm \sqrt{(-6)^2 - 4(2)(3)}}{2(2)} = \frac{6 \pm \sqrt{12}}{4} = \frac{6 \pm 2\sqrt{3}}{4} = \frac{3 \pm \sqrt{3}}{2}.$$

Hence, the solutions are $\dfrac{3 + \sqrt{3}}{2} \approx 2.37$ and $\dfrac{3 - \sqrt{3}}{2} \approx 0.63$.

An equation is of **quadratic type** if it can be written in the form

$$au^2 + bu + c = 0,$$

where $a \neq 0$ and u is an expression in some variable. If we find the solutions in terms of u, then the solutions of the given equation can be obtained by referring to the specific form of u. For example, suppose we want to solve the equation

$$3y^4 + 5y^2 - 1 = 0.$$

By letting $u = y^2$, we can solve the equation

$$3u^2 + 5u - 1 = 0$$

for u. Next we re-substitute y^2 for u and solve for y, obtaining the four solutions given by $y = \pm \sqrt{\dfrac{-5 \pm \sqrt{37}}{6}}$.

An **inequality** is a statement that two quantities or expressions are not equal. It may be the case that one quantity is less than ($<$), less than or equal to ($\leq$), greater than ($>$), or greater than or equal to ($\geq$) another quantity. Consider the inequality

$$2x + 3 > 11,$$

where x is a variable. If a true statement is obtained when a number b is substituted for x, then b is a **solution** of the inequality. Thus, $x = 5$ is a solution of $2x + 3 > 11$ since $13 > 11$ is true, but $x = 3$ is not a solution since $9 > 11$ is false. To **solve** an inequality means to find *all* solutions. Two inequalities are **equivalent** if they have exactly the same solutions.

Most inequalities have an infinite number of solutions. To illustrate, the solutions of the continued inequality

$$2 < x < 5$$

consist of *every* real number x between 2 and 5. We call this set of numbers an **open interval** and denote it by $(2, 5)$. The **graph** of the open interval $(2, 5)$ is the set of all points on a coordinate line that lie between—but do not include—the points corresponding to $x = 2$ and $x = 5$. The graph is represented by shading an appropriate part of the axis, as shown in Figure 6. We refer to this process as **sketching the graph** of the interval. The numbers 2 and 5 are called the **endpoints** of the interval $(2, 5)$. The parentheses in the notation $(2, 5)$ and in Figure 6 are used to indicate that endpoints of the interval are not included.

If we wish to include an endpoint, we use a bracket instead of a parenthesis. For example, the solutions of the inequality $2 \leq x \leq 5$ are denoted by $[2, 5]$ and are referred to as a **closed interval.** The graph of $[2, 5]$ is sketched in Figure 7, where brackets indicate that endpoints are included. We shall also consider **half-open intervals** $[a, b)$ and $(a, b]$ and **infinite intervals,** as described in the following chart. The symbol ∞ (read "infinity") used for infinite intervals is merely a notational device and does *not* represent a real number.

Figure 6

Figure 7

Intervals

Notation	Inequality	Graph
(1) (a, b)	$a < x < b$	(a to b, open-open)
(2) $[a, b]$	$a \le x \le b$	(a to b, closed-closed)
(3) $[a, b)$	$a \le x < b$	(a to b, closed-open)
(4) $(a, b]$	$a < x \le b$	(a to b, open-closed)
(5) (a, ∞)	$x > a$	(a to ∞, open)
(6) $[a, \infty)$	$x \ge a$	(a to ∞, closed)
(7) $(-\infty, b)$	$x < b$	(−∞ to b, open)
(8) $(-\infty, b]$	$x \le b$	(−∞ to b, closed)
(9) $(-\infty, \infty)$	$-\infty < x < \infty$	(all reals)

Some properties of inequalities are illustrated in the following example.

EXAMPLE 4   **Solving a continued inequality**

Solve the continued inequality $-5 \le \dfrac{4 - 3x}{2} < 1$.

Solution   A number x is a solution of the given inequality if and only if

$$-5 \le \frac{4 - 3x}{2} \quad \text{and} \quad \frac{4 - 3x}{2} < 1.$$

We can either work with each inequality separately or solve both inequalities simultaneously, as follows (keep in mind that our goal is to isolate x):

$$-5 \le \frac{4 - 3x}{2} < 1 \quad \text{given}$$

$$-10 \le 4 - 3x < 2 \quad \text{multiply by 2}$$

$$-14 \le -3x < -2 \quad \text{subtract 4}$$

$$\tfrac{14}{3} \ge x > \tfrac{2}{3} \quad \begin{array}{l}\text{divide by } -3\text{; reverse}\\\text{the inequality signs}\end{array}$$

$$\tfrac{2}{3} < x \le \tfrac{14}{3} \quad \text{equivalent inequality}$$

Figure 8

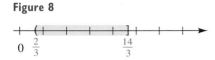

Thus, the solutions of the inequality are all numbers in the half-open interval $\left(\tfrac{2}{3}, \tfrac{14}{3}\right]$ sketched in Figure 8.

1.1 EXERCISES

Exer. 1–2: If $x < 0$ and $y > 0$, determine the sign of the real number.

1 (a) xy (b) x^2y (c) $\dfrac{x}{y} + x$ (d) $y - x$

2 (a) $\dfrac{x}{y}$ (b) xy^2 (c) $\dfrac{x - y}{xy}$ (d) $y(y - x)$

Exer. 3–6: Replace the symbol □ with either <, >, or = to make the resulting statement true.

3 (a) $-7 \;\square\; -4$ (b) $\dfrac{\pi}{2} \;\square\; 1.57$ (c) $\sqrt{225} \;\square\; 15$

4 (a) $-3 \;\square\; -5$ (b) $\dfrac{\pi}{4} \;\square\; 0.8$ (c) $\sqrt{289} \;\square\; 17$

5 (a) $\frac{1}{11} \;\square\; 0.09$ (b) $\frac{2}{3} \;\square\; 0.6666$ (c) $\frac{22}{7} \;\square\; \pi$

6 (a) $\frac{1}{7} \;\square\; 0.143$ (b) $\frac{5}{6} \;\square\; 0.833$ (c) $\sqrt{2} \;\square\; 1.4$

Exer. 7–8: Express the statement as an inequality.

7 (a) x is negative.

(b) y is nonnegative.

(c) q is less than or equal to π.

(d) d is between 4 and 2.

(e) t is not less than 5.

(f) The negative of z is not greater than 3.

(g) The quotient of p and q is at most 7.

(h) The reciprocal of w is at least 9.

(i) The absolute value of x is greater than 7.

8 (a) b is positive.

(b) s is nonpositive.

(c) w is greater than or equal to -4.

(d) c is between $\frac{1}{5}$ and $\frac{1}{3}$.

(e) p is not greater than -2.

(f) The negative of m is not less than -2.

(g) The quotient of r and s is at least $\frac{1}{5}$.

(h) The reciprocal of f is at most 14.

(i) The absolute value of x is less than 4.

Exer. 9–14: Rewrite the number without using the absolute value symbol, and simplify the result.

9 (a) $|-3 - 2|$ (b) $|-5| - |2|$ (c) $|7| + |-4|$

10 (a) $|-11 + 1|$ (b) $|6| - |-3|$ (c) $|8| + |-9|$

11 (a) $(-5)|3 - 6|$ (b) $|-6|/(-2)$ (c) $|-7| + |4|$

12 (a) $(4)|6 - 7|$ (b) $5/|-2|$ (c) $|-1| + |-9|$

13 (a) $|4 - \pi|$ (b) $|\pi - 4|$ (c) $|\sqrt{2} - 1.5|$

14 (a) $|\sqrt{3} - 1.7|$ (b) $|1.7 - \sqrt{3}|$ (c) $\left|\frac{1}{5} - \frac{1}{3}\right|$

Exer. 15–18: The given numbers are coordinates of points A, B, and C, respectively, on a coordinate line. Find the distance.

(a) $d(A, B)$ (b) $d(B, C)$

(c) $d(C, B)$ (d) $d(A, C)$

15 $3, 7, -5$

16 $-6, -2, 4$

17 $-9, 1, 10$

18 $8, -4, -1$

Exer. 19–24: The two given numbers are coordinates of points A and B, respectively, on a coordinate line. Express the indicated statement as an inequality involving the absolute value symbol.

19 x, 7; $d(A, B)$ is less than 5

20 x, $-\sqrt{2}$; $d(A, B)$ is greater than 1

21 x, -3; $d(A, B)$ is at least 8

22 x, 4; $d(A, B)$ is at most 2

23 4, x; $d(A, B)$ is not greater than 3

24 -2, x; $d(A, B)$ is not less than 2

Exer. 25–32: Rewrite the expression without using the absolute value symbol, and simplify the result.

25 $|3 + x|$ if $x < -3$ 26 $|5 - x|$ if $x > 5$

27 $|2 - x|$ if $x < 2$ 28 $|7 + x|$ if $x \geq -7$

29 $|a - b|$ if $a < b$ 30 $|a - b|$ if $a > b$

31 $|x^2 + 4|$ 32 $|-x^2 - 1|$

Exer. 33–38: Solve the equation.

33 $4(2x + 5) = 3(5x - 2)$

34 $6(2x + 3) - 3(x - 5) = 0$

35 $3x(x - 2)(4x + 3) = 0$

36 $-4x(x + 5)(3x - 7) = 0$

37 $8 - \dfrac{5}{x} = 2 + \dfrac{3}{x}$

38 $\dfrac{3}{x} + \dfrac{6}{x} - \dfrac{1}{x} = 11$

Exer. 39–42: Solve the equation by factoring.

39 $6x^2 + x - 12 = 0$ 40 $4x^2 + x - 14 = 0$

41 $15x^2 - 12 = -8x$ 42 $15x^2 - 14 = 29x$

Exer. 43–46: Solve the equation by using the special quadratic equation on page 8.

43 $25x^2 = 9$ 44 $16x^2 = 49$

45 $(x - 3)^2 = 17$ 46 $(x + 4)^2 = 31$

Exer. 47–50: Solve by using the quadratic formula.

47 $x^2 + 4x + 2 = 0$ 48 $x^2 - 6x - 3 = 0$

49 $2x^2 - 3x - 4 = 0$ 50 $3x^2 + 5x + 1 = 0$

Exer. 51–52: Solve the equation.

51 $x^4 - 25x^2 + 144 = 0$ 52 $2x^4 - 10x^2 + 8 = 0$

Exer. 53–62: Express the inequality as an interval, and sketch its graph.

53 $x < -2$ 54 $x \le 5$

55 $x \ge 4$ 56 $x > -3$

57 $-2 < x \le 4$ 58 $-3 \le x < 5$

59 $3 \le x \le 7$ 60 $-3 < x < -1$

61 $0 < x < \pi$ 62 $\dfrac{\pi}{2} \le x \le \dfrac{3\pi}{2}$

Exer. 63–72: Express the interval as an inequality in the variable x.

63 $(-5, 8]$ 64 $[0, 4)$

65 $[-4, -1]$ 66 $(3, 7)$

67 $[4, \infty)$ 68 $(-3, \infty)$

69 $(-\infty, -5)$ 70 $(-\infty, 2]$

71 $[0, 2\pi]$ 72 $\left(-\dfrac{\pi}{2}, \dfrac{\pi}{2}\right)$

Exer. 73–82: Solve the inequality, and express the solutions in terms of intervals whenever possible.

73 $2x + 5 < 3x - 7$ 74 $x - 8 > 5x + 3$

75 $3 \le \dfrac{2x - 3}{5} < 7$ 76 $-2 < \dfrac{4x + 1}{3} \le 0$

77 $\dfrac{4}{3x + 2} \ge 0$ 78 $\dfrac{3}{2x + 5} \le 0$

79 $\dfrac{-2}{4 - 3x} > 0$ 80 $\dfrac{-3}{2 - x} < 0$

81 $\dfrac{2}{(1 - x)^2} > 0$ 82 $\dfrac{4}{x^2 + 4} < 0$

83 The point on a coordinate line corresponding to $\sqrt{2}$ may be determined by constructing a right triangle with sides of length 1, as shown in the figure. Determine the points that correspond to $\sqrt{3}$ and $\sqrt{5}$, respectively. (*Hint:* Use the Pythagorean theorem.)

Exercise 83

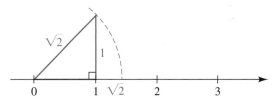

84 A circle of radius 1 rolls along a coordinate line in the positive direction, as shown in the figure. If point P is initially at the origin, find the coordinate of P after one, two, and ten complete revolutions.

Exercise 84

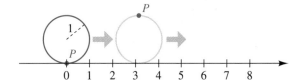

85 *Men's weight* The average weight W (in pounds) for men with height h between 64 and 79 inches can be

approximated using the formula $W = 0.1166h^{1.7}$. Construct a table for W by letting $h = 64, 65, \ldots, 79$. Round all weights to the nearest pound.

Height	Weight	Height	Weight
64		72	
65		73	
66		74	
67		75	
68		76	
69		77	
70		78	
71		79	

86 *Women's weight*   The mean weight W (in pounds) for women with height h between 60 and 75 inches can be approximated using the formula $W = 0.1049h^{1.7}$. Construct a table for W by letting $h = 60, 61, \ldots, 75$. Round all weights to the nearest pound.

Height	Weight	Height	Weight
60		68	
61		69	
62		70	
63		71	
64		72	
65		73	
66		74	
67		75	

87 *Business expenditure*   A construction firm is trying to decide which of two models of a crane to purchase. Model A costs $50,000 and requires $4000 per year to maintain. Model B has an initial cost of $40,000 and a maintenance cost of $5500 per year. For how many years must model A be used before it becomes more economical than B?

88 *Buying a car*   A consumer is trying to decide whether to purchase car A or car B. Car A costs $10,000 and has an mpg rating of 30, and insurance is $550 per year. Car B costs $12,000 and has an mpg rating of 50, and

insurance is $600 per year. Assume that the consumer drives 15,000 miles per year and that the price of gas remains constant at $1.25 per gallon. Based only on these facts, determine how long it will take for the total cost of car B to become less than that of car A.

89 *Calorie requirements*   The basal energy requirement for an individual indicates the minimum number of calories necessary to maintain essential life-sustaining processes such as circulation, body temperature, and respiration. Given a person's sex, weight w (in kilograms), height h (in centimeters), and age y (in years), we can estimate the basal energy requirement in calories using the following formulas, where C_f and C_m are the calories necessary for females and males, respectively:

$$C_f = 66.5 + 13.8w + 5h - 6.8y$$
$$C_m = 655 + 9.6w + 1.9h - 4.7y$$

(a) Determine the basal energy requirements first for a 25-year-old female who weighs 59 kilograms and is 163 centimeters tall and then for a 55-year-old male who weighs 75 kilograms and is 178 centimeters tall.

(b) Discuss why, in both formulas, the coefficient for y is negative but the other coefficients are positive.

C Exer. 90–91: Choose the equation that best describes the table of data. Some specific keystrokes for creating a table using the TI-82/83 are given in Example 1 of Appendix I.

90

x	y
1	0.8
2	−0.4
3	−1.6
4	−2.8
5	−4.0

(1) $y = -1.2x + 2$
(2) $y = -1.2x^2 + 2$
(3) $y = 0.8\sqrt{x}$
(4) $y = x^{3/4} - 0.2$

91

x	y
1	−9
2	−4
3	11
4	42
5	95

(1) $y = 13x - 22$
(2) $y = x^2 - 2x - 8$
(3) $y = 4\sqrt{x} - 13$
(4) $y = x^3 - x^2 + x - 10$

c 92 *Temperature-latitude relationships* The table contains average annual temperatures for the northern and southern hemispheres at various latitudes.

Latitude	N. Hem.	S. Hem.
85°	−8°F	−5°F
75°	13°F	10°F
65°	30°F	27°F
55°	41°F	42°F
45°	57°F	53°F
35°	68°F	65°F
25°	78°F	73°F
15°	80°F	78°F
5°	79°F	79°F

(a) Which of the following equations more accurately predicts the average annual temperature in the southern hemisphere at latitude L?

    (1) $T_1 = -1.09L + 96.01$

    (2) $T_2 = -0.011L^2 - 0.126L + 81.45$

(b) Approximate the average annual temperature in the southern hemisphere at latitude 50°.

c 93 *Daylight-latitude relationships* The table gives the numbers of minutes of daylight occurring at various latitudes in the northern hemisphere at the summer and winter solstices.

Latitude	Summer	Winter
0°	720	720
10°	755	685
20°	792	648
30°	836	604
40°	892	548
50°	978	462
60°	1107	333

(a) Which of the following equations more accurately predicts the length of day at the summer solstice at latitude L?

    (1) $D_1 = 6.096L + 685.7$

    (2) $D_2 = 0.00178L^3 - 0.072L^2 + 4.37L + 719$

(b) Approximate the length of daylight at 35° during the summer solstice.

1.2 RECTANGULAR COORDINATE SYSTEMS AND GRAPHS

*The term *Cartesian* is used in honor of the French mathematician and philosopher René Descartes (1596–1650), who was one of the first to employ such coordinate systems.

In Section 1.1 we discussed how to assign a real number (coordinate) to each point on a line. We shall now show how to assign an **ordered pair** (*a, b*) of real numbers to each point in a plane. Although we have also used the notation (*a, b*) to denote an open interval, there is little chance for confusion, since it should always be clear from our discussion whether (*a, b*) represents a point or an interval.

We introduce a **rectangular**, or **Cartesian,** coordinate system in a plane by means of two perpendicular coordinate lines, called **coordinate axes,** that intersect at the **origin** *O*, as shown in Figure 9. We often refer to the horizontal line as the **x-axis** and the vertical line as the **y-axis** and label them *x* and *y*, respectively. The plane is then a **coordinate plane,** or an **xy-plane.** The coordinate axes divide the plane into four parts called the **first, second, third,** and **fourth quadrants,** labeled I, II, III, and IV, respectively (see Figure 9). Points on the axes do not belong to any quadrant.

Each point *P* in an *xy*-plane may be assigned an ordered pair (*a, b*), as shown in Figure 9. We call *a* the **x-coordinate** (or **abscissa**) of *P*, and *b* the **y-coordinate** (or **ordinate**). We say that *P has coordinates* (*a, b*) and refer to the *point* (a, b) or the *point* $P(a, b)$. Conversely, every ordered pair (a, b) determines a point P with coordinates a and b. We **plot a point** by using a dot, as illustrated in Figure 10.

Figure 9   **Figure 10**

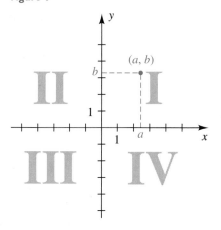 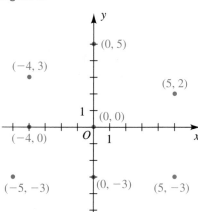

We may use the following formula to find the distance between two points in a coordinate plane.

Distance Formula

> The distance $d(P_1, P_2)$ between any two points $P_1(x_1, y_1)$ and $P_2(x_2, y_2)$ in a coordinate plane is
> $$d(P_1, P_2) = \sqrt{(x_2 - x_1)^2 + (y_2 - y_1)^2}.$$

Figure 11

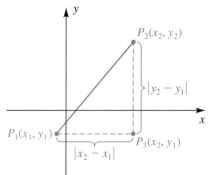

Proof   If $x_1 \neq x_2$ and $y_1 \neq y_2$, then, as illustrated in Figure 11, the points P_1, P_2, and $P_3(x_2, y_1)$ are vertices of a right triangle. By the Pythagorean theorem,

$$[d(P_1, P_2)]^2 = [d(P_1, P_3)]^2 + [d(P_3, P_2)]^2.$$

From the figure we see that

$$d(P_1, P_3) = |x_2 - x_1| \quad \text{and} \quad d(P_3, P_2) = |y_2 - y_1|.$$

Since $|a|^2 = a^2$ for every real number a, we may write

$$[d(P_1, P_2)]^2 = (x_2 - x_1)^2 + (y_2 - y_1)^2.$$

Taking the square root of each side of the last equation and using the fact that $d(P_1, P_2) \geq 0$ gives us the distance formula.

If $y_1 = y_2$, the points P_1 and P_2 lie on the same horizontal line, and

$$d(P_1, P_2) = |x_2 - x_1| = \sqrt{(x_2 - x_1)^2}.$$

Similarly, if $x_1 = x_2$, the points are on the same vertical line, and

$$d(P_1, P_2) = |y_2 - y_1| = \sqrt{(y_2 - y_1)^2}.$$

These are special cases of the distance formula.

Although we referred to the points shown in Figure 11, our proof is independent of the positions of P_1 and P_2.   ◢

When applying the distance formula, note that $d(P_1, P_2) = d(P_2, P_1)$ and, hence, the order in which we subtract the x-coordinates and the y-coordinates of the points is immaterial. We may think of the distance between two points as the length of the hypotenuse of a right triangle.

ILLUSTRATION

Finding the Distance Between Two Points A and B

▶   For $A(0, 0)$ and $B(4, 3)$, $d(A, B) = \sqrt{(4 - 0)^2 + (3 - 0)^2}$
$$= \sqrt{25} = 5.$$

▶   For $A(-3, 6)$ and $B(5, 1)$, $d(A, B) = \sqrt{(5 + 3)^2 + (1 - 6)^2}$
$$= \sqrt{89} \approx 9.43.$$

EXAMPLE 1   **Applying the distance formula**

Given $A(1, 7)$, $B(-3, 2)$, and $C\left(4, \frac{1}{2}\right)$, prove that C is on the perpendicular bisector of segment AB.

Figure 12

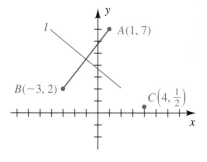

l

y

$A(1, 7)$

$B(-3, 2)$

$C\left(4, \frac{1}{2}\right)$

x

Solution   The points A, B, C and the *perpendicular bisector l* are illustrated in Figure 12. From plane geometry, l can be characterized by either of the following conditions:

(1) l is the line perpendicular to segment AB at its midpoint.

(2) l is the set of all points equidistant from the endpoints of segment AB.

We shall use condition 2 to show that C is on l by verifying that

$$d(A, C) = d(B, C).$$

We apply the distance formula:

$$d(A, C) = \sqrt{(4 - 1)^2 + \left(\tfrac{1}{2} - 7\right)^2} = \sqrt{3^2 + \left(-\tfrac{13}{2}\right)^2} = \sqrt{9 + \tfrac{169}{4}} = \sqrt{\tfrac{205}{4}}$$

$$d(B, C) = \sqrt{[4 - (-3)]^2 + \left(\tfrac{1}{2} - 2\right)^2} = \sqrt{7^2 + \left(-\tfrac{3}{2}\right)^2} = \sqrt{49 + \tfrac{9}{4}} = \sqrt{\tfrac{205}{4}}$$

Thus, C is equidistant from A and B, and the verification is complete.

In Example 9 of Appendix VI, we find a formula for the perpendicular bisector of a segment using condition 1 of Example 1.

We can find the midpoint of a line segment by using the following formula.

Midpoint Formula

> The midpoint M of the line segment from $P_1(x_1, y_1)$ to $P_2(x_2, y_2)$ is
>
> $$\left(\frac{x_1 + x_2}{2}, \frac{y_1 + y_2}{2} \right).$$

Figure 13

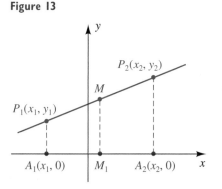

Proof   The lines through P_1 and P_2 parallel to the y-axis intersect the x-axis at $A_1(x_1, 0)$ and $A_2(x_2, 0)$. From plane geometry, the line through the midpoint M parallel to the y-axis bisects the segment $A_1 A_2$ at point M_1 (see Figure 13). If $x_1 < x_2$, then $x_2 - x_1 > 0$, and hence $d(A_1, A_2) = x_2 - x_1$. Since M_1 is halfway from A_1 to A_2, the x-coordinate of M_1 is equal to the x-coordinate of A_1 plus one-half the distance from A_1 to A_2; that is,

$$x\text{-coordinate of } M_1 = x_1 + \tfrac{1}{2}(x_2 - x_1).$$

The expression on the right side of the last equation simplifies to

$$\frac{x_1 + x_2}{2}.$$

This quotient is the *average* of the numbers x_1 and x_2. It follows that the x-coordinate of M is also $(x_1 + x_2)/2$. Similarly, the y-coordinate of M is $(y_1 + y_2)/2$. These formulas hold for all positions of P_1 and P_2. ◢

To apply the midpoint formula, it may suffice to remember that

the x-coordinate of the midpoint $=$ the *average* of the x-coordinates

and that

the y-coordinate of the midpoint $=$ the *average* of the y-coordinates.

ILLUSTRATION

Finding the Midpoint M of Two Points A and B

For $A(-2, 3)$ and $B(4, -2)$, $M = \left(\dfrac{-2 + 4}{2}, \dfrac{3 + (-2)}{2} \right) = \left(1, \dfrac{1}{2} \right).$

Graphs are often used to illustrate changes in quantities. A graph in the business section of a newspaper may show the fluctuation of the Dow Jones average during a given month; a meteorologist might use a graph to indicate how the air temperature varied throughout a day; a cardiologist employs graphs (electrocardiograms) to analyze heart irregularities; an engineer or physicist may turn to a graph to illustrate the manner in which the pressure of a confined gas increases as the gas is heated. Such visual aids usually reveal the behavior of quantities more readily than a long table of numerical values.

Two quantities are sometimes related by means of an equation or formula that involves two variables. In this section we discuss how to represent such an equation geometrically, by a graph in a coordinate plane. The graph may then be used to discover properties of the quantities that are not evident from the equation alone. The following chart introduces the basic concept of the graph of an equation in two variables x and y. Of course, other letters can also be used for the variables.

Terminology	Definition	Illustration
Solution of an equation in x and y	An ordered pair (a, b) that yields a true statement if $x = a$ and $y = b$	$(2, 3)$ is a solution of $y^2 = 5x - 1$, since substituting $x = 2$ and $y = 3$ gives us LS: $3^2 = 9$ RS: $5(2) - 1 = 10 - 1 = 9$

For each solution (a, b) of an equation in x and y there is a point $P(a, b)$ in a coordinate plane. The set of all such points is called the **graph** of the equation. To *sketch the graph of an equation,* we illustrate the significant features of the graph in a coordinate plane. In simple cases, a graph can be sketched by plotting few, if any, points. For a complicated equation, plotting points may give very little information about the graph. In such cases, methods of calculus or computer graphics are often employed. Let us begin with a simple example.

Figure 14

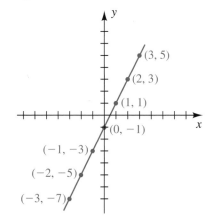

*Appendix VI contains a thorough review of lines. Please refer to this appendix for material on the following topics: slope and slopes of parallel and perpendicular lines; horizontal and vertical lines; and various forms of equations of lines (point-slope, slope-intercept, general, intercept).

EXAMPLE 2   Sketching a simple graph by plotting points

Sketch the graph of the equation $y = 2x - 1$.

Solution   We wish to find the points (x, y) in a coordinate plane that correspond to the solutions of the equation. It is convenient to list coordinates of several such points in a table, where for each x we obtain the value for y from $y = 2x - 1$:

x	-3	-2	-1	0	1	2	3
y	-7	-5	-3	-1	1	3	5

The points with these coordinates appear to lie on a line, and we can sketch the graph in Figure 14. Ordinarily, the few points we have plotted would not be enough to illustrate the graph of an equation; however, in this elementary case we can be reasonably sure that the graph is a line. In Appendix VI* we establish this fact.

It is impossible to sketch the entire graph in Example 2, because we can assign values to x that are numerically as large as desired. Nevertheless, we call the drawing in Figure 14 *the graph of the equation* or *a sketch of the graph*. In general, the sketch of a graph should illustrate its essential features so that the remaining (unsketched) parts are self-evident. For instance, in Figure 14, the **end behavior**—the pattern of the graph as x assumes large positive and negative values (that is, the shape of the right and left ends)—is apparent to the reader.

If a graph terminates at some point (as would be the case for a half-line or line segment), we place a dot at the appropriate *endpoint* of the graph. As a final general remark, *if ticks on the coordinate axes are not labeled* (as in Figure 14), *then each tick represents one unit.* We shall label ticks only when different units are used on the axes. For *arbitrary* graphs, where units of measurement are irrelevant, we omit ticks completely (see, for example, Figures 18 and 19).

EXAMPLE 3   Sketching the graph of an equation

Sketch the graph of the equation $y = x^2 - 3$.

Solution   Substituting values for x and finding the corresponding values of y using $y = x^2 - 3$, we obtain a table of coordinates for several points on the graph:

x	-3	-2	-1	0	1	2	3
y	6	1	-2	-3	-2	1	6

Larger values of $|x|$ produce larger values of y. For example, the points $(4, 13)$, $(5, 22)$, and $(6, 33)$ are on the graph, as are $(-4, 13)$, $(-5, 22)$, and $(-6, 33)$. Plotting the points given by the table and drawing a smooth curve through these points (in the order of increasing values of x) gives us the sketch in Figure 15.

Figure 15

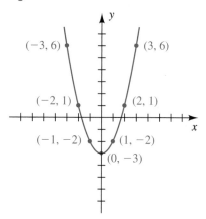

The graph in Figure 15 is a **parabola**, and the y-axis is the **axis of the parabola**. The lowest point $(0, -3)$ is the **vertex** of the parabola, and we say that the parabola *opens upward*. If we invert the graph, then the parabola *opens downward* and the vertex is the highest point on the graph. In general, the graph of *any* equation of the form $y = ax^2 + c$ with $a \neq 0$ is a parabola with vertex $(0, c)$, opening upward if $a > 0$ or downward if $a < 0$. If $c = 0$, the equation reduces to $y = ax^2$ and the vertex is at the origin $(0, 0)$. Parabolas may also open to the right or to the left (see Example 5) or in other directions.

We shall use the following terminology to describe where the graph of an equation in x and y intersects the x-axis or the y-axis.

Intercepts of the Graph of an Equation in x and y

Terminology	Definition	Graphical interpretation	How to find
x-intercepts	The x-coordinates of points where the graph intersects the x-axis		Let $y = 0$ and solve for x
y-intercepts	The y-coordinates of points where the graph intersects the y-axis		Let $x = 0$ and solve for y

An x-intercept is sometimes referred to as a *zero* of the graph of an equation or as a *root* of an equation. When using a graphing utility to find an x-intercept, we will say that we are using a *root feature*.

EXAMPLE 4   Finding x-intercepts and y-intercepts

Find the x- and y-intercepts of the graph of $y = x^2 - 3$.

Solution   The graph is sketched in Figure 15 (Example 3). We find the intercepts as stated in the preceding chart.

(1) *x-intercepts:*

$$
\begin{aligned}
y &= x^2 - 3 && \text{given} \\
0 &= x^2 - 3 && \text{let } y = 0 \\
x^2 &= 3 && \text{equivalent equation} \\
x &= \pm\sqrt{3} \approx \pm 1.73 && \text{take the square root}
\end{aligned}
$$

Thus, the x-intercepts are $-\sqrt{3}$ and $\sqrt{3}$. The points at which the graph crosses the x-axis are $(-\sqrt{3}, 0)$ and $(\sqrt{3}, 0)$.

(2) *y-intercepts:* $y = x^2 - 3$ given

$y = 0 - 3 = -3$ let $x = 0$

Thus, the y-intercept is -3, and the point at which the graph crosses the y-axis is $(0, -3)$.

If the coordinate plane in Figure 15 is folded along the y-axis, the graph that lies in the left half of the plane coincides with that in the right half, and we say that *the graph is symmetric with respect to the y-axis.* A graph is symmetric with respect to the y-axis provided that the point $(-x, y)$ is on the graph whenever (x, y) is on the graph. The graph of $y = x^2 - 3$ in Example 3 has this property, since substitution of $-x$ for x yields the same equation:

$$y = (-x)^2 - 3 = x^2 - 3$$

This substitution is an application of symmetry test 1 in the following chart. Two other types of symmetry and the appropriate tests are also listed. The graphs of $x = y^2$ and $4y = x^3$ in the illustration column are discussed in Examples 5 and 6, respectively.

Symmetries of Graphs of Equations in x and y

Terminology	Graphical interpretation	Test for symmetry	Illustration
The graph is symmetric with respect to the y-axis.	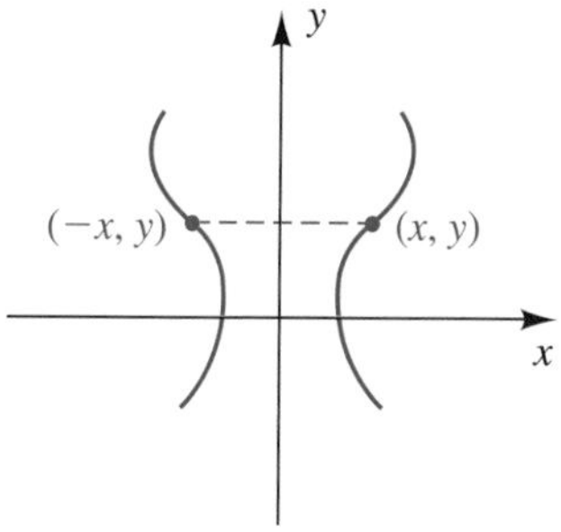	**(1)** Substitution of $-x$ for x leads to the same equation.	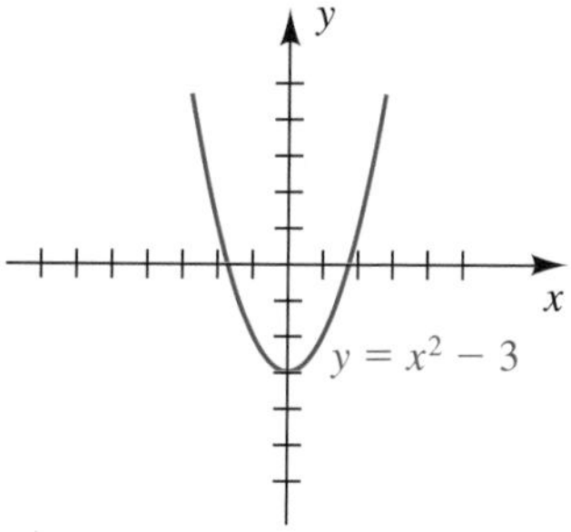
The graph is symmetric with respect to the x-axis.	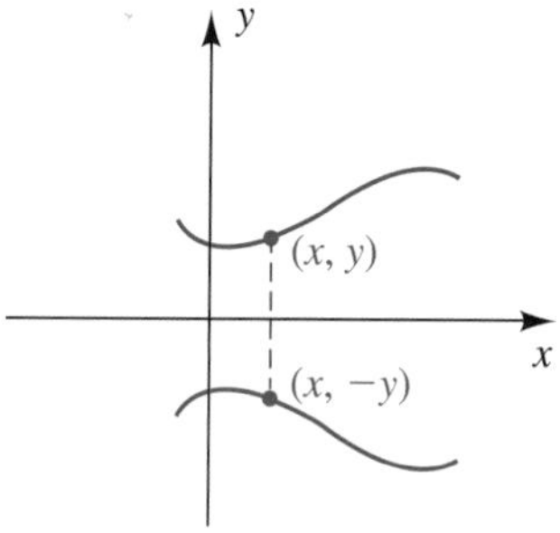	**(2)** Substitution of $-y$ for y leads to the same equation.	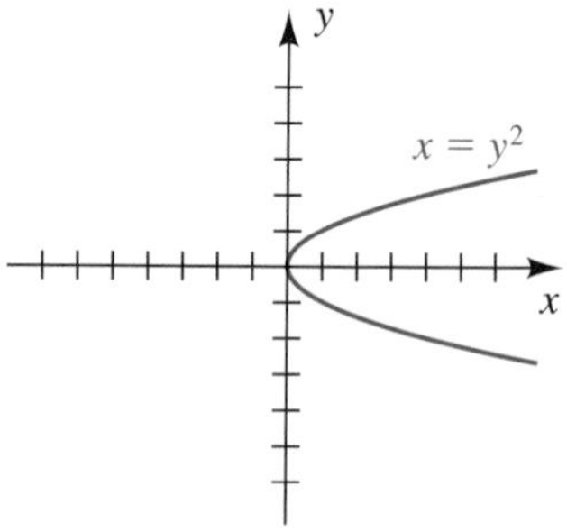
The graph is symmetric with respect to the origin.	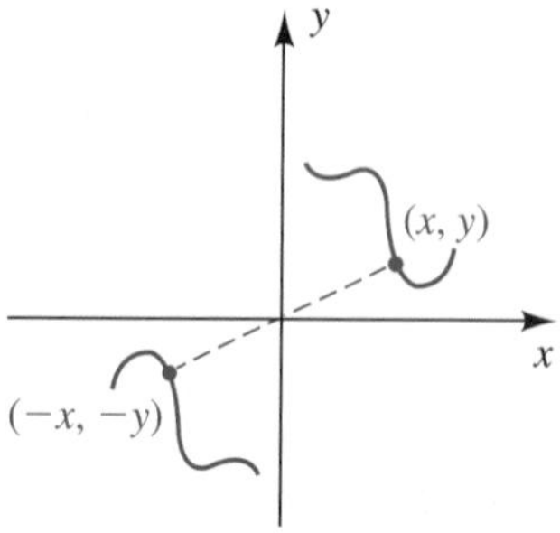	**(3)** Simultaneous substitution of $-x$ for x and $-y$ for y leads to the same equation.	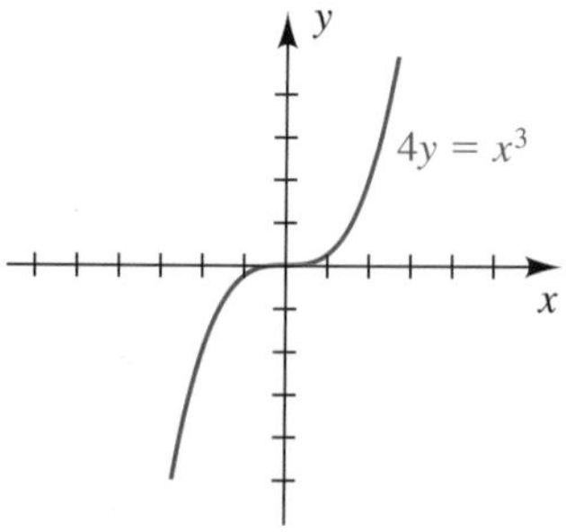

If a graph is symmetric with respect to an axis, it is sufficient to determine the graph in half of the coordinate plane, since we can sketch the remainder of the graph by taking a *mirror image,* or *reflection,* through the appropriate axis.

EXAMPLE 5   **A graph that is symmetric with respect to the x-axis**

Sketch the graph of the equation $y^2 = x$.

Solution   Since substitution of $-y$ for y does not change the equation, the graph is symmetric with respect to the x-axis (see symmetry test 2). Hence, if the point (x, y) is on the graph, then the point $(x, -y)$ is on the graph. Thus, it is sufficient to find points with nonnegative y-coordinates and then reflect through the x-axis. The equation $y^2 = x$ is equivalent to $y = \pm\sqrt{x}$. The y-coordinates of points *above* the x-axis (y is *positive*) are given by $y = \sqrt{x}$, whereas the y-coordinates of points *below* the x-axis (y is *negative*) are given by $y = -\sqrt{x}$. Coordinates of some points on the graph are listed below. The graph is sketched in Figure 16.

Figure 16

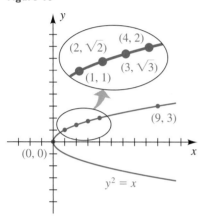

x	0	1	2	3	4	9
y	0	1	$\sqrt{2} \approx 1.4$	$\sqrt{3} \approx 1.7$	2	3

The graph is a parabola that opens to the right, with its vertex at the origin. In this case, the x-axis is the axis of the parabola.

EXAMPLE 6   **A graph that is symmetric with respect to the origin**

Sketch the graph of the equation $4y = x^3$.

Solution   If we simultaneously substitute $-x$ for x and $-y$ for y, then

$$4(-y) = (-x)^3 \quad \text{or, equivalently,} \quad -4y = -x^3.$$

Multiplying both sides by -1, we see that the last equation has the same solutions as the equation $4y = x^3$. Hence, from symmetry test 3, the graph is symmetric with respect to the origin—and if the point (x, y) is on the graph, then the point $(-x, -y)$ is on the graph. The following table lists coordinates of some points on the graph.

Figure 17

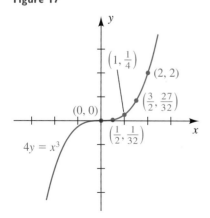

x	0	$\frac{1}{2}$	1	$\frac{3}{2}$	2	$\frac{5}{2}$
y	0	$\frac{1}{32}$	$\frac{1}{4}$	$\frac{27}{32}$	2	$\frac{125}{32}$

Because of the symmetry, we can see that the points $\left(-1, -\frac{1}{4}\right)$, $(-2, -2)$, and so on, are also on the graph. The graph is sketched in Figure 17.

If $C(h, k)$ is a point in a coordinate plane, then a circle with center C and radius $r > 0$ consists of all points in the plane that are r units from C. As shown in Figure 18, a point $P(x, y)$ is on the circle provided $d(C, P) = r$ or, by the distance formula,

$$\sqrt{(x - h)^2 + (y - k)^2} = r.$$

The above equation is equivalent to the following equation, which we will refer to as the **standard equation of a circle.**

Standard Equation of a Circle with Center (h, k) and Radius r

$$(x - h)^2 + (y - k)^2 = r^2$$

If $h = 0$ and $k = 0$, this equation reduces to $x^2 + y^2 = r^2$, which is an equation of a circle of radius r with center at the origin (see Figure 19). If $r = 1$, we call the graph a **unit circle.**

Figure 18

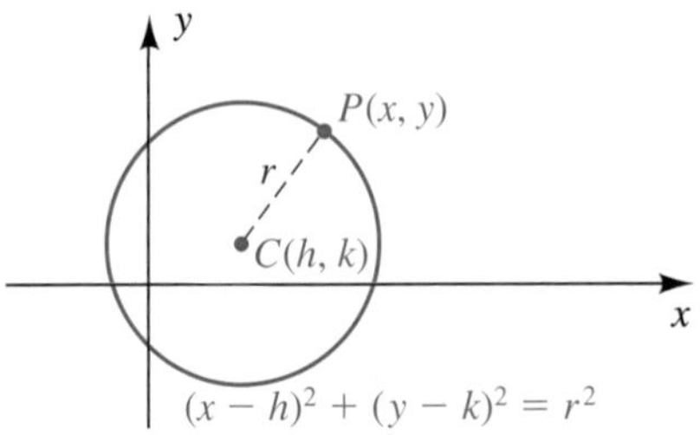

Figure 19

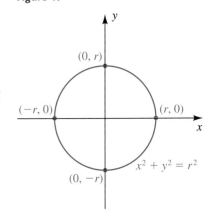

Figure 20

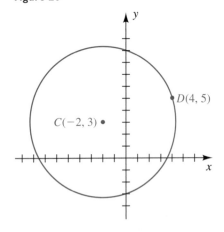

E X A M P L E 7   **Finding an equation of a circle**

Find an equation of the circle that has center $C(-2, 3)$ and contains the point $D(4, 5)$.

Solution   The circle is shown in Figure 20. Since D is on the circle, the radius r is $d(C, D)$. By the distance formula,

$$r = \sqrt{(4 + 2)^2 + (5 - 3)^2} = \sqrt{36 + 4} = \sqrt{40}.$$

Using the standard equation of a circle with $h = -2$, $k = 3$, and $r = \sqrt{40}$, we obtain

$$(x + 2)^2 + (y - 3)^2 = 40.$$

By squaring terms and simplifying the last equation, we may write it as

$$x^2 + y^2 + 4x - 6y - 27 = 0.$$

As in the solution to Example 7, squaring terms of an equation of the form $(x - h)^2 + (y - k)^2 = r^2$ and simplifying leads to an equation of the form

$$x^2 + y^2 + ax + by + c = 0,$$

where a, b, and c are real numbers. Conversely, if we begin with this equation, it is always possible, by *completing squares,* to obtain an equation of the form

$$(x - h)^2 + (y - k)^2 = d.$$

This method will be illustrated in Example 8. If $d > 0$, the graph is a circle with center (h, k) and radius $r = \sqrt{d}$. If $d = 0$, the graph consists of only the point (h, k). Finally, if $d < 0$, the equation has no real solutions, and hence there is no graph.

EXAMPLE 8   **Finding the center and radius of a circle**

Find the center and radius of the circle with equation

$$3x^2 + 3y^2 - 12x + 18y = 9.$$

Solution   Since it is easier to complete the square if the coefficients of x^2 and y^2 are 1, we begin by dividing the given equation by 3, obtaining

$$x^2 + y^2 - 4x + 6y = 3.$$

Next, we rewrite the equation as follows, where the underscored spaces represent numbers to be determined:

$$(x^2 - 4x + \underline{\ \ }) + (y^2 + 6y + \underline{\ \ }) = 3 + \underline{\ \ } + \underline{\ \ }$$

We then complete the squares for the expressions within parentheses, taking care to add the appropriate numbers to *both* sides of the equation. To complete the square for an expression of the form $x^2 + ax$, we add the square of half the coefficient of x (that is, $(a/2)^2$) to both sides of the equation. Similarly, for $y^2 + by$, we add $(b/2)^2$ to both sides. In this example, $a = -4$, $b = 6$, $(a/2)^2 = (-2)^2 = 4$, and $(b/2)^2 = 3^2 = 9$. These additions lead to

$$(x^2 - 4x + \underline{4}) + (y^2 + 6y + \underline{9}) = 3 + \underline{4} + \underline{9} \qquad \text{completing the squares}$$

$$(x - 2)^2 + (y + 3)^2 = 16. \qquad \text{equivalent equation}$$

Comparing the last equation with the standard equation of a circle, we see that $h = 2$ and $k = -3$ and conclude that the circle has center $(2, -3)$ and radius $\sqrt{16} = 4$.

In some applications it is necessary to work with only one-half of a circle—that is, a **semicircle**. The next example indicates how to find equations of semicircles for circles with centers at the origin.

EXAMPLE 9   **Finding equations of semicircles**

Find equations for the upper half, lower half, right half, and left half of the circle $x^2 + y^2 = 81$.

Solution   The graph of $x^2 + y^2 = 81$ is a circle of radius 9 with center at the origin (compare with Figure 19). To find equations for the upper and lower halves, we solve for y in terms of x:

$$x^2 + y^2 = 81 \qquad \text{given}$$
$$y^2 = 81 - x^2 \qquad \text{subtract } x^2$$
$$y = \pm\sqrt{81 - x^2} \qquad \text{take the square root}$$

Since $\sqrt{81 - x^2} \geq 0$, it follows that the upper half of the circle has the equation $y = \sqrt{81 - x^2}$ (y is positive) and the lower half is given by $y = -\sqrt{81 - x^2}$ (y is negative), as illustrated in Figure 21(a) and (b).

Figure 21

(a) $y = \sqrt{81 - x^2}$

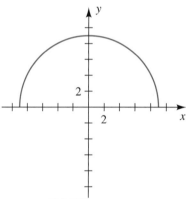

(b) $y = -\sqrt{81 - x^2}$

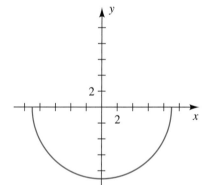

(c) $x = \sqrt{81 - y^2}$

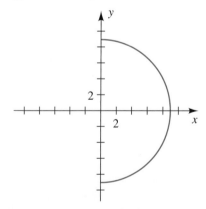

(d) $x = -\sqrt{81 - y^2}$

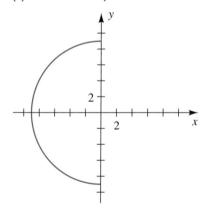

Similarly, to find equations for the right and left halves, we solve $x^2 + y^2 = 81$ for x in terms of y, obtaining

$$x = \pm\sqrt{81 - y^2}.$$

Since $\sqrt{81 - y^2} \geq 0$, it follows that the right half of the circle has the equation $x = \sqrt{81 - y^2}$ (x is positive) and the left half is given by $x = -\sqrt{81 - y^2}$ (x is negative), as illustrated in Figure 21(c) and (d).

Engineers, scientists, and mathematicians often use graphing calculators or computer software to obtain sketches of graphs. These devices have the capability to *zoom in* (or *zoom out*) on any part of a graph, allowing the *user* to approximate x-intercepts, y-intercepts, high points, low points, and other important aspects of the graph. Because of the great variety of graphing calculators and computer software, we shall not discuss methods of using any specific type. User's manuals usually give adequate instructions.

The term **graphing utility** refers to either a graphing calculator or a computer equipped with appropriate software packages. The **viewing rectangle** of a graphing utility is the portion of the xy-plane shown on the screen. The boundaries (sides) of the viewing rectangle can be manually set by assigning a minimum x value (Xmin), a maximum x value (Xmax), a minimum y value (Ymin), and a maximum y value (Ymax). In examples, we often use the standard (or default) values for the viewing rectangle. These values depend on the dimensions (measured in pixels) of the graphing utility screen. If we want a different view of the graph, we use the phrase "using [Xmin, Xmax] by [Ymin, Ymax]" to indicate the change in the viewing rectangle.

In many applications it is essential to find the points at which the graphs of two equations in x and y intersect. To approximate such points of intersection with a graphing utility, it is often necessary to solve each equation for y in terms of x. For example, suppose one equation is

$$4x^2 - 3x + 2y + 6 = 0.$$

Solving for y gives us

$$y = \frac{-4x^2 + 3x - 6}{2} = -2x^2 + \frac{3}{2}x - 3.$$

The graph of the equation is then found by *making the assignment*

$$Y_1 = -2x^2 + \tfrac{3}{2}x - 3$$

in the graphing utility. (The symbol Y_1 indicates the *first* equation, or the first y value.) We also solve the second equation for y in terms of x and make the assignment

$$Y_2 = \text{an expression in } x.$$

Pressing appropriate keys gives us sketches of the graphs, which we refer to as the graphs of Y_1 and Y_2. We then use a graphing utility feature such as *zoom* and *trace* (or simply zoom) or *intersect* to estimate the coordinates of the points of intersection. Both of these features are demonstrated for the TI-82/83 in Appendix I. When estimating coordinates in examples, we usually find one-decimal-place approximations unless otherwise specified.

In the next example we demonstrate this technique for the graphs discussed in Examples 2 and 3.

E X A M P L E   10   Estimating points of intersection of graphs

Use a graphing utility to estimate the points of intersection of the graphs of $y = 2x - 1$ and $y = x^2 - 3$.

Solution   Some specific keystrokes for the TI-82/83 are given in Example 5 of Appendix I. We first make the assignments

$$Y_1 = 2x - 1 \qquad \text{and} \qquad Y_2 = x^2 - 3.$$

Figure 22
[−15, 15] by [−10, 10]

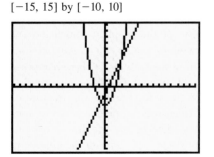

Using a standard viewing rectangle, [−15, 15] by [−10, 10], we see from the graphs of Y_1 and Y_2 in Figure 22 that there are two points of intersection: P_1 in quadrant I and P_2 in quadrant III.

We next either manually move the cursor or use a tracing feature (consult your user's manual for specific directions) to get close to P_1. Using the zoom or intersect feature, we estimate the coordinates of P_1 as (2.7, 4.5).

We can now either zoom out or redraw the original graphs to view P_2. Using the zoom or intersect feature, we obtain (−0.7, −2.5) as approximate coordinates of P_2.

E X A M P L E   11   Estimating points of intersection of graphs

Use a graphing utility to estimate the points of intersection of the circles $x^2 + y^2 = 25$ and $x^2 + y^2 - 4y = 12$.

Solution   Some specific keystrokes for the TI-82/83 are given in Example 8 of Appendix I. As in Example 10, we solve $x^2 + y^2 = 25$ for y in terms of x, obtaining

$$y = \pm\sqrt{25 - x^2}.$$

To display the entire circle on the graphing utility, we make the following assignments:

$$Y_1 = \sqrt{25 - x^2} \qquad \text{and} \qquad Y_2 = -Y_1$$

(We often assign Y_2 in terms of Y_1 to avoid repetitive keystroking.)

At this point you should graph Y_1 and Y_2 using a standard viewing rectangle. If the circle has an oval shape, consult your user's manual to find out which dimensions for the viewing rectangle will yield a more circular shape.

We may regard the equation of the second circle as a quadratic equation $ay^2 + by + c = 0$ in y by rearranging terms, as follows:

$$y^2 - 4y + (x^2 - 12) = 0$$

Applying the quadratic formula with $a = 1$, $b = -4$, and $c = x^2 - 12$ ($x^2 - 12$ is considered to be the constant term, since it does not contain the variable y) gives us

$$y = \frac{-(-4) \pm \sqrt{(-4)^2 - 4(1)(x^2 - 12)}}{2(1)}$$

$$= \frac{4 \pm \sqrt{16 - 4(x^2 - 12)}}{2} = \frac{4 \pm 2\sqrt{4 - (x^2 - 12)}}{2} = 2 \pm \sqrt{16 - x^2}.$$

(It is unnecessary to simplify the equation as much as we have, but the simplified form is easier to enter in a graphing utility.)

We now make the assignments

$$Y_3 = \sqrt{16 - x^2}, \qquad Y_4 = 2 + Y_3, \qquad \text{and} \qquad Y_5 = 2 - Y_3.$$

(If a Y_5 is not available on your graphing utility, consult the user's manual for information about displaying additional graphs on the screen.) We then select Y_1, Y_2, Y_4, and Y_5 to be graphed, obtaining a display similar to Figure 23. There are two points of intersection. Zooming in on the point in the first quadrant, we estimate its coordinates as (3.8, 3.25). Since both circles are symmetric with respect to the y-axis, the other point of intersection is approximately $(-3.8, 3.25)$.

Figure 23
[−15, 15] by [−10, 10]

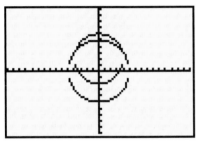

It should be noted that the approximate solutions found in Examples 10 and 11 do not satisfy the given equations because of the inaccuracy of the estimates made from the graph. To find the *exact* values for the points of intersection, we can *solve the system* using the *elimination* or *substitution* method, which you learned in a previous course.

1.2 EXERCISES

Exer. 1–2: Describe the set of all points $P(x, y)$ in a coordinate plane that satisfy the given condition.

1 (a) $x = -2$ (b) $y = 3$

(c) $x \geq 0$ (d) $xy > 0$

(e) $y < 0$ (f) $x = 0$

2 (a) $y = -2$ (b) $x = -4$

(c) $x/y < 0$ (d) $xy = 0$

(e) $y > 1$ (f) $y = 0$

Exer. 3–4: (a) Find the distance $d(A, B)$ between A and B. (b) Find the midpoint of the segment AB.

3 $A(4, -3)$, $B(6, 2)$ 4 $A(-2, -5)$, $B(4, 6)$

Exer. 5–6: Show that the triangle with vertices A, B, and C is a right triangle, and find its area.

5

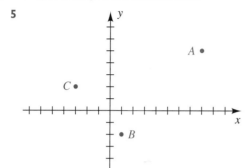

6

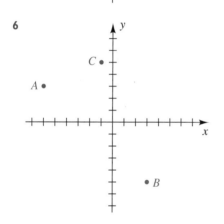

7 Show that $A(-4, 2)$, $B(1, 4)$, $C(3, -1)$, and $D(-2, -3)$ are vertices of a square.

8 Show that $A(-4, -1)$, $B(0, -2)$, $C(6, 1)$, and $D(2, 2)$ are vertices of a parallelogram.

9 Given $A(-3, 8)$, find the coordinates of the point B such that $C(5, -10)$ is the midpoint of segment AB.

10 Given $A(5, -8)$ and $B(-6, 2)$, find the point on segment AB that is three-fourths of the way from A to B.

Exer. 11–12: Prove that C is on the perpendicular bisector of segment AB.

11 $A(-4, -3)$, $B(6, 1)$, $C(5, -11)$

12 $A(-3, 2)$, $B(5, -4)$, $C(7, 7)$

13 Find all points on the y-axis that are a distance 6 from $P(5, 3)$.

14 Find all points on the x-axis that are a distance 5 from $P(-2, 4)$.

15 For what values of a is the distance between $P(a, 3)$ and $Q(5, 2a)$ greater than $\sqrt{26}$?

16 Find all points with coordinates of the form (a, a) that are a distance 3 from $P(-2, 1)$.

Exer. 17–36: Sketch the graph of the equation, and label the x- and y-intercepts.

17 $y = 2x - 3$ **18** $y = 3x + 2$

19 $y = -x + 1$ **20** $y = -2x - 3$

21 $y = -4x^2$ **22** $y = \frac{1}{3}x^2$

23 $y = 2x^2 - 1$ **24** $y = -x^2 + 2$

25 $x = \frac{1}{4}y^2$ **26** $x = -2y^2$

27 $x = -y^2 + 3$ **28** $x = 2y^2 - 4$

29 $y = -\frac{1}{2}x^3$ **30** $y = \frac{1}{2}x^3$

31 $y = x^3 - 8$ **32** $y = -x^3 + 1$

33 $y = \sqrt{x}$ **34** $y = \sqrt{-x}$

35 $y = \sqrt{x} - 4$ **36** $y = \sqrt{x - 4}$

Exer. 37–38: Use tests for symmetry to determine which graphs in the indicated exercises are symmetric with respect to (a) the y-axis, (b) the x-axis, and (c) the origin.

37 The odd-numbered exercises in 17–36

38 The even-numbered exercises in 17–36

Exer. 39–50: Sketch the graph of the circle or semicircle.

39 $x^2 + y^2 = 11$ **40** $x^2 + y^2 = 7$

41 $(x + 3)^2 + (y - 2)^2 = 9$

42 $(x - 4)^2 + (y + 2)^2 = 4$

43 $(x + 3)^2 + y^2 = 16$ **44** $x^2 + (y - 2)^2 = 25$

45 $4x^2 + 4y^2 = 25$ **46** $9x^2 + 9y^2 = 1$

47 $y = -\sqrt{16 - x^2}$ **48** $y = \sqrt{4 - x^2}$

49 $x = \sqrt{9 - y^2}$ **50** $x = -\sqrt{25 - y^2}$

Exer. 51–62: Find an equation of the circle that satisfies the stated conditions.

51 Center $C(2, -3)$, radius 5

52 Center $C(-4, 1)$, radius 3

53 Center $C\left(\frac{1}{4}, 0\right)$, radius $\sqrt{5}$

54 Center $C\left(\frac{3}{4}, -\frac{2}{3}\right)$, radius $3\sqrt{2}$

55 Center $C(-4, 6)$, passing through $P(1, 2)$

56 Center at the origin, passing through $P(4, -7)$

57 Center $C(-3, 6)$, tangent to the y-axis

58 Center $C(4, -1)$, tangent to the x-axis

59 Tangent to both axes, center in the second quadrant, radius 4

60 Tangent to both axes, center in the fourth quadrant, radius 3

61 Endpoints of a diameter $A(4, -3)$ and $B(-2, 7)$

62 Endpoints of a diameter $A(-5, 2)$ and $B(3, 6)$

Exer. 63–72: Find the center and radius of the circle with the given equation.

63 $x^2 + y^2 - 4x + 6y - 36 = 0$

64 $x^2 + y^2 + 8x - 10y + 37 = 0$

65 $x^2 + y^2 + 4y - 117 = 0$

66 $x^2 + y^2 - 10x + 18 = 0$

67 $2x^2 + 2y^2 - 12x + 4y - 15 = 0$

68 $9x^2 + 9y^2 + 12x - 6y + 4 = 0$

69 $x^2 + y^2 + 4x - 2y + 5 = 0$

70 $x^2 + y^2 - 6x + 4y + 13 = 0$

71 $x^2 + y^2 - 2x - 8y + 19 = 0$

72 $x^2 + y^2 + 4x + 6y + 16 = 0$

Exer. 73–76: Find equations for the upper half, lower half, right half, and left half of the circle.

73 $x^2 + y^2 = 36$ 74 $(x + 3)^2 + y^2 = 64$

75 $(x - 2)^2 + (y + 1)^2 = 49$

76 $(x - 3)^2 + (y - 5)^2 = 4$

Exer. 77–78: Determine whether the point P is inside, outside, or on the circle with center C and radius r.

77 (a) $P(2, 3)$, $C(4, 6)$, $r = 4$

   (b) $P(4, 2)$, $C(1, -2)$, $r = 5$

   (c) $P(-3, 5)$, $C(2, 1)$, $r = 6$

78 (a) $P(3, 8)$, $C(-2, -4)$, $r = 13$

   (b) $P(-2, 5)$, $C(3, 7)$, $r = 6$

   (c) $P(1, -2)$, $C(6, -7)$, $r = 7$

Exer. 79–80: For the given circle, find (a) the x-intercepts and (b) the y-intercepts.

79 $x^2 + y^2 - 4x - 6y + 4 = 0$

80 $x^2 + y^2 - 10x + 4y + 13 = 0$

81 Find an equation of the circle that is concentric with $x^2 + y^2 + 4x - 6y + 4 = 0$ and passes through the point $P(2, 6)$.

82 *Radio broadcasting ranges* The signal from a radio station has a circular range of 50 miles. A second radio station, located 100 miles east and 80 miles north of the first station, has a range of 80 miles. Are there locations where signals can be received from both radio stations? Explain your answer.

83 A circle C_1 of radius 5 has its center at the origin. Inside this circle is a first-quadrant circle C_2 of radius 2 that is tangent to C_1. The y-coordinate of the center of C_2 is 2. Find the x-coordinate of the center of C_2.

84 A circle C_1 of radius 5 has its center at the origin. Outside this circle is a first-quadrant circle C_2 of radius 2 that is tangent to C_1. The y-coordinate of the center of C_2 is 3. Find the x-coordinate of the center of C_2.

Exer. 85–88: The figure represents the screen of a graphing utility, where Y_1 and Y_2 are y-value assignments for two equations in x. Express, in interval form, the x-values such that $Y_1 < Y_2$ for the indicated viewing rectangle. Assume that the x- and y-values of each point of intersection are integers.

85
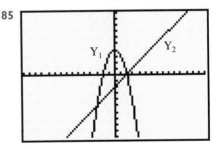

86
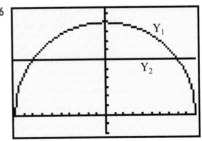

87

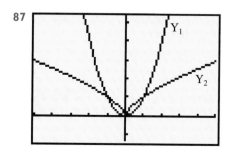

88

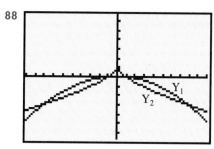

[c] 89 *Cable subscribers* The table lists the number of basic and pay cable television subscribers from 1990 to 1994. Some specific keystrokes for the TI-82/83 are given in Example 2 of Appendix I.

Year	Subscribers
1990	54,871,330
1991	55,786,390
1992	57,211,600
1993	58,834,440
1994	59,332,200

(a) Plot the data in the viewing rectangle [1988, 1996] by [54×10^6, 61×10^6].

(b) Discuss how the number of subscribers is changing.

[c] 90 *Published newspapers* The table lists the number of daily newspapers published in the United States for various years.

(a) Plot the data in the viewing rectangle [1895, 2000] by [0, 3000].

(b) Use the midpoint formula to estimate the number of newspapers in 1930. Compare your answer with the actual value of 1942.

Year	Newspapers
1900	2226
1920	2042
1940	1878
1960	1763
1980	1745
1993	1556

[c] 91 Graph the unit circle $x^2 + y^2 = 1$ using the equations $Y_1 = \sqrt{1 - x^2}$ and $Y_2 = -Y_1$ in the given viewing rectangles. Then discuss how the viewing rectangle affects the graph, and determine the viewing rectangle that results in a graph that most looks like a circle.

(1) [−2, 2] by [−2, 2]   (2) [−3, 3] by [−2, 2]
(3) [−2, 2] by [−5, 5]   (4) [−5, 5] by [−2, 2]

[c] 92 Graph the equation $|x| + |y| = 5$, using the equations $Y_1 = 5 - |x|$ and $Y_2 = -Y_1$ in the viewing rectangle [−5, 5] by [−5, 5].

(a) Find the number of x- and y-intercepts.

(b) Use the graph to determine the region where $|x| + |y| < 5$.

[c] Exer. 93–94: Graph the equation, and estimate the x-intercepts.

93 $y = x^3 - \frac{9}{10}x^2 - \frac{43}{25}x + \frac{24}{25}$

94 $y = x^4 + 0.85x^3 - 2.46x^2 - 1.07x + 0.51$

[c] Exer. 95–98: Graph the two equations on the same coordinate plane, and estimate the coordinates of their points of intersection.

95 $y = x^3 + x$; $x^2 + y^2 = 1$

96 $y = 3x^4 - \frac{3}{2}$; $x^2 + y^2 = 1$

97 $x^2 + (y - 1)^2 = 1$; $\left(x - \frac{5}{4}\right)^2 + y^2 = 1$

98 $(x + 1)^2 + (y - 1)^2 = \frac{1}{4}$; $\left(x + \frac{1}{2}\right)^2 + \left(y - \frac{1}{2}\right)^2 = 1$

[c] 99 *Distance between cars* The distance D (in miles) between two cars meeting on the same highway at time t (in minutes) is described by the equation $D = |2t - 4|$ on the interval [0, 4]. Graph D, and describe the motion of the cars.

c 100 *Water in a pool*   The amount of water A in a swimming pool on day x is given by $A = 12{,}000x - 2000x^2$, where A is in gallons and $x = 0$ corresponds to noon on Sunday. Graph A on the interval $[0, 6]$, and describe the amount of water in the pool.

c 101 *Speed of sound*   The speed of sound v in air varies with temperature. It can be calculated in ft/sec using the equation $v = 1087 \sqrt{\dfrac{T + 273}{273}}$, where T is temperature (in °C).

(a) Approximate v when $T = 20°C$.

(b) Determine the temperature to the nearest degree, both algebraically and graphically, when the speed of sound is 1000 ft/sec.

c 102 The area A of an equilateral triangle with a side of length s is $A = \dfrac{\sqrt{3}}{4}s^2$. Suppose that A must be equal to 100 ft^2 with an error of at most ± 1 ft^2. Determine graphically how accurately s must be measured in order to satisfy this error requirement. (*Hint:* Graph $y = A$, $y = 99$, and $y = 101$.)

1.3   FUNCTIONS

The notion of **correspondence** occurs frequently in everyday life. Some examples are given in the following illustration.

ILLUSTRATION

Correspondence

- To each book in a library there corresponds the number of pages in the book.
- To each human being there corresponds a birth date.
- If the temperature of the air is recorded throughout a day, then to each instant of time there corresponds a temperature.

Each correspondence in the previous illustration involves two sets, D and E. In the first illustration, D denotes the set of books in a library and E the set of positive integers. To each book x in D there corresponds a positive integer y in E—namely, the number of pages in the book.

We sometimes depict correspondences by diagrams of the type shown in Figure 24, where the sets D and E are represented by points within regions in a plane. The curved arrow indicates that the element y of E corresponds to the element x of D. The two sets may have elements in common. As a matter of fact, we often have $D = E$. It is important to note that *to each x in D there corresponds exactly one y in E.* However, the same element of E may correspond to different elements of D. For example, two books may have the same number of pages, two people may have the same birthday, and the temperature may be the same at different times.

In most of our work, D and E will be sets of numbers. To illustrate, let both D and E denote the set $\mathbb{R}$ of real numbers, and to each real number x let us assign its square x^2. This gives us a correspondence from $\mathbb{R}$ to $\mathbb{R}$.

Each of our illustrations of a correspondence is a *function,* which we define as follows.

Figure 24

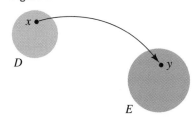

DEFINITION OF FUNCTION

> A **function** f from a set D to a set E is a correspondence that assigns to each element x of D exactly one element y of E.

The element y of E is the **value** of f at x (or the **image** of x under f) and is denoted by $f(x)$, read "f of x." The set D is the **domain** of the function. The **range** of f is the subset R of E consisting of all possible values $f(x)$ for x in D. Note that there may be elements in the set E that are not in the range R of f.

Consider the diagram in Figure 25. The curved arrows indicate that the elements $f(w)$, $f(z)$, $f(x)$, and $f(a)$ of E correspond to the elements w, z, x, and a of D. *To each element in D there is assigned exactly one function value in E;* however, different elements of D, such as w and z in Figure 25, may have the same value in E.

Figure 25

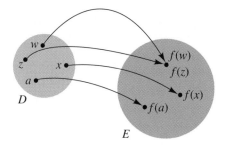

The symbols

$$D \xrightarrow{f} E, \qquad f: D \to E, \qquad \text{and}$$

signify that f is a function from D to E, and we say that f **maps** D *into* E. Initially, the notations f and $f(x)$ may be confusing. Remember that f is used to represent the function. It is neither in D nor in E. However, $f(x)$ is an element of the range R—the element that the function f assigns to the element x, which is in the domain D.

Two functions f and g from D to E are **equal**, and we write

$$f = g \quad \text{provided} \quad f(x) = g(x) \quad \text{for every } x \text{ in } D.$$

For example, if $g(x) = \frac{1}{2}(2x^2 - 6) + 3$ and $f(x) = x^2$ for every x in $\mathbb{R}$, then $g = f$.

E X A M P L E 1 Finding function values

Let f be the function with domain $\mathbb{R}$ such that $f(x) = x^2$ for every x in $\mathbb{R}$.

(a) Find $f(-6)$, $f(\sqrt{3})$, $f(a + b)$, and $f(a) + f(b)$, where a and b are real numbers.

(b) What is the range of f?

Solution

(a) We find values of f by substituting for x in the equation $f(x) = x^2$:

$$f(-6) = (-6)^2 = 36$$
$$f(\sqrt{3}) = (\sqrt{3})^2 = 3$$
$$f(a + b) = (a + b)^2 = a^2 + 2ab + b^2$$
$$f(a) + f(b) = a^2 + b^2$$

(b) By definition, the range of f consists of all numbers of the form $f(x) = x^2$ for x in $\mathbb{R}$. Since the square of every real number is nonnegative, the range is contained in the set of all nonnegative real numbers. Moreover, every nonnegative real number c is a value of f, since $f(\sqrt{c}) = (\sqrt{c})^2 = c$. Hence, the range of f is the set of all nonnegative real numbers.

If a function is defined as in Example 1, the symbols used for the function and variable are immaterial; that is, expressions such as $f(x) = x^2$, $f(s) = s^2$, $g(t) = t^2$, and $k(r) = r^2$ all define the same function. This is true because if a is any number in the domain, then the same value a^2 is obtained regardless of which expression is employed.

In the remainder of our work, the phrase f *is a function* will mean that the domain and range are sets of real numbers. If a function is defined by means of an expression, as in Example 1, and the domain D is not stated, then we will consider D to be the totality of real numbers x such that $f(x)$ is real. This is sometimes called the **implied domain** of f. To illustrate, if $f(x) = \sqrt{x - 2}$, then the implied domain is the set of real numbers x such that $\sqrt{x - 2}$ is real—that is, $x - 2 \geq 0$, or $x \geq 2$. Thus, the domain is the infinite interval $[2, \infty)$. If x is in the domain, we say that f *is defined at x* or that $f(x)$ *exists*. If a set S is contained in the domain, f *is defined on S*. The terminology f *is undefined at x* means that x is not in the domain of f.

E X A M P L E 2 Finding function values

Let $g(x) = \dfrac{\sqrt{4 + x}}{1 - x}$.

(a) Find the domain of g.

(b) Find $g(5)$, $g(-2)$, $g(-a)$, and $-g(a)$.

Solution

(a) The expression $\sqrt{4 + x}/(1 - x)$ is a real number if and only if the radicand $4 + x$ is nonnegative and the denominator $1 - x$ is not equal to 0. Thus, $g(x)$ exists if and only if

$$4 + x \geq 0 \qquad \text{and} \qquad 1 - x \neq 0$$

or, equivalently, $\qquad x \geq -4 \qquad$ and $\qquad x \neq 1.$

We may express the domain in terms of intervals as $[-4, 1) \cup (1, \infty)$.

(b) To find values of g, we substitute for x:

$$g(5) = \frac{\sqrt{4 + 5}}{1 - 5} = \frac{\sqrt{9}}{-4} = -\frac{3}{4}$$

$$g(-2) = \frac{\sqrt{4 + (-2)}}{1 - (-2)} = \frac{\sqrt{2}}{3}$$

$$g(-a) = \frac{\sqrt{4 + (-a)}}{1 - (-a)} = \frac{\sqrt{4 - a}}{1 + a}$$

$$-g(a) = -\frac{\sqrt{4 + a}}{1 - a} = \frac{\sqrt{4 + a}}{a - 1}$$

Of special interest in calculus is the expression

$$\frac{f(x + h) - f(x)}{h}$$

(with $h \neq 0$), which is commonly called a **difference quotient.**

E X A M P L E   3   **Simplifying a difference quotient**

Simplify the difference quotient

$$\frac{f(x + h) - f(x)}{h}$$

using the function $f(x) = x^2 + 6x - 4$.

Solution

$$\frac{f(x + h) - f(x)}{h} = \frac{[(x + h)^2 + 6(x + h) - 4] - [x^2 + 6x - 4]}{h}$$

definition of f

$$= \frac{(x^2 + 2xh + h^2 + 6x + 6h - 4) - (x^2 + 6x - 4)}{h}$$

expand numerator

$$= \frac{(x^2 + 2xh + h^2 + 6x + 6h - 4\!\!\!/) - (x^2 + 6x - 4\!\!\!/)}{h}$$

subtract terms

$$= \frac{2xh + h^2 + 6h}{h}$$

simplify

$$= \frac{h(2x + h + 6)}{h}$$

factor out h

$$= 2x + h + 6$$

cancel $h \neq 0$

Figure 26

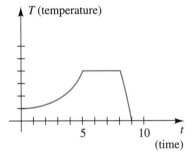

Graphs are often used to describe the variation of physical quantities. For example, a scientist may use the graph in Figure 26 to indicate the temperature T of a certain solution at various times t during an experiment. The sketch shows that the temperature increased gradually for time $t = 0$ to time $t = 5$, did not change between $t = 5$ and $t = 8$, and then decreased rapidly from $t = 8$ to $t = 9$.

Similarly, if f is a function, we may use a graph to indicate the change in $f(x)$ as x varies through the domain of f. Specifically, we have the following definition.

DEFINITION OF GRAPH OF A FUNCTION

The **graph of a function** f is the graph of the equation $y = f(x)$ for x in the domain of f.

We often attach the label $y = f(x)$ to a sketch of the graph. If $P(a, b)$ is a point on the graph, then the y-coordinate b is the function value $f(a)$, as illustrated in Figure 27. The figure displays the domain of f (the set of possible values of x) and the range of f (the corresponding values of y). Although we have pictured the domain and range as closed intervals, they may be infinite intervals or other sets of real numbers.

Since there is exactly one value $f(a)$ for each a in the domain of f, only *one* point on the graph of f has x-coordinate a. In general, we may use the following graphical test to determine whether a graph is the graph of a function.

Vertical Line Test

> The graph of a set of points in a coordinate plane is the graph of a function if every vertical line intersects the graph in at most one point.

Figure 27

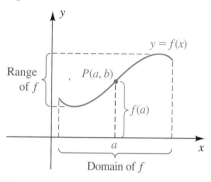

Thus, *every vertical line intersects the graph of a function in at most one point.* Consequently, the graph of a function cannot be a figure such as a circle, in which a vertical line may intersect the graph in more than one point.

The *x*-intercepts of the graph of a function *f* are the solutions of the equation $f(x) = 0$. These numbers are called the **zeros** of the function. The *y*-intercept of the graph is $f(0)$, if it exists.

EXAMPLE 4   Sketching the graph of a function

Let $f(x) = \sqrt{x - 1}$.

(a) Sketch the graph of *f*.

(b) Find the domain and range of *f*.

Figure 28

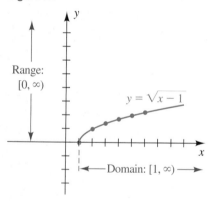

Solution

(a) By definition, the graph of *f* is the graph of the equation $y = \sqrt{x - 1}$. The following table lists coordinates of several points on the graph.

x	1	2	3	4	5	6
$y = f(x)$	0	1	$\sqrt{2} \approx 1.4$	$\sqrt{3} \approx 1.7$	2	$\sqrt{5} \approx 2.2$

Plotting points, we obtain the sketch shown in Figure 28. Note that the *x*-intercept is 1 and there is no *y*-intercept.

(b) Referring to Figure 28, note that the domain of *f* consists of all real numbers *x* such that $x \geq 1$ or, equivalently, the interval $[1, \infty)$. The range of *f* is the set of all real numbers *y* such that $y \geq 0$ or, equivalently, $[0, \infty)$.

The **square root function,** defined by $f(x) = \sqrt{x}$, has a graph similar to the one in Figure 28, but the endpoint is at $(0, 0)$. The *y*-value of a point on this graph is the number displayed on a calculator when a square root is requested. This graphical relationship may help you remember that $\sqrt{9}$ is 3 and that $\sqrt{9}$ is *not* ± 3. Similarly, $f(x) = x^2$, $f(x) = x^3$, and $f(x) = \sqrt[3]{x}$ are often referred to as the **squaring function**, the **cubing function**, and the **cube root function,** respectively.

In Example 4, as *x* increases, the function value $f(x)$ also increases, and we say that the graph of *f rises* (see Figure 28). A function of this type is said to be *increasing*. For certain functions, $f(x)$ decreases as *x* increases. In this case the graph *falls,* and *f* is a *decreasing* function. In general, we shall consider functions that increase or decrease on an interval *I*, as described in the following chart, where x_1 and x_2 denote numbers in *I*.

Increasing, Decreasing, and Constant Functions

Terminology	Definition	Graphical interpretation
f is **increasing** on an interval *I*	$f(x_1) < f(x_2)$ whenever $x_1 < x_2$	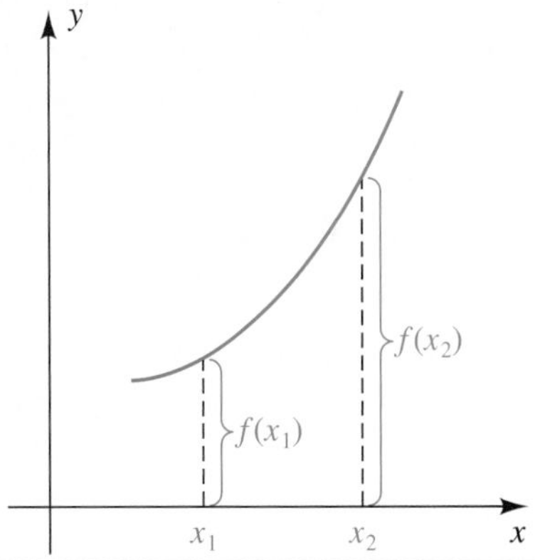
f is **decreasing** on an interval *I*	$f(x_1) > f(x_2)$ whenever $x_1 < x_2$	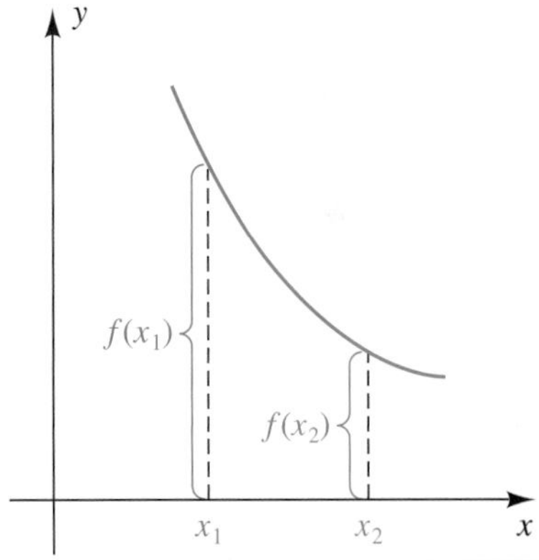
f is **constant** on an interval *I*	$f(x_1) = f(x_2)$ for every x_1 and x_2	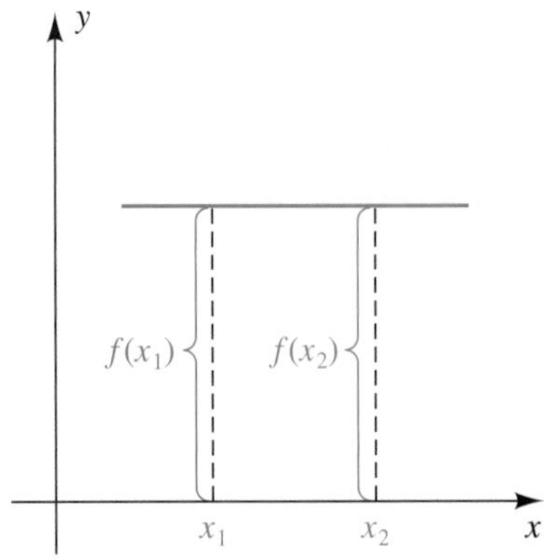

An example of an *increasing function* is the **identity function,** whose equation is $f(x) = x$ and whose graph is the line through the origin with slope 1. An example of a *decreasing function* is $f(x) = -x$, an equation of the line through the origin with slope -1. If $f(x) = c$ for every real number x, then f is called a *constant function.*

We shall use the phrases f *is increasing* and $f(x)$ *is increasing* interchangeably. We shall do the same with the terms *decreasing* and *constant.*

EXAMPLE 5   Using a graph to find domain, range, and where a function increases or decreases

Let $f(x) = \sqrt{9 - x^2}$.

(a) Sketch the graph of f.

(b) Find the domain and range of f.

(c) Find the intervals on which f is increasing or is decreasing.

Solution

(a) By definition, the graph of f is the graph of the equation $y = \sqrt{9 - x^2}$. We know from our work with circles in Section 1.2 that the graph of $x^2 + y^2 = 9$ is a circle of radius 3 with center at the origin. Solving the equation $x^2 + y^2 = 9$ for y gives us $y = \pm\sqrt{9 - x^2}$. It follows that the graph of f is the *upper half* of the circle, as illustrated in Figure 29.

(b) Referring to Figure 29, we see that the domain of f is the closed interval $[-3, 3]$, and the range of f is the interval $[0, 3]$.

(c) The graph rises as x increases from -3 to 0, so f is increasing on the closed interval $[-3, 0]$. Thus, as shown in the preceding chart, if $x_1 < x_2$ in $[-3, 0]$, then $f(x_1) < f(x_2)$ (note that *possibly* $x_1 = -3$ or $x_2 = 0$).

The graph falls as x increases from 0 to 3, so f is decreasing on the closed interval $[0, 3]$. In this case, the chart indicates that if $x_1 < x_2$ in $[0, 3]$, then $f(x_1) > f(x_2)$ (note that *possibly* $x_1 = 0$ or $x_2 = 3$).

Figure 29

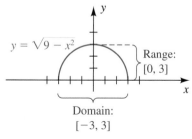

$y = \sqrt{9 - x^2}$

Range: [0, 3]

Domain: [−3, 3]

The following type of function is one of the most basic in algebra.

DEFINITION OF LINEAR FUNCTION

A function f is a **linear function** if

$$f(x) = ax + b,$$

where x is any real number and a and b are constants.

The graph of f in the preceding definition is the graph of $y = ax + b$, which, by the slope-intercept form, is a line with slope a and y-intercept b. Thus, *the graph of a linear function is a line.* Since $f(x)$ exists for every x, the domain of f is $\mathbb{R}$. As illustrated in the next example, if $a \neq 0$, then the range of f is also $\mathbb{R}$.

E X A M P L E   6   Sketching the graph of a linear function

Let $f(x) = 2x + 3$.

(a) Sketch the graph of f.

(b) Find the domain and range of f.

(c) Determine where f is increasing or is decreasing.

Solution

(a) Since $f(x)$ has the form $ax + b$, with $a = 2$ and $b = 3$, f is a linear function. The graph of $y = 2x + 3$ is the line with slope 2 and y-intercept 3, illustrated in Figure 30.

(b) We see from the graph that x and y may be any real numbers, so both the domain and the range of f are $\mathbb{R}$.

(c) Since the slope a is positive, the graph of f rises as x increases; that is, $f(x_1) < f(x_2)$ whenever $x_1 < x_2$. Thus, f is increasing throughout its domain.

Figure 30

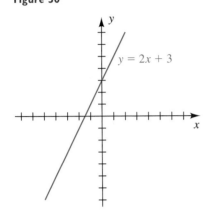

$y = 2x + 3$

In applications it is sometimes necessary to determine a specific linear function from given data, as in the next example.

E X A M P L E   7   Finding a linear function

If f is a linear function such that $f(-2) = 5$ and $f(6) = 3$, find $f(x)$, where x is any real number.

Solution   By the definition of linear function, $f(x) = ax + b$, where a and b are constants. Moreover, the given function values tell us that the points $(-2, 5)$ and $(6, 3)$ are on the graph of f—that is, on the line $y = ax + b$ illustrated in Figure 31. The slope a of this line is

$$a = \frac{5 - 3}{-2 - 6} = \frac{2}{-8} = -\frac{1}{4},$$

and hence $f(x)$ has the form

$$f(x) = -\tfrac{1}{4}x + b.$$

To find the value of b, we may use the fact that $f(6) = 3$, as follows:

$$f(6) = -\tfrac{1}{4}(6) + b \quad \text{let } x = 6 \text{ in } f(x) = -\tfrac{1}{4}x + b$$
$$3 = -\tfrac{3}{2} + b \quad\quad f(6) = 3$$
$$b = 3 + \tfrac{3}{2} = \tfrac{9}{2} \quad \text{solve for } b$$

Thus, the linear function satisfying $f(-2) = 5$ and $f(6) = 3$ is

$$f(x) = -\tfrac{1}{4}x + \tfrac{9}{2}.$$

Figure 31

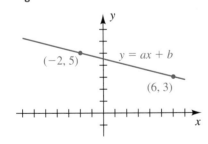

$(-2, 5)$ $y = ax + b$

$(6, 3)$

Many formulas that occur in mathematics and the sciences determine functions. For instance, the formula $A = \pi r^2$ for the area A of a circle of radius r assigns to each positive real number r exactly one value of A. This determines a function f such that $f(r) = \pi r^2$, and we may write $A = f(r)$. The letter r, which represents an arbitrary number from the domain of f, is called an **independent variable.** The letter A, which represents a number from the range of f, is a **dependent variable,** since its value depends on the number assigned to r. If two variables r and A are related in this manner, we say that A *is a function of r.* In applications, the independent variable and dependent variable are sometimes referred to as the **input variable** and **output variable,** respectively. As another example, if an automobile travels at a uniform rate of 50 mi/hr, then the distance d (miles) traveled in time t (hours) is given by $d = 50t$, and hence *the distance d is a function of time t.*

EXAMPLE 8   Expressing the volume of a tank as a function of its radius

A steel storage tank for propane gas is to be constructed in the shape of a right circular cylinder of altitude 10 feet with a hemisphere attached to each end. The radius r is yet to be determined. Express the volume V (in ft^3) of the tank as a function of r (in feet).

Solution   The tank is illustrated in Figure 32. We may find the volume of the cylindrical part of the tank by multiplying the altitude 10 by the area πr^2 of the base of the cylinder. This gives us

$$\text{volume of cylinder} = 10(\pi r^2) = 10\pi r^2.$$

The two hemispherical ends, taken together, form a sphere of radius r. Using the formula for the volume of a sphere, we obtain

$$\text{volume of the two ends} = \tfrac{4}{3}\pi r^3.$$

Thus, the volume V of the tank is

$$V = \tfrac{4}{3}\pi r^3 + 10\pi r^2.$$

This formula expresses V as a function of r. In factored form,

$$V = \tfrac{1}{3}\pi r^2(4r + 30) = \tfrac{2}{3}\pi r^2(2r + 15).$$

Figure 32

Figure 33

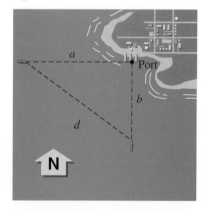

EXAMPLE 9   Expressing a distance as a function of time

Two ships leave port at the same time, one sailing west at a rate of 17 mi/hr and the other sailing south at 12 mi/hr. If t is the time (in hours) after their departure, express the distance d between the ships as a function of t.

Solution   To help visualize the problem, we begin by drawing a picture and labeling it, as in Figure 33. By the Pythagorean theorem,

$$d^2 = a^2 + b^2, \quad \text{or} \quad d = \sqrt{a^2 + b^2}.$$

Since distance = (rate)(time) and the rates are 17 and 12, respectively,

$$a = 17t \quad \text{and} \quad b = 12t.$$

Substitution in $d = \sqrt{a^2 + b^2}$ gives us

$$d = \sqrt{(17t)^2 + (12t)^2} = \sqrt{289t^2 + 144t^2} = \sqrt{433t^2} \approx (20.8)t.$$

Ordered pairs can be used to obtain an alternative approach to functions. We first observe that a function f from D to E determines the following set W of ordered pairs:

$$W = \{(x, f(x)): x \text{ is in } D\}$$

Thus, W consists of all ordered pairs such that the first number x is in D and the second number is the function value $f(x)$. In Example 1, where $f(x) = x^2$, W is the set of all ordered pairs of the form (x, x^2). It is important to note that, *for each x, there is exactly one ordered pair (x, y) in W having x in the first position.*

Conversely, if we begin with a set W of ordered pairs such that each x in D appears exactly once in the first position of an ordered pair, then W determines a function. Specifically, for each x in D there is exactly one pair (x, y) in W, and by letting y correspond to x, we obtain a function with domain D. The range consists of all real numbers y that appear in the second position of the ordered pairs.

It follows from the preceding discussion that the next statement could also be used as a definition of function.

ALTERNATIVE DEFINITION OF FUNCTION

A **function** with domain D is a set W of ordered pairs such that, for each x in D, there is exactly one ordered pair (x, y) in W having x in the first position.

In terms of the preceding definition, the ordered pairs $(x, \sqrt{x - 1})$ determine the function of Example 4 given by $f(x) = \sqrt{x - 1}$. Note, however, that if

$$W = \{(x, y): x^2 = y^2\},$$

then W is *not* a function, since for a given x there may be more than one pair in W with x in the first position. For example, if $x = 2$, then both $(2, 2)$ and $(2, -2)$ are in W.

In some scientific investigations, the terminology of *variation* or *proportion* is used to describe relationships between variable quantities. In the following chart, k is a nonzero real number called a **constant of variation** or a **constant of proportionality**.

Terminology	General formula	Illustration
y **varies directly** as x, or y is **directly proportional** to x	$y = kx$	$C = 2\pi r$, where C is the circumference of a circle, r is the radius, and $k = 2\pi$
y **varies inversely** as x, or y is **inversely proportional** to x	$y = \dfrac{k}{x}$	$I = \dfrac{110}{R}$, where I is the current in an electrical circuit, R is the resistance, and $k = 110$ is the voltage

The variable x in the chart can also represent a power. For example, the formula $A = \pi r^2$ states that the area A of a circle varies directly as the *square* of the radius r, where π is the constant of variation. Similarly, the formula $V = \frac{4}{3}\pi r^3$ states that the volume V of a sphere is directly proportional to the *cube* of the radius. In this case the constant of proportionality is $\frac{4}{3}\pi$.

E X A M P L E   10   Finding the support load of a rectangular beam

The weight W that can be safely supported by a beam with a rectangular cross section varies directly as the product of the width w and square of the depth d of the cross section and inversely as the length l of the beam. If a 2-inch by 4-inch beam that is 8 feet long safely supports a load of 500 pounds, what weight can be safely supported by a 2-inch by 8-inch beam that is 10 feet long? (Assume that the width is the *shorter* dimension of the cross section.)

Solution   A general formula for the weight W is

$$W = k\frac{wd^2}{l},$$

where k is a constant of variation. Using the given data—that is, $w = 2$, $d = 4$, $l = 8$, and $W = 500$—we substitute and determine the value of k as follows:

$$500 = k\frac{2(4^2)}{8}, \qquad \text{or, equivalently,} \qquad k = 125$$

To answer the question, we substitute $k = 125$, $w = 2$, $d = 8$, and $l = 10$ into the general formula for W, obtaining

$$W = 125\frac{2 \cdot 8^2}{10} = 1600 \text{ lb.}$$

In the next example we illustrate how some of the concepts presented in this section may be studied with the aid of a graphing utility. Hereafter, when making assignments on a graphing utility, we will frequently refer to variables such as Y_1 and Y_2 as the *functions* Y_1 and Y_2.

EXAMPLE 11   Analyzing the graph of a function

Let $f(x) = x^{2/3} - 3$.

(a) Find $f(-2)$.

(b) Sketch the graph of f.

(c) State the domain and range of f.

(d) State the intervals on which f is increasing or is decreasing.

(e) Estimate the x-intercepts of the graph to one-decimal-place accuracy.

Solution   Some specific keystrokes for parts (a) and (b) using the TI-82/83 are given in Example 9 of Appendix I.

(a) A representation of f on a computational device may have the form

$$X^\wedge(2/3) - 3 \quad \text{or} \quad (X^\wedge(1/3))^\wedge 2 - 3 \quad \text{or} \quad (X^\wedge 2)^\wedge(1/3) - 3.$$

We assign one of these expressions to the function Y_1. To find the value of Y_1 at $x = -2$, we first assign the value -2 to a memory location identified as X. This is usually done with a "store" or "assign" operation on a computational device. We next determine the value of Y_1 by requesting the computational device to indicate the contents of the memory location that contains the value of Y_1. We often refer to this process of finding a value of Y_1 as "querying Y_1." Upon querying Y_1, we see that its value is approximately -1.41 (that is, $f(-2) \approx -1.41$). It should be noted that not all computational devices process rational exponents in the same fashion. If you did not obtain this answer for $f(-2)$, change your representation of Y_1 before proceeding.

(b) Using the viewing rectangle $[-15, 15]$ by $[-10, 10]$ to graph Y_1 gives us a display similar to that of Figure 34. The v-shaped part of the graph of f at $x = 0$ is called a **cusp**.

(c) The domain of f is $\mathbb{R}$, since we may input any value for x. The figure indicates that $y \geq -3$, so we conclude that the range of f is $[-3, \infty)$.

(d) From the figure, we see that f is decreasing on $(-\infty, 0]$ and is increasing on $[0, \infty)$.

(e) Using the zoom or root feature, we find that the positive x-intercept in Figure 34 is approximately 5.2. Since f is symmetric with respect to the y-axis, the negative x-intercept is about -5.2.

Figure 34
$[-15, 15]$ by $[-10, 10]$

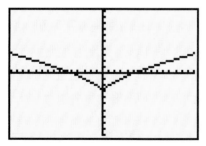

As a reference aid, some common graphs and their equations are listed in Appendix II. Many of these graphs are graphs of functions.

1.3 EXERCISES

1 If $f(x) = -x^2 - x - 4$, find $f(-2)$, $f(0)$, and $f(4)$.

2 If $f(x) = -x^3 - x^2 + 3$, find $f(-3)$, $f(0)$, and $f(2)$.

3 If $f(x) = \sqrt{x - 4} - 3x$, find $f(4)$, $f(8)$, and $f(13)$.

4 If $f(x) = \dfrac{x}{x - 3}$, find $f(-2)$, $f(0)$, and $f(3)$.

Exer. 5–8: If a and h are real numbers, find

(a) $f(a)$ (b) $f(-a)$ (c) $-f(a)$ (d) $f(a + h)$

(e) $f(a) + f(h)$ (f) $\dfrac{f(a + h) - f(a)}{h}$ if $h \neq 0$

5 $f(x) = 5x - 2$ 6 $f(x) = 3 - 4x$

7 $f(x) = x^2 - x + 3$ 8 $f(x) = 2x^2 + 3x - 7$

Exer. 9–10: Simplify the difference quotient
$\dfrac{f(x + h) - f(x)}{h}$ if $h \neq 0$.

9 $f(x) = x^2 + 5$ 10 $f(x) = 1/x^2$

Exer. 11–12: Simplify the difference quotient
$\dfrac{f(x) - f(a)}{x - a}$ if $x \neq a$.

11 $f(x) = \sqrt{x - 3}$ (*Hint:* Rationalize the numerator.)

12 $f(x) = x^3 - 2$

Exer. 13–16: If a is a positive real number, find

(a) $g\left(\dfrac{1}{a}\right)$ (b) $\dfrac{1}{g(a)}$ (c) $g(\sqrt{a})$ (d) $\sqrt{g(a)}$

13 $g(x) = 4x^2$ 14 $g(x) = 2x - 5$

15 $g(x) = \dfrac{2x}{x^2 + 1}$ 16 $g(x) = \dfrac{x^2}{x + 1}$

Exer. 17–18: For the graph of the function f sketched in the figure, determine

(a) the domain   (b) the range   (c) $f(1)$

(d) all x such that $f(x) = 1$

(e) all x such that $f(x) > 1$

17

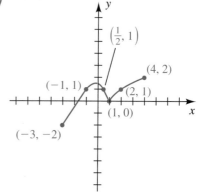

18

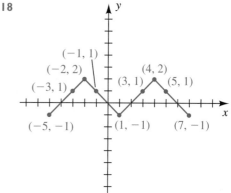

Exer. 19–30: Find the domain of f.

19 $f(x) = \sqrt{2x + 7}$ 20 $f(x) = \sqrt{8 - 3x}$

21 $f(x) = \sqrt{9 - x^2}$ 22 $f(x) = \sqrt{x^2 - 25}$

23 $f(x) = \dfrac{x + 1}{x^3 - 4x}$ 24 $f(x) = \dfrac{4x}{6x^2 + 13x - 5}$

25 $f(x) = \dfrac{\sqrt{2x - 3}}{x^2 - 5x + 4}$ 26 $f(x) = \dfrac{\sqrt{4x - 3}}{x^2 - 4}$

27 $f(x) = \dfrac{x - 4}{\sqrt{x - 2}}$ 28 $f(x) = \dfrac{1}{(x - 3)\sqrt{x + 3}}$

29 $f(x) = \sqrt{x + 2} + \sqrt{2 - x}$

30 $f(x) = \sqrt{(x - 2)(x - 6)}$

Exer. 31–40: (a) Sketch the graph of f. (b) Find the domain D and range R of f. (c) Find the intervals on which f is increasing, is decreasing, or is constant.

31 $f(x) = 3x - 2$

32 $f(x) = -2x + 3$

33 $f(x) = 4 - x^2$

34 $f(x) = x^2 - 1$

35 $f(x) = \sqrt{x + 4}$

36 $f(x) = \sqrt{4 - x}$

37 $f(x) = -2$

38 $f(x) = 3$

39 $f(x) = -\sqrt{36 - x^2}$

40 $f(x) = \sqrt{16 - x^2}$

Exer. 41–42: If a linear function f satisfies the given conditions, find f(x).

41 $f(-3) = 1$ and $f(3) = 2$

42 $f(-2) = 7$ and $f(4) = -2$

Exer. 43–52: Determine whether the set W of ordered pairs is a function in the sense of the alternative definition of function on page 43.

43 $W = \{(x, y): 2y = x^2 + 5\}$

44 $W = \{(x, y): x = 3y + 2\}$

45 $W = \{(x, y): x^2 + y^2 = 4\}$

46 $W = \{(x, y): y^2 - x^2 = 1\}$

47 $W = \{(x, y): y = 3\}$ **48** $W = \{(x, y): x = 3\}$

49 $W = \{(x, y): xy = 0\}$

50 $W = \{(x, y): x + y = 0\}$

51 $W = \{(x, y): |y| = |x|\}$ **52** $W = \{(x, y): y < x\}$

53 *Constructing a box* From a rectangular piece of cardboard having dimensions 20 inches × 30 inches, an open box is to be made by cutting out an identical square of area x^2 from each corner and turning up the sides (see the figure). Express the volume V of the box as a function of x.

Exercise 53

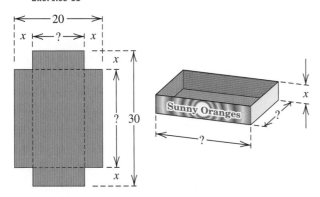

54 *Constructing a storage tank* Refer to Example 8. A steel storage tank for propane gas is to be constructed in the shape of a right circular cylinder of altitude 10 feet with a hemisphere attached to each end. The radius r is yet to be determined. Express the surface area S of the tank as a function of r.

55 *Dimensions of a building* A small office unit is to contain 500 ft² of floor space. A simplified model is shown in the figure.

(a) Express the length y of the building as a function of the width x.

(b) If the walls cost $100 per running foot, express the cost C of the walls as a function of the width x. (Disregard the wall space above the doors and the thickness of the walls.)

Exercise 55

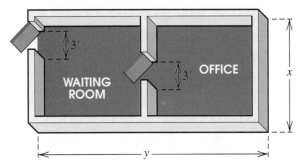

56 *Dimensions of an aquarium* An aquarium of height 1.5 feet is to have a volume of 6 ft³. Let x denote the length of the base and y the width (see the figure).

(a) Express y as a function of x.

(b) Express the total number S of square feet of glass needed as a function of x.

Exercise 56

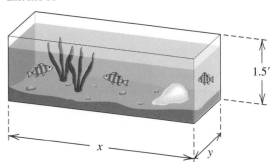

57 *Skyline ordinance* A city council is proposing a new skyline ordinance. It would require the setback S for any building from a residence to be a minimum of 100 feet, plus an additional 6 feet for each foot of height above 25 feet. Find a linear function for S in terms of h.

Exercise 57

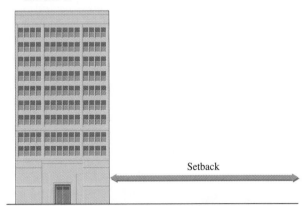

Setback

58 *Energy tax* A proposed energy tax T on gasoline, which would affect the cost of driving a vehicle, is to be computed by multiplying the number x of gallons of gasoline that you buy by 125,000 (the number of BTUs per gallon of gasoline) and then multiplying the total BTUs by the tax—34.2 cents per million BTUs. Find a linear function for T in terms of x.

59 *Childhood growth* For children between ages 6 and 10, height y (in inches) is frequently a linear function of age t (in years). The height of a certain child is 48 inches at age 6 and 50.5 inches at age 7.

(a) Express y as a function of t.

(b) Sketch the line in part (a), and interpret the slope.

(c) Predict the height of the child at age 10.

60 *Radioactive contamination* It has been estimated that 1000 curies of a radioactive substance introduced at a point on the surface of the open sea would spread over an area of 40,000 km² in 40 days. Assuming that the area covered by the radioactive substance is a linear function of time t and is always circular in shape, express the radius r of the contamination as a function of t.

61 *Distance to a hot-air balloon* A hot-air balloon is released at 1:00 P.M. and rises vertically at a rate of 2 m/sec. An observation point is situated 100 meters from a point on the ground directly below the balloon (see the figure). If t denotes the time (in seconds) after

1:00 P.M., express the distance d between the balloon and the observation point as a function of t.

Exercise 61

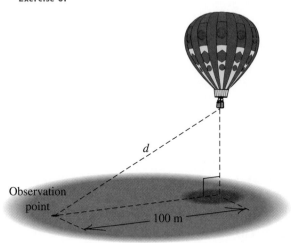

Observation point

100 m

62 Triangle ABC is inscribed in a semicircle of diameter 15 (see the figure).

(a) If x denotes the length of side AC, express the length y of side BC as a function of x. (*Hint:* Angle ACB is a right angle.)

(b) Express the area $\mathcal{A}$ of triangle ABC as a function of x, and state the domain of this function.

Exercise 62

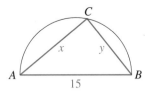

63 *Distance to the earth* From an exterior point P that is h units from a circle of radius r, a tangent line is drawn to the circle (see the figure). Let y denote the distance from the point P to the point of tangency T.

(a) Express y as a function of h. (*Hint:* If C is the center of the circle, then PT is perpendicular to CT.)

(b) If r is the radius of the earth and h is the altitude of a space shuttle, then y is the maximum distance to the earth that an astronaut can see from the shuttle. In particular, if $h = 200$ mi and $r \approx 4000$ mi, approximate y.

Exercise 63

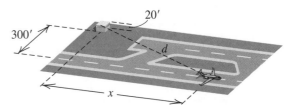

64 *Length of a tightrope* The figure illustrates the apparatus for a tightrope walker. Two poles are set 50 feet apart, but the point of attachment *P* for the rope is yet to be determined.

(a) Express the length *L* of the rope as a function of the distance *x* from *P* to the ground.

(b) If the total walk is to be 75 feet, determine the distance from *P* to the ground.

Exercise 64

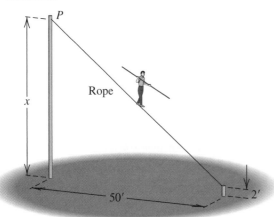

65 *Airport runway* The relative positions of an aircraft runway and a 20-foot-tall control tower are shown in the figure. The beginning of the runway is at a perpendicular distance of 300 feet from the base of the tower. If *x* denotes the distance an airplane has moved down the runway, express the distance *d* between the airplane and the control booth as a function of *x*.

Exercise 65

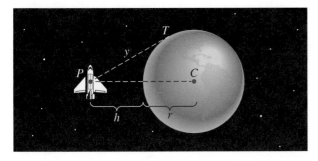

66 *Destination time* A man in a rowboat that is 2 miles from the nearest point *A* on a straight shoreline wishes to reach a house located at a point *B* that is 6 miles farther down the shoreline (see the figure). He plans to row to a point *P* that is between *A* and *B* and is *x* miles from the house, and then he will walk the remainder of the distance. Suppose he can row at a rate of 3 mi/hr and can walk at a rate of 5 mi/hr. If *T* is the total time required to reach the house, express *T* as a function of *x*.

Exercise 66

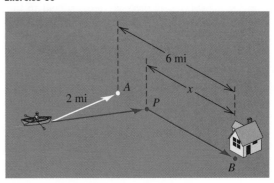

67 *Hooke's law* Hooke's law states that the force *F* required to stretch a spring *x* units beyond its natural length is directly proportional to *x*. If a weight of 4 pounds stretches a certain spring from its natural length of 10 inches to a length of 10.3 inches, what weight will stretch the spring to a length of 11.5 inches?

68 *Period of a planet* Kepler's third law states that the period *T* of a planet (the time needed to make one complete revolution about the sun) is directly proportional to the $\frac{3}{2}$ power of its average distance *d* (in millions of miles) from the sun. For the planet Earth, $T = 365$ days and $d = 93$. Estimate the period of Venus if its average distance from the sun is 67 million miles.

69 *Electrical resistance*   The electrical resistance R of a wire varies directly as its length l and inversely as the square of its diameter d. A wire 100 feet long of diameter 0.01 inch has a resistance of 25 ohms. Find the resistance of a wire made of the same material that has a diameter of 0.015 inch and is 50 feet long.

70 *Coulomb's law*   Coulomb's law in electrical theory states that the force F of attraction between two oppositely charged particles varies directly as the product of the magnitudes Q_1 and Q_2 of the charges and inversely as the square of the distance d between the particles. What is the effect of reducing the distance between the particles by a factor of one-fourth?

71 *Density at a point*   A thin flat plate is situated in an xy-plane such that the density d (in lb/ft^2) at the point $P(x, y)$ is inversely proportional to the square of the distance from the origin. What is the effect on the density at P if the x- and y-coordinates are each multiplied by $\frac{1}{3}$?

72 *Temperature at a point*   A flat metal plate is positioned in an xy-plane such that the temperature T (in °C) at the point (x, y) is inversely proportional to the distance from the origin. If the temperature at the point $P(3, 4)$ is 20°C, find the temperature at the point $Q(24, 7)$.

c **Exer. 73–76: (a) Sketch the graph of f on the given interval $[a, b]$. (b) Estimate the range of f on $[a, b]$. (c) Estimate the intervals on which f is increasing or is decreasing.**

73 $f(x) = \dfrac{x^{1/3}}{1 + x^4};$ $[-2, 2]$

74 $f(x) = x^4 - 0.4x^3 - 0.8x^2 + 0.2x + 0.1;$ $[-1, 1]$

75 $f(x) = x^5 - 3x^2 + 1;$ $[-0.7, 1.4]$

76 $f(x) = \dfrac{1 - x^3}{1 + x^4};$ $[-4, 4]$

c **Exer. 77–78: Solve the equation graphically by assigning the expression on the left side to Y_1 and the number on the right side to Y_2 and then finding the x-coordinates of all points of intersection of the two graphs.**

77 (a) $x^{5/3} = 32$ (b) $x^{4/3} = 16$ (c) $x^{2/3} = -36$

  (d) $x^{3/4} = 125$ (e) $x^{3/2} = -27$

78 (a) $x^{3/5} = -27$ (b) $x^{2/3} = 25$ (c) $x^{4/3} = -49$

  (d) $x^{3/2} = 27$ (e) $x^{3/4} = -8$

c **79** *New car prices*   In 1985 the average price paid for a new car was \$11,450. The average price increased linearly to \$20,021 in 1994.

  (a) Find a function f that models the average price paid for a new car. Graph f together with the two data points.

  (b) Interpret the slope of the graph of f.

  (c) Graphically approximate the year when the average price paid will be \$25,000.

80 *Calculator screen*   A particular graphing calculator screen is 95 pixels wide and 63 pixels high.

  (a) Find the total number of pixels in the screen.

  (b) If a function is graphed in dot mode, determine the maximum number of pixels that would typically be darkened on the calculator screen.

c **81** *Stopping distances*   The table lists the practical stopping distances D (in feet) for cars at speeds S (in miles per hour) on level surfaces, as used by the American Association of State Highway and Transportation Officials.

S	20	30	40	50	60	70
D	33	86	167	278	414	593

  (a) Plot the data.

  (b) Determine whether stopping distance is a linear function of speed.

  (c) Discuss the practical implications of these data for safely driving a car.

c **82** *Stopping distances*   Refer to Exercise 81. The distance D (in feet) required for a car to safely stop varies directly with its speed S (in mi/hr).

  (a) Use the table to determine an approximate value for k in the variation formula $D = kS^{2.3}$.

  (b) Check your approximation by graphing both the data and D on the same coordinate axes.

1.4 GRAPHS OF FUNCTIONS

In this section we discuss aids for sketching graphs of certain types of functions. In particular, a function f is called *even* if $f(-x) = f(x)$ for every x in its domain. In this case, the equation $y = f(x)$ is not changed if $-x$ is substituted for x, and hence, from symmetry test 1 of Section 1.2, the graph of an even function is symmetric with respect to the y-axis.

A function f is called *odd* if $f(-x) = -f(x)$ for every x in its domain. If we apply symmetry test 3 of Section 1.2 to the equation $y = f(x)$, we see that the graph of an odd function is symmetric with respect to the origin.

These facts are summarized in the first two columns of the next chart.

Even and Odd Functions

Terminology	Definition	Illustration	Symmetry of graph
f is an **even function.**	$f(-x) = f(x)$ for every x in the domain.	$y = f(x) = x^2$	y-axis
f is an **odd function.**	$f(-x) = -f(x)$ for every x in the domain.	$y = f(x) = x^3$	the origin

E X A M P L E   1   Determining whether a function is even or odd

Determine whether f is even, odd, or neither even nor odd.

(a) $f(x) = 3x^4 - 2x^2 + 5$ **(b)** $f(x) = 2x^5 - 7x^3 + 4x$

(c) $f(x) = x^3 + x^2$

Solution   In each case the domain of f is $\mathbb{R}$. To determine whether f is even or odd, we begin by examining $f(-x)$, where x is any real number.

(a) $f(-x) = 3(-x)^4 - 2(-x)^2 + 5$ substitute $-x$ for x in $f(x)$

$\qquad\qquad = 3x^4 - 2x^2 + 5$ simplify

$\qquad\qquad = f(x)$ definition of f

Since $f(-x) = f(x)$, f is an even function.

(b) $f(-x) = 2(-x)^5 - 7(-x)^3 + 4(-x)$ substitute $-x$ for x in $f(x)$

$\qquad\qquad = -2x^5 + 7x^3 - 4x$ simplify

$\qquad\qquad = -(2x^5 - 7x^3 + 4x)$ factor out -1

$\qquad\qquad = -f(x)$ definition of f

Since $f(-x) = -f(x)$, f is an odd function.

(continued)

(c) $f(-x) = (-x)^3 + (-x)^2$ substitute $-x$ for x in $f(x)$

  $= -x^3 + x^2$ simplify

Since $f(-x) \neq f(x)$, and $f(-x) \neq -f(x)$ (note that $-f(x) = -x^3 - x^2$), the function f is neither even nor odd.

A function f is a **polynomial function** if $f(x)$ is a polynomial—that is, if

$$f(x) = a_n x^n + a_{n-1} x^{n-1} + \cdots + a_1 x + a_0,$$

where the coefficients $a_0, a_1, \ldots, a_n$ are real numbers and the exponents are nonnegative integers. A polynomial function may be regarded as a sum of functions whose values are of the form cx^k, where c is a real number and k is a nonnegative integer. The three functions in Example 1 are polynomial functions. All polynomial functions are **continuous functions**—that is, their graphs can be drawn without any breaks.

An **algebraic function** is a function that can be expressed in terms of finite sums, differences, products, quotients, or roots of polynomial functions.

ILLUSTRATION

Algebraic Function

$$f(x) = 5x^4 - 2\sqrt[3]{x} + \frac{x(x^2 + 5)}{\sqrt{x^3 + \sqrt{x}}}$$

Functions that are not algebraic are **transcendental**. The exponential, logarithmic, and trigonometric functions considered later are examples of transcendental functions.

In the next example we consider the **absolute value function** f, defined by $f(x) = |x|$.

EXAMPLE 2   Sketching the graph of the absolute value function

Let $f(x) = |x|$.

(a) Determine whether f is even or odd.

(b) Sketch the graph of f.

(c) Find the intervals on which f is increasing or is decreasing.

Solution

(a) The domain of f is $\mathbb{R}$, because the absolute value of x exists for every real number x. If x is in $\mathbb{R}$, then

$$f(-x) = |-x| = |x| = f(x).$$

Thus, f is an even function, since $f(-x) = f(x)$.

Figure 35

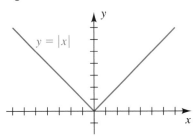

(b) Since f is even, its graph is symmetric with respect to the y-axis. If $x \geq 0$, then $|x| = x$, and therefore the first quadrant part of the graph coincides with the line $y = x$. Sketching this half-line and using symmetry gives us Figure 35.

(c) Referring to the graph, we see that f is decreasing on $(-\infty, 0]$ and is increasing on $[0, \infty)$.

A function f is a **rational function** if

$$f(x) = \frac{g(x)}{h(x)},$$

where $g(x)$ and $h(x)$ are polynomials. The domain of f consists of all real numbers *except* the zeros of the denominator $h(x)$. We next graph one of the simplest rational functions, the **reciprocal function.** The graph of the reciprocal function is the graph of a *hyperbola* (discussed in Section 6.3).

EXAMPLE 3   **Sketching the graph of $f(x) = 1/x$**

Sketch the graph of f if $f(x) = \dfrac{1}{x}$.

Solution   The domain of f is the set of all nonzero real numbers. The function is odd, since

$$f(-x) = \frac{1}{-x} = -\frac{1}{x} = -f(x).$$

Figure 36

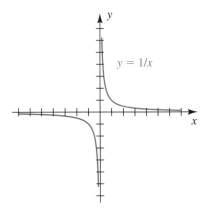

Hence, the graph is symmetric with respect to the origin. If x is positive, so is $f(x)$, and thus no part of the graph lies in quadrant IV. Coordinates of several points on the graph of $y = 1/x$ for $x > 0$ are listed in the following table.

x	$\frac{1}{4}$	$\frac{1}{2}$	1	2	3	4
y	4	2	1	$\frac{1}{2}$	$\frac{1}{3}$	$\frac{1}{4}$

These coordinates lead to the part of the graph in quadrant I (see Figure 36). Using symmetry with respect to the origin, we obtain the points $\left(-\frac{1}{4}, -4\right)$, $\left(-\frac{1}{2}, -2\right)$, $(-1, -1), \ldots,$ which give us the part of the graph in quadrant III.

In Example 3, if x is close to zero and $x > 0$, then $f(x) = 1/x$ is large, as shown in the following table.

x	0.1	0.01	0.001	0.0001	0.00001
$f(x) = 1/x$	10	100	1000	10,000	100,000

We can make $f(x)$ as large as we desire by choosing x close to 0 and $x > 0$. We denote this fact by writing

$$f(x) \to \infty \quad \text{as} \quad x \to 0^+.$$

We say that "$f(x)$ *increases without bound as x approaches* 0 *from the right*"—that is, through values of x greater than 0. The symbol ∞ (read "infinity") does not represent a real number, but is used to denote the variation of $f(x)$ that we have described.

If x is close to 0 and $x < 0$, then $|f(x)|$ is large, but $f(x)$ is negative, as shown in the next table.

x	-0.1	-0.01	-0.001	-0.0001	-0.00001
$f(x) = 1/x$	-10	-100	-1000	$-10,000$	$-100,000$

In this case we use the symbol $-\infty$ (read "minus infinity") and write

$$f(x) \to -\infty \quad \text{as} \quad x \to 0^-,$$

which may be read "$f(x)$ *decreases without bound as x approaches* 0 *from the left*"—that is, through values of x less than 0.

In general, the notation

$$x \to a^+$$

signifies that x approaches a from the *right*—that is, through values of x greater than a. The notation

$$x \to a^-$$

means that x approaches a from the *left,* through values of x *less* than a. Some illustrations of the manner in which a function f may increase or decrease without bound are shown in Figure 37, along with the corresponding notation. In the figure a is positive, but we can also have $a \leq 0$.

The dashed line $x = a$ in Figure 37 is called a *vertical asymptote,* as in the following definition.

Figure 37

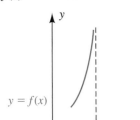

$f(x) \to \infty$ as $x \to a^-$

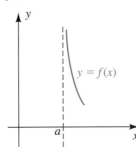

$f(x) \to \infty$ as $x \to a^+$

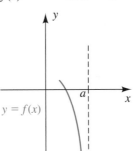

$f(x) \to -\infty$ as $x \to a^-$

$f(x) \to -\infty$ as $x \to a^+$

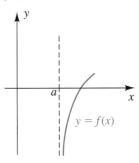

**DEFINITION OF VERTICAL
ASYMPTOTE**

> The line $x = a$ is a **vertical asymptote** for the graph of a function f if
>
> $$f(x) \to \infty \qquad \text{or} \qquad f(x) \to -\infty$$
>
> as x approaches a from either the left or the right.

In Figure 36 the line $x = 0$ (the y-axis) is a vertical asymptote. As we shall see in Chapter 2, the graphs of some of the trigonometric functions have vertical asymptotes.

If, for $f(x) = 1/x$, we assign larger and larger positive values to x, then points on the graph of f approach the x-axis, as indicated in Figure 36. The same is true if $x < 0$ and $|x|$ is large. We call the horizontal line $y = 0$ (the x-axis) a *horizontal asymptote* for the graph of $f(x) = 1/x$, since it satisfies the following definition.

**DEFINITION OF HORIZONTAL
ASYMPTOTE**

> The line $y = c$ is a **horizontal asymptote** for the graph of a function f if
>
> $$f(x) \to c \quad \text{as} \quad x \to \infty \quad \text{or as} \quad x \to -\infty.$$

If we know the graph of $y = f(x)$, it is easy to sketch the graphs of

$$y = f(x) + c \qquad \text{and} \qquad y = f(x) - c$$

for any positive real number c. As in the next chart, for $y = f(x) + c$, we add c to the y-coordinate of each point on the graph of $y = f(x)$. This *shifts* the graph of f *upward* a distance c. For $y = f(x) - c$ with $c > 0$, we subtract c from each y-coordinate, thereby shifting the graph of f a distance c *downward*. These are called **vertical shifts** of graphs.

**Vertically Shifting
the Graph of $y = f(x)$**

Equation	$y = f(x) + c$ with $c > 0$	$y = f(x) - c$ with $c > 0$
Effect on graph	The graph of f is shifted vertically upward a distance c.	The graph of f is shifted vertically downward a distance c.
Graphical interpretation		

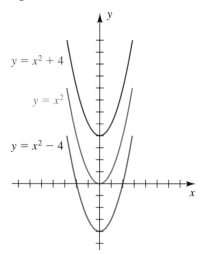

Figure 38

E X A M P L E 4   **Vertically shifting a graph**

Sketch the graph of f:

(a) $f(x) = x^2$ **(b)** $f(x) = x^2 + 4$ **(c)** $f(x) = x^2 - 4$

Solution   We shall sketch all graphs on the same coordinate plane.

(a) Since $f(-x) = (-x)^2 = x^2 = f(x),$

the function f is even, and hence its graph is symmetric with respect to the y-axis. Several points on the graph of $y = x^2$ are $(0, 0)$, $(1, 1)$, $(2, 4)$, and $(3, 9)$. Drawing a smooth curve through these points and reflecting through the y-axis gives us the sketch in Figure 38. The graph is a parabola with vertex at the origin and opening upward.

(b) To sketch the graph of $y = x^2 + 4$, we add 4 to the y-coordinate of each point on the graph of $y = x^2$; that is, we shift the graph in part (a) upward 4 units, as shown in the figure.

(c) To sketch the graph of $y = x^2 - 4$, we decrease the y-coordinates of $y = x^2$ by 4; that is, we shift the graph in part (a) downward 4 units.

The polynomial functions in Example 4 are called **quadratic functions** since they are of the form

$$f(x) = ax^2 + bx + c,$$

where a, b, and c are real numbers with $a \neq 0$.

We can also consider **horizontal shifts** of graphs. Specifically, if $c > 0$, consider the graphs of $y = f(x)$ and $y = g(x) = f(x - c)$ sketched on the same coordinate plane, as illustrated in the next chart. Since

$$g(a + c) = f([a + c] - c) = f(a)$$

we see that the point with x-coordinate a on the graph of $y = f(x)$ has the same y-coordinate as the point with x-coordinate $a + c$ on the graph of $y = g(x) = f(x - c)$. This implies that the graph of $y = g(x) = f(x - c)$ can be obtained by shifting the graph of $y = f(x)$ *to the right* a distance c. Similarly, the graph of $y = h(x) = f(x + c)$ can be obtained by shifting the graph of f *to the left* a distance c, as shown in the chart.

Horizontally Shifting the Graph of $y = f(x)$

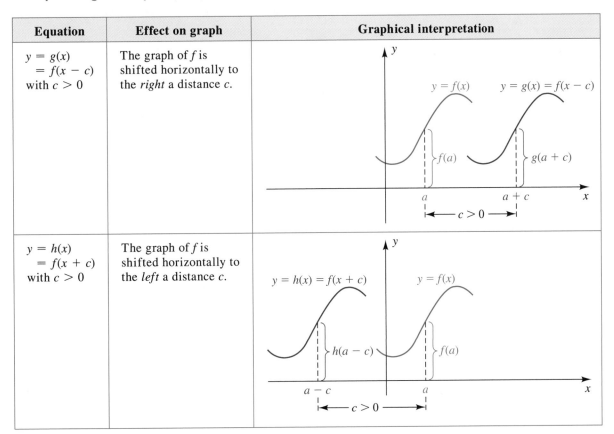

Equation	Effect on graph	Graphical interpretation
$y = g(x)$ $= f(x - c)$ with $c > 0$	The graph of f is shifted horizontally to the *right* a distance c.	
$y = h(x)$ $= f(x + c)$ with $c > 0$	The graph of f is shifted horizontally to the *left* a distance c.	

Horizontal and vertical shifts are also referred to as *translations*.

EXAMPLE 5   **Horizontally shifting a graph**

Sketch the graph of f:

(a) $f(x) = (x - 4)^2$ **(b)** $f(x) = (x + 2)^2$

Figure 39

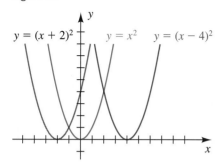

$y = (x + 2)^2$ $y = x^2$ $y = (x - 4)^2$

Solution   The graph of $y = x^2$ is sketched in Figure 39.

(a) Shifting the graph of $y = x^2$ to the right 4 units gives us the graph of $y = (x - 4)^2$, shown in the figure.

(b) Shifting the graph of $y = x^2$ to the left 2 units leads to the graph of $y = (x + 2)^2$, shown in the figure.

To obtain the graph of $y = cf(x)$ for some real number c, we may *multiply* the y-coordinates of points on the graph of $y = f(x)$ by c. For example, if $y = 2f(x)$, we double y-coordinates; or if $y = \frac{1}{2}f(x)$, we multiply each y-coordinate by $\frac{1}{2}$. This procedure is referred to as **vertically stretching** the graph of f (if $c > 1$) or **vertically compressing** the graph (if $0 < c < 1$) and is summarized in the following chart.

Vertically Stretching or Compressing the Graph of $y = f(x)$

Equation	$y = cf(x)$ with $c > 1$	$y = cf(x)$ with $0 < c < 1$
Effect on graph	The graph of f is stretched vertically by a factor c.	The graph of f is compressed vertically by a factor $1/c$.
Graphical interpretation	$y = cf(x)$ with $c > 1$ $y = f(x)$	$y = cf(x)$ with $0 < c < 1$ $y = f(x)$

E X A M P L E   6   **Vertically stretching or compressing a graph**

Sketch the graph of the equation:

(a) $y = 4x^2$ **(b)** $y = \frac{1}{4}x^2$

Solution

(a) To sketch the graph of $y = 4x^2$, we may refer to the graph of $y = x^2$ in Figure 40 and multiply the y-coordinate of each point by 4. This stretches the graph of $y = x^2$ vertically by a factor 4 and gives us a narrower parabola that is sharper at the vertex, as illustrated in the figure.

Figure 40

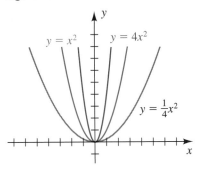

(b) The graph of $y = \frac{1}{4}x^2$ may be sketched by multiplying y-coordinates of points on the graph of $y = x^2$ by $\frac{1}{4}$. This compresses the graph of $y = x^2$ vertically by a factor $1/\frac{1}{4} = 4$ and gives us a wider parabola that is flatter at the vertex, as shown in Figure 40.

We may obtain the graph of $y = -f(x)$ by multiplying the y-coordinate of each point on the graph of $y = f(x)$ by -1. Thus, every point (a, b) on the graph of $y = f(x)$ that lies above the x-axis determines a point $(a, -b)$ on the graph of $y = -f(x)$ that lies below the x-axis. Similarly, if (c, d) lies below the x-axis (that is, $d < 0$), then $(c, -d)$ lies above the x-axis. The graph of $y = -f(x)$ is a **reflection** of the graph of $y = f(x)$ through the x-axis.

E X A M P L E   7   **Reflecting a graph through the x-axis**

Sketch the graph of $y = -x^2$.

Solution   The graph may be found by plotting points; however, since the graph of $y = x^2$ is familiar to us, we sketch it as in Figure 41 and then multiply y-coordinates of points by -1. This procedure gives us the reflection through the x-axis indicated in the figure.

Figure 41

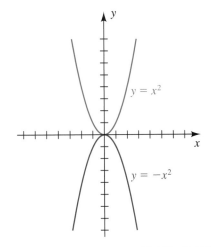

Sometimes it is useful to compare the graphs of $y = f(x)$ and $y = f(cx)$ if $c \neq 0$. In this case the function values $f(x)$ for

$$a \leq x \leq b$$

are the same as the function values $f(cx)$ for

$$a \le cx \le b \quad \text{or, equivalently,} \quad \frac{a}{c} \le x \le \frac{b}{c}.$$

This implies that the graph of f is **horizontally compressed** (if $c > 1$) or **horizontally stretched** (if $0 < c < 1$), as summarized in the following chart.

Horizontally Compressing or Stretching the Graph of $y = f(x)$

Equation	Effect on graph	Graphical interpretation
$y = f(cx)$ with $c > 1$	The graph of f is compressed horizontally by a factor c.	$y = f(x)$ $\qquad$ $y = f(cx)$ with $c > 1$
$y = f(cx)$ with $0 < c < 1$	The graph of f is stretched horizontally by a factor $1/c$.	$y = f(x)$ $\qquad$ $y = f(cx)$ with $0 < c < 1$

If $c < 0$, then the graph of $y = f(cx)$ may be obtained by reflecting the graph of $y = f(|c|x)$ through the y-axis. For example, to sketch the graph of $y = f(-2x)$, we reflect the graph of $y = f(2x)$ through the y-axis. As a special case, the graph of $y = f(-x)$ is a **reflection** of the graph of $y = f(x)$ through the y-axis.

EXAMPLE 8   Horizontally stretching or compressing a graph

If $f(x) = x^3 - 4x^2$, sketch the graphs of $y = f(x)$, $y = f(2x)$, and $y = f\left(\frac{1}{2}x\right)$.

Solution   We have the following:

$$y = f(x) = x^3 - 4x^2 = x^2(x - 4)$$
$$y = f(2x) = (2x)^3 - 4(2x)^2 = 8x^3 - 16x^2 = 8x^2(x - 2)$$
$$y = f\left(\tfrac{1}{2}x\right) = \left(\tfrac{1}{2}x\right)^3 - 4\left(\tfrac{1}{2}x\right)^2 = \tfrac{1}{8}x^3 - x^2 = \tfrac{1}{8}x^2(x - 8)$$

Figure 42
[−6, 15] by [−10, 4]

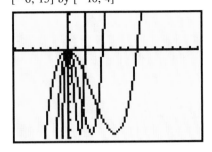

Note that the x-intercepts of the graph of $y = f(2x)$ are 0 and 2, which are $\frac{1}{2}$ the x-intercepts of 0 and 4 for $y = f(x)$. This indicates a horizontal compression by a factor 2.

The x-intercepts of the graph of $y = f\left(\frac{1}{2}x\right)$ are 0 and 8, which are 2 times the x-intercepts for $y = f(x)$. This indicates a horizontal stretching by a factor $1/\frac{1}{2} = 2$.

The graphs, obtained by using a graphing utility with viewing rectangle [−6, 15] by [−10, 4], are shown in Figure 42.

Functions are sometimes described by more than one expression, as in the next examples. We call such functions **piecewise-defined functions.**

EXAMPLE 9   Sketching the graph of a piecewise-defined function

Sketch the graph of the function f if

$$f(x) = \begin{cases} 2x + 3 & \text{if } x < 0 \\ x^2 & \text{if } 0 \le x < 2 \\ 1 & \text{if } x \ge 2 \end{cases}$$

Figure 43

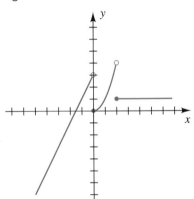

Solution   Some specific keystrokes for the TI-82/83 are given in Example 20 of Appendix I. If $x < 0$, then $f(x) = 2x + 3$ and the graph of f coincides with the line $y = 2x + 3$. This gives that portion of the graph to the left of the y-axis, sketched in Figure 43. The small circle indicates that the point $(0, 3)$ is *not* on the graph.

If $0 \le x < 2$, we use x^2 to find values of f, and therefore this part of the graph of f coincides with the parabola $y = x^2$, as indicated in the figure. Note that the point $(2, 4)$ is not on the graph.

Finally, if $x \ge 2$, the values of f are always 1. Thus, the graph of f for $x \ge 2$ is the horizontal half-line illustrated in Figure 43.

If x is a real number, we define the symbol $[\![x]\!]$ as follows:

$$[\![x]\!] = n, \qquad \text{where } n \text{ is the greatest integer such that } n \le x$$

If we identify $\mathbb{R}$ with points on a coordinate line, then n is the first integer to the *left* of (or *equal* to) x.

ILLUSTRATION

The Symbol $[\![x]\!]$

- $[\![0.5]\!] = 0$
- $[\![1.8]\!] = 1$
- $[\![\sqrt{5}]\!] = 2$
- $[\![3]\!] = 3$
- $[\![-3]\!] = -3$
- $[\![-2.7]\!] = -3$
- $[\![-\sqrt{3}]\!] = -2$
- $[\![-0.5]\!] = -1$

The **greatest integer function** f is defined by $f(x) = [\![x]\!]$.

EXAMPLE 10   Sketching the graph of the greatest integer function

Sketch the graph of the greatest integer function.

Solution   The x- and y-coordinates of some points on the graph may be listed as follows:

Figure 44

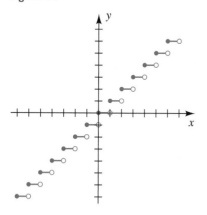

Values of x	$f(x) = [\![x]\!]$
$\vdots$	$\vdots$
$-2 \le x < -1$	-2
$-1 \le x < 0$	-1
$0 \le x < 1$	0
$1 \le x < 2$	1
$2 \le x < 3$	2
$\vdots$	$\vdots$

Whenever x is between successive integers, the corresponding part of the graph is a segment of a horizontal line. Part of the graph is sketched in Figure 44. The graph continues indefinitely to the right and to the left.

The next example involves absolute values.

Figure 45

(a)

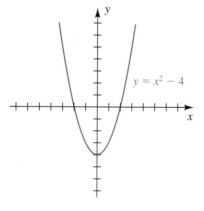

$y = x^2 - 4$

EXAMPLE 11   Sketching the graph of an equation containing an absolute value

Sketch the graph of $y = |x^2 - 4|$.

Solution   The graph of $y = x^2 - 4$ was sketched in Figure 38 and is re-sketched in Figure 45(a). We note the following facts:

(1) If $x \le -2$ or $x \ge 2$, then $x^2 - 4 \ge 0$, and hence $|x^2 - 4| = x^2 - 4$.

(2) If $-2 < x < 2$, then $x^2 - 4 < 0$, and hence $|x^2 - 4| = -(x^2 - 4)$.

It follows from (1) that the graphs of $y = |x^2 - 4|$ and $y = x^2 - 4$ coincide for $|x| \ge 2$. We see from (2) that if $|x| < 2$, then the graph of $y = |x^2 - 4|$ is the reflection of the graph of $y = x^2 - 4$ through the x-axis. This gives us the sketch in Figure 45(b).

(b)

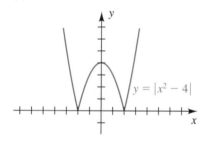

$y = |x^2 - 4|$

In general, if the graph of $y = f(x)$ contains a point $P(c, -d)$ with d positive, then the graph of $y = |f(x)|$ contains the point $Q(c, d)$—that is, Q is the reflection of P through the x-axis. Points with nonnegative y-values are the same for the graphs of $y = f(x)$ and $y = |f(x)|$.

In a previous course, you used algebraic methods to solve inequalities involving absolute values of polynomials of degree 1, such as

$$|2x - 5| < 7 \quad \text{and} \quad |5x + 2| \geq 3.$$

Much more complicated inequalities can be investigated using a graphing utility, as illustrated in the next example.

E X A M P L E 12   **Solving an absolute value inequality graphically**

Estimate the solutions of

$$|0.14x^2 - 13.72| > |0.58x| + 11.$$

Solution   To solve the inequality, we make the assignments

$$Y_1 = \text{ABS}(0.14x^2 - 13.72) \quad \text{and} \quad Y_2 = \text{ABS}(0.58x) + 11$$

and estimate the values of x for which the graph of Y_1 is *above* the graph of Y_2 (since we want Y_1 *greater* than Y_2). After perhaps several trials, we choose the viewing rectangle $[-30, 30]$ by $[0, 40]$, obtaining graphs similar to those in Figure 46, where each tick represents 5 units. Since there is symmetry with respect to the y-axis, it is sufficient to find the x-coordinates of the points of intersection of the graphs for $x > 0$. Using a zoom or intersect feature, we obtain $x \approx 2.80$ and $x \approx 15.52$. Referring to Figure 46, we obtain the (approximate) solution

$$(-\infty, -15.52) \cup (-2.80, 2.80) \cup (15.52, \infty).$$

Figure 46
$[-30, 30]$ by $[0, 40]$

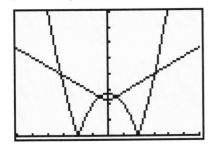

Later in this text and in calculus, you will encounter functions such as

$$g(x) = \ln |x| \quad \text{and} \quad h(x) = \sin |x|.$$

Both functions are of the form $y = f(|x|)$. The effect of substituting $|x|$ for x can be described as follows: If the graph of $y = f(x)$ contains a point $P(c, d)$ with c positive, then the graph of $y = f(|x|)$ contains the point $Q(-c, d)$— that is, Q is the reflection of P through the y-axis. Points on the y-axis ($x = 0$) are the same for the graphs of $y = f(x)$ and $y = f(|x|)$. Points with negative x-values on the graph of $y = f(x)$ are not on the graph of $y = f(|x|)$, since the result of the absolute value is always nonnegative.

The processes of shifting, stretching, compressing, and reflecting a graph may be collectively termed *transforming a graph,* and the resulting graph is called a **transformation** of the original graph. A graphical summary of the types of transformations encountered in this section appears in Appendix III.

1.4 EXERCISES

Exer. 1–10: Determine whether f is even, odd, or neither even nor odd.

1 $f(x) = 5x^3 + 2x$

2 $f(x) = |x| - 3$

3 $f(x) = 3x^4 + 2x^2 - 5$

4 $f(x) = 7x^5 - 4x^3$

5 $f(x) = 8x^3 - 3x^2$

6 $f(x) = 12$

7 $f(x) = \sqrt{x^2 + 4}$

8 $f(x) = 3x^2 - 5x + 1$

9 $f(x) = \sqrt[3]{x^3 - x}$

10 $f(x) = x^3 - \dfrac{1}{x}$

Exer. 11–26: Sketch, on the same coordinate plane, the graphs of f for the given values of c. (Make use of symmetry, shifting, stretching, compressing, or reflecting.)

11 $f(x) = |x| + c;$ $c = -3, 1, 3$

12 $f(x) = |x - c|;$ $c = -3, 1, 3$

13 $f(x) = -x^2 + c;$ $c = -4, 2, 4$

14 $f(x) = 2x^2 - c;$ $c = -4, 2, 4$

15 $f(x) = 2\sqrt{x} + c;$ $c = -3, 0, 2$

16 $f(x) = \sqrt{9 - x^2} + c;$ $c = -3, 0, 2$

17 $f(x) = \frac{1}{2}\sqrt{x - c};$ $c = -2, 0, 3$

18 $f(x) = -\frac{1}{2}(x - c)^2;$ $c = -2, 0, 3$

19 $f(x) = c\sqrt{4 - x^2};$ $c = -2, 1, 3$

20 $f(x) = (x + c)^3;$ $c = -2, 1, 2$

21 $f(x) = cx^3;$ $c = -\frac{1}{3}, 1, 2$

22 $f(x) = (cx)^3 + 1;$ $c = -1, 1, 4$

23 $f(x) = \sqrt{cx} - 1;$ $c = -1, \frac{1}{9}, 4$

24 $f(x) = -\sqrt{16 - (cx)^2};$ $c = 1, \frac{1}{2}, 4$

25 $f(x) = \dfrac{2}{x - c};$ $c = -3, 0, 2$

26 $f(x) = \dfrac{1}{(x + c)^2};$ $c = -1, 0, 2$

Exer. 27–28: If the point P is on the graph of a function f, find the corresponding point on the graph of the given function.

27 $P(3, -2);$ $y = 2f(x - 4) + 1$

28 $P(-2, 1);$ $y = -3f(2x) - 5$

Exer. 29–30: The graph of a function f with domain $[0, 4]$ is shown in the figure. Sketch the graph of the given equation.

29

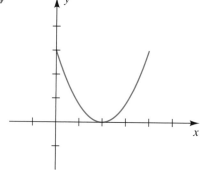

(a) $y = f(x + 3)$ (b) $y = f(x - 3)$

(c) $y = f(x) + 3$ (d) $y = f(x) - 3$

(e) $y = -3f(x)$ (f) $y = -\frac{1}{3}f(x)$

(g) $y = f\left(-\frac{1}{2}x\right)$ (h) $y = f(2x)$

(i) $y = -f(x + 2) - 3$ (j) $y = f(x - 2) + 3$

(k) $y = |f(x)|$ (l) $y = f(|x|)$

30

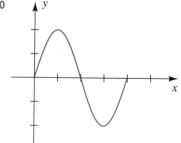

(a) $y = f(x - 2)$ (b) $y = f(x + 2)$

(c) $y = f(x) - 2$ (d) $y = f(x) + 2$

(e) $y = -2f(x)$ (f) $y = -\frac{1}{2}f(x)$

(g) $y = f(-2x)$ (h) $y = f\left(\frac{1}{2}x\right)$

(i) $y = -f(x + 4) - 2$ (j) $y = f(x - 4) + 2$

(k) $y = |f(x)|$ (l) $y = f(|x|)$

Exer. 31–34: The graph of a function f is shown, together with graphs of three other functions (a), (b), and (c). Use properties of symmetry, shifts, and reflecting to find equations for graphs (a), (b), and (c) in terms of f.

31
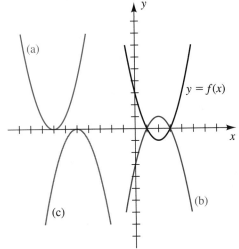

34
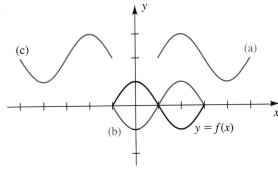

Exer. 35–40: Sketch the graph of f.

35 $f(x) = \begin{cases} 3 & \text{if } x \le -1 \\ -2 & \text{if } x > -1 \end{cases}$

36 $f(x) = \begin{cases} -1 & \text{if } x \text{ is an integer} \\ -2 & \text{if } x \text{ is not an integer} \end{cases}$

37 $f(x) = \begin{cases} 3 & \text{if } x < -2 \\ -x + 1 & \text{if } |x| \le 2 \\ -3 & \text{if } x > 2 \end{cases}$

38 $f(x) = \begin{cases} -2x & \text{if } x < -1 \\ x^2 & \text{if } -1 \le x < 1 \\ -2 & \text{if } x \ge 1 \end{cases}$

32
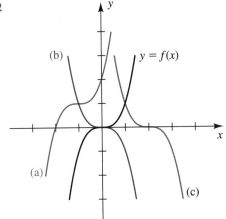

39 $f(x) = \begin{cases} x + 2 & \text{if } x \le -1 \\ x^3 & \text{if } |x| < 1 \\ -x + 3 & \text{if } x \ge 1 \end{cases}$

40 $f(x) = \begin{cases} x - 3 & \text{if } x \le -2 \\ -x^2 & \text{if } -2 < x < 1 \\ -x + 4 & \text{if } x \ge 1 \end{cases}$

33
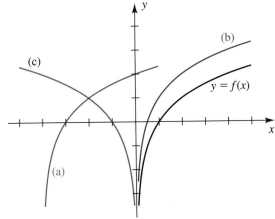

Exer. 41–42: The symbol $[\![x]\!]$ denotes values of the greatest integer function. Sketch the graph of f.

41 (a) $f(x) = [\![x - 3]\!]$ (b) $f(x) = [\![x]\!] - 3$
    (c) $f(x) = 2[\![x]\!]$ (d) $f(x) = [\![2x]\!]$
    (e) $f(x) = [\![-x]\!]$

42 (a) $f(x) = [\![x + 2]\!]$ (b) $f(x) = [\![x]\!] + 2$
    (c) $f(x) = \frac{1}{2}[\![x]\!]$ (d) $f(x) = [\![\frac{1}{2}x]\!]$
    (e) $f(x) = -[\![-x]\!]$

Exer. 43–44: Explain why the graph of the equation is not the graph of a function.

43 $x = y^2$ 44 $x = -|y|$

Exer. 45–46: For the graph of $y = f(x)$ shown in the figure, sketch the graph of $y = |f(x)|$.

45

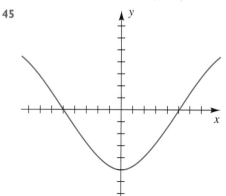

46

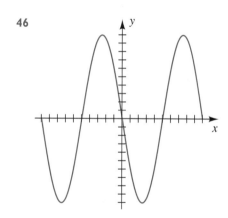

Exer. 47–50: Sketch the graph of the equation.

47 $y = |9 - x^2|$ 48 $y = |x^3 - 1|$

49 $y = |\sqrt{x} - 1|$ 50 $y = ||x| - 1|$

51 Let $y = f(x)$ be a function with domain $D = [-2, 6]$ and range $R = [-4, 8]$. Find the domain D and range R for each function. Assume $f(2) = 8$ and $f(6) = -4$.

(a) $y = -2f(x)$ (b) $y = f(\frac{1}{2}x)$

(c) $y = f(x - 3) + 1$ (d) $y = f(x + 2) - 3$

(e) $y = f(-x)$ (f) $y = -f(x)$

(g) $y = f(|x|)$ (h) $y = |f(x)|$

52 Let $y = f(x)$ be a function with domain $D = [-6, -2]$ and range $R = [-10, -4]$. Find the domain D and range R for each function.

(a) $y = \frac{1}{2}f(x)$ (b) $y = f(2x)$

(c) $y = f(x - 2) + 5$ (d) $y = f(x + 4) - 1$

(e) $y = f(-x)$ (f) $y = -f(x)$

(g) $y = f(|x|)$ (h) $y = |f(x)|$

53 *Income tax rates* A certain country taxes the first $20,000 of an individual's income at a rate of 15%, and all income over $20,000 is taxed at 20%. Find a piecewise-defined function T that specifies the total tax on an income of x dollars.

54 *Telephone rates* A telephone company charges 25 cents for a long-distance call that does not exceed one minute; for longer calls it charges 15 cents for each additional minute. Find a piecewise-defined function C that specifies the total cost of a long-distance call of x minutes.

55 *Royalty rates* A certain paperback sells for $12. The author is paid royalties of 10% on the first 10,000 copies sold, 12.5% on the next 5000 copies, and 15% on any additional copies. Find a piecewise-defined function R that specifies the total royalties if x copies are sold.

56 *Electricity rates* An electric company charges its customers $0.0577 per kilowatt-hour (kWh) for the first 1000 kWh used, $0.0532 for the next 4000 kWh, and $0.0511 for any kWh over 5000. Find a piecewise-defined function C for a customer's bill of x kWh.

C **Exer. 57–60: Estimate the solutions of the inequality.**

57 $|1.3x + 2.8| < 1.2x + 5$

58 $|0.3x| - 2 > 2.2 - 0.63x^2$

59 $|1.2x^2 - 10.8| > 1.36x + 4.08$

60 $|\sqrt{16 - x^2} - 3| < 0.12x^2 - 0.3$

C **Exer. 61–66: Graph f in the viewing rectangle $[-12, 12]$ by $[-8, 8]$. Use the graph of f to predict the graph of g. Verify your prediction by graphing g in the same viewing rectangle.**

61 $f(x) = 0.5x^3 - 4x - 5;$ $g(x) = 0.5x^3 - 4x - 1$

62 $f(x) = 0.25x^3 - 2x + 1;$ $g(x) = -0.25x^3 + 2x - 1$

63 $f(x) = x^2 - 5;$ $g(x) = \frac{1}{4}x^2 - 5$

64 $f(x) = |x + 2|;$ $g(x) = |x - 3| - 3$

65 $f(x) = x^3 - 5x;$ $g(x) = |x^3 - 5x|$

66 $f(x) = 0.5x^2 - 2x - 5;$ $g(x) = 0.5x^2 + 2x - 5$

C 67 *Car rental charges* There are two car rental options available for a four-day trip. Option I is $29.95 per day, with 200 free miles and $0.25 per mile for each addi-

(e) $y = f(-x)$ **(f)** $y = -f(x)$

(g) $y = f(|x|)$ **(h)** $y = |f(x)|$

tional mile. Option II is $39.95 per day, with a charge of $0.15 per mile.

(a) Determine the cost of a 500-mile trip for both options.

(b) Model the data with a cost function for each four-day option.

(c) Make a table that lists the mileage and the charge for each option for trips between 100 and 1200 miles, using increments of 100 miles.

(d) Use the table to determine the mileages at which each option is preferable.

1.5 INVERSE FUNCTIONS

A function f may have the same value for different numbers in its domain. For example, if $f(x) = x^2$, then $f(2) = 4$ and $f(-2) = 4$, but $2 \neq -2$. For *the inverse of a function* to be defined, it is essential that different numbers in the domain *always* give different values of f. Such functions are called *one-to-one functions*.

DEFINITION OF ONE-TO-ONE FUNCTION

A function f with domain D and range R is a **one-to-one function** if either of the following equivalent conditions is satisfied:

(1) Whenever $a \neq b$ in D, then $f(a) \neq f(b)$ in R.

(2) Whenever $f(a) = f(b)$ in R, then $a = b$ in D.

Figure 47

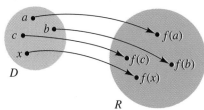

The arrow diagram in Figure 47 illustrates a one-to-one function. Note that each function value in the range R corresponds to *exactly one* element in the domain D. The function illustrated in Figure 25 of Section 1.3 is not one-to-one, since $f(w) = f(z)$, but $w \neq z$.

EXAMPLE 1   **Determining whether a function is one-to-one**

(a) If $f(x) = 3x + 2$, prove that f is one-to-one.

(b) If $g(x) = x^4 + 2x^2$, prove that g is not one-to-one.

Solution

(a) We shall use condition 2 of the preceding definition. Thus, suppose that $f(a) = f(b)$ for some numbers a and b in the domain of f. This gives us

$$3a + 2 = 3b + 2 \quad \text{definition of } f(x)$$
$$3a = 3b \quad \text{subtract 2}$$
$$a = b \quad \text{divide by 3}$$

Hence, f is one-to-one.

(b) Showing that a function *is* one-to-one requires a *general* proof, as in part (a). To show that g is *not* one-to-one we need only find two distinct real numbers in the domain that produce the same function value. For example, $-1 \neq 1$, but $g(-1) = g(1)$. In fact, since g is an even function, $f(-a) = f(a)$ for every real number a.

If we know the graph of a function f, it is easy to determine whether f is one-to-one. For example, the function whose graph is sketched in Figure 48 is not one-to-one, since $a \neq b$, but $f(a) = f(b)$. Note that the horizontal line $y = f(a)$ (or $y = f(b)$) intersects the graph in more than one point. In general, we may use the following graphical test to determine whether a function is one-to-one.

Figure 48

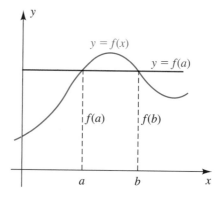

Horizontal Line Test

> A function f is one-to-one if and only if every horizontal line intersects the graph of f in at most one point.

Since every increasing function or decreasing function passes the horizontal line test, we obtain the following result.

THEOREM: INCREASING OR DECREASING FUNCTIONS ARE ONE-TO-ONE

(1) A function that is increasing throughout its domain is one-to-one.

(2) A function that is decreasing throughout its domain is one-to-one.

Let f be a one-to-one function with domain D and range R. Thus, for each number y in R, there is *exactly one* number x in D such that $y = f(x)$, as illustrated by the arrow in Figure 49(a). We may, therefore, define a function g from R to D by means of the following rule:

$$x = g(y)$$

As in Figure 49(b), *g reverses the correspondence given by f.* We call g the *inverse function* of f, as in the next definition.

Figure 49

(a) $y = f(x)$

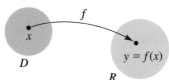

(b) $x = g(y)$

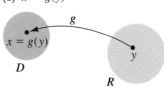

DEFINITION OF

INVERSE FUNCTION

Let f be a one-to-one function with domain D and range R. A function g with domain R and range D is the **inverse function** of f, provided the following condition is true for every x in D and every y in R:

$$y = f(x) \quad \text{if and only if} \quad x = g(y)$$

Remember that for the inverse of a function f to be defined, *it is absolutely essential that f be one-to-one.* The following theorem, stated without proof, is useful to verify that a function g is the inverse of f.

THEOREM ON

INVERSE FUNCTIONS

Let f be a one-to-one function with domain D and range R. If g is a function with domain R and range D, then g is the inverse function of f if and only if both of the following conditions are true:

(1) $g(f(x)) = x$ for every x in D

(2) $f(g(y)) = y$ for every y in R

Conditions 1 and 2 of the preceding theorem are illustrated in Figure 50(a) and (b), respectively, where the blue arrow indicates that f is a function from D to R and the red arrow indicates that g is a function from R to D.

Figure 50

(a) First f, then g

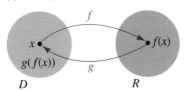

(b) First g, then f

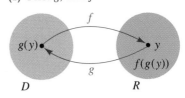

Note that in Figure 50(a) we first apply f to the number x in D, obtaining the function value $f(x)$ in R, and then apply g to $f(x)$, obtaining the number $g(f(x))$ in D. Condition 1 of the theorem states that $g(f(x)) = x$ for every x; that is, g *reverses* the correspondence given by f.

In Figure 50(b) we use the opposite order for the functions. We first apply g to the number y in R, obtaining the function value $g(y)$ in D, and then apply f to $g(y)$, obtaining the number $f(g(y))$ in R. Condition 2 of the theorem states that $f(g(y)) = y$ for every y; that is, f *reverses* the correspondence given by g.

If a function f has an inverse function g, we often denote g by f^{-1}. The -1 used in this notation should not be mistaken for an exponent; that is, $f^{-1}(y)$ *does not mean* $1/[f(y)]$. The reciprocal $1/[f(y)]$ may be denoted by $[f(y)]^{-1}$. It is important to remember the following facts about the domain and range of f and f^{-1}.

Domain and Range of f and f^{-1}

$$\boxed{\begin{array}{l} \text{domain of } f^{-1} = \text{range of } f \\ \text{range of } f^{-1} = \text{domain of } f \end{array}}$$

When we discuss functions, we often let x denote an arbitrary number in the domain. Thus, for the inverse function f^{-1}, we may wish to consider $f^{-1}(x)$, *where x is in the domain R of f^{-1}*. In this event, the two conditions in the theorem on inverse functions are written as follows:

(1) $f^{-1}(f(x)) = x$ for every x in the domain of f

(2) $f(f^{-1}(x)) = x$ for every x in the domain of f^{-1}

Figure 50 contains a hint for finding the inverse of a one-to-one function in certain cases: If possible, *we solve the equation $y = f(x)$ for x in terms of y*, obtaining an equation of the form $x = g(y)$. If the two conditions $g(f(x)) = x$ and $f(g(x)) = x$ are true for every x in the domains of f and g, respectively, then g is the required inverse function f^{-1}. The following guidelines summarize this procedure; in guideline 2, in anticipation of finding f^{-1}, we write $x = f^{-1}(y)$ instead of $x = g(y)$.

GUIDELINES FOR FINDING f^{-1} IN SIMPLE CASES

1 Verify that f is a one-to-one function throughout its domain.

2 Solve the equation $y = f(x)$ for x in terms of y, obtaining an equation of the form $x = f^{-1}(y)$.

3 Verify the following two conditions:

  (a) $f^{-1}(f(x)) = x$ for every x in the domain of f

  (b) $f(f^{-1}(x)) = x$ for every x in the domain of f^{-1}

The success of this method depends on the nature of the equation $y = f(x)$, since we must be able to solve for x in terms of y. For this reason, we include the phrase *in simple cases* in the title of the guidelines. We shall follow these guidelines in the next three examples.

Finding the inverse of a function

Let $f(x) = 3x - 5$. Find the inverse function of f.

Solution

Guideline 1 The graph of the linear function f is a line of slope 3, and hence f is increasing throughout $\mathbb{R}$. Thus, f is one-to-one and the inverse function f^{-1} exists. Moreover, since the domain and range of f is $\mathbb{R}$, the same is true for f^{-1}.

Guideline 2 Solve the equation $y = f(x)$ for x:

$$y = 3x - 5 \quad \text{let } y = f(x)$$

$$x = \frac{y + 5}{3} \quad \text{solve for } x \text{ in terms of } y$$

We now formally let $x = f^{-1}(y)$; that is,

$$f^{-1}(y) = \frac{y + 5}{3}.$$

Since the symbol used for the variable is immaterial, we may also write

$$f^{-1}(x) = \frac{x + 5}{3},$$

where x is in the domain of f^{-1}.

Guideline 3 Since the domain and range of both f and f^{-1} is $\mathbb{R}$, we must verify conditions a and b for every real number x. We proceed as follows:

(a) $f^{-1}(f(x)) = f^{-1}(3x - 5)$ definition of f

$$= \frac{(3x - 5) + 5}{3} \quad \text{definition of } f^{-1}$$

$$= x \quad \text{simplify}$$

(b) $f(f^{-1}(x)) = f\left(\dfrac{x + 5}{3}\right)$ definition of f^{-1}

$$= 3\left(\frac{x + 5}{3}\right) - 5 \quad \text{definition of } f$$

$$= x \quad \text{simplify}$$

These verifications prove that the inverse function of f is given by

$$f^{-1}(x) = \frac{x + 5}{3}.$$

Finding the inverse of a function

Let $f(x) = x^2 - 3$ for $x \geq 0$. Find the inverse function of f.

Figure 51

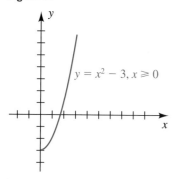

$y = x^2 - 3, \, x \geq 0$

Solution

Guideline 1   The graph of f is sketched in Figure 51. The domain of f is $[0, \infty)$, and the range is $[-3, \infty)$. Since f is increasing, it is one-to-one and hence has an inverse function f^{-1} with domain $[-3, \infty)$ and range $[0, \infty)$.

Guideline 2   We consider the equation

$$y = x^2 - 3$$

and solve for x, obtaining

$$x = \pm\sqrt{y + 3}.$$

Since x is nonnegative, we reject $x = -\sqrt{y + 3}$ and let

$$f^{-1}(y) = \sqrt{y + 3} \qquad \text{or, equivalently,} \qquad f^{-1}(x) = \sqrt{x + 3}.$$

(Note that if the function f had domain $x \leq 0$, we would choose the function $f^{-1}(x) = -\sqrt{x + 3}$.)

Guideline 3   We verify conditions a and b for x in the domains of f and f^{-1}, respectively.

(a) $f^{-1}(f(x)) = f^{-1}(x^2 - 3)$
$$= \sqrt{(x^2 - 3) + 3} = \sqrt{x^2} = x \text{ for } x \geq 0$$

(b) $f(f^{-1}(x)) = f(\sqrt{x + 3})$
$$= (\sqrt{x + 3})^2 - 3 = (x + 3) - 3 = x \text{ for } x \geq -3$$

Thus, the inverse function is given by

$$f^{-1}(x) = \sqrt{x + 3} \quad \text{for } x \geq -3.$$

There is an interesting relationship between the graph of a function f and the graph of its inverse function f^{-1}. We first note that $b = f(a)$ is equivalent to $a = f^{-1}(b)$. These equations imply that *the point (a, b) is on the graph of f if and only if the point (b, a) is on the graph of f^{-1}.*

As an illustration, in Example 3 we found that the functions f and f^{-1} given by

$$f(x) = x^2 - 3 \qquad \text{and} \qquad f^{-1}(x) = \sqrt{x + 3}$$

are inverse functions of each other, provided that x is suitably restricted. Some points on the graph of f are $(0, -3)$, $(1, -2)$, $(2, 1)$, and $(3, 6)$. Corresponding points on the graph of f^{-1} are $(-3, 0)$, $(-2, 1)$, $(1, 2)$, and $(6, 3)$. The graphs of f and f^{-1} are sketched on the same coordinate plane in Figure 52. If the page is folded along the line $y = x$ that bisects quadrants I and III (as indicated by the dashes in the figure), then the graphs of f and f^{-1} coincide. The two graphs are *reflections* of each other through the line $y = x$, or are *symmetric* with respect to this line. This is typical of the graph of every function f that has an inverse function f^{-1} (see Exercise 40).

Figure 52

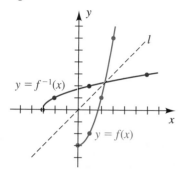

$y = f^{-1}(x)$

l

$y = f(x)$

Figure 53

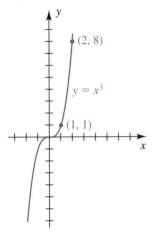

EXAMPLE 4   The relationship between the graphs of f and f^{-1}

Let $f(x) = x^3$. Find the inverse function f^{-1} of f, and sketch the graphs of f and f^{-1} on the same coordinate plane.

Solution   The graph of f is sketched in Figure 53. Note that f is an odd function, and hence the graph is symmetric with respect to the origin.

Guideline 1   Since f is increasing throughout its domain $\mathbb{R}$, it is one-to-one and hence has an inverse function f^{-1}.

Guideline 2   We consider the equation

$$y = x^3$$

and solve for x by taking the cube root of each side, obtaining

$$x = y^{1/3} = \sqrt[3]{y}.$$

Figure 54

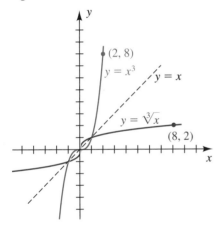

We now let

$$f^{-1}(y) = \sqrt[3]{y} \quad \text{or, equivalently,} \quad f^{-1}(x) = \sqrt[3]{x}.$$

Guideline 3   We verify conditions a and b:

(a) $f^{-1}(f(x)) = f^{-1}(x^3) = \sqrt[3]{x^3} = x$ for every x in $\mathbb{R}$

(b) $f(f^{-1}(x)) = f(\sqrt[3]{x}) = (\sqrt[3]{x})^3 = x$ for every x in $\mathbb{R}$

The graph of f^{-1} (that is, the graph of the equation $y = \sqrt[3]{x}$) may be obtained by reflecting the graph in Figure 53 through the line $y = x$, as shown in Figure 54. Three points on the graph of f^{-1} are $(0, 0)$, $(1, 1)$, and $(8, 2)$.

In the next example we illustrate how some of the concepts presented in this section are helpful in determining the sketch of a graph with the aid of a graphing utility.

EXAMPLE 5   Sketching the graph of the inverse of a function

(a) Sketch the graph of $f(x) = \dfrac{x}{\sqrt{x^2 + 1}}$.

(b) Explain why f is one-to-one, and sketch the graph of f^{-1}.

Figure 55
[−15, 15] by [−10, 10]

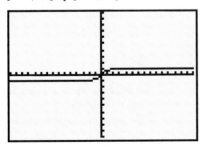

Solution

(a) We assign $x/\sqrt{x^2 + 1}$ to Y_1 and use a standard viewing rectangle to obtain a display similar to Figure 55. The y-coordinate is close to 1 if x is large positive and close to −1 if x is large negative. To improve the graph, we change the viewing rectangle to $[−15, 15]$ by $[−1, 1]$, obtaining a display similar to Figure 56.

(continued)

Figure 56
[−15, 15] by [−1, 1]

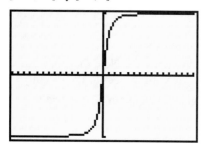

(b) By the horizontal line test, the function f appears to be one-to-one, with domain $\mathbb{R}$ and range $(−1, 1)$. (These facts can be proved algebraically; however, we shall not do so here.) Hence, f has an inverse function f^{-1} that has domain $(−1, 1)$ and range $\mathbb{R}$.

To find f^{-1}, we solve the equation $y = f(x)$ for x in terms of y, as follows:

$$y = \frac{x}{\sqrt{x^2 + 1}} \qquad \text{given}$$

$$y\sqrt{x^2 + 1} = x \qquad \text{multiply by } \sqrt{x^2 + 1}$$

$$y^2(x^2 + 1) = x^2 \qquad \text{square both sides}$$

$$y^2 x^2 + y^2 = x^2 \qquad \text{multiply}$$

$$(y^2 - 1)x^2 = -y^2 \qquad \text{combine } x^2 \text{ terms}$$

$$x^2 = \frac{y^2}{1 - y^2} \qquad \substack{\text{divide by } 1 - y^2 \text{ and} \\ \text{change signs}}$$

$$x = \pm\sqrt{\frac{y^2}{1 - y^2}} = \pm\frac{\sqrt{y^2}}{\sqrt{1 - y^2}} \qquad \text{take the square root}$$

$$x = \pm\frac{|y|}{\sqrt{1 - y^2}} = \pm\frac{y}{\sqrt{1 - y^2}} \qquad \sqrt{y^2} = |y|$$

Figure 57
[−1, 1] by [−15, 15]

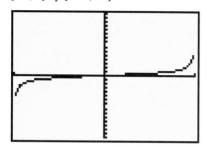

Referring to Figure 56, we see that x *and* y *are always both positive or both negative*, so the last equation can be simplified to

$$x = \frac{y}{\sqrt{1 - y^2}}.$$

Hence,

$$f^{-1}(y) = \frac{y}{\sqrt{1 - y^2}} \qquad \text{or, equivalently,} \qquad f^{-1}(x) = \frac{x}{\sqrt{1 - x^2}}.$$

Using the viewing rectangle $[−1, 1]$ by $[−15, 15]$ gives us a display similar to Figure 57.

1.5 EXERCISES

Exer. 1–12: Determine whether the function f is one-to-one.

1 $f(x) = 3x - 7$

2 $f(x) = \dfrac{1}{x - 2}$

3 $f(x) = x^2 - 9$

4 $f(x) = x^2 + 4$

5 $f(x) = \sqrt{x}$

6 $f(x) = \sqrt[3]{x}$

7 $f(x) = |x|$

8 $f(x) = 3$

9 $f(x) = \sqrt{4 - x^2}$

10 $f(x) = 2x^3 - 4$

11 $f(x) = \dfrac{1}{x}$

12 $f(x) = \dfrac{1}{x^2}$

Exer. 13–16: Use the theorem on inverse functions to prove that f and g are inverse functions of each other, and sketch the graphs of f and g on the same coordinate plane.

13 $f(x) = 3x - 2;$ $g(x) = \dfrac{x + 2}{3}$

14 $f(x) = x^2 + 5, x \le 0;$ $g(x) = -\sqrt{x - 5}, x \ge 5$

15 $f(x) = -x^2 + 3, x \ge 0;$ $g(x) = \sqrt{3 - x}, x \le 3$

16 $f(x) = x^3 - 4;$ $g(x) = \sqrt[3]{x + 4}$

Exer. 17–32: Find the inverse function of f.

17 $f(x) = 3x + 5$ 　　　　18 $f(x) = 7 - 2x$

19 $f(x) = \dfrac{1}{3x - 2}$ 　　　20 $f(x) = \dfrac{1}{x + 3}$

21 $f(x) = \dfrac{3x + 2}{2x - 5}$ 　　　22 $f(x) = \dfrac{4x}{x - 2}$

23 $f(x) = 2 - 3x^2, x \le 0$ 　　24 $f(x) = 5x^2 + 2, x \ge 0$

25 $f(x) = 2x^3 - 5$ 　　　　26 $f(x) = -x^3 + 2$

27 $f(x) = \sqrt{3 - x}$

　　(*Note:* Instructions on graphing the inverse of a function on the TI-82/83 are given in Example 11 of Appendix I.)

28 $f(x) = \sqrt{4 - x^2}, 0 \le x \le 2$

29 $f(x) = \sqrt[3]{x} + 1$ 　　　30 $f(x) = (x^3 + 1)^5$

31 $f(x) = x$ 　　　　　　32 $f(x) = -x$

Exer. 33–34: Find $f^{-1}(x)$ for the given function and condition.

33 $f(x) = x^2 - 4;$ 　　　　$f^{-1}(5) = -3$

34 $f(x) = x^2 - 4x + 3;$ 　　$f^{-1}(3) = 0$

Exer. 35–38: The graph of a one-to-one function f is shown. (a) Use the reflection property to sketch the graph of f^{-1}. (b) Find the domain D and range R of the function f. (c) Find the domain D_1 and range R_1 of the inverse function f^{-1}.

35

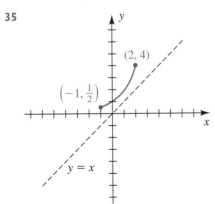

36

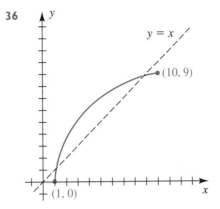

37

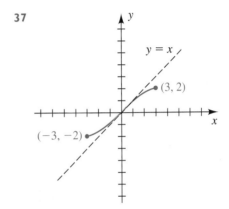

38

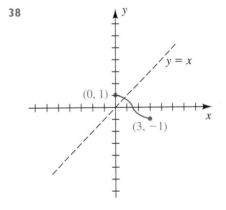

39 (a) Prove that the function defined by $f(x) = ax + b$ (a linear function) for $a \ne 0$ has an inverse function, and find $f^{-1}(x)$.

　　(b) Does a constant function have an inverse? Explain.

40 Show that the graph of f^{-1} is the reflection of the graph of f through the line $y = x$ by verifying the following conditions:

(1) If $P(a, b)$ is on the graph of f, then $Q(b, a)$ is on the graph of f^{-1}.

(2) The midpoint of line segment PQ is on the line $y = x$.

(3) The line PQ is perpendicular to the line $y = x$.

41 Verify that $f(x) = f^{-1}(x)$ if

(a) $f(x) = -x + b$ **(b)** $f(x) = \dfrac{ax + b}{cx - a}$ for $c \neq 0$

(c) $f(x)$ has the following graph:

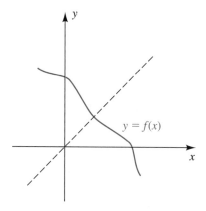

42 Let n be any positive integer. Find the inverse function of f if

(a) $f(x) = x^n$ for $x \geq 0$

(b) $f(x) = x^{m/n}$ for $x \geq 0$ and m any positive integer

[c] **Exer. 43–44: Use the graph of f to determine whether f is one-to-one.**

43 $f(x) = 0.4x^5 - 0.4x^4 + 1.2x^3 - 1.2x^2 + 0.8x - 0.8$

44 $f(x) = \dfrac{x - 8}{x^{2/3} + 4}$

[c] **Exer. 45–46: Graph f on the given interval. (a) Estimate the largest interval $[a, b]$ with $a < 0 < b$ on which f is one-to-one. (b) If g is the function with domain $[a, b]$ such that $g(x) = f(x)$ for $a \leq x \leq b$, estimate the domain and range of g^{-1}.**

45 $f(x) = 2.1x^3 - 2.98x^2 - 2.11x + 3$; $[-1, 2]$

46 $f(x) = 0.05x^4 - 0.24x^3 - 0.15x^2 + 1.18x + 0.24$; $[-2, 2]$

[c] **Exer. 47–48: Graph f in the given viewing rectangle. Use the graph of f to predict the shape of the graph of f^{-1}. Verify your prediction by graphing f^{-1} and the line $y = x$ in the same viewing rectangle.**

47 $f(x) = \sqrt[3]{x - 1}$; $[-12, 12]$ by $[-8, 8]$

48 $f(x) = 2(x - 2)^2 + 3$, $x \geq 2$; $[0, 12]$ by $[0, 8]$

49 *Ventilation requirements* Ventilation is an effective way to improve indoor air quality. In nonsmoking restaurants, air circulation requirements (in ft³/min) are given by the function $V(x) = 35x$, where x is the number of people in the dining area.

(a) Determine the ventilation requirements for 23 people.

(b) Find $V^{-1}(x)$. Explain the significance of V^{-1}.

(c) Use V^{-1} to determine the maximum number of people that should be in a restaurant having a ventilation capability of 2350 ft³/min.

[c] **50** *Radio stations* The table lists the total numbers of radio stations in the United States for certain years.

Year	Number
1950	2773
1960	4133
1970	6760
1980	8566
1990	10,819

(a) Plot the data.

(b) Determine a linear function $f(x) = ax + b$ that models these data, where x is the year. Plot f and the data on the same coordinate axes.

(c) Find $f^{-1}(x)$. Explain the significance of f^{-1}.

(d) Use f^{-1} to predict the year in which there were 7744 radio stations. Compare it with the true value of 1975.

1.6 EXPONENTIAL AND LOGARITHMIC FUNCTIONS

Several exercises in this text involve exponential or logarithmic functions. The brief survey in this section should provide sufficient information for working such exercises. The material in this section also appears in Chapter 5, which contains a more complete discussion of these functions.

If a is a positive real number different from 1, then it can be shown that to each real number x there corresponds exactly one positive number a^x such that the laws of exponents are true. Thus, as in the following chart, we may define a function f whose domain is $\mathbb{R}$ and range is the set of positive real numbers.

Terminology	Definition	Graph of f for $a > 1$	Graph of f for $0 < a < 1$
Exponential function f with base a	$f(x) = a^x$ for every x in $\mathbb{R}$, where $a > 0$ and $a \neq 1$		

The graphs in the chart show that if $a > 1$, then f is increasing on $\mathbb{R}$, and if $0 < a < 1$, then f is decreasing on $\mathbb{R}$. (These facts can be proved using calculus.) The graphs merely indicate the *general* appearance—the *exact* shape depends on the value of a. Note, however, that since $a^0 = 1$, the y-intercept is 1 for every a.

If $a > 1$, then as x *decreases* through negative values, the graph of f approaches the x-axis (see the third column in the chart). Thus, the x-axis is a *horizontal asymptote*. As x increases through positive values, the graph rises rapidly. This type of variation is characteristic of the **exponential law of growth,** and f is sometimes called a **growth function.**

If $0 < a < 1$, then as x *increases,* the graph of f approaches the x-axis asymptotically (see the last column of the chart). This type of variation is known as **exponential decay.**

In the next two examples we sketch the graphs of several different exponential functions.

EXAMPLE 1   Sketching graphs of exponential functions

If $f(x) = \left(\frac{3}{2}\right)^x$ and $g(x) = 3^x$, sketch the graphs of f and g on the same coordinate plane.

Solution   Since $\frac{3}{2} > 1$ and $3 > 1$, each graph *rises* as x increases. The following table displays coordinates for several points on the graphs.

(continued)

Figure 58

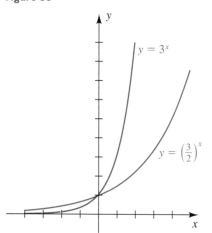

x	-2	-1	0	1	2	3	4
$y = \left(\frac{3}{2}\right)^x$	$\frac{4}{9} \approx 0.4$	$\frac{2}{3} \approx 0.7$	1	$\frac{3}{2}$	$\frac{9}{4} \approx 2.3$	$\frac{27}{8} \approx 3.4$	$\frac{81}{16} \approx 5.1$
$y = 3^x$	$\frac{1}{9} \approx 0.1$	$\frac{1}{3} \approx 0.3$	1	3	9	27	81

Plotting points and being familiar with the general graph of $y = a^x$ leads to the graphs in Figure 58.

Example 1 illustrates the fact that if $1 < a < b$, then $a^x < b^x$ for positive values of x and $b^x < a^x$ for negative values of x. In particular, since $\frac{3}{2} < 2 < 3$, the graph of $y = 2^x$ lies between the graphs of f and g in Figure 58.

EXAMPLE 2   **Sketching the graph of an exponential function**

Sketch the graph of the equation $y = \left(\frac{1}{2}\right)^x$.

Figure 59

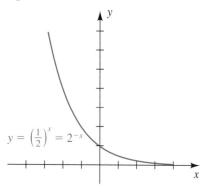

Solution   Since $0 < \frac{1}{2} < 1$, the graph *falls* as x increases. Coordinates of some points on the graph are listed in the following table.

x	-3	-2	-1	0	1	2	3
$y = \left(\frac{1}{2}\right)^x$	8	4	2	1	$\frac{1}{2}$	$\frac{1}{4}$	$\frac{1}{8}$

The graph is sketched in Figure 59. Since $\left(\frac{1}{2}\right)^x = 2^{-x}$, the graph is the same as the graph of the equation $y = 2^{-x}$.

In investigations involving physical phenomena such as the growth of a culture of bacteria or the decay of a radioactive substance, a certain irrational number, denoted by e, arises as the base of an exponential function. The following five-decimal-place approximation can be obtained using calculus.

Approximation to e

$$e \approx 2.71828$$

Since e occurs in the study of a large variety of situations that occur in nature, the exponential function with base e is called the *natural exponential function*, as in the following definition.

DEFINITION OF THE NATURAL EXPONENTIAL FUNCTION

The **natural exponential function** f is defined by

$$f(x) = e^x$$

for every real number x.

Figure 60

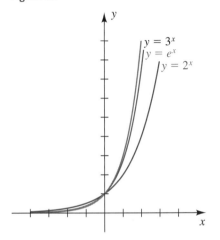

Since $2 < e < 3$, the graph of $y = e^x$ lies between the graphs of $y = 2^x$ and $y = 3^x$, as shown in Figure 60. Scientific calculators have an $\boxed{e^x}$ key for approximating values of the natural exponential function, and graphics calculators are useful for sketching its graph.

EXAMPLE 3   Sketching the graph of an exponential function

Sketch the graph of f if $f(x) = e^{-x}$.

Solution   Let us write

$$f(x) = e^{-x} = \frac{1}{e^x} = \left(\frac{1}{e}\right)^x$$

and note that since $2 < e < 3$, we have $\frac{1}{3} < 1/e < \frac{1}{2}$. Thus, the graph of f is decreasing on $\mathbb{R}$ and lies between the graphs of $y = \left(\frac{1}{3}\right)^x$ and $y = \left(\frac{1}{2}\right)^x$. More accurately, we can see in Figure 61 that the graph of f is the reflection of the graph of $y = e^x$ through the y-axis, for if (k, e^k) is a point on the graph of $y = e^x$, then $(-k, e^k)$ is a point on the graph of f.

Figure 61

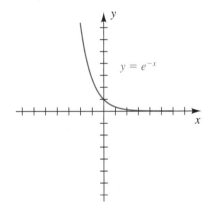

Logarithmic functions may be used to calculate the magnitude of an earthquake, find the half-life of a radioactive substance, or investigate a variety of other physical phenomena. These functions are defined in terms of exponential functions, as in the next definition.

DEFINITION OF $\log_a$

Let a be a positive real number different from 1. The **logarithm of x with base a** is defined by

$$y = \log_a x \quad \text{if and only if} \quad x = a^y$$

for every $x > 0$ and every real number y.

Note that the two equations in the definition are equivalent. We call the first equation the **logarithmic form** and the second the **exponential form.**

The following diagram may help you remember how to change one form into the other.

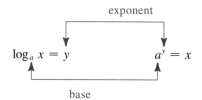

Logarithmic form   **Exponential form**

exponent

$$\log_a x = y \qquad\qquad a^y = x$$

base

Observe that when forms are changed, *the bases of the logarithmic and exponential forms are the same.* The number y (that is, $\log_a x$) corresponds to the exponent in the exponential form. In words, $\log_a x$ *is the exponent to which the base a must be raised to obtain x.*

The following illustration contains examples of equivalent forms.

ILLUSTRATION

Equivalent Forms

Logarithmic form	**Exponential form**
$\log_5 u = 2$	$5^2 = u$
$\log_b 8 = 3$	$b^3 = 8$
$r = \log_p q$	$p^r = q$
$w = \log_4 (2t + 3)$	$4^w = 2t + 3$

The next example contains an application that involves changing from an exponential form to a logarithmic form.

EXAMPLE 4   Changing exponential form to logarithmic form

The number N of bacteria in a certain culture after t hours is given by $N = (1000)2^t$. Express t as a logarithmic function of N with base 2.

Solution   If $N = (1000)2^t$, then

$$2^t = \frac{N}{1000}.$$

Changing to logarithmic form, we obtain

$$t = \log_2 \frac{N}{1000}.$$

The following general properties follow from the interpretation of $\log_a x$ as an exponent.

Property of $\log_a x$	Reason	Illustration
(1) $\log_a 1 = 0$	$a^0 = 1$	$\log_3 1 = 0$
(2) $\log_a a = 1$	$a^1 = a$	$\log_{10} 10 = 1$
(3) $\log_a a^x = x$	$a^x = a^x$	$\log_2 8 = \log_2 2^3 = 3$
(4) $a^{\log_a x} = x$	see below	$5^{\log_5 7} = 7$

The reason for property (4) follows directly from the definition of $\log_a$, since

$$\text{if} \quad y = \log_a x, \quad \text{then} \quad x = a^y, \quad \text{or} \quad x = a^{\log_a x}.$$

EXAMPLE 5   Sketching the graph of a logarithmic function

Sketch the graph of f if $f(x) = \log_3 x$.

Solution   If we write $y = \log_3 x$ and change to exponential form, we obtain

$$x = 3^y.$$

Substituting several values for y and finding the corresponding values of x gives us the following table.

y	-3	-2	-1	0	1	2	3
$x = 3^y$	$\frac{1}{27}$	$\frac{1}{9}$	$\frac{1}{3}$	1	3	9	27

These values lead to the sketch in Figure 62.

Figure 62

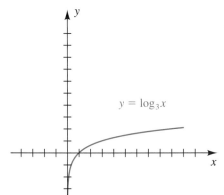

$y = \log_3 x$

If $a > 1$, the graph of $y = \log_a x$ has the general appearance of the graph in Figure 62; however, if a increases, the graph gets closer to the x-axis for $x > 1$. If a is close to 1 (and $a > 1$), the graph rises more rapidly for $x > 1$.

The base $0 < a < 1$ is seldom used in applications, and hence we will not discuss graphs for this case.

Before electronic calculators were invented, logarithms with base 10 were used for complicated numerical computations involving products, quotients, and powers of real numbers. Base 10 was employed because it is well suited for real numbers that are expressed in decimal form. Logarithms with base 10 are called **common logarithms.** The symbol $\log x$ is used as an abbreviation for $\log_{10} x$.

DEFINITION OF COMMON LOGARITHM

$$\log x = \log_{10} x \quad \text{for every } x > 0$$

Figure 63

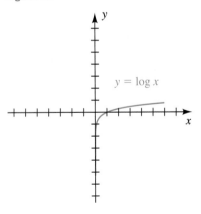

$y = \log x$

Since inexpensive calculators are now available, there is no need for common logarithms as a tool for computational work. Base 10 does occur in applications, however, and hence many calculators have a $\boxed{\text{LOG}}$ key, which can be used to approximate common logarithms. The graph of $y = \log x$ in Figure 63 can be obtained as in Example 5 by sketching $x = 10^y$.

The natural exponential function is given by $f(x) = e^x$. The logarithmic function with base e is called the **natural logarithmic function.** The symbol **ln x** (read "ell-en of x") is an abbreviation for $\log_e x$, and we refer to it as the **natural logarithm of x.**

DEFINITION OF NATURAL LOGARITHM

$$\ln x = \log_e x \quad \text{for every } x > 0$$

Many calculators have a key labeled $\boxed{\text{LN}}$, which can be used to approximate natural logarithms. Since $e \approx 2.7 \approx 3$, the graph of $y = \ln x$ is similar to the graph of $y = \log_3 x$ in Figure 62.

The following laws are fundamental for all work with logarithms. Proofs, based on the laws of exponents, can be found in Chapter 5.

Laws of Logarithms

If u and w denote positive real numbers, then

(1) $\log_a (uw) = \log_a u + \log_a w$

(2) $\log_a \dfrac{u}{w} = \log_a u - \log_a w$

(3) $\log_a (u^c) = c \log_a u \quad$ for every real number c

The following example illustrates one use of these laws.

EXAMPLE 6   Using laws of logarithms

Express $\log_a \dfrac{x^2 \sqrt[3]{y}}{z^4}$ in terms of logarithms of x, y, and z.

Solution   We write $\sqrt[3]{y}$ as $y^{1/3}$ and use laws of logarithms:

$$\log_a \frac{x^2 \sqrt[3]{y}}{z^4} = \log_a (x^2 y^{1/3}) - \log_a z^4 \qquad \text{law 2}$$

$$= \log_a x^2 + \log_a y^{1/3} - \log_a z^4 \qquad \text{law 1}$$

$$= 2 \log_a x + \tfrac{1}{3} \log_a y - 4 \log_a z \qquad \text{law 3}$$

We can solve the equation $3^x = 21$ by changing it from the given exponential form to the equivalent logarithmic form, obtaining

$$x = \log_3 21.$$

This is, in fact, the solution of the equation; however, since calculators typically have keys only for log and ln, we cannot approximate $\log_3 21$ directly. The next theorem gives us a simple *change of base formula* for finding $\log_b u$ if $u > 0$ and b is *any* logarithmic base.

THEOREM: CHANGE OF BASE FORMULA

If $u > 0$ and if a and b are positive real numbers different from 1, then

$$\log_b u = \frac{\log_a u}{\log_a b}.$$

The change of base formula enables us to evaluate $\log_3 21$ as $\log 21/\log 3$ (or $\ln 21/\ln 3$), to obtain an approximate value of 2.77. It also enables us to graph a logarithmic function such as $f(x) = \log_6 x$ by representing it as $f(x) = (\log x)/(\log 6)$.

1.6 EXERCISES

1 Sketch the graph of f if $a = 2$.

(a) $f(x) = a^x$ (b) $f(x) = -a^x$

(c) $f(x) = 3a^x$ (d) $f(x) = a^{x+3}$

(e) $f(x) = a^x + 3$ (f) $f(x) = a^{x-3}$

(g) $f(x) = a^x - 3$ (h) $f(x) = a^{-x}$

(i) $f(x) = \left(\dfrac{1}{a}\right)^x$ (j) $f(x) = a^{3-x}$

2 Work Exercise 1 if $a = \frac{1}{2}$.

Exer. 3–6: Sketch the graph of f.

3 $f(x) = \left(\frac{2}{5}\right)^{-x}$ **4** $f(x) = \left(\frac{2}{5}\right)^x$

5 $f(x) = -\left(\frac{1}{2}\right)^x + 4$ **6** $f(x) = -3^x + 9$

Exer. 7–8: Use the graph of $y = e^x$ to help sketch the graph of f.

7 (a) $f(x) = e^{x+4}$ (b) $f(x) = e^x + 4$

8 (a) $f(x) = e^{-2x}$ (b) $f(x) = -2e^x$

Exer. 9–10: Change to logarithmic form.

9 (a) $4^3 = 64$ (b) $4^{-3} = \frac{1}{64}$ (c) $t^r = s$

10 (a) $3^5 = 243$ (b) $3^{-4} = \frac{1}{81}$ (c) $c^p = d$

Exer. 11–12: Change to exponential form.

11 (a) $\log_2 32 = 5$ (b) $\log_3 \frac{1}{243} = -5$ (c) $\log_t r = p$

12 (a) $\log_3 81 = 4$ (b) $\log_4 \frac{1}{256} = -4$ (c) $\log_v w = q$

Exer. 13–14: Change to logarithmic form.

13 (a) $10^5 = 100{,}000$ (b) $10^{-3} = 0.001$ (c) $10^x = y + 1$

14 (a) $10^4 = 10{,}000$ (b) $10^{-2} = 0.01$ (c) $10^x = 38z$

Exer. 15–16: Change to exponential form.

15 (a) $\log x = 50$ (b) $\log x = 20t$ (c) $\ln x = 0.1$

16 (a) $\log x = -8$ (b) $\log x = y - 2$ (c) $\ln x = \frac{1}{2}$

Exer. 17–18: Find the number, if possible.

17 (a) $\log_5 1$ (b) $\log_3 3$ (c) $\log_4 (-2)$

(d) $\log_7 7^2$ (e) $3^{\log_3 8}$ (f) $\log_5 125$

(g) $\log_4 \frac{1}{16}$

18 (a) $\log_8 1$ (b) $\log_9 9$ (c) $\log_5 0$

(d) $\log_6 6^7$ (e) $5^{\log_5 4}$ (f) $\log_3 243$

(g) $\log_2 128$

Exer. 19–20: Find the number.

19 (a) $10^{\log 3}$ (b) $\log 10^5$ (c) $\log 100$

(d) $\log 0.0001$ (e) $e^{\ln 2}$ (f) $\ln e^{-3}$

(g) $e^{2+\ln 3}$

20 (a) $10^{\log 7}$ (b) $\log 10^{-6}$ (c) $\log 100{,}000$

(d) $\log 0.001$ (e) $e^{\ln 8}$ (f) $\ln e^{2/3}$

(g) $e^{1+\ln 5}$

21 Sketch the graph of f if $a = 4$.

(a) $f(x) = \log_a x$ (b) $f(x) = -\log_a x$

(c) $f(x) = 2 \log_a x$ (d) $f(x) = \log_a (x + 2)$

(e) $f(x) = (\log_a x) + 2$ (f) $f(x) = \log_a (x - 2)$

(g) $f(x) = (\log_a x) - 2$ (h) $f(x) = \log_a |x|$

(i) $f(x) = \log_a (-x)$ (j) $f(x) = \log_a (3 - x)$

(k) $f(x) = |\log_a x|$

22 Work Exercise 21 if $a = 5$.

Exer. 23–26: Sketch the graph of f.

23 $f(x) = \log x - 1$ 24 $f(x) = \log (-x)$

25 $f(x) = 2 - \ln x$ 26 $f(x) = \ln (x - 1)$

Exer. 27–30: Express in terms of logarithms of x, y, z, or w.

27 (a) $\log_4 (xz)$ (b) $\log_4 (y/x)$ (c) $\log_4 \sqrt[3]{z}$

28 (a) $\log_3 (xyz)$ (b) $\log_3 (xz/y)$ (c) $\log_3 \sqrt[5]{y}$

29 $\log_a \dfrac{x^3 w}{y^2 z^4}$ 30 $\log_a \dfrac{y^5 w^2}{x^4 z^3}$

Exer. 31–34: Evaluate using the change of base formula.

31 $\log_5 6$ 32 $\log_2 20$

33 $\log_9 0.2$ 34 $\log_6 \frac{1}{2}$

Exer. 35–36: Evaluate using the change of base formula (without a calculator).

35 $\dfrac{\log_5 16}{\log_5 4}$ 36 $\dfrac{\log_7 243}{\log_7 3}$

Exer. 37–38: Sketch the graph of f, and use the change of base formula to approximate the y-intercept.

37 $f(x) = \log_2 (x + 3)$ 38 $f(x) = \log_3 (x + 5)$

Exer. 39–40: Sketch the graph of f, and use the change of base formula to approximate the x-intercept.

39 $f(x) = 4^x - 3$ 40 $f(x) = 3^x - 6$

CHAPTER 1 REVIEW EXERCISES

1 Replace the symbol $\square$ with either $<$, $>$, or $=$ to make the resulting statement true.

(a) $-0.1 \square -0.001$ (b) $\sqrt{9} \square -3$ (c) $\frac{1}{6} \square 0.166$

2 Express as an inequality:

(a) x is negative.

(b) a is between $\frac{1}{2}$ and $\frac{1}{3}$.

(c) The absolute value of x is not greater than 4.

3 Rewrite without using the absolute value symbol, and simplify:

(a) $|-7|$ (b) $\dfrac{|-5|}{-5}$ (c) $|3^{-1} - 2^{-1}|$

4 If points A, B, and C on a coordinate line have coordinates -8, 4, and -3, respectively, find the distance:

(a) $d(A, C)$ (b) $d(C, A)$ (c) $d(B, C)$

Exer. 5–10: Solve the equation or inequality. Express the solutions in terms of intervals whenever possible.

5 $3x + 5 = -4(x - 2)$ 6 $-2(4x - 7) = 5(8 - x)$

7 $2x^2 + 5x - 12 = 0$ 8 $(x - 2)(x + 1) = 3$

9 $10 - 7x < 4 + 2x$ 10 $-\dfrac{1}{2} < \dfrac{2x + 3}{5} < \dfrac{3}{2}$

11 Express the inequality as an interval, and sketch its graph:

(a) $x < 3$ (b) $-3 \le x \le 3$ (c) $0 < x < \dfrac{\pi}{2}$

12 Express the interval as an inequality in the variable x:

(a) $[-5, \infty)$ (b) $(-2, 2]$ (c) $\left[-\dfrac{\pi}{2}, \dfrac{3\pi}{2} \right)$

13 Describe the set of all points (x, y) in a coordinate plane such that $y/x < 0$.

14 Given $P(-5, 9)$ and $Q(-8, -7)$, find

(a) the distance $d(P, Q)$

(b) the midpoint of the segment PQ

Exer. 15–24: Sketch the graph of the equation and label the x- and y-intercepts.

15 $x + 5 = 0$ 16 $2y - 7 = 0$

17 $2y + 5x - 8 = 0$ 18 $x = 3y + 4$

19 $9y + 2x^2 = 0$ 20 $3x - 7y^2 = 0$

21 $y = \sqrt{1 - x}$ 22 $y = (x - 1)^3$

23 $y^2 = 16 - x^2$

24 $x^2 + y^2 + 4x - 16y + 64 = 0$

25 Find an equation of the circle that has center $C(7, -4)$ and passes through $P(-3, 3)$.

26 Find an equation of the circle that has endpoints of a diameter $A(8, 10)$ and $B(-2, -14)$.

Exer. 27–28: Find the center and radius of the circle with the given equation.

27 $x^2 + y^2 - 12y + 31 = 0$

28 $4x^2 + 4y^2 + 24x - 16y + 39 = 0$

29 If $f(x) = \dfrac{x}{\sqrt{x + 3}}$, find

(a) $f(1)$ (b) $f(-1)$ (c) $f(0)$ (d) $f(-x)$
(e) $-f(x)$ (f) $f(x^2)$ (g) $[f(x)]^2$

30 Find the domain and range of f if

(a) $f(x) = \sqrt{3x - 4}$ (b) $f(x) = \dfrac{1}{(x + 3)^2}$

31 Find $\dfrac{f(a + h) - f(a)}{h}$ if $f(x) = -x^2 + x + 5$ and $h \ne 0$.

32 Determine whether the function f is one-to-one.

(a) $f(x) = 2x^3 - 5$ (b) $f(x) = 3x^2 + 1$

Exer. 33–36: (a) Sketch the graph of f. (b) Find the domain D and range R of f. (c) Find the intervals on which f is increasing, is decreasing, or is constant.

33 $f(x) = \dfrac{1 - 3x}{2}$ 34 $f(x) = 1000$

35 $f(x) = |x + 3|$ 36 $f(x) = 9 - x^2$

37 Determine whether f is even, odd, or neither even nor odd.

(a) $f(x) = \sqrt[3]{x^3 + 4x}$

(b) $f(x) = \sqrt[3]{3x^2 - x^3}$

(c) $f(x) = \sqrt[3]{x^4 + 3x^2 + 5}$

38 Sketch the graphs of the following equations, making use of shifting, stretching, or reflecting.

(a) $y = \sqrt{x}$ (b) $y = \sqrt{x + 4}$ (c) $y = \sqrt{x} + 4$
(d) $y = 4\sqrt{x}$ (e) $y = \frac{1}{4}\sqrt{x}$ (f) $y = -\sqrt{x}$

39 The graph of a function f with domain $[-3, 3]$ is shown in the figure. Sketch the graph of the given equation.

(a) $y = f(x - 2)$ (b) $y = f(x) - 2$
(c) $y = f(-x)$ (d) $y = f(2x)$
(e) $y = f\left(\frac{1}{2}x\right)$ (f) $y = f^{-1}(x)$
(g) $y = |f(x)|$ (h) $y = f(|x|)$

Exercise 39

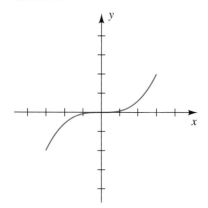

40 Find an equation for the graph shown in the figure.

Exercise 40

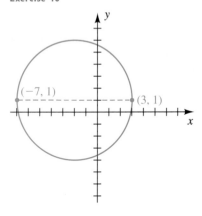

Exer. 41–42: (a) **Find** $f^{-1}(x)$. (b) **Sketch the graphs of** f **and** f^{-1} **on the same coordinate plane.**

41 $f(x) = 10 - 15x$ **42** $f(x) = 9 - 2x^2, \ x \le 0$

43 Refer to the figure to determine each of the following. [Recall that $(f \circ g)(x) = f(g(x))$.]

   (a) $f(1)$ (b) $(f \circ f)(1)$ (c) $f^{-1}(4)$

   (d) all x such that $f(x) = 4$

   (e) all x such that $f(x) > 4$

Exercise 43

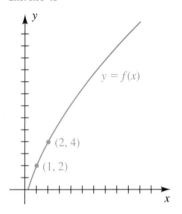

44 Suppose f and g are one-to-one functions such that $f(2) = 7$, $f(4) = 2$, and $g(2) = 5$. Find the value, if possible.

   (a) $(g \circ f^{-1})(7)$ (b) $(f \circ g^{-1})(5)$

   (c) $(f^{-1} \circ g^{-1})(5)$ (d) $(g^{-1} \circ f^{-1})(2)$

Exer. 45–49: Sketch the graph of f.

45 $f(x) = 3^{x+2}$

46 $f(x) = 3^{-2x}$

47 $f(x) = 1 - 3^{-x}$

48 $f(x) = e^{x/2}$

49 $f(x) = \log_2(x + 4)$

Exer. 50–51: Evaluate without using a calculator or table.

50 (a) $\log_2 \frac{1}{16}$ (b) $\log_\pi 1$ (c) $\ln e$

     (d) $6^{\log_6 4}$ (e) $\log 1{,}000{,}000$ (f) $10^{3\log 2}$

     (g) $\log_4 2$

51 (a) $\log_5 \sqrt[3]{5}$ (b) $\log_5 1$ (c) $\log 10$

     (d) $e^{\ln 5}$ (e) $\log \log 10^{10}$ (f) $e^{2\ln 5}$

     (g) $\log_{27} 3$

52 Express $\log x^4 \sqrt[3]{y^2/z}$ in terms of logarithms of x, y, and z.

Exer. 53–54: (a) **Find the domain and range of the function.** (b) **Find the inverse of the function and its domain and range.**

53 $y = \log_2(x + 1)$

54 $y = 2^{3-x} - 2$

55 *Wheelchair ramp* The Americans with Disabilities Act of 1990 guarantees all persons the right of accessibility of public accommodations. Providing access to a building often involves building a wheelchair ramp. Ramps should have approximately 1 inch of vertical rise for every 12–20 inches of horizontal run. If the base of an exterior door is located 3 feet above a sidewalk, determine the range of appropriate lengths for a wheelchair ramp.

56 *Discus throw* Based on Olympic records, the winning distance for the discus throw can be approximated by $d = 175 + 1.75t$, where d is in feet and $t = 0$ corresponds to the year 1948.

   (a) Predict the winning distance for the Summer Olympics in the year 2004.

   (b) Estimate the year in which the winning distance will be 280 feet.

57 *House appreciation* Six years ago a house was purchased for $89,000. This year it was appraised at

$125,000. Assume that the value V of the house after its purchase is a linear function of time t (in years).

(a) Express V in terms of t.

(b) How many years after the purchase date was the house worth $103,000?

58 *Temperature scales* The freezing point of water is $0°C$, or $32°F$, and the boiling point is $100°C$, or $212°F$.

(a) Express the Fahrenheit temperature F as a linear function of the Celsius temperature C.

(b) What temperature increase in $°F$ corresponds to an increase in temperature of $1°C$?

59 *Air temperature* Below the cloud base, the air temperature T (in $°F$) at height h (in feet) can be approximated by the equation $T = T_0 - \left(\frac{5.5}{1000}\right)h$, where T_0 is the temperature at ground level.

(a) Determine the air temperature at a height of 1 mile if the ground temperature is $70°F$.

(b) At what altitude is the temperature freezing?

60 *Height of a cloud* The height h (in feet) of the cloud base can be estimated using $h = 227(T - D)$, where T is the ground temperature and D is the dew point.

(a) If the temperature is $70°F$ and the dew point is $55°F$, approximate the height of the cloud base.

(b) If the dew point is $65°F$ and the cloud base is 3500 feet, estimate the ground temperature.

61 *A cloud's temperature* The temperature T within a cloud at height h (in feet) above the cloud base can be approximated using the equation $T = B - \left(\frac{3}{1000}\right)h$, where B is the temperature of the cloud at its base. Determine the temperature at 10,000 feet in a cloud with a base temperature of $55°F$ and a base height of 4000 feet. **Note:** For an interesting application involving the three preceding exercises, see Exercise 4 in the Discussion Exercises at the end of the chapter.

62 *Frustum of a cone* The shape of the first spacecraft in the Apollo program was a frustum of a right circular cone—a solid formed by truncating a cone by a plane parallel to its base. For the frustum shown in the figure, the radii a and b have already been determined.

(a) Use similar triangles to express y as a function of h.

(b) Derive a formula for the volume of the frustum as a function of h.

(c) If $a = 6$ ft and $b = 3$ ft, for what value of h is the volume of the frustum 600 ft^3?

Exercise 62

63 *Radioactive decay of radon gas* When uranium disintegrates into lead, one step in the process is the radioactive decay of radium into radon gas. Radon enters through the soil into home basements, where it presents a health hazard if inhaled. In the simplest case of radon detection, a sample of air with volume V is taken. After equilibrium has been established, the radioactive decay D of the radon gas is counted with efficiency E over time t. The radon concentration C present in the sample of air varies directly as the product of D and E and inversely as the product of V and t. For a fixed radon concentration C and time t, find the change in the radioactive decay count D if V is doubled and E is reduced by 20%.

64 Express the area A of an equilateral triangle as a function of the length s of a side.

65 Choose the equation that best describes the table of data.

x	y
1	2.1213
2	3.6742
3	4.7434
4	5.6125
5	6.3640

(1) $y = 1.5529x + 0.5684$

(2) $y = \dfrac{3}{x} + x^2 - 1$

(3) $y = 3\sqrt{x - 0.5}$

(4) $y = 3x^{1/3} + 1.1213$

CHAPTER 1 DISCUSSION EXERCISES

1 Surface area of a tank You know that a spherical tank holds 10,000 gallons of water. What do you need to know to determine the surface area of the tank? Estimate the surface area of the tank.

2 Determine the conditions under which $\sqrt{a^2 + b^2} = a + b$.

3 Refer to Appendix V for a review of complex numbers.

(a) Find an expression of the form $p + qi$ for the multiplicative inverse of $\dfrac{a + bi}{c + di}$, where a, b, c, and d are real numbers.

(b) Does this expression apply to real numbers of the form a/c?

(c) Are there any restrictions on your answer for part (a)?

4 Freezing level in a cloud Refer to Exercises 59–61 in the Chapter 1 Review Exercises.

(a) Approximate the height of the freezing level in a cloud if the ground temperature is 80°F and the dew point is 68°F.

(b) Find a formula for the height h of the freezing level in a cloud for ground temperature G and dew point D.

5 The midpoint formula could be considered to be the "halfway" formula since it gives us the point that is $\frac{1}{2}$ of the distance from the point $P(x_1, y_1)$ to the point $Q(x_2, y_2)$. Develop a "m-nth way" formula that gives the point $R(x_3, y_3)$ that is m/n of the distance from P to Q (assume m and n are positive integers with $m < n$).

6 Compare the graphs of $y = \sqrt[3]{x}$, $y = \sqrt{x}$, $y = x$, $y = x^2$, and $y = x^3$ on the interval $0 \le x \le 2$. Write a generalization based on what you find out about graphs of equations of the form $y = x^{p/q}$, where $x \ge 0$ and p and q are positive integers.

7 Simplify the difference quotient in Exercises 9 and 10 of Section 1.3 for an arbitrary quadratic function of the form $f(x) = ax^2 + bx + c$.

8 Write an expression for $g(x)$ if the graph of g is obtained from the graph of $f(x) = \frac{1}{2}x - 3$ by reflecting f about the

(a) x-axis   (b) y-axis   (c) line $y = 2$

(d) line $x = 3$ (e) line $y = x$

9 Billing for service A common method of billing for service calls is to charge a flat fee plus an additional fee for each quarter-hour spent on the call. Create a function for a washer repair company that charges $40 plus $20 for each quarter-hour or portion thereof—for example, a 30-minute repair call would cost $80, while a 31-minute repair call would cost $100. The input to your function is any positive integer. (*Hint:* See Exercise 42(e) of Section 1.4.)

C 10 Density of the ozone layer The density D (in 10^{-3} cm/km) of the ozone layer at altitudes x between 3 and 15 kilometers during winter at Edmonton, Canada, was determined experimentally to be

$$D = 0.0833x^2 - 0.4996x + 3.5491.$$

Express x as a function of D.

C 11 Precipitation in Minneapolis The average monthly precipitation in inches in Minneapolis is listed in the table.

Month	Precipitation
Jan.	0.7
Feb.	0.8
Mar.	1.5
Apr.	1.9
May	3.2
June	4.0
July	3.3
Aug.	3.2
Sept.	2.4
Oct.	1.6
Nov.	1.4
Dec.	0.9

(a) Plot the average monthly precipitation.

(b) Model these data with a piecewise function f that is first quadratic and then linear.

(c) Graph f together with the data.

C 12 You need an inverse function graphing feature on your graphing utility for part (a) of this exercise.

(a) Graph
$$f(x) = 0.0346x^3 - 0.0731x^2 - 0.7154x + 3.4462$$
and $y = f^{-1}(x)$ on $[-15, 15]$ by $[-10, 10]$.

(b) Discuss what happens to the graph of $y = f^{-1}(x)$ (in general) as the graph of $y = f(x)$ is increasing or is decreasing.

(c) What can you conclude about the intersection points of the graphs of a function and its inverse?

THE TRIGONOMETRIC FUNCTIONS

Trigonometry was invented over 2000 years ago by the Greeks, who needed precise methods for measuring angles and sides of triangles. In fact, the word *trigonometry* was derived from the two Greek words *trigonon* (triangle) and *metria* (measurement). The chapter begins with a discussion of angles and how they are measured. We next introduce the trigonometric functions by using a unit circle and then discuss their graphs. This modern approach to trigonometry is followed by classical methods that employ ratios of sides of right triangles. We then consider graphing techniques that make use of amplitudes, periods, and phase shifts. The chapter concludes with a section on applied problems.

2.1 ANGLES

Figure 1

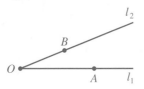

Figure 2
Coterminal angles

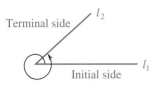

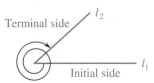

In geometry an **angle** is defined as the set of points determined by two rays, or half-lines, l_1 and l_2, having the same endpoint O. If A and B are points on l_1 and l_2, as in Figure 1, we refer to **angle *AOB*** (denoted $\angle AOB$). An angle may also be considered as two finite line segments with a common endpoint.

In trigonometry we often interpret angles as rotations of rays. Start with a fixed ray l_1, having endpoint O, and rotate it about O, in a plane, to a position specified by ray l_2. We call l_1 the **initial side,** l_2 the **terminal side,** and O the **vertex** of $\angle AOB$. The amount or direction of rotation is not restricted in any way. We might let l_1 make several revolutions in either direction about O before coming to position l_2, as illustrated by the curved arrows in Figure 2. Thus, many different angles have the same initial and terminal sides. Any two such angles are called **coterminal angles.** A **straight angle** is an angle whose sides lie on the same straight line but extend in opposite directions from its vertex.

If we introduce a rectangular coordinate system, then the **standard position** of an angle is obtained by taking the vertex at the origin and letting the initial side l_1 coincide with the positive *x*-axis. If l_1 is rotated in a *counterclockwise* direction to the terminal position l_2, then the angle is considered **positive.** If l_1 is rotated in a *clockwise* direction, the angle is **negative.** We often denote angles by lowercase Greek letters such as α (*alpha*), β (*beta*), γ (*gamma*), θ (*theta*), ϕ (*phi*), and so on. Figure 3 contains sketches of two positive angles, α and β, and a negative angle, γ. If the terminal side of an angle in standard position is in a certain quadrant, we say that the *angle* is in that quadrant. In Figure 3, α is in quadrant III, β is in quadrant I, and γ is in quadrant II. An angle is called a **quadrantal angle** if its terminal side lies on a coordinate axis.

Figure 3   Standard position of an angle

Positive angle   Positive angle   Negative angle

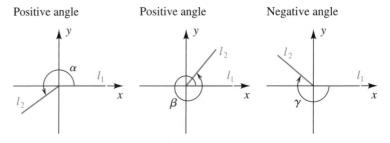

One unit of measurement for angles is the **degree**. The angle in standard position obtained by one complete revolution in the counterclockwise direction has measure 360 degrees, written 360°. Thus, an angle of measure 1 degree (1°) is obtained by $\frac{1}{360}$ of one complete counterclockwise revolution. In Figure 4, several angles measured in degrees are shown in standard

Figure 4

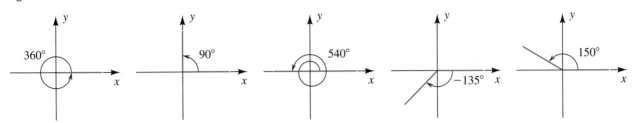

position on rectangular coordinate systems. Note that the first three are quadrantal angles.

Throughout our work, a notation such as $\theta = 60°$ specifies an angle θ whose measure is 60°. We also refer to *an angle of* 60° or *a* 60° *angle,* instead of using the more precise (but cumbersome) phrase *an angle having measure* 60°.

EXAMPLE 1   Finding coterminal angles

If $\theta = 60°$ is in standard position, find two positive angles and two negative angles that are coterminal with θ.

Solution   The angle θ is shown in standard position in the first sketch in Figure 5. To find positive coterminal angles, we may add 360° or 720° (or any other positive multiple of 360°) to θ, obtaining

$$60° + 360° = 420° \quad \text{and} \quad 60° + 720° = 780°.$$

These coterminal angles are also shown in Figure 5.

To find negative coterminal angles, we may add −360° or −720° (or any other negative multiple of 360°), obtaining

$$60° + (-360°) = -300° \quad \text{and} \quad 60° + (-720°) = -660°,$$

as shown in the last two sketches in Figure 5.

Figure 5

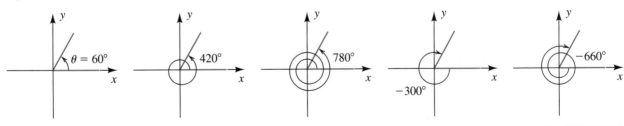

A **right angle** is half of a straight angle and has measure 90°. The following chart contains definitions of other special types of angles.

Terminology	Definition	Illustration
acute angle θ	$0° < \theta < 90°$	$12°$; $37°$
obtuse angle θ	$90° < \theta < 180°$	$95°$; $157°$
complementary angles α, β	$\alpha + \beta = 90°$	$20°, 70°$; $7°, 83°$
supplementary angles α, β	$\alpha + \beta = 180°$	$115°, 65°$; $18°, 162°$

If smaller measurements than the degree are required, we can use tenths, hundredths, or thousandths of degrees. Alternatively, we can divide the degree into 60 equal parts, called **minutes** (denoted by $'$), and each minute into 60 equal parts, called **seconds** (denoted by $''$). Thus, $1° = 60'$, and $1' = 60''$. The notation $\theta = 73°56'18''$ refers to an angle θ that has measure 73 degrees, 56 minutes, 18 seconds.

EXAMPLE 2   Finding complementary angles

Find the angle that is complementary to θ:

(a) $\theta = 25°43'37''$ **(b)** $\theta = 73.26°$

Solution   We wish to find $90° - \theta$. It is convenient to write $90°$ as an equivalent measure, $89°59'60''$.

(a)

$$90° = 89°59'60''$$
$$\underline{\quad\theta\quad = 25°43'37''}$$
$$90° - \theta = 64°16'23''$$

(b)

$$90° = 90.00°$$
$$\underline{\quad\theta\quad = 73.26°}$$
$$90° - \theta = 16.74°$$

Figure 6
Central angle θ

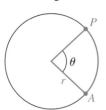

Degree measure for angles is used in applied areas such as surveying, navigation, and the design of mechanical equipment. In scientific applications that require calculus, it is customary to employ *radian measure*. To define an angle of radian measure 1, we consider a circle of any radius r. A **central angle** of a circle is an angle whose vertex is at the center of the circle. If θ is the central angle shown in Figure 6, we say that the **arc** AP (denoted $\overset{\frown}{AP}$) of the circle **subtends** θ or that θ **is subtended by** $\overset{\frown}{AP}$. If the length of $\overset{\frown}{AP}$ is equal to the radius r of the circle, then θ has a measure of one radian, as in the next definition.

DEFINITION OF
RADIAN MEASURE

One radian is the measure of the central angle of a circle subtended by an arc equal in length to the radius of the circle.

If we consider a circle of radius r, then an angle α whose measure is 1 radian intercepts an arc AP of length r, as illustrated in Figure 7(a). The angle β in Figure 7(b) has radian measure 2, since it is subtended by an arc of length $2r$. Similarly, γ in (c) of the figure has radian measure 3, since it is subtended by an arc of length $3r$.

Figure 7

(a) $\alpha = 1$ radian

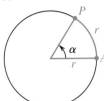

(b) $\beta = 2$ radians

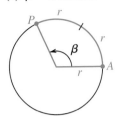

(c) $\gamma = 3$ radians

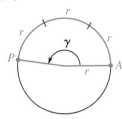

(d) $360° = 2\pi \approx 6.28$ radians

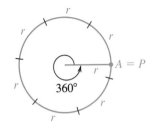

To find the radian measure corresponding to 360°, we must find the number of times that a circular arc of length r can be laid off along the circumference (see Figure 7(d)). This number is not an integer or even a rational number. Since the circumference of the circle is $2\pi r$, the number of times r units can be laid off is 2π. Thus, an angle of measure 2π radians corresponds to the degree measure 360°, and we write $360° = 2\pi$ radians. This result gives us the following relationships.

Relationships Between Degrees and Radians

> **(1)** $180° = \pi$ radians
>
> **(2)** $1° = \dfrac{\pi}{180}$ radian ≈ 0.0175 radian
>
> **(3)** 1 radian $= \left(\dfrac{180°}{\pi}\right) \approx 57.2958°$

When radian measure of an angle is used, no units will be indicated. Thus, if an angle has radian measure 5, we write $\theta = 5$ instead of $\theta = 5$ *radians.* There should be no confusion as to whether radian or degree measure is being used, since if θ has *degree* measure 5°, we write $\theta = 5°$, and *not* $\theta = 5$.

The next chart illustrates how to change from one angular measure to another.

Changing Angular Measures

To change	Multiply by	Illustration
degrees to radians	$\dfrac{\pi}{180°}$	$150° = 150°\left(\dfrac{\pi}{180°}\right) = \dfrac{5\pi}{6}$ $225° = 225°\left(\dfrac{\pi}{180°}\right) = \dfrac{5\pi}{4}$
radians to degrees	$\dfrac{180°}{\pi}$	$\dfrac{7\pi}{4} = \dfrac{7\pi}{4}\left(\dfrac{180°}{\pi}\right) = 315°$ $\dfrac{\pi}{3} = \dfrac{\pi}{3}\left(\dfrac{180°}{\pi}\right) = 60°$

We may use the technique illustrated in the preceding chart to obtain the following table, which displays the corresponding radian and degree measures of special angles.

Radians	0	$\dfrac{\pi}{6}$	$\dfrac{\pi}{4}$	$\dfrac{\pi}{3}$	$\dfrac{\pi}{2}$	$\dfrac{2\pi}{3}$	$\dfrac{3\pi}{4}$	$\dfrac{5\pi}{6}$	π	$\dfrac{7\pi}{6}$	$\dfrac{5\pi}{4}$	$\dfrac{4\pi}{3}$	$\dfrac{3\pi}{2}$	$\dfrac{5\pi}{3}$	$\dfrac{7\pi}{4}$	$\dfrac{11\pi}{6}$	2π
Degrees	0°	30°	45°	60°	90°	120°	135°	150°	180°	210°	225°	240°	270°	300°	315°	330°	360°

Several of these special angles, in radian measure, are shown in standard position in Figure 8.

Figure 8

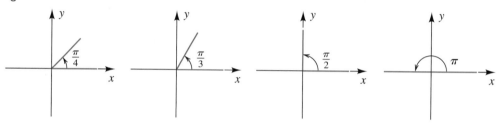

E X A M P L E   3   Changing radians to degrees, minutes, and seconds

If $\theta = 3$, approximate θ in terms of degrees, minutes, and seconds.

Solution

$$3 \text{ radians} = 3\left(\frac{180°}{\pi}\right) \qquad \text{multiply by } \frac{180°}{\pi}$$

$$\approx 171.8873° \qquad \text{approximate}$$

$$= 171° + (0.8873)(60') \qquad 1° = 60'$$

$$= 171° + 53.238' \qquad \text{multiply}$$

$$= 171° + 53' + (0.238)(60'') \qquad 1' = 60''$$

$$= 171°53' + 14.28'' \qquad \text{multiply}$$

$$\approx 171°53'14'' \qquad \text{approximate}$$

Some calculators have a key that can be used to convert radian measure to degrees, and vice versa. Calculators may also have a $\boxed{\text{DMS}}$ key for converting decimal degree measure to degrees, minutes, and seconds, and vice versa. Since entries are made in terms of decimals, angles that are expressed in terms of degrees, minutes, and seconds must be changed to decimal form. If a $\boxed{\text{DMS}}$ key is not available, the procedure given in the next example may be followed.

> **EXAMPLE 4** Expressing minutes and seconds
> as decimal degrees

Express $19°47'23''$ as a decimal, to the nearest ten-thousandth of a degree.

Solution Since $1' = \left(\frac{1}{60}\right)°$ and $1'' = \left(\frac{1}{60}\right)' = \left(\frac{1}{3600}\right)°$,

$$19°47'23'' = 19° + \left(\frac{47}{60}\right)° + \left(\frac{23}{3600}\right)°$$
$$\approx 19° + 0.7833° + 0.0064°$$
$$= 19.7897°.$$

The next result specifies the relationship between the length of a circular arc and the central angle that it subtends.

Formula for the Length of a Circular Arc

> If an arc of length s on a circle of radius r subtends a central angle of radian measure θ, then
>
> $$s = r\theta.$$

Figure 9

(a)

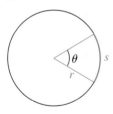

(b)

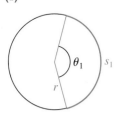

Proof A typical arc of length s and the corresponding central angle θ are shown in Figure 9(a). Figure 9(b) shows an arc of length s_1 and central angle θ_1. If radian measure is used, then, from plane geometry, the ratio of the lengths of the arcs is the same as the ratio of the angular measures; that is,

$$\frac{s}{s_1} = \frac{\theta}{\theta_1}.$$

If we consider the special case in which θ_1 has radian measure 1, then, from the definition of radian, $s_1 = r$ and

$$\frac{s}{r} = \frac{\theta}{1}, \quad \text{or} \quad s = r\theta.$$

The next formula is proved in a similar manner.

Formula for the Area of a Circular Sector

> If θ is the radian measure of a central angle of a circle of radius r and if A is the area of the circular sector determined by θ, then
>
> $$A = \tfrac{1}{2}r^2\theta.$$

Proof If A and A_1 are the areas of the sectors in Figures 9(a) and 9(b), respectively, then, from plane geometry,

$$\frac{A}{A_1} = \frac{\theta}{\theta_1}, \quad \text{or} \quad A = \frac{A_1}{\theta_1}\theta.$$

If we consider the special case $\theta_1 = 2\pi$, then $A_1 = \pi r^2$ and

$$A = \frac{\pi r^2}{2\pi}\theta = \frac{1}{2}r^2\theta.$$

When using the preceding formulas, it is important to remember to use the radian measure of θ rather than the degree measure, as illustrated in the next example.

EXAMPLE 5   **Using the circular arc and sector formulas**

A central angle θ is subtended by an arc 10 centimeters long on a circle of radius 4 centimeters.

(a) Approximate the measure of θ in degrees.

(b) Find the area of the circular sector determined by θ.

Solution   We proceed as follows:

(a) $s = r\theta$ length of a circular arc formula

$$\theta = \frac{s}{r} \qquad \text{solve for } \theta$$

$$= \tfrac{10}{4} = 2.5 \qquad \text{let } s = 10,\ r = 4$$

This is the *radian* measure of θ. Changing to degrees, we have

$$\theta = 2.5\left(\frac{180°}{\pi}\right) = \frac{450°}{\pi} \approx 143.24°.$$

(b) $A = \tfrac{1}{2}r^2\theta$ area of a circular sector formula

$$= \tfrac{1}{2}(4)^2(2.5) \quad \text{let } r = 4,\ \theta = 2.5 \text{ radians}$$

$$= 20 \text{ cm}^2 \qquad \text{multiply}$$

Figure 10

The **angular speed** of a wheel that is rotating at a constant rate is the angle generated in one unit of time by a line segment from the center of the wheel to a point P on the circumference (see Figure 10). The **linear speed** of a point P on the circumference is the distance that P travels per unit of time.

EXAMPLE 6   **Finding angular and linear speeds**

Suppose that a machine contains a wheel of diameter 3 feet, rotating at a rate of 1600 rpm (revolutions per minute).

(a) Find the angular speed of the wheel.

(b) Find the linear speed of a point P on the circumference of the wheel.

Solution

(a) Let O denote the center of the wheel, and let P be a point on the circumference. Because the number of revolutions per minute is 1600 and because each revolution generates an angle of 2π radians, the angle generated by the line segment OP in one minute has radian measure $(1600)(2\pi)$; that is,

$$\text{angular speed} = (1600)(2\pi) = 3200\pi \text{ radians per minute.}$$

Note that the diameter of the wheel is irrelevant in finding the angular speed.

(b) The linear speed of P is the distance it travels per minute. We may find this distance by using the formula $s = r\theta$, with $r = \frac{3}{2}$ ft and $\theta = 3200\pi$. Thus,

$$s = \tfrac{3}{2}(3200\pi) = 4800\pi \text{ ft,}$$

and consequently the linear speed of P is 4800π ft/min. To the nearest integer, this is approximately 15,080 ft/min, or 171.36 mi/hr. Unlike the angular speed, the linear speed *is* dependent on the diameter of the wheel.

2.1 EXERCISES

Exer. 1–4: If the given angle is in standard position, find two positive coterminal angles and two negative coterminal angles.

1 (a) $120°$ (b) $135°$ (c) $-30°$

2 (a) $240°$ (b) $315°$ (c) $-150°$

3 (a) $620°$ (b) $\dfrac{5\pi}{6}$ (c) $-\dfrac{\pi}{4}$

4 (a) $570°$ (b) $\dfrac{2\pi}{3}$ (c) $-\dfrac{5\pi}{4}$

Exer. 5–6: Find the angle that is complementary to θ.

5 (a) $\theta = 5°17'34''$ (b) $\theta = 32.5°$

6 (a) $\theta = 63°4'15''$ (b) $\theta = 82.73°$

Exer. 7–8: Find the angle that is supplementary to θ.

7 (a) $\theta = 48°51'37''$ (b) $\theta = 136.42°$

8 (a) $\theta = 152°12'4''$ (b) $\theta = 15.9°$

Exer. 9–12: Find the exact radian measure of the angle.

9 (a) $150°$ (b) $-60°$ (c) $225°$

10 (a) $120°$ (b) $-135°$ (c) $210°$

11 (a) $450°$ (b) $72°$ (c) $100°$

12 (a) $630°$ (b) $54°$ (c) $95°$

Exer. 13–16: Find the exact degree measure of the angle.

13 (a) $\dfrac{2\pi}{3}$ (b) $\dfrac{11\pi}{6}$ (c) $\dfrac{3\pi}{4}$

14 (a) $\dfrac{5\pi}{6}$ (b) $\dfrac{4\pi}{3}$ (c) $\dfrac{11\pi}{4}$

15 (a) $-\dfrac{7\pi}{2}$ (b) 7π (c) $\dfrac{\pi}{9}$

16 (a) $-\dfrac{5\pi}{2}$ (b) 9π (c) $\dfrac{\pi}{16}$

Exer. 17–20: Express θ in terms of degrees, minutes, and seconds, to the nearest second.

17 $\theta = 2$ 18 $\theta = 1.5$

19 $\theta = 5$ 20 $\theta = 4$

Exer. 21–24: Express the angle as a decimal, to the nearest ten-thousandth of a degree.

21 37°41′

22 83°17′

23 115°26′27″

24 258°39′52″

Exer. 25–28: Express the angle in terms of degrees, minutes, and seconds, to the nearest second.

25 63.169°

26 12.864°

27 310.6215°

28 81.7238°

Exer. 29–30: If a circular arc of the given length s subtends the central angle θ on a circle, find the radius of the circle.

29 $s = 10$ cm, $\theta = 4$

30 $s = 3$ km, $\theta = 20°$

Exer. 31–32: (a) Find the length of the arc of the colored sector in the figure. (b) Find the area of the sector.

31

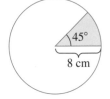

32

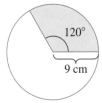

Exer. 33–34: (a) Find the radian and degree measures of the central angle θ subtended by the given arc of length s on a circle of radius r. (b) Find the area of the sector determined by θ.

33 $s = 7$ cm, $r = 4$ cm

34 $s = 3$ ft, $r = 20$ in.

Exer. 35–36: (a) Find the length of the arc that subtends the given central angle θ on a circle of diameter d. (b) Find the area of the sector determined by θ.

35 $\theta = 50°$, $d = 16$ m

36 $\theta = 2.2$, $d = 120$ cm

37 *Measuring distances on the earth* The distance between two points A and B on the earth is measured along a circle having center C at the center of the earth and radius equal to the distance from C to the surface (see the figure). If the diameter of the earth is approximately 8000 miles, approximate the distance between A and B if angle ACB has the indicated measure:

(a) 60°   (b) 45°   (c) 30°   (d) 10°   (e) 1°

Exercise 37

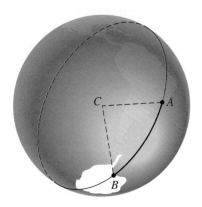

38 *Nautical miles* Refer to Exercise 37. If angle ACB has measure 1′, then the distance between A and B is a nautical mile. Approximate the number of land (statute) miles in a nautical mile.

39 *Measuring angles using distance* Refer to Exercise 37. If two points A and B are 500 miles apart, express angle ACB in radians and in degrees.

40 *A tornado's core* A simple model of the core of a tornado is a right circular cylinder that rotates about its axis. If a tornado has a core diameter of 200 feet and maximum wind speed of 180 mi/hr (or 264 ft/sec) at the perimeter of the core, approximate the number of revolutions the core makes each minute.

41 *The earth's rotation* The earth rotates about its axis once every 23 hours, 56 minutes, and 4 seconds. Approximate the number of radians the earth rotates in one second.

42 *The earth's rotation* Refer to Exercise 41. The equatorial radius of the earth is approximately 3963.3 miles. Find the linear speed of a point on the equator as a result of the earth's rotation.

Exer. 43–44: A wheel of the given radius is rotating at the indicated rate.

(a) Find the angular speed (in radians per minute).

(b) Find the linear speed of a point on the circumference (in ft/min).

43 radius 5 in., 40 rpm

44 radius 9 in., 2400 rpm

45 *Rotation of phonograph records* The two common types of phonograph records, LP albums and singles, have diameters of 12 inches and 7 inches, respectively. The album rotates at a rate of $33\frac{1}{3}$ rpm, and the single rotates at 45 rpm.

(a) Find the angular speed (in radians per minute) of the album and of the single.

(b) Find the linear speed (in ft/min) of a point on the circumference of the album and of the single.

46 *Tire revolutions* A typical tire for a compact car is 22 inches in diameter. If the car is traveling at a speed of 60 mi/hr, find the number of revolutions the tire makes per minute.

47 *Cargo winch* A large winch of diameter 3 feet is used to hoist cargo, as shown in the figure.

(a) Find the distance the cargo is lifted if the winch rotates through an angle of radian measure $7\pi/4$.

(b) Find the angle (in radians) through which the winch must rotate in order to lift the cargo d feet.

Exercise 47

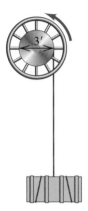

48 *Pendulum's swing* A pendulum in a grandfather clock is 4 feet long and swings back and forth along a 6-inch arc. Approximate the angle (in degrees) through which the pendulum passes during one swing.

49 *Pizza values* A vendor sells two sizes of pizza by the slice. The *small* slice is $\frac{1}{6}$ of a circular 18-inch-diameter pizza, and it sells for $2.00. The *large* slice is $\frac{1}{8}$ of a circular 26-inch-diameter pizza, and it sells for $3.00. Which slice provides more pizza per dollar?

50 *Bicycle mechanics* The sprocket assembly for a bicycle is shown in the figure. If the sprocket of radius r_1 rotates through an angle of θ_1 radians, find the corresponding angle of rotation for the sprocket of radius r_2.

Exercise 50

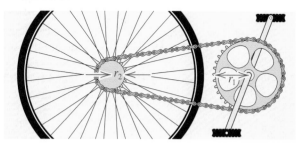

51 *Bicycle mechanics* An expert cyclist can attain a speed of 40 mi/hr. If in Exercise 50 the sprocket assembly has $r_1 = 5$ in., $r_2 = 2$ in. and the wheel has a diameter of 28 inches, approximately how many revolutions per minute of the front sprocket wheel will produce a speed of 40 mi/hr? (*Hint:* First change 40 mi/hr to in./sec.)

52 *Magnetic pole drift* The geographic and magnetic north poles have different locations. Currently, the magnetic north pole is drifting westward through 0.0017 radian per year, where the angle of drift has its vertex at the center of the earth. If this movement continues, approximately how many years will it take for the magnetic north pole to drift a total of 5°?

2.2 THE TRIGONOMETRIC FUNCTIONS

We can introduce trigonometry in two different ways. We can use a modern development that involves a unit circle, or we can use classical right triangle methods that were employed by the ancient Greeks. In this section we discuss the unit circle approach. Right triangle trigonometry will be considered in Section 2.4.

Let U be a unit circle—that is, a circle of radius 1, with center at the origin O of a rectangular coordinate system. Thus, U is the graph of the

Figure II

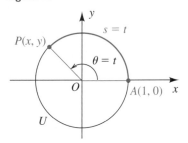

equation $x^2 + y^2 = 1$. Let t be a real number such that $0 < t < 2\pi$, and let θ denote the angle (in standard position) of radian measure t. One possibility is illustrated in Figure 11, where $P(x, y)$ is the point of intersection of the terminal side of θ and the unit circle U and where s is the length of the circular arc from $A(1, 0)$ to $P(x, y)$. Using the formula $s = r\theta$ for the length of a circular arc, with $\theta = t$ and $r = 1$, we see that

$$s = r\theta = 1(t) = t.$$

Thus, t may be regarded either as the radian measure of the angle θ or as the length of the circular arc AP on U.

Next consider *any* nonnegative real number t. If we regard the angle θ of radian measure t as having been generated by rotating the line segment OA about O in the counterclockwise direction, then t is the distance along U that A travels before reaching its final position $P(x, y)$. In Figure 12 we have illustrated a case for $t < 2\pi$; however, if $t > 2\pi$, then A may travel around U several times in a counterclockwise direction before reaching $P(x, y)$.

If $t < 0$, then the rotation of OA is in the *clockwise* direction, and the distance A travels before reaching $P(x, y)$ is $|t|$, as illustrated in Figure 13.

Figure 12
$\theta = t, t > 0$

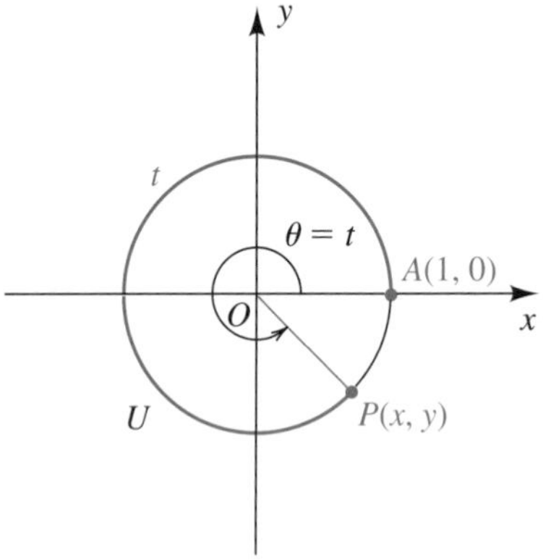

Figure 13
$\theta = t, t < 0$

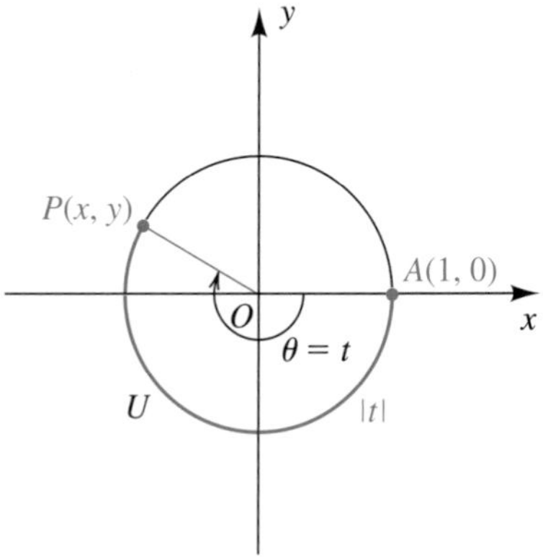

*There are other, less common trigonometric functions that we will not use in this text. Hence, we will refer to these six trigonometric functions as *the* trigonometric functions.

The preceding discussion indicates how *we may associate with each real number t a unique point $P(x, y)$ on U.* We shall call $P(x, y)$ the **point on the unit circle U that corresponds to t.** The coordinates (x, y) of P may be used to define the six **trigonometric functions.*** These functions are called the **sine, cosine, tangent, cotangent, secant,** and **cosecant** functions and are abbreviated **sin, cos, tan, cot, sec,** and **csc,** respectively. If t is a real number, then the real number that the sine function associates with t is designated by either $\sin (t)$ or $\sin t$. Similar notation is used for the other five functions.

DEFINITION OF THE TRIGONOMETRIC FUNCTIONS IN TERMS OF A UNIT CIRCLE

If t is a real number and $P(x, y)$ is the point on the unit circle U that corresponds to t, then

$$\sin t = y \qquad\qquad \cos t = x \qquad\qquad \tan t = \frac{y}{x} \ (\text{if } x \neq 0)$$

$$\csc t = \frac{1}{y} \ (\text{if } y \neq 0) \qquad \sec t = \frac{1}{x} \ (\text{if } x \neq 0) \qquad \cot t = \frac{x}{y} \ (\text{if } y \neq 0).$$

The formulas in this definition express function values in terms of coordinates of a point P on a unit circle. For this reason, the trigonometric functions are sometimes referred to as the **circular functions.**

EXAMPLE 1   Finding values of the trigonometric functions

A point $P(x, y)$ on the unit circle U corresponding to a real number t is shown in Figure 14, for $\pi < t < 3\pi/2$. Find the values of the trigonometric functions at t.

Solution   Referring to Figure 14, we see that the coordinates of $P(x, y)$ are

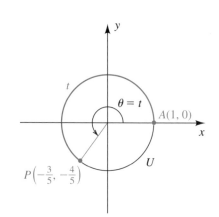

$$x = -\tfrac{3}{5}, \qquad y = -\tfrac{4}{5}.$$

Using the definition of the trigonometric functions in terms of a unit circle gives us the following:

$$\sin t = -\frac{4}{5} \qquad\qquad \cos t = -\frac{3}{5} \qquad\qquad \tan t = \frac{-\frac{4}{5}}{-\frac{3}{5}} = \frac{4}{3}$$

$$\csc t = \frac{1}{-\frac{4}{5}} = -\frac{5}{4} \qquad \sec t = \frac{1}{-\frac{3}{5}} = -\frac{5}{3} \qquad \cot t = \frac{-\frac{3}{5}}{-\frac{4}{5}} = \frac{3}{4}$$

EXAMPLE 2   Finding a point on U relative to a given point

Let $P(t)$ denote the point on the unit circle U that corresponds to t for $0 \leq t < 2\pi$. If $P(t) = \left(\tfrac{4}{5}, \tfrac{3}{5}\right)$, find

(a) $P(t + \pi)$ **(b)** $P(t - \pi)$ **(c)** $P(-t)$

Solution

(a) The point $P(t)$ on U is plotted in Figure 15(a), where we have also shown the arc AP of length t. To find $P(t + \pi)$, we travel a distance π in the *counterclockwise* direction along U from $P(t)$, as indicated by the blue arc in the figure. Since π is one-half the circumference of U, this gives us the point $P(t + \pi) = \left(-\tfrac{4}{5}, -\tfrac{3}{5}\right)$, diametrically opposite $P(t)$.

(continued)

Figure 15

(a)

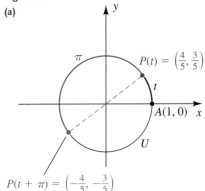

$$P(t + \pi) = \left(-\frac{4}{5}, -\frac{3}{5}\right)$$

(b)

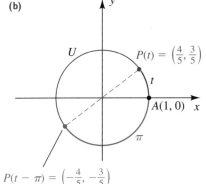

$$P(t - \pi) = \left(-\frac{4}{5}, -\frac{3}{5}\right)$$

(c)

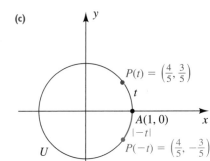

(b) To find $P(t - \pi)$, we travel a distance π in the *clockwise* direction along U from $P(t)$, as indicated in Figure 15(b). This gives us $P(t - \pi) = \left(-\frac{4}{5}, -\frac{3}{5}\right)$. Note that $P(t + \pi) = P(t - \pi)$.

(c) To find $P(-t)$, we travel along U a distance $|-t|$ in the *clockwise* direction from $A(1, 0)$, as indicated in Figure 15(c). This is equivalent to reflecting $P(t)$ through the x-axis. Thus, we merely change the sign of the y-coordinate of $P(t) = \left(\frac{4}{5}, \frac{3}{5}\right)$ to obtain $P(-t) = \left(\frac{4}{5}, -\frac{3}{5}\right)$.

Since $\sin t = y$ and $\csc t = 1/y$, we see that $\sin t$ and $\csc t$ are reciprocals of each other. This gives us the two identities in the left-hand column of the next box. Similarly, $\cos t$ and $\sec t$ are reciprocals of each other, as are $\tan t$ and $\cot t$.

Reciprocal Identities

$$\sin t = \frac{1}{\csc t} \qquad \cos t = \frac{1}{\sec t} \qquad \tan t = \frac{1}{\cot t}$$

$$\csc t = \frac{1}{\sin t} \qquad \sec t = \frac{1}{\cos t} \qquad \cot t = \frac{1}{\tan t}$$

Several other important identities involving the trigonometric functions will be discussed later in this section.

The domain of both the sine and the cosine function is $\mathbb{R}$, since $\sin t = y$ and $\cos t = x$ exist for every real number t.

For $\tan t = y/x$ and $\sec t = 1/x$, the number x is a denominator, and hence we must exclude values of t for which x is 0—that is, values of t that give us the points $(0, 1)$ and $(0, -1)$ on the y-axis. Thus, the domain of the tangent and secant functions consists of all real numbers *except* $\pm\pi/2$, $\pm3\pi/2$, $\pm5\pi/2$, and, in general, $(\pi/2) + \pi n$ for any integer n.

For $\cot t = x/y$ and $\csc t = 1/y$, the number y is a denominator, and hence we must exclude values of t that give us the points $(1, 0)$ and $(-1, 0)$ on the x-axis. Thus, the domain of the cotangent and cosecant functions consists of all real numbers *except* $0, \pm\pi, \pm2\pi, \pm3\pi$, and, in general, πn for any integer n.

The following chart summarizes this discussion.

Function	Domain
sine, cosine	all real numbers
tangent, secant	all real numbers except $(\pi/2) + \pi n$ for any integer n
cotangent, cosecant	all real numbers except πn for any integer n

Note that $P(x, y)$ is a point on the unit circle U, and hence $|x| \leq 1$ and $|y| \leq 1$. This implies that

$$|\sin t| \leq 1, \qquad |\cos t| \leq 1, \qquad |\csc t| \geq 1, \qquad \text{and} \qquad |\sec t| \geq 1$$

for every t in the domains of these functions.

In the next two examples we find some special values of the trigonometric functions.

Figure 16

(a)

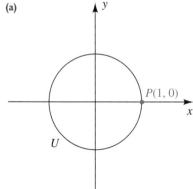

EXAMPLE 3 Finding special values of the trigonometric functions

Find the values of the trigonometric functions at t:

(a) $t = 0$ **(b)** $t = \dfrac{\pi}{4}$ **(c)** $t = \dfrac{\pi}{2}$

Solution

(a) The point P on the unit circle U that corresponds to $t = 0$ has coordinates $(1, 0)$, as shown in Figure 16(a). Thus, we let $x = 1$ and $y = 0$ in the definition of the trigonometric functions in terms of a unit circle, obtaining

$$\sin 0 = y = 0, \quad \cos 0 = x = 1,$$

$$\tan 0 = \frac{y}{x} = \frac{0}{1} = 0, \quad \sec 0 = \frac{1}{x} = \frac{1}{1} = 1.$$

Note that $\csc 0$ and $\cot 0$ are undefined, since $y = 0$ is a denominator.

(b)

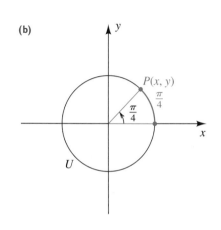

(b) If $t = \pi/4$, then the angle of radian measure $\pi/4$ shown in Figure 16(b) bisects the first quadrant and the point $P(x, y)$ lies on the line $y = x$. Since $P(x, y)$ is on the unit circle $x^2 + y^2 = 1$ and since $y = x$, we obtain

$$x^2 + x^2 = 1, \qquad \text{or} \qquad 2x^2 = 1.$$

(continued)

Figure 16   *(continued)*
(c)

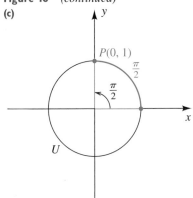

Solving for x and noting that $x > 0$ gives us

$$x = \frac{1}{\sqrt{2}} = \frac{\sqrt{2}}{2}.$$

Thus, P is the point $(\sqrt{2}/2, \sqrt{2}/2)$. Letting $x = \sqrt{2}/2$ and $y = \sqrt{2}/2$ in the definition of the trigonometric functions gives us the following:

$$\sin\frac{\pi}{4} = \frac{\sqrt{2}}{2} \qquad \cos\frac{\pi}{4} = \frac{\sqrt{2}}{2} \qquad \tan\frac{\pi}{4} = \frac{\sqrt{2}/2}{\sqrt{2}/2} = 1$$

$$\csc\frac{\pi}{4} = \frac{2}{\sqrt{2}} = \sqrt{2} \qquad \sec\frac{\pi}{4} = \frac{2}{\sqrt{2}} = \sqrt{2} \qquad \cot\frac{\pi}{4} = \frac{\sqrt{2}/2}{\sqrt{2}/2} = 1$$

(c) The point P on U that corresponds to $t = \pi/2$ has coordinates $(0, 1)$, as shown in Figure 16(c). Thus, we let $x = 0$ and $y = 1$ in the definition of the trigonometric functions, obtaining

$$\sin\frac{\pi}{2} = 1, \qquad \cos\frac{\pi}{2} = 0, \qquad \csc\frac{\pi}{2} = \frac{1}{1} = 1, \qquad \cot\frac{\pi}{2} = \frac{0}{1} = 0.$$

The tangent and secant functions are undefined, since $x = 0$ is a denominator in each case.

Figure 17
(a)

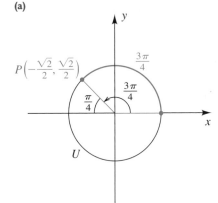

EXAMPLE 4   **Finding special values of the trigonometric functions**

Find the values of the trigonometric functions at t:

(a) $t = \dfrac{3\pi}{4}$ **(b)** $t = \pi$ **(c)** $t = -\dfrac{\pi}{4}$

Solution

(a) We can find the coordinates of the point P on U that corresponds to $t = 3\pi/4$ by symmetry from the point $(\sqrt{2}/2, \sqrt{2}/2)$, which we obtained for $t = \pi/4$ in Example 3. This gives us $P(-\sqrt{2}/2, \sqrt{2}/2)$, as shown in Figure 17(a).

Letting $x = -\sqrt{2}/2$ and $y = \sqrt{2}/2$ in the definition of the trigonometric functions leads to the values in the first row of the following table.

(b)

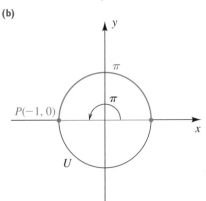

t	$\sin t$	$\cos t$	$\tan t$	$\cot t$	$\sec t$	$\csc t$
$\dfrac{3\pi}{4}$	$\dfrac{\sqrt{2}}{2}$	$-\dfrac{\sqrt{2}}{2}$	-1	-1	$-\sqrt{2}$	$\sqrt{2}$
π	0	-1	0	undefined	-1	undefined
$-\dfrac{\pi}{4}$	$-\dfrac{\sqrt{2}}{2}$	$\dfrac{\sqrt{2}}{2}$	-1	-1	$\sqrt{2}$	$-\sqrt{2}$

(c)

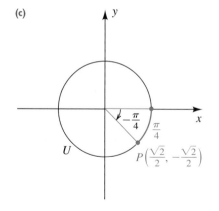

(b) If $t = \pi$, we obtain the point $P(-1, 0)$ on U, shown in Figure 17(b). Letting $x = -1$ and $y = 0$ in the definition of the trigonometric functions gives us the second row of the table in part (a).

(c) We can find the coordinates of the point P on U that corresponds to $t = -\pi/4$ by reflecting the point $P(x, y)$ in Figure 16(b) through the x-axis. Using the results of Example 3(b), we obtain $P(\sqrt{2}/2, -\sqrt{2}/2)$, as shown in Figure 17(c). Letting $x = \sqrt{2}/2$ and $y = -\sqrt{2}/2$ in the definition of the trigonometric functions gives us the third row of the table in part (a).

In Examples 3 and 4 we considered only multiples of $t = \pi/4$ in order to obtain the *exact* coordinates of the point P on U. Other special cases involving multiples of $t = \pi/6$ will be discussed in Section 2.4. For arbitrary values of t we can find only *approximate* values of the trigonometric functions. This will be discussed in later sections.

Let us determine the signs associated with values of the trigonometric functions for various values of t. Suppose that t determines a point $P(x, y)$ on U that lies in quadrant II. In this case, x is negative and y is positive; therefore, $\sin t = y$ and $\csc t = 1/y$ are positive. The other four trigonometric functions are negative. By checking the remaining quadrants in a similar fashion, we obtain the following table.

Signs of the Trigonometric Functions

Quadrant containing $P(x, y)$	Positive functions	Negative functions
I	all	none
II	sin, csc	cos, sec, tan, cot
III	tan, cot	sin, csc, cos, sec
IV	cos, sec	sin, csc, tan, cot

Figure 18

Positive trigonometric functions

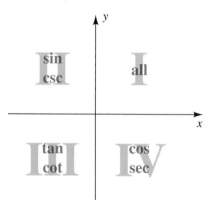

The diagram in Figure 18 may be useful for remembering quadrants in which trigonometric functions are *positive*. If a function is not listed (such as cos in quadrant II), then that function is negative.

EXAMPLE 5   **Determining the quadrant that contains a point on U**

Let $P(x, y)$ be the point on U corresponding to t. Find the quadrant containing P if $\cos t > 0$ and $\sin t < 0$.

Solution   Referring to the table of signs, we see that $\cos t > 0$ if P is in quadrant I or IV and $\sin t < 0$ if P is in quadrant III or IV. Hence, for both conditions to be satisfied, P must be in quadrant IV.

The formulas listed in the next box are, without doubt, the most important identities in trigonometry, because they may be used to simplify and unify many different aspects of the subject. Since the formulas are part of the foundation for work in trigonometry, they are called the *fundamental identities.*

Three of the fundamental identities involve squares, such as $(\sin t)^2$ and $(\cos t)^2$. In general, if n is an integer different from -1, then a power such as $(\cos t)^n$ is written $\cos^n t$. The symbols $\sin^{-1} t$ and $\cos^{-1} t$ are reserved for inverse trigonometric functions, which we will discuss in the next chapter. With this agreement on notation, we have, for example,

$$\cos^2 t = (\cos t)^2 = (\cos t)(\cos t)$$
$$\tan^3 t = (\tan t)^3 = (\tan t)(\tan t)(\tan t)$$
$$\sec^4 t = (\sec t)^4 = (\sec t)(\sec t)(\sec t)(\sec t).$$

Let us first list all the fundamental identities and then discuss the proofs. These identities are true for every allowable value of t, and t may take on various forms. For example, using the first Pythagorean identity with $t = 4\theta$, we know that

$$\sin^2 4\theta + \cos^2 4\theta = 1.$$

We shall see later that these identities also are true for angles.

The Fundamental Identities

---

(1) The reciprocal identities:

$$\csc t = \frac{1}{\sin t} \qquad \sec t = \frac{1}{\cos t} \qquad \cot t = \frac{1}{\tan t}$$

(2) The tangent and cotangent identities:

$$\tan t = \frac{\sin t}{\cos t} \qquad \cot t = \frac{\cos t}{\sin t}$$

(3) The Pythagorean identities:

$$\sin^2 t + \cos^2 t = 1 \quad 1 + \tan^2 t = \sec^2 t \quad 1 + \cot^2 t = \csc^2 t$$

Proof

(1) The reciprocal identities were proved earlier.

(2) To prove the tangent and cotangent identities, we apply the definition of the trigonometric functions as follows:

$$\tan t = \frac{y}{x} = \frac{\sin t}{\cos t}, \qquad \cot t = \frac{x}{y} = \frac{\cos t}{\sin t}$$

(3) If $P(x, y)$ is a point on the unit circle, we have the following:

$$y^2 + x^2 = 1 \quad \text{equation of } U$$

$$(\sin t)^2 + (\cos t)^2 = 1 \quad \text{definitions of } \sin t \text{ and } \cos t$$

$$\sin^2 t + \cos^2 t = 1 \quad \text{equivalent notation}$$

This proves the first Pythagorean identity.

If $\cos t \neq 0$, then we may prove the second identity as follows:

$$\sin^2 t + \cos^2 t = 1 \qquad \text{first Pythagorean identity}$$

$$\frac{\sin^2 t}{\cos^2 t} + \frac{\cos^2 t}{\cos^2 t} = \frac{1}{\cos^2 t} \qquad \text{divide by } \cos^2 t$$

$$\left(\frac{\sin t}{\cos t}\right)^2 + 1 = \left(\frac{1}{\cos t}\right)^2 \qquad \text{equivalent equation}$$

$$\tan^2 t + 1 = \sec^2 t \qquad \text{tangent and reciprocal identities}$$

The final identity, $1 + \cot^2 t = \csc^2 t$, can be proved in similar fashion by dividing the first Pythagorean identity by $\sin^2 t$ (provided $\sin t \neq 0$).   ◣

Figure 19

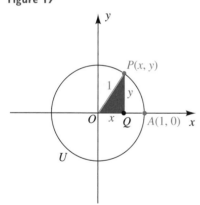

The Pythagorean identities are so named because if $0 < t < \pi/2$ and $P(x, y)$ is the point on U corresponding to t, then, as shown in Figure 19, OQP is a right triangle with hypotenuse 1 and sides of lengths x and y. Hence, by the Pythagorean theorem,

$$y^2 + x^2 = 1 \qquad \text{or, equivalently,} \qquad \sin^2 t + \cos^2 t = 1.$$

In the preceding proof we obtained this result using an equation for U.

We can use the fundamental identities to express each trigonometric function in terms of any other trigonometric function. Two illustrations are given in the next example.

> **EXAMPLE 6**   Using fundamental identities

Let t be a real number such that $0 < t < \pi/2$.

(a) Express $\sin t$ in terms of $\cos t$.

(b) Express $\tan t$ in terms of $\sin t$.

Solution

(a) We may proceed as follows:

$$\sin^2 t + \cos^2 t = 1 \qquad \text{Pythagorean identity}$$

$$\sin^2 t = 1 - \cos^2 t \qquad \text{isolate } \sin^2 t$$

$$\sin t = \pm\sqrt{1 - \cos^2 t} \qquad \text{take the square root}$$

$$\sin t = \sqrt{1 - \cos^2 t} \qquad \sin t > 0 \text{ if } 0 < t < \pi/2$$

(continued)

Later in this section (Example 10) we will consider a simplification involving a non-acute angle t.

(b) If we begin with the fundamental identity

$$\tan t = \frac{\sin t}{\cos t},$$

then all that remains is to express $\cos t$ in terms of $\sin t$. We can do this by solving $\sin^2 t + \cos^2 t = 1$ for $\cos t$, obtaining

$$\cos t = \sqrt{1 - \sin^2 t} \quad \text{for} \quad 0 < t < \frac{\pi}{2}.$$

Hence,

$$\tan t = \frac{\sin t}{\cos t} = \frac{\sin t}{\sqrt{1 - \sin^2 t}} \quad \text{for} \quad 0 < t < \frac{\pi}{2}.$$

EXAMPLE 7   **Finding values of trigonometric functions from prescribed conditions**

If $\sin t = \frac{3}{5}$ and $\tan t < 0$, use fundamental identities to find the values of the other five trigonometric functions.

Solution   Since $\sin t = \frac{3}{5} > 0$ and $\tan t < 0$, the point P on U that corresponds to t is in quadrant II. Since $\cos t$ is negative in quadrant II,

$$\cos t = -\sqrt{1 - \sin^2 t} = -\sqrt{1 - \left(\tfrac{3}{5}\right)^2} = -\sqrt{\tfrac{16}{25}} = -\tfrac{4}{5}.$$

Next we use the tangent identity to obtain

$$\tan t = \frac{\sin t}{\cos t} = \frac{\frac{3}{5}}{-\frac{4}{5}} = -\frac{3}{4}.$$

Finally, using the reciprocal identities gives us

$$\csc t = \frac{1}{\sin t} = \frac{1}{\frac{3}{5}} = \frac{5}{3}$$

$$\sec t = \frac{1}{\cos t} = \frac{1}{-\frac{4}{5}} = -\frac{5}{4}$$

$$\cot t = \frac{1}{\tan t} = \frac{1}{-\frac{3}{4}} = -\frac{4}{3}.$$

Fundamental identities are often used to simplify expressions involving trigonometric functions, as illustrated in the next example.

EXAMPLE 8   **Showing that an equation is an identity**

Show that the following equation is an identity by transforming the left side into the right side:

$$(\sec t + \tan t)(1 - \sin t) = \cos t$$

Solution We begin with the left side and proceed as follows:

$$(\sec t + \tan t)(1 - \sin t) = \left(\frac{1}{\cos t} + \frac{\sin t}{\cos t}\right)(1 - \sin t) \quad \text{reciprocal and tangent identities}$$

$$= \left(\frac{1 + \sin t}{\cos t}\right)(1 - \sin t) \quad \text{add fractions}$$

$$= \frac{1 - \sin^2 t}{\cos t} \quad \text{multiply}$$

$$= \frac{\cos^2 t}{\cos t} \quad \sin^2 t + \cos^2 t = 1$$

$$= \cos t \quad \text{cancel } \cos t$$

There are other ways to simplify the expression on the left side in Example 8. We could first multiply the two factors and then simplify and combine terms. The method we employed—changing all expressions to expressions that involve only sines and cosines—is often useful. However, that technique does not always lead to the shortest possible simplification.

Hereafter, we shall use the phrase *verify an identity* instead of *show that an equation is an identity*. When verifying an identity, we often use fundamental identities and algebraic manipulations to simplify expressions, as we did in the preceding example. We understand that, as with the fundamental identities, an identity containing fractions is valid for all values of the variables such that no denominator is zero.

EXAMPLE 9 **Verifying an identity**

Verify the following identity by transforming the left side into the right side:

$$\frac{\tan t + \cos t}{\sin t} = \sec t + \cot t$$

Solution We may transform the left side into the right side as follows:

$$\frac{\tan t + \cos t}{\sin t} = \frac{\tan t}{\sin t} + \frac{\cos t}{\sin t} \quad \text{divide numerator by } \sin t$$

$$= \frac{\left(\dfrac{\sin t}{\cos t}\right)}{\sin t} + \cot t \quad \text{tangent and cotangent identities}$$

$$= \frac{\sin t}{\cos t} \cdot \frac{1}{\sin t} + \cot t \quad \text{rule for quotients}$$

$$= \frac{1}{\cos t} + \cot t \quad \text{cancel } \sin t$$

$$= \sec t + \cot t \quad \text{reciprocal identity}$$

In Section 3.1, we will verify many other identities using methods similar to those used in the last two examples.

EXAMPLE 10   Using fundamental identities

Rewrite $\sqrt{\cos^2 t + \sin^2 t + \cot^2 t}$ in nonradical form without using absolute values for $\pi < t < 2\pi$.

Solution

$$\sqrt{\cos^2 t + \sin^2 t + \cot^2 t} = \sqrt{1 + \cot^2 t} \qquad \cos^2 t + \sin^2 t = 1$$
$$= \sqrt{\csc^2 t} \qquad 1 + \cot^2 t = \csc^2 t$$
$$= |\csc t| \qquad \sqrt{x^2} = |x|$$

Since $\pi < t < 2\pi$, we know that t is in quadrant III or IV. Thus, $\csc t$ is *negative,* and by the definition of absolute value, we have

$$|\csc t| = -\csc t.$$

2.2 EXERCISES

Exer. 1–4: A point $P(x, y)$ is shown on the unit circle U corresponding to a real number t. Find the values of the trigonometric functions at t.

1

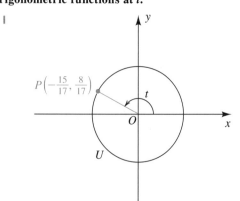

2

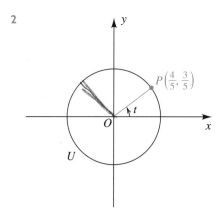

3

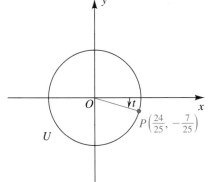

4

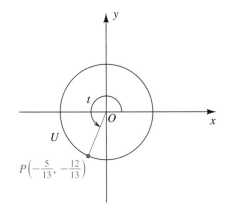

Exer. 5–8: Let $P(t)$ be the point on the unit circle U that corresponds to t. If $P(t)$ has the given rectangular coordinates, find

(a) $P(t + \pi)$ (b) $P(t - \pi)$

(c) $P(-t)$ (d) $P(-t - \pi)$

5 $\left(\frac{3}{5}, \frac{4}{5}\right)$ 6 $\left(-\frac{8}{17}, \frac{15}{17}\right)$

7 $\left(-\frac{12}{13}, -\frac{5}{13}\right)$ 8 $\left(\frac{7}{25}, -\frac{24}{25}\right)$

Exer. 9–16: Let P be the point on the unit circle U that corresponds to t. Find the coordinates of P and the exact values of the trigonometric functions at t, whenever possible.

9 (a) 2π (b) -3π

10 (a) $-\pi$ (b) 6π

11 (a) $3\pi/2$ (b) $-7\pi/2$

12 (a) $5\pi/2$ (b) $-\pi/2$

13 (a) $9\pi/4$ (b) $-5\pi/4$

14 (a) $3\pi/4$ (b) $-7\pi/4$

15 (a) $5\pi/4$ (b) $-\pi/4$

16 (a) $7\pi/4$ (b) $-3\pi/4$

Exer. 17–18: Find the quadrant containing the point P on the unit circle U for the given conditions.

17 (a) $\cos t > 0$ and $\sin t < 0$

(b) $\sin t < 0$ and $\cot t > 0$

(c) $\csc t > 0$ and $\sec t < 0$

(d) $\sec t < 0$ and $\tan t > 0$

18 (a) $\tan t < 0$ and $\cos t > 0$

(b) $\sec t > 0$ and $\tan t < 0$

(c) $\csc t > 0$ and $\cot t < 0$

(d) $\cos t < 0$ and $\csc t < 0$

Exer. 19–24: Use fundamental identities to write the first expression in terms of the second, if $0 < t < \pi/2$.

19 $\cot t, \sin t$ 20 $\tan t, \cos t$

21 $\sec t, \sin t$ 22 $\csc t, \cos t$

23 $\sin t, \sec t$ 24 $\cos t, \cot t$

Exer. 25–32: Use fundamental identities to find the values of the trigonometric functions for the given conditions.

25 $\tan t = -\frac{3}{4}$ and $\sin t > 0$

26 $\cot t = \frac{3}{4}$ and $\cos t < 0$

27 $\sin t = -\frac{5}{13}$ and $\sec t > 0$

28 $\cos t = \frac{1}{2}$ and $\sin t < 0$

29 $\cos t = -\frac{1}{3}$ and $\sin t < 0$

30 $\csc t = 5$ and $\cot t < 0$

31 $\sec t = -4$ and $\csc t > 0$

32 $\sin t = \frac{2}{5}$ and $\cos t < 0$

Exer. 33–36: Use the Pythagorean identities to write the expression as an integer.

33 (a) $\tan^2 4\beta - \sec^2 4\beta$ (b) $4 \tan^2 \beta - 4 \sec^2 \beta$

34 (a) $\csc^2 3\alpha - \cot^2 3\alpha$ (b) $3 \csc^2 \alpha - 3 \cot^2 \alpha$

35 (a) $5 \sin^2 \theta + 5 \cos^2 \theta$

(b) $5 \sin^2 (\theta/4) + 5 \cos^2 (\theta/4)$

36 (a) $7 \sec^2 \gamma - 7 \tan^2 \gamma$

(b) $7 \sec^2 (\gamma/3) - 7 \tan^2 (\gamma/3)$

Exer. 37–40: Simplify the expression.

37 $\dfrac{\sin^3 t + \cos^3 t}{\sin t + \cos t}$

38 $\dfrac{\cot^2 t - 4}{\cot^2 t - \cot t - 6}$

39 $\dfrac{2 - \tan t}{2 \csc t - \sec t}$

40 $\dfrac{\csc t + 1}{(1/\sin^2 t) + \csc t}$

Exer. 41–62: Verify the identity by transforming the left side into the right side.

41 $\cos t \sec t = 1$

42 $\tan t \cot t = 1$

43 $\sin t \sec t = \tan t$

44 $\sin t \cot t = \cos t$

45 $\dfrac{\csc t}{\sec t} = \cot t$

46 $\cot t \sec t = \csc t$

47 $(1 + \cos 2t)(1 - \cos 2t) = \sin^2 2t$

48 $\cos^2 2t - \sin^2 2t = 2 \cos^2 2t - 1$

49 $\cos^2 t(\sec^2 t - 1) = \sin^2 t$

50 $(\tan t + \cot t) \tan t = \sec^2 t$

51 $\dfrac{\sin (t/2)}{\csc (t/2)} + \dfrac{\cos (t/2)}{\sec (t/2)} = 1$

52 $1 - 2 \sin^2 (t/2) = 2 \cos^2 (t/2) - 1$

53 $(1 + \sin t)(1 - \sin t) = \dfrac{1}{\sec^2 t}$

54 $(1 - \sin^2 t)(1 + \tan^2 t) = 1$

55 $\sec t - \cos t = \tan t \sin t$

56 $\dfrac{\sin t + \cos t}{\cos t} = 1 + \tan t$

57 $(\cot t + \csc t)(\tan t - \sin t) = \sec t - \cos t$

58 $\cot t + \tan t = \csc t \sec t$

59 $\sec^2 3t \csc^2 3t = \sec^2 3t + \csc^2 3t$

60 $\dfrac{1 + \cos^2 3t}{\sin^2 3t} = 2 \csc^2 3t - 1$

61 $\log \csc t = -\log \sin t$

62 $\log \tan t = \log \sin t - \log \cos t$

Exer. 63–68: Rewrite the expression in nonradical form without using absolute values for the indicated values of t.

63 $\sqrt{\sec^2 t - 1}$; $\pi/2 < t < \pi$

64 $\sqrt{1 + \cot^2 t}$; $0 < t < \pi$

65 $\sqrt{1 + \tan^2 t}$; $3\pi/2 < t < 2\pi$

66 $\sqrt{\csc^2 t - 1}$; $3\pi/2 < t < 2\pi$

67 $\sqrt{\sin^2 (t/2)}$; $2\pi < t < 4\pi$

68 $\sqrt{\cos^2 (t/2)}$; $0 < t < \pi$

Exer. 69–72: Use the figure to approximate the following to one decimal place.

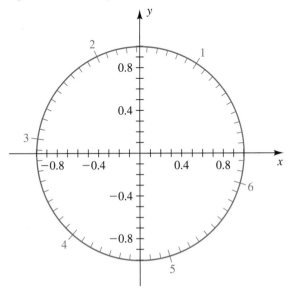

69 (a) $\sin 4$ (b) $\sin (-1.2)$

  (c) All numbers t between 0 and 2π such that $\sin t = 0.5$

70 (a) $\sin 2$ (b) $\sin (-2.3)$

  (c) All numbers t between 0 and 2π such that $\sin t = -0.2$

71 (a) $\cos 4$ (b) $\cos (-1.2)$

  (c) All numbers t between 0 and 2π such that $\cos t = -0.6$

72 (a) $\cos 2$ (b) $\cos (-2.3)$

  (c) All numbers t between 0 and 2π such that $\cos t = 0.2$

73 *Temperature-humidity relationship* On March 17, 1981, in Tucson, Arizona, the temperature in degrees Fahrenheit could be described by the equation

$$T(t) = -12 \cos \left(\frac{\pi}{12} t\right) + 60,$$

while the relative humidity in percent could be expressed by

$$H(t) = 20 \cos\left(\frac{\pi}{12}t\right) + 60,$$

where t is in hours and $t = 0$ corresponds to 6 A.M.

(a) Construct a table that lists the temperature and relative humidity every three hours, beginning at midnight.

(b) Determine the times when the maximums and minimums occurred for T and H.

(c) Discuss the relationship between the temperature and relative humidity on this day.

74 Robotic arm movement Trigonometric functions are used extensively in the design of industrial robots. Suppose that a robot's shoulder joint is motorized so that the angle θ increases at a constant rate of $\pi/12$ radian per second from an initial angle of $\theta = 0$. Assume that the elbow joint is always kept straight and that the arm has a constant length of 153 centimeters, as shown in the figure.

(a) Assume that $h = 50$ cm when $\theta = 0$. Construct a table that lists the angle θ and the height h of the robotic hand every second while $0 \leq \theta \leq \pi/2$.

(b) Determine whether or not a constant increase in the angle θ produces a constant increase in the height of the hand.

(c) Find the total distance that the hand moves.

Exercise 74

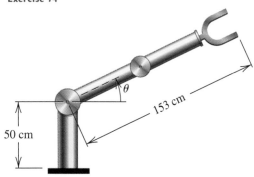

2.3 GRAPHS OF THE TRIGONOMETRIC FUNCTIONS

If t is a real number and $P(x, y)$ is the point on the unit circle U that corresponds to t, then by the definitions of the trigonometric functions in terms of a unit circle,

$$x = \cos t \quad \text{and} \quad y = \sin t.$$

Thus, as shown in Figure 20, we may denote $P(x, y)$ by

$$P(\cos t, \sin t).$$

If $t > 0$, the real number t may be interpreted either as the radian measure of the angle θ or as the length of arc AP.

If we let t increase from 0 to 2π radians, the point $P(\cos t, \sin t)$ travels around the unit circle U one time in the counterclockwise direction. By observing the variation of the x- and y-coordinates of P, we obtain the next table. The notation $0 \rightarrow \pi/2$ in the first row of the table means that t increases from 0 to $\pi/2$, and the notation $(1, 0) \rightarrow (0, 1)$ denotes the corresponding variation of $P(\cos t, \sin t)$ as it travels along U from $(1, 0)$ to $(0, 1)$. If t increases from 0 to $\pi/2$, then $\sin t$ increases from 0 to 1, which we denote by $0 \rightarrow 1$. Moreover, $\sin t$ takes on every value between 0 and 1. If t increases from $\pi/2$ to π, then $\sin t$ decreases from 1 to 0, which is denoted by $1 \rightarrow 0$. Other entries in the table may be interpreted in similar fashion.

Figure 20

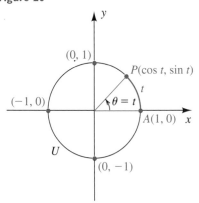

t	$P(\cos t, \sin t)$	$\cos t$	$\sin t$
$0 \to \dfrac{\pi}{2}$	$(1, 0) \to (0, 1)$	$1 \to 0$	$0 \to 1$
$\dfrac{\pi}{2} \to \pi$	$(0, 1) \to (-1, 0)$	$0 \to -1$	$1 \to 0$
$\pi \to \dfrac{3\pi}{2}$	$(-1, 0) \to (0, -1)$	$-1 \to 0$	$0 \to -1$
$\dfrac{3\pi}{2} \to 2\pi$	$(0, -1) \to (1, 0)$	$0 \to 1$	$-1 \to 0$

If t increases from 2π to 4π, the point $P(\cos t, \sin t)$ in Figure 20 traces the unit circle U again, and the patterns for $\sin t$ and $\cos t$ are repeated; that is,

$$\sin (t + 2\pi) = \sin t \quad \text{and} \quad \cos (t + 2\pi) = \cos t$$

for every t in the interval $[0, 2\pi]$. The same is true if t increases from 4π to 6π, from 6π to 8π, and so on. In general, we have the following theorem.

THEOREM ON REPEATED FUNCTION VALUES FOR sin AND cos

If n is any integer, then

$$\sin (t + 2\pi n) = \sin t \quad \text{and} \quad \cos (t + 2\pi n) = \cos t.$$

The repetitive variation of the sine and cosine functions is *periodic* in the sense of the following definition.

DEFINITION OF PERIODIC FUNCTION

A function f is **periodic** if there exists a positive real number k such that

$$f(t + k) = f(t)$$

for every t in the domain of f. The least such positive real number k, if it exists, is the **period** of f.

From the discussion preceding the previous theorem, we see that the period of the sine and cosine functions is 2π.

Let us sketch the graph of $y = \sin t$ on a ty-coordinate system, where t is a real number or the radian measure of an angle. The table in the margin lists coordinates of several points on the graph for $0 \le t \le 2\pi$. Some of these values were found in Examples 3 and 4 of the preceding section. In the next section we will obtain several other exact values, such as $\sin (\pi/6) = \frac{1}{2}$ and $\sin (\pi/3) = \sqrt{3}/2 \approx 0.87$.

t	$y = \sin t$
0	0
$\dfrac{\pi}{4}$	$\dfrac{\sqrt{2}}{2} \approx 0.7$
$\dfrac{\pi}{2}$	1
$\dfrac{3\pi}{4}$	$\dfrac{\sqrt{2}}{2} \approx 0.7$
π	0
$\dfrac{5\pi}{4}$	$-\dfrac{\sqrt{2}}{2} \approx -0.7$
$\dfrac{3\pi}{2}$	-1
$\dfrac{7\pi}{4}$	$-\dfrac{\sqrt{2}}{2} \approx -0.7$
2π	0

To sketch the graph for $0 \le t \le 2\pi$, we plot the points given by the table and remember that $\sin t$ increases on $[0, \pi/2]$, decreases on $[\pi/2, \pi]$ and $[\pi, 3\pi/2]$, and increases on $[3\pi/2, 2\pi]$. This gives us the sketch in Figure 21. Since the sine function is periodic, the pattern shown in Figure 21 is repeated to the right and to the left, in intervals of length 2π. This gives us the sketch in Figure 22.

Figure 21 $y = \sin t$; $0 \le t \le 2\pi$

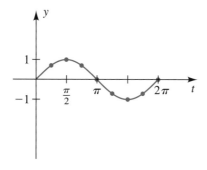

Figure 22 $y = \sin t$

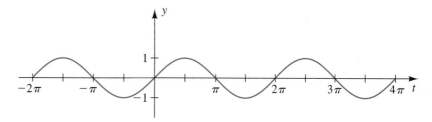

t	$y = \cos t$
0	1
$\dfrac{\pi}{4}$	$\dfrac{\sqrt{2}}{2} \approx 0.7$
$\dfrac{\pi}{2}$	0
$\dfrac{3\pi}{4}$	$-\dfrac{\sqrt{2}}{2} \approx -0.7$
π	-1
$\dfrac{5\pi}{4}$	$-\dfrac{\sqrt{2}}{2} \approx -0.7$
$\dfrac{3\pi}{2}$	0
$\dfrac{7\pi}{4}$	$\dfrac{\sqrt{2}}{2} \approx 0.7$
2π	1

We can use the same procedure to sketch the graph of $y = \cos t$. The table in the margin lists coordinates of several points on the graph for $0 \le t \le 2\pi$. Plotting these points leads to the part of the graph shown in Figure 23. Repeating this pattern to the right and to the left, in intervals of length 2π, we obtain the sketch in Figure 24 on the next page.

Figure 23 $y = \cos t$; $0 \le t \le 2\pi$

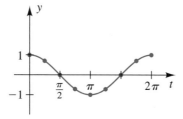

Figure 24 $y = \cos t$

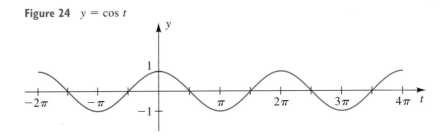

The part of the graph of the sine or cosine function corresponding to $0 \le t \le 2\pi$ is one **cycle**. We sometimes refer to a cycle as a **sine wave** or a **cosine wave**.

The range of the sine and cosine functions consists of all real numbers in the closed interval $[-1, 1]$. Since $\csc t = 1/\sin t$ and $\sec t = 1/\cos t$, it follows that the range of the cosecant and secant functions consists of all real numbers having absolute value greater than or equal to 1.

As we shall see, the range of the tangent and cotangent functions consists of all real numbers.

Before discussing graphs of the other trigonometric functions, let us establish formulas that involve functions of $-t$ for any t. Since a minus sign is involved, we call them *formulas for negatives.*

Formulas for Negatives

$$\begin{array}{lll} \sin(-t) = -\sin t & \cos(-t) = \cos t & \tan(-t) = -\tan t \\ \csc(-t) = -\csc t & \sec(-t) = \sec t & \cot(-t) = -\cot t \end{array}$$

Figure 25

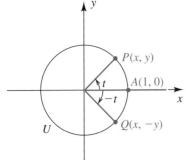

Proof   Consider the unit circle U in Figure 25. As t increases from 0 to 2π, the point $P(x, y)$ traces the unit circle U once in the counterclockwise direction and the point $Q(x, -y)$, corresponding to $-t$, traces U once in the clockwise direction. Applying the definition of trigonometric functions in terms of a unit circle, we have

$$\sin(-t) = -y = -\sin t$$
$$\cos(-t) = x = \cos t$$
$$\tan(-t) = \frac{-y}{x} = -\frac{y}{x} = -\tan t.$$

The proofs of the remaining three formulas are similar. ◂

In the following example, formulas for negatives are used to find an exact value for each trigonometric function.

E X A M P L E   1   Using formulas for negatives

Use formulas for negatives to find the exact values of

(a) $\sin\left(-\dfrac{\pi}{4}\right)$ **(b)** $\cos\left(-\dfrac{\pi}{4}\right)$ **(c)** $\tan\left(-\dfrac{\pi}{4}\right)$

Solution   Using the formulas for negatives and Example 3 of Section 2.2, we obtain the following:

(a) $\sin\left(-\dfrac{\pi}{4}\right) = -\sin\dfrac{\pi}{4} = -\dfrac{\sqrt{2}}{2}$

(b) $\cos\left(-\dfrac{\pi}{4}\right) = \cos\dfrac{\pi}{4} = \dfrac{\sqrt{2}}{2}$

(c) $\tan\left(-\dfrac{\pi}{4}\right) = -\tan\dfrac{\pi}{4} = -1$

E X A M P L E   2   Using formulas for negatives
to verify an identity

Verify the following identity by transforming the left side into the right side:

$$\sin(-t)\tan(-t) + \cos(-t) = \sec t$$

Solution   We may proceed as follows:

$$\sin(-t)\tan(-t) + \cos(-t) = (-\sin t)(-\tan t) + \cos t \qquad \text{formulas for negatives}$$

$$= \sin t\,\dfrac{\sin t}{\cos t} + \cos t \qquad \text{tangent identity}$$

$$= \dfrac{\sin^2 t}{\cos t} + \cos t \qquad \text{multiply}$$

$$= \dfrac{\sin^2 t + \cos^2 t}{\cos t} \qquad \text{add terms}$$

$$= \dfrac{1}{\cos t} \qquad \text{Pythagorean identity}$$

$$= \sec t \qquad \text{reciprocal identity}$$

We may use the formulas for negatives to prove the following theorem.

**THEOREM ON EVEN AND ODD
TRIGONOMETRIC FUNCTIONS**

(1) The cosine and secant functions are even.

(2) The sine, tangent, cotangent, and cosecant functions are odd.

t	$y = \tan t$
$-\dfrac{\pi}{3}$	$-\sqrt{3} \approx -1.7$
$-\dfrac{\pi}{4}$	-1
$-\dfrac{\pi}{6}$	$-\dfrac{\sqrt{3}}{3} \approx -0.6$
0	0
$\dfrac{\pi}{6}$	$\dfrac{\sqrt{3}}{3} \approx 0.6$
$\dfrac{\pi}{4}$	1
$\dfrac{\pi}{3}$	$\sqrt{3} \approx 1.7$

Proof   We shall prove the theorem for the cosine and sine functions. If $f(t) = \cos t$, then

$$f(-t) = \cos (-t) = \cos t = f(t),$$

which means that the cosine function is even.

If $f(t) = \sin t$, then

$$f(-t) = \sin (-t) = -\sin t = -f(t).$$

Thus, the sine function is odd.   ◢

Since the sine function is odd, its graph is symmetric with respect to the origin (see Figure 22). Since the cosine function is even, its graph is symmetric with respect to the y-axis (see Figure 24).

Scientific calculators have keys labeled $\boxed{\text{SIN}}$, $\boxed{\text{COS}}$, and $\boxed{\text{TAN}}$ that can be used to approximate values of these functions. The values of csc, sec, and cot may then be found by means of the reciprocal key. *Before using a calculator to find function values that correspond to a real number t, be sure that the calculator is in radian mode.* (Note that in Figure 20 the real number t determines a central angle of U of radian measure t.) In Section 2.5 we will discuss function values of *angles* that are measured in either radians or degrees.

As an illustration, to find $\sin (\pi/2)$ on a typical calculator, place the calculator in radian mode and use the $\boxed{\text{SIN}}$ key to obtain $\sin (\pi/2) = 1$, which is the exact value. Using the same procedure for $\pi/4$, we obtain a decimal approximation to $\sqrt{2}/2$, such as

$$\sin (\pi/4) \approx 0.7071.$$

Most calculators give eight- to ten-decimal-place accuracy for such function values; throughout the text, however, we will usually round off values to four decimal places.

To find a value such as $\cos 1.3$, we place the calculator in radian mode and use the $\boxed{\text{COS}}$ key, obtaining

$$\cos 1.3 \approx 0.2675.$$

For $\sec 1.3$, we could find $\cos 1.3$ and then use the reciprocal key, usually labeled $\boxed{1/x}$ or $\boxed{x^{-1}}$, to obtain

$$\sec 1.3 = \frac{1}{\cos 1.3} \approx 3.7383.$$

In Section 2.2 we calculated several exact values of $\tan t$. In approximating values of $\tan t$ for $0 < t < \pi/2$, the values near $t = \pi/2$ require special attention. If we consider $\tan t = \sin t/\cos t$, then as t increases toward $\pi/2$, the numerator $\sin t$ approaches 1 and the denominator $\cos t$

approaches 0. Consequently, tan t takes on increasingly large positive values. Following are some approximations of tan t for t close to $\pi/2 \approx 1.5708$:

$$\tan 1.57000 \approx \quad 1{,}255.8$$
$$\tan 1.57030 \approx \quad 2{,}014.8$$
$$\tan 1.57060 \approx \quad 5{,}093.5$$
$$\tan 1.57070 \approx \quad 10{,}381.3$$
$$\tan 1.57079 \approx \quad 158{,}057.9$$

Notice how rapidly tan t increases as t approaches $\pi/2$. We say that tan t *increases without bound* as t approaches $\pi/2$ through values *less* than $\pi/2$. Similarly, if t approaches $-\pi/2$ through values *greater* than $-\pi/2$, then tan t *decreases without bound*. We may denote this variation using the notation introduced on page 54:

$$\text{as } t \to (\pi/2)^-, \quad \tan t \to \infty$$
$$\text{as } t \to (-\pi/2)^+, \quad \tan t \to -\infty$$

This variation of tan t in the open interval $(-\pi/2, \pi/2)$ is illustrated in Figure 26. The lines $t = \pi/2$ and $t = -\pi/2$ are vertical asymptotes for the graph. The same pattern is repeated in the open intervals $(-3\pi/2, -\pi/2)$, $(\pi/2, 3\pi/2)$, and $(3\pi/2, 5\pi/2)$ and in similar intervals of length π, as shown in the figure. Thus, *the tangent function is periodic with period π.*

We may use the graphs of $y = \sin t$, $y = \cos t$, and $y = \tan t$ to help sketch the graphs of the remaining trigonometric functions. For example,

Figure 26 $y = \tan t$

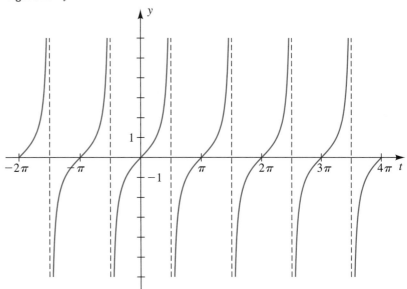

since csc $t = 1/\sin t$, we may find the y-coordinate of a point on the graph of the cosecant function by taking the reciprocal of the corresponding y-coordinate on the sine graph for every value of t except $t = \pi n$ for any integer n. (If $t = \pi n$, $\sin t = 0$, and hence $1/\sin t$ is undefined.) As an aid to sketching the graph of the cosecant function, it is convenient to sketch the graph of the sine function (shown in red in Figure 27) and then take reciprocals to obtain points on the cosecant graph.

Figure 27 $y = \csc t, \quad y = \sin t$

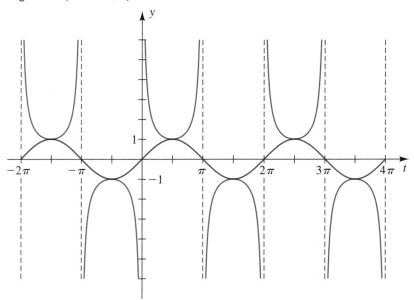

Notice the manner in which the cosecant function increases or decreases without bound as t approaches πn for any integer n. The graph has vertical asymptotes $t = \pi n$, as indicated in the figure.

Since sec $t = 1/\cos t$ and cot $t = 1/\tan t$, we may obtain the graphs of the secant and cotangent functions by taking reciprocals of y-coordinates of points on the graphs of the cosine and tangent functions, as illustrated in Figures 28 and 29.

A graphical summary of the six trigonometric functions and their inverses (discussed in Section 3.6) appears in Appendix IV.

We have considered many properties of the six trigonometric functions of t, where t is a real number or the radian measure of an angle. The following chart (on page 124) contains a summary of important features of these functions. As usual, x and y are coordinates of a point P on the unit circle U corresponding to a real number t.

Figure 28 $y = \sec t,$ $y = \cos t$

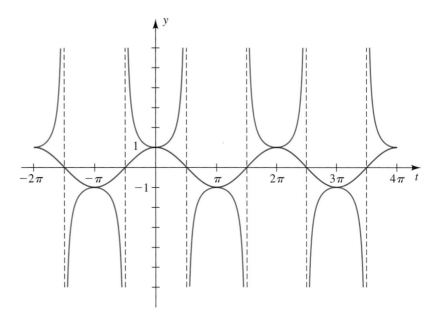

Figure 29 $y = \cot t,$ $y = \tan t$

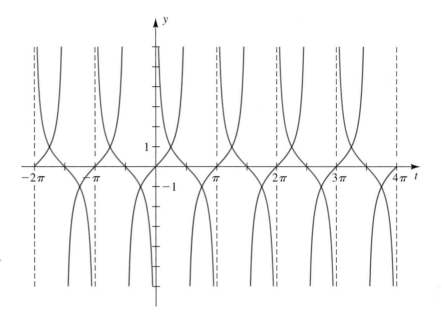

Summary of Features of the Trigonometric Functions and Their Graphs

Feature	$\sin t$	$\cos t$	$\tan t$	$\cot t$	$\sec t$	$\csc t$
Definition	y	x	$\dfrac{y}{x}$	$\dfrac{x}{y}$	$\dfrac{1}{x}$	$\dfrac{1}{y}$
Domain	$\mathbb{R}$	$\mathbb{R}$	$t \neq \dfrac{\pi}{2} + \pi n$	$t \neq \pi n$	$t \neq \dfrac{\pi}{2} + \pi n$	$t \neq \pi n$
Vertical asymptotes	none	none	$t = \dfrac{\pi}{2} + \pi n$	$t = \pi n$	$t = \dfrac{\pi}{2} + \pi n$	$t = \pi n$
Range	$[-1, 1]$	$[-1, 1]$	$\mathbb{R}$	$\mathbb{R}$	$(-\infty, -1] \cup [1, \infty)$	$(-\infty, -1] \cup [1, \infty)$
t-intercepts	πn	$\dfrac{\pi}{2} + \pi n$	πn	$\dfrac{\pi}{2} + \pi n$	none	none
y-intercept	0	1	0	none	1	none
Period	2π	2π	π	π	2π	2π
Even or odd	odd	even	odd	odd	even	odd
Symmetry	origin	y-axis	origin	origin	y-axis	origin

E X A M P L E 3   **Investigating the variation of** $\csc t$

Investigate the variation of $\csc t$ as

$$t \to \pi^-, \quad t \to \pi^+, \quad t \to \frac{\pi}{2}^-, \quad \text{and} \quad t \to \frac{\pi}{4}^+.$$

Solution   Referring to the graph of $y = \csc t$ in Figure 27 and using our knowledge of the special values of the cosecant function, we obtain the following:

$$\text{as} \quad t \to \pi^-, \quad \csc t \to \infty$$
$$\text{as} \quad t \to \pi^+, \quad \csc t \to -\infty$$
$$\text{as} \quad t \to \frac{\pi}{2}^-, \quad \csc t \to 1$$
$$\text{as} \quad t \to \frac{\pi}{4}^+, \quad \csc t \to \sqrt{2}$$

E X A M P L E 4   **Solving equations and inequalities that involve a trigonometric function**

Find all values of t in the interval $[-2\pi, 2\pi]$ such that

(a) $\cos t = \dfrac{\sqrt{2}}{2}$ **(b)** $\cos t > \dfrac{\sqrt{2}}{2}$ **(c)** $\cos t < \dfrac{\sqrt{2}}{2}$

Figure 30

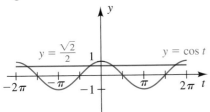

Solution   We shall find the solutions by referring to the graphs of the equations $y = \cos t$ and $y = \sqrt{2}/2$, sketched on the same ty-plane in Figure 30 for $-2\pi \le t \le 2\pi$.

(a) The values of t such that $\cos t = \sqrt{2}/2$ are the t-coordinates of the points at which the graphs intersect. From Example 3(b) in Section 2.2 we know that $\cos(\pi/4) = \sqrt{2}/2$. Hence, we see that the positive values are

$$t = \frac{\pi}{4} \quad \text{and} \quad t = 2\pi - \frac{\pi}{4} = \frac{7\pi}{4}.$$

By symmetry, the negative values are

$$t = -\frac{\pi}{4} \quad \text{and} \quad t = -\frac{7\pi}{4}.$$

(b) The values of t such that $\cos t > \sqrt{2}/2$ can be found by determining where the graph of $y = \cos t$ in Figure 30 lies *above* the line $y = \sqrt{2}/2$. This gives us the t-intervals

$$\left[-2\pi, -\frac{7\pi}{4}\right), \quad \left(-\frac{\pi}{4}, \frac{\pi}{4}\right), \quad \text{and} \quad \left(\frac{7\pi}{4}, 2\pi\right].$$

(c) To solve $\cos t < \sqrt{2}/2$, we again refer to Figure 30 and note where the graph of $y = \cos t$ lies *below* the line $y = \sqrt{2}/2$. This gives us the t-intervals

$$\left(-\frac{7\pi}{4}, -\frac{\pi}{4}\right) \quad \text{and} \quad \left(\frac{\pi}{4}, \frac{7\pi}{4}\right).$$

Another method of solving $\cos t < \sqrt{2}/2$ is to note that the solutions are the open subintervals of $[-2\pi, 2\pi]$ that are *not* included in the intervals obtained in part (b).

The result discussed in the next example plays an important role in advanced mathematics.

E X A M P L E   5   Sketching the graph of $f(t) = (\sin t)/t$

If $f(t) = (\sin t)/t$, sketch the graph of f on $[-\pi, \pi]$, and investigate the behavior of $f(t)$ as $t \to 0^-$ and as $t \to 0^+$.

Solution   Some specific keystrokes for the TI-82/83 graphing calculator are given in Example 21 of Appendix I. Note that f is undefined at $t = 0$, because substitution yields the meaningless expression $0/0$.

(continued)

Figure 3I
$[-\pi, \pi]$ by $[-2.1, 2.1]$

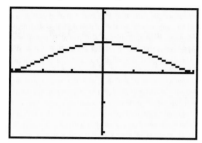

We assign $(\sin x)/x$ to Y_1. Because our screen has a $3:2$ (horizontal: vertical) proportion, we use the viewing rectangle $[-\pi, \pi]$ by $[-2.1, 2.1]$ $\left(\text{since } \frac{2}{3}\pi \approx 2.1\right)$, obtaining a sketch similar to Figure 31. Using tracing and zoom features, we find it appears that

$$\text{as} \quad t \to 0^-, \quad f(t) \to 1 \quad \text{and as} \quad t \to 0^+, \quad f(t) \to 1.$$

There is a hole in the graph at the point $(0, 1)$; however, most graphing utilities are not capable of showing this fact.

Our graphical technique does not *prove* that $f(t) \to 1$ as $t \to 0$, but it does make it appear highly probable. A rigorous proof, based on the definition of $\sin t$ and geometric considerations, can be found in calculus texts.

An interesting result obtained from Example 5 is that *if t is in radians* and

$$\text{if} \quad t \approx 0, \quad \text{then} \quad \frac{\sin t}{t} \approx 1, \quad \text{or} \quad \sin t \approx t.$$

The last statement gives us an *approximation formula* for $\sin t$ if t is close to 0. To illustrate, using a calculator we find that

$$\sin (0.06) \approx 0.0599640$$
$$\sin (0.05) \approx 0.0499792$$
$$\sin (0.04) \approx 0.0399893$$
$$\sin (0.03) \approx 0.0299955$$
$$\sin (0.02) \approx 0.0199987$$
$$\sin (0.01) \approx 0.0099998.$$

2.3 EXERCISES

Exer. 1–4: Use a formula for negatives to find the exact value.

1 (a) $\sin (-90°)$ (b) $\cos (-3\pi/4)$ (c) $\tan (-45°)$

2 (a) $\sin (-3\pi/2)$ (b) $\cos (-225°)$ (c) $\tan (-\pi)$

3 (a) $\cot (-3\pi/4)$ (b) $\sec (-180°)$ (c) $\csc (-3\pi/2)$

4 (a) $\cot (-225°)$ (b) $\sec (-\pi/4)$ (c) $\csc (-45°)$

Exer. 5–10: Verify the identity by transforming the left side into the right side.

5 $\sin (-t) \sec (-t) = -\tan t$

6 $\csc (-t) \cos (-t) = -\cot t$

7 $\dfrac{\cot (-t)}{\csc (-t)} = \cos t$

8 $\dfrac{\sec (-t)}{\tan (-t)} = -\csc t$

9 $\dfrac{1}{\cos (-t)} - \tan (-t) \sin (-t) = \cos t$

10 $\cot (-t) \cos (-t) + \sin (-t) = -\csc t$

Exer. 11–22: Complete the statement by referring to the graph of a trigonometric function.

11 (a) As $t \to 0^+$, $\sin t \to$ ___

    (b) As $t \to (-\pi/2)^-$, $\sin t \to$ ___

12 (a) As $t \to \pi^+$, $\sin t \to$ ___

   (b) As $t \to (\pi/4)^-$, $\sin t \to$ ___

13 (a) As $t \to (\pi/4)^+$, $\cos t \to$ ___

   (b) As $t \to \pi^-$, $\cos t \to$ ___

14 (a) As $t \to 0^+$, $\cos t \to$ ___

   (b) As $t \to (-3\pi/4)^-$, $\cos t \to$ ___

15 (a) As $t \to (\pi/4)^+$, $\tan t \to$ ___

   (b) As $t \to (\pi/2)^+$, $\tan t \to$ ___

16 (a) As $t \to 0^+$, $\tan t \to$ ___

   (b) As $t \to (-\pi/2)^-$, $\tan t \to$ ___

17 (a) As $t \to (-\pi/4)^-$, $\cot t \to$ ___

   (b) As $t \to 0^+$, $\cot t \to$ ___

18 (a) As $t \to (\pi/2)^+$, $\cot t \to$ ___

   (b) As $t \to \pi^-$, $\cot t \to$ ___

19 (a) As $t \to (\pi/2)^-$, $\sec t \to$ ___

   (b) As $t \to (\pi/4)^+$, $\sec t \to$ ___

20 (a) As $t \to (\pi/2)^+$, $\sec t \to$ ___

   (b) As $t \to 0^-$, $\sec t \to$ ___

21 (a) As $t \to 0^-$, $\csc t \to$ ___

   (b) As $t \to (\pi/2)^+$, $\csc t \to$ ___

22 (a) As $t \to \pi^+$, $\csc t \to$ ___

   (b) As $t \to (\pi/4)^-$, $\csc t \to$ ___

Exer. 23–30: Refer to the graph of $y = \sin t$ or $y = \cos t$ to find the exact values of t in the interval $[0, 4\pi]$ that satisfy the equation.

23 $\sin t = -1$

24 $\sin t = 1$

25 $\sin t = \sqrt{2}/2$

26 $\sin t = -\sqrt{2}/2$

27 $\cos t = 1$

28 $\cos t = -1$

29 $\cos t = -\sqrt{2}/2$

30 $\cos t = 0$

Exer. 31–32: Refer to the graph of $y = \tan t$ to find the exact values of t in the interval $(-\pi/2, 3\pi/2)$ that satisfy the equation.

31 $\tan t = 1$

32 $\tan t = -1$

Exer. 33–36: Refer to the graph of the equation on the specified interval. Find all values of t such that for the real number a, (a) $y = a$, (b) $y > a$, and (c) $y < a$.

33 $y = \sin t$; $[-2\pi, 2\pi]$; $a = \sqrt{2}/2$

34 $y = \cos t$; $[0, 4\pi]$; $a = \sqrt{2}/2$

35 $y = \cos t$; $[-2\pi, 2\pi]$; $a = -\sqrt{2}/2$

36 $y = \sin t$; $[0, 4\pi]$; $a = -\sqrt{2}/2$

Exer. 37–44: Use the graph of a trigonometric function to sketch the graph of the equation without plotting points.

37 $y = 2 + \sin t$

38 $y = 3 + \cos t$

39 $y = \cos t - 2$

40 $y = \sin t - 1$

41 $y = 1 + \tan t$

42 $y = \cot t - 1$

43 $y = \sec t - 2$

44 $y = 1 + \csc t$

Exer. 45–48: Find the intervals between -2π and 2π on which $f(t)$ is (a) increasing or (b) decreasing.

45 $f(t) = \sec t$

46 $f(t) = \csc t$

47 $f(t) = \tan t$

48 $f(t) = \cot t$

49 Practice sketching the graph of the sine function, taking different units of length on the horizontal and vertical axes. Practice sketching graphs of the cosine and tangent functions in the same manner. Continue practicing until you reach the stage at which, if you were awakened from a sound sleep in the middle of the night and asked to sketch one of these graphs, you could do so in less than thirty seconds.

50 Work Exercise 49 for the cosecant, secant, and cotangent functions.

C Exer. 51–52: Graph the equation, and estimate the values of t within the specified interval that correspond to the given value of y.

51 $y = \sin(t^2)$, $[-\pi, \pi]$; $y = 0.5$

52 $y = \tan(\sqrt{t})$, $[0, 25]$; $y = 5$

C Exer. 53–54: Graph f on the interval $[-2\pi, 2\pi]$, and estimate the coordinates of the high and low points.

53 $f(t) = t \sin t$

54 $f(t) = \sin^2 t \cos t$

C Exer. 55–60: As $t \to 0^+$, $f(t) \to L$ for some real number L. Use a graph to predict L.

55 $f(t) = \dfrac{1 - \cos t}{t}$

56 $f(t) = \dfrac{6t - 6\sin t}{t^3}$

57 $f(t) = t \cot t$

58 $f(t) = \dfrac{t + \tan t}{\sin t}$

59 $f(t) = \dfrac{\tan t}{t}$

60 $f(t) = \dfrac{\cos\left(t + \frac{1}{2}\pi\right)}{t}$

2.4 TRIGONOMETRIC FUNCTIONS OF ANGLES

For certain applied problems it is convenient to change the domains of the trigonometric functions from sets of real numbers to sets of angles. If θ is an angle, then a natural way to assign the values sin θ, cos θ, and so on, is to use the radian measure of θ, as in the following definition.

DEFINITION OF TRIGONOMETRIC FUNCTIONS OF ANGLES

> If θ is an angle with radian measure t, then the value of each trigonometric function at θ is its value at the real number t.

By the preceding definition, if t is the radian measure of an angle θ, then

$$\sin \theta = \sin t, \qquad \cos \theta = \cos t, \qquad \tan \theta = \tan t,$$

and so on. We shall use the terminology *trigonometric functions* if the domains consist of either angles or numbers. We use notation such as sin 65° or tan 150° whenever the angle is measured in degrees. Numerals without any symbol attached—such as cos 3 and csc $(\pi/4)$—will indicate that radian measure is to be used. This is not in conflict with our previous work where, for example, cos 3 meant the value of the cosine function at the real number 3, since by definition the cosine of an angle of measure 3 radians is the same as the cosine of the real number 3. Note, however, that

$$\sin 3° \neq \sin 3,$$

since 3° ≠ 3 radians.

EXAMPLE 1   Finding trigonometric function values of special angles

Find sin 90°, cos 45°, and tan 720°.

Solution   The angles are shown in standard position in Figure 32.

Figure 32

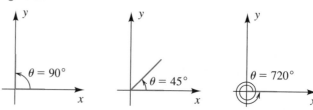

We first convert the degree measure to radian measure:

$$90° = 90°\left(\frac{\pi}{180°}\right) = \frac{\pi}{2}, \qquad 45° = 45°\left(\frac{\pi}{180°}\right) = \frac{\pi}{4}, \qquad 720° = 720°\left(\frac{\pi}{180°}\right) = 4\pi$$

Using the definition of trigonometric functions of angles and Example 3 of Section 2.2, we obtain the function values:

$$\sin 90° = \sin \frac{\pi}{2} = 1$$

$$\cos 45° = \cos \frac{\pi}{4} = \frac{\sqrt{2}}{2}$$

$$\tan 720° = \tan 4\pi = \tan 0 = 0$$

In the next section we shall introduce techniques for approximating trigonometric function values corresponding to *any* angle θ.

The values of the trigonometric functions at an angle θ may be determined by means of a point on the terminal side of θ, as follows. Let θ be an angle in standard position, and let $Q(x, y)$ be any point on the terminal side of θ other than the origin O, as shown in Figure 33. Let r denote the distance between O and Q; that is, $r = d(O, Q)$. By the distance formula,

$$r = \sqrt{x^2 + y^2}.$$

The point $Q(x, y)$ is not necessarily a point on the unit circle U, since r may be different from 1. If we let $P(x_1, y_1)$ be the point on the terminal side of θ such that $d(O, P) = 1$, then P is on the unit circle U. If t is the radian measure of θ, then, by definition,

$$\sin \theta = \sin t = y_1 \qquad \text{and} \qquad \cos \theta = \cos t = x_1.$$

As shown in Figure 34, vertical lines through Q and P intersect the x-axis at $Q'(x, 0)$ and $P'(x_1, 0)$, respectively. Since triangles OQQ' and OPP' are similar, ratios of corresponding sides are equal. In particular,

$$\frac{d(P', P)}{d(O, P)} = \frac{d(Q', Q)}{d(O, Q)} \qquad \text{or, equivalently,} \qquad \frac{|y_1|}{1} = \frac{|y|}{r}.$$

Since y and y_1 have the same sign, this gives us

$$y_1 = \frac{y}{r} \qquad \text{and hence} \qquad \sin \theta = y_1 = \frac{y}{r}.$$

In similar fashion we obtain

$$\cos \theta = \frac{x}{r}.$$

Figure 33

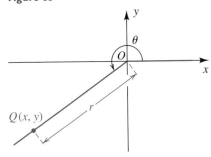

Figure 34

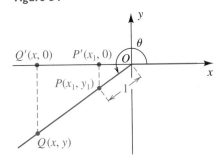

Using the tangent identity, we have

$$\tan \theta = \frac{\sin \theta}{\cos \theta} = \frac{y/r}{x/r} = \frac{y}{x}.$$

The remaining three function values can be obtained by taking reciprocals. This gives us the following theorem.

THEOREM ON TRIGONOMETRIC FUNCTIONS AS RATIOS

Let θ be an angle in standard position on a rectangular coordinate system, and let $Q(x, y)$ be any point other than the origin O on the terminal side of θ. If $d(O, Q) = r = \sqrt{x^2 + y^2}$, then

$$\sin \theta = \frac{y}{r} \qquad\qquad \cos \theta = \frac{x}{r} \qquad\qquad \tan \theta = \frac{y}{x} \ \text{(if } x \neq 0)$$

$$\csc \theta = \frac{r}{y} \ \text{(if } y \neq 0) \qquad \sec \theta = \frac{r}{x} \ \text{(if } x \neq 0) \qquad \cot \theta = \frac{x}{y} \ \text{(if } y \neq 0).$$

Let us reiterate that, in this theorem, x and y are not necessarily the coordinates of a point on the unit circle U; however, if $r = 1$, then the formulas reduce to the unit circle definition of the trigonometric functions given in Section 2.2. The preceding theorem has many applications. In Section 2.8 we shall use it to solve applied problems involving right triangles. It is also useful for finding values of trigonometric functions, as illustrated in the following examples.

EXAMPLE 2 Finding trigonometric function values of an angle in standard position

If θ is an angle in standard position on a rectangular coordinate system and if $P(-15, 8)$ is on the terminal side of θ, find the values of the six trigonometric functions of θ.

Figure 35

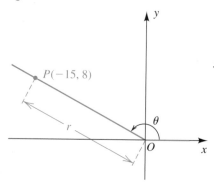

Solution The point $P(-15, 8)$ is shown in Figure 35. Applying the theorem on trigonometric functions as ratios with

$$x = -15, \quad y = 8, \quad \text{and} \quad r = \sqrt{x^2 + y^2} = \sqrt{(-15)^2 + 8^2} = \sqrt{289} = 17,$$

we obtain the following:

$$\sin \theta = \frac{y}{r} = \frac{8}{17} \qquad \cos \theta = \frac{x}{r} = -\frac{15}{17} \qquad \tan \theta = \frac{y}{x} = -\frac{8}{15}$$

$$\csc \theta = \frac{r}{y} = \frac{17}{8} \qquad \sec \theta = \frac{r}{x} = -\frac{17}{15} \qquad \cot \theta = \frac{x}{y} = -\frac{15}{8}$$

Figure 36

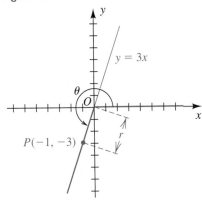

E X A M P L E 3 Finding trigonometric function values
of an angle in standard position

An angle θ is in standard position, and its terminal side lies in quadrant III
on the line $y = 3x$. Find the values of the trigonometric functions of θ.

Solution The graph of $y = 3x$ is sketched in Figure 36, together with
the initial and terminal sides of θ. Since the terminal side of θ is in quadrant III, we begin by choosing a convenient negative value for x, say
$x = -1$. Substituting for x in $y = 3x$ gives us $y = 3(-1) = -3$, and hence
$P(-1, -3)$ is on the terminal side. Applying the theorem on trigonometric
functions as ratios with

$$x = -1, \quad y = -3, \quad \text{and} \quad r = \sqrt{x^2 + y^2} = \sqrt{(-1)^2 + (-3)^2} = \sqrt{10}$$

gives us

$$\sin \theta = -\frac{3}{\sqrt{10}} \qquad \cos \theta = -\frac{1}{\sqrt{10}} \qquad \tan \theta = \frac{-3}{-1} = 3$$

$$\csc \theta = -\frac{\sqrt{10}}{3} \qquad \sec \theta = -\frac{\sqrt{10}}{1} \qquad \cot \theta = \frac{-1}{-3} = \frac{1}{3}.$$

The theorem on trigonometric functions as ratios may be applied if θ is
a quadrantal angle. This is illustrated by the next example.

E X A M P L E 4 Finding trigonometric function
values of a quadrantal angle

If $\theta = 3\pi/2$, find the values of the trigonometric functions of θ.

Solution Note that $3\pi/2 = 270°$. Placing θ in standard position, we find
that the terminal side of θ coincides with the negative y-axis, as shown in
Figure 37. To use the theorem on trigonometric functions as ratios, we may
choose *any* point P on the terminal side of θ. For simplicity, we use
$P(0, -1)$. In this case, $x = 0$, $y = -1$, $r = 1$, and hence

$$\sin \frac{3\pi}{2} = \frac{-1}{1} = -1 \qquad \cos \frac{3\pi}{2} = \frac{0}{1} = 0$$

$$\csc \frac{3\pi}{2} = \frac{1}{-1} = -1 \qquad \cot \frac{3\pi}{2} = \frac{0}{-1} = 0.$$

Figure 37

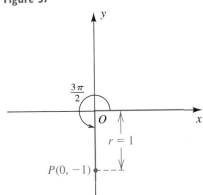

The tangent and secant functions are undefined because the meaningless
expressions $\tan \theta = (-1)/0$ and $\sec \theta = 1/0$ occur when we substitute in
the appropriate formulas.

Figure 38

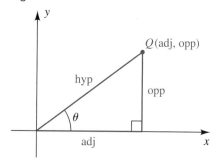

We shall conclude this section by showing that for acute angles, values of the trigonometric functions can be interpreted as ratios of the lengths of the sides of a right triangle. A triangle is a **right triangle** if one of its angles is a right angle. We sometimes use the right angle symbol ⌐ to specify the angle that has measure 90° (see Figure 38). If θ is an acute angle, then it can be regarded as an angle of a right triangle, and we may refer to the lengths of the **hypotenuse,** the **opposite side,** and the **adjacent side.** We abbreviate these lengths **hyp, opp,** and **adj,** respectively. Introducing a rectangular coordinate system as in Figure 38, we see that the lengths of the adjacent side and the opposite side for θ are the x-coordinate and the y-coordinate, respectively, of a point Q on the terminal side of θ. Using the theorem on trigonometric functions as ratios gives us the following result.

Trigonometric Functions of an Acute Angle of a Right Triangle

$$\sin \theta = \frac{\text{opp}}{\text{hyp}} \qquad \cos \theta = \frac{\text{adj}}{\text{hyp}} \qquad \tan \theta = \frac{\text{opp}}{\text{adj}}$$

$$\csc \theta = \frac{\text{hyp}}{\text{opp}} \qquad \sec \theta = \frac{\text{hyp}}{\text{adj}} \qquad \cot \theta = \frac{\text{adj}}{\text{opp}}$$

These formulas will be useful in future applications.

E X A M P L E   5   **Finding trigonometric function values of an acute angle**

If θ is an acute angle and $\cos \theta = \frac{3}{4}$, find the values of the trigonometric functions of θ.

Figure 39

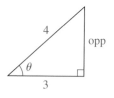

Solution   We begin by sketching a right triangle having an acute angle θ with adj $= 3$ and hyp $= 4$, as shown in Figure 39, and proceed as follows:

$$3^2 + (\text{opp})^2 = 4^2 \qquad \text{Pythagorean theorem}$$

$$(\text{opp})^2 = 16 - 9 = 7 \qquad \text{isolate (opp)}^2$$

$$\text{opp} = \sqrt{7} \qquad \text{take the square root}$$

Applying the definition of trigonometric functions of an acute angle of a right triangle, we obtain the following:

$$\sin \theta = \frac{\text{opp}}{\text{hyp}} = \frac{\sqrt{7}}{4} \qquad \cos \theta = \frac{\text{adj}}{\text{hyp}} = \frac{3}{4} \qquad \tan \theta = \frac{\text{opp}}{\text{adj}} = \frac{\sqrt{7}}{3}$$

$$\csc \theta = \frac{\text{hyp}}{\text{opp}} = \frac{4}{\sqrt{7}} \qquad \sec \theta = \frac{\text{hyp}}{\text{adj}} = \frac{4}{3} \qquad \cot \theta = \frac{\text{adj}}{\text{opp}} = \frac{3}{\sqrt{7}}$$

In Example 5 we could have rationalized the denominators for $\csc \theta$ and $\cot \theta$, writing

$$\csc \theta = \frac{4\sqrt{7}}{7} \qquad \text{and} \qquad \cot \theta = \frac{3\sqrt{7}}{7}.$$

However, in most examples and exercises we will leave expressions in unrationalized form. An exception to this practice is the special trigonometric function values corresponding to 60°, 30°, and 45°, which are obtained in the following example.

E X A M P L E   6   **Finding trigonometric function values of 60°, 30°, and 45°**

Find the values of the trigonometric functions that correspond to θ:

(a) $\theta = 60°$ **(b)** $\theta = 30°$ **(c)** $\theta = 45°$

Solution   Consider an equilateral triangle with sides of length 2. The median from one vertex to the opposite side bisects the angle at that vertex, as illustrated by the dashes in Figure 40. By the Pythagorean theorem, the side opposite 60° in the shaded right triangle has length $\sqrt{3}$. Using the formulas for the trigonometric functions of an acute angle of a right triangle, we obtain the values corresponding to 60° and 30° as follows:

Figure 40

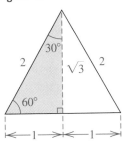

(a) $\sin 60° = \dfrac{\sqrt{3}}{2}$ $\cos 60° = \dfrac{1}{2}$ $\tan 60° = \dfrac{\sqrt{3}}{1} = \sqrt{3}$

$\csc 60° = \dfrac{2}{\sqrt{3}} = \dfrac{2\sqrt{3}}{3}$ $\sec 60° = \dfrac{2}{1} = 2$ $\cot 60° = \dfrac{1}{\sqrt{3}} = \dfrac{\sqrt{3}}{3}$

(b) $\sin 30° = \dfrac{1}{2}$ $\cos 30° = \dfrac{\sqrt{3}}{2}$ $\tan 30° = \dfrac{1}{\sqrt{3}} = \dfrac{\sqrt{3}}{3}$

$\csc 30° = \dfrac{2}{1} = 2$ $\sec 30° = \dfrac{2}{\sqrt{3}} = \dfrac{2\sqrt{3}}{3}$ $\cot 30° = \dfrac{\sqrt{3}}{1} = \sqrt{3}$

(c) To find the values for $\theta = 45°$, we may consider an isosceles right triangle whose two equal sides have length 1, as illustrated in Figure 41. By the Pythagorean theorem, the length of the hypotenuse is $\sqrt{2}$. Hence, the values corresponding to 45° are as follows:

Figure 41

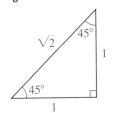

$$\sin 45° = \dfrac{1}{\sqrt{2}} = \dfrac{\sqrt{2}}{2} = \cos 45°   \qquad \tan 45° = \dfrac{1}{1} = 1$$

$$\csc 45° = \dfrac{\sqrt{2}}{1} = \sqrt{2} = \sec 45°   \qquad \cot 45° = \dfrac{1}{1} = 1$$

For reference, we list the values found in Example 6, together with the radian measures of the angles, in the following table. Two reasons for stressing these values are that they are exact and that they occur frequently in work involving trigonometry. Because of the importance of these special values, it is a good idea either to memorize the table or to learn to find the values quickly by using triangles, as in Example 6.

Special Values of the Trigonometric Functions

θ (radians)	θ (degrees)	$\sin \theta$	$\cos \theta$	$\tan \theta$	$\cot \theta$	$\sec \theta$	$\csc \theta$
$\dfrac{\pi}{6}$	$30°$	$\dfrac{1}{2}$	$\dfrac{\sqrt{3}}{2}$	$\dfrac{\sqrt{3}}{3}$	$\sqrt{3}$	$\dfrac{2\sqrt{3}}{3}$	2
$\dfrac{\pi}{4}$	$45°$	$\dfrac{\sqrt{2}}{2}$	$\dfrac{\sqrt{2}}{2}$	1	1	$\sqrt{2}$	$\sqrt{2}$
$\dfrac{\pi}{3}$	$60°$	$\dfrac{\sqrt{3}}{2}$	$\dfrac{1}{2}$	$\sqrt{3}$	$\dfrac{\sqrt{3}}{3}$	2	$\dfrac{2\sqrt{3}}{3}$

The next example illustrates a practical use for trigonometric functions of acute angles. Additional applications involving right triangles will be considered in Section 2.8.

EXAMPLE 7   **Finding the height of a flagpole**

A surveyor observes that at a point A, located on level ground a distance 25.0 feet from the base B of a flagpole, the angle between the ground and the top of the pole is 30°. Approximate the height h of the pole to the nearest tenth of a foot.

Figure 42

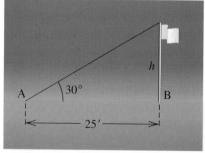

Solution   Referring to Figure 42, we see that we want to relate the opposite side h and the adjacent side, of length 25, to the 30° angle. This suggests that we use a trigonometric function involving those two sides—either tan or cot. It is usually easier to solve the problem if we select the function for which the variable is in the numerator. Hence, we have

$$\tan 30° = \frac{h}{25} \quad \text{or, equivalently,} \quad h = 25 \tan 30°.$$

We use the value of tan 30° from Example 6 to find h:

$$h = 25\left(\frac{\sqrt{3}}{3}\right) \approx 14.4 \text{ ft}$$

We have now discussed two different approaches to the trigonometric functions. The definition in terms of a unit circle, introduced in Section 2.2, emphasizes the fact that the trigonometric functions have domains consisting of real numbers. Such functions are the building blocks for calculus. In addition, the unit circle approach is useful for discussing graphs and deriving trigonometric identities. The development in terms of angles and ratios, considered in this section, has many applications in the sciences and engineering. You should work to become proficient in the use of both formulations of the trigonometric functions, since each will reinforce the other and thus facilitate your mastery of more advanced aspects of trigonometry.

2.4 EXERCISES

Exer. 1–4: Find the exact values of the six trigonometric functions of θ if θ is in standard position and P is on the terminal side.

1 $P(4, -3)$

2 $P(-8, -15)$

3 $P(-2, -5)$

4 $P(-1, 2)$

Exer. 5–10: Find the exact values of the six trigonometric functions of θ if θ is in standard position and the terminal side of θ is in the specified quadrant and satisfies the given condition.

5 II;   on the line $y = -4x$

6 IV;   on the line $3y + 5x = 0$

7 I;   on a line having slope $\frac{4}{3}$

8 III;   bisects the quadrant

9 III;   parallel to the line $2y - 7x + 2 = 0$

10 II;   parallel to the line through $A(1, 4)$ and $B(3, -2)$

Exer. 11–12: Find the exact values of the six trigonometric functions of the angle, whenever possible.

11 **(a)** 90°   **(b)** 0°   **(c)** $7\pi/2$ **(d)** 3π

12 **(a)** 180°   **(b)** −90°   **(c)** 2π **(d)** $5\pi/2$

Exer. 13–18: Find the exact values of the trigonometric functions if θ is an acute angle.

13 $\sin \theta = \frac{3}{5}$

14 $\cos \theta = \frac{8}{17}$

15 $\tan \theta = \frac{5}{12}$

16 $\cot \theta = \frac{7}{24}$

17 $\sec \theta = \frac{6}{5}$

18 $\csc \theta = 4$

Exer. 19–26: Find the values of the six trigonometric functions for the angle θ.

19

20
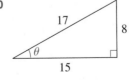

21
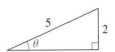

22

23
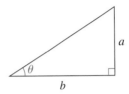

24
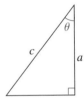

25
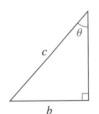

26
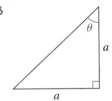

Exer. 27–32: Find the exact values of x and y.

27
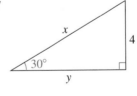

28
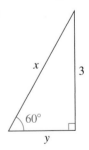

29

30
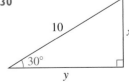

31

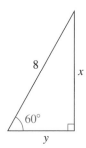

32

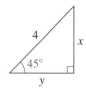

33 Height of a tree A forester, 200 feet from the base of a redwood tree, observes that the angle between the ground and the top of the tree is 60°. Estimate the height of the tree.

34 Distance to Mt. Fuji Mt. Fuji in Japan is approximately 12,400 feet high. A trigonometry student, several miles away, notes that the angle between level ground and the top of the mountain is 30°. Estimate the distance to the base of the mountain.

35 Stonehenge blocks Stonehenge in Salisbury Plains, England, was constructed using solid stone blocks weighing over 99,000 pounds each. Lifting a single stone required 550 people, who pulled the stone up a ramp inclined at an angle of 9°. Approximate the distance that a stone was moved in order to raise it to a height of 30 feet.

36 Advertising sign height The highest advertising sign in the world is a large letter I situated at the top of the 73-story First Interstate World Center building in Los Angeles. At a distance of 200 feet from a point directly below the sign, the angle between the ground and the top of the sign is 78.87°. Approximate the height of the top of the sign.

37 Telescope resolution Two stars that are very close may appear to be one. The ability of a telescope to separate their images is called its resolution. The smaller the resolution, the better a telescope's ability to separate images in the sky. In a refracting telescope, resolution θ (see the figure) can be improved by using a lens with a larger diameter D. The relationship between θ in degrees and D in meters is $\sin \theta = 1.22\lambda/D$, where λ is the wavelength of light in meters. The largest refracting

telescope in the world is at the University of Chicago. At a wavelength of $\lambda = 550 \times 10^{-9}$ meter, its resolution is 0.00003769°. Approximate the diameter of the lens.

Exercise 37

38 Moon phases The phases of the moon can be described using the phase angle θ, determined by the sun, the moon, and the earth, as shown in the figure. Because the moon orbits the earth, θ changes during the course of a month. The area of the region A of the moon, which appears illuminated to an observer on the earth, is given by $A = \frac{1}{2}\pi R^2(1 + \cos \theta)$, where $R = 1080$ mi is the radius of the moon. Approximate A for the following positions of the moon:

(a) $\theta = 0°$ (full moon)

(b) $\theta = 180°$ (new moon)

(c) $\theta = 90°$ (first quarter)

(d) $\theta = 103°$

Exercise 38

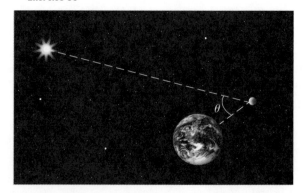

2.5 VALUES OF THE TRIGONOMETRIC FUNCTIONS

In previous sections we calculated special values of the trigonometric functions by using either the unit circle definition or the theorem on trigonometric functions as ratios. In practice we most often use a calculator to approximate function values.

We will next show how the values of any trigonometric function at a real number t or at an angle of θ degrees can be found from its values in the t-interval $(0, \pi/2)$ or the θ-interval $(0°, 90°)$, respectively. This technique is sometimes necessary when a calculator is used to find all angles or real numbers that correspond to a given function value.

We shall make use of the following concept.

DEFINITION OF
REFERENCE ANGLE

> Let θ be a nonquadrantal angle in standard position. The **reference angle** for θ is the acute angle θ_R that the terminal side of θ makes with the x-axis.

Figure 43 illustrates the reference angle θ_R for a nonquadrantal angle θ, with $0° < \theta < 360°$ or $0 < \theta < 2\pi$, in each of the four quadrants.

Figure 43 Reference angles

(a) Quadrant I   (b) Quadrant II   (c) Quadrant III   (d) Quadrant IV

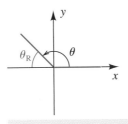

$$\theta_R = \theta$$

$$\theta_R = 180° - \theta$$
$$= \pi - \theta$$

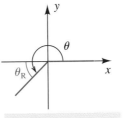

$$\theta_R = \theta - 180°$$
$$= \theta - \pi$$

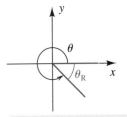

$$\theta_R = 360° - \theta$$
$$= 2\pi - \theta$$

The formulas below the axes in Figure 43 may be used to find the degree or radian measure of θ_R when θ is in degrees or radians, respectively. *For a nonquadrantal angle greater than 360° or less than 0°*, first find the coterminal angle θ with $0° < \theta < 360°$ or $0 < \theta < 2\pi$, and then use the formulas in Figure 43.

Figure 44

(a)

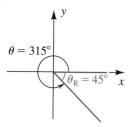

(b)

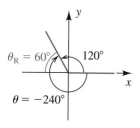

(c)

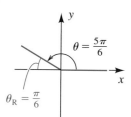

(d)

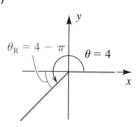

Finding reference angles

Find the reference angle θ_R for θ, and sketch θ and θ_R in standard position on the same coordinate plane.

(a) $\theta = 315°$ **(b)** $\theta = -240°$ **(c)** $\theta = \dfrac{5\pi}{6}$ **(d)** $\theta = 4$

Solution

(a) The angle $\theta = 315°$ is in quadrant IV, and hence, as in Figure 43(d),

$$\theta_R = 360° - 315° = 45°.$$

The angles θ and θ_R are sketched in Figure 44(a).

(b) The angle between $0°$ and $360°$ that is coterminal with $\theta = -240°$ is

$$-240° + 360° = 120°,$$

which is in quadrant II. Using the formula in Figure 43(b) gives

$$\theta_R = 180° - 120° = 60°.$$

The angles θ and θ_R are sketched in Figure 44(b).

(c) Since the angle $\theta = 5\pi/6$ is in quadrant II, we have

$$\theta_R = \pi - \frac{5\pi}{6} = \frac{\pi}{6},$$

as shown in Figure 44(c).

(d) Since $\pi < 4 < 3\pi/2$, the angle $\theta = 4$ is in quadrant III. Using the formula in Figure 43(c), we obtain

$$\theta_R = 4 - \pi.$$

The angles are sketched in Figure 44(d).

We shall next show how reference angles can be used to find values of the trigonometric functions.

If θ is a nonquadrantal angle with reference angle θ_R, then $0° < \theta_R < 90°$ or $0 < \theta_R < \pi/2$. Let $P(x, y)$ be a point on the terminal side of θ, and consider the point $Q(x, 0)$ on the x-axis. Figure 45 illustrates a typical situation for θ in each quadrant. In each case, the lengths of the sides of triangle OQP are

$$d(O, Q) = |x|, \qquad d(Q, P) = |y|, \qquad \text{and} \qquad d(O, P) = \sqrt{x^2 + y^2} = r.$$

Figure 45

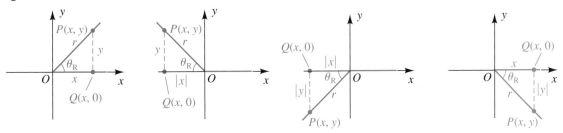

We may apply the theorem on trigonometric functions as ratios and also use triangle OQP to obtain the following formulas:

$$|\sin \theta| = \left| \frac{y}{r} \right| = \frac{|y|}{|r|} = \frac{|y|}{r} = \sin \theta_R$$

$$|\cos \theta| = \left| \frac{x}{r} \right| = \frac{|x|}{|r|} = \frac{|x|}{r} = \cos \theta_R$$

$$|\tan \theta| = \left| \frac{y}{x} \right| = \frac{|y|}{|x|} = \tan \theta_R$$

These formulas lead to the next theorem. If θ is a quadrantal angle, the theorem on trigonometric functions should be used to find values.

THEOREM ON REFERENCE ANGLES

If θ is a nonquadrantal angle in standard position, then to find the value of a trigonometric function at θ, find its value for the reference angle θ_R and prefix the appropriate sign.

The "appropriate sign" referred to in the theorem can be determined from the table of signs of the trigonometric functions given on page 107.

EXAMPLE 2   **Using reference angles**

Use reference angles to find the exact values of $\sin \theta$, $\cos \theta$, and $\tan \theta$ if

(a) $\theta = \dfrac{5\pi}{6}$ **(b)** $\theta = 315°$

Solution

(a) The angle $\theta = 5\pi/6$ and its reference angle $\theta_R = \pi/6$ are sketched in Figure 44(c). Since θ is in quadrant II, $\sin \theta$ is positive and both $\cos \theta$ and

(continued)

$\tan \theta$ are negative. Hence, by the theorem on reference angles and known results about special angles (see page 134), we obtain the following values:

$$\sin \frac{5\pi}{6} = \; + \quad \sin \frac{\pi}{6} = \frac{1}{2}$$

$$\cos \frac{5\pi}{6} = \; - \quad \cos \frac{\pi}{6} = -\frac{\sqrt{3}}{2}$$

$$\tan \frac{5\pi}{6} = \; - \quad \tan \frac{\pi}{6} = -\frac{\sqrt{3}}{3}$$

(b) The angle $\theta = 315°$ and its reference angle $\theta_R = 45°$ are sketched in Figure 44(a). Since θ is in quadrant IV, $\sin \theta < 0$, $\cos \theta > 0$, and $\tan \theta < 0$. Hence, by the theorem on reference angles, we obtain

$$\sin 315° = \; - \quad \sin 45° = -\frac{\sqrt{2}}{2}$$

$$\cos 315° = \; + \quad \cos 45° = \frac{\sqrt{2}}{2}$$

$$\tan 315° = \; - \quad \tan 45° = -1.$$

If we use a calculator to approximate function values, reference angles are usually unnecessary (see Discussion Exercise 2 at the end of the chapter). As an illustration, to find $\sin 210°$, we place the calculator in degree mode and obtain $\sin 210° = -0.5$, which is the exact value. Using the same procedure for $240°$, we obtain a decimal representation:

$$\sin 240° \approx -0.8660$$

A calculator should not be used to find the *exact* value of $\sin 240°$. In this case, we find the reference angle $60°$ of $240°$ and use the theorem on reference angles, together with known results about special angles, to obtain

$$\sin 240° = -\sin 60° = -\frac{\sqrt{3}}{2}.$$

Let us next consider the problem of solving an equation of the following type:

Problem   If θ is an acute angle and $\sin \theta = 0.6635$, approximate θ.

Most calculators have a key labeled $\boxed{\text{SIN}^{-1}}$ that can be used to help solve the equation. With other calculators, it may be necessary to use another key or a keystroke sequence such as $\boxed{\text{INV}}$ $\boxed{\text{SIN}}$ (refer to the user manual for the calculator). We shall use the following notation when finding θ, where $0 \le k \le 1$:

$$\text{if} \quad \sin \theta = k, \quad \text{then} \quad \theta = \sin^{-1} k$$

This notation is similar to that used for the inverse function f^{-1} of a function f in Section 1.5, where we saw that under certain conditions,

$$\text{if} \quad f(x) = y, \quad \text{then} \quad x = f^{-1}(y).$$

For the given problem, $\sin \theta = 0.6635$, f is the sine function, $x = \theta$, and $y = 0.6635$. The notation $\sin^{-1}$ is based on the *inverse trigonometric functions* discussed in Section 3.6. At this stage of our work, *we shall regard* $\sin^{-1}$ *simply as an entry made on a calculator using a* $\boxed{\text{SIN}^{-1}}$ *key.* Thus, for the stated problem, we obtain

$$\theta = \sin^{-1}(0.6635) \approx 41.57° \approx 0.7255.$$

As indicated, when finding an angle we will usually round off degree measure to the nearest $0.01°$ and radian measure to four decimal places.

Similarly, given $\cos \theta = k$ or $\tan \theta = k$, we write

$$\theta = \cos^{-1} k \qquad \text{or} \qquad \theta = \tan^{-1} k$$

to indicate the use of a $\boxed{\text{COS}^{-1}}$ or $\boxed{\text{TAN}^{-1}}$ key on a calculator.

Given $\csc \theta$, $\sec \theta$, or $\cot \theta$, we use a reciprocal relationship to find θ, as indicated in the following illustration.

ILLUSTRATION

Finding Acute Angle Solutions of Equations with a Calculator

Equation	Calculator solution (degree and radian)
$\sin \theta = 0.5$	$\theta = \sin^{-1}(0.5) = 30° \approx 0.5236$
$\cos \theta = 0.5$	$\theta = \cos^{-1}(0.5) = 60° \approx 1.0472$
$\tan \theta = 0.5$	$\theta = \tan^{-1}(0.5) \approx 26.57° \approx 0.4636$
$\csc \theta = 2$	$\theta = \sin^{-1}\left(\frac{1}{2}\right) = 30° \approx 0.5236$
$\sec \theta = 2$	$\theta = \cos^{-1}\left(\frac{1}{2}\right) = 60° \approx 1.0472$
$\cot \theta = 2$	$\theta = \tan^{-1}\left(\frac{1}{2}\right) \approx 26.57° \approx 0.4636$

The same technique may be employed if θ is *any* angle or real number. Thus, using the $\boxed{\text{SIN}^{-1}}$ key, we obtain, in degree or radian mode,

$$\theta = \sin^{-1}(0.6635) \approx 41.57° \approx 0.7255,$$

which is the reference angle for θ. If $\sin \theta$ is *negative,* then a calculator gives us the *negative* of the reference angle. For example,

$$\sin^{-1}(-0.6635) \approx -41.57° \approx -0.7255.$$

Similarly, given $\cos \theta$ or $\tan \theta$, we find θ with a calculator by using $\boxed{\text{COS}^{-1}}$ or $\boxed{\text{TAN}^{-1}}$, respectively. The interval containing θ is listed in the next chart. It is important to note that if $\cos \theta$ is negative, then θ is *not* the negative of the reference angle, but instead is in the interval $\pi/2 < \theta \leq \pi$, or $90° < \theta \leq 180°$. The reasons for using these intervals are explained in Section 3.6. We may use reciprocal relationships to solve similar equations involving $\csc \theta$, $\sec \theta$, and $\cot \theta$.

Equation	Values of k	Calculator solution	Interval containing θ if a calculator is used
$\sin \theta = k$	$-1 \leq k \leq 1$	$\theta = \sin^{-1} k$	$-\dfrac{\pi}{2} \leq \theta \leq \dfrac{\pi}{2}$, or $-90° \leq \theta \leq 90°$
$\cos \theta = k$	$-1 \leq k \leq 1$	$\theta = \cos^{-1} k$	$0 \leq \theta \leq \pi$, or $0° \leq \theta \leq 180°$
$\tan \theta = k$	any k	$\theta = \tan^{-1} k$	$-\dfrac{\pi}{2} < \theta < \dfrac{\pi}{2}$, or $-90° < \theta < 90°$

The following illustration contains some specific examples for both degree and radian modes.

ILLUSTRATION

Finding Angles with a Calculator

Equation	Calculator solution (degree and radian)
$\sin \theta = -0.5$	$\theta = \sin^{-1}(-0.5) = -30° \approx -0.5236$
$\cos \theta = -0.5$	$\theta = \cos^{-1}(-0.5) = 120° \approx 2.0944$
$\tan \theta = -0.5$	$\theta = \tan^{-1}(-0.5) \approx -26.57° \approx -0.4636$

When using a calculator to find θ, be sure to keep the restrictions on θ in mind. If other values are desired, then reference angles or other methods may be employed, as illustrated in the next example.

EXAMPLE 3   **Approximating an angle with a calculator**

If $\tan \theta = -0.4623$ and $0° \leq \theta < 360°$, find θ to the nearest 0.1°.

Solution   As pointed out in the preceding discussion, if we use a calculator (in degree mode) to find θ when $\tan \theta$ is negative, then the degree measure is in the interval $(-90°, 0°)$. In particular, we obtain the following:

$$\theta = \tan^{-1}(-0.4623) \approx -24.8°$$

Since we wish to find values of θ between $0°$ and $360°$, we use the (approximate) reference angle $\theta_R \approx 24.8°$. There are two possible values of θ such that $\tan \theta$ is negative—one in quadrant II, the other in quadrant IV. If θ is in quadrant II and $0° \leq \theta < 360°$, we have the situation shown in Figure 46, and

$$\theta = 180° - \theta_R \approx 180° - 24.8° = 155.2°.$$

If θ is in quadrant IV and $0° \leq \theta < 360°$, then, as in Figure 47,

$$\theta = 360° - \theta_R \approx 360° - 24.8° = 335.2°.$$

Figure 46

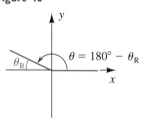

Figure 47

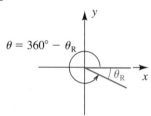

2.5 EXERCISES

Exer. 1–6: Find the reference angle θ_R if θ has the given measure.

1 (a) 240°   (b) 340°   (c) −202°   (d) −660°

2 (a) 165°   (b) 275°   (c) −110°   (d) 400°

3 (a) $3\pi/4$ (b) $4\pi/3$ (c) $-\pi/6$ (d) $9\pi/4$

4 (a) $7\pi/4$ (b) $2\pi/3$ (c) $-3\pi/4$ (d) $-23\pi/6$

5 (a) 3   (b) −2   (c) 5.5   (d) 100

6 (a) 6   (b) −4   (c) 4.5   (d) 80

Exer. 7–18: Find the exact value.

7 (a) $\sin (2\pi/3)$ (b) $\sin (-5\pi/4)$

8 (a) $\sin 210°$ (b) $\sin (-315°)$

9 (a) $\cos 150°$ (b) $\cos (-60°)$

10 (a) $\cos (5\pi/4)$ (b) $\cos (-11\pi/6)$

11 (a) $\tan (5\pi/6)$ (b) $\tan (-\pi/3)$

12 (a) $\tan 330°$ (b) $\tan (-225°)$

13 (a) $\cot 120°$ (b) $\cot (-150°)$

14 (a) $\cot (3\pi/4)$ (b) $\cot (-2\pi/3)$

15 (a) $\sec (2\pi/3)$ (b) $\sec (-\pi/6)$

16 (a) $\sec 135°$ (b) $\sec (-210°)$

17 (a) $\csc 240°$ (b) $\csc (-330°)$

18 (a) $\csc (3\pi/4)$ (b) $\csc (-2\pi/3)$

Exer. 19–24: Approximate to three decimal places.

19 (a) $\sin 73°20'$ (b) $\cos 0.68$

20 (a) $\cos 38°30'$ (b) $\sin 1.48$

21 (a) $\tan 21°10'$ (b) $\cot 1.13$

22 (a) $\cot 9°10'$ (b) $\tan 0.75$

23 (a) $\sec 67°50'$ (b) $\csc 0.32$

24 (a) $\csc 43°40'$ (b) $\sec 0.26$

Exer. 25–32: Approximate the acute angle θ to the nearest (a) 0.01° and (b) 1′.

25 $\cos \theta = 0.8620$ 26 $\sin \theta = 0.6612$

27 $\tan \theta = 3.7$ 28 $\cos \theta = 0.8$

29 $\sin \theta = 0.4217$ 30 $\tan \theta = 4.91$

31 $\sec \theta = 4.246$ 32 $\csc \theta = 11$

Exer. 33–34: Approximate to four decimal places.

33 (a) $\sin 98°10'$ (b) $\cos 623.7°$ (c) $\tan 3$

   (d) $\cot 231°40'$ (e) $\sec 1175.1°$ (f) $\csc 0.82$

34 (a) $\sin 496.4°$ (b) $\cos 0.65$ (c) $\tan 105°40'$

   (d) $\cot 1030.2°$ (e) $\sec 1.46$ (f) $\csc 320°50'$

Exer. 35–36: Approximate, to the nearest 0.1°, all angles θ in the interval [0°, 360°) that satisfy the equation.

35 (a) $\sin \theta = -0.5640$ (b) $\cos \theta = 0.7490$

   (c) $\tan \theta = 2.798$ (d) $\cot \theta = -0.9601$

   (e) $\sec \theta = -1.116$ (f) $\csc \theta = 1.485$

36 (a) $\sin \theta = 0.8225$ (b) $\cos \theta = -0.6604$

   (c) $\tan \theta = -1.5214$ (d) $\cot \theta = 1.3752$

   (e) $\sec \theta = 1.4291$ (f) $\csc \theta = -2.3179$

Exer. 37–38: Approximate, to the nearest 0.01 radian, all angles θ in the interval [0, 2π) that satisfy the equation.

37 (a) $\sin \theta = 0.4195$ (b) $\cos \theta = -0.1207$

   (c) $\tan \theta = -3.2504$ (d) $\cot \theta = 2.6815$

   (e) $\sec \theta = 1.7452$ (f) $\csc \theta = -4.8521$

38 (a) $\sin \theta = -0.0135$ (b) $\cos \theta = 0.9235$

   (c) $\tan \theta = 0.42$ (d) $\cot \theta = -2.731$

   (e) $\sec \theta = -3.51$ (f) $\csc \theta = 1.258$

39 *Thickness of the ozone layer* The thickness of the ozone layer can be estimated using the formula

$$\ln I_0 - \ln I = kx \sec \theta,$$

where I_0 is the intensity of a particular wavelength of light from the sun before it reaches the atmosphere, I is the intensity of the same wavelength after passing through a layer of ozone x centimeters thick, k is the absorption constant of ozone for that wavelength, and θ is the acute angle that the sunlight makes with the vertical. Suppose that for a wavelength of 3055×10^{-8} centimeter with $k \approx 1.88$, I_0/I is measured as 1.72 and $\theta = 12°$. Approximate the thickness of the ozone layer to the nearest 0.01 centimeter.

40 Ozone calculations Refer to Exercise 39. If the ozone layer is estimated to be 0.31 centimeter thick and, for a wavelength of 3055×10^{-8} centimeter, I_0/I is measured as 2.05, approximate the angle the sun made with the vertical at the time of the measurement.

41 Solar radiation The amount of sunshine illuminating a wall of a building can greatly affect the energy efficiency of the building. The solar radiation striking a vertical wall that faces east is given by

$$R = R_0 \cos \theta \sin \phi,$$

where R_0 is the maximum solar radiation possible, θ is the angle that the sun makes with the horizontal, and ϕ is the direction of the sun in the sky, with $\phi = 90°$ when the sun is in the east and $\phi = 0°$ when the sun is in the south.

(a) When does the maximum solar radiation R_0 strike the wall?

(b) What percentage of R_0 is striking the wall when $\theta = 60°$ and the sun is in the southeast?

42 Meteorological calculations In the mid-latitudes it is sometimes possible to estimate the distance between consecutive regions of low pressure. If ϕ is the latitude (in degrees), R is the earth's radius (in kilometers), and v is the horizontal wind velocity (in km/hr), then the distance d (in kilometers) from one low pressure area to the next can be estimated using the formula

$$d = 2\pi \left(\frac{vR}{0.52 \cos \phi} \right)^{1/3}.$$

(a) At a latitude of 48°, the earth's radius is approximately 6369 kilometers. Approximate d if the wind speed is 45 km/hr.

(b) If v and R are constant, how does d vary as the latitude increases?

43 Robot's arm Points on the terminal sides of angles play an important part in the design of arms for robots. Suppose a robot has a straight arm 18 inches long that can rotate about the origin in a coordinate plane. If the robot's hand is located at (18, 0) and then rotates through an angle of 60°, what is the new location of the hand?

44 Robot's arm Suppose the robot's arm in Exercise 43 can change its length in addition to rotating about the origin. If the hand is initially at (12, 12), approximately how many degrees should the arm be rotated and how much should its length be changed to move the hand to (−16, 10)?

2.6 TRIGONOMETRIC GRAPHS

In Section 2.3 we used sin t to denote the value of the sine function at the real number t, and we sketched the graph of $y = \sin t$ on a ty-coordinate system. Since we now wish to sketch graphs on an xy-coordinate system, we shall consider equations such as $y = \sin x$ instead of $y = \sin t$. *The variables x and y used here should not be confused with those used in Section 2.4 for a point P(x, y) on the terminal side of an angle.* We shall regard x as the radian measure of an angle or as a real number. These are equivalent points of view, since the sine of an angle of x radians is the same as the sine of the real number x. The number y is the function value that corresponds to x.

In this section we consider graphs of the equations

$$y = a \sin (bx + c) \quad \text{and} \quad y = a \cos (bx + c)$$

for real numbers a, b, and c. Our goal is to sketch such graphs without plotting many points. To do so we shall use facts about the graphs of the sine and cosine functions discussed in Section 2.3.

Let us begin by considering the special case $c = 0$ and $b = 1$—that is,

$$y = a \sin x \quad \text{and} \quad y = a \cos x.$$

We can find y-coordinates of points on the graphs by multiplying y-coordinates of points on the graphs of $y = \sin x$ and $y = \cos x$ by a. To illustrate, if $y = 2 \sin x$, we multiply the y-coordinate of each point on the graph of $y = \sin x$ by 2. This gives us Figure 48, where for comparison we also show the graph of $y = \sin x$. The procedure is the same as that for vertically stretching the graph of a function, discussed in Section 1.4.

Figure 48

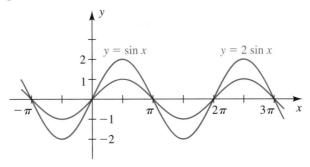

As another illustration, if $y = \frac{1}{2} \sin x$, we multiply y-coordinates of points on the graph of $y = \sin x$ by $\frac{1}{2}$. This multiplication vertically compresses the graph of $y = \sin x$ by a factor of 2, as illustrated in Figure 49.

Figure 49

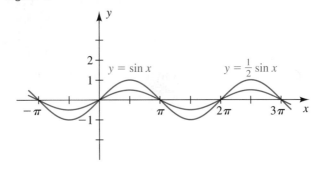

The following example illustrates a graph of $y = a \sin x$ with a negative.

E X A M P L E   1   Sketching the graph of an equation involving $\sin x$

Sketch the graph of the equation $y = -2 \sin x$.

Solution   The graph of $y = -2 \sin x$ sketched in Figure 50 can be obtained by first sketching the graph of $y = \sin x$ (shown in the figure) and then multiplying y-coordinates by -2. An alternative method is to reflect the graph of $y = 2 \sin x$ (see Figure 48) through the x-axis.

(continued)

Figure 50

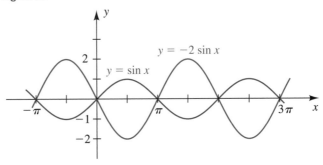

For any $a \neq 0$, the graph of $y = a \sin x$ has the general appearance of one of the graphs illustrated in Figures 48, 49, and 50. The amount of stretching of the graph of $y = \sin x$ and whether the graph is reflected are determined by the absolute value of a and the sign of a, respectively. The largest y-coordinate $|a|$ is the **amplitude of the graph** or, equivalently, the **amplitude of the function** f given by $f(x) = a \sin x$. In Figures 48 and 50 the amplitude is 2. In Figure 49 the amplitude is $\frac{1}{2}$. Similar remarks and techniques apply if $y = a \cos x$.

EXAMPLE 2   Sketching the graph of an equation involving $\cos x$

Find the amplitude and sketch the graph of $y = 3 \cos x$.

Solution   By the preceding discussion, the amplitude is 3. As indicated in Figure 51, we first sketch the graph of $y = \cos x$ and then multiply y-coordinates by 3.

Figure 51

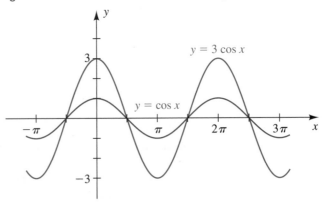

Let us next consider $y = a \sin bx$ and $y = a \cos bx$ for nonzero real numbers a and b. As before, the amplitude is $|a|$. If $b > 0$, then exactly one cycle occurs as bx increases from 0 to 2π or, equivalently, as x increases from 0 to $2\pi/b$. If $b < 0$, then $-b > 0$ and one cycle occurs as x increases from 0 to $2\pi/(-b)$. Thus, the period of the function f given by $f(x) = a \sin bx$ or $f(x) = a \cos bx$ is $2\pi/|b|$. For convenience, we shall also refer to $2\pi/|b|$ as the period of the *graph of f*. The next theorem summarizes our discussion.

THEOREM ON AMPLITUDES AND PERIODS

> If $y = a \sin bx$ or $y = a \cos bx$ for nonzero real numbers a and b, then the graph has amplitude $|a|$ and period $\dfrac{2\pi}{|b|}$.

We can also relate the role of b to the discussion of horizontal compressing and stretching of a graph in Section 1.4. If $|b| > 1$, the graph of $y = \sin bx$ or $y = \cos bx$ can be considered to be compressed horizontally by a factor b. If $0 < |b| < 1$, the graphs are stretched horizontally by a factor $1/b$. This concept is illustrated in the next two examples.

EXAMPLE 3 Finding an amplitude and a period

Find the amplitude and the period and sketch the graph of $y = 3 \sin 2x$.

Solution Using the theorem on amplitudes and periods with $a = 3$ and $b = 2$, we obtain the following:

$$\text{amplitude:} \quad |a| = |3| = 3$$

$$\text{period:} \quad \frac{2\pi}{|b|} = \frac{2\pi}{2} = \pi$$

Thus, there is exactly one sine wave of amplitude 3 on the x-interval $[0, \pi]$. Sketching this wave and then extending the graph to the right and left gives us Figure 52.

Figure 52

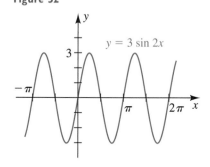

$y = 3 \sin 2x$

EXAMPLE 4 Finding an amplitude and a period

Find the amplitude and the period and sketch the graph of $y = 2 \sin \frac{1}{2}x$.

Solution Using the theorem on amplitudes and periods with $a = 2$ and $b = \frac{1}{2}$, we obtain the following:

$$\text{amplitude:} \quad |a| = |2| = 2$$

$$\text{period:} \quad \frac{2\pi}{|b|} = \frac{2\pi}{\frac{1}{2}} = 4\pi$$

Thus, there is one sine wave of amplitude 2 on the interval $[0, 4\pi]$. Sketching this wave and extending it left and right gives us the graph in Figure 53.

Figure 53

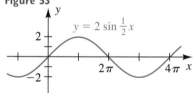

$y = 2 \sin \frac{1}{2}x$

If $y = a \sin bx$ and if b is a large positive number, then the period $2\pi/b$ is small and the sine waves are close together, with b sine waves on the interval $[0, 2\pi]$. For example, in Figure 52, $b = 2$ and we have two sine waves on $[0, 2\pi]$. If b is a small positive number, then the period $2\pi/b$ is large and the waves are far apart. To illustrate, if $y = \sin \frac{1}{10}x$, then one-tenth of a sine wave occurs on $[0, 2\pi]$ and an interval 20π units long is required for one complete cycle. (See also Figure 53—for $y = 2 \sin \frac{1}{2}x$, one-half of a sine wave occurs on $[0, 2\pi]$.)

If $b < 0$, we can use the fact that $\sin(-x) = -\sin x$ to obtain the graph of $y = a \sin bx$. To illustrate, the graph of $y = \sin(-2x)$ is the same as the graph of $y = -\sin 2x$.

EXAMPLE 5   **Finding an amplitude and a period**

Find the amplitude and the period and sketch the graph of the equation $y = 2 \sin(-3x)$.

Solution   Since $\sin(-3x) = -\sin 3x$, we may write $y = -2 \sin 3x$. The amplitude is $|-2| = 2$, and the period is $2\pi/3$. Thus, there is one cycle on an interval of length $2\pi/3$. The negative sign indicates a reflection through the x-axis. If we consider the interval $[0, 2\pi/3]$ and sketch a sine wave of amplitude 2 (reflected through the x-axis), the shape of the graph is apparent. The part of the graph in the interval $[0, 2\pi/3]$ is repeated periodically, as illustrated in Figure 54.

Figure 54

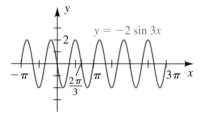

EXAMPLE 6   **Finding an amplitude and a period**

Find the amplitude and the period and sketch the graph of $y = 4 \cos \pi x$.

Solution   The amplitude is $|4| = 4$, and the period is $2\pi/\pi = 2$. Thus, there is exactly one cosine wave of amplitude 4 on the interval $[0, 2]$. Sketching this wave and extending it left and right gives us the graph in Figure 55.

Figure 55

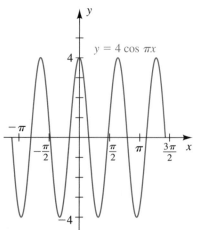

As in Section 1.4, if f is a function and c is a positive real number, then the graph of $y = f(x) + c$ can be obtained by shifting the graph of $y = f(x)$ vertically upward a distance c. For the graph of $y = f(x) - c$, we shift the graph of $y = f(x)$ vertically downward a distance c. In the next example we use this technique for a trigonometric graph.

EXAMPLE 7   **Vertically shifting a trigonometric graph**

Sketch the graph of $y = 2 \sin x + 3$.

Figure 56

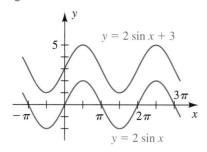

Solution It is important to note that $y \neq 2 \sin (x + 3)$. The graph of $y = 2 \sin x$ is sketched in Figure 56. If we shift this graph vertically upward a distance 3, we obtain the graph of $y = 2 \sin x + 3$.

Let us next consider the graph of

$$y = a \sin (bx + c).$$

As before, the amplitude is $|a|$ and the period is $2\pi/|b|$. One cycle occurs if $bx + c$ increases from 0 to 2π. Hence, we can find an interval containing exactly one sine wave by solving the equations

$$bx + c = 0 \qquad \text{and} \qquad bx + c = 2\pi.$$

The solutions are

$$x = -\frac{c}{b} \qquad \text{and} \qquad x = -\frac{c}{b} + \frac{2\pi}{b}.$$

Equivalently, we could solve the following inequality for x:

$$0 \leq bx + c \leq 2\pi$$
$$-c \leq bx \leq 2\pi - c$$
$$-\frac{c}{b} \leq x \leq \frac{2\pi}{b} - \frac{c}{b}$$

The number $-c/b$ is the **phase shift** associated with the graph. The graph of $y = a \sin (bx + c)$ may be obtained by shifting the graph of $y = a \sin bx$ to the left if the phase shift is negative or to the right if the phase shift is positive.

Analogous results are true for $y = a \cos (bx + c)$. The next theorem summarizes our discussion.

THEOREM ON AMPLITUDES, PERIODS, AND PHASE SHIFTS

If $y = a \sin (bx + c)$ or $y = a \cos (bx + c)$ for nonzero real numbers a and b, then

(1) the amplitude is $|a|$, the period is $\dfrac{2\pi}{|b|}$, and the phase shift is $-c/b$;

(2) an interval containing exactly one cycle can be found by solving the two equations

$$bx + c = 0 \qquad \text{and} \qquad bx + c = 2\pi$$

or the inequality

$$0 \leq bx + c \leq 2\pi.$$

EXAMPLE 8   Finding an amplitude, a period, and a phase shift

Find the amplitude, the period, and the phase shift and sketch the graph of

$$y = 3 \sin \left(2x + \frac{\pi}{2} \right).$$

Solution   The equation is of the form $y = a \sin(bx + c)$ with $a = 3$, $b = 2$, and $c = \pi/2$. Thus, the amplitude is $|a| = 3$, and the period is $2\pi/|b| = 2\pi/2 = \pi$.

By part (2) of the theorem on amplitudes, periods, and phase shifts, the phase shift and an interval containing one sine wave can be found by solving the two equations

$$2x + \frac{\pi}{2} = 0 \quad \text{and} \quad 2x + \frac{\pi}{2} = 2\pi.$$

This gives us

$$x = -\frac{\pi}{4} \quad \text{and} \quad x = \frac{3\pi}{4}.$$

Thus, the phase shift is $-\pi/4$, and one sine wave of amplitude 3 occurs on the interval $[-\pi/4, 3\pi/4]$. Sketching that wave and then repeating it to the right and left gives us the graph in Figure 57.

Figure 57

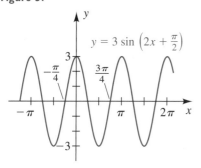

$y = 3 \sin \left(2x + \frac{\pi}{2} \right)$

EXAMPLE 9   Finding an amplitude, a period, and a phase shift

Find the amplitude, the period, and the phase shift and sketch the graph of $y = 2 \cos(3x - \pi)$.

Solution   The equation has the form $y = a \cos(bx + c)$ with $a = 2$, $b = 3$, and $c = -\pi$. Thus, the amplitude is $|a| = 2$, and the period is $2\pi/|b| = 2\pi/3$.

By part (2) of the theorem on amplitudes, periods, and phase shifts, the phase shift and an interval containing one cycle can be found by solving the following inequality:

$$0 \le 3x - \pi \le 2\pi$$
$$\pi \le 3x \quad\quad \le 3\pi$$
$$\frac{\pi}{3} \le x \quad\quad \le \pi$$

Hence, the phase shift is $\pi/3$, and one cosine-type cycle of amplitude 2 occurs on the interval $[\pi/3, \pi]$. Sketching that part of the graph and then repeating it to the right and left gives us the sketch in Figure 58.

Figure 58

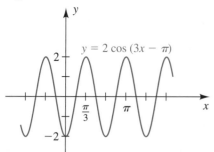

$y = 2 \cos(3x - \pi)$

If we solve the inequality

$$-\frac{\pi}{2} \le 3x - \pi \le \frac{3\pi}{2} \quad \text{instead of} \quad 0 \le 3x - \pi \le 2\pi,$$

we obtain the interval $\pi/6 \le x \le 5\pi/6$, which gives us a cycle between *x*-intercepts rather than a cycle between maximums.

E X A M P L E 10   **Finding an equation for a sine wave**

Express the equation for the sine wave shown in Figure 59 in the form

$$y = a \sin (bx + c)$$

for $a > 0$, $b > 0$, and the least positive real number c.

Figure 59

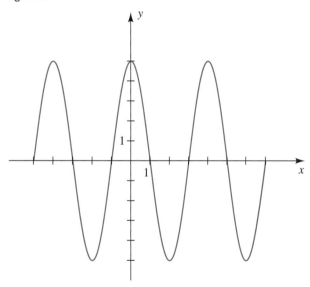

Solution   The largest and smallest *y*-coordinates of points on the graph are 5 and -5, respectively. Hence, the amplitude is $a = 5$.

Since one sine wave occurs on the interval $[-1, 3]$, the period has value $3 - (-1) = 4$. Hence, by the theorem on amplitudes, periods, and phase shifts (with $b > 0$),

$$\frac{2\pi}{b} = 4 \quad \text{or, equivalently,} \quad b = \frac{\pi}{2}.$$

(continued)

The phase shift is $-c/b = -c/(\pi/2)$. Since c is to be positive, the phase shift must be *negative;* that is, the graph in Figure 59 must be obtained by shifting the graph of $y = 5 \sin [(\pi/2)x]$ to the *left*. Since we want c to be as small as possible, we choose the phase shift -1. Hence,

$$-\frac{c}{\pi/2} = -1 \qquad \text{or, equivalently,} \qquad c = \frac{\pi}{2}.$$

Thus, the desired equation is

$$y = 5 \sin \left(\frac{\pi}{2} x + \frac{\pi}{2} \right).$$

There are many other equations for the graph. For example, we could use the phase shifts -5, -9, -13, and so on, but these would not give us the *least* positive value for c. Two other equations for the graph are

$$y = 5 \sin \left(\frac{\pi}{2} x - \frac{3\pi}{2} \right) \qquad \text{and} \qquad y = -5 \sin \left(\frac{\pi}{2} x + \frac{3\pi}{2} \right).$$

However, neither of these equations satisfies the given criteria for a, b, and c, since in the first, $c < 0$, and in the second, $a < 0$ and c does not have its least positive value.

As an alternative solution, we could write

$$y = a \sin (bx + c) \qquad \text{as} \qquad y = a \sin \left[b \left(x + \frac{c}{b} \right) \right].$$

As before, we find $a = 5$ and $b = \pi/2$. Now since the graph has an x-intercept at $x = -1$, we can consider this graph to be a horizontal shift of the graph of $y = 5 \sin [(\pi/2)x]$ to the left by 1 unit—that is, replace x with $x + 1$. Thus, an equation is

$$y = 5 \sin \left[\frac{\pi}{2} (x + 1) \right], \qquad \text{or} \qquad y = 5 \sin \left(\frac{\pi}{2} x + \frac{\pi}{2} \right).$$

Many phenomena that occur in nature vary in a cyclic or rhythmic manner. It is sometimes possible to represent such behavior by means of trigonometric functions, as illustrated in the next two examples.

EXAMPLE 11　　**Analyzing the process of breathing**

The rhythmic process of breathing consists of alternating periods of inhaling and exhaling. One complete cycle normally takes place every 5 seconds. If $F(t)$ denotes the air flow rate at time t (in liters per second) and if the maximum flow rate is 0.6 liter per second, find a formula of the form $F(t) = a \sin bt$ that fits this information.

Solution   If $F(t) = a \sin bt$ for some $b > 0$, then the period of F is $2\pi/b$. In this application the period is 5 seconds, and hence

$$\frac{2\pi}{b} = 5, \quad \text{or} \quad b = \frac{2\pi}{5}.$$

Since the maximum flow rate corresponds to the amplitude a of F, we let $a = 0.6$. This gives us the formula

$$F(t) = 0.6 \sin\left(\frac{2\pi}{5}t\right).$$

E X A M P L E   12   **Approximating the number of hours of daylight in a day**

The number of hours of daylight $D(t)$ at a particular time of the year can be approximated by

$$D(t) = \frac{K}{2} \sin \frac{2\pi}{365}(t - 79) + 12$$

for t in days and $t = 0$ corresponding to January 1. The constant K determines the total variation in day length and depends on the latitude of the locale.

(a) For Boston, $K \approx 6$. Sketch the graph of D for $0 \le t \le 365$.

(b) When is the day length the longest? the shortest?

Solution

(a) If $K = 6$, then $K/2 = 3$, and we may write $D(t)$ in the form

$$D(t) = f(t) + 12,$$

where $$f(t) = 3 \sin \frac{2\pi}{365}(t - 79).$$

We shall sketch the graph of f and then apply a vertical shift through a distance 12.

As in part (2) of the theorem on amplitudes, periods, and phase shifts, we can obtain a t-interval containing exactly one cycle by solving the following inequality:

$$0 \le \frac{2\pi}{365}(t - 79) \le 2\pi$$

$$0 \le \quad t - 79 \quad \le 365$$

$$79 \le \quad t \quad \le 444$$

(continued)

Figure 60

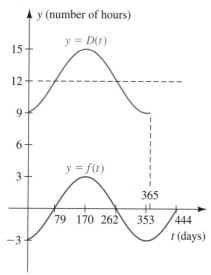

Hence, one sine wave occurs on the interval [79, 444]. Dividing this interval into four equal parts, we obtain the following table of values, which indicates the familiar sine wave pattern of amplitude 3.

t	79	170.25	261.5	352.75	444
$f(t)$	0	3	0	−3	0

If $t = 0$,

$$f(0) = 3 \sin \frac{2\pi}{365}(-79) \approx 3 \sin (-1.36) \approx -2.9.$$

Since the period of f is 365, this implies that $f(365) \approx -2.9$.

The graph of f for the interval [0, 444] is sketched in Figure 60, with different scales on the axes and t rounded off to the nearest day.

Applying a vertical shift of 12 units gives us the graph of D for $0 \le t \le 365$ shown in Figure 60.

(b) The longest day—that is, the largest value of $D(t)$—occurs 170 days after January 1. Except for leap year, this corresponds to June 20. The shortest day occurs 353 days after January 1, or December 20.

In the next example we use a graphing utility to approximate the solution of an inequality that involves trigonometric expressions.

 EXAMPLE 13   **Approximating solutions of a trigonometric inequality**

Approximate the solution of the inequality

$$\sin 3x < x + \sin x.$$

Solution   The given inequality is equivalent to

$$\sin 3x - x - \sin x < 0.$$

If we assign $\sin 3x - x - \sin x$ to Y_1, then the given problem is equivalent to finding where the graph of Y_1 is below the x-axis. Using the standard viewing rectangle gives us a sketch similar to Figure 61(a), where we see that the graph of Y_1 has an x-intercept c between -1 and 0. It appears that the graph is below the x-axis on the interval (c, ∞); however, this fact is not perfectly clear because of the small scale on the axes.

Using the viewing rectangle $[-1.5, 1.5]$ by $[-1, 1]$ with Xscl and Yscl both equal to 0.25, we obtain Figure 61(b), where we see that the x-intercepts are approximately $-0.5, 0$, and 0.5. Using a zoom or root feature yields the more accurate positive value 0.51. Since the function involved is odd, the

Figure 61

(a) [−15, 15] by [−10, 10]   (b) [−1.5, 1.5] by [−1, 1]

 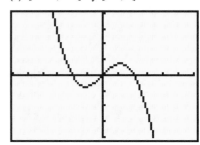

negative value is approximately −0.51. Hence, the solutions of the inequality are in the (approximate) intervals

$$(-0.51, 0) \cup (0.51, \infty).$$

EXAMPLE 14   **Investigating alternating current in an electrical circuit**

The current I (in amperes) in an alternating current circuit at time t (in seconds) is given by

$$I = 30 \sin \left(50\pi t - \frac{7\pi}{3} \right).$$

Approximate the smallest value of t for which $I = 15$.

Solution   Letting $I = 15$ in the given formula, we obtain

$$15 = 30 \sin \left(50\pi t - \frac{7\pi}{3} \right)$$

or, equivalently,

$$\sin \left(50\pi t - \frac{7\pi}{3} \right) - \frac{1}{2} = 0.$$

If we assign $\sin (50\pi x - 7\pi/3) - \frac{1}{2}$ to Y_1, then the given problem is equivalent to approximating the smallest x-intercept of the graph.

Since the period of Y_1 is

$$\frac{2\pi}{b} = \frac{2\pi}{50\pi} = \frac{1}{25} = 0.04$$

and since $-\frac{3}{2} \le Y_1 \le \frac{1}{2}$, we select the viewing rectangle [0, 0.04] by [−1.5, 0.5] with Xscl = 0.01 and Yscl = 0.25, obtaining a sketch similar to Figure 62. Using a zoom or root feature gives us $t \approx 0.01$ sec.

Figure 62
[0, 0.04] by [−1.5, 0.5]

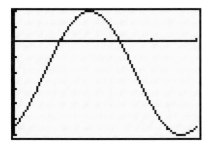

We will rework the preceding example in Section 3.2 and show how to find the exact value of t without the aid of a graphing utility.

2.6 EXERCISES

1 Find the amplitude and the period and sketch the graph of the equation:

(a) $y = 4 \sin x$ (b) $y = \sin 4x$

(c) $y = \frac{1}{4} \sin x$ (d) $y = \sin \frac{1}{4}x$

(e) $y = 2 \sin \frac{1}{4}x$ (f) $y = \frac{1}{2} \sin 4x$

(g) $y = -4 \sin x$ (h) $y = \sin (-4x)$

2 Sketch the graphs of the equations that involve the cosine and are analogous to those in (a)–(h) of Exercise 1.

3 Find the amplitude and the period and sketch the graph of the equation:

(a) $y = 3 \cos x$ (b) $y = \cos 3x$

(c) $y = \frac{1}{3} \cos x$ (d) $y = \cos \frac{1}{3}x$

(e) $y = 2 \cos \frac{1}{3}x$ (f) $y = \frac{1}{2} \cos 3x$

(g) $y = -3 \cos x$ (h) $y = \cos (-3x)$

4 Sketch the graphs of the equations that involve the sine and are analogous to those in (a)–(h) of Exercise 3.

Exer. 5–40: Find the amplitude, the period, and the phase shift and sketch the graph of the equation.

5 $y = \sin \left(x - \dfrac{\pi}{2} \right)$ 6 $y = \sin \left(x + \dfrac{\pi}{4} \right)$

7 $y = 3 \sin \left(x + \dfrac{\pi}{6} \right)$ 8 $y = 2 \sin \left(x - \dfrac{\pi}{3} \right)$

9 $y = \cos \left(x + \dfrac{\pi}{2} \right)$ 10 $y = \cos \left(x - \dfrac{\pi}{3} \right)$

11 $y = 4 \cos \left(x - \dfrac{\pi}{4} \right)$ 12 $y = 3 \cos \left(x + \dfrac{\pi}{6} \right)$

13 $y = \sin (2x - \pi) + 1$ 14 $y = -\sin (3x + \pi) - 1$

15 $y = -\cos (3x + \pi) - 2$ 16 $y = \cos (2x - \pi) + 2$

17 $y = -2 \sin (3x - \pi)$ 18 $y = 3 \cos (3x - \pi)$

19 $y = \sin \left(\dfrac{1}{2}x - \dfrac{\pi}{3} \right)$ 20 $y = \sin \left(\dfrac{1}{2}x + \dfrac{\pi}{4} \right)$

21 $y = 6 \sin \pi x$ 22 $y = 3 \cos \dfrac{\pi}{2}x$

23 $y = 2 \cos \dfrac{\pi}{2}x$ 24 $y = 4 \sin 3\pi x$

25 $y = \dfrac{1}{2} \sin 2\pi x$ 26 $y = \dfrac{1}{2} \cos \dfrac{\pi}{2}x$

27 $y = 5 \sin \left(3x - \dfrac{\pi}{2} \right)$ 28 $y = -4 \cos \left(2x + \dfrac{\pi}{3} \right)$

29 $y = 3 \cos \left(\dfrac{1}{2}x - \dfrac{\pi}{4} \right)$ 30 $y = -2 \sin \left(\dfrac{1}{2}x + \dfrac{\pi}{2} \right)$

31 $y = -5 \cos \left(\dfrac{1}{3}x + \dfrac{\pi}{6} \right)$ 32 $y = 4 \sin \left(\dfrac{1}{3}x - \dfrac{\pi}{3} \right)$

33 $y = 3 \cos (\pi x + 4\pi)$ 34 $y = -2 \sin (2\pi x + \pi)$

35 $y = -\sqrt{2} \sin \left(\dfrac{\pi}{2}x - \dfrac{\pi}{4} \right)$

36 $y = \sqrt{3} \cos \left(\dfrac{\pi}{4}x - \dfrac{\pi}{2} \right)$

37 $y = -2 \sin (2x - \pi) + 3$

38 $y = 3 \cos (x + 3\pi) - 2$

39 $y = 5 \cos (2x + 2\pi) + 2$

40 $y = -4 \sin (3x - \pi) - 3$

Exer. 41–44: The graph of an equation is shown in the figure. (a) Find the amplitude, period, and phase shift. (b) Write the equation in the form $y = a \sin (bx + c)$ for $a > 0, b > 0$, and the least positive real number c.

41

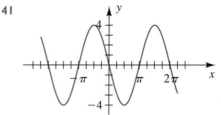

42

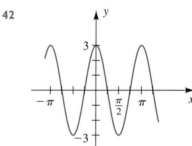

43

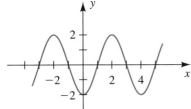

44

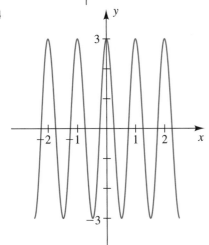

45 Electroencephalography Shown in the figure is an electroencephalogram of human brain waves during deep sleep. If we use $W = a \sin(bt + c)$ to represent these waves, what is the value of b?

Exercise 45

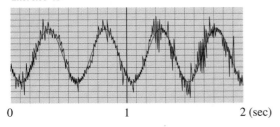

0   1   2 (sec)

46 Intensity of daylight On a certain spring day with 12 hours of daylight, the light intensity I takes on its largest value of 510 calories/cm² at midday. If $t = 0$ corresponds to sunrise, find a formula $I = a \sin bt$ that fits this information.

47 Heart action The pumping action of the heart consists of the systolic phase, in which blood rushes from the left ventricle into the aorta, and the diastolic phase, during which the heart muscle relaxes. The function whose

graph is shown in the figure is sometimes used to model one complete cycle of this process. For a particular individual, the systolic phase lasts $\frac{1}{4}$ second and has a maximum flow rate of 8 liters per minute. Find a and b.

Exercise 47

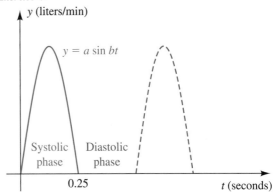

48 Biorhythms The popular biorhythm theory uses the graphs of three simple sine functions to make predictions about an individual's physical, emotional, and intellectual potential for a particular day. The graphs are given by $y = a \sin bt$ for t in days, with $t = 0$ corresponding to birth and $a = 1$ denoting 100% potential.

(a) Find the value of b for the physical cycle, which has a period of 23 days; for the emotional cycle (period 28 days); and for the intellectual cycle (period 33 days).

(b) Evaluate the biorhythm cycles for a person who has just become 21 years of age and is exactly 7670 days old.

49 Tidal components The height of the tide at a particular point on shore can be predicted by using seven trigonometric functions (called tidal components) of the form $f(t) = a \cos(bt + c)$. The principal lunar component may be approximated by

$$f(t) = a \cos\left(\frac{\pi}{6}t - \frac{11\pi}{12}\right),$$

where t is in hours and $t = 0$ corresponds to midnight. Sketch the graph of f if $a = 0.5$ m.

50 Tidal components Refer to Exercise 49. The principal solar diurnal component may be approximated by

$$f(t) = a \cos\left(\frac{\pi}{12}t - \frac{7\pi}{12}\right).$$

Sketch the graph of f if $a = 0.2$ m.

51 *Hours of daylight in Fairbanks* If the formula for $D(t)$ in Example 12 is used for Fairbanks, Alaska, then $K \approx 12$. Sketch the graph of D in this case for $0 \le t \le 365$.

52 *Low temperature in Fairbanks* Based on years of weather data, the expected low temperature T (in °F) in Fairbanks, Alaska, can be approximated by

$$T = 36 \sin \frac{2\pi}{365}(t - 101) + 14,$$

where t is in days and $t = 0$ corresponds to January 1.

(a) Sketch the graph of T for $0 \le t \le 365$.

(b) Predict when the coldest day of the year will occur.

Exer. 53–54: Graph the equation $y = f(t)$ on the interval [0, 24]. Let y represent the outdoor temperature (in °F) at time t (in hours), where $t = 0$ corresponds to 9 A.M. Describe the temperature during the 24-hour interval.

53 $y = 20 + 15 \sin \dfrac{\pi}{12}t$

54 $y = 80 + 22 \cos \dfrac{\pi}{12}(t - 3)$

Exer. 55–58: Scientists sometimes use the formula

$$f(t) = a \sin (bt + c) + d$$

to simulate temperature variations during the day, with time t in hours, temperature $f(t)$ in °C, and $t = 0$ corresponding to midnight. Assume that $f(t)$ is decreasing at midnight.

(a) **Determine values of a, b, c, and d that fit the information.**

(b) **Sketch the graph of f for $0 \le t \le 24$.**

55 The high temperature is 10°C, and the low temperature of -10°C occurs at 4 A.M.

56 The temperature at midnight is 15°C, and the high and low temperatures are 20°C and 10°C.

57 The temperature varies between 10°C and 30°C, and the average temperature of 20°C first occurs at 9 A.M.

58 The high temperature of 28°C occurs at 2 P.M., and the average temperature of 20°C occurs 6 hours later.

59 *Precipitation at South Lake Tahoe* The average monthly precipitation P (in inches) at South Lake Tahoe, California, is listed in the table.

Month	P	Month	P	Month	P
Jan.	6.1	May	1.2	Sept.	0.5
Feb.	5.4	June	0.6	Oct.	2.8
March	3.9	July	0.3	Nov.	3.1
April	2.2	Aug.	0.2	Dec.	5.4

(a) Let t be time in months, with $t = 1$ corresponding to January, $t = 2$ to February, . . . , $t = 12$ to December, $t = 13$ to January, and so on. Plot the data points for a two-year period.

(b) Find a function $P(t) = a \sin (bt + c) + d$ that approximates the average monthly precipitation. Plot the data and the function P on the same coordinate axes.

60 *Thames River depth* When a river flows into an ocean, the depth of the river varies near its mouth as a result of tides. Information about this change in depth is critical for safety. The following table gives the depth D (in feet) of the Thames River in London for a 24-hour period.

Time	D	Time	D	Time	D
12 A.M.	27.1	8 A.M.	20.0	4 P.M.	34.0
1 A.M.	30.1	9 A.M.	18.0	5 P.M.	32.4
2 A.M.	33.0	10 A.M.	18.3	6 P.M.	29.1
3 A.M.	34.3	11 A.M.	20.6	7 P.M.	25.2
4 A.M.	33.7	12 P.M.	24.2	8 P.M.	21.9
5 A.M.	31.1	1 P.M.	28.1	9 P.M.	19.6
6 A.M.	27.1	2 P.M.	31.7	10 P.M.	18.6
7 A.M.	23.2	3 P.M.	33.7	11 P.M.	19.6

(a) Plot the data, with time on the horizontal axis and depth on the vertical axis. Let $t = 0$ correspond to 12.00 A.M.

(b) Determine a function $D(t) = a \sin (bt + c) + d$, where $D(t)$ represents the depth of the water in the harbor at time t. Graph the function D with the data. (*Hint:* To determine b, find the time between maximum depths.)

(c) If a ship requires at least 24 feet of water to navigate the Thames safely, graphically determine the time interval(s) when navigation is *not* safe.

61 *Hours of daylight* The number of daylight hours D at a particular location varies with both the month and the latitude. The table lists the number of daylight hours on the first day of each month at 60°N latitude.

Month	D	Month	D	Month	D
Jan.	6.03	May	15.97	Sept.	14.18
Feb.	7.97	June	18.28	Oct.	11.50
March	10.43	July	18.72	Nov.	8.73
April	13.27	Aug.	16.88	Dec.	5.88

(a) Let t be time in months, with $t = 1$ corresponding to January, $t = 2$ to February, $\ldots$, $t = 12$ to December, $t = 13$ to January, and so on. Plot the data for a two-year period.

(b) Find a function $D(t) = a \sin (bt + c) + d$ that approximates the number of daylight hours. Graph the function D with the data.

62 *Hours of daylight* Refer to Exercise 61. The maximum number of daylight hours at 40°N is 15.02 hours and occurs on June 21. The minimum number of daylight hours is 9.32 hours and occurs on December 22.

(a) Determine a function $D(t) = a \sin (bt + c) + d$ that models the number of daylight hours, where t is in months and $t = 1$ corresponds to January 1.

(b) Graph the function D using the viewing rectangle [0.5, 24.5] by [0, 20].

(c) Predict the number of daylight hours on February 1 and September 1. Compare your answers to the true values of 10.17 and 13.08 hours, respectively.

C Exer. 63–66: Graph the equation on the interval $[-2, 2]$, and describe the behavior of y as $x \to 0^-$ and as $x \to 0^+$.

63 $y = \sin \dfrac{1}{x}$ **64** $y = |x| \sin \dfrac{1}{x}$

65 $y = \dfrac{\sin 2x}{x}$ **66** $y = \dfrac{1 - \cos 3x}{x}$

C Exer. 67–68: Graph the equation on the interval $[-20, 20]$, and estimate the horizontal asymptote.

67 $y = x^2 \sin^2 \left(\dfrac{2}{x}\right)$ **68** $y = \dfrac{1 - \cos^2 (2/x)}{\sin (1/x)}$

C Exer. 69–70: Use a graph to solve the inequality on the interval $[-\pi, \pi]$.

69 $\cos 3x \geq \frac{1}{2}x - \sin x$

70 $\frac{1}{4} \tan \left(\frac{1}{3}x^2\right) < \frac{1}{2} \cos 2x + \frac{1}{5}x^2$

2.7 ADDITIONAL TRIGONOMETRIC GRAPHS

Methods we developed in Section 2.6 for the sine and cosine can be applied to the other four trigonometric functions; however, there are several differences. Since the tangent, cotangent, secant, and cosecant functions have no largest values, the notion of amplitude has no meaning. Moreover, we do not refer to cycles. For some tangent and cotangent graphs, we begin by sketching the portion between successive vertical asymptotes and then repeat that pattern to the right and to the left.

The graph of $y = a \tan x$ for $a > 0$ can be obtained by stretching or compressing the graph of $y = \tan x$. If $a < 0$, then we also use a reflection about the x-axis. Since the tangent function has period π, it is sufficient to sketch the graph between the two successive vertical asymptotes $x = -\pi/2$ and $x = \pi/2$. The same pattern occurs to the right and to the left, as in the next example.

EXAMPLE 1   Sketching the graph of an equation involving tan x

Sketch the graph of the equation:

(a) $y = 2 \tan x$ **(b)** $y = \frac{1}{2} \tan x$

Solution   We begin by sketching the graph of $y = \tan x$, as shown in red in Figures 63 and 64, between the vertical asymptotes $x = -\pi/2$ and $x = \pi/2$.

(a) For $y = 2 \tan x$, we multiply the y-coordinate of each point by 2 and then extend the resulting graph to the right and left, as shown in Figure 63.

Figure 63 $y = 2 \tan x$

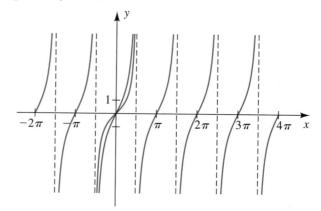

(b) For $y = \frac{1}{2} \tan x$, we multiply the y-coordinates by $\frac{1}{2}$, obtaining the sketch in Figure 64.

Figure 64 $y = \frac{1}{2} \tan x$

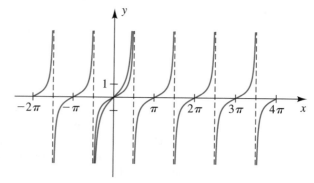

The method used in Example 1 can be applied to other functions. Thus, to sketch the graph of the equation $y = 3 \sec x$, we could first sketch the graph of $y = \sec x$ and then multiply the y-coordinate of each point by 3.

The next theorem is an analogue of the theorem stated in Section 2.6 for the sine and cosine functions.

THEOREM ON THE GRAPH OF $y = a \tan (bx + c)$

If $y = a \tan (bx + c)$ for nonzero real numbers a and b, then

(1) the period is $\dfrac{\pi}{|b|}$ and the phase shift is $-\dfrac{c}{b}$;

(2) successive vertical asymptotes for the graph may be found by solving the equations

$$bx + c = -\frac{\pi}{2} \quad \text{and} \quad bx + c = \frac{\pi}{2}$$

or the inequality

$$-\frac{\pi}{2} < bx + c < \frac{\pi}{2}.$$

E X A M P L E   2 **Sketching the graph of an equation of the form** $y = a \tan (bx + c)$

Find the period and sketch the graph of $y = \dfrac{1}{2} \tan \left(x + \dfrac{\pi}{4} \right)$.

Figure 65

$$y = \frac{1}{2} \tan \left(x + \frac{\pi}{4} \right)$$

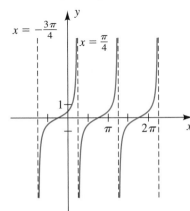

Solution The equation has the form given in the preceding theorem with $a = \frac{1}{2}$, $b = 1$, and $c = \pi/4$. Hence, by part (1), the period is given by $\pi/|b| = \pi/1 = \pi$.

As in part (2), to find successive vertical asymptotes we solve the two equations

$$x + \frac{\pi}{4} = -\frac{\pi}{2} \quad \text{and} \quad x + \frac{\pi}{4} = \frac{\pi}{2},$$

obtaining

$$x = -\frac{3\pi}{4} \quad \text{and} \quad x = \frac{\pi}{4}.$$

Because $a = \frac{1}{2}$, the graph of the equation on the interval $[-3\pi/4, \pi/4]$ has the shape of the graph of $y = \frac{1}{2} \tan x$ (see Figure 64). Sketching that part of the graph and extending it to the right and left gives us Figure 65.

Note that since $c = \pi/4$ and $b = 1$, the phase shift is $-c/b = -\pi/4$. Hence, the graph can also be obtained by shifting the graph of $y = \frac{1}{2} \tan x$ in Figure 64 to the left a distance $\pi/4$.

If $y = a \cot (bx + c)$, we have a situation similar to that stated in the previous theorem. The only difference is part (2). Since successive vertical asymptotes for the graph of $y = \cot x$ are $x = 0$ and $x = \pi$ (see Figure 29), we obtain successive vertical asymptotes for the graph of an equation of the form $y = a \cot (bx + c)$ by solving the equations

$$bx + c = 0 \quad \text{and} \quad bx + c = \pi$$

or the inequality

$$0 < bx + c < \pi.$$

EXAMPLE 3   Sketching the graph of an equation of the form $y = a \cot (bx + c)$

Find the period and sketch the graph of $y = \cot \left(2x - \dfrac{\pi}{2} \right)$.

Figure 66 $y = \cot \left(2x - \dfrac{\pi}{2} \right)$

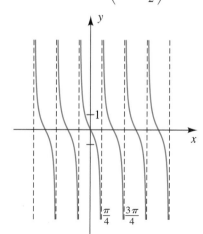

Solution   Using the usual notation, we see that $a = 1$, $b = 2$, and $c = -\pi/2$. The period is $\pi/|b| = \pi/2$. Hence, the graph repeats itself in intervals of length $\pi/2$.

As in the discussion preceding this example, to find two successive vertical asymptotes for the graph we solve the equations

$$2x - \frac{\pi}{2} = 0 \quad \text{and} \quad 2x - \frac{\pi}{2} = \pi,$$

obtaining

$$x = \frac{\pi}{4} \quad \text{and} \quad x = \frac{3\pi}{4}.$$

Since a is positive, we sketch a cotangent-shaped graph on the interval $[\pi/4, 3\pi/4]$ and then repeat it to the right and left in intervals of length $\pi/2$, as shown in Figure 66.

Graphs involving the secant and cosecant functions can be obtained by using methods similar to those for the tangent and cotangent or by taking reciprocals of corresponding graphs of the cosine and sine functions.

EXAMPLE 4   Sketching the graph of an equation of the form $y = a \sec (bx + c)$

Sketch the graph of the equation:

(a) $y = \sec \left(x - \dfrac{\pi}{4} \right)$ **(b)** $y = 2 \sec \left(x - \dfrac{\pi}{4} \right)$

Solution

(a) The graph of $y = \sec x$ is sketched (without asymptotes) in red in Figure 67. We can obtain the graph of $y = \sec\left(x - \dfrac{\pi}{4}\right)$ by shifting this graph to the right a distance $\pi/4$, as shown in blue in Figure 67.

(b) We can sketch this graph by multiplying the y-coordinates of the graph in part (a) by 2. This gives us Figure 68.

Figure 67 $y = \sec\left(x - \dfrac{\pi}{4}\right)$

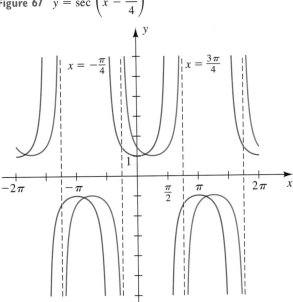

Figure 68 $y = 2 \sec\left(x - \dfrac{\pi}{4}\right)$

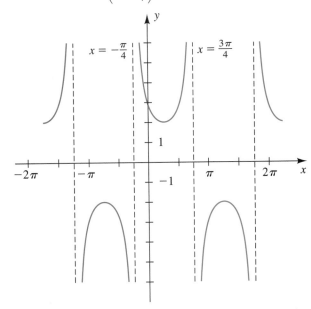

EXAMPLE 5   **Sketching the graph of an equation of the form $y = a \csc (bx + c)$**

Sketch the graph of $y = \csc (2x + \pi)$.

Solution   Since $\csc \theta = 1/\sin \theta$, we may write the given equation as

$$y = \frac{1}{\sin (2x + \pi)}.$$

Thus, we may obtain the graph of $y = \csc (2x + \pi)$ by finding the graph of $y = \sin (2x + \pi)$ and then taking the reciprocal of the y-coordinate of each point. Using $a = 1$, $b = 2$, and $c = \pi$, we see that the amplitude of

(continued)

Figure 69 $y = \csc(2x + \pi)$

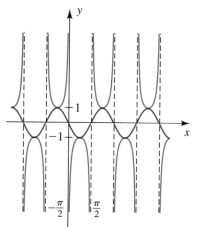

$y = \sin(2x + \pi)$ is 1 and the period is $2\pi/|b| = 2\pi/2 = \pi$. To find an interval containing one cycle, we solve the inequality

$$0 \le 2x + \pi \le 2\pi$$
$$-\pi \le 2x \qquad \le \pi$$
$$-\frac{\pi}{2} \le \; x \quad \le \frac{\pi}{2}.$$

This leads to the graph in red in Figure 69. Taking reciprocals gives us the graph of $y = \csc(2x + \pi)$ shown in blue in the figure. Note that the zeros of the sine curve correspond to the asymptotes of the cosecant graph.

The next example involves the absolute value of a trigonometric function.

Figure 70

(a)

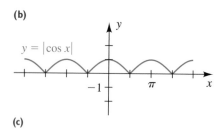

EXAMPLE 6   **Sketching the graph of an equation involving an absolute value**

Sketch the graph of $y = |\cos x| + 1$.

Solution   We shall sketch the graph in three stages. First, we sketch the graph of $y = \cos x$, as in Figure 70(a).

Next, we obtain the graph of $y = |\cos x|$ by reflecting the negative y-coordinates in Figure 70(a) through the x-axis. This gives us Figure 70(b).

Finally, we shift the graph in (b) upward 1 unit to obtain Figure 70(c).

We have used three graphs for clarity. In practice, we could sketch the graphs successively on one coordinate plane.

(b)

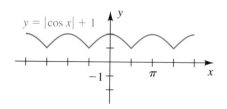

(c)

Mathematical applications often involve a function f that is a sum of two or more other functions. To illustrate, suppose

$$f(x) = g(x) + h(x),$$

where f, g, and h have the same domain D. Before graphing utilities were invented, a technique known as **addition of y-coordinates** was sometimes used to sketch the graph of f. The method is illustrated in Figure 71, where for each x_1, the y-coordinate $f(x_1)$ of a point on the graph of f is the *sum* $g(x_1) + h(x_1)$ of y-coordinates of points on the graphs of g and h. The graph of f is obtained by *graphically adding* a sufficient number of such y-coordinates using a ruler and compass.

Figure 71

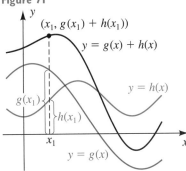

This cumbersome method is no longer necessary, since the graphs can be readily sketched with the aid of a graphing utility. However, it is sometimes useful to compare the graph of a sum of functions with the individual functions, as illustrated in the next example.

EXAMPLE 7 Sketching the graph of a sum of two trigonometric functions

Sketch the graph of $y = \cos x$, $y = \sin x$, and $y = \cos x + \sin x$ on the same coordinate plane for $0 \le x \le 3\pi$.

Figure 72

(a) $[0, 3\pi]$ by $[-\pi, \pi]$

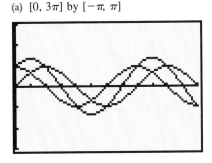

(b) $[0, 3\pi]$ by $[-1.5, 1.5]$

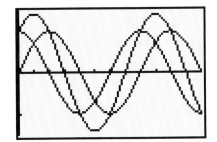

Solution Some specific keystrokes for the TI-82/83 graphing calculator are given in Example 22 of Appendix I. We make the following assignments:

$$Y_1 = \cos x, \qquad Y_2 = \sin x, \qquad \text{and} \qquad Y_3 = Y_1 + Y_2$$

Since we desire a 3:2 (horizontal:vertical) screen proportion, we choose the viewing rectangle $[0, 3\pi]$ by $[-\pi, \pi]$, obtaining Figure 72(a). The clarity of the graph can be enhanced by changing the viewing rectangle to $[0, 3\pi]$ by $[-1.5, 1.5]$, as in Figure 72(b).

Note that the graph of Y_3 intersects the graph of Y_1 when $Y_2 = 0$, and the graph of Y_2 when $Y_1 = 0$. The x-intercepts for Y_3 correspond to the solutions of $Y_2 = -Y_1$. Finally, we see that the maximum and minimum values of Y_3 occur when $Y_1 = Y_2$ (that is, when $x = \pi/4$, $5\pi/4$, and $9\pi/4$). These y-values are

$$\sqrt{2}/2 + \sqrt{2}/2 = \sqrt{2} \qquad \text{and} \qquad -\sqrt{2}/2 + (-\sqrt{2}/2) = -\sqrt{2}.$$

The graph of an equation of the form

$$y = f(x) \sin (ax + b) \qquad \text{or} \qquad y = f(x) \cos (ax + b),$$

where f is a function and a and b are real numbers, is called a **damped sine wave** or **damped cosine wave,** respectively, and $f(x)$ is called the **damping factor.** The next example illustrates a method for graphing such equations.

Figure 73

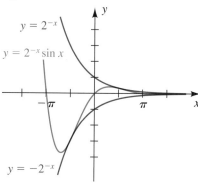

$y = 2^{-x}$

$y = 2^{-x} \sin x$

$y = -2^{-x}$

EXAMPLE 8   **Sketching the graph of a damped sine wave**

Sketch the graph of f if $f(x) = 2^{-x} \sin x$.

Solution   Using properties of absolute values, we see that

$$|f(x)| = |2^{-x}| \, |\sin x|.$$

Since $|\sin x| \le 1$ and $2^{-x} > 0$, it follows that $|f(x)| < 2^{-x}$, and hence

$$-2^{-x} \le f(x) \le 2^{-x}.$$

The last inequality implies that the graph of f lies between the graphs of the equations $y = -2^{-x}$ and $y = 2^{-x}$. The graph of f will coincide with one of these graphs if $|\sin x| = 1$—that is, if $x = (\pi/2) + \pi n$ for some integer n.

Since $2^{-x} > 0$, the x-intercepts on the graph of f occur at $\sin x = 0$—that is, at $x = \pi n$. Because there are an infinite number of x-intercepts, this is an example of a function that intersects its horizontal asymptote an infinite number of times. With this information, we obtain the sketch shown in Figure 73.

The damping factor in Example 8 is 2^{-x}. By using different damping factors, we can obtain other compressed or expanded variations of sine waves. The analysis of such graphs is important in physics and engineering.

2.7 EXERCISES

Exer. 1–52: Find the period and sketch the graph of the equation. Show the asymptotes.

1   $y = 4 \tan x$ 2 $y = \frac{1}{4} \tan x$

3   $y = 3 \cot x$ 4 $y = \frac{1}{3} \cot x$

5   $y = 2 \csc x$ 6 $y = \frac{1}{2} \csc x$

7   $y = 3 \sec x$ 8 $y = \frac{1}{4} \sec x$

9   $y = \tan \left(x - \dfrac{\pi}{4} \right)$ 10 $y = \tan \left(x + \dfrac{3\pi}{4} \right)$

11   $y = \tan 2x$ 12 $y = \tan \frac{1}{2} x$

13   $y = \tan \frac{1}{4} x$ 14 $y = \tan 4x$

15   $y = 2 \tan \left(2x + \dfrac{\pi}{2} \right)$ 16 $y = \dfrac{1}{3} \tan \left(2x - \dfrac{\pi}{4} \right)$

17   $y = -\dfrac{1}{4} \tan \left(\dfrac{1}{2} x + \dfrac{\pi}{3} \right)$

18   $y = -3 \tan \left(\dfrac{1}{3} x - \dfrac{\pi}{3} \right)$

19   $y = \cot \left(x - \dfrac{\pi}{2} \right)$ 20 $y = \cot \left(x + \dfrac{\pi}{4} \right)$

21   $y = \cot 2x$ 22 $y = \cot \frac{1}{2} x$

23   $y = \cot \frac{1}{3} x$ 24 $y = \cot 3x$

25   $y = 2 \cot \left(2x + \dfrac{\pi}{2} \right)$ 26 $y = -\frac{1}{3} \cot (3x - \pi)$

27   $y = -\dfrac{1}{2} \cot \left(\dfrac{1}{2} x + \dfrac{\pi}{4} \right)$ 28 $y = 4 \cot \left(\dfrac{1}{3} x - \dfrac{\pi}{6} \right)$

29   $y = \sec \left(x - \dfrac{\pi}{2} \right)$ 30 $y = \sec \left(x - \dfrac{3\pi}{4} \right)$

31   $y = \sec 2x$ 32 $y = \sec \frac{1}{2} x$

33   $y = \sec \frac{1}{3} x$ 34 $y = \sec 3x$

35 $y = 2 \sec \left(2x - \dfrac{\pi}{2} \right)$ **36** $y = \dfrac{1}{2} \sec \left(2x - \dfrac{\pi}{2} \right)$

37 $y = -\dfrac{1}{3} \sec \left(\dfrac{1}{2}x + \dfrac{\pi}{4} \right)$

38 $y = -3 \sec \left(\dfrac{1}{3}x + \dfrac{\pi}{3} \right)$

39 $y = \csc \left(x - \dfrac{\pi}{2} \right)$ **40** $y = \csc \left(x + \dfrac{3\pi}{4} \right)$

41 $y = \csc 2x$ **42** $y = \csc \frac{1}{2}x$

43 $y = \csc \frac{1}{3}x$ **44** $y = \csc 3x$

45 $y = 2 \csc \left(2x + \dfrac{\pi}{2} \right)$ **46** $y = -\dfrac{1}{2} \csc (2x - \pi)$

47 $y = -\dfrac{1}{4} \csc \left(\dfrac{1}{2}x + \dfrac{\pi}{2} \right)$ **48** $y = 4 \csc \left(\dfrac{1}{2}x - \dfrac{\pi}{4} \right)$

49 $y = \tan \dfrac{\pi}{2} x$ **50** $y = \cot \pi x$

51 $y = \csc 2\pi x$ **52** $y = \sec \dfrac{\pi}{8} x$

53 Find an equation using the cotangent function that has the same graph as $y = \tan x$.

54 Find an equation using the cosecant function that has the same graph as $y = \sec x$.

Exer. 55–60: Use the graph of a trigonometric function to aid in sketching the graph of the equation without plotting points.

55 $y = |\sin x|$ **56** $y = |\cos x|$

57 $y = |\sin x| + 2$ **58** $y = |\cos x| - 3$

59 $y = -|\cos x| + 1$ **60** $y = -|\sin x| - 2$

Exer. 61–66: Sketch the graph of the equation.

61 $y = x + \cos x$ **62** $y = x - \sin x$

63 $y = 2^{-x} \cos x$ **64** $y = e^x \sin x$

65 $y = |x| \sin x$ **66** $y = |x| \cos x$

C Exer. 67–72: Graph f in the viewing rectangle $[-2\pi, 2\pi]$ by $[-4, 4]$. Use the graph of f to predict the graph of g. Verify your prediction by graphing g in the same viewing rectangle.

67 $f(x) = \tan 0.5x;$ $g(x) = \tan \left[0.5 \left(x + \dfrac{\pi}{2} \right) \right]$

68 $f(x) = 0.5 \csc 0.5x;$ $g(x) = 0.5 \csc 0.5x - 2$

69 $f(x) = 0.5 \sec 0.5x;$ $g(x) = 0.5 \sec \left[0.5 \left(x - \dfrac{\pi}{2} \right) \right] - 1$

70 $f(x) = \tan x - 1;$ $g(x) = -\tan x + 1$

71 $f(x) = 3 \cos 2x;$ $g(x) = |3 \cos 2x| - 1$

72 $f(x) = 1.2^{-x} \cos x;$ $g(x) = 1.2^x \cos x$

C Exer. 73–74: Identify the damping factor $f(x)$ for the damped wave. Sketch graphs of $y = \pm f(x)$ and the equation on the same coordinate plane for $-2\pi \le x \le 2\pi$.

73 $y = e^{-x/4} \sin 4x$ **74** $y = 3^{-x/5} \cos 2x$

C Exer. 75–76: Graph the function f on $[-\pi, \pi]$, and estimate the high and low points.

75 $f(x) = \cos 2x + 2 \sin 4x - \sin x$

76 $f(x) = \tan \frac{1}{4} x - 2 \sin 2x$

C Exer. 77–78: Use a graph to estimate the largest interval $[a, b]$, with $a < 0$ and $b > 0$, on which f is one-to-one.

77 $f(x) = \sin (2x + 2) \cos (1.5x - 1)$

78 $f(x) = 1.5 \cos \left(\frac{1}{2}x - 0.3 \right) + \sin (1.5x + 0.5)$

C Exer. 79–80: Use a graph to solve the inequality on the interval $[-\pi, \pi]$.

79 $\cos (2x - 1) + \sin 3x \ge \sin \frac{1}{3}x + \cos x$

80 $\frac{1}{2} \cos 2x + 2 \cos (x - 2) <$
$$2 \cos (1.5x + 1) + \sin (x - 1)$$

81 *Radio signal intensity* Radio stations often have more than one broadcasting tower because federal guidelines do not usually permit a radio station to broadcast its signal in all directions with equal power. Since radio waves can travel over long distances, it is important to control their directional patterns so that radio stations do not interfere with one another. Suppose that a radio station has two broadcasting towers located along a north-south line, as shown in the figure. If the radio station is broadcasting at a wavelength λ and the distance between the two radio towers is equal to $\frac{1}{2}\lambda$, then the intensity I of the signal in the direction θ is given by
$$I = \tfrac{1}{2} I_0 [1 + \cos (\pi \sin \theta)],$$

where I_0 is the maximum intensity. Approximate I in terms of I_0 for each θ.

(a) $\theta = 0$ (b) $\theta = \pi/3$ (c) $\theta = \pi/7$

Exercise 81

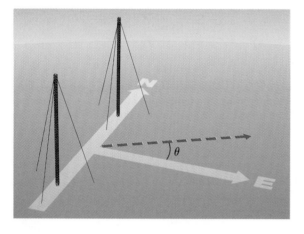

C 82 *Radio signal intensity* Refer to Exercise 81.

(a) Determine the directions in which I has maximum or minimum values.

(b) Graph I on the interval $[0, 2\pi)$. Graphically approximate θ to three decimal places, when I is equal to $\frac{1}{3}I_0$. (*Hint:* Let $I_0 = 1$.)

83 *The earth's magnetic field* The strength of the earth's magnetic field varies with the depth below the surface. The strength at depth z and time t can sometimes be approximated using the damped sine wave

$$S = A_0 e^{-\alpha z} \sin (kt - \alpha z),$$

where A_0, α, and k are constants.

(a) What is the damping factor?

(b) Find the phase shift at depth z_0.

(c) At what depth is the amplitude of the wave one-half the amplitude of the surface strength?

2.8 APPLIED PROBLEMS

Trigonometry was developed to help solve problems involving angles and lengths of sides of triangles. Problems of that type are no longer the most important applications; however, questions about triangles still arise in physical situations. When considering such questions in this section, we shall restrict our discussion to right triangles. Triangles that do not contain a right angle will be considered in Chapter 4.

We shall often use the following notation. The vertices of a triangle will be denoted by A, B, and C; the angles at A, B, and C will be denoted by α, β, and γ, respectively; and the lengths of the sides opposite these angles by a, b, and c, respectively. The triangle itself will be referred to as *triangle ABC* (or denoted $\triangle ABC$). If a triangle is a right triangle and if one of the acute angles and a side are known or if two sides are given, then we may find the remaining parts by using the formulas in Section 2.4 that express the trigonometric functions as ratios of sides of a triangle. We can refer to the process of finding the remaining parts as **solving the triangle.**

In all examples *it is assumed that you know how to find trigonometric function values and angles by using either a calculator or results about special angles.*

EXAMPLE 1 Solving a right triangle

Solve $\triangle ABC$, given $\gamma = 90°$, $\alpha = 34°$, and $b = 10.5$.

Solution Since the sum of the three interior angles in a triangle is $180°$, we have $\alpha + \beta + \gamma = 180°$. Solving for the unknown angle β gives us

$$\beta = 180° - \alpha - \gamma = 180° - 34° - 90° = 56°.$$

Referring to Figure 74, we obtain

$$\tan 34° = \frac{a}{10.5} \qquad\qquad \tan \alpha = \frac{\text{opp}}{\text{adj}}$$

$$a = (10.5)\tan 34° \approx 7.1. \qquad \text{solve for } a; \text{ approximate}$$

To find side c, we can use either the cosine or the secant function, as in (1) or (2), respectively:

$$\textbf{(1)} \qquad \cos 34° = \frac{10.5}{c} \qquad\qquad \cos \alpha = \frac{\text{adj}}{\text{hyp}}$$

$$c = \frac{10.5}{\cos 34°} \approx 12.7 \qquad \text{solve for } c; \text{ approximate}$$

$$\textbf{(2)} \qquad \sec 34° = \frac{c}{10.5} \qquad\qquad \sec \alpha = \frac{\text{hyp}}{\text{adj}}$$

$$c = (10.5)\sec 34° \approx 12.7 \qquad \text{solve for } c; \text{ approximate}$$

Figure 74

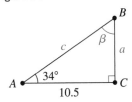

As illustrated in Example 1, when working with triangles we usually round off answers. One reason for doing so is that in most applications the lengths of sides of triangles and measures of angles are found by mechanical devices and hence are only approximations to the exact values. Consequently, a number such as 10.5 in Example 1 is assumed to have been rounded off to the nearest tenth. We cannot expect more accuracy in the calculated values for the remaining sides, and therefore they should also be rounded off to the nearest tenth.

In finding angles, answers should be rounded off as indicated in the following table.

Number of significant figures for sides	Round off degree measure of angles to the nearest
2	1°
3	0.1°, or 10′
4	0.01°, or 1′

Justification of this table requires a careful analysis of problems that involve approximate data.

EXAMPLE 2   **Solving a right triangle**

Solve $\triangle ABC$, given $\gamma = 90°$, $a = 12.3$, and $b = 31.6$.

Figure 75

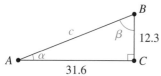

Solution   Referring to the triangle illustrated in Figure 75 gives us

$$\tan \alpha = \frac{12.3}{31.6}.$$

Since the sides are given with three significant figures, the rule stated in the preceding table tells us that α should be rounded off to the nearest 0.1°, or the nearest multiple of 10′. Using the degree mode on a calculator, we have

$$\alpha = \tan^{-1} \frac{12.3}{31.6} \approx 21.3° \qquad \text{or, equivalently,} \qquad \alpha \approx 21°20'.$$

Since α and β are complementary angles,

$$\beta = 90° - \alpha \approx 90° - 21.3° = 68.7°.$$

The only remaining part to find is c. We could use several relationships involving c to determine its value. Among these are

$$\cos \alpha = \frac{31.6}{c}, \quad \sec \beta = \frac{c}{12.3}, \qquad \text{and} \qquad a^2 + b^2 = c^2.$$

Whenever possible, it is best to use a relationship that involves only given information, since it doesn't depend on any previously calculated value. Hence, with $a = 12.3$ and $b = 31.6$, we have

$$c = \sqrt{a^2 + b^2} = \sqrt{(12.3)^2 + (31.6)^2} = \sqrt{1149.85} \approx 33.9.$$

Figure 76

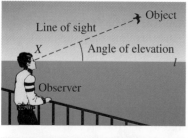

As illustrated in Figure 76, if an observer at point X sights an object, then the angle that the line of sight makes with the horizontal line l is the **angle of elevation** of the object, if the object is above the horizontal line, or the **angle of depression** of the object, if the object is below the horizontal line. We use this terminology in the next two examples.

EXAMPLE 3   **Using an angle of elevation**

From a point on level ground 135 feet from the base of a tower, the angle of elevation of the top of the tower is 57°20′. Approximate the height of the tower.

Solution   If we let d denote the height of the tower, then the given facts are represented by the triangle in Figure 77. Referring to the figure, we obtain

$$\tan 57°20' = \frac{d}{135} \qquad\qquad \tan 57°20' = \frac{\text{opp}}{\text{adj}}$$

$$d = 135 \tan 57°20' \approx 211. \quad \text{solve for } d; \text{ approximate}$$

The tower is approximately 211 feet high.

Figure 77

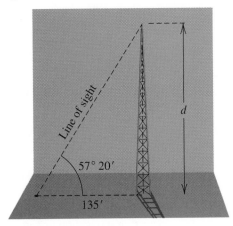

Line of sight

57° 20′

135′

d

E X A M P L E   4   **Using angles of depression**

From the top of a building that overlooks an ocean, an observer watches a boat sailing directly toward the building. If the observer is 100 feet above sea level and if the angle of depression of the boat changes from 25° to 40° during the period of observation, approximate the distance that the boat travels.

Solution   As in Figure 78, let A and B be the positions of the boat that correspond to the 25° and 40° angles, respectively. Suppose that the ob-

Figure 78

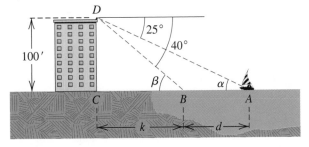

D

25°

40°

100′

β

α

C

B

A

k

d

(continued)

server is at point D and that C is the point 100 feet directly below. Let d denote the distance the boat travels, and let k denote the distance from B to C. If α and β denote angles DAC and DBC, respectively, then it follows from geometry (alternate interior angles) that $\alpha = 25°$ and $\beta = 40°$.

From triangle BCD:

$$\cot \beta = \cot 40° = \frac{k}{100} \qquad \cot \beta = \frac{\text{adj}}{\text{opp}}$$

$$k = 100 \cot 40° \qquad \text{solve for } k$$

From triangle DAC:

$$\cot \alpha = \cot 25° = \frac{d + k}{100} \qquad \cot \alpha = \frac{\text{adj}}{\text{opp}}$$

$$d + k = 100 \cot 25° \qquad \text{multiply by lcd}$$

$$d = 100 \cot 25° - k \qquad \text{solve for } d$$

$$= 100 \cot 25° - 100 \cot 40° \qquad k = 100 \cot 40°$$

$$= 100(\cot 25° - \cot 40°) \qquad \text{factor out 100}$$

$$\approx 100(2.145 - 1.192) \approx 95 \qquad \text{approximate}$$

Hence, the boat travels approximately 95 feet.

In certain navigation or surveying problems, the **direction**, or **bearing**, from a point P to a point Q is specified by stating the acute angle that segment PQ makes with the north-south line through P. We also state whether Q is north or south and east or west of P. Figure 79 illustrates four possibilities. The bearing from P to Q_1 is $25°$ east of north and is denoted by N25°E. We also refer to the **direction** N25°E, meaning the direction from P to Q_1. The bearings from P to Q_2, to Q_3, and to Q_4 are represented in a similar manner in the figure. Note that when this notation is used for bear-

Figure 79

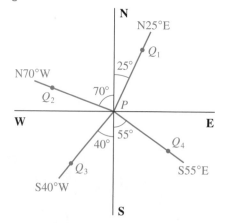

Figure 80

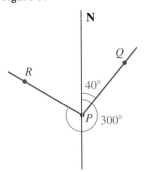

ings or directions, N or S always appears to the *left* of the angle and W or E to the *right*.

In air navigation, directions and bearings are specified by measuring from the north in a *clockwise* direction. In this case, a positive measure is assigned to the angle instead of the negative measure to which we are accustomed for clockwise rotations. Referring to Figure 80, we see that the direction of *PQ* is 40° and the direction of *PR* is 300°.

EXAMPLE 5 Using bearings

Two ships leave port at the same time, one ship sailing in the direction N23°E at a speed of 11 mi/hr and the second ship sailing in the direction S67°E at 15 mi/hr. Approximate the bearing from the second ship to the first, one hour later.

Figure 81

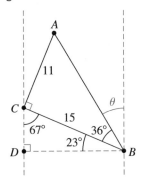

Solution The sketch in Figure 81 indicates the positions of the first and second ships at points *A* and *B*, respectively, after one hour. Point *C* represents the port. We wish to find the bearing from *B* to *A*. Note that

$$\angle ACB = 180° - 23° - 67° = 90°,$$

and hence triangle *ACB* is a right triangle. Thus,

$$\tan \beta = \frac{11}{15} \qquad\qquad \tan \beta = \frac{\text{opp}}{\text{adj}}$$

$$\beta = \tan^{-1} \tfrac{11}{15} \approx 36°. \quad \text{solve for } \beta; \text{ approximate}$$

We have rounded β to the nearest degree because the sides of the triangles are given with two significant figures.

Referring to Figure 82, we obtain the following:

$$\angle CBD = 90° - \angle BCD = 90° - 67° = 23°$$

$$\angle ABD = \angle ABC + \angle CBD \approx 36° + 23° = 59°$$

$$\theta = 90° - \angle ABD \approx 90° - 59° = 31°$$

Thus, the bearing from *B* to *A* is approximately N31°W.

Figure 82

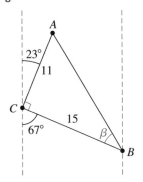

Trigonometric functions are useful in the investigation of vibratory or oscillatory motion, such as the motion of a particle in a vibrating guitar string or in a spring that has been compressed or elongated and then released to oscillate back and forth. The fundamental type of particle displacement in these illustrations is *harmonic motion*.

DEFINITION OF SIMPLE HARMONIC MOTION

A point moving on a coordinate line is in **simple harmonic motion** if its distance *d* from the origin at time *t* is given by either

$$d = a \cos \omega t \qquad \text{or} \qquad d = a \sin \omega t,$$

where *a* and ω are constants, with $\omega > 0$.

Figure 83

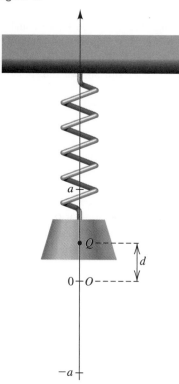

In the preceding definition, the **amplitude** of the motion is the maximum displacement $|a|$ of the point from the origin. The **period** is the time $2\pi/\omega$ required for one complete oscillation. The reciprocal of the period, $\omega/(2\pi)$, is the number of oscillations per unit of time and is called the **frequency**.

A physical interpretation of simple harmonic motion can be obtained by considering a spring with an attached weight that is oscillating vertically relative to a coordinate line, as illustrated in Figure 83. The number d represents the coordinate of a fixed point Q in the weight, and we assume that the amplitude a of the motion is constant. In this case no frictional force is retarding the motion. If friction is present, then the amplitude decreases with time, and the motion is said to be *damped.*

E X A M P L E 6   **Describing harmonic motion**

Suppose that the oscillation of the weight shown in Figure 83 is given by

$$d = 10 \cos\left(\frac{\pi}{6}t\right),$$

with t measured in seconds and d in centimeters. Discuss the motion of the weight.

Solution   By definition, the motion is simple harmonic with amplitude $a = 10$ cm. Since $\omega = \pi/6$, we obtain the following:

$$\text{period} = \frac{2\pi}{\omega} = \frac{2\pi}{\pi/6} = 12$$

Thus, in 12 seconds the weight makes one complete oscillation. The frequency is $\frac{1}{12}$, which means that one-twelfth of an oscillation takes place each second. The following table indicates the position of Q at various times.

t	0	1	2	3	4	5	6
$\dfrac{\pi}{6}t$	0	$\dfrac{\pi}{6}$	$\dfrac{\pi}{3}$	$\dfrac{\pi}{2}$	$\dfrac{2\pi}{3}$	$\dfrac{5\pi}{6}$	π
$\cos\left(\dfrac{\pi}{6}t\right)$	1	$\dfrac{\sqrt{3}}{2}$	$\dfrac{1}{2}$	0	$-\dfrac{1}{2}$	$-\dfrac{\sqrt{3}}{2}$	-1
d	10	$5\sqrt{3} \approx 8.7$	5	0	-5	$-5\sqrt{3} \approx -8.7$	-10

The initial position of Q is 10 centimeters above the origin O. It moves downward, gaining speed until it reaches O. Note that Q travels approximately $10 - 8.7 = 1.3$ cm during the first second, $8.7 - 5 = 3.7$ cm during the next second, and $5 - 0 = 5$ cm during the third second. It then slows down until it reaches a point 10 centimeters below O at the end of

6 seconds. The direction of motion is then reversed, and the weight moves upward, gaining speed until it reaches O. Once it reaches O, it slows down until it returns to its original position at the end of 12 seconds. The direction of motion is then reversed again, and the same pattern is repeated indefinitely.

2.8 EXERCISES

Exer. 1–8: Given the indicated parts of triangle ABC with $\gamma = 90°$, find the exact values of the remaining parts.

1	$\alpha = 30°,$	$b = 20$	2 $\beta = 45°,$	$b = 35$
3	$\beta = 45°,$	$c = 30$	4 $\alpha = 60°,$	$c = 6$
5	$a = 5,$	$b = 5$	6 $a = 4\sqrt{3},$	$c = 8$
7	$b = 5\sqrt{3},$	$c = 10\sqrt{3}$	8 $b = 7\sqrt{2},$	$c = 14$

Exer. 9–16: Given the indicated parts of triangle ABC with $\gamma = 90°$, approximate the remaining parts.

9	$\alpha = 37°,$	$b = 24$	10 $\beta = 64°20',$	$a = 20.1$
11	$\beta = 71°51',$	$b = 240.0$	12 $\alpha = 31°10',$	$a = 510$
13	$a = 25,$	$b = 45$	14 $a = 31,$	$b = 9.0$
15	$c = 5.8,$	$b = 2.1$	16 $a = 0.42,$	$c = 0.68$

Exer. 17–24: Given the indicated parts of triangle ABC with $\gamma = 90°$, express the third part in terms of the first two.

17	$\alpha, c;$	b
18	$\beta, c;$	b
19	$\beta, b;$	a
20	$\alpha, b;$	a
21	$\alpha, a;$	c
22	$\beta, a;$	c
23	$a, c;$	b
24	$a, b;$	c

25 *Height of a kite* A person flying a kite holds the string 4 feet above ground level. The string of the kite is taut and makes an angle of 60° with the horizontal (see the figure). Approximate the height of the kite above level ground if 500 feet of string is payed out.

Exercise 25

26 *Surveying* From a point 15 meters above level ground, a surveyor measures the angle of depression of an object on the ground at 68°. Approximate the distance from the object to the point on the ground directly beneath the surveyor.

27 *Airplane landing* A pilot, flying at an altitude of 5000 feet, wishes to approach the numbers on a runway at an angle of 10°. Approximate, to the nearest 100 feet, the distance from the airplane to the numbers at the beginning of the descent.

28 *Radio antenna* A guy wire is attached to the top of a radio antenna and to a point on horizontal ground that is 40.0 meters from the base of the antenna. If the wire makes an angle of 58°20' with the ground, approximate the length of the wire.

29 *Surveying* To find the distance d between two points P and Q on opposite shores of a lake, a surveyor locates a point R that is 50.0 meters from P such that RP is

perpendicular to PQ, as shown in the figure. Next, using a transit, the surveyor measures angle PRQ as 72°40′. Find d.

Exercise 29

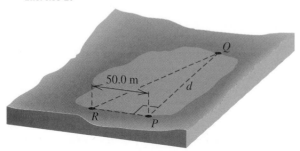

30 *Meteorological calculations* To measure the height h of a cloud cover, a meteorology student directs a spotlight vertically upward from the ground. From a point P on level ground that is d meters from the spotlight, the angle of elevation θ of the light image on the clouds is then measured (see the figure).

(a) Express h in terms of d and θ.

(b) Approximate h if $d = 1000$ m and $\theta = 59°$.

Exercise 30

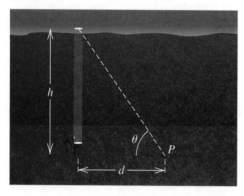

31 *Altitude of a rocket* A rocket is fired at sea level and climbs at a constant angle of 75° through a distance of 10,000 feet. Approximate its altitude to the nearest foot.

32 *Airplane takeoff* An airplane takes off at a 10° angle and travels at the rate of 250 ft/sec. Approximately how long does it take the airplane to reach an altitude of 15,000 feet?

33 *Designing a drawbridge* A drawbridge is 150 feet long when stretched across a river. As shown in the figure, the two sections of the bridge can be rotated upward through an angle of 35°.

(a) If the water level is 15 feet below the closed bridge, find the distance d between the end of a section and the water level when the bridge is fully open.

(b) Approximately how far apart are the ends of the two sections when the bridge is fully opened, as shown in the figure?

Exercise 33

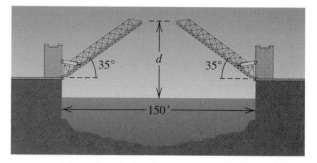

34 *Designing a water slide* Shown in the figure is part of a design for a water slide. Find the total length of the slide to the nearest foot.

Exercise 34

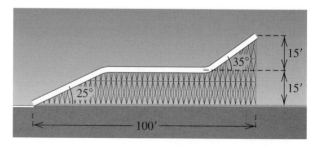

35 *Sun's elevation* Approximate the angle of elevation α of the sun if a person 5.0 feet tall casts a shadow 4.0 feet long on level ground (see the figure).

Exercise 35

36 *Constructing a ramp* A builder wishes to construct a ramp 24 feet long that rises to a height of 5.0 feet above level ground. Approximate the angle that the ramp should make with the horizontal.

37 *Video game* Shown in the figure is the screen for a simple video arcade game in which ducks move from A to B at the rate of 7 cm/sec. Bullets fired from point O travel 25 cm/sec. If a player shoots as soon as a duck appears at A, at which angle φ should the gun be aimed in order to score a direct hit?

Exercise 37

38 *Conveyor belt* A conveyor belt 9 meters long can be hydraulically rotated up to an angle of 40° to unload cargo from airplanes (see the figure).

(a) Find, to the nearest degree, the angle through which the conveyor belt should be rotated to reach a door that is 4 meters above the platform supporting the belt.

(b) Approximate the maximum height above the platform that the belt can reach.

Exercise 38

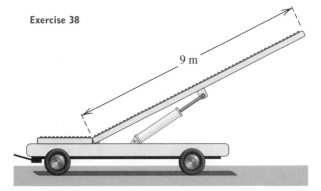

9 m

39 *Tallest structure* The tallest man-made structure in the world is a television transmitting tower located near Fargo, North Dakota. From a distance of 1 mile on level ground, its angle of elevation is 21°20′24″. Determine its height to the nearest foot.

40 *Elongation of Venus* The *elongation* of the planet Venus is defined to be the angle θ determined by the sun, the earth, and Venus, as shown in the figure. Maximum elongation of Venus occurs when the earth is at its minimum distance D_e from the sun and Venus is at its maximum distance D_v from the sun. If $D_e = 91,500,000$ mi and $D_v = 68,000,000$ mi, approximate the maximum elongation θ_{max} of Venus. Assume that the orbit of Venus is circular.

Exercise 40

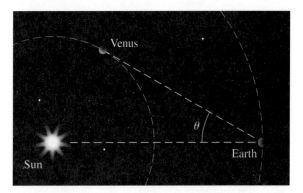

Venus

θ

Earth

Sun

41 *The Pentagon's ground area* The Pentagon is the largest office building in the world in terms of ground area. The perimeter of the building has the shape of a regular pentagon with each side of length 921 feet. Find the area enclosed by the perimeter of the building.

42 A regular octagon is inscribed in a circle of radius 12.0 centimeters. Approximate the perimeter of the octagon.

43 A rectangular box has dimensions 8″ × 6″ × 4″. Approximate, to the nearest tenth of a degree, the angle θ formed by a diagonal of the base and the diagonal of the box, as shown in the figure.

Exercise 43

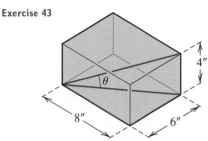

θ

8″

6″

4″

44 *Volume of a conical cup*   A conical paper cup has a radius of 2 inches. Approximate, to the nearest degree, the angle β (see the figure) so that the cone will have a volume of 20 in³.

Exercise 44

45 *Height of a tower*   From a point P on level ground, the angle of elevation of the top of a tower is 26°50′. From a point 25.0 meters closer to the tower and on the same line with P and the base of the tower, the angle of elevation of the top is 53°30′. Approximate the height of the tower.

46 *Ladder calculations*   A ladder 20 feet long leans against the side of a building, and the angle between the ladder and the building is 22°.

  (a) Approximate the distance from the bottom of the ladder to the building.

  (b) If the distance from the bottom of the ladder to the building is increased by 3.0 feet, approximately how far does the top of the ladder move down the building?

47 *Ascent of a hot-air balloon*   As a hot-air balloon rises vertically, its angle of elevation from a point P on level

Exercise 47

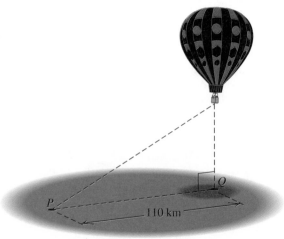

ground 110 kilometers from the point Q directly underneath the balloon changes from 19°20′ to 31°50′ (see the figure). Approximately how far does the balloon rise during this period?

48 *Height of a building*   From a point A that is 8.20 meters above level ground, the angle of elevation of the top of the building is 31°20′ and the angle of depression of the base of the building is 12°50′. Approximate the height of the building.

49 *Radius of the earth*   A spacelab circles the earth at an altitude of 380 miles. When an astronaut views the horizon of the earth, the angle θ shown in the figure is 65.8°. Use this information to estimate the radius of the earth.

Exercise 49

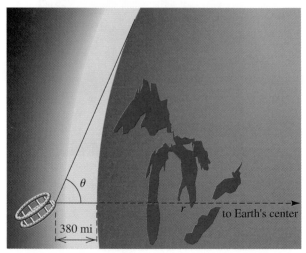

50 *Length of an antenna*   A CB antenna is located on the top of a garage that is 16 feet tall. From a point on level ground that is 100 feet from a point directly below the antenna, the antenna subtends an angle of 12°, as shown in the figure. Approximate the length of the antenna.

Exercise 50

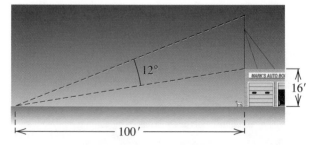

51 *Speed of an airplane* An airplane flying at an altitude of 10,000 feet passes directly over a fixed object on the ground. One minute later, the angle of depression of the object is 42°. Approximate the speed of the airplane to the nearest mile per hour.

52 *Height of a mountain* A motorist, traveling along a level highway at a speed of 60 km/hr directly toward a mountain, observes that between 1:00 P.M. and 1:10 P.M. the angle of elevation of the top of the mountain changes from 10° to 70°. Approximate the height of the mountain.

53 *Communications satellite* Shown in the figure on the left is a communications satellite with an equatorial orbit—that is, a nearly circular orbit in the plane determined by the earth's equator. If the satellite circles the earth at an altitude of $a = 22{,}300$ mi, its speed is the same as the rotational speed of earth; to an observer on the equator, the satellite appears to be stationary—that is, its orbit is synchronous.

(a) Using $R = 4000$ mi for the radius of the earth, determine the percentage of the equator that is within signal range of such a satellite.

(b) As shown in the figure on the right, three satellites are equally spaced in equatorial synchronous orbits. Use the value of θ obtained in part (a) to explain why all points on the equator are within signal range of at least one of the three satellites.

Exercise 53

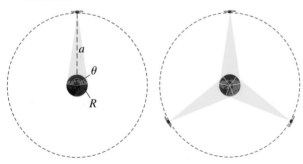

54 *Communications satellite* Refer to Exercise 53. Shown in the figure is the area served by a communications satellite circling a planet of radius R at an altitude a. The portion of the planet's surface within range of the satellite is a spherical cap of depth d and surface area $A = 2\pi R d$.

(a) Express d in terms of R and θ.

(b) Estimate the percentage of the planet's surface that is within signal range of a single satellite in equatorial synchronous orbit.

Exercise 54

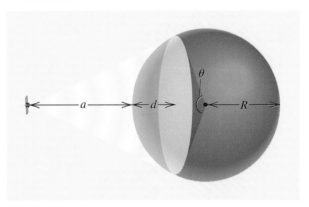

55 *Height of a kite* Generalize Exercise 25 to the case where the angle is α, the number of feet of string payed out is d, and the end of the string is held c feet above the ground. Express the height h of the kite in terms of α, d, and c.

56 *Surveying* Generalize Exercise 26 to the case where the point is d meters above level ground and the angle of depression is α. Express the distance x in terms of d and α.

57 *Height of a tower* Generalize Exercise 45 to the case where the first angle is α, the second angle is β, and the distance between the two points is d. Express the height h of the tower in terms of d, α, and β.

58 Generalize Exercise 42 to the case of an n-sided polygon inscribed in a circle of radius r. Express the perimeter P in terms of n and r.

59 *Ascent of a hot-air balloon* Generalize Exercise 47 to the case where the distance from P to Q is d kilometers and the angle of elevation changes from α to β.

60 *Height of a building* Generalize Exercise 48 to the case where point A is d meters above ground and the angles of elevation and depression are α and β, respectively. Express the height h of the building in terms of d, α, and β.

Exer. 61–62: Find the bearing from *P* to each of the points *A*, *B*, *C*, and *D*.

61

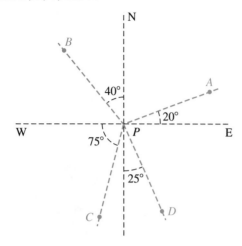

62

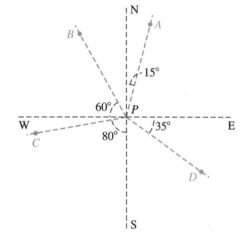

63 *Ships' bearings* A ship leaves port at 1:00 P.M. and sails in the direction N34°W at a rate of 24 mi/hr. Another ship leaves port at 1:30 P.M. and sails in the direction N56°E at a rate of 18 mi/hr.

(a) Approximately how far apart are the ships at 3:00 P.M.?

(b) What is the bearing, to the nearest degree, from the first ship to the second?

64 *Pinpointing a forest fire* From an observation point *A*, a forest ranger sights a fire in the direction S35°50′W (see the figure). From a point *B*, 5 miles due west of *A*, another ranger sights the same fire in the direction S54°10′E. Approximate, to the nearest tenth of a mile, the distance of the fire from *A*.

Exercise 64

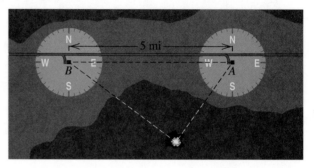

65 *Airplane flight* An airplane flying at a speed of 360 mi/hr flies from a point *A* in the direction 137° for 30 minutes and then flies in the direction 227° for 45 minutes. Approximate, to the nearest mile, the distance from the airplane to *A*.

66 *Airplane flight plan* An airplane flying at a speed of 400 mi/hr flies from a point *A* in the direction 153° for 1 hour and then flies in the direction 63° for 1 hour.

(a) In what direction does the plane need to fly in order to get back to point *A*?

(b) How long will it take to get back to point *A*?

Exer. 67–70: The formula specifies the position of a point *P* that is moving harmonically on a vertical axis, where *t* is in seconds and *d* is in centimeters. Determine the amplitude, period, and frequency, and describe the motion of the point during one complete oscillation (starting at *t* = 0).

67 $d = 10 \sin 6\pi t$

68 $d = \dfrac{1}{3} \cos \dfrac{\pi}{4} t$

69 $d = 4 \cos \dfrac{3\pi}{2} t$

70 $d = 6 \sin \dfrac{2\pi}{3} t$

71 A point *P* in simple harmonic motion has a period of 3 seconds and an amplitude of 5 centimeters. Express the motion of *P* by means of an equation of the form $d = a \cos \omega t$.

72 A point *P* in simple harmonic motion has a frequency of $\frac{1}{2}$ oscillation per minute and an amplitude of 4 feet. Express the motion of *P* by means of an equation of the form $d = a \sin \omega t$.

73 *Tsunamis* A tsunami is a tidal wave caused by an earthquake beneath the sea. These waves can be more than 100 feet in height and can travel at great speeds. Engineers sometimes represent such waves by trigonometric expressions of the form $y = a \cos bt$ and use

these representations to estimate the effectiveness of sea walls. Suppose that a wave has height $h = 50$ ft and period 30 minutes and is traveling at the rate of 180 ft/sec.

(a) Let (x, y) be a point on the wave represented in the figure. Express y as a function of t if $y = 25$ ft when $t = 0$.

(b) The wave length L is the distance between two successive crests of the wave. Approximate L in feet.

Exercise 73

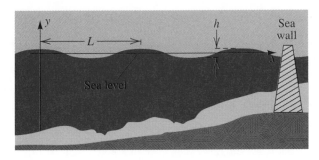

Sea level

74 *Some Hawaiian tsunamis* For an interval of 45 minutes, the tsunamis near Hawaii caused by the Chilean earthquake of 1960 could be modeled by the equation

$$y = 8 \sin \frac{\pi}{6} t,$$ where y is in feet and t is in minutes.

(a) Find the amplitude and period of the waves.

(b) If the distance from one crest of the wave to the next was 21 kilometers, what was the velocity of the wave? (Tidal waves can have velocities of more than 700 km/hr in deep sea water.)

CHAPTER 2 REVIEW EXERCISES

1 Find the radian measure that corresponds to each degree measure: $330°$, $405°$, $-150°$, $240°$, $36°$.

2 Find the degree measure that corresponds to each radian measure: $\dfrac{9\pi}{2}$, $-\dfrac{2\pi}{3}$, $\dfrac{7\pi}{4}$, 5π, $\dfrac{\pi}{5}$.

3 A central angle θ is subtended by an arc 20 centimeters long on a circle of radius 2 meters.

(a) Find the radian measure of θ.

(b) Find the area of the sector determined by θ.

4 (a) Find the length of the arc that subtends an angle of measure $70°$ on a circle of diameter 15 centimeters.

(b) Find the area of the sector in part (a).

Exer. 5–6: $P(t)$ **denotes the point on the unit circle** U **that corresponds to the real number** t.

5 Find the rectangular coordinates of $P(7\pi)$, $P(-5\pi/2)$, $P(9\pi/2)$, $P(-3\pi/4)$, $P(18\pi)$, and $P(\pi/6)$.

6 If $P(t)$ has coordinates $\left(-\frac{3}{5}, -\frac{4}{5}\right)$, find the coordinates of $P(t + 3\pi)$, $P(t - \pi)$, $P(-t)$, and $P(2\pi - t)$.

7 Find the exact values of the remaining trigonometric functions if

(a) $\sin t = -\frac{4}{5}$ and $\cos t = \frac{3}{5}$

(b) $\csc t = \dfrac{\sqrt{13}}{2}$ and $\cot t = -\dfrac{3}{2}$

8 Find the quadrant containing the point P on the unit circle U for the given conditions.

(a) $\sec t < 0$ and $\sin t > 0$

(b) $\cot t > 0$ and $\csc t < 0$

(c) $\cos t > 0$ and $\tan t < 0$

Exer. 9–10: Use fundamental identities to write the first expression in terms of the second, if $0 < t < \pi/2$.

9 $\tan t$, $\sec t$ **10** $\cot t$, $\csc t$

Exer. 11–20: Verify the identity by transforming the left side into the right side.

11 $\sin t (\csc t - \sin t) = \cos^2 t$

12 $\cos t (\tan t + \cot t) = \csc t$

13 $(\cos^2 t - 1)(\tan^2 t + 1) = 1 - \sec^2 t$

14 $\dfrac{\sec t - \cos t}{\tan t} = \dfrac{\tan t}{\sec t}$

15 $\dfrac{1 + \tan^2 t}{\tan^2 t} = \csc^2 t$

16 $\dfrac{\sec t + \csc t}{\sec t - \csc t} = \dfrac{\sin t + \cos t}{\sin t - \cos t}$

17 $\dfrac{\cot t - 1}{1 - \tan t} = \cot t$

18 $\dfrac{1 + \sec t}{\tan t + \sin t} = \csc t$

19 $\dfrac{\tan (-t) + \cot (-t)}{\tan t} = -\csc^2 t$

20 $-\dfrac{1}{\csc (-t)} - \dfrac{\cot (-t)}{\sec (-t)} = \csc t$

21 Find, whenever possible, the exact values of the six trigonometric functions of θ if θ is in standard position and satisfies the stated condition.

  (a) The point $(30, -40)$ is on the terminal side of θ.

  (b) The terminal side of θ is in quadrant II and is parallel to the line $2x + 3y + 6 = 0$.

  (c) The terminal side of θ is on the negative y-axis.

22 If θ is an acute angle of a right triangle and if the adjacent side and hypotenuse have lengths 4 and 7, respectively, find the values of the trigonometric functions of θ.

Exer. 23–24: Find the exact values of x and y.

23

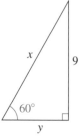

24

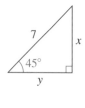

25 (a) Find the reference angle for each radian measure:
$$\frac{5\pi}{4}, \quad -\frac{5\pi}{6}, \quad -\frac{9\pi}{8}.$$

  (b) Find the reference angle for each degree measure: 245°, 137°, 892°.

26 Without the use of a calculator, find the exact values of the trigonometric functions corresponding to each real number, whenever possible.

  (a) $\dfrac{9\pi}{2}$ **(b)** $-\dfrac{5\pi}{4}$ **(c)** 0 **(d)** $\dfrac{11\pi}{6}$

27 Find the exact value.

  (a) $\cos 225°$ **(b)** $\tan 150°$ **(c)** $\sin \left(-\dfrac{\pi}{6} \right)$

  (d) $\sec \dfrac{4\pi}{3}$ **(e)** $\cot \dfrac{7\pi}{4}$ **(f)** $\csc 300°$

28 If $\sin \theta = -0.7604$ and $\sec \theta$ is positive, approximate θ to the nearest 0.1° for $0° \le \theta < 360°$.

Exer. 29–36: Find the amplitude and period and sketch the graph of the equation.

29 $y = 5 \cos x$ **30** $y = \frac{2}{3} \sin x$

31 $y = \frac{1}{3} \sin 3x$ **32** $y = -\frac{1}{2} \cos \frac{1}{3}x$

33 $y = -3 \cos \frac{1}{2}x$ **34** $y = 4 \sin 2x$

35 $y = 2 \sin \pi x$ **36** $y = 4 \cos \dfrac{\pi}{2}x - 2$

Exer. 37–40: The graph of an equation is shown in the figure. (a) Find the amplitude and period. (b) Express the equation in the form $y = a \sin bx$ or in the form $y = a \cos bx$.

37

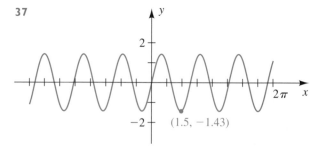

38

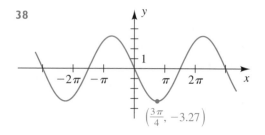

39

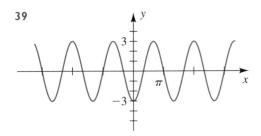

40

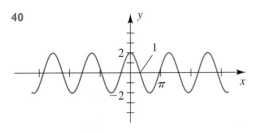

Exer. 41–52: Sketch the graph of the equation.

41 $y = 2 \sin \left(x - \dfrac{2\pi}{3} \right)$

42 $y = -3 \sin \left(\dfrac{1}{2}x - \dfrac{\pi}{4} \right)$

43 $y = -4 \cos \left(x + \dfrac{\pi}{6} \right)$ **44** $y = 5 \cos \left(2x + \dfrac{\pi}{2} \right)$

45 $y = 2 \tan \left(\dfrac{1}{2}x - \pi \right)$ **46** $y = -3 \tan \left(2x + \dfrac{\pi}{3} \right)$

47 $y = -4 \cot \left(2x - \dfrac{\pi}{2} \right)$ **48** $y = 2 \cot \left(\dfrac{1}{2}x + \dfrac{\pi}{4} \right)$

49 $y = \sec \left(\dfrac{1}{2}x + \pi \right)$ **50** $y = \sec \left(2x - \dfrac{\pi}{2} \right)$

51 $y = \csc \left(2x - \dfrac{\pi}{4} \right)$ **52** $y = \csc \left(\dfrac{1}{2}x + \dfrac{\pi}{4} \right)$

Exer. 53–56: Given the indicated parts of triangle *ABC* with $\gamma = 90°$, approximate the remaining parts.

53 $\beta = 60°$, $b = 40$ **54** $\alpha = 54°40'$, $b = 220$

55 $a = 62$, $b = 25$ **56** $a = 9.0$, $c = 41$

57 *Airplane propeller* The length of the largest airplane propeller ever used was 22 feet 7.5 inches. The plane was powered by four engines that turned the propeller at 545 revolutions per minute.

(a) What was the angular speed of the propeller in radians per second?

(b) Approximately how fast (in mi/hr) did the tip of the propeller travel along the circle it generated?

58 *The Eiffel Tower* When the top of the Eiffel Tower is viewed at a distance of 200 feet from the base, the angle of elevation is 79.2°. Estimate the height of the tower.

59 *Lasers and velocities* Lasers are used to accurately measure velocities of objects. Laser light produces an oscillating electromagnetic field E with a constant frequency f that can be described by

$$E = E_0 \cos (2\pi f t).$$

If a laser beam is pointed at an object moving toward the laser, light will be reflected toward the laser at a slightly higher frequency, in much the same way as a train whistle sounds higher when it is moving toward you. If Δf is this change in frequency and v is the object's velocity, then the equation

$$\Delta f = \frac{2fv}{c}$$

can be used to determine v, where $c = 186{,}000$ mi/sec is the velocity of the light. Approximate the velocity v of an object if $\Delta f = 10^8$ and $f = 10^{14}$.

60 *The Great Pyramid* The Great Pyramid of Egypt is 147 meters high, with a square base of side 230 meters (see the figure). Approximate, to the nearest degree, the angle φ formed when an observer stands at the midpoint of one of the sides and views the apex of the pyramid.

Exercise 60

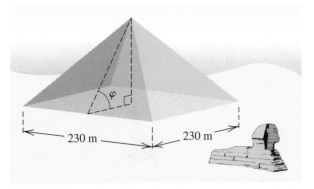

61 *Venus* When viewed from the earth over a period of time, the planet Venus appears to move back and forth along a line segment with the sun at its midpoint (see the figure). At the maximum apparent distance from the sun, angle *SEV* is approximately 47°. If *ES* is approximately 92,900,000 miles, estimate the distance of Venus from the sun.

Exercise 61

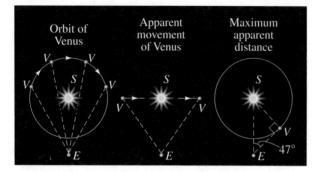

62 **Constructing a conical cup** A conical paper cup is constructed by removing a sector from a circle of radius 5 inches and attaching edge *OA* to *OB* (see the figure). Find angle *AOB* so that the cup has a depth of 4 inches.

Exercise 62

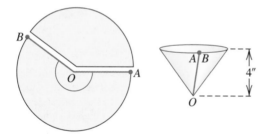

63 **Length of a tunnel** A tunnel for a new highway is to be cut through a mountain that is 260 feet high. At a distance of 200 feet from the base of the mountain, the angle of elevation is 36° (see the figure). From a distance of 150 feet on the other side, the angle of elevation is 47°. Approximate the length of the tunnel to the nearest foot.

Exercise 63

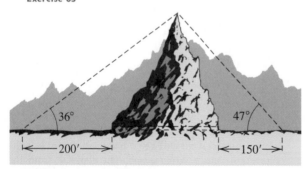

64 **Height of a skyscraper** When a certain skyscraper is viewed from the top of a building 50 feet tall, the angle of elevation is 59° (see the figure). When viewed from the street next to the shorter building, the angle of elevation is 62°.

(a) Approximately how far apart are the two structures?

(b) Approximate the height of the skyscraper to the nearest tenth of a foot.

Exercise 64

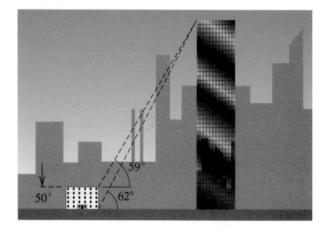

65 **Height of a mountain** When a mountaintop is viewed from the point *P* shown in the figure, the angle of elevation is α. From a point *Q*, which is *d* miles closer to the mountain, the angle of elevation increases to β.

(a) Show that the height *h* of the mountain is given by

$$h = \frac{d}{\cot \alpha - \cot \beta}.$$

(b) If $d = 2$ mi, $\alpha = 15°$, and $\beta = 20°$, approximate the height of the mountain.

Exercise 65

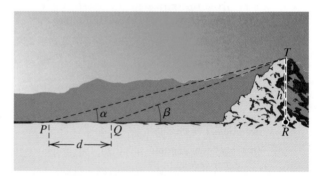

66 *Height of a building* An observer of height h stands on an incline at a distance d from the base of a building of height T, as shown in the figure. The angle of elevation from the observer to the top of the building is θ, and the incline makes an angle of α with the horizontal.

(a) Express T in terms of h, d, α, and θ.

(b) If $h = 6$ ft, $d = 50$ ft, $\alpha = 15°$, and $\theta = 31.4°$, estimate the height of the building.

Exercise 66

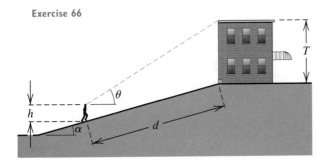

67 *Illuminance* A spotlight with intensity 5000 candles is located 15 feet above a stage. If the spotlight is rotated through an angle θ as shown in the figure, the illuminance E (in foot-candles) in the lighted area of the stage is given by

$$E = \frac{5000 \cos \theta}{s^2},$$

where s is the distance (in feet) that the light must travel.

(a) Find the illuminance if the spotlight is rotated through an angle of 30°.

(b) The maximum illuminance occurs when $\theta = 0°$. For what value of θ is the illuminance one-half the maximum value?

Exercise 67

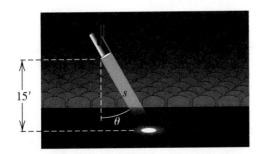

68 *Height of a mountain* If a mountaintop is viewed from a point P due south of the mountain, the angle of

elevation is α (see the figure). If viewed from a point Q that is d miles east of P, the angle of elevation is β.

(a) Show that the height h of the mountain is given by

$$h = \frac{d \sin \alpha \sin \beta}{\sqrt{\sin^2 \alpha - \sin^2 \beta}}.$$

(b) If $\alpha = 30°$, $\beta = 20°$, and $d = 10$ mi, approximate h to the nearest hundredth of a mile.

Exercise 68

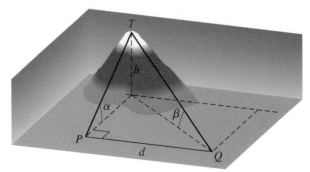

69 *Mounting a projection unit* The manufacturer of a computerized projection system recommends that a projection unit be mounted on the ceiling as shown in the figure. The distance from the end of the mounting bracket to the center of the screen is 85.5 inches, and the angle of depression is 30°.

(a) If the thickness of the screen is disregarded, how far from the wall should the bracket be mounted?

(b) If the bracket is 18 inches long and the screen is 6 feet high, determine the distance from the ceiling to the top edge of the screen.

Exercise 69

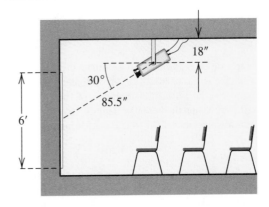

70 Pyramid relationships A pyramid has a square base and congruent triangular faces. Let θ be the angle that the altitude a of a triangular face makes with the altitude y of the pyramid, and let x be the length of a side (see the figure).

(a) Express the total surface area S of the four faces in terms of a and θ.

(b) The volume V of the pyramid equals one-third the area of the base times the altitude. Express V in terms of a and θ.

Exercise 70

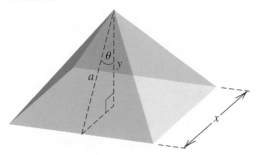

71 Surveying a bluff A surveyor, using a transit, sights the edge B of a bluff, as shown in the figure on the left (not drawn to scale). Because of the curvature of the earth, the true elevation h of the bluff is larger than that measured by the surveyor. A cross-sectional schematic view of the earth is shown in the figure on the right.

(a) If s is the length of arc PQ and R is the distance from P to the center C of the earth, express h in terms of R and s.

(b) If $R = 4000$ mi and $s = 50$ mi, estimate the elevation of the bluff in feet.

Exercise 71

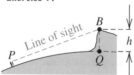

72 Earthquake response To simulate the response of a structure to an earthquake, an engineer must choose a shape for the initial displacement of the beams in the building. When the beam has length L feet and the maximum displacement is a feet, the equation

$$y = a - a \cos \frac{\pi}{2L} x$$

has been used by engineers to estimate the displacement y (see the figure). If $a = 1$ and $L = 10$, sketch the graph of the equation for $0 \le x \le 10$.

Exercise 72

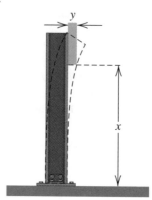

73 Circadian rhythms The variation in body temperature is an example of a circadian rhythm, a biological process that repeats itself approximately every 24 hours. Body temperature is highest about 5 P.M. and lowest at 5 A.M. Let y denote the body temperature (in °F), and let $t = 0$ correspond to midnight. If the low and high body temperatures are 98.3° and 98.9°, respectively, find an equation of the form $y = 98.6 + a \sin (bt + c)$ that fits this information.

74 Temperature variation in Ottawa The annual variation in temperature T (in °C) in Ottawa, Canada, may be approximated by

$$T(t) = 15.8 \sin \left[\frac{\pi}{6}(t - 3) \right] + 5,$$

where t is the time in months and $t = 0$ corresponds to January 1.

(a) Sketch the graph of T for $0 \le t \le 12$.

(b) Find the highest temperature of the year and the date on which it occurs.

75 Water demand A reservoir supplies water to a community. During the summer months, the demand $D(t)$ for water (in ft^3/day) is given by

$$D(t) = 2000 \sin \frac{\pi}{90}t + 4000,$$

where t is time in days and $t = 0$ corresponds to the beginning of summer.

(a) Sketch the graph of D for $0 \le t \le 90$.

(b) When is the demand for water the greatest?

76 Bobbing cork A cork bobs up and down in a lake. The distance from the bottom of the lake to the center of the cork at time $t \geq 0$ is given by $s(t) = 12 + \cos \pi t$, where $s(t)$ is in feet and t is in seconds.

(a) Describe the motion of the cork for $0 \leq t \leq 2$.

(b) During what time intervals is the cork rising?

CHAPTER 2 DISCUSSION EXERCISES

c **1** Graph $y = \sin(ax)$ on $[-2\pi, 2\pi]$ by $[-1, 1]$ for $a = 15$, 30, and 45. Discuss the accuracy of the graphs and the graphing capabilities (in terms of precision) of your graphing calculator. (*Note:* If something strange doesn't occur for $a = 45$, keep increasing a until it does.)

2 Find the maximum integer k on your calculator such that $\sin(10^k)$ can be evaluated. Now discuss how you can evaluate $\sin(10^{k+1})$ on the same calculator, and then actually find that value.

3 Determine the number of solutions of the equation

$$\cos x + \cos 2x + \cos 3x = \pi.$$

4 Discuss the relationships among periodic functions, one-to-one functions, and inverse functions. With these relationships in mind, discuss what must happen for the trigonometric functions to have inverses.

c **5** Graph $y_1 = x$, $y_2 = \sin x$, and $y_3 = \tan x$ on $[-0.1, 0.1]$ by $[-0.1, 0.1]$. Create a table of values for these three functions, with small positive values (on the order of 10^{-10} or so). What conclusion can you draw from the graph and the table?

6 Racetrack coordinates Shown in the figure is a circular racetrack of diameter 2 kilometers. All races begin at S and proceed in a counterclockwise direction. Approximate, to four decimal places, the coordinates of the point at which the following races end relative to a rectangular coordinate system with origin at the center of the track and S on the positive x-axis.

(a) A drag race of length 2 kilometers

(b) An endurance race of length 500 kilometers

Exercise 6

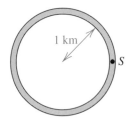

7 Racetrack coordinates Work Exercise 6 for the track shown in the figure, if the origin of the rectangular coordinate system is at the center of the track and S is on the negative y-axis.

Exercise 7

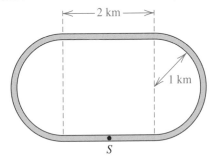

c **8 Outboard motor propeller** A 90-horsepower outboard motor at full throttle will rotate its propeller at 5000 revolutions per minute.

(a) Find the angular speed ω of the propeller in radians per second.

(b) The center of a 10-inch-diameter propeller is located 18 inches below the surface of the water. Express the depth $D(t) = a \cos(\omega t + c) + d$ of a point on the edge of a propeller blade as a function of time t, where t is in seconds. Assume that the point is initially at a depth of 23 inches.

(c) Graphically determine the number of times the propeller rotates in 0.12 second.

ANALYTIC TRIGONOMETRY

In advanced mathematics, the natural sciences, and engineering, it is sometimes necessary to simplify complicated trigonometric expressions and to solve equations that involve trigonometric functions. These topics are discussed in the first two sections of this chapter. We then derive many useful formulas with respect to sums, differences, and multiples; for reference they are listed on the inside back cover of the text. In addition to formal manipulations, we also consider numerous applications of these formulas. The last section contains the definitions and properties of the inverse trigonometric functions.

3.1 VERIFYING TRIGONOMETRIC IDENTITIES

A **trigonometric expression** contains symbols involving trigonometric functions.

ILLUSTRATION

Trigonometric Expressions

▸ $x + \sin x$ ▸ $\dfrac{\sqrt{\theta} + 2^{\sin \theta}}{\cot \theta}$ ▸ $\dfrac{\cos (3t + 1)}{t^2 + \tan^2 (2 - t^2)}$

We assume that the domain of each variable in a trigonometric expression is the set of real numbers or angles for which the expression is meaningful. To provide manipulative practice in simplifying complicated trigonometric expressions, we shall use the fundamental identities (see page 108) and algebraic manipulations, as we did in Examples 8 and 9 of Section 2.2. In the first three examples our method consists of transforming the left side of a given identity into the right side, or vice versa.

EXAMPLE 1   Verifying an identity

Verify the identity $\sec \alpha - \cos \alpha = \sin \alpha \tan \alpha$.

Solution   We transform the left side into the right side:

$$\sec \alpha - \cos \alpha = \frac{1}{\cos \alpha} - \cos \alpha \qquad \text{reciprocal identity}$$

$$= \frac{1 - \cos^2 \alpha}{\cos \alpha} \qquad \text{add expressions}$$

$$= \frac{\sin^2 \alpha}{\cos \alpha} \qquad \sin^2 \alpha + \cos^2 \alpha = 1$$

$$= \sin \alpha \left(\frac{\sin \alpha}{\cos \alpha} \right) \qquad \text{equivalent expression}$$

$$= \sin \alpha \tan \alpha \qquad \text{tangent identity}$$

EXAMPLE 2   Verifying an identity

Verify the identity $\sec \theta = \sin \theta (\tan \theta + \cot \theta)$.

Solution   Since the expression on the right side is more complicated than that on the left side, we transform the right side into the left side:

$$\sin\theta\,(\tan\theta + \cot\theta) = \sin\theta\left(\frac{\sin\theta}{\cos\theta} + \frac{\cos\theta}{\sin\theta}\right) \qquad \text{tangent and cotangent identities}$$

$$= \sin\theta\left(\frac{\sin^2\theta + \cos^2\theta}{\cos\theta\,\sin\theta}\right) \qquad \text{add fractions}$$

$$= \sin\theta\left(\frac{1}{\cos\theta\,\sin\theta}\right) \qquad \text{Pythagorean identity}$$

$$= \frac{1}{\cos\theta} \qquad \text{cancel } \sin\theta$$

$$= \sec\theta \qquad \text{reciprocal identity}$$

EXAMPLE 3   **Verifying an identity**

Verify the identity $\dfrac{\cos x}{1 - \sin x} = \dfrac{1 + \sin x}{\cos x}$.

Solution   Since the denominator of the left side is a binomial and the denominator of the right side is a monomial, we change the form of the fraction on the left side by multiplying the numerator and denominator by the conjugate of the denominator and then use one of the Pythagorean identities.

$$\frac{\cos x}{1 - \sin x} = \frac{\cos x}{1 - \sin x} \cdot \frac{1 + \sin x}{1 + \sin x} \qquad \text{multiply numerator and denominator by } 1 + \sin x$$

$$= \frac{\cos x\,(1 + \sin x)}{1 - \sin^2 x} \qquad \text{property of quotients}$$

$$= \frac{\cos x\,(1 + \sin x)}{\cos^2 x} \qquad \sin^2 x + \cos^2 x = 1$$

$$= \frac{1 + \sin x}{\cos x} \qquad \text{cancel } \cos x$$

Another technique for showing that an equation $p = q$ is an identity is to begin by transforming the left side p into another expression s, making sure that each step is *reversible*—that is, making sure it is possible to transform s back into p by reversing the procedure used in each step. In this case, the equation $p = s$ is an identity. Next, as a *separate* exercise, we show that the right side q can also be transformed into the expression s by means of reversible steps and, therefore, that $q = s$ is an identity. It then follows that $p = q$ is an identity. This method is illustrated in the next example.

EXAMPLE 4   **Verifying an identity**

Verify the identity $(\tan\theta - \sec\theta)^2 = \dfrac{1 - \sin\theta}{1 + \sin\theta}$.

Solution   We shall verify the identity by showing that each side of the equation can be transformed into the same expression. First we work only with the left side:

$$(\tan\theta - \sec\theta)^2 = \tan^2\theta - 2\tan\theta\sec\theta + \sec^2\theta \qquad \text{square expression}$$

$$= \left(\frac{\sin\theta}{\cos\theta}\right)^2 - 2\left(\frac{\sin\theta}{\cos\theta}\right)\left(\frac{1}{\cos\theta}\right) + \left(\frac{1}{\cos\theta}\right)^2$$

$$\text{tangent and reciprocal identities}$$

$$= \frac{\sin^2\theta}{\cos^2\theta} - \frac{2\sin\theta}{\cos^2\theta} + \frac{1}{\cos^2\theta} \qquad \text{equivalent expression}$$

$$= \frac{\sin^2\theta - 2\sin\theta + 1}{\cos^2\theta} \qquad \text{add fractions}$$

At this point it may not be obvious how we can obtain the right side of the given equation from the last expression. Thus, we next work with only the right side and try to obtain the last expression. Multiplying numerator and denominator by the conjugate of the denominator gives us the following:

$$\frac{1 - \sin\theta}{1 + \sin\theta} = \frac{1 - \sin\theta}{1 + \sin\theta} \cdot \frac{1 - \sin\theta}{1 - \sin\theta} \qquad \begin{array}{l}\text{multiply numerator and} \\ \text{denominator by } 1 - \sin\theta\end{array}$$

$$= \frac{1 - 2\sin\theta + \sin^2\theta}{1 - \sin^2\theta} \qquad \text{property of quotients}$$

$$= \frac{1 - 2\sin\theta + \sin^2\theta}{\cos^2\theta} \qquad \sin^2\theta + \cos^2\theta = 1$$

The last expression is the same as that obtained from $(\tan\theta - \sec\theta)^2$. Since all steps are reversible, the given equation is an identity.

In calculus it is sometimes convenient to change the form of certain algebraic expressions by making a **trigonometric substitution,** as illustrated in the following example.

EXAMPLE 5   **Making a trigonometric substitution**

Express $\sqrt{a^2 - x^2}$ in terms of a trigonometric function of θ, without radicals, by making the substitution $x = a\sin\theta$ for $-\pi/2 \le \theta \le \pi/2$ and $a > 0$.

Solution We proceed as follows:

$$\sqrt{a^2 - x^2} = \sqrt{a^2 - (a \sin \theta)^2} \qquad \text{let } x = a \sin \theta$$
$$= \sqrt{a^2 - a^2 \sin^2 \theta} \qquad \text{law of exponents}$$
$$= \sqrt{a^2(1 - \sin^2 \theta)} \qquad \text{factor out } a^2$$
$$= \sqrt{a^2 \cos^2 \theta} \qquad \sin^2 \theta + \cos^2 \theta = 1$$
$$= a \cos \theta \qquad \text{see below}$$

The last equality is true because (1) if $a > 0$, then $\sqrt{a^2} = a$, and (2) if $-\pi/2 \le \theta \le \pi/2$, then $\cos \theta \ge 0$ and hence $\sqrt{\cos^2 \theta} = \cos \theta$.

We may also use a geometric solution. If $x = a \sin \theta$, then $\sin \theta = x/a$, and the triangle in Figure 1 illustrates the problem for $0 < \theta < \pi/2$. The third side of the triangle, $\sqrt{a^2 - x^2}$, can be found by using the Pythagorean theorem. From the figure we can see that

$$\cos \theta = \frac{\sqrt{a^2 - x^2}}{a} \qquad \text{or, equivalently,} \qquad \sqrt{a^2 - x^2} = a \cos \theta.$$

Figure 1

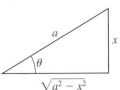

3.1 EXERCISES

Exer. 1–50: Verify the identity.

1 $\csc \theta - \sin \theta = \cot \theta \cos \theta$

2 $\sin x + \cos x \cot x = \csc x$

3 $\dfrac{\sec^2 2u - 1}{\sec^2 2u} = \sin^2 2u$

4 $\tan t + 2 \cos t \csc t = \sec t \csc t + \cot t$

5 $\dfrac{\csc^2 \theta}{1 + \tan^2 \theta} = \cot^2 \theta$

6 $(\tan u + \cot u)(\cos u + \sin u) = \csc u + \sec u$

7 $\dfrac{1 + \cos 3t}{\sin 3t} + \dfrac{\sin 3t}{1 + \cos 3t} = 2 \csc 3t$

8 $\tan^2 \alpha - \sin^2 \alpha = \tan^2 \alpha \sin^2 \alpha$

9 $\dfrac{1}{1 - \cos \gamma} + \dfrac{1}{1 + \cos \gamma} = 2 \csc^2 \gamma$

10 $\dfrac{1 + \csc 3\beta}{\sec 3\beta} - \cot 3\beta = \cos 3\beta$

11 $(\sec u - \tan u)(\csc u + 1) = \cot u$

12 $\dfrac{\cot \theta - \tan \theta}{\sin \theta + \cos \theta} = \csc \theta - \sec \theta$

13 $\csc^4 t - \cot^4 t = \csc^2 t + \cot^2 t$

14 $\cos^4 2\theta + \sin^2 2\theta = \cos^2 2\theta + \sin^4 2\theta$

15 $\dfrac{\cos \beta}{1 - \sin \beta} = \sec \beta + \tan \beta$

16 $\dfrac{1}{\csc y - \cot y} = \csc y + \cot y$

17 $\dfrac{\tan^2 x}{\sec x + 1} = \dfrac{1 - \cos x}{\cos x}$

18 $\dfrac{\cot x}{\csc x + 1} = \dfrac{\csc x - 1}{\cot x}$

19 $\dfrac{\cot 4u - 1}{\cot 4u + 1} = \dfrac{1 - \tan 4u}{1 + \tan 4u}$

20 $\dfrac{1 + \sec 4x}{\sin 4x + \tan 4x} = \csc 4x$

21 $\sin^4 r - \cos^4 r = \sin^2 r - \cos^2 r$

22 $\sin^4 \theta + 2 \sin^2 \theta \cos^2 \theta + \cos^4 \theta = 1$

23 $\tan^4 k - \sec^4 k = 1 - 2 \sec^2 k$

24 $\sec^4 u - \sec^2 u = \tan^2 u + \tan^4 u$

25 $(\sec t + \tan t)^2 = \dfrac{1 + \sin t}{1 - \sin t}$

26 $\sec^2 \gamma + \tan^2 \gamma = (1 - \sin^4 \gamma) \sec^4 \gamma$

27 $(\sin^2 \theta + \cos^2 \theta)^3 = 1$

28 $\dfrac{\sin t}{1 - \cos t} = \csc t + \cot t$

29 $\dfrac{1 + \csc \beta}{\cot \beta + \cos \beta} = \sec \beta$

30 $\dfrac{\cos^3 x - \sin^3 x}{\cos x - \sin x} = 1 + \sin x \cos x$

31 $(\csc t - \cot t)^4 (\csc t + \cot t)^4 = 1$

32 $(a \cos t - b \sin t)^2 + (a \sin t + b \cos t)^2 = a^2 + b^2$

33 $\dfrac{\sin \alpha \cos \beta + \cos \alpha \sin \beta}{\cos \alpha \cos \beta - \sin \alpha \sin \beta} = \dfrac{\tan \alpha + \tan \beta}{1 - \tan \alpha \tan \beta}$

34 $\dfrac{\tan u - \tan v}{1 + \tan u \tan v} = \dfrac{\cot v - \cot u}{\cot u \cot v + 1}$

35 $\dfrac{\tan \alpha}{1 + \sec \alpha} + \dfrac{1 + \sec \alpha}{\tan \alpha} = 2 \csc \alpha$

36 $\dfrac{\csc x}{1 + \csc x} - \dfrac{\csc x}{1 - \csc x} = 2 \sec^2 x$

37 $\dfrac{1}{\tan \beta + \cot \beta} = \sin \beta \cos \beta$

38 $\dfrac{\cot y - \tan y}{\sin y \cos y} = \csc^2 y - \sec^2 y$

39 $\sec \theta + \csc \theta - \cos \theta - \sin \theta = \sin \theta \tan \theta + \cos \theta \cot \theta$

40 $\sin^3 t + \cos^3 t = (1 - \sin t \cos t)(\sin t + \cos t)$

41 $(1 - \tan^2 \phi)^2 = \sec^4 \phi - 4 \tan^2 \phi$

42 $\cos^4 w + 1 - \sin^4 w = 2 \cos^2 w$

43 $\dfrac{\cot (-t) + \tan (-t)}{\cot t} = -\sec^2 t$

44 $\dfrac{\csc (-t) - \sin (-t)}{\sin (-t)} = \cot^2 t$

45 $\log 10^{\tan t} = \tan t$ **46** $10^{\log |\sin t|} = |\sin t|$

47 $\ln \cot x = -\ln \tan x$ **48** $\ln \sec \theta = -\ln \cos \theta$

49 $\ln |\sec \theta + \tan \theta| = -\ln |\sec \theta - \tan \theta|$

50 $\ln |\csc x - \cot x| = -\ln |\csc x + \cot x|$

Exer. 51–62: Show that the equation is *not* an identity. (*Hint:* Find one number for which the equation is false.)

51 $\cos t = \sqrt{1 - \sin^2 t}$

52 $\sqrt{\sin^2 t + \cos^2 t} = \sin t + \cos t$

53 $\sqrt{\sin^2 t} = \sin t$ **54** $\sec t = \sqrt{\tan^2 t + 1}$

55 $(\sin \theta + \cos \theta)^2 = \sin^2 \theta + \cos^2 \theta$

56 $\log \left(\dfrac{1}{\sin t} \right) = \dfrac{1}{\log \sin t}$

57 $\cos (-t) = -\cos t$ **58** $\sin (t + \pi) = \sin t$

59 $\cos (\sec t) = 1$ **60** $\cot (\tan \theta) = 1$

61 $\sin^2 t - 4 \sin t - 5 = 0$

62 $3 \cos^2 \theta + \cos \theta - 2 = 0$

Exer. 63–66: Refer to Example 5. Make the trigonometric substitution $x = a \sin \theta$ for $-\pi/2 < \theta < \pi/2$ and $a > 0$. Use fundamental identities to simplify the resulting expression.

63 $(a^2 - x^2)^{3/2}$ **64** $\dfrac{\sqrt{a^2 - x^2}}{x}$

65 $\dfrac{x^2}{\sqrt{a^2 - x^2}}$ **66** $\dfrac{1}{x \sqrt{a^2 - x^2}}$

Exer. 67–70: Make the trigonometric substitution $x = a \tan \theta$ for $-\pi/2 < \theta < \pi/2$ and $a > 0$. Simplify the resulting expression.

67 $\sqrt{a^2 + x^2}$ **68** $\dfrac{1}{\sqrt{a^2 + x^2}}$

69 $\dfrac{1}{x^2 + a^2}$ **70** $\dfrac{(x^2 + a^2)^{3/2}}{x}$

Exer. 71–74: Make the trigonometric substitution $x = a \sec \theta$ for $0 < \theta < \pi/2$ and $a > 0$. Simplify the resulting expression.

71 $\sqrt{x^2 - a^2}$ **72** $\dfrac{1}{x^2 \sqrt{x^2 - a^2}}$

73 $x^3 \sqrt{x^2 - a^2}$ **74** $\dfrac{\sqrt{x^2 - a^2}}{x^2}$

c **Exer. 75–78:** Use the graph of f to find the simplest expression $g(x)$ such that the equation $f(x) = g(x)$ is an identity. Verify this identity.

75 $f(x) = \dfrac{\sin^2 x - \sin^4 x}{(1 - \sec^2 x)\cos^4 x}$

76 $f(x) = \dfrac{\sin x - \sin^3 x}{\cos^4 x + \cos^2 x \sin^2 x}$

77 $f(x) = \sec x \,(\sin x \cos x + \cos^2 x) - \sin x$

78 $f(x) = \dfrac{\sin^3 x + \sin x \cos^2 x}{\csc x} + \dfrac{\cos^3 x + \cos x \sin^2 x}{\sec x}$

3.2 TRIGONOMETRIC EQUATIONS

A **trigonometric equation** is an equation that contains trigonometric expressions. Each identity considered in the preceding section is an example of a trigonometric equation with every number (or angle) in the domain of the variable a solution of the equation. If a trigonometric equation is not an identity, we often find solutions by using techniques similar to those used for algebraic equations. The main difference is that we first solve the trigonometric equation for $\sin x$, $\cos \theta$, and so on, and then find values of x or θ that satisfy the equation. Solutions may be expressed either as real numbers or as angles. Throughout our work we shall use the following rule: *If degree measure is not specified, then solutions of a trigonometric equation should be expressed in radian measure (or as real numbers).* If solutions in degree measure are desired, an appropriate statement will be included in the example or exercise.

EXAMPLE 1 Solving a trigonometric equation involving the sine function

Find the solutions of the equation $\sin \theta = \frac{1}{2}$ if

(a) θ is in the interval $[0, 2\pi)$

(b) θ is any real number

Solution

(a) If $\sin \theta = \frac{1}{2}$, then the reference angle for θ is $\theta_R = \pi/6$. If we regard θ as an angle in standard position, then, since $\sin \theta > 0$, the terminal side is in either quadrant I or quadrant II, as illustrated in Figure 2. Thus, there are two solutions for $0 \le \theta < 2\pi$:

$$\theta = \frac{\pi}{6} \quad \text{and} \quad \theta = \pi - \frac{\pi}{6} = \frac{5\pi}{6}$$

(b) Since the sine function has period 2π, we may obtain all solutions by adding multiples of 2π to $\pi/6$ and $5\pi/6$. This gives us

$$\theta = \frac{\pi}{6} + 2\pi n \quad \text{and} \quad \theta = \frac{5\pi}{6} + 2\pi n \quad \text{for every integer } n.$$

Figure 2

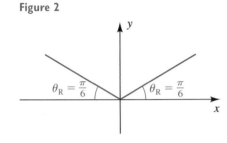

(continued)

Figure 3

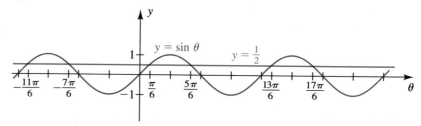

An alternative (graphical) solution involves determining where the graph of $y = \sin \theta$ intersects the horizontal line $y = \frac{1}{2}$, as illustrated in Figure 3.

EXAMPLE 2   **Solving a trigonometric equation involving the tangent function**

Find the solutions of the equation $\tan u = -1$.

Solution   Since the tangent function has period π, it is sufficient to find one real number u such that $\tan u = -1$ and then add multiples of π.

A portion of the graph of $y = \tan u$ is sketched in Figure 4. Since $\tan (3\pi/4) = -1$, one solution is $3\pi/4$; hence,

$$\text{if} \quad \tan u = -1, \quad \text{then} \quad u = \frac{3\pi}{4} + \pi n \quad \text{for every integer } n.$$

Figure 4 $y = \tan u$

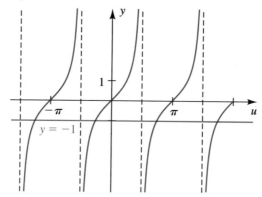

We could also have chosen $-\pi/4$ (or some other number u such that $\tan u = -1$) for the initial solution and written

$$u = -\frac{\pi}{4} + \pi n \quad \text{for every integer } n.$$

EXAMPLE 3 Solving a trigonometric equation involving multiple angles

(a) Solve the equation $\cos 2x = 0$, and express the solutions both in radians and in degrees.

(b) Find the solutions that are in the intervals $[0, 2\pi)$ and $[0°, 360°)$.

Solution

(a) We proceed as follows, where n denotes any integer:

$$\cos 2x = 0 \qquad \text{given}$$

$$\cos \theta = 0 \qquad \text{let } \theta = 2x$$

$$\theta = \frac{\pi}{2} + \pi n \qquad \text{refer to graph of } y = \cos t \text{ (page 118)}$$

$$2x = \frac{\pi}{2} + \pi n \qquad \theta = 2x$$

$$x = \frac{\pi}{4} + \frac{\pi}{2} n \qquad \text{divide by 2}$$

In degrees, we have $x = 45° + 90°n$.

(b) We may find particular solutions of the equation by substituting integers for n in either of the formulas for x obtained in part (a). Several such solutions are listed in the following table.

n	$\dfrac{\pi}{4} + \dfrac{\pi}{2}n$	$45° + 90°n$
-1	$\dfrac{\pi}{4} + \dfrac{\pi}{2}(-1) = -\dfrac{\pi}{4}$	$45° + 90°(-1) = -45°$
0	$\dfrac{\pi}{4} + \dfrac{\pi}{2}(0) = \dfrac{\pi}{4}$	$45° + 90°(0) = 45°$
1	$\dfrac{\pi}{4} + \dfrac{\pi}{2}(1) = \dfrac{3\pi}{4}$	$45° + 90°(1) = 135°$
2	$\dfrac{\pi}{4} + \dfrac{\pi}{2}(2) = \dfrac{5\pi}{4}$	$45° + 90°(2) = 225°$
3	$\dfrac{\pi}{4} + \dfrac{\pi}{2}(3) = \dfrac{7\pi}{4}$	$45° + 90°(3) = 315°$
4	$\dfrac{\pi}{4} + \dfrac{\pi}{2}(4) = \dfrac{9\pi}{4}$	$45° + 90°(4) = 405°$

(continued)

Note that the solutions in the interval $[0, 2\pi)$, or $[0°, 360°)$, are given by $n = 0$, $n = 1$, $n = 2$, and $n = 3$. These solutions are

$$\frac{\pi}{4}, \frac{3\pi}{4}, \frac{5\pi}{4}, \frac{7\pi}{4}, \quad \text{or} \quad 45°, 135°, 225°, 315°.$$

EXAMPLE 4   **Solving a trigonometric equation by factoring**

Solve the equation $\sin \theta \tan \theta = \sin \theta$.

Solution

$$
\begin{array}{lll}
\sin \theta \tan \theta = \sin \theta & \text{given} \\
\sin \theta \tan \theta - \sin \theta = 0 & \text{make one side 0} \\
\sin \theta (\tan \theta - 1) = 0 & \text{factor out } \sin \theta \\
\sin \theta = 0, \quad \tan \theta - 1 = 0 & \text{zero factor theorem} \\
\sin \theta = 0, \quad \tan \theta = 1 & \text{solve for } \sin \theta \text{ and } \tan \theta
\end{array}
$$

The solutions of the equation $\sin \theta = 0$ are $0, \pm\pi, \pm 2\pi, \ldots$. Thus,

$$\text{if} \quad \sin \theta = 0, \quad \text{then} \quad \theta = \pi n \quad \text{for every integer } n.$$

The tangent function has period π, and hence we find the solutions of the equation $\tan \theta = 1$ that are in the interval $(-\pi/2, \pi/2)$ and then add multiples of π. Since the only solution of $\tan \theta = 1$ in $(-\pi/2, \pi/2)$ is $\pi/4$, we see that

$$\text{if} \quad \tan \theta = 1, \quad \text{then} \quad \theta = \frac{\pi}{4} + \pi n \quad \text{for every integer } n.$$

Thus, the solutions of the given equation are

$$\pi n \quad \text{and} \quad \frac{\pi}{4} + \pi n \quad \text{for every integer } n.$$

Some *particular* solutions, obtained by letting $n = 0$, $n = 1$, $n = 2$, and $n = -1$, are

$$0, \quad \frac{\pi}{4}, \quad \pi, \quad \frac{5\pi}{4}, \quad 2\pi, \quad \frac{9\pi}{4}, \quad -\pi, \quad -\frac{3\pi}{4}.$$

In Example 4 it would have been incorrect to begin by dividing both sides by $\sin \theta$, since we would have lost the solutions of $\sin \theta = 0$.

EXAMPLE 5 Solving a trigonometric equation by factoring

Solve the equation $2 \sin^2 t - \cos t - 1 = 0$, and express the solutions both in radians and in degrees.

Solution It appears that we have a quadratic equation in either $\sin t$ or $\cos t$. We do not have a simple substitution for $\cos t$ in terms of $\sin t$, but we do have one for $\sin^2 t$ in terms of $\cos^2 t$ ($\sin^2 t = 1 - \cos^2 t$), so we shall first express the equation in terms of $\cos t$ alone and then solve by factoring.

$$2 \sin^2 t - \cos t - 1 = 0 \qquad \text{given}$$
$$2(1 - \cos^2 t) - \cos t - 1 = 0 \qquad \sin^2 t + \cos^2 t = 1$$
$$-2 \cos^2 t - \cos t + 1 = 0 \qquad \text{simplify}$$
$$2 \cos^2 t + \cos t - 1 = 0 \qquad \text{multiply by } -1$$
$$(2 \cos t - 1)(\cos t + 1) = 0 \qquad \text{factor}$$
$$2 \cos t - 1 = 0, \quad \cos t + 1 = 0 \qquad \text{zero factor theorem}$$
$$\cos t = \tfrac{1}{2}, \qquad \cos t = -1 \qquad \text{solve for } \cos t$$

Since the cosine function has period 2π, we may find all solutions of these equations by adding multiples of 2π to the solutions that are in the interval $[0, 2\pi)$.

If $\cos t = \tfrac{1}{2}$, the reference angle is $\pi/3$ (or $60°$). Since $\cos t$ is positive, the angle of radian measure t is in either quadrant I or quadrant IV. Hence, in the interval $[0, 2\pi)$, we see that

$$\text{if} \quad \cos t = \frac{1}{2}, \quad \text{then} \quad t = \frac{\pi}{3} \quad \text{or} \quad t = 2\pi - \frac{\pi}{3} = \frac{5\pi}{3}.$$

Referring to the graph of the cosine function, we see that

$$\text{if} \quad \cos t = -1, \quad \text{then} \quad t = \pi.$$

Thus, the solutions of the given equation are the following, where n is any integer:

$$\frac{\pi}{3} + 2\pi n, \quad \frac{5\pi}{3} + 2\pi n, \quad \pi + 2\pi n$$

In degree measure, we have

$$60° + 360° n, \quad 300° + 360° n, \quad 180° + 360° n.$$

EXAMPLE 6 Solving a trigonometric equation by factoring

Find the solutions of $4 \sin^2 x \tan x - \tan x = 0$ that are in the interval $[0, 2\pi)$.

Solution

$$4 \sin^2 x \tan x - \tan x = 0 \qquad \text{given}$$

$$\tan x \,(4 \sin^2 x - 1) = 0 \qquad \text{factor out } \tan x$$

$$\tan x = 0, \quad 4 \sin^2 x - 1 = 0 \qquad \text{zero factor theorem}$$

$$\tan x = 0, \qquad \sin^2 x = \tfrac{1}{4} \qquad \text{solve for } \tan x, \ \sin^2 x$$

$$\tan x = 0, \qquad \sin x = \pm \tfrac{1}{2} \qquad \text{solve for } \sin x$$

Figure 5

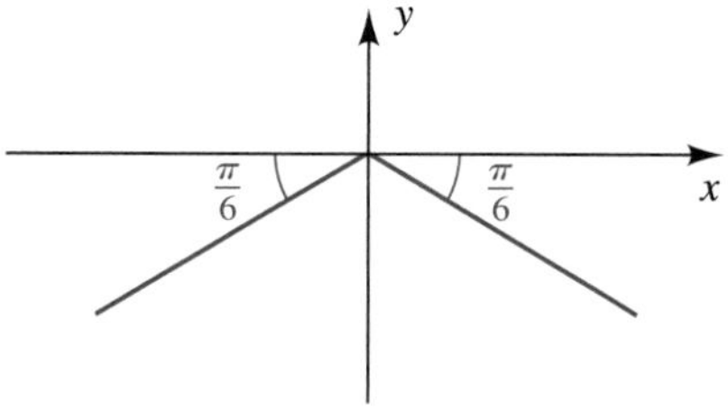

The reference angle $\pi/6$ for the third and fourth quadrants is shown in Figure 5. These angles, $7\pi/6$ and $11\pi/6$, are the solutions of the equation $\sin x = -\tfrac{1}{2}$ for $0 \le x < 2\pi$. The solutions of all three equations are listed in the following table.

Equation	Solutions in $[0, 2\pi)$	Refer to
$\tan x = 0$	$0, \pi$	Figure 4
$\sin x = \dfrac{1}{2}$	$\dfrac{\pi}{6}, \dfrac{5\pi}{6}$	Example 1
$\sin x = -\dfrac{1}{2}$	$\dfrac{7\pi}{6}, \dfrac{11\pi}{6}$	Figure 5 (use reference angle)

Thus, the given equation has the six solutions listed in the second column of the table.

E X A M P L E   7   **Solving a trigonometric equation involving multiple angles**

Find the solutions of $\csc^4 2u - 4 = 0$.

Solution

$$\csc^4 2u - 4 = 0 \qquad \text{given}$$

$$(\csc^2 2u - 2)(\csc^2 2u + 2) = 0 \qquad \text{difference of two squares}$$

$$\csc^2 2u - 2 = 0, \qquad \csc^2 2u + 2 = 0 \qquad \text{zero factor theorem}$$

$$\csc^2 2u = 2, \qquad \csc^2 2u = -2 \qquad \text{solve for } \csc^2 2u$$

$$\csc 2u = \pm\sqrt{2}, \qquad \csc 2u = \pm\sqrt{-2} \qquad \text{take square roots}$$

The second equation has no solution because $\sqrt{-2}$ is not a real number. The first equation is equivalent to

$$\sin 2u = \pm \frac{1}{\sqrt{2}} = \pm \frac{\sqrt{2}}{2}.$$

Since the reference angle for $2u$ is $\pi/4$, we obtain the following table, in which n denotes any integer.

Equation	Solution for $2u$	Solution for u
$\sin 2u = \dfrac{\sqrt{2}}{2}$	$2u = \dfrac{\pi}{4} + 2\pi n$	$u = \dfrac{\pi}{8} + \pi n$
	$2u = \dfrac{3\pi}{4} + 2\pi n$	$u = \dfrac{3\pi}{8} + \pi n$
$\sin 2u = -\dfrac{\sqrt{2}}{2}$	$2u = \dfrac{5\pi}{4} + 2\pi n$	$u = \dfrac{5\pi}{8} + \pi n$
	$2u = \dfrac{7\pi}{4} + 2\pi n$	$u = \dfrac{7\pi}{8} + \pi n$

The solutions of the given equation are listed in the last column. Note that *all* of these solutions can be written in the one form

$$u = \frac{\pi}{8} + \frac{\pi}{4}n.$$

The next example illustrates the use of a calculator in solving a trigonometric equation.

EXAMPLE 8   **Approximating the solutions of a trigonometric equation**

Approximate, to the nearest degree, the solutions of the following equation in the interval $[0, 360°)$:

$$5 \sin \theta \tan \theta - 10 \tan \theta + 3 \sin \theta - 6 = 0$$

Solution

$5 \sin \theta \tan \theta - 10 \tan \theta + 3 \sin \theta - 6 = 0$	given
$(5 \sin \theta \tan \theta - 10 \tan \theta) + (3 \sin \theta - 6) = 0$	group terms
$5 \tan \theta (\sin \theta - 2) + 3(\sin \theta - 2) = 0$	factor each group
$(5 \tan \theta + 3)(\sin \theta - 2) = 0$	factor out $(\sin \theta - 2)$
$5 \tan \theta + 3 = 0, \quad \sin \theta - 2 = 0$	zero factor theorem
$\tan \theta = -\tfrac{3}{5}, \qquad \sin \theta = 2$	solve for $\tan \theta$ and $\sin \theta$

(continued)

The equation $\sin \theta = 2$ has no solution, since $\sin \theta \le 1$ for every θ. For $\tan \theta = -\frac{3}{5}$, we use a calculator in degree mode, obtaining

$$\theta = \tan^{-1}\left(-\tfrac{3}{5}\right) \approx -31°.$$

Hence, the reference angle is $\theta_R \approx 31°$. Since θ is in either quadrant II or quadrant IV, we obtain the following solutions:

$$\theta = 180° - \theta_R \approx 180° - 31° = 149°$$
$$\theta = 360° - \theta_R \approx 360° - 31° = 329°$$

EXAMPLE 9   **Investigating the number of hours of daylight**

In Boston, the number of hours of daylight $D(t)$ at a particular time of the year may be approximated by

$$D(t) = 3 \sin\left[\frac{2\pi}{365}(t - 79)\right] + 12,$$

with t in days and $t = 0$ corresponding to January 1. How many days of the year have more than 10.5 hours of daylight?

Figure 6

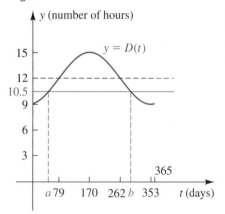

Solution   The graph of D was discussed in Example 12 of Section 2.6 and is resketched in Figure 6. As illustrated in the figure, if we can find two numbers a and b with $D(a) = 10.5$, $D(b) = 10.5$, and $0 < a < b < 365$, then there will be more than 10.5 hours of daylight in the tth day of the year if $a < t < b$.

Let us solve the equation $D(t) = 10.5$ as follows:

$$3 \sin\left[\frac{2\pi}{365}(t - 79)\right] + 12 = 10.5 \qquad \text{let } D(t) = 10.5$$

$$3 \sin\left[\frac{2\pi}{365}(t - 79)\right] = -1.5 \qquad \text{subtract 12}$$

$$\sin\left[\frac{2\pi}{365}(t - 79)\right] = -0.5 = -\frac{1}{2} \qquad \text{divide by 3}$$

If $\sin \theta = -\frac{1}{2}$, then the reference angle is $\pi/6$ and the angle θ is in either quadrant III or quadrant IV. Thus, we can find the numbers a and b by solving the equations

$$\frac{2\pi}{365}(t - 79) = \frac{7\pi}{6} \qquad \text{and} \qquad \frac{2\pi}{365}(t - 79) = \frac{11\pi}{6}.$$

From the first of these equations we obtain

$$t - 79 = \frac{7\pi}{6} \cdot \frac{365}{2\pi} = \frac{2555}{12} \approx 213,$$

and hence $\qquad t \approx 213 + 79, \qquad \text{or} \qquad t \approx 292.$

Similarly, the second equation gives us $t \approx 414$. Since the period of the function D is 365 days (see Figure 6), we obtain

$$t \approx 414 - 365, \quad \text{or} \quad t \approx 49.$$

Thus, there will be at least 10.5 hours of daylight from $t = 49$ to $t = 292$—that is, for 243 days of the year.

A graphical solution of the next example was given in Example 14 of Section 2.6.

EXAMPLE 10   **Finding the minimum current in an electrical circuit**

The current I (in amperes) in an alternating current circuit at time t (in seconds) is given by

$$I = 30 \sin \left(50\pi t - \frac{7\pi}{3} \right).$$

Find the smallest exact value of t for which $I = 15$.

Solution   Letting $I = 15$ in the given formula, we obtain

$$15 = 30 \sin \left(50\pi t - \frac{7\pi}{3} \right) \quad \text{or, equivalently,} \quad \sin \left(50\pi t - \frac{7\pi}{3} \right) = \frac{1}{2}.$$

Thus, the reference angle is $\pi/6$, and consequently

$$50\pi t - \frac{7\pi}{3} = \frac{\pi}{6} + 2\pi n \quad \text{or} \quad 50\pi t - \frac{7\pi}{3} = \frac{5\pi}{6} + 2\pi n,$$

where n is any integer. Solving for t gives us

$$t = \frac{\frac{15}{6} + 2n}{50} \quad \text{or} \quad t = \frac{\frac{19}{6} + 2n}{50}.$$

The smallest positive value of t will occur when one of the numerators of these two fractions has its least positive value. Since $\frac{15}{6} = 2.5$, $\frac{19}{6} \approx 3.17$, and $2(-1) = -2$, we see that the smallest positive value of t occurs when $n = -1$ in the first fraction—that is, when

$$t = \frac{\frac{15}{6} + 2(-1)}{50} = \frac{1}{100}.$$

The next example illustrates how a graphing utility can aid in solving a complicated trigonometric equation.

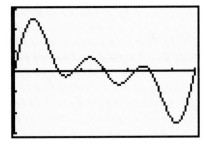

EXAMPLE 11   **Using a graph to determine solutions of a trigonometric equation**

Find the solutions of the following equation that are in the interval $[0, 2\pi)$:

$$\sin x + \sin 2x + \sin 3x = 0$$

Solution   We assign $\sin x + \sin 2x + \sin 3x$ to Y_1. Since $|\sin \theta| \le 1$ for $\theta = x, 2x,$ and $3x$, the left side of the equation is between -3 and 3, and we choose the viewing rectangle $[0, 2\pi]$ by $[-3, 3]$ and obtain a sketch similar to Figure 7. Using a zoom or root feature, we obtain the following approximations for the x-intercepts—that is, the approximate solutions of the given equation in $[0, 2\pi)$:

$$0, \quad 1.57, \quad 2.09, \quad 3.14, \quad 4.19, \quad 4.71$$

Changing to degree measure and rounding off to the nearest degree, we obtain

$$0°, \quad 90°, \quad 120°, \quad 180°, \quad 240°, \quad 270°.$$

Converting these degree measures to radian measures gives us

$$0, \quad \frac{\pi}{2}, \quad \frac{2\pi}{3}, \quad \pi, \quad \frac{4\pi}{3}, \quad \frac{3\pi}{2}.$$

Checking these values in the given equation, we see that all six are solutions. Figure 7 suggests that the graph has period 2π. After studying Section 3.4, you will be able to change the form of Y_1 and *prove* that the period is 2π and, therefore, that *all* solutions of the given equation can be obtained by adding integer multiples of 2π.

Figure 7
$[0, 2\pi]$ by $[-3, 3]$

In the preceding example we were able to use a graphing utility to help us find the *exact* solutions of the equation. For many equations that occur in applications, however, it is only possible to approximate the solutions.

3.2 EXERCISES

Exer. 1–36: Find all solutions of the equation.

1 $\sin x = -\dfrac{\sqrt{2}}{2}$

2 $\cos t = -1$

3 $\tan \theta = \sqrt{3}$

4 $\cot \alpha = -\dfrac{1}{\sqrt{3}}$

5 $\sec \beta = 2$

6 $\csc \gamma = \sqrt{2}$

7 $\sin x = \dfrac{\pi}{2}$

8 $\cos x = -\dfrac{\pi}{3}$

9 $\cos \theta = \dfrac{1}{\sec \theta}$

10 $\csc \theta \sin \theta = 1$

11 $2 \cos 2\theta - \sqrt{3} = 0$

12 $2 \sin 3\theta + \sqrt{2} = 0$

13 $\sqrt{3} \tan \dfrac{1}{3}t = 1$

14 $\cos \dfrac{1}{4}x = -\dfrac{\sqrt{2}}{2}$

15 $\sin \left(\theta + \dfrac{\pi}{4} \right) = \dfrac{1}{2}$

16 $\cos \left(x - \dfrac{\pi}{3} \right) = -1$

17 $\sin \left(2x - \dfrac{\pi}{3} \right) = \dfrac{1}{2}$

18 $\cos \left(4x - \dfrac{\pi}{4} \right) = \dfrac{\sqrt{2}}{2}$

19 $2 \cos t + 1 = 0$

20 $\cot \theta + 1 = 0$

21 $\tan^2 x = 1$

22 $4 \cos \theta - 2 = 0$

23 $(\cos \theta - 1)(\sin \theta + 1) = 0$

24 $2 \cos x = \sqrt{3}$

25 $\sec^2 \alpha - 4 = 0$

26 $3 - \tan^2 \beta = 0$

27 $\sqrt{3} + 2 \sin \beta = 0$

28 $4 \sin^2 x - 3 = 0$

29 $\cot^2 x - 3 = 0$

30 $(\sin t - 1) \cos t = 0$

31 $(2 \sin \theta + 1)(2 \cos \theta + 3) = 0$

32 $(2 \sin u - 1)(\cos u - \sqrt{2}) = 0$

33 $\sin 2x (\csc 2x - 2) = 0$

34 $\tan \alpha + \tan^2 \alpha = 0$

35 $\cos (\ln x) = 0$

36 $\ln (\sin x) = 0$

Exer. 37–60: Find the solutions of the equation that are in the interval $[0, 2\pi)$.

37 $\cos \left(2x - \dfrac{\pi}{4} \right) = 0$

38 $\sin \left(3x - \dfrac{\pi}{4} \right) = 1$

39 $2 - 8 \cos^2 t = 0$

40 $\cot^2 \theta - \cot \theta = 0$

41 $2 \sin^2 u = 1 - \sin u$

42 $2 \cos^2 t + 3 \cos t + 1 = 0$

43 $\tan^2 x \sin x = \sin x$

44 $\sec \beta \csc \beta = 2 \csc \beta$

45 $2 \cos^2 \gamma + \cos \gamma = 0$

46 $\sin x - \cos x = 0$

47 $\sin^2 \theta + \sin \theta - 6 = 0$

48 $2 \sin^2 u + \sin u - 6 = 0$

49 $1 - \sin t = \sqrt{3} \cos t$

50 $\cos \theta - \sin \theta = 1$

51 $\cos \alpha + \sin \alpha = 1$

52 $\sqrt{3} \sin t + \cos t = 1$

53 $2 \tan t - \sec^2 t = 0$

54 $\tan \theta + \sec \theta = 1$

55 $\cot \alpha + \tan \alpha = \csc \alpha \sec \alpha$

56 $\sin x + \cos x \cot x = \csc x$

57 $2 \sin^3 x + \sin^2 x - 2 \sin x - 1 = 0$

58 $\sec^5 \theta = 4 \sec \theta$

59 $2 \tan t \csc t + 2 \csc t + \tan t + 1 = 0$

60 $2 \sin v \csc v - \csc v = 4 \sin v - 2$

Exer. 61–66: Approximate, to the nearest 10′, the solutions of the equation in the interval $[0°, 360°)$.

61 $\sin^2 t - 4 \sin t + 1 = 0$

62 $\cos^2 t - 4 \cos t + 2 = 0$

63 $\tan^2 \theta + 3 \tan \theta + 2 = 0$

64 $2 \tan^2 x - 3 \tan x - 1 = 0$

65 $12 \sin^2 u - 5 \sin u - 2 = 0$

66 $5 \cos^2 \alpha + 3 \cos \alpha - 2 = 0$

67 *Tidal waves* A tidal wave of height 50 feet and period 30 minutes is approaching a sea wall that is 12.5 feet above sea level (see the figure). From a particular point on shore, the distance y from sea level to the top of the wave is given by

$$y = 25 \cos \dfrac{\pi}{15}t,$$

with t in minutes. For approximately how many minutes of each 30-minute period is the top of the wave above the level of the top of the sea wall?

Exercise 67

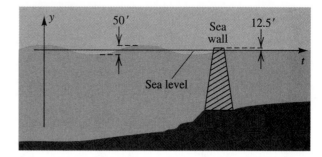

68 *Temperature in Fairbanks* The expected low temperature T (in °F) in Fairbanks, Alaska, may be approximated by

$$T = 36 \sin \left[\dfrac{2\pi}{365}(t - 101) \right] + 14,$$

where t is in days, with $t = 0$ corresponding to January 1. For how many days during the year is the low temperature expected to be below -4°F?

c **69** *Temperature in Chicago* The average monthly high temperature T (in °F) in Chicago, Illinois, can be approximated using the function

$$T(t) = 26.5 \sin\left(\frac{\pi}{6}t - \frac{2\pi}{3}\right) + 56.5,$$

where t is in months and $t = 1$ corresponds to January.

(a) Graph T over the two-year interval $[1, 25]$.

(b) Calculate the average high temperature in July and in October.

(c) Graphically approximate the months when the average high temperature is 69°F or higher.

(d) Discuss why a sine function is an appropriate function to approximate these temperatures.

c **70** *Temperature in Augusta* The average monthly high temperature T (in °F) in Augusta, Georgia, can be approximated using the function

$$T(t) = 17 \cos\left(\frac{\pi}{6}t - \frac{7\pi}{6}\right) + 75,$$

where t is in months and $t = 1$ corresponds to January.

(a) Graph T over the two-year interval $[1, 25]$.

(b) Calculate the average high temperature in April and in December.

(c) Graphically approximate the months when the average high temperature is 67°F or lower.

71 *Intensity of sunlight* On a clear day with D hours of daylight, the intensity of sunlight I (in calories/cm²) may be approximated by

$$I = I_\text{M} \sin^3 \frac{\pi t}{D} \quad \text{for} \quad 0 \le t \le D,$$

where $t = 0$ corresponds to sunrise and I_M is the maximum intensity. If $D = 12$, approximately how many hours after sunrise is $I = \frac{1}{2}I_\text{M}$?

72 *Intensity of sunlight* Refer to Exercise 71. On cloudy days, a better approximation of the sun intensity I is given by

$$I = I_\text{M} \sin^2 \frac{\pi t}{D}.$$

If $D = 12$, approximately how many hours after sunrise is $I = \frac{1}{2}I_\text{M}$?

73 *Protection from sunlight* Refer to Exercises 71 and 72. A dermatologist recommends protection from the sun when the intensity I exceeds 75% of the maximum intensity. If $D = 12$ hours, approximate the number of hours for which protection is required on

(a) a clear day   (b) a cloudy day

74 *Highway engineering* In the study of frost penetration problems in highway engineering, the temperature T at time t hours and depth x feet is given by

$$T = T_0 e^{-\lambda x} \sin(\omega t - \lambda x),$$

where T_0, ω, and λ are constants and the period of T is 24 hours.

(a) Find a formula for the temperature at the surface.

(b) At what times is the surface temperature a minimum?

(c) If $\lambda = 2.5$, find the times when the temperature is a minimum at a depth of 1 foot.

75 *Rabbit population* Many animal populations, such as that of rabbits, fluctuate over ten-year cycles. Suppose that the number of rabbits at time t (in years) is given by

$$N(t) = 1000 \cos\frac{\pi}{5}t + 4000.$$

(a) Sketch the graph of N for $0 \le t \le 10$.

(b) For what values of t in part (a) does the rabbit population exceed 4500?

76 *River flow rate* The flow rate (or water discharge rate) at the mouth of the Orinoco River in South America may be approximated by

$$F(t) = 26,000 \sin\left[\frac{\pi}{6}(t - 5.5)\right] + 34,000,$$

where t is the time in months and $F(t)$ is the flow rate in m³/sec. For approximately how many months each year does the flow rate exceed 55,000 m³/sec?

77 Shown in the figure is a graph of $y = \frac{1}{2}x + \sin x$ for $-2\pi \le x \le 2\pi$. Using calculus, it can be shown that the x-coordinates of the turning points A, B, C, and D on the graph are solutions of the equation $\frac{1}{2} + \cos x = 0$. Determine the coordinates of these points.

Exercise 77

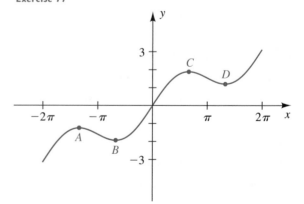

78 Shown in the figure is the graph of the equation

$$y = e^{-x/2} \sin 2x.$$

The x-coordinates of the turning points on the graph are solutions of $4 \cos 2x - \sin 2x = 0$. Approximate the x-coordinates of these points for $x > 0$.

Exercise 78

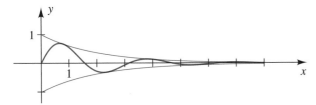

Exer. 79–80: If $I(t)$ is the current (in amperes) in an alternating current circuit at time t (in seconds), find the smallest exact value of t for which $I(t) = k$.

79 $I(t) = 20 \sin (60\pi t - 6\pi); \quad k = -10$

80 $I(t) = 40 \sin (100\pi t - 4\pi); \quad k = 20$

c **Exer. 81–84: Approximate the solution to each inequality on the interval $[0, 2\pi]$.**

81 $\cos x \geq 0.3$

82 $\sin x < -0.6$

83 $\cos 3x < \sin x$

84 $\tan x \leq \sin 2x$

c **Exer. 85–86: Graph f in the viewing rectangle $[0, 3]$ by $[-1.5, 1.5]$.**

(a) Approximate to within two decimal places the largest solution of $f(x) = 0$ on $[0, 3]$.

(b) Discuss what happens to the graph of f as x becomes large.

(c) Examine graphs of f on the interval $[0, c]$, where $c = 0.1, 0.01, 0.001$. How many zeros does f appear to have on the interval $[0, c]$, where $c > 0$?

85 $f(x) = \cos \dfrac{1}{x}$

86 $f(x) = \sin \dfrac{1}{x^2}$

c **Exer. 87–90: Because planets do not move in precisely circular orbits, the computation of the position of a planet requires the solution of Kepler's equation. Kepler's equation cannot be solved algebraically. It has the form $M = \theta + e \sin \theta$, where M is the mean anomaly, e is the eccentricity of the orbit, and θ is an angle called the eccentric anomaly. For the specified values of M and e, use graphical techniques to solve Kepler's equation for θ to three decimal places.**

87 *Position of Mercury* $M = 5.241, \qquad e = 0.206$

88 *Position of Mars* $M = 4.028, \qquad e = 0.093$

89 *Position of Earth* $M = 3.611, \qquad e = 0.0167$

90 *Position of Pluto* $M = 0.09424, \quad e = 0.255$

c **Exer. 91–96: Estimate the solutions of the equation in the interval $[-\pi, \pi]$.**

91 $\sin 2x = 2 - x^2$

92 $\cos^3 x + \cos 3x - \sin^3 x = 0$

93 $\ln (1 + \sin^2 x) = \cos x$

94 $e^{\sin x} = \sec \left(\tfrac{1}{3}x - \tfrac{1}{2}\right)$

95 $3 \cos^4 x - 2 \cos^3 x + \cos x - 1 = 0$

96 $\cos 2x + \sin 3x - \tan \tfrac{1}{3}x = 0$

97 *Weight at various latitudes*   The weight W of a person on the earth's surface is directly proportional to the force of gravity g (in m/sec^2). Because of rotation, the earth is flattened at the poles, and as a result weight will vary at different latitudes. If θ is the latitude, then g can be approximated by $g = 9.8066(1 - 0.00264 \cos 2\theta)$.

(a) At what latitude is $g = 9.8$?

(b) If a person weighs 150 pounds at the equator $(\theta = 0°)$, at what latitude will the person weigh 150.5 pounds?

3.3 THE ADDITION AND SUBTRACTION FORMULAS

In this section we derive formulas that involve trigonometric functions of $u + v$ or $u - v$ for any real numbers or angles u and v. These formulas are known as *addition* and *subtraction formulas*, respectively, or as *sum* and *difference identities.* The first formula that we will consider may be stated as follows.

Subtraction Formula for Cosine

$$\cos (u - v) = \cos u \cos v + \sin u \sin v$$

Figure 8

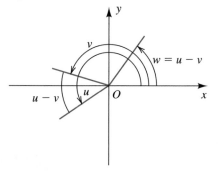

Proof   Let u and v be any real numbers, and consider angles of radian measure u and v. Let $w = u - v$. Figure 8 illustrates one possibility with the angles in standard position. For convenience we have assumed that both u and v are positive and that $0 \le u - v < v$.

As in Figure 9, let $P(u_1, u_2)$, $Q(v_1, v_2)$, and $R(w_1, w_2)$ be the points on the terminal sides of the indicated angles that are each a distance 1 from the origin. In this case P, Q, and R are on the unit circle U with center at the origin. From the definition of trigonometric functions in terms of a unit circle,

(∗)
$$\cos u = u_1 \qquad \cos v = v_1 \qquad \cos (u - v) = w_1$$
$$\sin u = u_2 \qquad \sin v = v_2 \qquad \sin (u - v) = w_2.$$

Figure 9

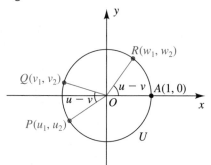

We next observe that the distance between $A(1, 0)$ and R must equal the distance between Q and P, because angles AOR and QOP have the same measure, $u - v$. Using the distance formula yields

$$d(A, R) = d(Q, P)$$
$$\sqrt{(w_1 - 1)^2 + (w_2 - 0)^2} = \sqrt{(u_1 - v_1)^2 + (u_2 - v_2)^2}.$$

Squaring both sides and simplifying the expressions under the radicals gives us

$$w_1^2 - 2w_1 + 1 + w_2^2 = u_1^2 - 2u_1v_1 + v_1^2 + u_2^2 - 2u_2v_2 + v_2^2.$$

Since the points (u_1, u_2), (v_1, v_2), and (w_1, w_2) are on the unit circle U and since an equation for U is $x^2 + y^2 = 1$, we may substitute 1 for each of $u_1^2 + u_2^2$, $v_1^2 + v_2^2$, and $w_1^2 + w_2^2$. Doing this and simplifying, we obtain

$$2 - 2w_1 = 2 - 2u_1v_1 - 2u_2v_2,$$

which reduces to

$$w_1 = u_1v_1 + u_2v_2.$$

Substituting from the formulas stated in (∗) gives us

$$\cos (u - v) = \cos u \cos v + \sin u \sin v,$$

which is what we wished to prove. It is possible to extend our discussion to all values of u and v. ◢

The next example demonstrates the use of the subtraction formula in finding the *exact* value of cos 15°. Of course, if only an approximation were desired, we could use a calculator.

EXAMPLE 1   Using a subtraction formula

Find the exact value of cos 15° by using the fact that $15° = 60° - 45°$.

Solution   We use the subtraction formula for cosine with $u = 60°$ and $v = 45°$:

$$\cos 15° = \cos (60° - 45°)$$
$$= \cos 60° \cos 45° + \sin 60° \sin 45°$$
$$= \frac{1}{2} \frac{\sqrt{2}}{2} + \frac{\sqrt{3}}{2} \frac{\sqrt{2}}{2}$$
$$= \frac{\sqrt{2} + \sqrt{6}}{4}$$

It is relatively easy to obtain a formula for cos $(u + v)$. We begin by writing $u + v$ as $u - (-v)$ and then use the subtraction formula for cosine:

$$\cos (u + v) = \cos [u - (-v)]$$
$$= \cos u \cos (-v) + \sin u \sin (-v)$$

Using the formulas for negatives, $\cos (-v) = \cos v$ and $\sin (-v) = -\sin v$, gives us the following addition formula for cosine.

Addition Formula for Cosine

$$\cos (u + v) = \cos u \cos v - \sin u \sin v$$

EXAMPLE 2   Using an addition formula

Find the exact value of $\cos \dfrac{7\pi}{12}$ by using the fact that $\dfrac{7\pi}{12} = \dfrac{\pi}{3} + \dfrac{\pi}{4}$.

Solution   We apply the addition formula for cosine:

$$\cos \frac{7\pi}{12} = \cos \left(\frac{\pi}{3} + \frac{\pi}{4} \right)$$
$$= \cos \frac{\pi}{3} \cos \frac{\pi}{4} - \sin \frac{\pi}{3} \sin \frac{\pi}{4}$$
$$= \frac{1}{2} \frac{\sqrt{2}}{2} - \frac{\sqrt{3}}{2} \frac{\sqrt{2}}{2}$$
$$= \frac{\sqrt{2} - \sqrt{6}}{4}$$

We refer to the sine and cosine functions as **cofunctions** of each other. Similarly, the tangent and cotangent functions are cofunctions, as are the

Figure 10

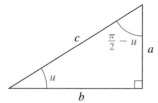

secant and cosecant. If u is the radian measure of an acute angle, then the angle with radian measure $\pi/2 - u$ is complementary to u, and we may consider the right triangle shown in Figure 10. Using ratios, we see that

$$\sin u = \frac{a}{c} = \cos\left(\frac{\pi}{2} - u\right)$$

$$\cos u = \frac{b}{c} = \sin\left(\frac{\pi}{2} - u\right)$$

$$\tan u = \frac{a}{b} = \cot\left(\frac{\pi}{2} - u\right).$$

These three formulas and their analogues for $\sec u$, $\csc u$, and $\cot u$ state that *the function value of u equals the cofunction of the complementary angle $\pi/2 - u$.*

In the following formulas we use subtraction formulas to extend these relationships to any real number u, provided the function values are defined.

Cofunction Formulas

> If u is a real number or the radian measure of an angle, then
>
> **(1)** $\cos\left(\dfrac{\pi}{2} - u\right) = \sin u$ **(2)** $\sin\left(\dfrac{\pi}{2} - u\right) = \cos u$
>
> **(3)** $\tan\left(\dfrac{\pi}{2} - u\right) = \cot u$ **(4)** $\cot\left(\dfrac{\pi}{2} - u\right) = \tan u$
>
> **(5)** $\sec\left(\dfrac{\pi}{2} - u\right) = \csc u$ **(6)** $\csc\left(\dfrac{\pi}{2} - u\right) = \sec u$

Proof   Using the subtraction formula for cosine, we have

$$\cos\left(\frac{\pi}{2} - u\right) = \cos\frac{\pi}{2}\cos u + \sin\frac{\pi}{2}\sin u$$

$$= (0)\cos u + (1)\sin u = \sin u.$$

This gives us formula 1.

If we substitute $\pi/2 - v$ for u in the first formula, we obtain

$$\cos\left[\frac{\pi}{2} - \left(\frac{\pi}{2} - v\right)\right] = \sin\left(\frac{\pi}{2} - v\right),$$

or $$\cos v = \sin\left(\frac{\pi}{2} - v\right).$$

Since the symbol v is arbitrary, this equation is equivalent to the second cofunction formula:

$$\sin\left(\frac{\pi}{2} - u\right) = \cos u$$

Using the tangent identity, cofunction formulas 1 and 2, and the cotangent identity, we obtain a proof for the third formula:

$$\tan\left(\frac{\pi}{2} - u\right) = \frac{\sin\left(\dfrac{\pi}{2} - u\right)}{\cos\left(\dfrac{\pi}{2} - u\right)} = \frac{\cos u}{\sin u} = \cot u.$$

The proofs of the remaining three formulas are similar.   ◢

An easy way to remember the cofunction formulas is to refer to the triangle in Figure 10.

We may now prove the following identities.

Addition and Subtraction Formulas for Sine and Tangent

> **(1)** $\sin(u + v) = \sin u \cos v + \cos u \sin v$
>
> **(2)** $\sin(u - v) = \sin u \cos v - \cos u \sin v$
>
> **(3)** $\tan(u + v) = \dfrac{\tan u + \tan v}{1 - \tan u \tan v}$
>
> **(4)** $\tan(u - v) = \dfrac{\tan u - \tan v}{1 + \tan u \tan v}$

Proof   We shall prove formulas 1 and 3. Using the cofunction formulas and the subtraction formula for cosine, we can verify formula 1:

$$\sin(u + v) = \cos\left[\frac{\pi}{2} - (u + v)\right]$$

$$= \cos\left[\left(\frac{\pi}{2} - u\right) - v\right]$$

$$= \cos\left(\frac{\pi}{2} - u\right)\cos v + \sin\left(\frac{\pi}{2} - u\right)\sin v$$

$$= \sin u \cos v + \cos u \sin v$$

To verify formula 3, we begin as follows:

$$\tan(u + v) = \frac{\sin(u + v)}{\cos(u + v)}$$

$$= \frac{\sin u \cos v + \cos u \sin v}{\cos u \cos v - \sin u \sin v}$$

If $\cos u \cos v \neq 0$, then we may divide the numerator and the denominator by $\cos u \cos v$, obtaining

$$\tan(u+v) = \frac{\left(\dfrac{\sin u}{\cos u}\right)\left(\dfrac{\cos v}{\cos v}\right) + \left(\dfrac{\cos u}{\cos u}\right)\left(\dfrac{\sin v}{\cos v}\right)}{\left(\dfrac{\cos u}{\cos u}\right)\left(\dfrac{\cos v}{\cos v}\right) - \left(\dfrac{\sin u}{\cos u}\right)\left(\dfrac{\sin v}{\cos v}\right)}$$

$$= \frac{\tan u + \tan v}{1 - \tan u \tan v}.$$

If $\cos u \cos v = 0$, then either $\cos u = 0$ or $\cos v = 0$. In this case, either $\tan u$ or $\tan v$ is undefined and the formula is invalid. Proofs of formulas 2 and 4 are left as exercises. ◢

EXAMPLE 3   **Using addition formulas to find the quadrant containing an angle**

Suppose $\sin \alpha = \frac{4}{5}$ and $\cos \beta = -\frac{12}{13}$, where α is in quadrant I and β is in quadrant II.

(a) Find the exact values of $\sin(\alpha + \beta)$ and $\tan(\alpha + \beta)$.

(b) Find the quadrant containing $\alpha + \beta$.

Figure 11

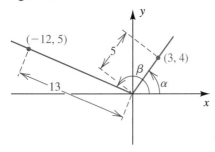

Solution   Angles α and β are illustrated in Figure 11. There is no loss of generality in regarding α and β as positive angles between 0 and 2π, as we have done in the figure. Since $\sin \alpha = \frac{4}{5}$, we may choose the point $(3, 4)$ on the terminal side of α. Similarly, since $\cos \beta = -\frac{12}{13}$, the point $(-12, 5)$ is on the terminal side of β. Referring to Figure 11 and using the theorem on trigonometric functions as ratios, we have

$$\cos \alpha = \tfrac{3}{5}, \quad \tan \alpha = \tfrac{4}{3}, \quad \sin \beta = \tfrac{5}{13}, \quad \tan \beta = -\tfrac{5}{12}.$$

(a) Addition formulas give us

$$\sin(\alpha + \beta) = \sin \alpha \cos \beta + \cos \alpha \sin \beta = \left(\tfrac{4}{5}\right)\left(-\tfrac{12}{13}\right) + \left(\tfrac{3}{5}\right)\left(\tfrac{5}{13}\right) = -\tfrac{33}{65}$$

$$\tan(\alpha + \beta) = \frac{\tan \alpha + \tan \beta}{1 - \tan \alpha \tan \beta} = \frac{\tfrac{4}{3} + \left(-\tfrac{5}{12}\right)}{1 - \left(\tfrac{4}{3}\right)\left(-\tfrac{5}{12}\right)} \cdot \frac{36}{36} = \frac{33}{56}.$$

(b) Since $\sin(\alpha + \beta)$ is negative and $\tan(\alpha + \beta)$ is positive, the angle $\alpha + \beta$ must be in quadrant III.

The next example illustrates a type of simplification of the difference quotient (introduced in Section 1.3) with the sine function. The resulting form is useful in calculus.

E X A M P L E   4   A formula used in calculus

If $f(x) = \sin x$ and $h \neq 0$, show that

$$\frac{f(x + h) - f(x)}{h} = \sin x \left(\frac{\cos h - 1}{h} \right) + \cos x \left(\frac{\sin h}{h} \right).$$

Solution   We use the definition of f and the addition formula for sine:

$$\frac{f(x + h) - f(x)}{h} = \frac{\sin (x + h) - \sin x}{h}$$

$$= \frac{\sin x \cos h + \cos x \sin h - \sin x}{h}$$

$$= \frac{\sin x (\cos h - 1) + \cos x \sin h}{h}$$

$$= \sin x \left(\frac{\cos h - 1}{h} \right) + \cos x \left(\frac{\sin h}{h} \right)$$

Addition formulas may also be used to derive **reduction formulas.** Reduction formulas may be used to change expressions such as

$$\sin \left(\theta + \frac{\pi}{2} n \right) \quad \text{and} \quad \cos \left(\theta + \frac{\pi}{2} n \right) \quad \text{for any integer } n$$

to expressions involving only $\sin \theta$ or $\cos \theta$. Similar formulas are true for the other trigonometric functions. Instead of deriving general reduction formulas, we shall illustrate two special cases in the next example.

E X A M P L E   5   Obtaining reduction formulas

Express in terms of a trigonometric function of θ alone:

(a) $\sin \left(\theta - \dfrac{3\pi}{2} \right)$ **(b)** $\cos (\theta + \pi)$

Solution   Using subtraction and addition formulas, we obtain the following:

(a) $\sin \left(\theta - \dfrac{3\pi}{2} \right) = \sin \theta \cos \dfrac{3\pi}{2} - \cos \theta \sin \dfrac{3\pi}{2}$

$$= \sin \theta \cdot (0) - \cos \theta \cdot (-1) = \cos \theta$$

(b) $\cos (\theta + \pi) = \cos \theta \cos \pi - \sin \theta \sin \pi$

$$= \cos \theta \cdot (-1) - \sin \theta \cdot (0) = -\cos \theta$$

EXAMPLE 6   **Combining a sum involving the sine and cosine functions**

Let a and b be real numbers with $a > 0$. Show that for every x,

$$a \cos Bx + b \sin Bx = A \cos (Bx - C),$$

where $A = \sqrt{a^2 + b^2}$ and $\tan C = \dfrac{b}{a}$ with $-\dfrac{\pi}{2} < C < \dfrac{\pi}{2}$.

Solution   Given $a \cos Bx + b \sin Bx$, let us consider $\tan C = b/a$ with $-\pi/2 < C < \pi/2$. Thus, $b = a \tan C$, and we may write

$$a \cos Bx + b \sin Bx = a \cos Bx + (a \tan C) \sin Bx$$

$$= a \cos Bx + a \frac{\sin C}{\cos C} \sin Bx$$

$$= \frac{a}{\cos C}(\cos C \cos Bx + \sin C \sin Bx)$$

$$= (a \sec C) \cos (Bx - C).$$

We shall complete the proof by showing that $a \sec C = \sqrt{a^2 + b^2}$. Since $-\pi/2 < C < \pi/2$, it follows that $\sec C$ is positive, and hence

$$a \sec C = a\sqrt{1 + \tan^2 C}.$$

Using $\tan C = b/a$ and $a > 0$, we obtain

$$a \sec C = a\sqrt{1 + \frac{b^2}{a^2}} = \sqrt{a^2\left(1 + \frac{b^2}{a^2}\right)} = \sqrt{a^2 + b^2}.$$

EXAMPLE 7   **An application of Example 6**

If $f(x) = \cos x + \sin x$, use the formulas given in Example 6 to express $f(x)$ in the form $A \cos (Bx - C)$, and then sketch the graph of f.

Solution   Letting $a = 1$, $b = 1$, and $B = 1$ in the formulas from Example 6, we have

$$A = \sqrt{a^2 + b^2} = \sqrt{1 + 1} = \sqrt{2} \quad \text{and} \quad \tan C = \frac{b}{a} = \frac{1}{1} = 1.$$

Since $\tan C = 1$ and $-\pi/2 < C < \pi/2$, we have $C = \pi/4$. Substituting for a, b, A, B, and C in the formula

$$a \cos Bx + b \sin Bx = A \cos (Bx - C)$$

gives us

$$f(x) = \cos x + \sin x = \sqrt{2} \cos \left(x - \frac{\pi}{4}\right).$$

Comparing the last formula with the equation $y = a \cos (bx + c)$, which we discussed in Section 2.6, we see that the amplitude of the graph is $\sqrt{2}$, the period is 2π, and the phase shift is $\pi/4$. The graph of f is sketched in Figure 12, where we have also shown the graphs of $y = \sin x$ and $y = \cos x$. Our sketch agrees with that obtained in Section 2.7 using a graphing utility. (See Figure 72 in Chapter 2.)

Figure 12

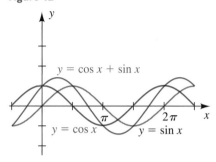

3.3 EXERCISES

Exer. 1–4: Express as a cofunction of a complementary angle.

1 (a) $\sin 46°37'$ (b) $\cos 73°12'$

   (c) $\tan \dfrac{\pi}{6}$ (d) $\sec 17.28°$

2 (a) $\tan 24°12'$ (b) $\sin 89°41'$

   (c) $\cos \dfrac{\pi}{3}$ (d) $\cot 61.87°$

3 (a) $\cos \dfrac{7\pi}{20}$ (b) $\sin \dfrac{1}{4}$

   (c) $\tan 1$ (d) $\csc 0.53$

4 (a) $\sin \dfrac{\pi}{12}$ (b) $\cos 0.64$

   (c) $\tan \sqrt{2}$ (d) $\sec 1.2$

Exer. 5–10: Find the exact values.

5 (a) $\cos \dfrac{\pi}{4} + \cos \dfrac{\pi}{6}$

   (b) $\cos \dfrac{5\pi}{12}$ $\left(\text{use } \dfrac{5\pi}{12} = \dfrac{\pi}{4} + \dfrac{\pi}{6}\right)$

6 (a) $\sin \dfrac{2\pi}{3} + \sin \dfrac{\pi}{4}$

   (b) $\sin \dfrac{11\pi}{12}$ $\left(\text{use } \dfrac{11\pi}{12} = \dfrac{2\pi}{3} + \dfrac{\pi}{4}\right)$

7 (a) $\tan 60° + \tan 225°$

   (b) $\tan 285°$ (use $285° = 60° + 225°$)

8 (a) $\cos 135° - \cos 60°$

   (b) $\cos 75°$ (use $75° = 135° - 60°$)

9 (a) $\sin \dfrac{3\pi}{4} - \sin \dfrac{\pi}{6}$

   (b) $\sin \dfrac{7\pi}{12}$ $\left(\text{use } \dfrac{7\pi}{12} = \dfrac{3\pi}{4} - \dfrac{\pi}{6}\right)$

10 (a) $\tan \dfrac{3\pi}{4} - \tan \dfrac{\pi}{6}$

   (b) $\tan \dfrac{7\pi}{12}$ $\left(\text{use } \dfrac{7\pi}{12} = \dfrac{3\pi}{4} - \dfrac{\pi}{6}\right)$

Exer. 11–16: Express as a trigonometric function of one angle.

11 $\cos 48° \cos 23° + \sin 48° \sin 23°$

12 $\cos 13° \cos 50° - \sin 13° \sin 50°$

13 $\cos 10° \sin 5° - \sin 10° \cos 5°$

14 $\sin 57° \cos 4° + \cos 57° \sin 4°$

15 $\cos 3 \sin (-2) - \cos 2 \sin 3$

16 $\sin (-5) \cos 2 + \cos 5 \sin (-2)$

17 If α and β are acute angles such that $\cos \alpha = \frac{4}{5}$ and $\tan \beta = \frac{8}{15}$, find
   (a) $\sin (\alpha + \beta)$ (b) $\cos (\alpha + \beta)$
   (c) the quadrant containing $\alpha + \beta$

18 If α and β are acute angles such that $\csc \alpha = \frac{13}{12}$ and $\cot \beta = \frac{4}{3}$, find
   (a) $\sin (\alpha + \beta)$ (b) $\tan (\alpha + \beta)$
   (c) the quadrant containing $\alpha + \beta$

19 If $\sin \alpha = -\frac{4}{5}$ and $\sec \beta = \frac{5}{3}$ for a third-quadrant angle α and a first-quadrant angle β, find
   (a) $\sin (\alpha + \beta)$ (b) $\tan (\alpha + \beta)$
   (c) the quadrant containing $\alpha + \beta$

20 If $\tan \alpha = -\frac{7}{24}$ and $\cot \beta = \frac{3}{4}$ for a second-quadrant angle α and a third-quadrant angle β, find
   (a) $\sin (\alpha + \beta)$ (b) $\cos (\alpha + \beta)$ (c) $\tan (\alpha + \beta)$
   (d) $\sin (\alpha - \beta)$ (e) $\cos (\alpha - \beta)$ (f) $\tan (\alpha - \beta)$

21 If α and β are third-quadrant angles such that $\cos \alpha = -\frac{2}{5}$ and $\cos \beta = -\frac{3}{5}$, find
   (a) $\sin (\alpha - \beta)$ (b) $\cos (\alpha - \beta)$
   (c) the quadrant containing $\alpha - \beta$

22 If α and β are second-quadrant angles such that $\sin \alpha = \frac{2}{3}$ and $\cos \beta = -\frac{1}{3}$, find
   (a) $\sin (\alpha + \beta)$ (b) $\tan (\alpha + \beta)$
   (c) the quadrant containing $\alpha + \beta$

Exer. 23–34: Verify the reduction formula.

23 $\sin (\theta + \pi) = -\sin \theta$ 24 $\sin \left(x + \frac{\pi}{2} \right) = \cos x$

25 $\sin \left(x - \frac{5\pi}{2} \right) = -\cos x$ 26 $\sin \left(\theta - \frac{3\pi}{2} \right) = \cos \theta$

27 $\cos (\theta - \pi) = -\cos \theta$ 28 $\cos \left(x + \frac{\pi}{2} \right) = -\sin x$

29 $\cos \left(x + \frac{3\pi}{2} \right) = \sin x$ 30 $\cos \left(\theta - \frac{5\pi}{2} \right) = \sin \theta$

31 $\tan \left(x - \frac{\pi}{2} \right) = -\cot x$ 32 $\tan (\pi - \theta) = -\tan \theta$

33 $\tan \left(\theta + \frac{\pi}{2} \right) = -\cot \theta$ 34 $\tan (x + \pi) = \tan x$

Exer. 35–44: Verify the identity.

35 $\sin \left(\theta + \frac{\pi}{4} \right) = \frac{\sqrt{2}}{2} (\sin \theta + \cos \theta)$

36 $\cos \left(\theta + \frac{\pi}{4} \right) = \frac{\sqrt{2}}{2} (\cos \theta - \sin \theta)$

37 $\tan \left(u + \frac{\pi}{4} \right) = \frac{1 + \tan u}{1 - \tan u}$

38 $\tan \left(x - \frac{\pi}{4} \right) = \frac{\tan x - 1}{\tan x + 1}$

39 $\cos (u + v) + \cos (u - v) = 2 \cos u \cos v$

40 $\sin (u + v) + \sin (u - v) = 2 \sin u \cos v$

41 $\sin (u + v) \cdot \sin (u - v) = \sin^2 u - \sin^2 v$

42 $\cos (u + v) \cdot \cos (u - v) = \cos^2 u - \sin^2 v$

43 $\dfrac{1}{\cot \alpha - \cot \beta} = \dfrac{\sin \alpha \sin \beta}{\sin (\beta - \alpha)}$

44 $\dfrac{1}{\tan \alpha + \tan \beta} = \dfrac{\cos \alpha \cos \beta}{\sin (\alpha + \beta)}$

45 Express $\sin (u + v + w)$ in terms of trigonometric functions of u, v, and w. (*Hint:* Write
$$\sin (u + v + w) \quad \text{as} \quad \sin [(u + v) + w]$$
and use addition formulas.)

46 Express $\tan (u + v + w)$ in terms of trigonometric functions of u, v, and w.

47 Derive the formula $\cot (u + v) = \dfrac{\cot u \cot v - 1}{\cot u + \cot v}$.

48 If α and β are complementary angles, show that
$$\sin^2 \alpha + \sin^2 \beta = 1.$$

49 Derive the subtraction formula for the sine function.

50 Derive the subtraction formula for the tangent function.

51 If $f(x) = \cos x$, show that

$$\frac{f(x + h) - f(x)}{h} = \cos x \left(\frac{\cos h - 1}{h}\right) - \sin x \left(\frac{\sin h}{h}\right).$$

52 If $f(x) = \tan x$, show that

$$\frac{f(x + h) - f(x)}{h} = \sec^2 x \left(\frac{\sin h}{h}\right) \frac{1}{\cos h - \sin h \tan x}.$$

Exer. 53–58: Use an addition or subtraction formula to find the solutions of the equation that are in the interval $[0, \pi)$.

53 $\sin 4t \cos t = \sin t \cos 4t$

54 $\cos 5t \cos 3t = \frac{1}{2} + \sin(-5t) \sin 3t$

55 $\cos 5t \cos 2t = -\sin 5t \sin 2t$

56 $\sin 3t \cos t + \cos 3t \sin t = -\frac{1}{2}$

57 $\tan 2t + \tan t = 1 - \tan 2t \tan t$

58 $\tan t - \tan 4t = 1 + \tan 4t \tan t$

Exer. 59–62: (a) Use the formula from Example 6 to express f in terms of the cosine function. (b) Determine the amplitude, period, and phase shift of f. (c) Sketch the graph of f.

59 $f(x) = \sqrt{3} \cos 2x + \sin 2x$

60 $f(x) = \cos 4x + \sqrt{3} \sin 4x$

61 $f(x) = 2 \cos 3x - 2 \sin 3x$

62 $f(x) = 5 \cos 10x - 5 \sin 10x$

Exer. 63–64: For certain applications in electrical engineering, the sum of several voltage signals or radio waves of the same frequency is expressed in the compact form $y = A \cos(Bt - C)$. Express the given signal in this form.

63 $y = 50 \sin 60\pi t + 40 \cos 60\pi t$

64 $y = 10 \sin\left(120\pi t - \frac{\pi}{2}\right) + 5 \sin 120\pi t$

65 *Motion of a mass* If a mass that is attached to a spring is raised y_0 feet and released with an initial vertical velocity of v_0 ft/sec, then the subsequent position y of the mass is given by

$$y = y_0 \cos \omega t + \frac{v_0}{\omega} \sin \omega t,$$

where t is time in seconds and ω is a positive constant.

(a) If $\omega = 1$, $y_0 = 2$ ft, and $v_0 = 3$ ft/sec, express y in the form $A \cos(Bt - C)$, and find the amplitude and period of the resulting motion.

(b) Determine the times when $y = 0$—that is, the times when the mass passes through the equilibrium position.

66 *Motion of a mass* Refer to Exercise 65. If $y_0 = 1$ and $\omega = 2$, find the initial velocities that result in an amplitude of 4 feet.

67 *Pressure on the eardrum* If a tuning fork is struck and then held a certain distance from the eardrum, the pressure $p_1(t)$ on the outside of the eardrum at time t may be represented by $p_1(t) = A \sin \omega t$, where A and ω are positive constants. If a second identical tuning fork is struck with a possibly different force and held a different distance from the eardrum (see the figure), its effect may be represented by $p_2(t) = B \sin(\omega t + \tau)$, where B is a positive constant and $0 \le \tau \le 2\pi$. The total pressure $p(t)$ on the eardrum is given by

$$p(t) = A \sin \omega t + B \sin(\omega t + \tau).$$

(a) Show that $p(t) = a \cos \omega t + b \sin \omega t$, where

$$a = B \sin \tau \quad \text{and} \quad b = A + B \cos \tau.$$

(b) Show that the amplitude C of p is given by

$$C^2 = A^2 + B^2 + 2AB \cos \tau.$$

Exercise 67

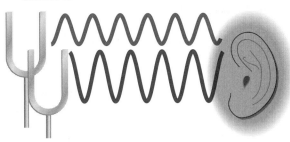

68 *Destructive interference* Refer to Exercise 67. Destructive interference occurs if the amplitude of the resulting sound wave is less than A. Suppose that the two tuning forks are struck with the same force—that is, $A = B$.

(a) When total destructive interference occurs, the amplitude of p is zero and no sound is heard. Find the least positive value of τ for which this occurs.

(b) Determine the τ-interval (a, b) for which destructive interference occurs and a has its least positive value.

69 Constructive interference Refer to Exercise 67. When two tuning forks are struck, constructive interference occurs if the amplitude C of the resulting sound wave is larger than either A or B (see the figure).

(a) Show that $C \leq A + B$.

(b) Find the values of τ such that $C = A + B$.

(c) If $A \geq B$, determine a condition under which constructive interference will occur.

Exercise 69

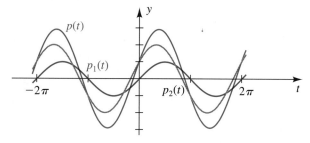

c 70 Pressure on the eardrum Refer to Exercise 67. If two tuning forks with different pitches are struck simultaneously with different forces, then the total pressure $p(t)$ on the eardrum at time t is given by

$$p(t) = p_1(t) + p_2(t) = A \sin \omega_1 t + B \sin (\omega_2 t + \tau),$$

where A, B, ω_1, ω_2, and τ are constants.

(a) Graph p for $-2\pi \leq t \leq 2\pi$ if $A = B = 2$, $\omega_1 = 1$, $\omega_2 = 20$, and $\tau = 3$.

(b) Use the graph to describe the variation of the tone that is produced.

c Exer. 71–72: Refer to Exercise 69. Graph the equation for $-\pi \leq t \leq \pi$, and estimate the intervals on which constructive interference occurs.

71 $y = 3 \sin 2t + 2 \sin (4t + 1)$

72 $y = 2 \sin t + 2 \sin (3t + 3)$

3.4 MULTIPLE-ANGLE FORMULAS

We refer to the formulas considered in this section as **multiple-angle formulas.** In particular, the following identities are **double-angle formulas,** because they contain the expression $2u$.

Double-Angle Formulas

$$
\begin{aligned}
\textbf{(1)} \quad & \underline{\sin 2u = 2 \sin u \cos u} \\
\textbf{(2)} \quad & \cos 2u = \cos^2 u - \sin^2 u \\
& \qquad\quad = 1 - 2 \sin^2 u \\
& \qquad\quad = 2 \cos^2 u - 1 \\
\textbf{(3)} \quad & \tan 2u = \frac{2 \tan u}{1 - \tan^2 u}
\end{aligned}
$$

Proof Each of these formulas may be proved by letting $v = u$ in the appropriate addition formulas. If we use the formula for $\sin (u + v)$, then

$$
\begin{aligned}
\sin 2u &= \sin (u + u) \\
&= \sin u \cos u + \cos u \sin u \\
&= 2 \sin u \cos u.
\end{aligned}
$$

Using the formula for $\cos (u + v)$, we have

$$
\begin{aligned}
\cos 2u &= \cos (u + u) \\
&= \cos u \cos u - \sin u \sin u \\
&= \cos^2 u - \sin^2 u.
\end{aligned}
$$

To obtain the other two forms for cos $2u$, we use the fundamental identity $\sin^2 u + \cos^2 u = 1$. Thus,

$$\cos 2u = \cos^2 u - \sin^2 u$$
$$= (1 - \sin^2 u) - \sin^2 u$$
$$= 1 - 2 \sin^2 u.$$

Similarly, if we substitute for $\sin^2 u$ instead of $\cos^2 u$, we obtain

$$\cos 2u = \cos^2 u - (1 - \cos^2 u)$$
$$= 2 \cos^2 u - 1.$$

Formula 3 for tan $2u$ may be obtained by letting $v = u$ in the formula for tan $(u + v)$.   ◀

EXAMPLE 1   Using double-angle formulas

If $\sin \alpha = \frac{4}{5}$ and α is an acute angle, find the exact values of sin 2α and cos 2α.

Solution   If we regard α as an acute angle of a right triangle, as shown in Figure 13, we obtain $\cos \alpha = \frac{3}{5}$. We next substitute in double-angle formulas:

$$\sin 2\alpha = 2 \sin \alpha \cos \alpha = 2\left(\tfrac{4}{5}\right)\left(\tfrac{3}{5}\right) = \tfrac{24}{25}$$
$$\cos 2\alpha = \cos^2 \alpha - \sin^2 \alpha = \left(\tfrac{3}{5}\right)^2 - \left(\tfrac{4}{5}\right)^2 = \tfrac{9}{25} - \tfrac{16}{25} = -\tfrac{7}{25}$$

Figure 13

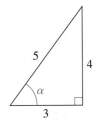

The next example demonstrates how to change a multiple-angle expression to a single-angle expression.

EXAMPLE 2   Changing the form of cos 3θ

Express cos 3θ in terms of cos θ.

Solution

$\cos 3\theta = \cos (2\theta + \theta)$	$3\theta = 2\theta + \theta$
$= \cos 2\theta \cos \theta - \sin 2\theta \sin \theta$	addition formula
$= (2 \cos^2 \theta - 1) \cos \theta - (2 \sin \theta \cos \theta) \sin \theta$	double-angle formulas
$= 2 \cos^3 \theta - \cos \theta - 2 \cos \theta \sin^2 \theta$	multiply
$= 2 \cos^3 \theta - \cos \theta - 2 \cos \theta (1 - \cos^2 \theta)$	$\sin^2 \theta + \cos^2 \theta = 1$
$= 4 \cos^3 \theta - 3 \cos \theta$	simplify

We call each of the next three formulas a **half-angle identity,** because the number u is one-half the number $2u$.

Half-Angle Identities

$$(1) \ \sin^2 u = \frac{1 - \cos 2u}{2} \qquad (2) \ \cos^2 u = \frac{1 + \cos 2u}{2}$$

$$(3) \ \tan^2 u = \frac{1 - \cos 2u}{1 + \cos 2u}$$

Proof The first identity may be verified as follows:

$$\cos 2u = 1 - 2 \sin^2 u \qquad \text{double-angle formula}$$
$$2 \sin^2 u = 1 - \cos 2u \qquad \text{isolate } 2 \sin^2 u$$
$$\sin^2 u = \frac{1 - \cos 2u}{2} \qquad \text{divide by 2}$$

The second identity can be derived in similar fashion by starting with

$$\cos 2u = 2 \cos^2 u - 1.$$

The third identity can be obtained from identities 1 and 2 by noting that

$$\tan^2 u = (\tan u)^2 = \left(\frac{\sin u}{\cos u}\right)^2 = \frac{\sin^2 u}{\cos^2 u}.$$

◢

Half-angle identities may be used to express even powers of trigonometric functions in terms of functions with exponent 1, as illustrated in the next two examples.

EXAMPLE 3   **Using half-angle identities to verify an identity**
Verify the identity $\sin^2 x \cos^2 x = \frac{1}{8}(1 - \cos 4x)$.

Solution

$$\sin^2 x \cos^2 x = \left(\frac{1 - \cos 2x}{2}\right)\left(\frac{1 + \cos 2x}{2}\right) \qquad \text{half-angle identities}$$
$$= \tfrac{1}{4}(1 - \cos^2 2x) \qquad \text{multiply}$$
$$= \tfrac{1}{4}(\sin^2 2x) \qquad \sin^2 2x + \cos^2 2x = 1$$
$$= \frac{1}{4}\left(\frac{1 - \cos 4x}{2}\right) \qquad \begin{array}{l}\text{half-angle identity}\\ \text{with } u = 2x\end{array}$$
$$= \tfrac{1}{8}(1 - \cos 4x) \qquad \text{multiply}$$

E X A M P L E 4   Using half-angle identities
to reduce a power of cos t

Express $\cos^4 t$ in terms of values of the cosine function with exponent 1.

Solution

$$\cos^4 t = (\cos^2 t)^2 \qquad\qquad\qquad \text{law of exponents}$$

$$= \left(\frac{1 + \cos 2t}{2}\right)^2 \qquad\qquad \text{half-angle identity}$$

$$= \tfrac{1}{4}(1 + 2\cos 2t + \cos^2 2t) \qquad \text{square}$$

$$= \frac{1}{4}\left(1 + 2\cos 2t + \frac{1 + \cos 4t}{2}\right) \qquad \text{half-angle identity with } u = 2t$$

$$= \tfrac{3}{8} + \tfrac{1}{2}\cos 2t + \tfrac{1}{8}\cos 4t \qquad\qquad \text{simplify}$$

Substituting $v/2$ for u in the three half-angle identities gives us

$$\sin^2\frac{v}{2} = \frac{1 - \cos v}{2} \qquad \cos^2\frac{v}{2} = \frac{1 + \cos v}{2} \qquad \tan^2\frac{v}{2} = \frac{1 - \cos v}{1 + \cos v}.$$

Taking the square roots of both sides of each of these equations, we obtain the following, which we call the *half-angle formulas* in order to distinguish them from the half-angle identities.

Half-Angle Formulas

$$\textbf{(1)} \ \sin\frac{v}{2} = \pm\sqrt{\frac{1 - \cos v}{2}} \qquad \textbf{(2)} \ \cos\frac{v}{2} = \pm\sqrt{\frac{1 + \cos v}{2}}$$

$$\textbf{(3)} \ \tan\frac{v}{2} = \pm\sqrt{\frac{1 - \cos v}{1 + \cos v}}$$

When using a half-angle formula, we choose either the + or the −, depending on the quadrant containing the angle of radian measure v/2. Thus, for sin (v/2) we use + if v/2 is an angle in quadrant I or II and − if v/2 is in quadrant III or IV. For cos (v/2) we use + if v/2 is in quadrant I or IV, and so on.

E X A M P L E 5   Using half-angle formulas for the sine and cosine

Find exact values for

(a) $\sin 22.5°$ **(b)** $\cos 112.5°$

Solution

(a) We choose the positive sign because $22.5°$ is in quadrant I, and hence $\sin 22.5° > 0$.

$$\sin 22.5° = +\sqrt{\frac{1 - \cos 45°}{2}} \qquad \text{half-angle formula for sine with } v = 45°$$

$$= \sqrt{\frac{1 - \sqrt{2}/2}{2}} \qquad \cos 45° = \frac{\sqrt{2}}{2}$$

$$= \frac{\sqrt{2 - \sqrt{2}}}{2} \qquad \text{multiply radicand by } \frac{2}{2} \text{ and simplify}$$

(b) Similarly, we choose the negative sign because $112.5°$ is in quadrant II, and so $\cos 112.5° < 0$.

$$\cos 112.5° = -\sqrt{\frac{1 + \cos 225°}{2}} \qquad \text{half-angle formula for cosine with } v = 225°$$

$$= -\sqrt{\frac{1 - \sqrt{2}/2}{2}} \qquad \cos 225° = -\frac{\sqrt{2}}{2}$$

$$= -\frac{\sqrt{2 - \sqrt{2}}}{2} \qquad \text{multiply radicand by } \frac{2}{2} \text{ and simplify}$$

We can obtain an alternative form for the half-angle formula for $\tan (v/2)$. Multiplying the numerator and denominator of the radicand in the third half-angle formula by $1 - \cos v$ gives us

$$\tan \frac{v}{2} = \pm\sqrt{\frac{1 - \cos v}{1 + \cos v} \cdot \frac{1 - \cos v}{1 - \cos v}}$$

$$= \pm\sqrt{\frac{(1 - \cos v)^2}{1 - \cos^2 v}}$$

$$= \pm\sqrt{\frac{(1 - \cos v)^2}{\sin^2 v}} = \pm\frac{1 - \cos v}{\sin v}.$$

We can eliminate the $\pm$ sign in the preceding formula. First note that the numerator $1 - \cos v$ is never negative. We can show that $\tan (v/2)$ and $\sin v$ always have the same sign. For example, if $0 < v < \pi$, then $0 < v/2 < \pi/2$, and consequently both $\sin v$ and $\tan (v/2)$ are positive. If $\pi < v < 2\pi$, then $\pi/2 < v/2 < \pi$, and hence both $\sin v$ and $\tan (v/2)$ are negative, which gives us the first of the next two identities. The second identity for $\tan (v/2)$ may be obtained by multiplying the numerator and denominator of the radicand in the third half-angle formula by $1 + \cos v$.

Half-Angle Formulas for the Tangent

$$\textbf{(1)} \ \tan\frac{v}{2} = \frac{1-\cos v}{\sin v} \qquad \textbf{(2)} \ \tan\frac{v}{2} = \frac{\sin v}{1+\cos v}$$

Figure 14

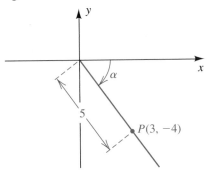

EXAMPLE 6   Using a half-angle formula for the tangent

If $\tan\alpha = -\dfrac{4}{3}$ and α is in quadrant IV, find $\tan\dfrac{\alpha}{2}$.

Solution   If we choose the point $(3, -4)$ on the terminal side of α, as illustrated in Figure 14, then $\sin\alpha = -\frac{4}{5}$ and $\cos\alpha = \frac{3}{5}$. Applying the first half-angle formula for the tangent, we obtain

$$\tan\frac{\alpha}{2} = \frac{1-\cos\alpha}{\sin\alpha} = \frac{1-\frac{3}{5}}{-\frac{4}{5}} = -\frac{1}{2}.$$

EXAMPLE 7   Finding the x-intercepts of a graph

Figure 15

A graph of the equation $y = \cos 2x + \cos x$ for $0 \le x \le 2\pi$ is sketched in Figure 15. The x-intercepts appear to be approximately 1.1, 3.1, and 5.2. Find their exact values and three-decimal-place approximations.

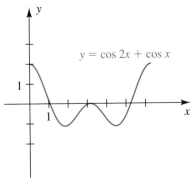

Solution   To find the x-intercepts, we proceed as follows:

$$\cos 2x + \cos x = 0 \qquad \text{let } y = 0$$
$$(2\cos^2 x - 1) + \cos x = 0 \qquad \text{double-angle formula}$$
$$2\cos^2 x + \cos x - 1 = 0 \qquad \text{equivalent equation}$$
$$(2\cos x - 1)(\cos x + 1) = 0 \qquad \text{factor}$$
$$2\cos x - 1 = 0, \quad \cos x + 1 = 0 \qquad \text{zero factor theorem}$$
$$\cos x = \tfrac{1}{2}, \qquad \cos x = -1 \qquad \text{solve for } \cos x$$

The solutions of the last two equations in the interval $[0, 2\pi]$ give us the following exact and approximate x-intercepts:

$$\frac{\pi}{3} \approx 1.047, \quad \frac{5\pi}{3} \approx 5.236, \quad \pi \approx 3.142$$

EXAMPLE 8   Deriving a formula for the area of an isosceles triangle

Figure 16

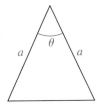

An **isosceles triangle** has two equal sides of length a, and the angle between them is θ (see Figure 16). Express the area A of the triangle in terms of a and θ.

Figure 17

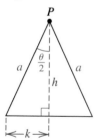

Solution   From Figure 17 we see that the altitude from point P bisects θ and that $A = \frac{1}{2}(2k)h = kh$. Thus, we have the following, where $\theta/2$ is an acute angle:

$$\sin \frac{\theta}{2} = \frac{k}{a} \qquad\qquad \cos \frac{\theta}{2} = \frac{h}{a} \qquad \text{see Figure 16}$$

$$k = a \sin \frac{\theta}{2} \qquad\qquad h = a \cos \frac{\theta}{2} \qquad \text{solve for } k \text{ and } h$$

We next find the area:

$$(*) \qquad A = a^2 \sin \frac{\theta}{2} \cos \frac{\theta}{2} \qquad\qquad \text{substitute in } A = kh$$

$$= a^2 \sqrt{\frac{1 - \cos \theta}{2}} \sqrt{\frac{1 + \cos \theta}{2}} \qquad \begin{array}{l}\text{half-angle formulas with} \\ \theta/2 \text{ in quadrant I}\end{array}$$

$$= a^2 \sqrt{\frac{1 - \cos^2 \theta}{4}} \qquad\qquad \text{law of radicals}$$

$$= a^2 \sqrt{\frac{\sin^2 \theta}{4}} \qquad\qquad \sin^2 \theta + \cos^2 \theta = 1$$

$$= \tfrac{1}{2} a^2 \, |\sin \theta| \qquad\qquad \text{take the square root}$$

$$= \tfrac{1}{2} a^2 \sin \theta \qquad\qquad \sin \theta > 0 \text{ for } 0° < \theta < 180°$$

Another method for simplifying $(*)$ is to write the double-angle formula for the sine, $\sin 2u = 2 \sin u \cos u$, as

$$(**) \qquad\qquad \sin u \cos u = \tfrac{1}{2} \sin 2u$$

and proceed as follows:

$$A = a^2 \sin \frac{\theta}{2} \cos \frac{\theta}{2} \qquad\qquad \text{substitute in } A = kh$$

$$= a^2 \cdot \frac{1}{2} \sin \left(2 \cdot \frac{\theta}{2} \right) \qquad \text{let } u = \frac{\theta}{2} \text{ in } (**)$$

$$= \tfrac{1}{2} a^2 \sin \theta \qquad\qquad \text{simplify}$$

3.4 EXERCISES

Exer. 1–4: Find the exact values of sin 2θ, cos 2θ, and tan 2θ for the given values of θ.

1 $\cos \theta = \frac{3}{5}$; $0° < \theta < 90°$

2 $\cot \theta = \frac{4}{3}$; $180° < \theta < 270°$

3 $\sec \theta = -3$; $90° < \theta < 180°$

4 $\sin \theta = -\frac{4}{5}$; $270° < \theta < 360°$

Exer. 5–8: Find the exact values of sin (θ/2), cos (θ/2), and tan (θ/2) for the given conditions.

5 $\sec \theta = \frac{5}{4}$; $0° < \theta < 90°$

6 $\csc \theta = -\frac{5}{3}$; $-90° < \theta < 0°$

7 $\tan \theta = 1$; $\quad -180° < \theta < -90°$

8 $\sec \theta = -4$; $\quad 180° < \theta < 270°$

Exer. 9–10: Use half-angle formulas to find the exact values.

9 (a) $\cos 67°30'$ (b) $\sin 15°$ (c) $\tan \dfrac{3\pi}{8}$

10 (a) $\cos 165°$ (b) $\sin 157°30'$ (c) $\tan \dfrac{\pi}{8}$

Exer. 11–28: Verify the identity.

11 $\sin 10\theta = 2 \sin 5\theta \cos 5\theta$

12 $\cos^2 3x - \sin^2 3x = \cos 6x$

13 $4 \sin \dfrac{x}{2} \cos \dfrac{x}{2} = 2 \sin x$

14 $\dfrac{\sin^2 2\alpha}{\sin^2 \alpha} = 4 - 4 \sin^2 \alpha$

15 $(\sin t + \cos t)^2 = 1 + \sin 2t$

16 $\csc 2u = \tfrac{1}{2} \csc u \sec u$

17 $\sin 3u = \sin u \, (3 - 4 \sin^2 u)$

18 $\sin 4t = 4 \sin t \cos t \, (1 - 2 \sin^2 t)$

19 $\cos 4\theta = 8 \cos^4 \theta - 8 \cos^2 \theta + 1$

20 $\cos 6t = 32 \cos^6 t - 48 \cos^4 t + 18 \cos^2 t - 1$

21 $\sin^4 t = \tfrac{3}{8} - \tfrac{1}{2} \cos 2t + \tfrac{1}{8} \cos 4t$

22 $\cos^4 x - \sin^4 x = \cos 2x$

23 $\sec 2\theta = \dfrac{\sec^2 \theta}{2 - \sec^2 \theta}$ 24 $\cot 2u = \dfrac{\cot^2 u - 1}{2 \cot u}$

25 $2 \sin^2 2t + \cos 4t = 1$

26 $\tan \theta + \cot \theta = 2 \csc 2\theta$

27 $\tan 3u = \dfrac{\tan u \, (3 - \tan^2 u)}{1 - 3 \tan^2 u}$

28 $\dfrac{1 + \sin 2v + \cos 2v}{1 + \sin 2v - \cos 2v} = \cot v$

Exer. 29–32: Express in terms of the cosine function with exponent 1.

29 $\cos^4 \dfrac{\theta}{2}$ 30 $\cos^4 2x$

31 $\sin^4 2x$ 32 $\sin^4 \dfrac{\theta}{2}$

Exer. 33–40: Find the solutions of the equation that are in the interval $[0, 2\pi)$.

33 $\sin 2t + \sin t = 0$ 34 $\cos t - \sin 2t = 0$

35 $\cos u + \cos 2u = 0$ 36 $\cos 2\theta - \tan \theta = 1$

37 $\tan 2x = \tan x$ 38 $\tan 2t - 2 \cos t = 0$

39 $\sin \tfrac{1}{2} u + \cos u = 1$ 40 $2 - \cos^2 x = 4 \sin^2 \tfrac{1}{2} x$

41 If $a > 0$, $b > 0$, and $0 < u < \pi/2$, show that

$$a \sin u + b \cos u = \sqrt{a^2 + b^2} \, \sin (u + v)$$

for $0 < v < \pi/2$, with

$$\sin v = \frac{b}{\sqrt{a^2 + b^2}} \quad \text{and} \quad \cos v = \frac{a}{\sqrt{a^2 + b^2}}.$$

42 Use Exercise 41 to express $8 \sin u + 15 \cos u$ in the form $c \sin (u + v)$.

43 A graph of $y = \cos 2x + 2 \cos x$ for $0 \le x \le 2\pi$ is shown in the figure.

  (a) Approximate the x-intercepts to two decimal places.

  (b) The x-coordinates of the turning points P, Q, and R on the graph are solutions of $\sin 2x + \sin x = 0$. Find the coordinates of these points.

Exercise 43

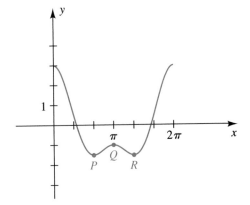

44 A graph of $y = \cos x - \sin 2x$ for $-2\pi \le x \le 2\pi$ is shown in the figure.

(a) Find the *x*-intercepts.

(b) The *x*-coordinates of the eight turning points on the graph are solutions of $\sin x + 2 \cos 2x = 0$. Approximate these *x*-coordinates to two decimal places.

Exercise 44

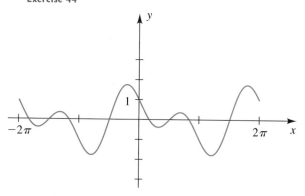

45 A graph of $y = \cos 3x - 3 \cos x$ for $-2\pi \le x \le 2\pi$ is shown in the figure.

(a) Find the *x*-intercepts. (*Hint:* Use the formula for $\cos 3\theta$ given in Example 2.)

(b) The *x*-coordinates of the 13 turning points on the graph are solutions of $\sin 3x - \sin x = 0$. Find these *x*-coordinates. (*Hint:* Use the formula for $\sin 3u$ in Exercise 17.)

Exercise 45

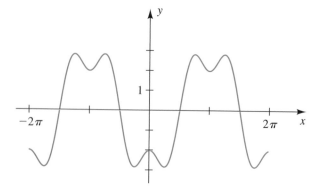

46 A graph of $y = \sin 4x - 4 \sin x$ for $-2\pi \le x \le 2\pi$ is shown in the figure. Find the *x*-intercepts. (*Hint:* Use the formula for $\sin 4t$ in Exercise 18.)

Exercise 46

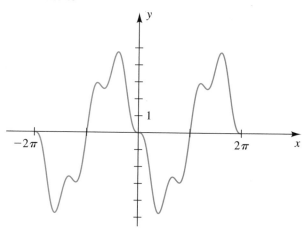

47 *Planning a railroad route* Shown in the figure is a proposed railroad route through three towns located at points *A*, *B*, and *C*. The track will branch out from *B* toward *C* at an angle θ.

(a) Show that the total distance *d* from *A* to *C* is given by $d = 20 \tan \frac{1}{2}\theta + 40$.

(b) Because of mountains between *A* and *C*, the branching point *B* must be at least 20 miles from *A*. Is there a route that avoids the mountains and measures exactly 50 miles?

Exercise 47

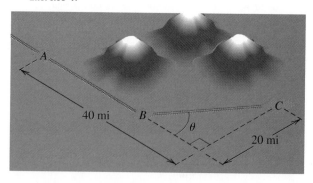

48 *Projectile's range* If a projectile is fired from ground level with an initial velocity of *v* ft/sec and at an angle of θ degrees with the horizontal, the range *R* of the projectile is given by

$$R = \frac{v^2}{16} \sin \theta \cos \theta.$$

If $v = 80$ ft/sec, approximate the angles that result in a range of 150 feet.

49 *Constructing a rain gutter* Shown in the figure is a design for a rain gutter.

(a) Express the volume V as a function of θ. (*Hint:* See Example 8.)

(b) Approximate the acute angle θ that results in a volume of 2 ft³.

Exercise 49

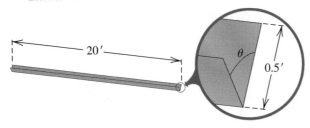

50 *Designing curbing* A highway engineer is designing curbing for a street at an intersection where two highways meet at an angle ϕ, as shown in the figure. The curbing between points A and B is to be constructed using a circle that is tangent to the highway at these two points.

(a) Show that the relationship between the radius R of the circle and the distance d in the figure is given by $d = R \tan (\phi/2)$.

(b) If $\phi = 45°$ and $d = 20$ ft, approximate R and the length of the curbing.

Exercise 50

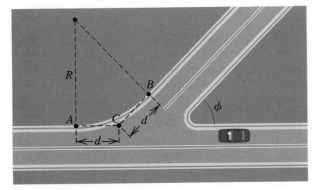

51 *Arterial bifurcation* A common form of cardiovascular branching is bifurcation, in which an artery splits into two smaller blood vessels. The bifurcation angle θ is the angle formed by the two smaller arteries. In the figure, the line through A and D bisects θ and is perpendicular to the line through B and C.

(a) Show that the length l of the artery from A to B is given by $l = a + \dfrac{b}{2} \tan \dfrac{\theta}{4}$.

(b) Estimate l from the measurements $a = 10$ mm, $b = 6$ mm, and $\theta = 156°$.

Exercise 51

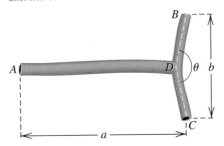

52 *Heat production in an AC circuit* By definition, the average value of $f(t) = c + a \cos bt$ for one or more complete cycles is c (see the figure).

(a) Use a double-angle formula to find the average value of $f(t) = \sin^2 \omega t$ for $0 \le t \le 2\pi/\omega$, with t in seconds.

(b) In an electrical circuit with an alternating current $I = I_0 \sin \omega t$, the rate r (in calories/sec) at which heat is produced in an R-ohm resistor is given by $r = RI^2$. Find the average rate at which heat is produced for one complete cycle.

Exercise 52

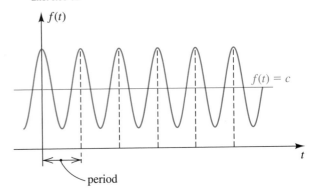

c **Exer. 53–54:** Use the graph of f to find the simplest expression $g(x)$ such that the equation $f(x) = g(x)$ is an identity. Verify this identity.

53 $f(x) = \dfrac{\sin 2x + \sin x}{\cos 2x + \cos x + 1}$

54 $f(x) = \dfrac{\sin x \,(1 + \cos 2x)}{\sin 2x}$

56 $\sec (2x + 1) = \cos \frac{1}{2}x + 1;$ $[-\pi/2, \pi/2]$

57 $\csc \left(\frac{1}{4}x + 1\right) = 1.5 - \cos 2x;$ $[-\pi, \pi]$

58 $3 \sin (2x) + 0.5 = 2 \sin \left(\frac{1}{2}x + 1\right);$ $[-\pi, \pi]$

[c] **Exer. 55–60:** Graphically solve the trigonometric equation on the indicated interval to two decimal places.

59 $2 \cot \frac{1}{4}x = 1 - \sec \frac{1}{2}x;$ $[-2\pi, 2\pi]$

55 $\tan \left(\frac{1}{2}x + 1\right) = \sin \frac{1}{2}x;$ $[-2\pi, 2\pi]$

60 $\tan \left(\frac{3}{2}x + \frac{1}{2}\right) = \frac{3}{2} \sin 2x;$ $[-\pi, \pi]$

3.5   PRODUCT-TO-SUM AND SUM-TO-PRODUCT FORMULAS

The following formulas may be used to change the form of certain trigonometric expressions from products to sums. We refer to these as **product-to-sum formulas** even though two of the formulas express a product as a difference, because any difference $x - y$ of two real numbers is also a sum $x + (-y)$.

Product-to-Sum Formulas

> **(1)** $\sin u \cos v = \frac{1}{2} [\sin (u + v) + \sin (u - v)]$
>
> **(2)** $\cos u \sin v = \frac{1}{2} [\sin (u + v) - \sin (u - v)]$
>
> **(3)** $\cos u \cos v = \frac{1}{2} [\cos (u + v) + \cos (u - v)]$
>
> **(4)** $\sin u \sin v = \frac{1}{2} [\cos (u - v) - \cos (u + v)]$

Proof   Let us add the left and right sides of the addition and subtraction formulas for the sine function, as follows:

$$\sin (u + v) = \quad \sin u \cos v + \cos u \sin v$$
$$\underline{\sin (u - v) = \quad \sin u \cos v - \cos u \sin v}$$
$$\sin (u + v) + \sin (u - v) = 2 \sin u \cos v$$

Dividing both sides of the last equation by 2 gives us formula 1.

Formula 2 is obtained by *subtracting* the left and right sides of the addition and subtraction formulas for the sine function. Formulas 3 and 4 are developed in a similar fashion, using the addition and subtraction formulas for the cosine function. ◂

EXAMPLE 1   Using product-to-sum formulas

Express as a sum:

(a) $\sin 4\theta \cos 3\theta$ **(b)** $\sin 3x \sin x$

Solution

(a) We use product-to-sum formula 1 with $u = 4\theta$ and $v = 3\theta$.

$$\sin 4\theta \cos 3\theta = \tfrac{1}{2} [\sin (4\theta + 3\theta) + \sin (4\theta - 3\theta)]$$
$$= \tfrac{1}{2} (\sin 7\theta + \sin \theta)$$

We can also obtain this relationship by using product-to-sum formula 2.

(b) We use product-to-sum formula 4 with $u = 3x$ and $v = x$:

$$\sin 3x \sin x = \tfrac{1}{2}[\cos (3x - x) - \cos (3x + x)]$$
$$= \tfrac{1}{2}(\cos 2x - \cos 4x)$$

We may use the product-to-sum formulas to express a sum or difference as a product. To obtain forms that can be applied more easily, we shall change the notation as follows. If we let

$$u + v = a \quad \text{and} \quad u - v = b,$$

then $(u + v) + (u - v) = a + b$, which simplifies to

$$u = \frac{a + b}{2}.$$

Similarly, since $(u + v) - (u - v) = a - b$, we obtain

$$v = \frac{a - b}{2}.$$

We now substitute for $u + v$ and $u - v$ on the right-hand sides of the product-to-sum formulas and for u and v on the left-hand sides. If we then multiply by 2, we obtain the following formulas.

Sum-to-Product Formulas

$$\textbf{(1)} \ \sin a + \sin b = 2 \sin \frac{a + b}{2} \cos \frac{a - b}{2}$$

$$\textbf{(2)} \ \sin a - \sin b = 2 \cos \frac{a + b}{2} \sin \frac{a - b}{2}$$

$$\textbf{(3)} \ \cos a + \cos b = 2 \cos \frac{a + b}{2} \cos \frac{a - b}{2}$$

$$\textbf{(4)} \ \cos a - \cos b = -2 \sin \frac{a + b}{2} \sin \frac{a - b}{2}$$

EXAMPLE 2   Using a sum-to-product formula

Express $\sin 5x - \sin 3x$ as a product.

Solution   We use sum-to-product formula 2 with $a = 5x$ and $b = 3x$:

$$\sin 5x - \sin 3x = 2 \cos \frac{5x + 3x}{2} \sin \frac{5x - 3x}{2}$$
$$= 2 \cos 4x \sin x$$

EXAMPLE 3   Using sum-to-product formulas to verify an identity

Verify the identity $\dfrac{\sin 3t + \sin 5t}{\cos 3t - \cos 5t} = \cot t$.

Solution   We first use a sum-to-product formula for the numerator and one for the denominator:

$$\frac{\sin 3t + \sin 5t}{\cos 3t - \cos 5t} = \frac{2 \sin \dfrac{3t + 5t}{2} \cos \dfrac{3t - 5t}{2}}{-2 \sin \dfrac{3t + 5t}{2} \sin \dfrac{3t - 5t}{2}} \qquad \text{sum-to-product formulas 1 and 4}$$

$$= \frac{2 \sin 4t \cos (-t)}{-2 \sin 4t \sin (-t)} \qquad \text{simplify}$$

$$= \frac{\cos (-t)}{-\sin (-t)} \qquad \text{cancel } 2 \sin 4t$$

$$= \frac{\cos t}{\sin t} \qquad \text{formulas for negatives}$$

$$= \cot t \qquad \text{cotangent identity}$$

EXAMPLE 4   Using a sum-to-product formula to solve an equation

Find the solutions of $\sin 5x + \sin x = 0$.

Solution

$$\sin 5x + \sin x = 0 \qquad \text{given}$$

$$2 \sin \frac{5x + x}{2} \cos \frac{5x - x}{2} = 0 \qquad \text{sum-to-product formula 1}$$

$$\sin 3x \cos 2x = 0 \qquad \text{simplify and divide by 2}$$

$$\sin 3x = 0, \quad \cos 2x = 0 \qquad \text{zero factor theorem}$$

The solutions of the last two equations are

$$3x = \pi n \quad \text{and} \quad 2x = \frac{\pi}{2} + \pi n \quad \text{for every integer } n.$$

Dividing by 3 and 2, respectively, we obtain

$$\frac{\pi}{3} n \quad \text{and} \quad \frac{\pi}{4} + \frac{\pi}{2} n \quad \text{for every integer } n.$$

E X A M P L E   5   Finding the x-intercepts of a graph

A graph of the equation $y = \cos x - \cos 3x - \sin 2x$ is shown in Figure 18. Find the 13 x-intercepts that are in the interval $[-2\pi, 2\pi]$.

Figure 18

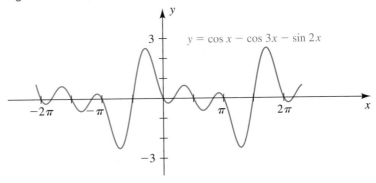

Solution   To find the x-intercepts, we proceed as follows:

$$\cos x - \cos 3x - \sin 2x = 0 \qquad \text{let } y = 0$$

$$(\cos x - \cos 3x) - \sin 2x = 0 \qquad \text{group the first two terms}$$

$$-2 \sin \frac{x + 3x}{2} \sin \frac{x - 3x}{2} - \sin 2x = 0 \qquad \text{sum-to-product formula 4}$$

$$-2 \sin 2x \sin (-x) - \sin 2x = 0 \qquad \text{simplify}$$

$$2 \sin 2x \sin x - \sin 2x = 0 \qquad \text{formula for negatives}$$

$$\sin 2x (2 \sin x - 1) = 0 \qquad \text{factor out } \sin 2x$$

$$\sin 2x = 0, \quad 2 \sin x - 1 = 0 \qquad \text{zero factor theorem}$$

$$\sin 2x = 0, \qquad \sin x = \tfrac{1}{2} \qquad \text{solve for } \sin x$$

The equation $\sin 2x = 0$ has solutions $2x = \pi n$, or, dividing by 2,

$$x = \frac{\pi}{2} n \quad \text{for every integer } n.$$

If we let $n = 0, \pm 1, \pm 2, \pm 3,$ and ± 4, we obtain nine x-intercepts in $[-2\pi, 2\pi]$:

$$0, \quad \pm \frac{\pi}{2}, \quad \pm \pi, \quad \pm \frac{3\pi}{2}, \quad \pm 2\pi$$

The solutions of the equation $\sin x = \tfrac{1}{2}$ are

$$\frac{\pi}{6} + 2\pi n \quad \text{and} \quad \frac{5\pi}{6} + 2\pi n \quad \text{for every integer } n.$$

(continued)

The four solutions in $[-2\pi, 2\pi]$ are obtained by letting $n = 0$ and $n = -1$:

$$\frac{\pi}{6}, \quad \frac{5\pi}{6}, \quad -\frac{11\pi}{6}, \quad -\frac{7\pi}{6}$$

3.5 EXERCISES

Exer. 1–8: Express as a sum or difference.

1 $\sin 7t \sin 3t$

2 $\sin(-4x)\cos 8x$

3 $\cos 6u \cos(-4u)$

4 $\cos 4t \sin 6t$

5 $2 \sin 9\theta \cos 3\theta$

6 $2 \sin 7\theta \sin 5\theta$

7 $3 \cos x \sin 2x$

8 $5 \cos u \cos 5u$

Exer. 9–16: Express as a product.

9 $\sin 6\theta + \sin 2\theta$

10 $\sin 4\theta - \sin 8\theta$

11 $\cos 5x - \cos 3x$

12 $\cos 5t + \cos 6t$

13 $\sin 3t - \sin 7t$

14 $\cos\theta - \cos 5\theta$

15 $\cos x + \cos 2x$

16 $\sin 8t + \sin 2t$

Exer. 17–24: Verify the identity.

17 $\dfrac{\sin 4t + \sin 6t}{\cos 4t - \cos 6t} = \cot t$

18 $\dfrac{\sin\theta + \sin 3\theta}{\cos\theta + \cos 3\theta} = \tan 2\theta$

19 $\dfrac{\sin u + \sin v}{\cos u + \cos v} = \tan\dfrac{1}{2}(u + v)$

20 $\dfrac{\sin u - \sin v}{\cos u - \cos v} = -\cot\dfrac{1}{2}(u + v)$

21 $\dfrac{\sin u - \sin v}{\sin u + \sin v} = \dfrac{\tan\frac{1}{2}(u - v)}{\tan\frac{1}{2}(u + v)}$

22 $\dfrac{\cos u - \cos v}{\cos u + \cos v} = -\tan\dfrac{1}{2}(u + v)\tan\dfrac{1}{2}(u - v)$

23 $4 \cos x \cos 2x \sin 3x = \sin 2x + \sin 4x + \sin 6x$

24 $\dfrac{\cos t + \cos 4t + \cos 7t}{\sin t + \sin 4t + \sin 7t} = \cot 4t$

Exer. 25–26: Express as a sum.

25 $(\sin ax)(\cos bx)$

26 $(\cos au)(\cos bu)$

Exer. 27–34: Use sum-to-product formulas to find the solutions of the equation.

27 $\sin 5t + \sin 3t = 0$

28 $\sin t + \sin 3t = \sin 2t$

29 $\cos x = \cos 3x$

30 $\cos 4x - \cos 3x = 0$

31 $\cos 3x + \cos 5x = \cos x$

32 $\cos 3x = -\cos 6x$

33 $\sin 2x - \sin 5x = 0$

34 $\sin 5x - \sin x = 2 \cos 3x$

Exer. 35–36: Shown in the figure is a graph of the function f for $0 \le x \le 2\pi$. Use a sum-to-product formula to help find the x-intercepts.

35 $f(x) = \cos x + \cos 3x$

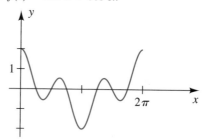

36 $f(x) = \sin 4x - \sin x$

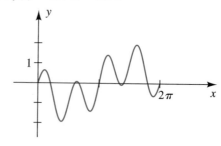

37 Refer to Exercise 45 of Section 3.4. The graph of the equation $y = \cos 3x - 3 \cos x$ has 13 turning points for $-2\pi \le x \le 2\pi$. The x-coordinates of these points are solutions of the equation $\sin 3x - \sin x = 0$. Use a sum-to-product formula to find these x-coordinates.

38 Refer to Exercise 46 of Section 3.4. The x-coordinates of the turning points on the graph of the equation $y = \sin 4x - 4 \sin x$ are solutions of $\cos 4x - \cos x = 0$. Use a sum-to-product formula to find these x-coordinates for $-2\pi \le x \le 2\pi$.

39 *Vibration of a violin string*   Mathematical analysis of a vibrating violin string of length l involves functions such that

$$f(x) = \sin\left(\frac{\pi n}{l}x\right) \cos\left(\frac{k \pi n}{l}t\right),$$

where n is an integer, k is a constant, and t is time. Express f as a sum of two sine functions.

40 *Pressure on the eardrum*   If two tuning forks are struck simultaneously with the same force and are then held at the same distance from the eardrum, the pressure on the outside of the eardrum at time t is given by

$$p(t) = a \cos \omega_1 t + a \cos \omega_2 t,$$

where a, ω_1, and ω_2 are constants. If ω_1 and ω_2 are almost equal, a tone is produced that alternates between loudness and virtual silence. This phenomenon is known as beats.

(a) Use a sum-to-product formula to express $p(t)$ as a product.

(b) Show that $p(t)$ may be considered as a cosine wave with approximate period $2\pi/\omega_1$ and variable amplitude $f(t) = 2a \cos \frac{1}{2}(\omega_1 - \omega_2)t$. Find the maximum amplitude.

(c) Shown in the figure is a graph of the equation

$$p(t) = \cos 4.5t + \cos 3.5t.$$

Near-silence occurs at points A and B, where the variable amplitude $f(t)$ in part (b) is zero. Find the coordinates of these points, and determine how frequently near-silence occurs.

(d) Use the graph to show that the function p in part (c) has period 4π. Conclude that the maximum amplitude of 2 occurs every 4π units of time.

Exercise 40

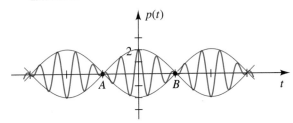

c **Exer. 41–42: Graph f on the interval $[-\pi, \pi]$. (a) Estimate the x-intercepts. (b) Use sum-to-product formulas to find the exact values of the x-intercepts.**

41 $f(x) = \sin 4x + \sin 2x$ **42** $f(x) = \cos 3x - \cos 2x$

c **Exer. 43–44: Use the graph of f to find the simplest expression $g(x)$ such that the equation $f(x) = g(x)$ is an identity. Verify this identity.**

43 $f(x) = \dfrac{\sin x + \sin 2x + \sin 3x}{\cos x + \cos 2x + \cos 3x}$

44 $f(x) = \dfrac{\cos x - \cos 2x + \cos 3x}{\sin x - \sin 2x + \sin 3x}$

3.6   THE INVERSE TRIGONOMETRIC FUNCTIONS

Recall from Section 1.5 that to define the inverse function f^{-1} of a function f, it is essential that f be one-to-one; that is, if $a \ne b$ in the domain of f, then $f(a) \ne f(b)$. The inverse function f^{-1} *reverses* the correspondence given by f; that is,

$$u = f(v) \qquad \text{if and only if} \qquad v = f^{-1}(u).$$

The following general relationships involving f and f^{-1} were discussed in Section 1.5.

Relationships Between f^{-1} and f

> **(1)** $y = f^{-1}(x)$ if and only if $x = f(y)$, where x is in the domain of f^{-1} and y is in the domain of f.
>
> **(2)** domain of f^{-1} = range of f
>
> **(3)** range of f^{-1} = domain of f
>
> **(4)** $f(f^{-1}(x)) = x$ for every x in the domain of f^{-1}
>
> **(5)** $f^{-1}(f(y)) = y$ for every y in the domain of f
>
> **(6)** The point (a, b) is on the graph of f if and only if the point (b, a) is on the graph of f^{-1}.
>
> **(7)** The graphs of f^{-1} and f are reflections of each other through the line $y = x$.

We shall use relationship 1 to define each of the inverse trigonometric functions.

The sine function is not one-to-one, since different numbers, such as $\pi/6$, $5\pi/6$, and $-7\pi/6$, yield the same function value $\left(\frac{1}{2}\right)$. If we restrict the domain to $[-\pi/2, \pi/2]$, then, as illustrated by the blue portion of the graph of $y = \sin x$ in Figure 19, we obtain a one-to-one (increasing) function that takes on every value of the sine function once and only once. We use this *new* function with domain $[-\pi/2, \pi/2]$ and range $[-1, 1]$ to define the *inverse sine function*.

Figure 19

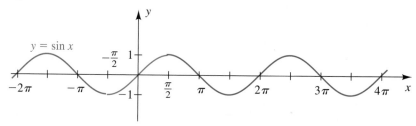

DEFINITION OF THE INVERSE SINE FUNCTION

> The **inverse sine function,** denoted by $\sin^{-1}$, is defined by
>
> $$y = \sin^{-1} x \quad \text{if and only if} \quad x = \sin y$$
>
> for $-1 \le x \le 1$ and $-\dfrac{\pi}{2} \le y \le \dfrac{\pi}{2}$.

The domain of the inverse sine function is $[-1, 1]$, and the range is $[-\pi/2, \pi/2]$.

The notation $y = \sin^{-1} x$ is sometimes read "y is the inverse sine of x." The equation $x = \sin y$ in the definition allows us to regard y as an angle, so $y = \sin^{-1} x$ may also be read *"y is the angle whose sine is x"* (with $-\pi/2 \le y \le \pi/2$).

The inverse sine function is also called the **arcsine function,** and arcsin x may be used in place of $\sin^{-1} x$. If $t = \arcsin x$, then $\sin t = x$, and t may be interpreted as an *arc length* on the unit circle U with center at the origin. We will use both notations—$\sin^{-1}$ and arcsin—throughout our work.

Several values of the inverse sine function are listed in the next chart.

Warning

> It is *essential* to choose the value y in the range $[-\pi/2, \pi/2]$ of $\sin^{-1}$. Thus, even though $\sin (5\pi/6) = \frac{1}{2}$, the number $y = 5\pi/6$ is not the inverse function value $\sin^{-1} \frac{1}{2}$.

Equation	Equivalent statement		Solution
$y = \sin^{-1} \left(\dfrac{1}{2} \right)$	$\sin y = \dfrac{1}{2}$ and	$-\dfrac{\pi}{2} \le y \le \dfrac{\pi}{2}$	$y = \dfrac{\pi}{6}$
$y = \sin^{-1} \left(-\dfrac{1}{2} \right)$	$\sin y = -\dfrac{1}{2}$ and	$-\dfrac{\pi}{2} \le y \le \dfrac{\pi}{2}$	$y = -\dfrac{\pi}{6}$
$y = \sin^{-1} (1)$	$\sin y = 1$ and	$-\dfrac{\pi}{2} \le y \le \dfrac{\pi}{2}$	$y = \dfrac{\pi}{2}$
$y = \arcsin (0)$	$\sin y = 0$ and	$-\dfrac{\pi}{2} \le y \le \dfrac{\pi}{2}$	$y = 0$
$y = \arcsin \left(-\dfrac{\sqrt{3}}{2} \right)$	$\sin y = -\dfrac{\sqrt{3}}{2}$ and	$-\dfrac{\pi}{2} \le y \le \dfrac{\pi}{2}$	$y = -\dfrac{\pi}{3}$

We have now justified the method of solving an equation of the form $\sin \theta = k$ as discussed in Chapter 2. We see that the calculator key $\boxed{\text{SIN}^{-1}}$ used to obtain $\theta = \sin^{-1} k$ gives us the value of the inverse sine function.

Relationship 7 for the graphs of f and f^{-1} tells us that we can sketch the graph of $y = \sin^{-1} x$ by reflecting the blue portion of Figure 19 through the line $y = x$. We can also use the equation $x = \sin y$ with the restriction $-\pi/2 \le y \le \pi/2$ to find points on the graph. This gives us Figure 20.

Relationship 4, $f(f^{-1}(x)) = x$, and relationship 5, $f^{-1}(f(y)) = y$, which hold for any inverse function f^{-1}, give us the following properties.

Figure 20

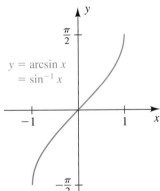

$y = \arcsin x$
$= \sin^{-1} x$

Properties of $\sin^{-1}$

> **(1)** $\sin (\sin^{-1} x) = \sin (\arcsin x) = x$ if $-1 \le x \le 1$
>
> **(2)** $\sin^{-1} (\sin y) = \arcsin (\sin y) = y$ if $-\dfrac{\pi}{2} \le y \le \dfrac{\pi}{2}$

E X A M P L E   1   Using properties of $\sin^{-1}$

Find the exact value:

(a) $\sin\left(\sin^{-1}\dfrac{1}{2}\right)$ **(b)** $\sin^{-1}\left(\sin\dfrac{\pi}{4}\right)$ **(c)** $\sin^{-1}\left(\sin\dfrac{2\pi}{3}\right)$

Solution

(a) The *difficult* way to find the value of this expression is to first find the angle $\sin^{-1}\frac{1}{2}$, namely $\pi/6$, and then evaluate $\sin(\pi/6)$, obtaining $\frac{1}{2}$. The *easy* way is to use property 1 of $\sin^{-1}$:

$$\text{since} \quad -1 \le \tfrac{1}{2} \le 1, \quad \sin\left(\sin^{-1}\tfrac{1}{2}\right) = \tfrac{1}{2}$$

(b) Since $-\pi/2 \le \pi/4 \le \pi/2$, we can use property 2 of $\sin^{-1}$ to obtain

$$\sin^{-1}\left(\sin\frac{\pi}{4}\right) = \frac{\pi}{4}.$$

(c) Be careful! Since $2\pi/3$ is *not* between $-\pi/2$ and $\pi/2$, we cannot use property 2 of $\sin^{-1}$. Instead, we first evaluate the inner expression, $\sin(2\pi/3)$, and then use the definition of $\sin^{-1}$, as follows:

$$\sin^{-1}\left(\sin\frac{2\pi}{3}\right) = \sin^{-1}\left(\frac{\sqrt{3}}{2}\right) = \frac{\pi}{3}$$

E X A M P L E   2   Finding a value of $\sin^{-1}$

Find the exact value of y if $y = \sin^{-1}\left(\tan\dfrac{3\pi}{4}\right)$.

Figure 21

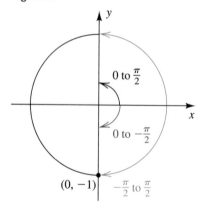

Solution   We first evaluate the inner expression—$\tan(3\pi/4)$—and then find the inverse sine of that number:

$$y = \sin^{-1}\left(\tan\frac{3\pi}{4}\right) = \sin^{-1}(-1)$$

In words, we have "y is the angle whose sine is -1." It may be helpful to recall the arcsine values by associating them with the angles corresponding to the blue portion of the unit circle shown in Figure 21. From the figure we see that $-\pi/2$ is the angle whose sine is -1. It follows that $y = -\pi/2$, and hence

$$y = \sin^{-1}\left(\tan\frac{3\pi}{4}\right) = -\frac{\pi}{2}.$$

The other trigonometric functions may also be used to introduce inverse trigonometric functions. The procedure is first to determine a convenient subset of the domain in order to obtain a one-to-one function. If the domain of the cosine function is restricted to the interval $[0, \pi]$, as illustrated by the blue portion of the graph of $y = \cos x$ in Figure 22, we obtain a one-to-one (decreasing) function that takes on every value of the cosine function once and only once. Then, we use this *new* function with domain $[0, \pi]$ and range $[-1, 1]$ to define the *inverse cosine function*.

Figure 22

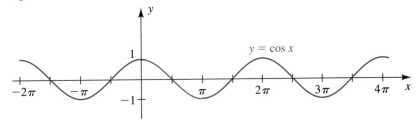

DEFINITION OF THE INVERSE COSINE FUNCTION

The **inverse cosine function,** denoted by $\cos^{-1}$, is defined by

$$y = \cos^{-1} x \quad \text{if and only if} \quad x = \cos y$$

for $-1 \le x \le 1$ and $0 \le y \le \pi$.

The domain of the inverse cosine function is $[-1, 1]$, and the range is $[0, \pi]$. Note that the range of $\cos^{-1}$ is *not* the same as the range of $\sin^{-1}$ but their domains are equal.

The notation $y = \cos^{-1} x$ may be read "y is the inverse cosine of x" or "y is the angle whose cosine is x" (with $0 \le y \le \pi$).

The inverse cosine function is also called the **arccosine function,** and the notation arccos x is used interchangeably with $\cos^{-1} x$.

Several values of the inverse cosine function are listed in the next chart.

Warning

It is *essential* to choose the value y in the range $[0, \pi]$ of $\cos^{-1}$.

Equation	Equivalent statement		Solution
$y = \cos^{-1}\left(\dfrac{1}{2}\right)$	$\cos y = \dfrac{1}{2}$	and $0 \le y \le \pi$	$y = \dfrac{\pi}{3}$
$y = \cos^{-1}\left(-\dfrac{1}{2}\right)$	$\cos y = -\dfrac{1}{2}$	and $0 \le y \le \pi$	$y = \dfrac{2\pi}{3}$
$y = \cos^{-1}(1)$	$\cos y = 1$	and $0 \le y \le \pi$	$y = 0$
$y = \arccos(0)$	$\cos y = 0$	and $0 \le y \le \pi$	$y = \dfrac{\pi}{2}$
$y = \arccos\left(-\dfrac{\sqrt{3}}{2}\right)$	$\cos y = -\dfrac{\sqrt{3}}{2}$	and $0 \le y \le \pi$	$y = \dfrac{5\pi}{6}$

We can sketch the graph of $y = \cos^{-1} x$ by reflecting the blue portion of Figure 22 through the line $y = x$. This gives us the sketch in Figure 23. We could also use the equation $x = \cos y$, with $0 \le y \le \pi$, to find points on the graph. As indicated by the graph, *the values of the inverse cosine function are never negative.*

As in Example 2 and Figure 21 for the arcsine, it may be helpful to associate the arccosine values with the angles corresponding to the blue arc in Figure 24.

Figure 23   **Figure 24**

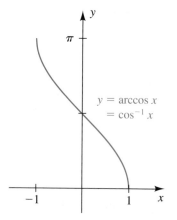
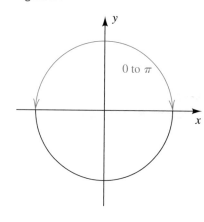

Using relationships 4 and 5 for general inverse functions f and f^{-1}, we obtain the following properties.

Properties of $\cos^{-1}$

(1) $\cos(\cos^{-1} x) = \cos(\arccos x) = x$ if $-1 \le x \le 1$
(2) $\cos^{-1}(\cos y) = \arccos(\cos y) = y$ if $0 \le y \le \pi$

E X A M P L E 3   **Using properties of $\cos^{-1}$**

Find the exact value:

(a) $\cos[\cos^{-1}(-0.5)]$ **(b)** $\cos^{-1}(\cos 3.14)$ **(c)** $\cos^{-1}\left[\sin\left(-\dfrac{\pi}{6}\right)\right]$

Solution   For parts (a) and (b), we may use properties 1 and 2 of $\cos^{-1}$, respectively.

(a) Since $-1 \le -0.5 \le 1$, $\cos[\cos^{-1}(-0.5)] = -0.5$.

(b) Since $0 \le 3.14 \le \pi$, $\cos^{-1}(\cos 3.14) = 3.14$.

(c) We first find $\sin(-\pi/6)$ and then use the definition of $\cos^{-1}$, as follows:

$$\cos^{-1}\left[\sin\left(-\frac{\pi}{6}\right)\right] = \cos^{-1}\left(-\frac{1}{2}\right) = \frac{2\pi}{3}$$

EXAMPLE 4   Finding a trigonometric function value

Find the exact value of $\sin \left[\arccos \left(-\frac{2}{3} \right) \right]$.

Solution   If we let $\theta = \arccos \left(-\frac{2}{3} \right)$, then, using the definition of the inverse cosine function, we have

$$\cos \theta = -\frac{2}{3} \quad \text{and} \quad 0 \le \theta \le \pi.$$

Figure 25

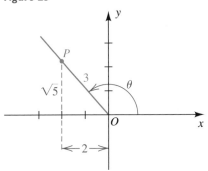

Hence, θ is in quadrant II, as illustrated in Figure 25. If we choose the point P on the terminal side with x-coordinate -2, the hypotenuse of the triangle in the figure must have length 3, since $\cos \theta = -\frac{2}{3}$. Thus, by the Pythagorean theorem, the y-coordinate of P is

$$\sqrt{3^2 - 2^2} = \sqrt{9 - 4} = \sqrt{5},$$

and therefore

$$\sin \left[\arccos \left(-\frac{2}{3} \right) \right] = \sin \theta = \frac{\sqrt{5}}{3}.$$

If we restrict the domain of the tangent function to the open interval $(-\pi/2, \pi/2)$, we obtain a one-to-one (increasing) function (see Figure 4 on page 196). We use this *new* function to define the *inverse tangent function*.

DEFINITION OF THE INVERSE TANGENT FUNCTION

The **inverse tangent function,** or **arctangent function,** denoted by $\tan^{-1}$ or arctan, is defined by

$$y = \tan^{-1} x = \arctan x \quad \text{if and only if} \quad x = \tan y$$

for any real number x and for $-\dfrac{\pi}{2} < y < \dfrac{\pi}{2}$.

Figure 26

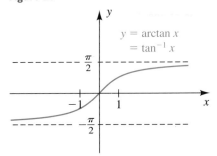

The domain of the arctangent function is $\mathbb{R}$, and the range is the open interval $(-\pi/2, \pi/2)$.

We can obtain the graph of $y = \tan^{-1} x$ in Figure 26 by sketching the graph of $x = \tan y$ for $-\pi/2 < y < \pi/2$. Note that the *vertical* asymptotes, $x = \pm\pi/2$, of the tangent function correspond to the *horizontal* asymptotes, $y = \pm\pi/2$, of the arctangent function.

As with $\sin^{-1}$ and $\cos^{-1}$, we have the following properties for $\tan^{-1}$.

Properties of $\tan^{-1}$

(1) $\tan (\tan^{-1} x) = \tan (\arctan x) = x$ for every x

(2) $\tan^{-1} (\tan y) = \arctan (\tan y) = y$ if $-\dfrac{\pi}{2} < y < \dfrac{\pi}{2}$

EXAMPLE 5   Using properties of $\tan^{-1}$

Find the exact value:

(a) $\tan (\tan^{-1} 1000)$ **(b)** $\tan^{-1} \left(\tan \dfrac{\pi}{4} \right)$ **(c)** $\arctan (\tan \pi)$

Solution

(a) By property 1 of $\tan^{-1}$,

$$\tan (\tan^{-1} 1000) = 1000.$$

(b) Since $-\pi/2 < \pi/4 < \pi/2$, we have, by property 2 of $\tan^{-1}$,

$$\tan^{-1} \left(\tan \dfrac{\pi}{4} \right) = \dfrac{\pi}{4}.$$

(c) Since $\pi > \pi/2$, we cannot use the second property of $\tan^{-1}$. Thus, we first find $\tan \pi$ and then evaluate, as follows:

$$\arctan (\tan \pi) = \arctan 0 = 0$$

EXAMPLE 6   Finding a trigonometric function value

Find the exact value of $\sec \left(\arctan \frac{2}{3} \right)$.

Figure 27

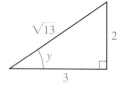

Solution   If we let $y = \arctan \frac{2}{3}$, then $\tan y = \frac{2}{3}$. We wish to find $\sec y$. Since $-\pi/2 < \arctan x < \pi/2$ for every x and $\tan y > 0$, it follows that $0 < y < \pi/2$. Thus, we may regard y as the radian measure of an angle of a right triangle such that $\tan y = \frac{2}{3}$, as illustrated in Figure 27. By the Pythagorean theorem, the hypotenuse is $\sqrt{3^2 + 2^2} = \sqrt{13}$. Referring to the triangle, we obtain

$$\sec \left(\arctan \dfrac{2}{3} \right) = \sec y = \dfrac{\sqrt{13}}{3}.$$

Figure 28

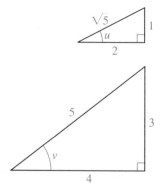

EXAMPLE 7   Finding a trigonometric function value

Find the exact value of $\sin \left(\arctan \frac{1}{2} - \arccos \frac{4}{5} \right)$.

Solution   If we let

$$u = \arctan \tfrac{1}{2} \qquad \text{and} \qquad v = \arccos \tfrac{4}{5},$$

then $\tan u = \tfrac{1}{2}$ and $\cos v = \tfrac{4}{5}$.

We wish to find $\sin (u - v)$. Since u and v are in the interval $(0, \pi/2)$, they can be considered as the radian measures of positive acute angles, and we may refer to the right triangles in Figure 28. This gives us

$$\sin u = \dfrac{1}{\sqrt{5}}, \quad \cos u = \dfrac{2}{\sqrt{5}}, \quad \sin v = \dfrac{3}{5}, \quad \cos v = \dfrac{4}{5}.$$

By the subtraction formula for sine,

$$\sin (u - v) = \sin u \cos v - \cos u \sin v$$

$$= \frac{1}{\sqrt{5}} \frac{4}{5} - \frac{2}{\sqrt{5}} \frac{3}{5}$$

$$= \frac{-2}{5\sqrt{5}}, \quad \text{or} \quad \frac{-2\sqrt{5}}{25}.$$

EXAMPLE 8   **Changing an expression involving $\sin^{-1} x$ to an algebraic expression**

If $-1 \le x \le 1$, rewrite $\cos (\sin^{-1} x)$ as an algebraic expression in x.

Solution   Let

$$y = \sin^{-1} x \quad \text{or, equivalently,} \quad \sin y = x.$$

We wish to express $\cos y$ in terms of x. Since $-\pi/2 \le y \le \pi/2$, it follows that $\cos y \ge 0$, and hence (from $\sin^2 y + \cos^2 y = 1$)

$$\cos y = \sqrt{1 - \sin^2 y} = \sqrt{1 - x^2}.$$

Consequently, $\cos (\sin^{-1} x) = \sqrt{1 - x^2}.$

Figure 29

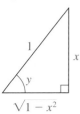

The last identity is also evident geometrically if $0 < x < 1$. In this case $0 < y < \pi/2$, and we may regard y as the radian measure of an angle of a right triangle such that $\sin y = x$, as illustrated in Figure 29. (The side of length $\sqrt{1 - x^2}$ is found by the Pythagorean theorem.) Referring to the triangle, we have

$$\cos (\sin^{-1} x) = \cos y = \frac{\sqrt{1 - x^2}}{1} = \sqrt{1 - x^2}.$$

Most of the trigonometric equations we considered in Section 3.2 had solutions that were rational multiples of π, such as $\pi/3$, $3\pi/4$, π, and so on. If solutions of trigonometric equations are not of that type, we can sometimes use inverse functions to express them in exact form, as illustrated in the next example.

EXAMPLE 9   **Using inverse trigonometric functions to solve an equation**

Find the solutions of $5 \sin^2 t + 3 \sin t - 1 = 0$ in $[0, 2\pi)$.

Figure 30

(a)

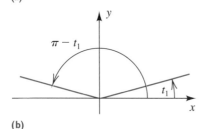

(b)

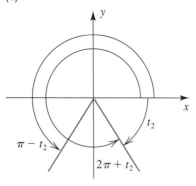

Figure 31

$[0, 2\pi]$ by $[-3, 8]$

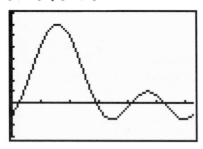

Solution   The equation may be regarded as a quadratic equation in $\sin t$. Applying the quadratic formula gives us

$$\sin t = \frac{-3 \pm \sqrt{3^2 - 4(5)(-1)}}{2(5)} = \frac{-3 \pm \sqrt{29}}{10}.$$

Using the definition of the inverse sine function, we obtain the following solutions:

$$t_1 = \sin^{-1} \tfrac{1}{10}(-3 + \sqrt{29}) \approx 0.2408$$
$$t_2 = \sin^{-1} \tfrac{1}{10}(-3 - \sqrt{29}) \approx -0.9946$$

Since the range of arcsin is $[-\pi/2, \pi/2]$, we know that t_1 is in $[0, \pi/2]$ and t_2 is in $[-\pi/2, 0]$. Using t_1 as a reference angle, we also have $\pi - t_1$ as a solution in quadrant II, as shown in Figure 30(a). We can add 2π to t_2 to obtain a solution in quadrant IV, as shown in Figure 30(b). The solution in quadrant III is $\pi - t_2$, not $\pi + t_2$, because t_2 is negative.

Hence, with t_1 and t_2 as previously defined, the four exact solutions are

$$t_1, \quad \pi - t_1, \quad \pi - t_2, \quad \text{and} \quad 2\pi + t_2,$$

and the four approximate solutions are

$$0.2408, \quad 2.9008, \quad 4.1361, \quad \text{and} \quad 5.2886.$$

If only approximate solutions are required, we may use a graphing utility to find the x-intercepts of $Y_1 = 5 \sin^2 x + 3 \sin x - 1$. Graphing Y_1 as shown in Figure 31 and using a root or zoom feature, we obtain the same four approximate solutions as listed above.

The next example illustrates one of many identities that are true for inverse trigonometric functions.

EXAMPLE 10   **Verifying an identity involving inverse trigonometric functions**

Verify the identity $\sin^{-1} x + \cos^{-1} x = \dfrac{\pi}{2}$ for $-1 \le x \le 1$.

Solution   Let

$$\alpha = \sin^{-1} x \quad \text{and} \quad \beta = \cos^{-1} x.$$

We wish to show that $\alpha + \beta = \pi/2$. From the definitions of $\sin^{-1}$ and $\cos^{-1}$,

$$\sin \alpha = x \quad \text{for} \quad -\frac{\pi}{2} \le \alpha \le \frac{\pi}{2}$$

and
$$\cos \beta = x \quad \text{for} \quad 0 \le \beta \le \pi.$$

Adding the two inequalities on the right, we see that

$$-\frac{\pi}{2} \le \alpha + \beta \le \frac{3\pi}{2}.$$

Note also that

$$\cos \alpha = \sqrt{1 - \sin^2 \alpha} = \sqrt{1 - x^2}$$

and
$$\sin \beta = \sqrt{1 - \cos^2 \beta} = \sqrt{1 - x^2}.$$

Using the addition formula for sine, we obtain

$$\sin (\alpha + \beta) = \sin \alpha \cos \beta + \cos \alpha \sin \beta$$
$$= x \cdot x + \sqrt{1 - x^2} \sqrt{1 - x^2}$$
$$= x^2 + (1 - x^2) = 1.$$

Since $\alpha + \beta$ is in the interval $[-\pi/2, 3\pi/2]$, the equation $\sin (\alpha + \beta) = 1$ has only one solution, $\alpha + \beta = \pi/2$, which is what we wished to show.

We may interpret the identity geometrically if $0 < x < 1$. If we construct a right triangle with one side of length x and hypotenuse of length 1, as illustrated in Figure 32, then angle β at B is an angle whose cosine is x; that is, $\beta = \cos^{-1} x$. Similarly, angle α at A is an angle whose sine is x; that is, $\alpha = \sin^{-1} x$. Since the acute angles of a right triangle are complementary, $\alpha + \beta = \pi/2$ or, equivalently,

$$\sin^{-1} x + \cos^{-1} x = \frac{\pi}{2}.$$

Figure 32

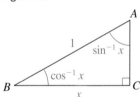

Each of the remaining inverse trigonometric functions is defined in the same manner as the first three—by choosing a domain D in which the corresponding trigonometric function is one-to-one and then using the usual technique (where y is in D):

$$y = \cot^{-1} x \quad \text{if and only if} \quad x = \cot y$$
$$y = \sec^{-1} x \quad \text{if and only if} \quad x = \sec y$$
$$y = \csc^{-1} x \quad \text{if and only if} \quad x = \csc y$$

The function $\sec^{-1}$ is used in calculus; however, $\cot^{-1}$ and $\csc^{-1}$ are seldom used. Because of their limited use in applications, we will not consider examples or exercises pertaining to these functions. We will merely summarize typical domains, ranges, and graphs in the following chart. A similar summary for the six trigonometric functions and their inverses appears in Appendix IV.

Summary of Features of $\cot^{-1}$, $\sec^{-1}$, **and** $\csc^{-1}$

Feature	$y = \cot^{-1} x$	$y = \sec^{-1} x$	$y = \csc^{-1} x$
Domain	$\mathbb{R}$	$\lvert x \rvert \geq 1$	$\lvert x \rvert \geq 1$
Range	$(0, \pi)$	$\left[0, \dfrac{\pi}{2} \right) \cup \left[\pi, \dfrac{3\pi}{2} \right)$	$\left(-\pi, -\dfrac{\pi}{2} \right] \cup \left(0, \dfrac{\pi}{2} \right]$
Graph			

It is often difficult to verify an identity involving inverse trigonometric functions, as we saw in Example 10. A graphing utility can be extremely helpful in determining whether an equation involving inverse trigonometric functions is an identity and, if it is not an identity, finding any solutions of the equation. The next example illustrates this process.

E X A M P L E   11   **Investigating an equation**

We know that $\tan x = (\sin x)/\cos x$ is an identity. Determine whether the equation

$$\arctan x = \frac{\arcsin x}{\arccos x}$$

is an identity. If it is not an identity, approximate the values of x for which the equation is true—that is, solve the equation.

Figure 33
$[-1, 1]$ by $[-\pi/2, \pi/2]$

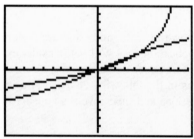

Solution   Some specific keystrokes for the TI-82/83 are given in Example 24 of Appendix I. We begin by making the assignments

$$Y_1 = \tan^{-1} x \qquad \text{and} \qquad Y_2 = \sin^{-1} x/\cos^{-1} x.$$

Since the domain of $\sin^{-1}$ and $\cos^{-1}$ is $[-1, 1]$ and the range of $\tan^{-1}$ is $(-\pi/2, \pi/2)$, we choose the viewing rectangle dimensions $[-1, 1]$ by $[-\pi/2, \pi/2]$, with Xscl = 0.1 and Yscl = 0.2, as shown in Figure 33.

Since the graphs representing Y_1 and Y_2 are not the same, we know that *the given equation is not an identity.* Because the graphs intersect twice, however, we know that the equation has two solutions. It appears that $x = 0$ is a solution, and a quick check in the given equation verifies that this is true. To estimate the point of intersection in the first quadrant, we use a zoom or intersect feature to determine that the point has the approximate coordinates (0.450, 0.423). Hence,

$$x = 0 \quad \text{and} \quad x \approx 0.450$$

are the values of x for which the given equation is true.

3.6 EXERCISES

Exer. 1–22: Find the exact value of the expression whenever it is defined.

1 (a) $\sin^{-1}\left(-\dfrac{\sqrt{2}}{2}\right)$ (b) $\cos^{-1}\left(-\dfrac{1}{2}\right)$

(c) $\tan^{-1}(-\sqrt{3})$

2 (a) $\sin^{-1}\left(-\dfrac{1}{2}\right)$ (b) $\cos^{-1}\left(-\dfrac{\sqrt{2}}{2}\right)$

(c) $\tan^{-1}(-1)$

3 (a) $\arcsin\dfrac{\sqrt{3}}{2}$ (b) $\arccos\dfrac{\sqrt{2}}{2}$ (c) $\arctan\dfrac{1}{\sqrt{3}}$

4 (a) $\arcsin 0$ (b) $\arccos(-1)$ (c) $\arctan 0$

5 (a) $\sin^{-1}\dfrac{\pi}{3}$ (b) $\cos^{-1}\dfrac{\pi}{2}$ (c) $\tan^{-1} 1$

6 (a) $\arcsin\dfrac{\pi}{2}$ (b) $\arccos\dfrac{\pi}{3}$ (c) $\arctan\left(-\dfrac{\sqrt{3}}{3}\right)$

7 (a) $\sin\left[\arcsin\left(-\dfrac{3}{10}\right)\right]$ (b) $\cos\left(\arccos\dfrac{1}{2}\right)$

(c) $\tan(\arctan 14)$

8 (a) $\sin\left(\sin^{-1}\dfrac{2}{3}\right)$ (b) $\cos\left[\cos^{-1}\left(-\dfrac{1}{5}\right)\right]$

(c) $\tan[\tan^{-1}(-9)]$

9 (a) $\sin^{-1}\left(\sin\dfrac{\pi}{3}\right)$ (b) $\cos^{-1}\left[\cos\left(\dfrac{5\pi}{6}\right)\right]$

(c) $\tan^{-1}\left[\tan\left(-\dfrac{\pi}{6}\right)\right]$

10 (a) $\arcsin\left[\sin\left(-\dfrac{\pi}{2}\right)\right]$ (b) $\arccos(\cos 0)$

(c) $\arctan\left(\tan\dfrac{\pi}{4}\right)$

11 (a) $\arcsin\left(\sin\dfrac{5\pi}{4}\right)$ (b) $\arccos\left(\cos\dfrac{5\pi}{4}\right)$

(c) $\arctan\left(\tan\dfrac{7\pi}{4}\right)$

12 (a) $\sin^{-1}\left(\sin\dfrac{2\pi}{3}\right)$ (b) $\cos^{-1}\left(\cos\dfrac{4\pi}{3}\right)$

(c) $\tan^{-1}\left(\tan\dfrac{7\pi}{6}\right)$

13 (a) $\sin\left[\cos^{-1}\left(-\dfrac{1}{2}\right)\right]$ (b) $\cos(\tan^{-1} 1)$

(c) $\tan[\sin^{-1}(-1)]$

14 (a) $\sin(\tan^{-1}\sqrt{3})$ (b) $\cos(\sin^{-1} 1)$

(c) $\tan(\cos^{-1} 0)$

15 (a) $\cot\left(\sin^{-1}\dfrac{2}{3}\right)$ (b) $\sec\left[\tan^{-1}\left(-\dfrac{3}{5}\right)\right]$

(c) $\csc\left[\cos^{-1}\left(-\dfrac{1}{4}\right)\right]$

16 (a) $\cot\left[\sin^{-1}\left(-\dfrac{2}{5}\right)\right]$ (b) $\sec\left(\tan^{-1}\dfrac{7}{4}\right)$

(c) $\csc\left(\cos^{-1}\dfrac{1}{5}\right)$

17 (a) $\sin\left(\arcsin\dfrac{1}{2} + \arccos 0\right)$

(b) $\cos\left[\arctan\left(-\dfrac{3}{4}\right) - \arcsin\dfrac{4}{5}\right]$

(c) $\tan\left(\arctan\dfrac{4}{3} + \arccos\dfrac{8}{17}\right)$

18 (a) $\sin \left[\sin^{-1} \frac{5}{13} - \cos^{-1} \left(-\frac{3}{5}\right)\right]$

(b) $\cos \left(\sin^{-1} \frac{4}{5} + \tan^{-1} \frac{3}{4}\right)$

(c) $\tan \left[\cos^{-1} \frac{1}{2} - \sin^{-1} \left(-\frac{1}{2}\right)\right]$

19 (a) $\sin \left[2 \arccos \left(-\frac{3}{5}\right)\right]$ (b) $\cos \left(2 \sin^{-1} \frac{15}{17}\right)$

(c) $\tan \left(2 \tan^{-1} \frac{3}{4}\right)$

20 (a) $\sin \left(2 \tan^{-1} \frac{5}{12}\right)$ (b) $\cos \left(2 \arccos \frac{9}{41}\right)$

(c) $\tan \left[2 \arcsin \left(-\frac{8}{17}\right)\right]$

21 (a) $\sin \left[\frac{1}{2} \sin^{-1} \left(-\frac{7}{25}\right)\right]$ (b) $\cos \left(\frac{1}{2} \tan^{-1} \frac{8}{15}\right)$

(c) $\tan \left(\frac{1}{2} \cos^{-1} \frac{3}{5}\right)$

22 (a) $\sin \left[\frac{1}{2} \cos^{-1} \left(-\frac{3}{5}\right)\right]$ (b) $\cos \left(\frac{1}{2} \sin^{-1} \frac{12}{13}\right)$

(c) $\tan \left(\frac{1}{2} \tan^{-1} \frac{40}{9}\right)$

Exer. 23–30: Write the expression as an algebraic expression in x for $x > 0$.

23 $\sin (\tan^{-1} x)$ 24 $\tan (\arccos x)$

25 $\sec \left(\sin^{-1} \dfrac{x}{\sqrt{x^2 + 4}}\right)$ 26 $\cot \left(\sin^{-1} \dfrac{\sqrt{x^2 - 9}}{x}\right)$

27 $\sin (2 \sin^{-1} x)$ 28 $\cos (2 \tan^{-1} x)$

29 $\cos \left(\frac{1}{2} \arccos x\right)$ 30 $\tan \left(\dfrac{1}{2} \cos^{-1} \dfrac{1}{x}\right)$

Exer. 31–32: Complete the statements.

31 (a) As $x \to -1^+$, $\sin^{-1} x \to$ ___

(b) As $x \to 1^-$, $\cos^{-1} x \to$ ___

(c) As $x \to \infty$, $\tan^{-1} x \to$ ___

32 (a) As $x \to 1^-$, $\sin^{-1} x \to$ ___

(b) As $x \to -1^+$, $\cos^{-1} x \to$ ___

(c) As $x \to -\infty$, $\tan^{-1} x \to$ ___

Exer. 33–42: Sketch the graph of the equation.

33 $y = \sin^{-1} 2x$ 34 $y = \frac{1}{2} \sin^{-1} x$

35 $y = \sin^{-1} (x + 1)$ 36 $y = \sin^{-1} (x - 2) + \dfrac{\pi}{2}$

37 $y = \cos^{-1} \frac{1}{2} x$ 38 $y = 2 \cos^{-1} x$

39 $y = 2 + \tan^{-1} x$ 40 $y = \tan^{-1} 2x$

41 $y = \sin (\arccos x)$ 42 $y = \sin (\sin^{-1} x)$

Exer. 43–46: The given equation has the form $y = f(x)$.
(a) Find the domain of f. (b) Find the range of f. (c) Solve for x in terms of y.

43 $y = \frac{1}{2} \sin^{-1} (x - 3)$ 44 $y = 3 \tan^{-1} (2x + 1)$

45 $y = 4 \cos^{-1} \frac{2}{3} x$ 46 $y = 2 \sin^{-1} (3x - 4)$

Exer. 47–50: Solve the equation for x in terms of y if x is restricted to the given interval.

47 $y = -3 - \sin x$; $\left[-\dfrac{\pi}{2}, \dfrac{\pi}{2}\right]$

48 $y = 2 + 3 \sin x$; $\left[-\dfrac{\pi}{2}, \dfrac{\pi}{2}\right]$

49 $y = 15 - 2 \cos x$; $[0, \pi]$

50 $y = 6 - 3 \cos x$; $[0, \pi]$

Exer. 51–52: Solve the equation for x in terms of y if $0 < x < \pi$ and $0 < y < \pi$.

51 $\dfrac{\sin x}{3} = \dfrac{\sin y}{4}$ 52 $\dfrac{4}{\sin x} = \dfrac{7}{\sin y}$

Exer. 53–64: Use inverse trigonometric functions to find the solutions of the equation that are in the given interval, and approximate the solutions to four decimal places.

53 $\cos^2 x + 2 \cos x - 1 = 0$; $[0, 2\pi)$

54 $\sin^2 x - \sin x - 1 = 0$; $[0, 2\pi)$

55 $2 \tan^2 t + 9 \tan t + 3 = 0$; $\left(-\dfrac{\pi}{2}, \dfrac{\pi}{2}\right)$

56 $3 \sin^2 t + 7 \sin t + 3 = 0$; $\left[-\dfrac{\pi}{2}, \dfrac{\pi}{2}\right]$

57 $15 \cos^4 x - 14 \cos^2 x + 3 = 0$; $[0, \pi]$

58 $3 \tan^4 \theta - 19 \tan^2 \theta + 2 = 0$; $\left(-\dfrac{\pi}{2}, \dfrac{\pi}{2}\right)$

59 $6 \sin^3 \theta + 18 \sin^2 \theta - 5 \sin \theta - 15 = 0$; $\left(-\dfrac{\pi}{2}, \dfrac{\pi}{2}\right)$

60 $6 \sin 2x - 8 \cos x + 9 \sin x - 6 = 0$; $\left(-\dfrac{\pi}{2}, \dfrac{\pi}{2}\right)$

61 $(\cos x)(15 \cos x + 4) = 3$; $[0, 2\pi)$

62 $6 \sin^2 x = \sin x + 2$; $[0, 2\pi)$

63 $3 \cos 2x - 7 \cos x + 5 = 0$; $[0, 2\pi)$

64 $\sin 2x = -1.5 \cos x$; $[0, 2\pi)$

Exer. 65–66: If an earthquake has a total horizontal displacement of S meters along its fault line, then the horizontal movement M of a point on the surface of the earth d kilometers from the fault line can be estimated

using the formula

$$M = \frac{S}{2}\left(1 - \frac{2}{\pi}\tan^{-1}\frac{d}{D}\right),$$

where D is the depth (in kilometers) below the surface of the focal point of the earthquake.

65 Earthquake movement For the San Francisco earthquake of 1906, S was 4 meters and D was 3.5 kilometers. Approximate M for the stated values of d.

(a) 1 kilometer   (b) 4 kilometers

(c) 10 kilometers

66 Earthquake movement Approximate the depth D of the focal point of an earthquake with $S = 3$ m if a point on the surface of the earth 5 kilometers from the fault line moved 0.6 meter horizontally.

67 A golfer's drive A golfer, centered in a 30-yard-wide straight fairway, hits a ball 280 yards. Approximate the largest angle the drive can have from the center of the fairway if the ball is to stay in the fairway (see the figure).

Exercise 67

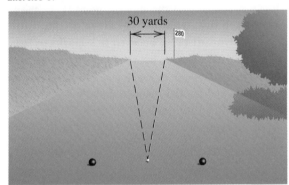

68 Placing a wooden brace A 14-foot piece of lumber is to be placed as a brace, as shown in the figure. Assuming all the lumber is 2 inches by 4 inches, find α and β.

Exercise 68

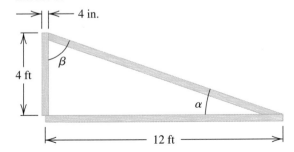

69 Tracking a sailboat As shown in the figure, a sailboat is following a straight-line course l. The shortest distance from a tracking station T to the course is d miles. As the boat sails, the tracking station records its distance k from T and its direction θ with respect to T. Angle α specifies the direction of the sailboat.

(a) Express α in terms of d, k, and θ.

(b) Estimate α to the nearest degree if $d = 50$ mi, $k = 210$ mi, and $\theta = 53.4°$.

Exercise 69

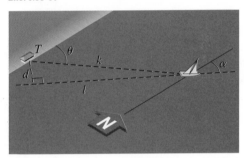

70 Calculating viewing angles An art critic whose eye level is 6 feet above the floor views a painting that is 10 feet in height and is mounted 4 feet above the floor, as shown in the figure.

(a) If the critic is standing x feet from the wall, express the viewing angle θ in terms of x.

(b) Use the addition formula for tangent to show that

$$\theta = \tan^{-1}\left(\frac{10x}{x^2 - 16}\right).$$

(c) For what value of x is $\theta = 45°$?

Exercise 70

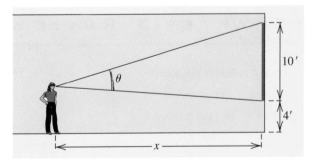

Exer. 71–76: Verify the identity.

71 $\sin^{-1} x = \tan^{-1}\dfrac{x}{\sqrt{1 - x^2}}$

72 $\arccos x + \arccos \sqrt{1 - x^2} = \dfrac{\pi}{2}, 0 \le x \le 1$

73 $\arcsin (-x) = -\arcsin x$

74 $\arccos (-x) = \pi - \arccos x$

75 $\arctan x + \arctan \dfrac{1}{x} = \dfrac{\pi}{2}, x > 0$

76 $2 \cos^{-1} x = \cos^{-1} (2x^2 - 1), 0 \le x \le 1$

c **Exer. 77–78: Graph f, and determine its domain and range.**

77 $f(x) = 2 \sin^{-1} (x - 1) + \cos^{-1} \frac{1}{2} x$

78 $f(x) = \frac{1}{2} \tan^{-1} (1 - 2x) + 3 \tan^{-1} \sqrt{x + 2}$

c **Exer. 79–80: Use a graph to estimate the solutions of the equation.**

79 $\sin^{-1} 2x = \tan^{-1} (1 - x)$

80 $\cos^{-1} \left(x - \frac{1}{5}\right) = 2 \sin^{-1} \left(\frac{1}{2} - x\right)$

c 81 *Designing a solar collector*   In designing a collector for solar power, an important consideration is the amount of sunlight that is transmitted through the glass into the water being heated. If the angle of incidence θ of the sun's rays is measured from a line perpendicular to the surface of the glass, then the fraction $f(\theta)$ of sunlight reflected off the glass can be approximated by

$$f(\theta) = \frac{1}{2} \left(\frac{\sin^2 \alpha}{\sin^2 \beta} + \frac{\tan^2 \alpha}{\tan^2 \beta} \right),$$

where

$$\alpha = \theta - \gamma, \quad \beta = \theta + \gamma, \quad \text{and} \quad \gamma = \sin^{-1} \left(\frac{\sin \theta}{1.52}\right).$$

Graph f for $0 < \theta < \pi/2$, and estimate θ when $f(\theta) = 0.2$.

c 82 *Designing a solar collector*   The altitude of the sun is the angle ϕ that the sun's rays make with the horizon at a given time and place. Determining ϕ is important in tilting a solar collector to obtain maximum efficiency. On June 21 at a latitude of 51.7°N, the altitude of the sun can be approximated using the formula

$$\sin \phi = \sin 23.5° \sin 51.7° + \cos 23.5° \cos 51.7° \cos H,$$

where H is called the hour angle, with $H = -\pi/2$ at 6 A.M., $H = 0$ at noon, and $H = \pi/2$ at 6 P.M.

(a) Solve the formula for ϕ, and graph the resulting equation for $-\pi/2 \le H \le \pi/2$.

(b) Estimate the times when $\phi = 45°$.

c **Exer. 83–86: Many calculators have viewing screens that are wider than they are high. The approximate ratio of the height to the width is often 2:3. Let the actual height of the calculator screen along the y-axis be 2 units, the actual width of the calculator screen along the x-axis be 3 units, and Xscl = Yscl = 1. Since the line y = x must pass through the point (1, 1), the actual slope m_A of this line on the calculator screen is given by**

$$m_A = \frac{\text{actual distance between tick marks on y-axis}}{\text{actual distance between tick marks on x-axis}}.$$

Using this information, graph y = x in the given viewing rectangle and predict the actual angle θ that the graph makes with the x-axis on the viewing screen.

83 [0, 3] by [0, 2]

84 [0, 6] by [0, 2]

85 [0, 3] by [0, 4]

86 [0, 2] by [0, 2]

CHAPTER 3   REVIEW   EXERCISES

Exer. 1–22: Verify the identity.

1 $(\cot^2 x + 1)(1 - \cos^2 x) = 1$

2 $\cos \theta + \sin \theta \tan \theta = \sec \theta$

3 $\dfrac{(\sec^2 \theta - 1) \cot \theta}{\tan \theta \sin \theta + \cos \theta} = \sin \theta$

4 $(\tan x + \cot x)^2 = \sec^2 x \csc^2 x$

5 $\dfrac{1}{1 + \sin t} = (\sec t - \tan t) \sec t$

6 $\dfrac{\sin (\alpha - \beta)}{\cos (\alpha + \beta)} = \dfrac{\tan \alpha - \tan \beta}{1 - \tan \alpha \tan \beta}$

7 $\tan 2u = \dfrac{2 \cot u}{\csc^2 u - 2}$

8 $\cos^2 \dfrac{v}{2} = \dfrac{1 + \sec v}{2 \sec v}$

9 $\dfrac{\tan^3 \phi - \cot^3 \phi}{\tan^2 \phi + \csc^2 \phi} = \tan \phi - \cot \phi$

10 $\dfrac{\sin u + \sin v}{\csc u + \csc v} = \dfrac{1 - \sin u \sin v}{-1 + \csc u \csc v}$

11 $\left(\dfrac{\sin^2 x}{\tan^4 x}\right)^3 \left(\dfrac{\csc^3 x}{\cot^6 x}\right)^2 = 1$

12 $\dfrac{\cos \gamma}{1 - \tan \gamma} + \dfrac{\sin \gamma}{1 - \cot \gamma} = \cos \gamma + \sin \gamma$

13 $\dfrac{\cos (-t)}{\sec (-t) + \tan (-t)} = 1 + \sin t$

14 $\dfrac{\cot (-t) + \csc (-t)}{\sin (-t)} = \dfrac{1}{1 - \cos t}$

15 $\sqrt{\dfrac{1 - \cos t}{1 + \cos t}} = \dfrac{1 - \cos t}{|\sin t|}$

16 $\sqrt{\dfrac{1 - \sin \theta}{1 + \sin \theta}} = \dfrac{|\cos \theta|}{1 + \sin \theta}$

17 $\cos \left(x - \dfrac{5\pi}{2}\right) = \sin x$

18 $\tan \left(x + \dfrac{3\pi}{4}\right) = \dfrac{\tan x - 1}{1 + \tan x}$

19 $\frac{1}{4} \sin 4\beta = \sin \beta \cos^3 \beta - \cos \beta \sin^3 \beta$

20 $\tan \frac{1}{2}\theta = \csc \theta - \cot \theta$

21 $\sin 8\theta = 8 \sin \theta \cos \theta (1 - 2 \sin^2 \theta)(1 - 8 \sin^2 \theta \cos^2 \theta)$

22 $\arctan x = \dfrac{1}{2} \arctan \dfrac{2x}{1 - x^2}$, $-1 < x < 1$

Exer. 23–40: Find the solutions of the equation that are in the interval $[0, 2\pi)$.

23 $2 \cos^3 \theta - \cos \theta = 0$

24 $2 \cos \alpha + \tan \alpha = \sec \alpha$

25 $\sin \theta = \tan \theta$

26 $\csc^5 \theta - 4 \csc \theta = 0$

27 $2 \cos^3 t + \cos^2 t - 2 \cos t - 1 = 0$

28 $\cos x \cot^2 x = \cos x$

29 $\sin \beta + 2 \cos^2 \beta = 1$

30 $\cos 2x + 3 \cos x + 2 = 0$

31 $2 \sec u \sin u + 2 = 4 \sin u + \sec u$

32 $\tan 2x \cos 2x = \sin 2x$

33 $2 \cos 3x \cos 2x = 1 - 2 \sin 3x \sin 2x$

34 $\sin x \cos 2x + \cos x \sin 2x = 0$

35 $\cos \pi x + \sin \pi x = 0$

36 $\sin 2u = \sin u$

37 $2 \cos^2 \frac{1}{2}\theta - 3 \cos \theta = 0$

38 $\sec 2x \csc 2x = 2 \csc 2x$

39 $\sin 5x = \sin 3x$

40 $\cos 3x = -\cos 2x$

Exer. 41–44: Find the exact value.

41 $\cos 75°$

42 $\tan 285°$

43 $\sin 195°$

44 $\csc \dfrac{\pi}{8}$

Exer. 45–56: If θ and ϕ are acute angles such that $\csc \theta = \frac{5}{3}$ and $\cos \phi = \frac{8}{17}$, find the exact value.

45 $\sin (\theta + \phi)$

46 $\cos (\theta + \phi)$

47 $\tan (\phi + \theta)$

48 $\tan (\theta - \phi)$

49 $\sin (\phi - \theta)$

50 $\sin (\theta - \phi)$

51 $\sin 2\phi$

52 $\cos 2\phi$

53 $\tan 2\theta$

54 $\sin \frac{1}{2}\theta$

55 $\tan \frac{1}{2}\theta$

56 $\cos \frac{1}{2}\phi$

57 Express as a sum or difference:

(a) $\sin 7t \sin 4t$

(b) $\cos \frac{1}{4}u \cos \left(-\frac{1}{6}u\right)$

(c) $6 \cos 5x \sin 3x$

(d) $4 \sin 3\theta \cos 7\theta$

58 Express as a product:

(a) $\sin 8u + \sin 2u$

(b) $\cos 3\theta - \cos 8\theta$

(c) $\sin \frac{1}{4}t - \sin \frac{1}{5}t$

(d) $3 \cos 2x + 3 \cos 6x$

Exer. 59–70: Find the exact value of the expression whenever it is defined.

59 $\cos^{-1} \left(\dfrac{\sqrt{3}}{2}\right)$

60 $\arcsin \left(\dfrac{\sqrt{2}}{2}\right)$

61 $\arctan \sqrt{3}$

62 $\arccos \left(\tan \dfrac{3\pi}{4}\right)$

63 $\arcsin \left(\sin \dfrac{5\pi}{4}\right)$

64 $\cos^{-1} \left(\cos \dfrac{5\pi}{4}\right)$

65 $\sin \left[\arccos \left(-\dfrac{\sqrt{3}}{2}\right)\right]$

66 $\tan (\tan^{-1} 2)$

67 $\sec \left(\sin^{-1} \frac{3}{2}\right)$

68 $\cos^{-1} (\sin 0)$

69 $\cos \left(\sin^{-1} \frac{15}{17} - \sin^{-1} \frac{8}{17}\right)$

70 $\cos \left(2 \sin^{-1} \frac{4}{5}\right)$

Exer. 71–74: Sketch the graph of the equation.

71 $y = \cos^{-1} 3x$

72 $y = 4 \sin^{-1} x$

73 $y = 1 - \sin^{-1} x$

74 $y = \sin \left(\frac{1}{2} \cos^{-1} x\right)$

75 Express $\cos (\alpha + \beta + \gamma)$ in terms of trigonometric functions of α, β, and γ.

76 *Force of a foot* When an individual is walking, the magnitude F of the vertical force of one foot on the ground (see the figure) can be described by

$$F = A(\cos bt - a \cos 3bt),$$

where t is time in seconds, $A > 0$, $b > 0$, and $0 < a < 1$.

(a) Show that $F = 0$ when $t = -\pi/(2b)$ and $t = \pi/(2b)$. (The time $t = -\pi/(2b)$ corresponds to the moment when the foot first touches the ground and the weight of the body is being supported by the other foot.)

(b) The maximum force occurs when

$$3a \sin 3bt = \sin bt.$$

If $a = \frac{1}{3}$, find the solutions of this equation for the interval $-\pi/(2b) < t < \pi/(2b)$.

(c) If $a = \frac{1}{3}$, express the maximum force in terms of A.

Exercise 76

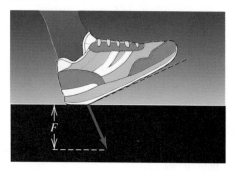

77 Shown in the figure is a graph of the equation

$$y = \sin x - \tfrac{1}{2}\sin 2x + \tfrac{1}{3}\sin 3x.$$

The x-coordinates of the turning points are solutions of the equation $\cos x - \cos 2x + \cos 3x = 0$. Use a sum-to-product formula to find these x-coordinates.

Exercise 77

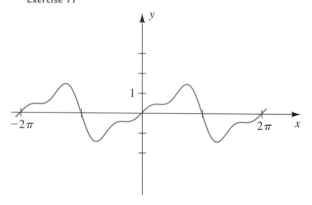

78 *Visual distinction* The human eye can distinguish between two distant points P and Q provided the angle of resolution θ is not too small. Suppose P and Q are x units apart and are d units from the eye, as illustrated in the figure.

(a) Express x in terms of d and θ.

(b) For a person with normal vision, the smallest distinguishable angle of resolution is about 0.0005 radian. If a pen 6 inches long is viewed by such an individual at a distance of d feet, for what values of d will the end points of the pen be distinguishable?

Exercise 78

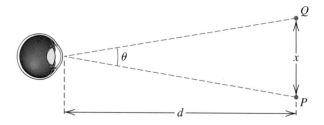

79 *Satellites* A satellite S circles a planet at a distance d miles from the planet's surface. The portion of the planet's surface that is visible from the satellite is determined by the angle θ indicated in the figure.

(a) Assuming that the planet is spherical in shape, express d in terms of θ and the radius r of the planet.

(b) Approximate θ for a satellite 300 miles from the surface of the earth, using $r = 4000$ mi.

Exercise 79

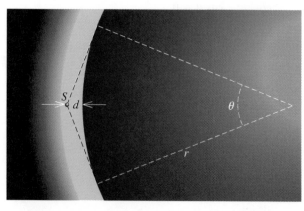

80 *Urban canyons* Because of the tall buildings and relatively narrow streets in some inner cities, the amount of

sunlight illuminating these "canyons" is greatly reduced. If h is the average height of the buildings and w
is the width of the street, the narrowness N of the street
is defined by $N = h/w$. The angle θ of the horizon is defined by $\tan \theta = N$. (The value $\theta = 63°$ may result in an

85% loss of illumination.) Approximate the angle of the
horizon for the following values of h and w.

(a) $h = 400$ ft, $w = 80$ ft

(b) $h = 55$ m, $w = 30$ m

CHAPTER 3   DISCUSSION   EXERCISES

1 Verify the following identity:

$$\frac{\tan x}{1 - \cot x} + \frac{\cot x}{1 - \tan x} = 1 + \sec x \csc x$$

(*Hint:* At some point, consider a special factoring.)

2 Refer to Example 5 of Section 3.1. Suppose $0 \le \theta < 2\pi$,
and rewrite the conclusion using a piecewise-defined
function.

3 How many solutions does the following equation have
on $[0, 2\pi)$? Find the largest one.

$$3 \cos 45x + 4 \sin 45x = 5$$

4 Graph the difference quotient for $f(x) = \sin x$ and $h =$
0.5, 0.1, and 0.001 on the viewing rectangle $[0, 2\pi]$ by
$[-2, 2]$. What generalization can you make from these
graphs? Show that this quotient can be written as

$$\sin x \left(\frac{\cos h - 1}{h} \right) + \cos x \left(\frac{\sin h}{h} \right).$$

5 There are several interesting exact relationships between
π and inverse trigonometric functions such as

$$\frac{\pi}{4} = 4 \tan^{-1} \left(\tfrac{1}{5} \right) - \tan^{-1} \left(\tfrac{1}{239} \right).$$

Use trigonometric identities to prove that this relationship is true. Another such relationship is

$$\frac{\pi}{4} = \tan^{-1} \left(\tfrac{1}{2} \right) + \tan^{-1} \left(\tfrac{1}{5} \right) + \tan^{-1} \left(\tfrac{1}{8} \right).$$

6 Shown in the figure is a function called a *sawtooth
function.*

(a) Define an inverse sawtooth function (**arcsaw**), including its domain and range.

(b) Find arcsaw (1.7) and arcsaw (−0.8).

(c) Formulate two properties of arcsaw (similar to the
sin (sin^{-1}) property).

(d) Graph the arcsaw function.

Exercise 6

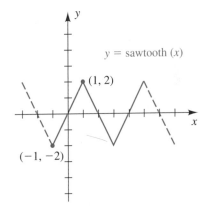

4

APPLICATIONS OF TRIGONOMETRY

In the first two sections of this chapter we consider methods of solving oblique triangles using the law of sines and the law of cosines. We next introduce the trigonometric form for complex numbers and use it to find all n solutions of equations of the form $w^n = z$, where n is any positive integer and w and z are complex numbers. The last two sections contain an introduction to vectors, a topic that has many applications in engineering, the natural sciences, and advanced mathematics.

4.1 THE LAW OF SINES

An **oblique triangle** is a triangle that does not contain a right angle. We shall use the letters A, B, C, a, b, c, α, β, and γ for parts of triangles, as we did in Chapter 2. Given triangle ABC, let us place angle α in standard position so that B is on the positive x-axis. The case for α obtuse is illustrated in Figure 1; however, the following discussion is also valid if α is acute.

Consider the line through C parallel to the y-axis and intersecting the x-axis at point D. If we let $d(C, D) = h$, then the y-coordinate of C is h. From the theorem on trigonometric functions as ratios,

Figure 1

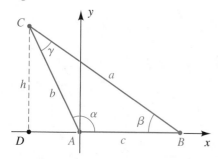

$$\sin \alpha = \frac{h}{b}, \quad \text{so} \quad h = b \sin \alpha.$$

Referring to right triangle BDC, we see that

$$\sin \beta = \frac{h}{a}, \quad \text{so} \quad h = a \sin \beta.$$

Equating the two expressions for h gives us

$$b \sin \alpha = a \sin \beta,$$

which we may write as

$$\frac{\sin \alpha}{a} = \frac{\sin \beta}{b}.$$

If we place α in standard position with C on the positive x-axis, then by the same reasoning,

$$\frac{\sin \alpha}{a} = \frac{\sin \gamma}{c}.$$

The last two equalities give us the following result.

The Law of Sines

> If ABC is an oblique triangle labeled in the usual manner, then
>
> $$\frac{\sin \alpha}{a} = \frac{\sin \beta}{b} = \frac{\sin \gamma}{c}.$$

Note that the law of sines consists of the following three formulas:

(1) $\dfrac{\sin \alpha}{a} = \dfrac{\sin \beta}{b}$ **(2)** $\dfrac{\sin \alpha}{a} = \dfrac{\sin \gamma}{c}$ **(3)** $\dfrac{\sin \beta}{b} = \dfrac{\sin \gamma}{c}$

To apply any one of these formulas to a specific triangle, we must know the values of three of the four variables. If we substitute these three values into the appropriate formula, we can then solve for the value of the fourth variable. It follows that the law of sines can be used to find the remaining parts of an oblique triangle whenever we know either of the following (the three letters in parentheses are used to denote the known parts, with S representing a side and A an angle):

(1) two sides and an angle *opposite* one of them (SSA)

(2) two angles and any side (AAS or ASA)

In the next section we will discuss the law of cosines and show how it can be used to find the remaining parts of an oblique triangle when given the following:

(1) two sides and the angle *between* them (SAS)

(2) three sides (SSS)

The law of sines cannot be applied directly to the last two cases.

The law of sines can also be written in the form

$$\frac{a}{\sin \alpha} = \frac{b}{\sin \beta} = \frac{c}{\sin \gamma}.$$

Instead of memorizing the three formulas associated with the law of sines, it may be more convenient to remember the following statement, which takes all of them into account.

The Law of Sines (General Form)

> In any triangle, the ratio of the sine of an angle to the side opposite that angle is equal to the ratio of the sine of another angle to the side opposite that angle.

In examples and exercises involving triangles, we shall assume that known lengths of sides and angles have been obtained by measurement and hence are approximations to exact values. Unless directed otherwise, when finding parts of triangles we will round off answers according to the following rule: *If known sides or angles are stated to a certain accuracy, then unknown sides or angles should be calculated to the same accuracy.* To illustrate, if known sides are stated to the nearest 0.1, then unknown sides should be calculated to the nearest 0.1. If known angles are stated to the nearest 10′, then unknown angles should be calculated to the nearest 10′. Similar remarks hold for accuracy to the nearest 0.01, 0.1°, and so on.

EXAMPLE 1 Using the law of sines (ASA)

Solve $\triangle ABC$, given $\alpha = 48°$, $\gamma = 57°$, and $b = 47$.

Figure 2

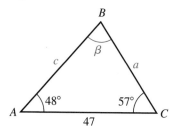

Solution   The triangle is sketched in Figure 2. Since the sum of the angles of a triangle is 180°,

$$\beta = 180° - 57° - 48° = 75°.$$

Since side b and all three angles are known, we can find a by using a form of the law of sines that involves a, α, b, and β:

$$\frac{a}{\sin \alpha} = \frac{b}{\sin \beta} \qquad \text{law of sines}$$

$$a = \frac{b \sin \alpha}{\sin \beta} \qquad \text{solve for } a$$

$$= \frac{47 \sin 48°}{\sin 75°} \qquad \text{substitute for } b, \alpha, \text{ and } \beta$$

$$\approx 36 \qquad \text{approximate to the nearest integer}$$

To find c, we merely replace $\dfrac{a}{\sin \alpha}$ with $\dfrac{c}{\sin \gamma}$ in the preceding solution for a, obtaining

$$c = \frac{b \sin \gamma}{\sin \beta} = \frac{47 \sin 57°}{\sin 75°} \approx 41.$$

Data such as those in Example 1 lead to exactly one triangle ABC. However, if two sides and an angle *opposite* one of them are given, a unique triangle is not always determined. To illustrate, suppose that a and b are to be lengths of sides of triangle ABC and that a given angle α is to be opposite the side of length a. Let us examine the case for α acute. Place α in standard position and consider the line segment AC of length b on the terminal side of α, as shown in Figure 3. The third vertex, B, should be somewhere on the x-axis. Since the length a of the side opposite α is given, we may find B by striking off a circular arc of length a with center at C. The four possible outcomes are illustrated in Figure 4 (without the coordinate axes).

Figure 3

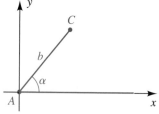

Figure 4

(a)   (b)   (c)   (d)

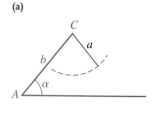 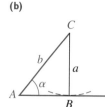 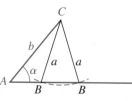 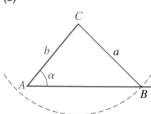

The four possibilities in the figure may be described as follows:

(a) The arc does not intersect the x-axis, and no triangle is formed.

(b) The arc is tangent to the x-axis, and a right triangle is formed.

(c) The arc intersects the positive x-axis in two distinct points, and two triangles are formed.

(d) The arc intersects both the positive and the nonpositive parts of the x-axis, and one triangle is formed.

Figure 5

(a) $a < b$

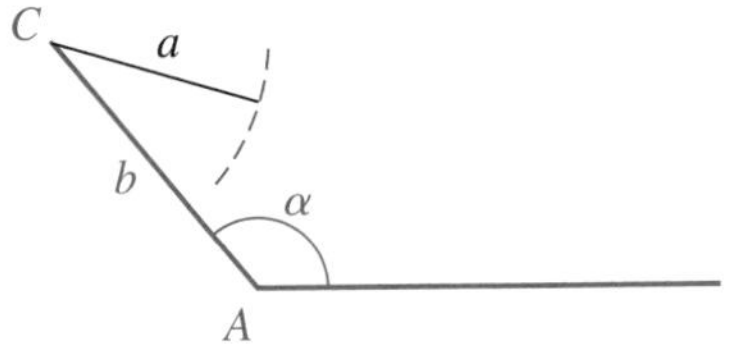

(b) $a > b$

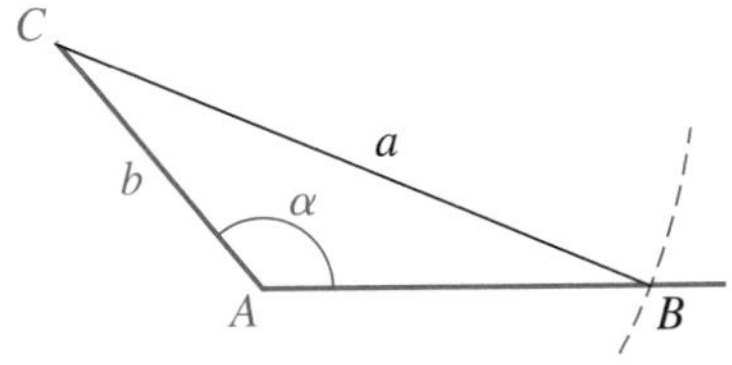

The particular case that occurs in a given problem will become evident when the solution is attempted. For example, if we solve the equation

$$\frac{\sin \alpha}{a} = \frac{\sin \beta}{b}$$

and obtain $\sin \beta > 1$, then no triangle exists and we have case (a). If we obtain $\sin \beta = 1$, then $\beta = 90°$ and hence (b) occurs. If $\sin \beta < 1$, then there are two possible choices for the angle β. By checking both possibilities, we may determine whether (c) or (d) occurs.

If the measure of α is greater than $90°$, then a triangle exists if and only if $a > b$ (see Figure 5). Since we may have more than one possibility when two sides and an angle opposite one of them are given, this situation is sometimes called the **ambiguous case.**

EXAMPLE 2   Using the law of sines (SSA)

Solve $\triangle ABC$, given $\alpha = 67°$, $a = 100$, and $c = 125$.

Solution   Since we know α, a, and c, we can find γ by using a form of the law of sines that involves a, α, c, and γ:

$$\frac{\sin \gamma}{c} = \frac{\sin \alpha}{a} \qquad \text{law of sines}$$

$$\sin \gamma = \frac{c \sin \alpha}{a} \qquad \text{solve for } \sin \gamma$$

$$= \frac{125 \sin 67°}{100} \qquad \text{substitute for } c, \alpha, \text{ and } a$$

$$\approx 1.1506 \qquad \text{approximate}$$

Since $\sin \gamma$ *cannot* be greater than 1, no triangle can be constructed with the given parts.

EXAMPLE 3   Using the law of sines (SSA)

Solve $\triangle ABC$, given $a = 12.4$, $b = 8.7$, and $\beta = 36.7°$.

Solution   To find α, we proceed as follows:

$$\frac{\sin \alpha}{a} = \frac{\sin \beta}{b} \qquad \text{law of sines}$$

$$\sin \alpha = \frac{a \sin \beta}{b} \qquad \text{solve for } \sin \alpha$$

$$= \frac{12.4 \sin 36.7°}{8.7} \qquad \text{substitute for } a, \beta, \text{ and } b$$

$$\approx 0.8518 \qquad \text{approximate}$$

There are two possible angles α between $0°$ and $180°$ such that $\sin \alpha$ is approximately 0.8518. The reference angle α_R is

$$\alpha_R \approx \sin^{-1}(0.8518) \approx 58.4°.$$

Consequently, the two possibilities for α are

$$\alpha_1 \approx 58.4° \qquad \text{and} \qquad \alpha_2 = 180° - \alpha_1 \approx 121.6°.$$

The angle $\alpha_1 \approx 58.4°$ gives us triangle $A_1 BC$ in Figure 6, and the angle $\alpha_2 \approx 121.6°$ gives us triangle $A_2 BC$.

If we let γ_1 and γ_2 denote the third angles of the triangles $A_1 BC$ and $A_2 BC$ corresponding to the angles α_1 and α_2, respectively, then

$$\gamma_1 = 180° - \alpha_1 - \beta \approx 180° - 58.4° - 36.7° \approx 84.9°$$
$$\gamma_2 = 180° - \alpha_2 - \beta \approx 180° - 121.6° - 36.7° \approx 21.7°.$$

If $c_1 = \overline{BA_1}$ is the side opposite γ_1 in triangle $A_1 BC$, then

$$\frac{c_1}{\sin \gamma_1} = \frac{a}{\sin \alpha_1} \qquad \text{law of sines}$$

$$c_1 = \frac{a \sin \gamma_1}{\sin \alpha_1} \qquad \text{solve for } c_1$$

$$\approx \frac{12.4 \sin 84.9°}{\sin 58.4°} \approx 14.5. \qquad \text{substitute and approximate}$$

Thus, the remaining parts of triangle $A_1 BC$ are

$$\alpha_1 \approx 58.4°, \quad \gamma_1 \approx 84.9°, \quad \text{and} \quad c_1 \approx 14.5.$$

Similarly, if $c_2 = \overline{BA_2}$ is the side opposite γ_2 in $\triangle A_2 BC$, then

$$c_2 = \frac{a \sin \gamma_2}{\sin \alpha_2} \approx \frac{12.4 \sin 21.7°}{\sin 121.6°} \approx 5.4,$$

and the remaining parts of triangle $A_2 BC$ are

$$\alpha_2 \approx 121.6°, \quad \gamma_2 \approx 21.7°, \quad \text{and} \quad c_2 \approx 5.4.$$

Figure 6

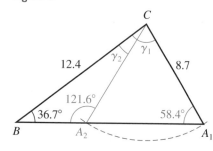

Figure 7

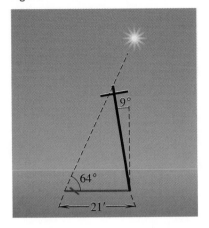

Figure 8

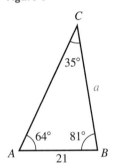

E X A M P L E   4   Using an angle of elevation

When the angle of elevation of the sun is 64°, a telephone pole that is tilted at an angle of 9° directly away from the sun casts a shadow 21 feet long on level ground. Approximate the length of the pole.

Solution   The problem is illustrated in Figure 7. Triangle *ABC* in Figure 8 also displays the given facts. Note that in Figure 8 we have calculated the following angles:

$$\beta = 90° - 9° = 81°$$
$$\gamma = 180° - 64° - 81° = 35°$$

To find the length of the pole—that is, side *a* of triangle *ABC*—we proceed as follows:

$$\frac{a}{\sin 64°} = \frac{21}{\sin 35°} \qquad \text{law of sines}$$

$$a = \frac{21 \sin 64°}{\sin 35°} \approx 33 \quad \text{solve for } a \text{ and approximate}$$

Thus, the telephone pole is approximately 33 feet in length.

E X A M P L E   5   Using bearings

A point *P* on level ground is 3.0 kilometers due north of a point *Q*. A runner proceeds in the direction N25°E from *Q* to a point *R*, and then from *R* to *P* in the direction S70°W. Approximate the distance run.

Solution   The notation used to specify directions was introduced in Section 2.8. The arrows in Figure 9 show the path of the runner, together with a north-south (dashed) line from *R* to another point *S*.

Figure 9

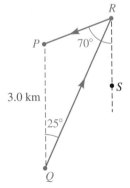

(continued)

Figure 10

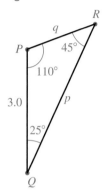

Since the lines through PQ and RS are parallel, it follows from geometry that the alternate interior angles PQR and QRS both have measure 25°. Hence,

$$\angle PRQ = \angle PRS - \angle QRS = 70° - 25° = 45°.$$

These observations give us triangle PQR in Figure 10 with

$$\angle QPR = 180° - 25° - 45° = 110°.$$

We apply the law of sines to find both q and p:

$$\frac{q}{\sin 25°} = \frac{3.0}{\sin 45°} \qquad \text{and} \qquad \frac{p}{\sin 110°} = \frac{3.0}{\sin 45°}$$

$$q = \frac{3.0 \sin 25°}{\sin 45°} \approx 1.8 \quad \text{and} \quad p = \frac{3.0 \sin 110°}{\sin 45°} \approx 4.0$$

The distance run, $p + q$, is approximately $4.0 + 1.8 = 5.8$ km.

EXAMPLE 6   **Locating a school of fish**

A commercial fishing boat uses sonar equipment to detect a school of fish 2 miles east of the boat and traveling in the direction N51°W at a rate of 8 mi/hr (see Figure 11).

Figure 11

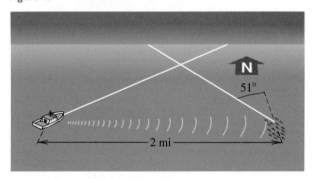

(a) If the boat travels at 20 mi/hr, approximate, to the nearest 0.1°, the direction it should head to intercept the school of fish.

(b) Find, to the nearest minute, the time it will take the boat to reach the fish.

Solution

(a) The problem is illustrated by the triangle in Figure 12, with the school of fish at A, the boat at B, and the point of interception at C. Note that $\alpha = 90° - 51° = 39°$. To obtain β, we begin as follows:

$$\frac{\sin \beta}{b} = \frac{\sin 39°}{a} \qquad \text{law of sines}$$

Figure 12

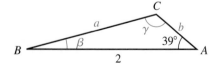

$(*)$ $\sin \beta = \dfrac{b}{a} \sin 39°$ solve for $\sin \beta$

We next find b/a, letting t denote the amount of time required for the boat and fish to meet at C:

$$a = 20t, \qquad b = 8t \qquad \text{(distance)} = \text{(rate)(time)}$$

$$\frac{b}{a} = \frac{8t}{20t} = \frac{2}{5} \qquad \text{divide } b \text{ by } a$$

$$\sin \beta = \tfrac{2}{5} \sin 39° \qquad \text{substitute for } b/a \text{ in } (*)$$

$$\beta = \sin^{-1}\left(\tfrac{2}{5} \sin 39°\right) \approx 14.6° \qquad \text{approximate}$$

Since $90° - 14.6° = 75.4°$, the boat should travel in the (approximate) direction N75.4°E.

(b) We can find t using the relationship $a = 20t$. Let us first find the distance a from B to C. Since the only known side is 2, we need to find the angle γ opposite the side of length 2 in order to use the law of sines. We begin by noting that

$$\gamma \approx 180° - 39° - 14.6° = 126.4°.$$

To find side a, we have

$$\frac{a}{\sin \alpha} = \frac{c}{\sin \gamma} \qquad \text{law of sines}$$

$$a = \frac{c \sin \alpha}{\sin \gamma} \qquad \text{solve for } a$$

$$\approx \frac{2 \sin 39°}{\sin 126.4°} \approx 1.56 \text{ mi.} \qquad \text{substitute and approximate}$$

Using $a = 20t$, we find the time t for the boat to reach C:

$$t = \frac{a}{20} \approx \frac{1.56}{20} \approx 0.08 \text{ hr} \approx 5 \text{ min}$$

4.1 EXERCISES

Exer. 1–16: Solve $\triangle ABC$.

1 $\alpha = 41°,$	$\gamma = 77°,$	$a = 10.5$
2 $\beta = 20°,$	$\gamma = 31°,$	$b = 210$
3 $\alpha = 27°40',$	$\beta = 52°10',$	$a = 32.4$

4 $\beta = 50°50',$	$\gamma = 70°30',$	$c = 537$
5 $\alpha = 42°10',$	$\gamma = 61°20',$	$b = 19.7$
6 $\alpha = 103.45°,$	$\gamma = 27.19°,$	$b = 38.84$
7 $\gamma = 81°,$	$c = 11,$	$b = 12$

8 $\alpha = 32.32°$, $\quad c = 574.3$, $\quad a = 263.6$

9 $\gamma = 53°20'$, $\quad a = 140$, $\quad c = 115$

10 $\alpha = 27°30'$, $\quad c = 52.8$, $\quad a = 28.1$

11 $\gamma = 47.74°$, $\quad a = 131.08$, $\quad c = 97.84$

12 $\alpha = 42.17°$, $\quad a = 5.01$, $\quad b = 6.12$

13 $\alpha = 65°10'$, $\quad a = 21.3$, $\quad b = 18.9$

14 $\beta = 113°10'$, $\quad b = 248$, $\quad c = 195$

15 $\beta = 121.624°$, $\quad b = 0.283$, $\quad c = 0.178$

16 $\gamma = 73.01°$, $\quad a = 17.31$, $\quad c = 20.24$

17 *Surveying* To find the distance between two points A and B that lie on opposite banks of a river, a surveyor lays off a line segment AC of length 240 yards along one bank and determines that the measures of $\angle BAC$ and $\angle ACB$ are $63°20'$ and $54°10'$, respectively (see the figure). Approximate the distance between A and B.

Exercise 17

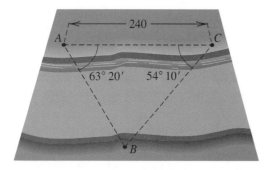

18 *Surveying* To determine the distance between two points A and B, a surveyor chooses a point C that is 375 yards from A and 530 yards from B. If $\angle BAC$ has measure $49°30'$, approximate the distance between A and B.

19 *Cable car route* As shown in the figure, a cable car carries passengers from a point A, which is 1.2 miles from a point B at the base of a mountain, to a point P at the top of the mountain. The angles of elevation of P from A and B are $21°$ and $65°$, respectively.

(a) Approximate the distance between A and P.

(b) Approximate the height of the mountain.

Exercise 19

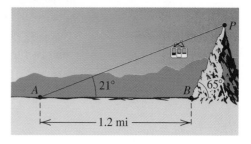

20 *Length of a shadow* A straight road makes an angle of $15°$ with the horizontal. When the angle of elevation of the sun is $57°$, a vertical pole at the side of the road casts a shadow 75 feet long directly down the road, as shown in the figure. Approximate the length of the pole.

Exercise 20

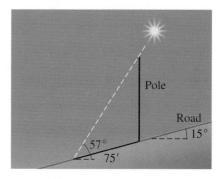

21 *Height of a hot-air balloon* The angles of elevation of a balloon from two points A and B on level ground are $24°10'$ and $47°40'$, respectively. As shown in the figure, points A and B are 8.4 miles apart and the balloon is between the points, in the same vertical plane. Approximate the height of the balloon above the ground.

Exercise 21

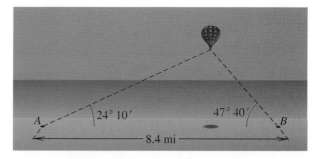

22 *Installing a solar panel* Shown in the figure is a solar panel 10 feet in width, which is to be attached to a roof that makes an angle of 25° with the horizontal. Approximate the length d of the brace that is needed for the panel to make an angle of 45° with the horizontal.

Exercise 22

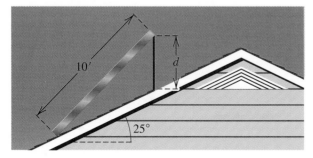

23 *Distance to an airplane* A straight road makes an angle of 22° with the horizontal. From a certain point P on the road, the angle of elevation of an airplane at point A is 57°. At the same instant, from another point Q, 100 meters farther up the road, the angle of elevation is 63°. As indicated in the figure, the points P, Q, and A lie in the same vertical plane. Approximate the distance from P to the airplane.

Exercise 23

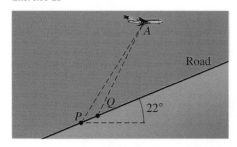

24 *Surveying* A surveyor notes that the direction from point A to point B is S63°W and the direction from A to C is S38°W. The distance from A to B is 239 yards, and the distance from B to C is 374 yards. Approximate the distance from A to C.

25 *Sighting a forest fire* A forest ranger at an observation point A sights a fire in the direction N27°10′E. Another ranger at an observation point B, 6.0 miles due east of A, sights the same fire at N52°40′W. Approximate the distance from each of the observation points to the fire.

26 *Leaning tower of Pisa* The leaning tower of Pisa was originally perpendicular to the ground and 179 feet tall. Because of sinking into the earth, it now leans at a certain angle θ from the perpendicular, as shown in the figure. When the top of the tower is viewed from a point 150 feet from the center of its base, the angle of elevation is 53.3°.

(a) Approximate the angle θ.

(b) Approximate the distance d that the center of the top of the tower has moved from the perpendicular.

Exercise 26

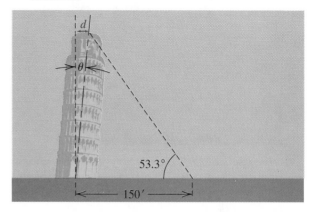

27 *Height of a cathedral* A cathedral is located on a hill, as shown in the figure. When the top of the spire is viewed from the base of the hill, the angle of elevation is 48°. When it is viewed at a distance of 200 feet from the base of the hill, the angle of elevation is 41°. The hill rises at an angle of 32°. Approximate the height of the cathedral.

Exercise 27

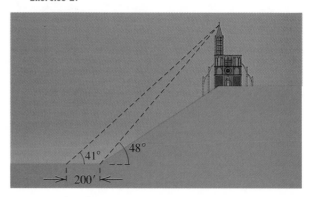

28 *Sighting from a helicopter*   A helicopter hovers at an altitude that is 1000 feet above a mountain peak of altitude 5210 feet, as shown in the figure. A second, taller peak is viewed from both the mountaintop and the helicopter. From the helicopter, the angle of depression is 43°, and from the mountaintop, the angle of elevation is 18°.

(a) Approximate the distance from peak to peak.

(b) Approximate the altitude of the taller peak.

Exercise 28

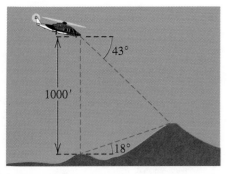

29 *Prism relationships*   The volume V of the right triangular prism shown in the figure is $\frac{1}{3}Bh$, where B is the area of the base and h is the height of the prism. Approximate

(a) h (b) V

Exercise 29

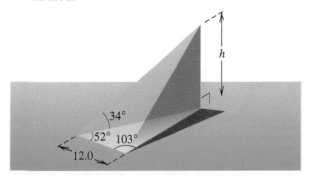

30 *Design for a jet fighter*   Shown in the figure is a plan for the top of a wing of a jet fighter.

(a) Approximate angle ϕ.

(b) If the fuselage is 4.80 feet wide, approximate the wing span CC'.

(c) Approximate the area of triangle ABC.

Exercise 30

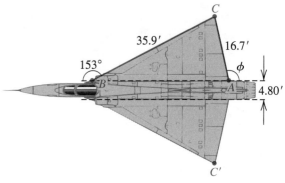

31 *Software for surveyors*   Computer software for surveyors makes use of coordinate systems to locate geographic positions. An offshore oil well at point R in the figure is viewed from points P and Q, and $\angle QPR$ and $\angle RQP$ are found to be 55°50′ and 65°22′, respectively. If points P and Q have coordinates (1487.7, 3452.8) and (3145.8, 5127.5), respectively, approximate the coordinates of R.

Exercise 31

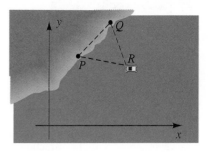

4.2 THE LAW OF COSINES

In the preceding section we stated that the law of sines cannot be applied directly to find the remaining parts of an oblique triangle given either of the following:

(1) two sides and the angle *between* them (SAS)

(2) three sides (SSS)

For these cases we may apply the *law of cosines,* which follows.

The Law of Cosines

If *ABC* is a triangle labeled in the usual manner, then

(**1**) $a^2 = b^2 + c^2 - 2bc \cos \alpha$

(**2**) $b^2 = a^2 + c^2 - 2ac \cos \beta$

(**3**) $c^2 = a^2 + b^2 - 2ab \cos \gamma$

Figure 13

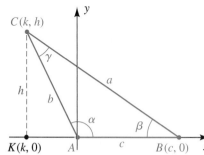

Proof   Let us prove the first formula. Given triangle *ABC*, place α in standard position, as illustrated in Figure 13. We have pictured α as obtuse; however, our discussion is also valid if α is acute. Consider the dashed line through *C*, parallel to the *y*-axis and intersecting the *x*-axis at the point *K*(*k*, 0). If we let $d(C, K) = h$, then *C* has coordinates (*k*, *h*). By the theorem on trigonometric functions as ratios,

$$\cos \alpha = \frac{k}{b} \quad \text{and} \quad \sin \alpha = \frac{h}{b}.$$

Solving for *k* and *h* gives us

$$k = b \cos \alpha \quad \text{and} \quad h = b \sin \alpha.$$

Since the segment *AB* has length *c*, the coordinates of *B* are (*c*, 0), and we obtain the following:

$$
\begin{aligned}
a^2 = [d(B, C)]^2 &= (k - c)^2 + (h - 0)^2 & \text{distance formula} \\
&= (b \cos \alpha - c)^2 + (b \sin \alpha)^2 & \text{substitute for } k \text{ and } h \\
&= b^2 \cos^2 \alpha - 2bc \cos \alpha + c^2 + b^2 \sin^2 \alpha & \text{square} \\
&= b^2(\cos^2 \alpha + \sin^2 \alpha) + c^2 - 2bc \cos \alpha & \text{factor the first and last terms} \\
&= b^2 + c^2 - 2bc \cos \alpha & \text{Pythagorean identity}
\end{aligned}
$$

Our result is the first formula stated in the law of cosines. The second and third formulas may be obtained by placing β and γ, respectively, in standard position on a coordinate system.   ◢

Note that if $\alpha = 90°$ in Figure 13, then $\cos \alpha = 0$ and the law of cosines reduces to $a^2 = b^2 + c^2$. This shows that the Pythagorean theorem is a special case of the law of cosines.

Instead of memorizing each of the three formulas of the law of cosines, it is more convenient to remember the following statement, which takes all of them into account.

The Law of Cosines (General Form)

The square of the length of any side of a triangle equals the sum of the squares of the lengths of the other two sides minus twice the product of the lengths of the other two sides and the cosine of the angle between them.

Given two sides and the included angle of a triangle, we can use the law of cosines to find the third side. We may then use the law of sines to find another angle of the triangle. Whenever this procedure is followed, it is best to find the angle opposite the shortest side, since that angle is always acute. In this way, we avoid the possibility of obtaining two solutions when solving a trigonometric equation involving that angle, as illustrated in the following example.

EXAMPLE 1   **Using the law of cosines (SAS)**

Solve $\triangle ABC$, given $a = 5.0$, $c = 8.0$, and $\beta = 77°$.

Solution   The triangle is sketched in Figure 14. Since β is the angle *between* sides a and c, we begin by approximating b (the side opposite β) as follows:

Figure 14

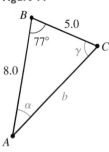

$$
\begin{aligned}
b^2 &= a^2 + c^2 - 2ac \cos \beta & \text{law of cosines} \\
&= (5.0)^2 + (8.0)^2 - 2(5.0)(8.0) \cos 77° & \text{substitute for } a, c, \text{ and } \beta \\
&= 89 - 80 \cos 77° \approx 71.0 & \text{simplify and approximate} \\
b &\approx \sqrt{71.0} \approx 8.4 & \text{take the square root}
\end{aligned}
$$

Let us find another angle of the triangle using the law of sines. In accordance with the remarks preceding this example, we will apply the law of sines and find α, since it is the angle opposite the shortest side a.

$$
\begin{aligned}
\frac{\sin \alpha}{a} &= \frac{\sin \beta}{b} & \text{law of sines} \\
\sin \alpha &= \frac{a \sin \beta}{b} & \text{solve for } \sin \alpha \\
&= \frac{5.0 \sin 77°}{\sqrt{71.0}} \approx 0.5782 & \text{substitute and approximate}
\end{aligned}
$$

Since α is acute,

$$\alpha = \sin^{-1}(0.5782) \approx 35.3° \approx 35°.$$

Finally, since $\alpha + \beta + \gamma = 180°$, we have

$$\gamma = 180° - \alpha - \beta \approx 180° - 35° - 77° = 68°.$$

Given the three sides of a triangle, we can use the law of cosines to find *any* of the three angles. We shall always find the largest angle first—that is, *the angle opposite the longest side*—since this practice will guarantee that the remaining angles are acute. We may then find another angle of the triangle by using either the law of sines or the law of cosines. Note that when an angle is found by means of the law of cosines, there is no ambiguous case, since we always obtain a unique angle between $0°$ and $180°$.

EXAMPLE 2   Using the law of cosines (SSS)

If triangle ABC has sides $a = 90$, $b = 70$, and $c = 40$, approximate angles α, β, and γ to the nearest degree.

Solution   In accordance with the remarks preceding this example, we first find the angle opposite the longest side a. Thus, we choose the form of the law of cosines that involves α and proceed as follows:

$$a^2 = b^2 + c^2 - 2bc \cos \alpha \qquad \text{law of cosines}$$

$$\cos \alpha = \frac{b^2 + c^2 - a^2}{2bc} \qquad \text{solve for } \cos \alpha$$

$$= \frac{70^2 + 40^2 - 90^2}{2(70)(40)} = -\frac{2}{7} \qquad \text{substitute and simplify}$$

$$\alpha = \cos^{-1}\left(-\tfrac{2}{7}\right) \approx 106.6° \approx 107° \qquad \text{approximate } \alpha$$

We may now use either the law of sines or the law of cosines to find β. Let us use the law of cosines in this case:

$$b^2 = a^2 + c^2 - 2ac \cos \beta \qquad \text{law of cosines}$$

$$\cos \beta = \frac{a^2 + c^2 - b^2}{2ac} \qquad \text{solve for } \cos \beta$$

$$= \frac{90^2 + 40^2 - 70^2}{2(90)(40)} = \frac{2}{3} \qquad \text{substitute and simplify}$$

$$\beta = \cos^{-1}\left(\tfrac{2}{3}\right) \approx 48.2° \approx 48° \qquad \text{approximate } \beta$$

Lastly, since $\alpha + \beta + \gamma = 180°$, we have

$$\gamma = 180° - \alpha - \beta \approx 180° - 107° - 48° = 25°.$$

EXAMPLE 3   Approximating the diagonals of a parallelogram

A parallelogram has sides of lengths 30 centimeters and 70 centimeters and one angle of measure 65°. Approximate the length of each diagonal to the nearest centimeter.

Figure 15

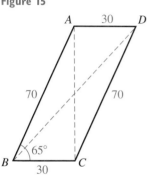

Solution   The parallelogram $ABCD$ and its diagonals AC and BD are shown in Figure 15. Using triangle ABC with $\angle ABC = 65°$, we may approximate AC as follows:

$$(AC)^2 = 30^2 + 70^2 - 2(30)(70) \cos 65° \qquad \text{law of cosines}$$

$$\approx 900 + 4900 - 1775 = 4025 \qquad \text{approximate}$$

$$AC \approx \sqrt{4025} \approx 63 \text{ cm} \qquad \text{take the square root}$$

Similarly, using triangle BAD and $\angle BAD = 180° - 65° = 115°$, we may approximate BD as follows:

$$(BD)^2 = 30^2 + 70^2 - 2(30)(70) \cos 115° \approx 7575 \qquad \text{law of cosines}$$

$$BD \approx \sqrt{7575} \approx 87 \text{ cm} \qquad \text{take the square root}$$

Figure 16

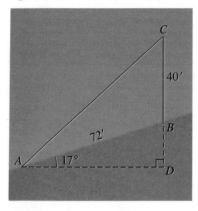

E X A M P L E   4   **Finding the length of a cable**

A vertical pole 40 feet tall stands on a hillside that makes an angle of 17° with the horizontal. Approximate the minimal length of cable that will reach from the top of the pole to a point 72 feet downhill from the base of the pole.

Solution   The sketch in Figure 16 displays the given data. We wish to find AC. Referring to the figure, we see that

$$\angle ABD = 90° - 17° = 73° \quad \text{and} \quad \angle ABC = 180° - 73° = 107°.$$

Using triangle ABC, we may approximate AC as follows:

$(AC)^2 = 72^2 + 40^2 - 2(72)(40) \cos 107° \approx 8468$ law of cosines

$AC \approx \sqrt{8468} \approx 92$ ft   take the square root

---

The law of cosines can be used to derive a formula for the area of a triangle. Let us first prove a preliminary result.

Given triangle ABC, place angle α in standard position (see Figure 13). As shown in the proof of the law of cosines, the altitude h from vertex C is $h = b \sin \alpha$. Since the area $\mathcal{A}$ of the triangle is given by $\mathcal{A} = \frac{1}{2}ch$, we see that

$$\mathcal{A} = \frac{1}{2}bc \sin \alpha.$$

Our argument is independent of the specific angle that is placed in standard position. By taking β and γ in standard position, we obtain the formulas

$$\mathcal{A} = \frac{1}{2}ac \sin \beta \quad \text{and} \quad \mathcal{A} = \frac{1}{2}ab \sin \gamma.$$

All three formulas are covered in the following statement.

Area of a Triangle

> The area of a triangle equals one-half the product of the lengths of any two sides and the sine of the angle between them.

The next two examples illustrate uses of this result.

E X A M P L E   5   **Approximating the area of a triangle**

Approximate the area of triangle ABC if $a = 2.20$ cm, $b = 1.30$ cm, and $\gamma = 43.2°$.

Solution   Since γ is the angle between sides a and b, we may use the preceding result directly, as follows:

$\mathcal{A} = \frac{1}{2}ab \sin \gamma$ area of a triangle formula

$\quad = \frac{1}{2}(2.20)(1.30) \sin 43.2° \approx 0.98$ cm^2 substitute and approximate

---

E X A M P L E   6   Approximating the area of a triangle

Approximate the area of triangle ABC if $a = 5.0$ cm, $b = 3.0$ cm, and $\alpha = 37°$.

Solution   To apply the formula for the area of a triangle, we must find the angle γ between known sides a and b. Since we are given a, b, and α, let us first find β as follows:

$$\frac{\sin \beta}{b} = \frac{\sin \alpha}{a} \qquad \text{law of sines}$$

$$\sin \beta = \frac{b \sin \alpha}{a} \qquad \text{solve for } \sin \beta$$

$$= \frac{3.0 \sin 37°}{5.0} \qquad \text{substitute for } b, \alpha, \text{ and } a$$

$$\beta_R = \sin^{-1}\left(\frac{3.0 \sin 37°}{5.0}\right) \approx 21° \qquad \text{reference angle for } \beta$$

$$\beta \approx 21° \quad \text{or} \quad \beta \approx 159° \qquad \beta_R \text{ or } 180° - \beta_R$$

We reject $\beta \approx 159°$, because then $\alpha + \beta = 196° \geq 180°$. Hence, $\beta \approx 21°$ and

$$\gamma = 180° - \alpha - \beta \approx 180° - 37° - 21° = 122°.$$

Finally, we approximate the area of the triangle as follows:

$$\mathcal{A} = \tfrac{1}{2}ab \sin \gamma \qquad \text{area of a triangle formula}$$

$$\approx \tfrac{1}{2}(5.0)(3.0) \sin 122° \approx 6.4 \text{ cm}^2 \qquad \text{substitute and approximate}$$

We will use the preceding result for the area of a triangle to derive *Heron's formula,* which expresses the area of a triangle in terms of the lengths of its sides.

Heron's Formula

> The area $\mathcal{A}$ of a triangle with sides a, b, and c is given by
> $$\mathcal{A} = \sqrt{s(s - a)(s - b)(s - c)},$$
> where s is one-half the perimeter; that is, $s = \tfrac{1}{2}(a + b + c)$.

Proof   The following equations are equivalent:

$$\mathcal{A} = \tfrac{1}{2}bc \sin \alpha$$

$$= \sqrt{\tfrac{1}{4}b^2c^2 \sin^2 \alpha}$$

$$= \sqrt{\tfrac{1}{4}b^2c^2(1 - \cos^2 \alpha)}$$

$$= \sqrt{\tfrac{1}{2}bc(1 + \cos \alpha) \cdot \tfrac{1}{2}bc(1 - \cos \alpha)}$$

We shall obtain Heron's formula by replacing the expressions under the radical sign by expressions involving only a, b, and c. We solve formula 1 of the law of cosines for $\cos \alpha$ and then substitute, as follows:

$$\begin{aligned}
\frac{1}{2}bc(1 + \cos \alpha) &= \frac{1}{2}bc\left(1 + \frac{b^2 + c^2 - a^2}{2bc}\right) \\
&= \frac{1}{2}bc\left(\frac{2bc + b^2 + c^2 - a^2}{2bc}\right) \\
&= \frac{2bc + b^2 + c^2 - a^2}{4} \\
&= \frac{(b + c)^2 - a^2}{4} \\
&= \frac{(b + c) + a}{2} \cdot \frac{(b + c) - a}{2}
\end{aligned}$$

We use the same type of manipulations on the second expression under the radical sign:

$$\frac{1}{2}bc(1 - \cos \alpha) = \frac{a - b + c}{2} \cdot \frac{a + b + c}{2}$$

If we now substitute for the expressions under the radical sign, we obtain

$$\mathcal{A} = \sqrt{\frac{b + c + a}{2} \cdot \frac{b + c - a}{2} \cdot \frac{a - b + c}{2} \cdot \frac{a + b - c}{2}}.$$

Letting $s = \frac{1}{2}(a + b + c)$, we see that

$$s - a = \frac{b + c - a}{2}, \quad s - b = \frac{a - b + c}{2}, \quad s - c = \frac{a + b - c}{2}.$$

Substitution in the last formula for $\mathcal{A}$ gives us Heron's formula. ◢

E X A M P L E 7   **Using Heron's formula**

A triangular field has sides of lengths 125 yards, 160 yards, and 225 yards. Approximate the number of acres in the field. (One acre is equivalent to 4840 square yards.)

Solution   We first find one-half the perimeter of the field with $a = 125$, $b = 160$, and $c = 225$, as well as the values of $s - a$, $s - b$, and $s - c$:

$$s = \tfrac{1}{2}(125 + 160 + 225) = \tfrac{1}{2}(510) = 255$$
$$s - a = 255 - 125 = 130$$
$$s - b = 255 - 160 = 95$$
$$s - c = 255 - 225 = 30$$

Substituting in Heron's formula gives us

$$\mathcal{A} = \sqrt{(255)(130)(95)(30)} \approx 9720 \text{ yd}^2.$$

Since there are 4840 square yards in one acre, the number of acres is $\frac{9720}{4840}$, or approximately 2.

4.2 EXERCISES

Exer. 1–10: Solve $\triangle ABC$.

1	$\alpha = 60°$,	$b = 20$,	$c = 30$
2	$\gamma = 45°$,	$b = 10.0$,	$a = 15.0$
3	$\beta = 150°$,	$a = 150$,	$c = 30$
4	$\beta = 73°50'$,	$c = 14.0$,	$a = 87.0$
5	$\gamma = 115°10'$,	$a = 1.10$,	$b = 2.10$
6	$\alpha = 23°40'$,	$c = 4.30$,	$b = 70.0$
7	$a = 2.0$,	$b = 3.0$,	$c = 4.0$
8	$a = 10$,	$b = 15$,	$c = 12$
9	$a = 25.0$,	$b = 80.0$,	$c = 60.0$
10	$a = 20.0$,	$b = 20.0$,	$c = 10.0$

11 *Dimensions of a triangular plot* The angle at one corner of a triangular plot of ground is 73°40′, and the sides that meet at this corner are 175 feet and 150 feet long. Approximate the length of the third side.

12 *Surveying* To find the distance between two points A and B, a surveyor chooses a point C that is 420 yards from A and 540 yards from B. If angle ACB has measure 63°10′, approximate the distance between A and B.

13 *Distance between automobiles* Two automobiles leave a city at the same time and travel along straight highways that differ in direction by 84°. If their speeds are 60 mi/hr and 45 mi/hr, respectively, approximately how far apart are the cars at the end of 20 minutes?

14 *Angles of a triangular plot* A triangular plot of land has sides of lengths 420 feet, 350 feet, and 180 feet. Approximate the smallest angle between the sides.

15 *Distance between ships* A ship leaves port at 1:00 P.M. and travels S35°E at the rate of 24 mi/hr. Another ship leaves the same port at 1:30 P.M. and travels S20°W at

18 mi/hr. Approximately how far apart are the ships at 3:00 P.M.?

16 *Flight distance* An airplane flies 165 miles from point A in the direction 130° and then travels in the direction 245° for 80 miles. Approximately how far is the airplane from A?

17 *Jogger's course* A jogger runs at a constant speed of one mile every 8 minutes in the direction S40°E for 20 minutes and then in the direction N20°E for the next 16 minutes. Approximate, to the nearest tenth of a mile, the distance from the endpoint to the starting point of the jogger's course.

18 *Surveying* Two points P and Q on level ground are on opposite sides of a building. To find the distance between the points, a surveyor chooses a point R that is 300 feet from P and 438 feet from Q and then determines that angle PRQ has measure 37°40′ (see the figure). Approximate the distance between P and Q.

Exercise 18

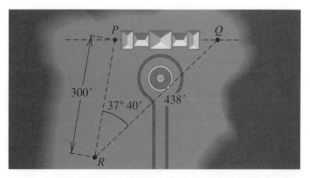

19 *Motorboat's course* A motorboat traveled along a triangular course having sides of lengths 2 kilometers, 4 kilometers, and 3 kilometers, respectively. The first side was traversed in the direction N20°W and the second in a direction Sθ°W, where θ° is the degree measure of an acute angle. Approximate, to the nearest minute, the direction in which the third side was traversed.

20 Angle of a box The rectangular box shown in the figure has dimensions 8″ × 6″ × 4″. Approximate the angle θ formed by a diagonal of the base and a diagonal of the 6″ × 4″ side.

Exercise 20

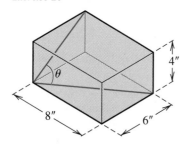

21 Distances in a baseball diamond A baseball diamond has four bases (forming a square) that are 90 feet apart; the pitcher's mound is 60.5 feet from home plate. Approximate the distance from the pitcher's mound to each of the other three bases.

22 A rhombus has sides of length 100 centimeters, and the angle at one of the vertices is 70°. Approximate the lengths of the diagonals to the nearest tenth of a centimeter.

23 Reconnaissance A reconnaissance airplane P, flying at 10,000 feet above a point R on the surface of the water, spots a submarine S at an angle of depression of 37° and a tanker T at an angle of depression of 21°, as shown in the figure. In addition, $\angle SPT$ is found to be 110°. Approximate the distance between the submarine and the tanker.

Exercise 23

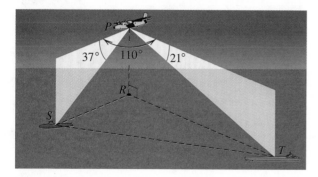

24 Correcting a ship's course A cruise ship sets a course N47°E from an island to a port on the mainland, which is 150 miles away. After moving through strong currents, the ship is off course at a position P that is N33°E and 80 miles from the island, as illustrated in the figure.

(a) Approximately how far is the ship from the port?

(b) In what direction should the ship head to correct its course?

Exercise 24

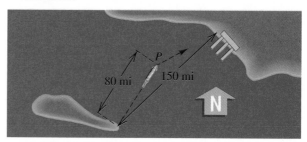

25 Seismology Seismologists investigate the structure of the earth's interior by analyzing seismic waves caused by earthquakes. If the interior of the earth is assumed to be homogeneous, then these waves will travel in straight lines at a constant velocity v. The figure shows a cross-sectional view of the earth, with the epicenter at E and an observation station at S. Use the law of cosines to show that the time t for a wave to travel through the earth's interior from E to S is given by

$$t = \frac{2R}{v} \sin \frac{\theta}{2},$$

where R is the radius of the earth and θ is the indicated angle with vertex at the center of the earth.

Exercise 25

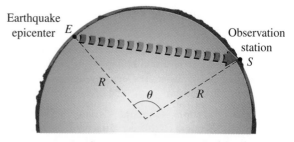

26 Calculating distances The distance across the river shown in the figure can be found without measuring angles. Two points B and C on the opposite shore are selected, and line segments AB and AC are extended as shown. Points D and E are chosen as indicated, and distances BC, BD, BE, CD, and CE are then measured. Suppose that $BC = 184$ ft, $BD = 102$ ft, $BE = 218$ ft, $CD = 236$ ft, and $CE = 80$ ft.

(a) Approximate the distances AB and AC.

(b) Approximate the shortest distance across the river from point A.

Exercise 26

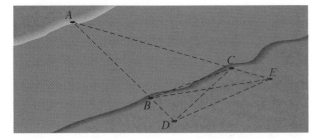

27 *Penrose tiles* Penrose tiles are formed from a rhombus $ABCD$ having sides of length 1 and an interior angle of $72°$. First a point P is located that lies on the diagonal AC and is a distance 1 from vertex C, and then segments PB and PD are drawn to the other vertices of the diagonal, as shown in the figure. The two tiles formed are called a dart and a kite. Three-dimensional counterparts of these tiles have been applied in molecular chemistry.

(a) Find the degree measures of $\angle BPC$, $\angle APB$, and $\angle ABP$.

(b) Approximate, to the nearest 0.01, the length of segment BP.

(c) Approximate, to the nearest 0.01, the area of a kite and the area of a dart.

Exercise 27

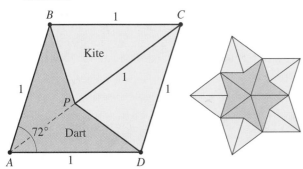

28 *Automotive design* The rear hatchback door of an automobile is 42 inches long. A strut with a fully

extended length of 24 inches is to be attached to the door and the body of the car so that when the door is opened completely, the strut is vertical and the rear clearance is 32 inches, as shown in the figure. Approximate the lengths of segments TQ and TP.

Exercise 28

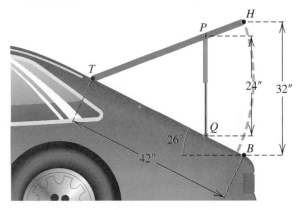

Exer. 29–36: Approximate the area of triangle ABC.

29 $\alpha = 60°,$	$b = 20,$	$c = 30$
30 $\gamma = 45°,$	$b = 10.0,$	$a = 15.0$
31 $\alpha = 40.3°,$	$\beta = 62.9°,$	$b = 5.63$
32 $\alpha = 35.7°,$	$\gamma = 105.2°,$	$b = 17.2$
33 $\alpha = 80.1°,$	$a = 8.0,$	$b = 3.4$
34 $\gamma = 32.1°,$	$a = 14.6,$	$c = 15.8$
35 $a = 25.0,$	$b = 80.0,$	$c = 60.0$
36 $a = 20.0,$	$b = 20.0,$	$c = 10.0$

Exer. 37–38: A triangular field has sides of lengths a, b, and c (in yards). Approximate the number of acres in the field (1 acre = 4840 yd^2).

37 $a = 115,$	$b = 140,$	$c = 200$
38 $a = 320,$	$b = 350,$	$c = 500$

Exer. 39–40: Approximate the area of a parallelogram that has sides of lengths a and b (in feet) if one angle at a vertex has measure θ.

39 $a = 12.0,$	$b = 16.0,$	$\theta = 40°$
40 $a = 40.3,$	$b = 52.6,$	$\theta = 100°$

4.3 TRIGONOMETRIC FORM FOR COMPLEX NUMBERS

*See Appendix V for a review of complex numbers.

In Section 1.1 we represented real numbers geometrically by using points on a coordinate line. We can obtain geometric representations for complex numbers by using points in a coordinate plane.* Specifically, each complex number $a + bi$ determines a unique ordered pair (a, b). The corresponding point $P(a, b)$ in a coordinate plane is the **geometric representation** of $a + bi$. To emphasize that we are assigning complex numbers to points in a plane, we may label the point $P(a, b)$ as $a + bi$. A coordinate plane with a complex number assigned to each point is referred to as a **complex plane** instead of an xy-plane. The x-axis is the **real axis** and the y-axis the **imaginary axis.** In Figure 17 we have represented several complex numbers geometrically. Note that to obtain the point corresponding to the conjugate $a - bi$ of any complex number $a + bi$, we simply reflect through the real axis.

Figure 17

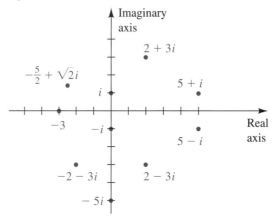

The absolute value of a real number a (denoted $|a|$) is the distance between the origin and the point on the x-axis that corresponds to a. Thus, it is natural to interpret the absolute value of a complex number as the distance between the origin of a complex plane and the point (a, b) that corresponds to $a + bi$.

DEFINITION OF THE ABSOLUTE VALUE OF A COMPLEX NUMBER

If $z = a + bi$ is a complex number, then its **absolute value,** denoted by $|a + bi|$, is

$$\sqrt{a^2 + b^2}.$$

E X A M P L E   1   Finding the absolute value of a complex number

Find

(a) $|2 - 6i|$ **(b)** $|3i|$

Solution   We use the previous definition:

(a) $|2 - 6i| = \sqrt{2^2 + (-6)^2} = \sqrt{40} = 2\sqrt{10} \approx 6.3$

(b) $|3i| = |0 + 3i| = \sqrt{0^2 + 3^2} = \sqrt{9} = 3$

The points corresponding to all complex numbers that have a fixed absolute value k are on a circle of radius k with center at the origin in the complex plane. For example, the points corresponding to the complex numbers z with $|z| = 1$ are on a unit circle.

Let us consider a nonzero complex number $z = a + bi$ and its geometric representation $P(a, b)$, as illustrated in Figure 18. Let θ be any angle in standard position whose terminal side lies on the segment OP, and let $r = |z| = \sqrt{a^2 + b^2}$. Since $\cos \theta = a/r$ and $\sin \theta = b/r$, we see that $a = r \cos \theta$ and $b = r \sin \theta$. Substituting for a and b in $z = a + bi$, we obtain

$$z = a + bi = (r \cos \theta) + (r \sin \theta)i = r(\cos \theta + i \sin \theta).$$

This expression is called the **trigonometric form for the complex number $a + bi$.** A common abbreviation is

$$r(\underline{c}os\ \theta + \underline{i}\ \underline{s}in\ \theta) = r\ \underline{cis}\ \theta.$$

The trigonometric form for $z = a + bi$ is not unique, since there are an unlimited number of different choices for the angle θ. When the trigonometric form is used, the absolute value r of z is sometimes referred to as the **modulus** of z and an angle θ associated with z as an **argument** (or **amplitude**) of z.

We may summarize our discussion as follows.

Figure 18

$z = a + bi = r(\cos \theta + i \sin \theta)$

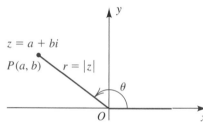

Trigonometric Form for a Complex Number

> Let $z = a + bi$. If $r = |z| = \sqrt{a^2 + b^2}$ and if θ is an argument of z, then
>
> $$z = r(\cos \theta + i \sin \theta) = r\ cis\ \theta.$$

E X A M P L E   2   Expressing a complex number in trigonometric form

Express the complex number in trigonometric form with $0 \le \theta < 2\pi$:

(a) $-4 + 4i$ **(b)** $2\sqrt{3} - 2i$ **(c)** $2 + 7i$ **(d)** $-2 + 7i$

Solution   We begin by representing each complex number geometrically and labeling its modulus r and argument θ, as in Figure 19.

Figure 19

(a)

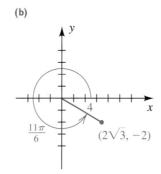

(b)

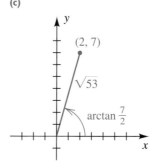

(c)

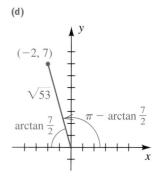

(d)

We next substitute for r and θ in the trigonometric form:

(a) $-4 + 4i = 4\sqrt{2}\left(\cos\dfrac{3\pi}{4} + i\sin\dfrac{3\pi}{4}\right) = 4\sqrt{2}\text{ cis }\dfrac{3\pi}{4}$

(b) $2\sqrt{3} - 2i = 4\left(\cos\dfrac{11\pi}{6} + i\sin\dfrac{11\pi}{6}\right) = 4\text{ cis }\dfrac{11\pi}{6}$

(c) $2 + 7i = \sqrt{53}\left[\cos\left(\arctan\tfrac{7}{2}\right) + i\sin\left(\arctan\tfrac{7}{2}\right)\right] = \sqrt{53}\text{ cis }\left(\arctan\tfrac{7}{2}\right)$

(d) $-2 + 7i = \sqrt{53}\left[\cos\left(\pi - \arctan\tfrac{7}{2}\right) + i\sin\left(\pi - \arctan\tfrac{7}{2}\right)\right]$
$\qquad = \sqrt{53}\text{ cis }\left(\pi - \arctan\tfrac{7}{2}\right)$

If we allow arbitrary values for θ, there are many other trigonometric forms for the complex numbers in Example 2. Thus, for $-4 + 4i$ in part (a) we could use

$$\theta = \frac{3\pi}{4} + 2\pi n \quad \text{for any integer } n.$$

If, for example, we let $n = 1$ and $n = -1$, we obtain

$$4\sqrt{2}\text{ cis }\frac{11\pi}{4} \quad \text{and} \quad 4\sqrt{2}\text{ cis }\left(-\frac{5\pi}{4}\right),$$

respectively. In general, arguments for the same complex number always differ by a multiple of 2π.

If complex numbers are expressed in trigonometric form, then multiplication and division may be performed as indicated in the next theorem.

THEOREM ON PRODUCTS AND QUOTIENTS OF COMPLEX NUMBERS

If trigonometric forms for two complex numbers z_1 and z_2 are

$$z_1 = r_1(\cos \theta_1 + i \sin \theta_1) \quad \text{and} \quad z_2 = r_2(\cos \theta_2 + i \sin \theta_2),$$

then

(1) $z_1 z_2 = r_1 r_2 [\cos (\theta_1 + \theta_2) + i \sin (\theta_1 + \theta_2)]$

(2) $\dfrac{z_1}{z_2} = \dfrac{r_1}{r_2} [\cos (\theta_1 - \theta_2) + i \sin (\theta_1 - \theta_2)], z_2 \neq 0$

Proof   We may prove (1) as follows:

$$z_1 z_2 = r_1(\cos \theta_1 + i \sin \theta_1) \cdot r_2(\cos \theta_2 + i \sin \theta_2)$$
$$= r_1 r_2 [(\cos \theta_1 \cos \theta_2 - \sin \theta_1 \sin \theta_2)$$
$$+ i(\sin \theta_1 \cos \theta_2 + \cos \theta_1 \sin \theta_2)]$$

Applying the addition formulas for $\cos (\theta_1 + \theta_2)$ and $\sin (\theta_1 + \theta_2)$ gives us (1). We leave the proof of (2) as an exercise.   ◄

Part (1) of the preceding theorem states that *the modulus of a product of two complex numbers is the product of their moduli, and an argument is the sum of their arguments.* An analogous statement can be made for (2).

EXAMPLE 3   Using trigonometric forms to find products and quotients

If $z_1 = 2\sqrt{3} - 2i$ and $z_2 = -1 + \sqrt{3}\,i$, use trigonometric forms to find **(a)** $z_1 z_2$ and **(b)** z_1/z_2. Check by using algebraic methods.

Solution   The complex number $2\sqrt{3} - 2i$ is represented geometrically in Figure 19(b). If we use $\theta = -\pi/6$ in the trigonometric form, then

$$z_1 = 2\sqrt{3} - 2i = 4\left[\cos \left(-\frac{\pi}{6} \right) + i \sin \left(-\frac{\pi}{6} \right) \right].$$

The complex number $z_2 = -1 + \sqrt{3}\,i$ is represented geometrically in Figure 20. A trigonometric form is

$$z_2 = -1 + \sqrt{3}\,i = 2\left(\cos \frac{2\pi}{3} + i \sin \frac{2\pi}{3} \right).$$

(a) We apply part (1) of the theorem on products and quotients of complex numbers:

$$z_1 z_2 = 4 \cdot 2\left[\cos \left(-\frac{\pi}{6} + \frac{2\pi}{3} \right) + i \sin \left(-\frac{\pi}{6} + \frac{2\pi}{3} \right) \right]$$

$$= 8\left(\cos \frac{\pi}{2} + i \sin \frac{\pi}{2} \right) = 8(0 + i) = 8i$$

(continued)

Figure 20

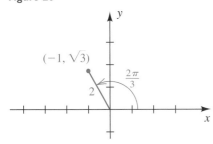

Figure 21

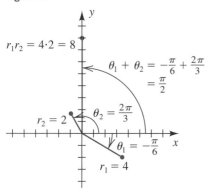

Using algebraic methods to check our result, we have

$$z_1 z_2 = (2\sqrt{3} - 2i)(-1 + \sqrt{3}\,i)$$
$$= (-2\sqrt{3} + 2\sqrt{3}) + (2 + 6)i = 0 + 8i = 8i.$$

Figure 21 gives a geometric interpretation of the product $z_1 z_2$.

(b) We apply part (2) of the theorem:

$$\frac{z_1}{z_2} = \frac{4}{2}\left[\cos\left(-\frac{\pi}{6} - \frac{2\pi}{3}\right) + i\sin\left(-\frac{\pi}{6} - \frac{2\pi}{3}\right)\right]$$
$$= 2\left[\cos\left(-\frac{5\pi}{6}\right) + i\sin\left(-\frac{5\pi}{6}\right)\right]$$
$$= 2\left[-\frac{\sqrt{3}}{2} + i\left(-\frac{1}{2}\right)\right] = -\sqrt{3} - i$$

Figure 22

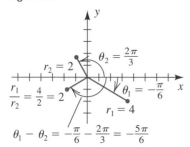

Using algebraic methods, we multiply the numerator and denominator by the conjugate of the denominator to obtain

$$\frac{z_1}{z_2} = \frac{2\sqrt{3} - 2i}{-1 + \sqrt{3}\,i} \cdot \frac{-1 - \sqrt{3}\,i}{-1 - \sqrt{3}\,i}$$
$$= \frac{(-2\sqrt{3} - 2\sqrt{3}) + (2 - 6)i}{(-1)^2 + (\sqrt{3})^2}$$
$$= \frac{-4\sqrt{3} - 4i}{4} = -\sqrt{3} - i.$$

Figure 22 gives a geometric interpretation of the quotient z_1/z_2.

4.3 EXERCISES

Exer. 1–10: Find the absolute value.

1 $|3 - 4i|$ **2** $|5 + 8i|$

3 $|-6 - 7i|$ **4** $|1 - i|$

5 $|8i|$ **6** $|i^7|$

7 $|i^{500}|$ **8** $|-15i|$

9 $|0|$ **10** $|-15|$

Exer. 11–20: Represent the complex number geometrically.

11 $4 + 2i$ **12** $-5 + 3i$

13 $3 - 5i$ **14** $-2 - 6i$

15 $-(3 - 6i)$ **16** $(1 + 2i)^2$

17 $2i(2 + 3i)$ **18** $(-3i)(2 - i)$

19 $(1 + i)^2$ **20** $4(-1 + 2i)$

Exer. 21–46: Express the complex number in trigonometric form with $0 \leq \theta < 2\pi$.

21 $1 - i$ **22** $\sqrt{3} + i$

23 $-4\sqrt{3} + 4i$ **24** $-2 - 2i$

25 $2\sqrt{3} + 2i$ **26** $3 - 3\sqrt{3}\,i$

27 $-4 - 4i$ **28** $-10 + 10i$

29 $-20i$ **30** $-6i$

31 12 **32** 15

33 -7 **34** -5

35 $6i$

36 $4i$

37 $-5 - 5\sqrt{3}\,i$

38 $\sqrt{3} - i$

39 $2 + i$

40 $3 + 2i$

41 $-3 + i$

42 $-4 + 2i$

43 $-5 - 3i$

44 $-2 - 7i$

45 $4 - 3i$

46 $1 - 3i$

Exer. 47–56: Express in the form $a + bi$, where a and b are real numbers.

47 $4\left(\cos \dfrac{\pi}{4} + i \sin \dfrac{\pi}{4}\right)$

48 $8\left(\cos \dfrac{7\pi}{4} + i \sin \dfrac{7\pi}{4}\right)$

49 $6\left(\cos \dfrac{2\pi}{3} + i \sin \dfrac{2\pi}{3}\right)$

50 $12\left(\cos \dfrac{4\pi}{3} + i \sin \dfrac{4\pi}{3}\right)$

51 $5(\cos \pi + i \sin \pi)$

52 $3\left(\cos \dfrac{3\pi}{2} + i \sin \dfrac{3\pi}{2}\right)$

53 $\sqrt{34}\ \text{cis}\left(\tan^{-1} \tfrac{3}{5}\right)$

54 $\sqrt{53}\ \text{cis}\left[\tan^{-1}\left(-\tfrac{2}{7}\right)\right]$

55 $\sqrt{5}\ \text{cis}\left[\tan^{-1}\left(-\tfrac{1}{2}\right)\right]$

56 $\sqrt{10}\ \text{cis}\,(\tan^{-1} 3)$

Exer. 57–64: Use trigonometric forms to find $z_1 z_2$ and z_1/z_2.

57 $z_1 = -1 + i$, $z_2 = 1 + i$

58 $z_1 = \sqrt{3} - i$, $z_2 = -\sqrt{3} - i$

59 $z_1 = -2 - 2\sqrt{3}\,i$, $z_2 = 5i$

60 $z_1 = -5 + 5i$, $z_2 = -3i$

61 $z_1 = -10$, $z_2 = -4$

62 $z_1 = 2i$, $z_2 = -3i$

63 $z_1 = 4$, $z_2 = 2 - i$

64 $z_1 = -3$, $z_2 = 5 + 2i$

65 Prove (2) of the theorem on products and quotients of complex numbers.

66 (a) Extend (1) of the theorem on products and quotients of complex numbers to three complex numbers.

(b) Generalize (1) of the theorem to n complex numbers.

Exer. 67–70: The trigonometric form of complex numbers is often used by electrical engineers to describe the current I, voltage V, and impedance Z in electrical circuits with alternating current. Impedance is the opposition to the flow of current in a circuit. Most common electrical devices operate on 115-volt, alternating current. The relationship among these three quantities is $I = V/Z$. Approximate the unknown quantity, and express the answer in rectangular form to two decimal places.

67 *Finding voltage* $I = 10 \text{ cis } 35°$, $Z = 3 \text{ cis } 20°$

68 *Finding voltage* $I = 12 \text{ cis } 5°$, $Z = 100 \text{ cis } 90°$

69 *Finding impedance* $I = 8 \text{ cis } 5°$, $V = 115 \text{ cis } 45°$

70 *Finding current* $Z = 78 \text{ cis } 61°$, $V = 163 \text{ cis } 17°$

71 *Modulus of impedance* The modulus of the impedance Z represents the total opposition to the flow of electricity in a circuit and is measured in ohms. If $Z = 14 - 13i$, compute $|Z|$.

72 *Resistance and reactance* The absolute value of the real part of Z represents the resistance in an electrical circuit; the absolute value of the complex part represents the reactance. Both quantities are measured in ohms. If $V = 220 \text{ cis } 34°$ and $I = 5 \text{ cis } 90°$, approximate

(a) the resistance   **(b)** the reactance

73 *Actual voltage* The real part of V represents the actual voltage delivered to an electrical appliance in volts. Approximate this voltage when $I = 4 \text{ cis } 90°$ and $Z = 18 \text{ cis } (-78°)$.

74 *Actual current* The real part of I represents the actual current delivered to an electrical appliance in amps. Approximate this current when $V = 163 \text{ cis } 43°$ and $Z = 100 \text{ cis } 17°$.

4.4 DE MOIVRE'S THEOREM AND nTH ROOTS OF COMPLEX NUMBERS

If z is a complex number and n is a positive integer, then a complex number w is an n**th root** of z if $w^n = z$. We will show that every nonzero complex number has n different nth roots. Since $\mathbb{R}$ is contained in $\mathbb{C}$, it will also follow that every nonzero real number has n different nth (complex) roots. If a is a positive real number and $n = 2$, then we already know that the roots are $\sqrt{a}$ and $-\sqrt{a}$.

If, in the theorem on products and quotients of complex numbers, we let both z_1 and z_2 equal the complex number $z = r(\cos \theta + i \sin \theta)$, we obtain

$$z^2 = r \cdot r[\cos (\theta + \theta) + i \sin (\theta + \theta)]$$
$$= r^2(\cos 2\theta + i \sin 2\theta).$$

Applying the same theorem to z^2 and z gives us

$$z^2 \cdot z = (r^2 \cdot r)[\cos (2\theta + \theta) + i \sin (2\theta + \theta)],$$

or
$$z^3 = r^3(\cos 3\theta + i \sin 3\theta).$$

Applying the theorem to z^3 and z, we obtain

$$z^4 = r^4(\cos 4\theta + i \sin 4\theta).$$

In general, we have the following result, named after the French mathematician Abraham De Moivre (1667–1754).

DE MOIVRE'S THEOREM

For every integer n,

$$[r(\cos \theta + i \sin \theta)]^n = r^n(\cos n\theta + i \sin n\theta).$$

We will use only positive integers for n in examples and exercises involving De Moivre's theorem. However, for completeness, the theorem holds for $n = 0$ and n negative if we use the respective real number exponent definitions—that is, $z^0 = 1$ and $z^{-n} = 1/z^n$, where z is a nonzero complex number and n is a positive integer.

E X A M P L E   1   **Using De Moivre's theorem**

Use De Moivre's theorem to change $(1 + i)^{20}$ to the form $a + bi$, where a and b are real numbers.

Figure 23

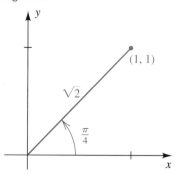

Solution   It would be tedious to change $(1 + i)^{20}$ using algebraic methods. Let us therefore introduce a trigonometric form for $1 + i$. Referring to Figure 23, we see that

$$1 + i = \sqrt{2}\left(\cos \frac{\pi}{4} + i \sin \frac{\pi}{4}\right).$$

We now apply De Moivre's theorem:

$$(1 + i)^{20} = (2^{1/2})^{20}\left[\cos \left(20 \cdot \frac{\pi}{4}\right) + i \sin \left(20 \cdot \frac{\pi}{4}\right)\right]$$
$$= 2^{10}(\cos 5\pi + i \sin 5\pi) = 2^{10}(-1 + 0i) = -1024$$

The number -1024 is of the form $a + bi$ with $a = -1024$ and $b = 0$.

If a nonzero complex number z has an nth root w, then $w^n = z$. If trigonometric forms for w and z are

$(*)$ $w = s(\cos \alpha + i \sin \alpha)$ and $z = r(\cos \theta + i \sin \theta)$,

then applying De Moivre's theorem to $w^n = z$ yields

$$s^n(\cos n\alpha + i \sin n\alpha) = r(\cos \theta + i \sin \theta).$$

If two complex numbers are equal, then so are their absolute values. Consequently, $s^n = r$, and since s and r are nonnegative, $s = \sqrt[n]{r}$. Substituting s^n for r in the last displayed equation and dividing both sides by s^n, we obtain

$$\cos n\alpha + i \sin n\alpha = \cos \theta + i \sin \theta.$$

Since the arguments of equal complex numbers differ by a multiple of 2π, there is an integer k such that $n\alpha = \theta + 2\pi k$. Dividing both sides of the last equation by n, we see that

$$\alpha = \frac{\theta + 2\pi k}{n} \quad \text{for some integer } k.$$

Substituting in the trigonometric form for w (see $(*)$) gives us the formula

$$w = \sqrt[n]{r}\left[\cos\left(\frac{\theta + 2\pi k}{n}\right) + i \sin\left(\frac{\theta + 2\pi k}{n}\right) \right].$$

If we substitute $k = 0, 1, \ldots, n - 1$ successively, we obtain n different nth roots of z. No other value of k will produce a new nth root. For example, if $k = n$, we obtain the angle $(\theta + 2\pi n)/n$, or $(\theta/n) + 2\pi$, which gives us the same nth root as $k = 0$. Similarly, $k = n + 1$ yields the same nth root as $k = 1$, and so on. The same is true for negative values of k. We have proved the following theorem.

THEOREM ON nTH ROOTS

If $z = r(\cos \theta + i \sin \theta)$ is any nonzero complex number and if n is any positive integer, then z has exactly n different nth roots $w_0, w_1, w_2, \ldots, w_{n-1}$. These roots, for θ in radians, are

$$w_k = \sqrt[n]{r}\left[\cos\left(\frac{\theta + 2\pi k}{n}\right) + i \sin\left(\frac{\theta + 2\pi k}{n}\right) \right]$$

or, equivalently, for θ in degrees,

$$w_k = \sqrt[n]{r}\left[\cos\left(\frac{\theta + 360°k}{n}\right) + i \sin\left(\frac{\theta + 360°k}{n}\right) \right],$$

where $k = 0, 1, \ldots, n - 1$.

The nth roots of z in this theorem all have absolute value $\sqrt[n]{r}$, and hence their geometric representations lie on a circle of radius $\sqrt[n]{r}$ with center at O.

Moreover, they are equispaced on this circle, since the difference in the arguments of successive nth roots is $2\pi/n$ (or $360°/n$).

EXAMPLE 2   **Finding the fourth roots of a complex number**

(a) Find the four fourth roots of $-8 - 8\sqrt{3}\,i$.

(b) Represent the roots geometrically.

Figure 24

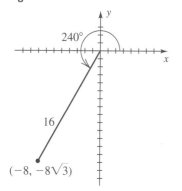

Solution

(a) The geometric representation of $-8 - 8\sqrt{3}\,i$ is shown in Figure 24. Introducing trigonometric form, we have

$$-8 - 8\sqrt{3}\,i = 16(\cos 240° + i \sin 240°).$$

Using the theorem on nth roots with $n = 4$ and noting that $\sqrt[4]{16} = 2$, we find that the fourth roots are

$$w_k = 2\left[\cos\left(\frac{240° + 360°k}{4}\right) + i \sin\left(\frac{240° + 360°k}{4}\right)\right]$$

for $k = 0, 1, 2, 3$. This formula may be written

$$w_k = 2[\cos(60° + 90°k) + i \sin(60° + 90°k)].$$

Substituting 0, 1, 2, and 3 for k in $(60° + 90°k)$ gives us the four fourth roots:

$$w_0 = 2(\cos 60° + i \sin 60°) = 1 + \sqrt{3}\,i$$
$$w_1 = 2(\cos 150° + i \sin 150°) = -\sqrt{3} + i$$
$$w_2 = 2(\cos 240° + i \sin 240°) = -1 - \sqrt{3}\,i$$
$$w_3 = 2(\cos 330° + i \sin 330°) = \sqrt{3} - i$$

Figure 25

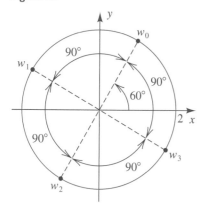

(b) By the comments preceding this example, all roots lie on a circle of radius $\sqrt[4]{16} = 2$ with center at O. The first root, w_0, has an argument of $60°$, and successive roots are spaced apart $360°/4 = 90°$, as shown in Figure 25.

EXAMPLE 3   **Finding the sixth roots of a real number**

(a) Find the six sixth roots of -1.

(b) Represent the roots geometrically.

Solution

(a) Writing $-1 = 1(\cos \pi + i \sin \pi)$ and using the theorem on nth roots with $n = 6$, we find that the sixth roots of -1 are given by

$$w_k = 1\left[\cos\left(\frac{\pi + 2\pi k}{6}\right) + i \sin\left(\frac{\pi + 2\pi k}{6}\right)\right]$$

for $k = 0, 1, 2, 3, 4, 5$. Substituting 0, 1, 2, 3, 4, 5 for k, we obtain the six sixth roots of -1:

$$w_0 = \cos \frac{\pi}{6} + i \sin \frac{\pi}{6} = \frac{\sqrt{3}}{2} + \frac{1}{2}i$$

$$w_1 = \cos \frac{\pi}{2} + i \sin \frac{\pi}{2} = i$$

$$w_2 = \cos \frac{5\pi}{6} + i \sin \frac{5\pi}{6} = -\frac{\sqrt{3}}{2} + \frac{1}{2}i$$

$$w_3 = \cos \frac{7\pi}{6} + i \sin \frac{7\pi}{6} = -\frac{\sqrt{3}}{2} - \frac{1}{2}i$$

$$w_4 = \cos \frac{3\pi}{2} + i \sin \frac{3\pi}{2} = -i$$

$$w_5 = \cos \frac{11\pi}{6} + i \sin \frac{11\pi}{6} = \frac{\sqrt{3}}{2} - \frac{1}{2}i$$

Figure 26

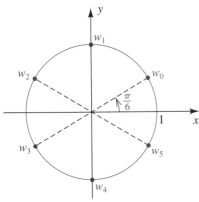

(b) Since $\sqrt[6]{1} = 1$, the points that represent the roots of -1 all lie on the unit circle shown in Figure 26. Moreover, they are equispaced on this circle by $\pi/3$ radians, or $60°$.

The special case in which $z = 1$ is of particular interest. The n distinct nth roots of 1 are called the **nth roots of unity.** In particular, if $n = 3$, we call these roots the **cube roots of unity.**

EXAMPLE 4   Finding the cube roots of unity

Find the three cube roots of unity.

Solution   Writing $1 = 1(\cos 0 + i \sin 0)$ and using the theorem on nth roots with $n = 3$, we obtain

$$w_k = 1\left[\cos \frac{2\pi k}{3} + i \sin \frac{2\pi k}{3}\right]$$

for $k = 0, 1, 2$. Substituting for k gives us the three roots:

$$w_0 = \cos 0 + i \sin 0 = 1$$

$$w_1 = \cos \frac{2\pi}{3} + i \sin \frac{2\pi}{3} = -\frac{1}{2} + \frac{\sqrt{3}}{2}i$$

$$w_2 = \cos \frac{4\pi}{3} + i \sin \frac{4\pi}{3} = -\frac{1}{2} - \frac{\sqrt{3}}{2}i$$

Note that finding the nth roots of a complex number c, as we did in Examples 2–4, is equivalent to finding all the solutions of the equation

$$x^n = c, \quad \text{or} \quad x^n - c = 0.$$

(See Exercises 23–30.)

4.4 EXERCISES

Exer. 1–12: Use De Moivre's theorem to change the given complex number to the form $a + bi$, where a and b are real numbers.

1 $(3 + 3i)^5$

2 $(1 + i)^{12}$

3 $(1 - i)^{10}$

4 $(-1 + i)^8$

5 $(1 - \sqrt{3}\,i)^3$

6 $(1 - \sqrt{3}\,i)^5$

7 $\left(-\dfrac{\sqrt{2}}{2} + \dfrac{\sqrt{2}}{2}i\right)^{15}$

8 $\left(\dfrac{\sqrt{2}}{2} + \dfrac{\sqrt{2}}{2}i\right)^{25}$

9 $\left(-\dfrac{\sqrt{3}}{2} - \dfrac{1}{2}i\right)^{20}$

10 $\left(-\dfrac{\sqrt{3}}{2} - \dfrac{1}{2}i\right)^{50}$

11 $(\sqrt{3} + i)^7$

12 $(-2 - 2i)^{10}$

13 Find the two square roots of $1 + \sqrt{3}\,i$.

14 Find the two square roots of $-9i$.

15 Find the four fourth roots of $-1 - \sqrt{3}\,i$.

16 Find the four fourth roots of $-8 + 8\sqrt{3}\,i$.

17 Find the three cube roots of $-27i$.

18 Find the three cube roots of $64i$.

Exer. 19–22: Find the indicated roots, and represent them geometrically.

19 The six sixth roots of unity

20 The eight eighth roots of unity

21 The five fifth roots of $1 + i$

22 The five fifth roots of $-\sqrt{3} - i$

Exer. 23–30: Find the solutions of the equation.

23 $x^4 - 16 = 0$

24 $x^6 - 64 = 0$

25 $x^6 + 64 = 0$

26 $x^5 + 1 = 0$

27 $x^3 + 8i = 0$

28 $x^3 - 64i = 0$

29 $x^5 - 243 = 0$

30 $x^4 + 81 = 0$

4.5 VECTORS

Figure 27
Equal vectors

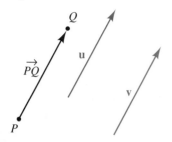

Quantities such as area, volume, length, temperature, and time have magnitude only and can be completely characterized by a single real number (with an appropriate unit of measurement such as in², ft³, cm, deg, or sec). A quantity of this type is a **scalar quantity,** and the corresponding real number is a **scalar**. A concept such as velocity or force has both magnitude and direction and is often represented by a **directed line segment**—that is, a line segment to which a direction has been assigned. Another name for a directed line segment is a **vector**.

As shown in Figure 27, we use $\overrightarrow{PQ}$ to denote the vector with **initial point** P and **terminal point** Q, and we indicate the direction of the vector by placing the arrowhead at Q. The **magnitude** of $\overrightarrow{PQ}$ is the length of the segment PQ and is denoted by $\|\overrightarrow{PQ}\|$. As in the figure, we use boldface letters such as **u** and **v** to denote vectors whose endpoints are not specified. In handwritten work, a notation such as $\vec{u}$ or $\vec{v}$ is often used.

Vectors that have the same magnitude and direction are said to be **equivalent**. In mathematics, a vector is determined only by its magnitude and direction, not by its location. Thus, we regard equivalent vectors, such as those in Figure 27, as **equal** and write

$$\mathbf{u} = \overrightarrow{PQ}, \quad \mathbf{v} = \overrightarrow{PQ}, \quad \text{and} \quad \mathbf{u} = \mathbf{v}.$$

Thus, *a vector may be translated from one location to another, provided neither the magnitude nor the direction is changed.*

We can represent many physical concepts by vectors. To illustrate, suppose an airplane is descending at a constant speed of 100 mi/hr and the line of flight makes an angle of 20° with the horizontal. Both of these facts are represented by the vector **v** of magnitude 100 in Figure 28. The vector **v** is a **velocity vector.**

Figure 28 Velocity vector

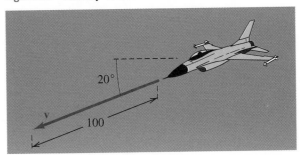

Figure 29
Force vector

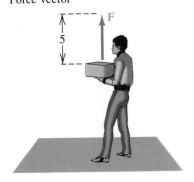

A vector that represents a pull or push of some type is a **force vector.** The force exerted when a person holds a 5-pound weight is illustrated by the vector **F** of magnitude 5 in Figure 29. This force has the same magnitude as the force exerted on the weight by gravity, but it acts in the opposite direction. As a result, there is no movement upward or downward.

We sometimes use $\overrightarrow{AB}$ to represent the path of a point (or particle) as it moves along the line segment from A to B. We then refer to $\overrightarrow{AB}$ as a **displacement** of the point (or particle). As in Figure 30, a displacement $\overrightarrow{AB}$ followed by a displacement $\overrightarrow{BC}$ leads to the same point as the single displacement $\overrightarrow{AC}$. By definition, the vector AC is the **sum** of $\overrightarrow{AB}$ and $\overrightarrow{BC}$, and we write

Figure 30
Sum of vectors

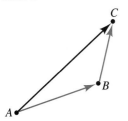

$$\overrightarrow{AC} = \overrightarrow{AB} + \overrightarrow{BC}.$$

Since vectors may be translated from one location to another, *any* two vectors may be added by placing the initial point of the second vector on the terminal point of the first and then drawing the line segment from the initial point of the first to the terminal point of the second, as in Figure 30.

Figure 31
Resultant force

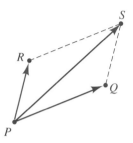

Another way to find the sum is to choose vector PQ and vector PR that are equal to $\overrightarrow{AB}$ and $\overrightarrow{BC}$, respectively, and have the same initial point P, as shown in Figure 31. If we construct parallelogram $RPQS$, then, since $\overrightarrow{PR} = \overrightarrow{QS}$, it follows that $\overrightarrow{PS} = \overrightarrow{PQ} + \overrightarrow{PR}$. If $\overrightarrow{PQ}$ and $\overrightarrow{PR}$ are two forces acting at P, then $\overrightarrow{PS}$ is the **resultant force**—that is, the single force that produces the same effect as the two combined forces.

If m is a scalar and $\mathbf{v}$ is a vector, then $m\mathbf{v}$ is defined as a vector whose magnitude is $|m|$ times $\|\mathbf{v}\|$ (the magnitude of $\mathbf{v}$) and whose direction is either the same as that of $\mathbf{v}$ (if $m > 0$) or opposite that of $\mathbf{v}$ (if $m < 0$). Illustrations are given in Figure 32. We refer to $m\mathbf{v}$ as a **scalar multiple** of $\mathbf{v}$.

Figure 32   Scalar multiples

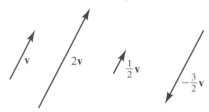

Figure 33

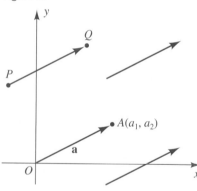

Throughout the remainder of this section we shall restrict our discussion to vectors that lie in an xy-plane. If $\overrightarrow{PQ}$ is such a vector, then, as indicated in Figure 33, there are many vectors that are equivalent to $\overrightarrow{PQ}$; however, there is exactly *one* equivalent vector $\mathbf{a} = \overrightarrow{OA}$ with initial point at the origin. In this sense, *each vector determines a unique ordered pair of real numbers,* the coordinates (a_1, a_2) of the terminal point A. Conversely, every ordered pair (a_1, a_2) determines the vector OA, where A has coordinates (a_1, a_2). Thus, *there is a one-to-one correspondence between vectors in an xy-plane and ordered pairs of real numbers.* This correspondence allows us to interpret a vector as both a directed line segment *and* an ordered pair of real numbers. To avoid confusion with the notation for open intervals or points, we use the symbol $\langle a_1, a_2 \rangle$ for an ordered pair that represents a vector, and we denote it by a boldface letter—for example, $\mathbf{a} = \langle a_1, a_2 \rangle$. The numbers a_1 and a_2 are the **components** of the vector $\langle a_1, a_2 \rangle$. If A is the point (a_1, a_2), as in Figure 33, we call $\overrightarrow{OA}$ the **position vector** for $\langle a_1, a_2 \rangle$ or for the *point A*.

The preceding discussion shows that vectors have two different natures, one geometric and the other algebraic. Often we do not distinguish between the two. It should always be clear from our discussion whether we are referring to ordered pairs or directed line segments.

The *magnitude* of the vector $\mathbf{a} = \langle a_1, a_2 \rangle$ is, by definition, the length of its position vector OA, as illustrated in Figure 34.

Figure 34
Magnitude $\|\mathbf{a}\|$

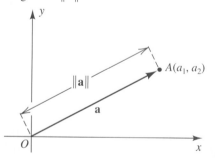

DEFINITION OF THE
MAGNITUDE OF A VECTOR

> The **magnitude** of the vector $\mathbf{a} = \langle a_1, a_2 \rangle$, denoted by $\|\mathbf{a}\|$, is given by
>
> $$\|\mathbf{a}\| = \|\langle a_1, a_2 \rangle\| = \sqrt{a_1^2 + a_2^2}.$$

Figure 35

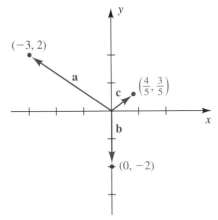

EXAMPLE 1   **Finding the magnitude of a vector**

Sketch the vectors for

$$\mathbf{a} = \langle -3, 2 \rangle, \quad \mathbf{b} = \langle 0, -2 \rangle, \quad \mathbf{c} = \left\langle \tfrac{4}{5}, \tfrac{3}{5} \right\rangle$$

on a coordinate plane, and find the magnitude of each vector.

Solution   The vectors are sketched in Figure 35. By the definition of the magnitude of a vector,

$$\|\mathbf{a}\| = \|\langle -3, 2 \rangle\| = \sqrt{(-3)^2 + 2^2} = \sqrt{13}$$
$$\|\mathbf{b}\| = \|\langle 0, -2 \rangle\| = \sqrt{0^2 + (-2)^2} = \sqrt{4} = 2$$
$$\|\mathbf{c}\| = \left\|\left\langle \tfrac{4}{5}, \tfrac{3}{5} \right\rangle\right\| = \sqrt{\left(\tfrac{4}{5}\right)^2 + \left(\tfrac{3}{5}\right)^2} = \sqrt{\tfrac{25}{25}} = 1.$$

Consider the vector OA and the vector OB corresponding to $\mathbf{a} = \langle a_1, a_2 \rangle$ and $\mathbf{b} = \langle b_1, b_2 \rangle$, respectively, as illustrated in Figure 36. If $\overrightarrow{OC}$ corresponds to $\mathbf{c} = \langle a_1 + b_1, a_2 + b_2 \rangle$, we can show, using slopes, that the points O, A, C, and B are vertices of a parallelogram; that is,

$$\overrightarrow{OA} + \overrightarrow{OB} = \overrightarrow{OC}.$$

Figure 36

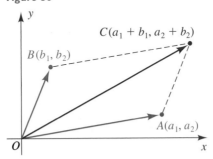

Expressing this equation in terms of ordered pairs leads to the following.

DEFINITION OF
ADDITION OF VECTORS

> $$\langle a_1, a_2 \rangle + \langle b_1, b_2 \rangle = \langle a_1 + b_1, a_2 + b_2 \rangle$$

Note that to add two vectors, we add corresponding components.

ILLUSTRATION

Addition of Vectors

$\langle 3, -4 \rangle + \langle 2, 7 \rangle = \langle 3 + 2, -4 + 7 \rangle = \langle 5, 3 \rangle$

$\langle 5, 1 \rangle + \langle -5, 1 \rangle = \langle 5 + (-5), 1 + 1 \rangle = \langle 0, 2 \rangle$

It can also be shown that if m is a scalar and $\overrightarrow{OA}$ corresponds to $\mathbf{a} = \langle a_1, a_2 \rangle$, then the ordered pair determined by $m\overrightarrow{OA}$ is (ma_1, ma_2), as illustrated in Figure 37 for $m > 1$. This leads to the next definition.

Figure 37

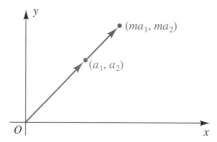

DEFINITION OF A SCALAR MULTIPLE OF A VECTOR

$$m\langle a_1, a_2 \rangle = \langle ma_1, ma_2 \rangle$$

Thus, to find a scalar multiple of a vector, we multiply each component by the scalar.

ILLUSTRATION

Scalar Multiple of a Vector

$2\langle -3, 4 \rangle = \langle 2(-3), 2(4) \rangle = \langle -6, 8 \rangle$

$-2\langle -3, 4 \rangle = \langle (-2)(-3), (-2)(4) \rangle = \langle 6, -8 \rangle$

$1\langle 5, 2 \rangle = \langle 1 \cdot 5, 1 \cdot 2 \rangle = \langle 5, 2 \rangle$

EXAMPLE 2   Finding a scalar multiple of a vector

If $\mathbf{a} = \langle 2, 1 \rangle$, find $3\mathbf{a}$ and $-2\mathbf{a}$, and sketch each vector in a coordinate plane.

Solution   Using the definition of scalar multiples of vectors, we find

$$3\mathbf{a} = 3\langle 2, 1 \rangle = \langle 3 \cdot 2, 3 \cdot 1 \rangle = \langle 6, 3 \rangle$$
$$-2\mathbf{a} = -2\langle 2, 1 \rangle = \langle (-2) \cdot 2, (-2) \cdot 1 \rangle = \langle -4, -2 \rangle.$$

The vectors are sketched in Figure 38.

Figure 38

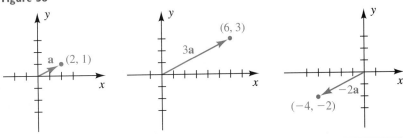

The **zero vector 0** and the **negative** $-\mathbf{a}$ of a vector $\mathbf{a} = \langle a_1, a_2 \rangle$ are defined as follows.

DEFINITION OF 0 AND $-\mathbf{a}$

$$\mathbf{0} = \langle 0, 0 \rangle \quad \text{and} \quad -\mathbf{a} = -\langle a_1, a_2 \rangle = \langle -a_1, -a_2 \rangle$$

ILLUSTRATION

The Zero Vector and the Negative of a Vector

- $\langle 3, 5 \rangle + \mathbf{0} = \langle 3, 5 \rangle + \langle 0, 0 \rangle = \langle 3 + 0, 5 + 0 \rangle = \langle 3, 5 \rangle$
- $-\langle 3, -5 \rangle = \langle -3, -(-5) \rangle = \langle -3, 5 \rangle$
- $\langle 3, -5 \rangle + \langle -3, 5 \rangle = \langle 3 + (-3), -5 + 5 \rangle = \langle 0, 0 \rangle = \mathbf{0}$
- $0\langle 2, 3 \rangle = \langle 0 \cdot 2, 0 \cdot 3 \rangle = \langle 0, 0 \rangle = \mathbf{0}$
- $5 \cdot \mathbf{0} = 5\langle 0, 0 \rangle = \langle 5 \cdot 0, 5 \cdot 0 \rangle = \langle 0, 0 \rangle = \mathbf{0}$

We next state properties of addition and scalar multiples of vectors for any vectors $\mathbf{a}$, $\mathbf{b}$, $\mathbf{c}$ and scalars m, n. You should have little difficulty in remembering these properties, since they resemble familiar properties of real numbers.

Properties of Addition and Scalar Multiples of Vectors

(1) $\mathbf{a} + \mathbf{b} = \mathbf{b} + \mathbf{a}$	**(5)** $m(\mathbf{a} + \mathbf{b}) = m\mathbf{a} + m\mathbf{b}$
(2) $\mathbf{a} + (\mathbf{b} + \mathbf{c}) = (\mathbf{a} + \mathbf{b}) + \mathbf{c}$	**(6)** $(m + n)\mathbf{a} = m\mathbf{a} + n\mathbf{a}$
(3) $\mathbf{a} + \mathbf{0} = \mathbf{a}$	**(7)** $(mn)\mathbf{a} = m(n\mathbf{a}) = n(m\mathbf{a})$
(4) $\mathbf{a} + (-\mathbf{a}) = \mathbf{0}$	**(8)** $1\mathbf{a} = \mathbf{a}$
	(9) $0\mathbf{a} = \mathbf{0} = m\mathbf{0}$

Proof   Let $\mathbf{a} = \langle a_1, a_2 \rangle$ and $\mathbf{b} = \langle b_1, b_2 \rangle$. To prove property 1, we note that

$$\mathbf{a} + \mathbf{b} = \langle a_1 + b_1, a_2 + b_2 \rangle = \langle b_1 + a_1, b_2 + a_2 \rangle = \mathbf{b} + \mathbf{a}.$$

The proof of property 5 is as follows:

$$
\begin{aligned}
m(\mathbf{a} + \mathbf{b}) &= m\langle a_1 + b_1, a_2 + b_2 \rangle && \text{definition of addition} \\
&= \langle m(a_1 + b_1), m(a_2 + b_2) \rangle && \text{definition of scalar multiple} \\
&= \langle ma_1 + mb_1, ma_2 + mb_2 \rangle && \text{distributive property} \\
&= \langle ma_1, ma_2 \rangle + \langle mb_1, mb_2 \rangle && \text{definition of addition} \\
&= m\mathbf{a} + m\mathbf{b} && \text{definition of scalar multiple}
\end{aligned}
$$

Proofs of the remaining properties are similar.

Vector subtraction (denoted by $-$) is defined by $\mathbf{a} - \mathbf{b} = \mathbf{a} + (-\mathbf{b})$. If we use the ordered pair notation for $\mathbf{a}$ and $\mathbf{b}$, then $-\mathbf{b} = \langle -b_1, -b_2 \rangle$, and we obtain the following.

DEFINITION OF SUBTRACTION OF VECTORS

$$
\mathbf{a} - \mathbf{b} = \langle a_1, a_2 \rangle - \langle b_1, b_2 \rangle = \langle a_1 - b_1, a_2 - b_2 \rangle
$$

Thus, to find $\mathbf{a} - \mathbf{b}$, we merely subtract the components of $\mathbf{b}$ from the corresponding components of $\mathbf{a}$.

ILLUSTRATION

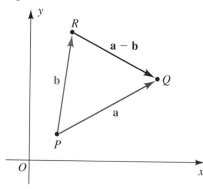

Figure 39

Subtraction of Vectors If $\mathbf{a} = \langle 5, -4 \rangle$ and $\mathbf{b} = \langle -3, 2 \rangle$

$$
\begin{aligned}
\mathbf{a} - \mathbf{b} &= \langle 5, -4 \rangle - \langle -3, 2 \rangle \\
&= \langle 5 - (-3), -4 - 2 \rangle = \langle 8, -6 \rangle
\end{aligned}
$$

$$
\begin{aligned}
2\mathbf{a} - 3\mathbf{b} &= 2\langle 5, -4 \rangle - 3\langle -3, 2 \rangle \\
&= \langle 10, -8 \rangle - \langle -9, 6 \rangle = \langle 19, -14 \rangle
\end{aligned}
$$

If $\mathbf{a}$ and $\mathbf{b}$ are arbitrary vectors, then

$$
\mathbf{b} + (\mathbf{a} - \mathbf{b}) = \mathbf{a};
$$

that is, $\mathbf{a} - \mathbf{b}$ *is the vector that, when added to* $\mathbf{b}$, *gives us* $\mathbf{a}$. If we represent $\mathbf{a}$ and $\mathbf{b}$ by vector PQ and vector PR *with the same initial point*, as in Figure 39, then $\overrightarrow{RQ}$ represents $\mathbf{a} - \mathbf{b}$.

The special vectors $\mathbf{i}$ and $\mathbf{j}$ are defined as follows.

DEFINITION OF i AND j

$$
\mathbf{i} = \langle 1, 0 \rangle, \qquad \mathbf{j} = \langle 0, 1 \rangle
$$

A **unit vector** is a vector of magnitude 1. The vectors $\mathbf{i}$ and $\mathbf{j}$ are unit vectors, as is the vector $\mathbf{c} = \langle \frac{4}{5}, \frac{3}{5} \rangle$ in Example 1.

The vectors **i** and **j** can be used to obtain an alternative way of denoting vectors. Specifically, if $\mathbf{a} = \langle a_1, a_2 \rangle$, then

$$\mathbf{a} = \langle a_1, 0 \rangle + \langle 0, a_2 \rangle = a_1 \langle 1, 0 \rangle + a_2 \langle 0, 1 \rangle.$$

This result gives us the following.

i, j Form for Vectors

$$\boxed{\mathbf{a} = \langle a_1, a_2 \rangle = a_1 \mathbf{i} + a_2 \mathbf{j}}$$

ILLUSTRATION

i, j Form

- $\langle 5, 2 \rangle = 5\mathbf{i} + 2\mathbf{j}$
- $\langle -3, 4 \rangle = -3\mathbf{i} + 4\mathbf{j}$
- $\langle 0, -6 \rangle = 0\mathbf{i} + (-6)\mathbf{j} = -6\mathbf{j}$

Vectors corresponding to **i, j**, and an arbitrary vector **a** are illustrated in Figure 40. Since **i** and **j** are unit vectors, $a_1\mathbf{i}$ and $a_2\mathbf{j}$ may be represented by horizontal and vertical vectors of magnitudes $\|a_1\|$ and $\|a_2\|$, respectively, as illustrated in Figure 41. For this reason we call a_1 the **horizontal component** and a_2 the **vertical component** of the vector **a**.

Figure 40 $\mathbf{a} = \langle a_1, a_2 \rangle$

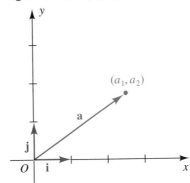

Figure 41 $\mathbf{a} = a_1\mathbf{i} + a_2\mathbf{j}$

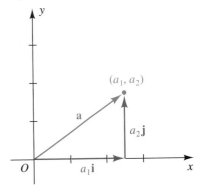

The vector sum $a_1\mathbf{i} + a_2\mathbf{j}$ is a **linear combination** of **i** and **j**. Rules for addition, subtraction, and multiplication by a scalar m may be written as follows, with $\mathbf{b} = \langle b_1, b_2 \rangle = b_1\mathbf{i} + b_2\mathbf{j}$:

$$(a_1\mathbf{i} + a_2\mathbf{j}) + (b_1\mathbf{i} + b_2\mathbf{j}) = (a_1 + b_1)\mathbf{i} + (a_2 + b_2)\mathbf{j}$$
$$(a_1\mathbf{i} + a_2\mathbf{j}) - (b_1\mathbf{i} + b_2\mathbf{j}) = (a_1 - b_1)\mathbf{i} + (a_2 - b_2)\mathbf{j}$$
$$m(a_1\mathbf{i} + a_2\mathbf{j}) = (ma_1)\mathbf{i} + (ma_2)\mathbf{j}$$

These formulas show that we may regard linear combinations of **i** and **j** as algebraic sums.

EXAMPLE 3   **Expressing a vector as a linear combination of i and j**

If $\mathbf{a} = 5\mathbf{i} + \mathbf{j}$ and $\mathbf{b} = 4\mathbf{i} - 7\mathbf{j}$, express $3\mathbf{a} - 2\mathbf{b}$ as a linear combination of $\mathbf{i}$ and $\mathbf{j}$.

Solution

$$
\begin{aligned}
3\mathbf{a} - 2\mathbf{b} &= 3(5\mathbf{i} + \mathbf{j}) - 2(4\mathbf{i} - 7\mathbf{j}) \\
&= (15\mathbf{i} + 3\mathbf{j}) - (8\mathbf{i} - 14\mathbf{j}) \\
&= 7\mathbf{i} + 17\mathbf{j}
\end{aligned}
$$

Figure 42

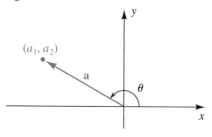

Let θ be an angle in standard position, measured from the positive x-axis to the vector $\mathbf{a} = \langle a_1, a_2 \rangle = a_1\mathbf{i} + a_2\mathbf{j}$, as illustrated in Figure 42. Since

$$
\cos \theta = \frac{a_1}{\|\mathbf{a}\|} \quad \text{and} \quad \sin \theta = \frac{a_2}{\|\mathbf{a}\|},
$$

we obtain the following formulas.

Formulas for Horizontal and Vertical Components of a = $\langle a_1, a_2 \rangle$

If the vector $\mathbf{a}$ and the angle θ are defined as above, then

$$
a_1 = \|\mathbf{a}\| \cos \theta \quad \text{and} \quad a_2 = \|\mathbf{a}\| \sin \theta.
$$

Using these formulas, we have

$$
\begin{aligned}
\mathbf{a} = \langle a_1, a_2 \rangle &= \langle \|\mathbf{a}\| \cos \theta, \|\mathbf{a}\| \sin \theta \rangle \\
&= \|\mathbf{a}\| \cos \theta \, \mathbf{i} + \|\mathbf{a}\| \sin \theta \, \mathbf{j} \\
&= \|\mathbf{a}\|(\cos \theta \, \mathbf{i} + \sin \theta \, \mathbf{j}).
\end{aligned}
$$

This is analogous to the trigonometric form of a complex number.

EXAMPLE 4   **Expressing wind velocity as a vector**

Figure 43

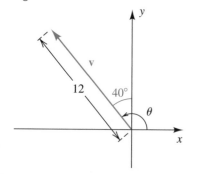

If the wind is blowing at 12 mi/hr in the direction N40°W, express its velocity as a vector $\mathbf{v}$.

Solution   The vector $\mathbf{v}$ is illustrated in Figure 43, where the angle $\theta = 90° + 40° = 130°$. Using the formulas for horizontal and vertical components with $\mathbf{v} = \langle v_1, v_2 \rangle$ gives us

$$
v_1 = \|\mathbf{v}\| \cos \theta = 12 \cos 130°, \qquad v_2 = \|\mathbf{v}\| \sin \theta = 12 \sin 130°.
$$

Hence,

$$
\begin{aligned}
\mathbf{v} &= v_1\mathbf{i} + v_2\mathbf{j} \\
&= (12 \cos 130°)\mathbf{i} + (12 \sin 130°)\mathbf{j} \\
&\approx (-7.7)\mathbf{i} + (9.2)\mathbf{j}.
\end{aligned}
$$

EXAMPLE 5   Finding a resultant vector

Two forces $\overrightarrow{PQ}$ and $\overrightarrow{PR}$ of magnitudes 5.0 kilograms and 8.0 kilograms, respectively, act at a point P. The direction of $\overrightarrow{PQ}$ is N20°E, and the direction of $\overrightarrow{PR}$ is N65°E. Approximate the magnitude and direction of the resultant $\overrightarrow{PS}$.

Figure 44

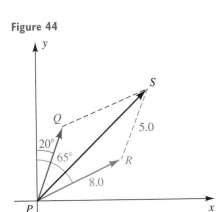

Solution   The forces are represented geometrically in Figure 44. Note that the angles from the positive x-axis to $\overrightarrow{PQ}$ and $\overrightarrow{PR}$ have measures 70° and 25°, respectively. Using the formulas for horizontal and vertical components, we obtain the following:

$$\overrightarrow{PQ} = (5 \cos 70°)\mathbf{i} + (5 \sin 70°)\mathbf{j}$$
$$\overrightarrow{PR} = (8 \cos 25°)\mathbf{i} + (8 \sin 25°)\mathbf{j}$$

Since $\overrightarrow{PS} = \overrightarrow{PQ} + \overrightarrow{PR}$,

$$\overrightarrow{PS} = (5 \cos 70° + 8 \cos 25°)\mathbf{i} + (5 \sin 70° + 8 \sin 25°)\mathbf{j}$$
$$\approx 8.9606\mathbf{i} + 8.0794\mathbf{j} \approx (9.0)\mathbf{i} + (8.1)\mathbf{j}.$$

Consequently,

$$\| \overrightarrow{PS} \| \approx \sqrt{(9.0)^2 + (8.1)^2} \approx 12.1.$$

We can also find $\| \overrightarrow{PS} \|$ by using the law of cosines (see Example 3 of Section 4.2). Since $\angle QPR = 45°$, it follows that $\angle PRS = 135°$, and hence

$$\| \overrightarrow{PS} \|^2 = (8.0)^2 + (5.0)^2 - 2(8.0)(5.0) \cos 135° \approx 145.6$$

and

$$\| \overrightarrow{PS} \| \approx \sqrt{145.6} \approx 12.1.$$

If θ is the angle from the positive x-axis to the resultant PS, then using the (approximate) coordinates (8.9606, 8.0794) of S, we obtain the following:

$$\tan \theta \approx \frac{8.0794}{8.9606} \approx 0.9017$$

$$\theta \approx \tan^{-1}(0.9017) \approx 42°$$

Hence, the direction of $\overrightarrow{PS}$ is approximately N48°E.

4.5 EXERCISES

Exer. 1–6: Find a + b, a − b, 4a + 5b, and 4a − 5b.

1 $\mathbf{a} = \langle 2, -3 \rangle$, $\mathbf{b} = \langle 1, 4 \rangle$

2 $\mathbf{a} = \langle -2, 6 \rangle$, $\mathbf{b} = \langle 2, 3 \rangle$

3 $\mathbf{a} = -\langle 7, -2 \rangle$, $\mathbf{b} = 4\langle -2, 1 \rangle$

4 $\mathbf{a} = 2\langle 5, -4 \rangle$, $\mathbf{b} = -\langle 6, 0 \rangle$

5 $\mathbf{a} = \mathbf{i} + 2\mathbf{j}$, $\mathbf{b} = 3\mathbf{i} - 5\mathbf{j}$

6 $\mathbf{a} = -3\mathbf{i} + \mathbf{j}$, $\mathbf{b} = -3\mathbf{i} + \mathbf{j}$

Exer. 7–10: Sketch vectors corresponding to a, b, a + b, 2a, and −3b.

7 $a = 3i + 2j$, $b = -i + 5j$

8 $a = -5i + 2j$, $b = i - 3j$

9 $a = \langle -4, 6 \rangle$, $b = \langle -2, 3 \rangle$

10 $a = \langle 2, 0 \rangle$, $b = \langle -2, 0 \rangle$

Exer. 11–16: Use components to express the sum or difference as a scalar multiple of one of the vectors a, b, c, d, e, or f shown in the figure.

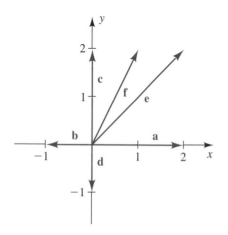

11 a + b   **12** c − d

13 b + e   **14** f − b

15 b + d   **16** e + c

Exer. 17–26: If $a = \langle a_1, a_2 \rangle$, $b = \langle b_1, b_2 \rangle$, $c = \langle c_1, c_2 \rangle$, and m and n are real numbers, prove the stated property.

17 $a + (b + c) = (a + b) + c$

18 $a + 0 = a$

19 $a + (-a) = 0$ **20** $(m + n)a = ma + na$

21 $(mn)a = m(na) = n(ma)$ **22** $1a = a$

23 $0a = 0 = m0$ **24** $(-m)a = -ma$

25 $-(a + b) = -a - b$ **26** $m(a - b) = ma - mb$

27 If $v = \langle a, b \rangle$, prove that the magnitude of 2v is twice the magnitude of v.

28 If $v = \langle a, b \rangle$ and k is any real number, prove that the magnitude of kv is $|k|$ times the magnitude of v.

Exer. 29–36: Find the magnitude of a and the smallest positive angle θ from the positive x-axis to the vector OP that corresponds to a.

29 $a = \langle 3, -3 \rangle$ **30** $a = \langle -2, -2\sqrt{3} \rangle$

31 $a = \langle -5, 0 \rangle$ **32** $a = \langle 0, 10 \rangle$

33 $a = -4i + 5j$ **34** $a = 10i - 10j$

35 $a = -18j$ **36** $a = 2i - 3j$

Exer. 37–40: The vectors a and b represent two forces acting at the same point, and θ is the smallest positive angle between a and b. Approximate the magnitude of the resultant force.

37 $a = 40$ lb, $b = 70$ lb, $\theta = 45°$

38 $a = 5.5$ lb, $b = 6.2$ lb, $\theta = 60°$

39 $a = 2.0$ kg, $b = 8.0$ kg, $\theta = 120°$

40 $a = 30$ kg, $b = 50$ kg, $\theta = 150°$

Exer. 41–44: The magnitudes and directions of two forces acting at a point P are given in **(a)** and **(b)**. Approximate the magnitude and direction of the resultant vector.

41 (a) 90 kg,   N75°W   **(b)** 60 kg,   S5°E

42 (a) 20 kg,   S17°W   **(b)** 50 kg,   N82°W

43 (a) 6.0 lb,   110°   **(b)** 2.0 lb,   215°

44 (a) 70 lb,   320°   **(b)** 40 lb,   30°

Exer. 45–48: Approximate the horizontal and vertical components of the vector that is described.

45 *Releasing a football*   A quarterback releases a football with a velocity of 50 ft/sec at an angle of 35° with the horizontal.

46 *Pulling a sled*   A child pulls a sled through the snow by exerting a force of 20 pounds at an angle of 40° with the horizontal.

47 *Biceps muscle*   The biceps muscle, in supporting the forearm and a weight held in the hand, exerts a force of 20 pounds. As shown in the figure, the muscle makes an angle of 108° with the forearm.

Exercise 47

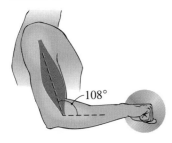

48 **Jet's approach** A jet airplane approaches a runway at an angle of 7.5° with the horizontal, traveling at a velocity of 160 mi/hr.

Exer. 49–52: If forces $F_1, F_2, \ldots, F_n$ act at a point P, the net (or resultant) force F is the sum $F_1 + F_2 + \cdots + F_n$. If $F = 0$, the forces are said to be in equilibrium. The given forces act at the origin O of an xy-plane.

(a) **Find the net force F.**

(b) **Find an additional force G such that equilibrium occurs.**

49 $F_1 = \langle 4, 3 \rangle$, $\quad F_2 = \langle -2, -3 \rangle$, $\quad F_3 = \langle 5, 2 \rangle$

50 $F_1 = \langle -3, -1 \rangle$, $\quad F_2 = \langle 0, -3 \rangle$, $\quad F_3 = \langle 3, 4 \rangle$

51

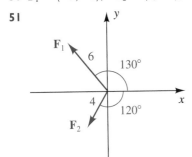

52

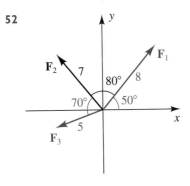

53 **Tugboat force** Two tugboats are towing a large ship into port, as shown in the figure. The larger tug exerts a force of 4000 pounds on its cable, and the smaller tug exerts a force of 3200 pounds on its cable. If the ship is to travel on a straight line l, approximate the angle θ that the larger tug must make with l.

Exercise 53

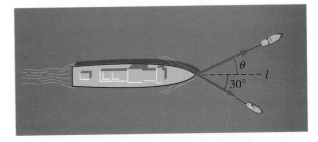

54 **Gravity simulation** Shown in the figure is a simple apparatus that may be used to simulate gravity conditions on other planets. A rope is attached to an astronaut who maneuvers on an inclined plane that makes an angle of θ degrees with the horizontal.

(a) If the astronaut weighs 160 pounds, find the x- and y-components of the downward force (see the figure for axes).

(b) The y-component in part (a) is the weight of the astronaut relative to the inclined plane. The astronaut would weigh 27 pounds on the moon and 60 pounds on Mars. Approximate the angles θ (to the nearest 0.01°) so that the inclined-plane apparatus will simulate walking on these surfaces.

Exercise 54

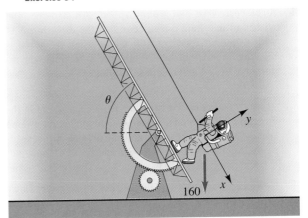

55 *Airplane course and ground speed*   An airplane with an airspeed of 200 mi/hr is flying in the direction 50°, and a 40 mi/hr wind is blowing directly from the west. As shown in the figure, these facts may be represented by vectors **p** and **w** of magnitudes 200 and 40, respectively. The direction of the resultant **p** + **w** gives the true course of the airplane relative to the ground, and the magnitude ∥**p** + **w**∥ is the ground speed of the airplane. Approximate the true course and ground speed.

Exercise 55

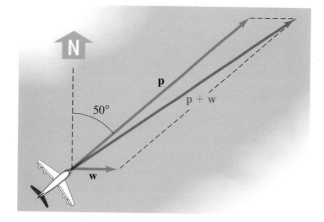

56 *Airplane course and ground speed*   Refer to Exercise 55. An airplane is flying in the direction 140° with an airspeed of 500 mi/hr, and a 30 mi/hr wind is blowing in the direction 65°. Approximate the true course and ground speed of the airplane.

57 *Airplane course and ground speed*   An airplane pilot wishes to maintain a true course in the direction 250° with a ground speed of 400 mi/hr when the wind is blowing directly north at 50 mi/hr. Approximate the required airspeed and compass heading.

58 *Wind direction and speed*   An airplane is flying in the direction 20° with an airspeed of 300 mi/hr. Its ground speed and true course are 350 mi/hr and 30°, respectively. Approximate the direction and speed of the wind.

59 *Rowboat navigation*   The current in a river flows directly from the west at a rate of 1.5 ft/sec. A person who rows a boat at a rate of 4 ft/sec in still water wishes

to row directly north across the river. Approximate, to the nearest degree, the direction in which the person should row.

60 *Motorboat navigation*   For a motorboat moving at a speed of 30 mi/hr to travel directly north across a river, it must aim at a point that has the bearing N15°E. If the current is flowing directly west, approximate the rate at which it flows.

61 *Flow of ground water*   Ground-water contaminants can enter a community's drinking water by migrating through porous rock into the aquifer. If underground water flows with a velocity v_1 through an interface between one type of rock and a second type of rock, its velocity changes to v_2, and both the direction and the speed of the flow can be obtained using the formula

$$\frac{\| v_1 \|}{\| v_2 \|} = \frac{\tan \theta_1}{\tan \theta_2},$$

where the angles θ_1 and θ_2 are as shown in the figure. For sandstone, $\| v_1 \| = 8.2$ cm/day; for limestone, $\| v_2 \| = 3.8$ cm/day. If $\theta_1 = 30°$, approximate the vectors v_1 and v_2 in **i**, **j** form.

Exercise 61

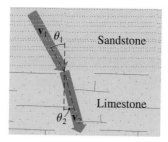

62 *Flow of ground water*   Refer to Exercise 61. Contaminated ground water is flowing through silty sand with the direction of flow θ_1 and speed (in cm/day) given by the vector $v_1 = 20\mathbf{i} - 82\mathbf{j}$. When the flow enters a region of clean sand, its rate increases to 725 cm/day. Find the new direction of flow by approximating θ_2.

63 *Robotic movement*   Vectors are useful for describing movement of robots.

(a) The robot's arm illustrated in the first figure can rotate at the joint connections P and Q. The upper arm, represented by **a**, is 15 inches long, and the forearm (including the hand), represented by **b**, is 17 inches long. Approximate the coordinates of the point R in the hand by using $\mathbf{a} + \mathbf{b}$.

Exercise 63(a)

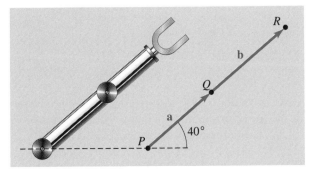

(b) If the upper arm is rotated 85° and the forearm is rotated an additional 35°, as illustrated in the second figure, approximate the new coordinates of R by using $\mathbf{c} + \mathbf{d}$.

Exercise 63(b)

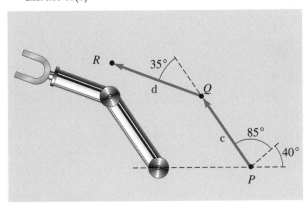

64 Robotic movement Refer to Exercise 63.

(a) Suppose the wrist joint of the robot's arm is allowed to rotate at the joint connection S and the arm is located as shown in the first figure. The upper arm has a length of 15 inches; the forearm, without the hand, has a length of 10 inches; and the hand has a length of 7 inches. Approximate the coordinates of R by using $\mathbf{a} + \mathbf{b} + \mathbf{c}$.

Exercise 64(a)

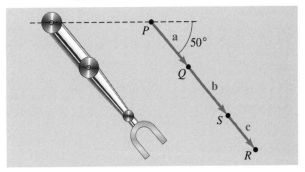

(b) Suppose the robot's upper arm is rotated 75°, and then the forearm is rotated −80°, and finally the hand is rotated an additional 40°, as shown in the second figure. Approximate the new coordinates of R by using $\mathbf{d} + \mathbf{e} + \mathbf{f}$.

Exercise 64(b)

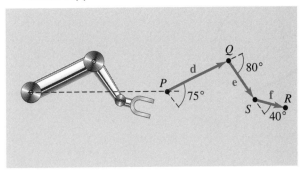

65 Stonehenge forces Refer to Exercise 35 in Section 2.4. In the construction of Stonehenge, groups of 550 people were used to pull 99,000-pound blocks of stone up ramps inclined at 9°. Ignoring friction, determine the force that each person had to contribute in order to move the stone up the ramp.

Exercise 65

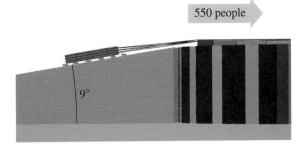

4.6 THE DOT PRODUCT

The *dot product* of two vectors has many applications. We begin with an algebraic definition.

**DEFINITION OF
THE DOT PRODUCT**

> Let $\mathbf{a} = \langle a_1, a_2 \rangle = a_1 \mathbf{i} + a_2 \mathbf{j}$ and $\mathbf{b} = \langle b_1, b_2 \rangle = b_1 \mathbf{i} + b_2 \mathbf{j}$. The **dot product** of $\mathbf{a}$ and $\mathbf{b}$, denoted $\mathbf{a} \cdot \mathbf{b}$, is
>
> $$\mathbf{a} \cdot \mathbf{b} = \langle a_1, a_2 \rangle \cdot \langle b_1, b_2 \rangle = a_1 b_1 + a_2 b_2.$$

The symbol $\mathbf{a} \cdot \mathbf{b}$ is read "**a** dot **b**." We also refer to the dot product as the **scalar product** or the **inner product**. Note that $\mathbf{a} \cdot \mathbf{b}$ *is a real number and not a vector,* as illustrated in the following example.

EXAMPLE 1   **Finding the dot product of two vectors**

Find $\mathbf{a} \cdot \mathbf{b}$.

(a) $\mathbf{a} = \langle -5, 3 \rangle$, $\mathbf{b} = \langle 2, 6 \rangle$ (b) $\mathbf{a} = 4\mathbf{i} + 6\mathbf{j}$, $\mathbf{b} = 3\mathbf{i} - 7\mathbf{j}$

Solution

(a) $\langle -5, 3 \rangle \cdot \langle 2, 6 \rangle = (-5)(2) + (3)(6) = -10 + 18 = 8$

(b) $(4\mathbf{i} + 6\mathbf{j}) \cdot (3\mathbf{i} - 7\mathbf{j}) = (4)(3) + (6)(-7) = 12 - 42 = -30$

Properties of the Dot Product

> If $\mathbf{a}, \mathbf{b}, \mathbf{c}$ are vectors and m is a real number, then
>
> **(1)** $\mathbf{a} \cdot \mathbf{a} = \|\mathbf{a}\|^2$
>
> **(2)** $\mathbf{a} \cdot \mathbf{b} = \mathbf{b} \cdot \mathbf{a}$
>
> **(3)** $\mathbf{a} \cdot (\mathbf{b} + \mathbf{c}) = \mathbf{a} \cdot \mathbf{b} + \mathbf{a} \cdot \mathbf{c}$
>
> **(4)** $(m\mathbf{a}) \cdot \mathbf{b} = m(\mathbf{a} \cdot \mathbf{b}) = \mathbf{a} \cdot (m\mathbf{b})$
>
> **(5)** $\mathbf{0} \cdot \mathbf{a} = 0$

Proof   The proof of each property follows from the definition of the dot product and the properties of real numbers. Thus, if $\mathbf{a} = \langle a_1, a_2 \rangle$, $\mathbf{b} = \langle b_1, b_2 \rangle$, and $\mathbf{c} = \langle c_1, c_2 \rangle$, then

$$\begin{aligned}
\mathbf{a} \cdot (\mathbf{b} + \mathbf{c}) &= \langle a_1, a_2 \rangle \cdot \langle b_1 + c_1, b_2 + c_2 \rangle && \text{definition of addition} \\
&= a_1(b_1 + c_1) + a_2(b_2 + c_2) && \text{definition of dot product} \\
&= (a_1 b_1 + a_2 b_2) + (a_1 c_1 + a_2 c_2) && \text{real number properties} \\
&= \mathbf{a} \cdot \mathbf{b} + \mathbf{a} \cdot \mathbf{c}, && \text{definition of dot product}
\end{aligned}$$

which proves property 3. The proofs of the remaining properties are left as exercises.

Any two nonzero vectors $\mathbf{a} = \langle a_1, a_2 \rangle$ and $\mathbf{b} = \langle b_1, b_2 \rangle$ may be represented in a coordinate plane by directed line segments from the origin O to the points $A(a_1, a_2)$ and $B(b_1, b_2)$, respectively. The **angle θ between a and b** is, by definition, $\angle AOB$ (see Figure 45). Note that $0 \le \theta \le \pi$ and that $\theta = 0$ if $\mathbf{a}$ and $\mathbf{b}$ have the same direction or $\theta = \pi$ if $\mathbf{a}$ and $\mathbf{b}$ have opposite directions.

Figure 45

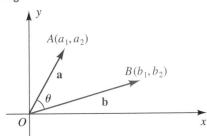

DEFINITION OF PARALLEL AND ORTHOGONAL VECTORS

Let θ be the angle between two nonzero vectors $\mathbf{a}$ and $\mathbf{b}$.

(1) $\mathbf{a}$ and $\mathbf{b}$ are **parallel** if $\theta = 0$ or $\theta = \pi$.

(2) $\mathbf{a}$ and $\mathbf{b}$ are **orthogonal** if $\theta = \dfrac{\pi}{2}$.

The vectors $\mathbf{a}$ and $\mathbf{b}$ in Figure 45 are parallel if and only if they lie on the same line that passes through the origin. In this case, $\mathbf{b} = m\mathbf{a}$ for some real number m. The vectors are orthogonal if and only if they lie on mutually perpendicular lines that pass through the origin. We assume that the zero vector $\mathbf{0}$ is parallel and orthogonal to *every* vector $\mathbf{a}$.

The next theorem shows the close relationship between the angle between two vectors and their dot product.

THEOREM ON THE DOT PRODUCT

If θ is the angle between two nonzero vectors $\mathbf{a}$ and $\mathbf{b}$, then

$$\mathbf{a} \cdot \mathbf{b} = \|\mathbf{a}\|\|\mathbf{b}\| \cos \theta.$$

Proof If $\mathbf{a}$ and $\mathbf{b}$ are not parallel, we have a situation similar to that illustrated in Figure 45. We may then apply the law of cosines to triangle AOB. Since the lengths of the three sides of the triangle are $\|\mathbf{a}\|$, $\|\mathbf{b}\|$, and $d(A, B)$,

$$[d(A, B)]^2 = \|\mathbf{a}\|^2 + \|\mathbf{b}\|^2 - 2\|\mathbf{a}\|\|\mathbf{b}\| \cos \theta.$$

Using the distance formula and the definition of the magnitude of a vector, we obtain

$$(b_1 - a_1)^2 + (b_2 - a_2)^2 = (a_1^2 + a_2^2) + (b_1^2 + b_2^2) - 2\|\mathbf{a}\|\|\mathbf{b}\| \cos \theta,$$

which reduces to

$$-2a_1b_1 - 2a_2b_2 = -2\|\mathbf{a}\|\|\mathbf{b}\| \cos \theta.$$

Dividing both sides of the last equation by -2 gives us

$$a_1b_1 + a_2b_2 = \|\mathbf{a}\|\|\mathbf{b}\| \cos \theta,$$

which is equivalent to what we wished to prove.

If $\mathbf{a}$ and $\mathbf{b}$ are parallel, then either $\theta = 0$ or $\theta = \pi$, and therefore $\mathbf{b} = m\mathbf{a}$ for some real number m with $m > 0$ if $\theta = 0$ and $m < 0$ if $\theta = \pi$. We can show, using properties of the dot product, that $\mathbf{a} \cdot (m\mathbf{a}) = \|\mathbf{a}\|\|m\mathbf{a}\| \cos \theta$, and hence the theorem is true for all nonzero vectors $\mathbf{a}$ and $\mathbf{b}$.

THEOREM ON THE COSINE OF THE ANGLE BETWEEN VECTORS

> If θ is the angle between two nonzero vectors $\mathbf{a}$ and $\mathbf{b}$, then
> $$\cos \theta = \frac{\mathbf{a} \cdot \mathbf{b}}{\|\mathbf{a}\|\|\mathbf{b}\|}.$$

Figure 46

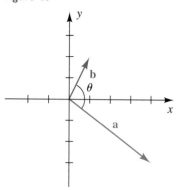

EXAMPLE 2   Finding the angle between two vectors

Find the angle between $\mathbf{a} = \langle 4, -3 \rangle$ and $\mathbf{b} = \langle 1, 2 \rangle$.

Solution   The vectors are sketched in Figure 46. We apply the preceding theorem:

$$\cos \theta = \frac{\mathbf{a} \cdot \mathbf{b}}{\|\mathbf{a}\|\|\mathbf{b}\|} = \frac{(4)(1) + (-3)(2)}{\sqrt{16 + 9}\sqrt{1 + 4}} = \frac{-2}{5\sqrt{5}}, \quad \text{or} \quad \frac{-2\sqrt{5}}{25}$$

Hence,

$$\theta = \arccos\left(\frac{-2\sqrt{5}}{25}\right) \approx 100.3°.$$

EXAMPLE 3   Showing that two vectors are parallel

Let $\mathbf{a} = \frac{1}{2}\mathbf{i} - 3\mathbf{j}$ and $\mathbf{b} = -2\mathbf{i} + 12\mathbf{j}$.

(a) Show that $\mathbf{a}$ and $\mathbf{b}$ are parallel.

(b) Find the scalar m such that $\mathbf{b} = m\mathbf{a}$.

Solution

(a) By definition, the vectors **a** and **b** are parallel if and only if the angle θ between them is either 0 or π. Since

$$\cos \theta = \frac{\mathbf{a} \cdot \mathbf{b}}{\|\mathbf{a}\|\|\mathbf{b}\|} = \frac{(\frac{1}{2})(-2) + (-3)(12)}{\sqrt{\frac{1}{4} + 9}\sqrt{4 + 144}} = \frac{-37}{37} = -1,$$

we conclude that

$$\theta = \arccos(-1) = \pi.$$

(b) Since **a** and **b** are parallel, there *is* a scalar m such that $\mathbf{b} = m\mathbf{a}$; that is,

$$-2\mathbf{i} + 12\mathbf{j} = m(\tfrac{1}{2}\mathbf{i} - 3\mathbf{j}) = \tfrac{1}{2}m\mathbf{i} - 3m\mathbf{j}.$$

Equating **i** and **j** components gives us

$$-2 = \tfrac{1}{2}m \quad \text{and} \quad 12 = -3m.$$

Thus, $m = -4$; that is, $\mathbf{b} = -4\mathbf{a}$. Note that **a** and **b** have opposite directions, since $m < 0$.

Using the formula $\mathbf{a} \cdot \mathbf{b} = \|\mathbf{a}\|\|\mathbf{b}\| \cos \theta$, together with the fact that two vectors are orthogonal if and only if the angle between them is $\pi/2$ (or one of the vectors is **0**), gives us the following result.

THEOREM ON ORTHOGONAL VECTORS

Two vectors **a** and **b** are orthogonal if and only if $\mathbf{a} \cdot \mathbf{b} = 0$.

EXAMPLE 4   Showing that two vectors are orthogonal

Show that the pair of vectors is orthogonal:

(a) **i**, **j**   **(b)** $2\mathbf{i} + 3\mathbf{j}$, $6\mathbf{i} - 4\mathbf{j}$

Solution   We may use the theorem on orthogonal vectors to prove orthogonality by showing that the dot product of each pair is zero:

(a) $\mathbf{i} \cdot \mathbf{j} = \langle 1, 0 \rangle \cdot \langle 0, 1 \rangle = (1)(0) + (0)(1) = 0 + 0 = 0$

(b) $(2\mathbf{i} + 3\mathbf{j}) \cdot (6\mathbf{i} - 4\mathbf{j}) = (2)(6) + (3)(-4) = 12 - 12 = 0$

DEFINITION OF $\text{comp}_b\, \mathbf{a}$

Let θ be the angle between two nonzero vectors **a** and **b**. The **component of a along b,** denoted by $\text{comp}_b\, \mathbf{a}$, is given by

$$\text{comp}_b\, \mathbf{a} = \|\mathbf{a}\| \cos \theta.$$

The geometric significance of the preceding definition with θ acute or obtuse is illustrated in Figure 47, where the x- and y-axes are not shown.

Figure 47 $\text{comp}_b\ \mathbf{a} = \|\mathbf{a}\|\cos\theta$

(a)   (b)

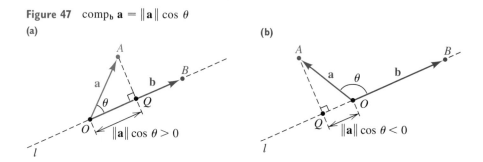

If angle θ is acute, then, as in Figure 47(a), we can form a right triangle by constructing a line segment AQ perpendicular to the line l through O and B. Note that $\overrightarrow{OQ}$ has the same direction as $\overrightarrow{OB}$. Referring to (a) of the figure, we see that

$$\cos\theta = \frac{d(O, Q)}{\|\mathbf{a}\|} \qquad \text{or, equivalently,} \qquad \|\mathbf{a}\|\cos\theta = d(O, Q).$$

If θ is obtuse, then, as in Figure 47(b), we again construct AQ perpendicular to l. In this case, the direction of $\overrightarrow{OQ}$ is opposite that of $\overrightarrow{OB}$, and since $\cos\theta$ is negative,

$$\cos\theta = \frac{-d(O, Q)}{\|\mathbf{a}\|} \qquad \text{or, equivalently,} \qquad \|\mathbf{a}\|\cos\theta = -d(O, Q).$$

special cases for
the component
of a along b

(1) If $\theta = \pi/2$, then $\mathbf{a}$ is orthogonal to $\mathbf{b}$ and $\text{comp}_b\ \mathbf{a} = 0$.

(2) If $\theta = 0$, then $\mathbf{a}$ has the same direction as $\mathbf{b}$ and $\text{comp}_b\ \mathbf{a} = \|\mathbf{a}\|$.

(3) If $\theta = \pi$, then $\mathbf{a}$ and $\mathbf{b}$ have opposite directions and $\text{comp}_b\ \mathbf{a} = -\|\mathbf{a}\|$.

The preceding discussion shows that the component of $\mathbf{a}$ along $\mathbf{b}$ may be found by *projecting* the endpoint of $\mathbf{a}$ onto the line l containing $\mathbf{b}$. For this reason, $\|\mathbf{a}\|\cos\theta$ is sometimes called the **projection of a on b** and is denoted by $\text{proj}_b\ \mathbf{a}$. The following formula shows how to compute this projection *without* knowing the angle θ.

Formula for $\text{comp}_b\ a$

If $\mathbf{a}$ and $\mathbf{b}$ are nonzero vectors, then

$$\text{comp}_b\ \mathbf{a} = \frac{\mathbf{a}\cdot\mathbf{b}}{\|\mathbf{b}\|}.$$

Proof   If θ is the angle between **a** and **b**, then, from the theorem on the dot product,

$$\mathbf{a} \cdot \mathbf{b} = \|\mathbf{a}\| \|\mathbf{b}\| \cos \theta.$$

Dividing both sides of this equation by $\|\mathbf{b}\|$ gives us

$$\frac{\mathbf{a} \cdot \mathbf{b}}{\|\mathbf{b}\|} = \|\mathbf{a}\| \cos \theta = \text{comp}_b \, \mathbf{a}. \quad \blacktriangleleft$$

EXAMPLE 5   **Finding the components of one vector along another**

If $\mathbf{c} = 10\mathbf{i} + 4\mathbf{j}$ and $\mathbf{d} = 3\mathbf{i} - 2\mathbf{j}$, find $\text{comp}_d \, \mathbf{c}$ and $\text{comp}_c \, \mathbf{d}$, and illustrate these numbers graphically.

Figure 48

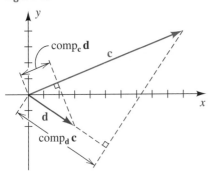

Solution   The vectors **c** and **d** and the desired components are illustrated in Figure 48. We use the formula for $\text{comp}_b \, \mathbf{a}$, as follows:

$$\text{comp}_d \, \mathbf{c} = \frac{\mathbf{c} \cdot \mathbf{d}}{\|\mathbf{d}\|} = \frac{(10)(3) + (4)(-2)}{\sqrt{9 + 4}} = \frac{22}{\sqrt{13}} \approx 6.10$$

$$\text{comp}_c \, \mathbf{d} = \frac{\mathbf{d} \cdot \mathbf{c}}{\|\mathbf{c}\|} = \frac{(3)(10) + (-2)(4)}{\sqrt{100 + 16}} = \frac{22}{\sqrt{116}} \approx 2.04$$

We shall conclude this section with a physical application of the dot product. First let us briefly discuss the scientific concept of *work*.

A **force** may be thought of as the physical entity that is used to describe a push or pull on an object. For example, a force is needed to push or pull an object along a horizontal plane, to lift an object off the ground, or to move a charged particle through an electromagnetic field. Forces are often measured in pounds. If an object weighs 10 pounds, then, by definition, the force required to lift it (or hold it off the ground) is 10 pounds. A force of this type is a **constant force,** since its magnitude does not change while it is applied to the given object.

If a constant force F is applied to an object, moving it a distance d in the direction of the force, then, by definition, the **work** W done is

$$W = Fd.$$

If F is measured in pounds and d in feet, then the units for W are foot-pounds (ft-lb). In the cgs system a **dyne** is used as the unit of force. If F is expressed in dynes and d in centimeters, then the unit for W is the dyne-centimeter, or **erg**. In the mks system, the **newton** is used as the unit of force. If F is in newtons and d is in meters, then the unit for W is the newton-meter, or **joule**.

EXAMPLE 6   Finding the work done by a constant force

Find the work done in pushing an automobile along a level road from a point A to another point B, 40 feet from A, while exerting a constant force of 90 pounds.

Solution   The problem is illustrated in Figure 49, where we have pictured the road as part of a line l. Since the constant force is $F = 90$ lb and the distance the automobile moves is $d = 40$ feet, the work done is

$$W = (90)(40) = 3600 \text{ ft-lb.}$$

Figure 49

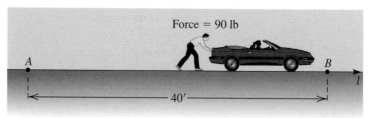

The formula $W = Fd$ is very restrictive, since it can be used only if the force is applied along the line of motion. More generally, suppose that a vector $\mathbf{a}$ represents a force and that its point of application moves along a vector $\mathbf{b}$. This is illustrated in Figure 50, where the force $\mathbf{a}$ is used to pull an object along a level path from O to B, and $\mathbf{b} = \overrightarrow{OB}$.

Figure 50

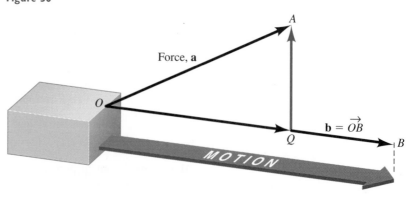

The vector $\mathbf{a}$ is the sum of the vectors $\overrightarrow{OQ}$ and $\overrightarrow{QA}$, where $\overrightarrow{QA}$ is orthogonal to $\mathbf{b}$. Since $\overrightarrow{QA}$ does not contribute to the horizontal movement, we may assume that the motion from O to B is caused by $\overrightarrow{OQ}$ alone. Applying $W = Fd$, we know that the work is the product of $\|\overrightarrow{OQ}\|$ and $\|\mathbf{b}\|$. Since $\|\overrightarrow{OQ}\| = \mathrm{comp}_\mathbf{b}\,\mathbf{a}$, we obtain

$$W = (\mathrm{comp}_\mathbf{b}\,\mathbf{a})\|\mathbf{b}\| = (\|\mathbf{a}\|\cos\theta)\|\mathbf{b}\| = \mathbf{a}\cdot\mathbf{b},$$

where θ represents $\angle AOQ$. This leads to the following definition.

DEFINITION OF WORK

The **work** W done by a constant force $\mathbf{a}$ as its point of application moves along a vector $\mathbf{b}$ is $W = \mathbf{a}\cdot\mathbf{b}$.

Figure 51

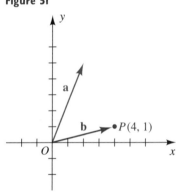

EXAMPLE 7   Finding the work done by a constant force

The magnitude and direction of a constant force are given by $\mathbf{a} = 2\mathbf{i} + 5\mathbf{j}$. Find the work done if the point of application of the force moves from the origin to the point $P(4, 1)$.

Solution   The force $\mathbf{a}$ and the vector $\mathbf{b} = \overrightarrow{OP}$ are sketched in Figure 51. Since $\mathbf{b} = \langle 4, 1\rangle = 4\mathbf{i} + \mathbf{j}$, we have, from the preceding definition,

$$W = \mathbf{a}\cdot\mathbf{b} = (2\mathbf{i} + 5\mathbf{j})\cdot(4\mathbf{i} + \mathbf{j})$$
$$= (2)(4) + (5)(1) = 13.$$

If, for example, the unit of length is feet and the magnitude of the force is measured in pounds, then the work done is 13 ft-lb.

EXAMPLE 8   Finding the work done against gravity

A small cart weighing 100 pounds is pushed up an incline that makes an angle of 30° with the horizontal, as shown in Figure 52. Find the work done against gravity in pushing the cart a distance of 80 feet.

Figure 52

Figure 53

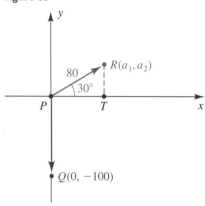

Solution   Let us introduce an *xy*-coordinate system, as shown in Figure 53. The vector *PQ* represents the force of gravity acting vertically downward with a magnitude of 100 pounds. The corresponding vector **F** is $0\mathbf{i} - 100\mathbf{j}$. The point of application of this force moves along the vector *PR* of magnitude 80. If $\overrightarrow{PR}$ corresponds to $\mathbf{a} = a_1\mathbf{i} + a_2\mathbf{j}$, then, referring to triangle *PTR*, we see that

$$a_1 = 80 \cos 30° = 40\sqrt{3}$$
$$a_2 = 80 \sin 30° = 40,$$

and hence $$\mathbf{a} = 40\sqrt{3}\,\mathbf{i} + 40\mathbf{j}.$$

Applying the definition, we find that the work done *by* gravity is

$$\mathbf{F} \cdot \mathbf{a} = (0\mathbf{i} - 100\mathbf{j}) \cdot (40\sqrt{3}\,\mathbf{i} + 40\mathbf{j}) = 0 - 4000 = -4000 \text{ ft-lb}.$$

The work done *against* gravity is

$$-\mathbf{F} \cdot \mathbf{a} = 4000 \text{ ft-lb}.$$

4.6 EXERCISES

Exer. 1–8: Find (a) the dot product of the two vectors and (b) the angle between the two vectors.

1 $\langle -2, 5 \rangle$, $\langle 3, 6 \rangle$

2 $\langle 4, -7 \rangle$, $\langle -2, 3 \rangle$

3 $4\mathbf{i} - \mathbf{j}$, $-3\mathbf{i} + 2\mathbf{j}$

4 $8\mathbf{i} - 3\mathbf{j}$, $2\mathbf{i} - 7\mathbf{j}$

5 $9\mathbf{i}$, $5\mathbf{i} + 4\mathbf{j}$

6 $6\mathbf{j}$, $-4\mathbf{i}$

7 $\langle 10, 7 \rangle$, $\langle -2, -\frac{7}{5} \rangle$

8 $\langle -3, 6 \rangle$, $\langle -1, 2 \rangle$

Exer. 9–12: Show that the vectors are orthogonal.

9 $\langle 4, -1 \rangle$, $\langle 2, 8 \rangle$

10 $\langle 3, 6 \rangle$, $\langle 4, -2 \rangle$

11 $-4\mathbf{j}$, $-7\mathbf{i}$

12 $8\mathbf{i} - 4\mathbf{j}$, $-6\mathbf{i} - 12\mathbf{j}$

Exer. 13–16: Show that the vectors are parallel, and determine whether they have the same direction or opposite directions.

13 $\mathbf{a} = 3\mathbf{i} - 5\mathbf{j}$, $\mathbf{b} = -\frac{12}{7}\mathbf{i} + \frac{20}{7}\mathbf{j}$

14 $\mathbf{a} = -\frac{5}{2}\mathbf{i} + 6\mathbf{j}$, $\mathbf{b} = -10\mathbf{i} + 24\mathbf{j}$

15 $\mathbf{a} = \langle \frac{2}{3}, \frac{1}{2} \rangle$, $\mathbf{b} = \langle 8, 6 \rangle$

16 $\mathbf{a} = \langle 6, 18 \rangle$, $\mathbf{b} = \langle -4, -12 \rangle$

Exer. 17–20: Determine *m* such that the two vectors are orthogonal.

17 $3\mathbf{i} - 2\mathbf{j}$, $4\mathbf{i} + 5m\mathbf{j}$

18 $4m\mathbf{i} + \mathbf{j}$, $9m\mathbf{i} - 25\mathbf{j}$

19 $9\mathbf{i} - 16m\mathbf{j}$, $\mathbf{i} + 4m\mathbf{j}$

20 $5m\mathbf{i} + 3\mathbf{j}$, $2\mathbf{i} + 7\mathbf{j}$

Exer. 21–28: Given that $\mathbf{a} = \langle 2, -3 \rangle$, $\mathbf{b} = \langle 3, 4 \rangle$, and $\mathbf{c} = \langle -1, 5 \rangle$, find the number.

21 (a) $\mathbf{a} \cdot (\mathbf{b} + \mathbf{c})$

(b) $\mathbf{a} \cdot \mathbf{b} + \mathbf{a} \cdot \mathbf{c}$

22 (a) $\mathbf{b} \cdot (\mathbf{a} - \mathbf{c})$

(b) $\mathbf{b} \cdot \mathbf{a} - \mathbf{b} \cdot \mathbf{c}$

23 $(2\mathbf{a} + \mathbf{b}) \cdot (3\mathbf{c})$

24 $(\mathbf{a} - \mathbf{b}) \cdot (\mathbf{b} + \mathbf{c})$

25 $\text{comp}_\mathbf{c}\,\mathbf{b}$

26 $\text{comp}_\mathbf{b}\,\mathbf{c}$

27 $\text{comp}_\mathbf{b}\,(\mathbf{a} + \mathbf{c})$

28 $\text{comp}_\mathbf{c}\,\mathbf{c}$

Exer. 29–32: If c represents a constant force, find the work done if the point of application of c moves along the line segment from *P* to *Q*.

29 $\mathbf{c} = 3\mathbf{i} + 4\mathbf{j}$; $P(0, 0)$, $Q(5, -2)$

30 $\mathbf{c} = -10\mathbf{i} + 12\mathbf{j}$; $P(0, 0)$, $Q(4, 7)$

31 $\mathbf{c} = 6\mathbf{i} + 4\mathbf{j}$; $P(2, -1)$, $Q(4, 3)$
    (*Hint:* Find a vector $\mathbf{b} = \langle b_1, b_2 \rangle$ such that $\mathbf{b} = \overrightarrow{PQ}$.)

32 $\mathbf{c} = -\mathbf{i} + 7\mathbf{j}$; $P(-2, 5)$, $Q(6, 1)$

33 A constant force of magnitude 4 has the same direction as **j**. Find the work done if its point of application moves from $P(0, 0)$ to $Q(8, 3)$.

34 A constant force of magnitude 10 has the same direction as $-\mathbf{i}$. Find the work done if its point of application moves from $P(0, 1)$ to $Q(1, 0)$.

Exer. 35–40: Prove the property if a and b are vectors and *m* is a real number.

35 $\mathbf{a} \cdot \mathbf{a} = \|\mathbf{a}\|^2$

36 $\mathbf{a} \cdot \mathbf{b} = \mathbf{b} \cdot \mathbf{a}$

37 $(m\mathbf{a}) \cdot \mathbf{b} = m(\mathbf{a} \cdot \mathbf{b})$

38 $m(\mathbf{a} \cdot \mathbf{b}) = \mathbf{a} \cdot (m\mathbf{b})$

39 $\mathbf{0} \cdot \mathbf{a} = 0$

40 $(\mathbf{a} + \mathbf{b}) \cdot (\mathbf{a} - \mathbf{b}) = \mathbf{a} \cdot \mathbf{a} - \mathbf{b} \cdot \mathbf{b}$

41 *Pulling a wagon* A child pulls a wagon along level ground by exerting a force of 20 pounds on a handle that makes an angle of 30° with the horizontal, as shown in the figure. Find the work done in pulling the wagon 100 feet.

Exercise 41

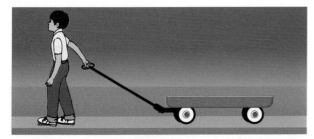

42 *Pulling a wagon* Refer to Exercise 41. Find the work done if the wagon is pulled, with the same force, 100 feet up an incline that makes an angle of 30° with the horizontal, as shown in the figure.

Exercise 42

43 *The sun's rays* The sun has a radius of 432,000 miles, and its center is 93,000,000 miles from the center of the earth. Let **v** and **w** be the vectors illustrated in the figure.

(a) Express **v** and **w** in **i**, **j** form.

(b) Approximate the angle between **v** and **w**.

Exercise 43

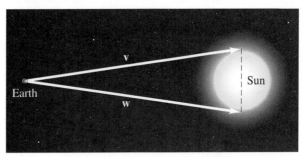

44 *July sunlight* The intensity I of sunlight (in watts/m²) can be calculated using the formula $I = ke^{-c/\sin\phi}$, where k and c are positive constants and ϕ is the angle between the sun's rays and the horizon. The amount of sunlight striking a vertical wall facing the sun is equal to the component of the sun's rays along the horizontal. If, during July, $\phi = 30°$, $k = 978$, and $c = 0.136$, approximate the total amount of sunlight striking a vertical wall that has an area of 160 m².

Exer. 45–46: Vectors are used extensively in computer graphics to perform shading. When light strikes a flat surface, it is reflected, and that area should not be shaded. Suppose that an incoming ray of light is represented by a vector L and that N is a vector orthogonal to the flat surface, as shown in the figure. The ray of reflected light can be represented by the vector R and is calculated using the formula R = 2(N · L)N − L. Compute R for the vectors L and N.

45 *Reflected light* $\mathbf{L} = \left\langle -\frac{4}{5}, \frac{3}{5} \right\rangle$, $\mathbf{N} = \langle 0, 1 \rangle$

46 *Reflected light* $\mathbf{L} = \left\langle \frac{12}{13}, -\frac{5}{13} \right\rangle$, $\mathbf{N} = \left\langle \frac{1}{2}\sqrt{2}, \frac{1}{2}\sqrt{2} \right\rangle$

Exercises 45–46

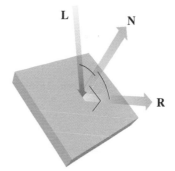

Exer. 47–48: Vectors are used in computer graphics to calculate the length of shadows over flat surfaces. Lengths of objects can sometimes be represented by a vector **a**. If a single light source is shining down on the object, then the length of its shadow on the ground will be equal to the absolute value of the component of the vector **a** along the direction of the ground, as shown in the figure. Compute the length of the shadow for the specified vector **a** if the ground is level.

47 *Shadow on level ground* $\mathbf{a} = \langle 2.6, 4.5 \rangle$

48 *Shadow on level ground* $\mathbf{a} = \langle -3.1, 7.9 \rangle$

Exercises 47–48

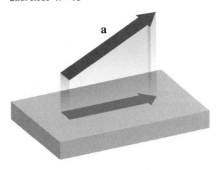

Exer. 49–50: Refer to Exercises 47 and 48. An object represented by a vector **a** is held over a flat surface inclined at an angle *θ*, as shown in the figure. If a light is shining directly downward, approximate the length of the shadow to two decimal places for the specified values of the vector **a** and *θ*.

49 *Shadow on inclined plane*
$$\mathbf{a} = \langle 25.7, -3.9 \rangle, \quad \theta = 12°$$

50 *Shadow on inclined plane*
$$\mathbf{a} = \langle -13.8, 19.4 \rangle, \quad \theta = -17°$$

Exercises 49–50

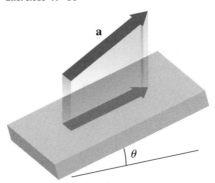

51 *Determining horsepower* The amount of horsepower *P* produced by an engine can be determined by using the formula $P = \frac{1}{550}(\mathbf{F} \cdot \mathbf{v})$, where **F** is the force (in pounds) exerted by the engine and **v** is the velocity (in ft/sec) of an object moved by the engine. An engine pulls with a force of 2200 pounds on a cable that makes an angle *θ* with the horizontal, moving a cart horizontally, as shown in the figure. Find the horsepower of the engine if the speed of the cart is 8 ft/sec when *θ* = 30°.

Exercise 51

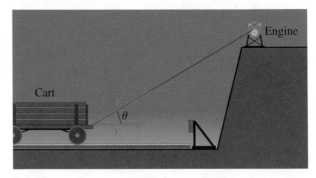

CHAPTER 4 REVIEW EXERCISES

Exer. 1–4: Find the exact values of the remaining parts of triangle *ABC*.

1 $\alpha = 60°, \quad b = 6, \quad c = 7$

2 $\gamma = 30°, \quad a = 2\sqrt{3}, \quad c = 2$

3 $\alpha = 60°, \quad \beta = 45°, \quad b = 100$

4 $a = 2, \quad b = 3, \quad c = 4$

Exer. 5–8: Approximate the remaining parts of triangle *ABC*.

5 $\beta = 67°, \quad \gamma = 75°, \quad b = 12$

6 $\alpha = 23°30', \quad c = 125, \quad a = 152$

7 $\beta = 115°$, $\quad a = 4.6$, $\quad c = 7.3$

8 $a = 37$, $\quad\quad b = 55$, $\quad c = 43$

Exer. 9–10: Approximate the area of triangle ABC to the nearest 0.1 square unit.

9 $\alpha = 75°$, $\quad b = 20$, $\quad c = 30$

10 $a = 4$, $\quad\quad b = 7$, $\quad\quad c = 10$

Exer. 11–16: Express the complex number in trigonometric form with $0 \le \theta < 2\pi$.

11 $-10 + 10i$ 12 $2 - 2\sqrt{3}\,i$

13 -17 14 $-12i$

15 $-5\sqrt{3} - 5i$ 16 $4 + 5i$

Exer. 17–18: Express in the form $a + bi$, where a and b are real numbers.

17 $20\left(\cos \dfrac{11\pi}{6} + i \sin \dfrac{11\pi}{6}\right)$

18 $13 \operatorname{cis}\left(\tan^{-1}\frac{5}{12}\right)$

Exer. 19–20: Use trigonometric forms to find $z_1 z_2$ and z_1/z_2.

19 $z_1 = -3\sqrt{3} - 3i$, $\quad z_2 = 2\sqrt{3} + 2i$

20 $z_1 = 2\sqrt{2} + 2\sqrt{2}\,i$, $\quad z_2 = -1 - i$

Exer. 21–24: Use De Moivre's theorem to change the given complex number to the form $a + bi$, where a and b are real numbers.

21 $(-\sqrt{3} + i)^9$ 22 $\left(\dfrac{\sqrt{2}}{2} - \dfrac{\sqrt{2}}{2}i\right)^{30}$

23 $(3 - 3i)^5$ 24 $(2 + 2\sqrt{3}\,i)^{10}$

25 Find the three cube roots of -27.

26 Let $z = 1 - \sqrt{3}\,i$.

(a) Find z^{24}.   (b) Find the three cube roots of z.

27 Find the solutions of the equation $x^5 - 32 = 0$.

28 If $\mathbf{a} = \langle -4, 5\rangle$ and $\mathbf{b} = \langle 2, -8\rangle$, sketch vectors corresponding to

(a) $\mathbf{a} + \mathbf{b}$ (b) $\mathbf{a} - \mathbf{b}$ (c) $2\mathbf{a}$ (d) $-\frac{1}{2}\mathbf{b}$

29 If $\mathbf{a} = 2\mathbf{i} + 5\mathbf{j}$ and $\mathbf{b} = 4\mathbf{i} - \mathbf{j}$, find the vector or number corresponding to

(a) $4\mathbf{a} + \mathbf{b}$ (b) $2\mathbf{a} - 3\mathbf{b}$

(c) $\|\mathbf{a} - \mathbf{b}\|$ (d) $\|\mathbf{a}\| - \|\mathbf{b}\|$

30 If $\mathbf{a} = \langle a_1, a_2\rangle$, $\mathbf{r} = \langle x, y\rangle$, and $c > 0$, describe the set of all points $P(x, y)$ such that $\|\mathbf{r} - \mathbf{a}\| = c$.

31 If $\mathbf{a}$ and $\mathbf{b}$ are vectors with the same initial point and angle θ between them, prove that

$$\|\mathbf{a} - \mathbf{b}\|^2 = \|\mathbf{a}\|^2 + \|\mathbf{b}\|^2 - 2\|\mathbf{a}\|\|\mathbf{b}\| \cos \theta.$$

32 *A ship's course* A ship is sailing at a speed of 14 mi/hr in the direction S50°E. Express its velocity $\mathbf{v}$ as a vector.

33 The magnitudes and directions of two forces are 72 kg, S60°E and 46 kg, N74°E, respectively. Approximate the magnitude and direction of the resultant force.

34 *Wind speed and direction* An airplane is flying in the direction 80° with an airspeed of 400 mi/hr. Its ground speed and true course are 390 mi/hr and 90°, respectively. Approximate the direction and speed of the wind.

35 If $\mathbf{a} = \langle 2, -3\rangle$ and $\mathbf{b} = \langle -1, -4\rangle$, find each of the following:

(a) $\mathbf{a} \cdot \mathbf{b}$ (b) the angle between $\mathbf{a}$ and $\mathbf{b}$

(c) $\operatorname{comp}_{\mathbf{a}} \mathbf{b}$

36 If $\mathbf{a} = 6\mathbf{i} - 2\mathbf{j}$ and $\mathbf{b} = \mathbf{i} + 3\mathbf{j}$, find each of the following:

(a) $(2\mathbf{a} - 3\mathbf{b}) \cdot \mathbf{a}$

(b) the angle between $\mathbf{a}$ and $\mathbf{a} + \mathbf{b}$

(c) $\operatorname{comp}_{\mathbf{a}} (\mathbf{a} + \mathbf{b})$

37 A constant force has the magnitude and direction of $\mathbf{a} = 7\mathbf{i} + 4\mathbf{j}$. Find the work done when the point of application of $\mathbf{a}$ moves along the x-axis from $P(-5, 0)$ to $Q(3, 0)$.

38 *Skateboard racecourse* A course for a skateboard race consists of a 200-meter downhill run and a 150-meter level portion. The angle of elevation of the starting point of the race from the finish line is 27.4°. What angle does the hill make with the horizontal?

39 *Distances to planets* The distances between the earth and nearby planets can be approximated using the phase angle α, as shown in the figure. Suppose that the distance between the earth and the sun is 93,000,000 miles and the distance between Venus and the sun is 67,000,000 miles. Approximate the distance between

the earth and Venus to the nearest million miles when $\alpha = 34°$.

Exercise 39

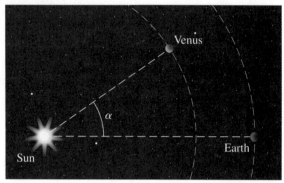

40 *Height of a skyscraper* If a skyscraper is viewed from the top of a 50-foot building, the angle of elevation is 59°. If it is viewed from street level, the angle of elevation is 62° (see the figure).

(a) Use the law of sines to approximate the shortest distance between the tops of the two buildings.

(b) Approximate the height of the skyscraper.

Exercise 40

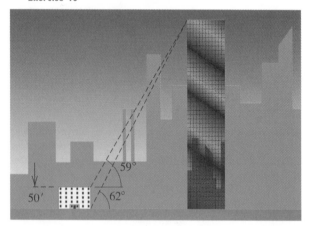

41 *Distances between cities* The beach communities of San Clemente and Long Beach are 41 miles apart, along a fairly straight stretch of coastline. Shown in the figure is the triangle formed by the two cities and the town of Avalon at the southeast corner of Santa Catalina Island. Angles *ALS* and *ASL* are found to be 66.4° and 47.2°, respectively.

(a) Approximate the distance from Avalon to each of the two cities.

(b) Approximate the shortest distance from Avalon to the coast.

Exercise 41

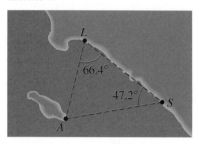

42 *Surveying* A surveyor wishes to find the distance between two inaccessible points *A* and *B*. As shown in the figure, two points *C* and *D* are selected from which it is possible to view both *A* and *B*. The distance *CD* and the angles *ACD*, *ACB*, *BDC*, and *BDA* are then measured. If $CD = 120$ ft, $\angle ACD = 115°$, $\angle ACB = 92°$, $\angle BDC = 125°$, and $\angle BDA = 100°$, approximate the distance *AB*.

Exercise 42

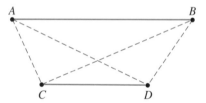

43 *Radio contact* Two girls with two-way radios are at the intersection of two country roads that meet at a 105° angle (see the figure). One begins walking in a northerly direction along one road at the rate of 5 mi/hr; at the same time the other walks east along the other road at the same rate. If each radio has a range of 10 miles, how long will the girls maintain contact?

Exercise 43

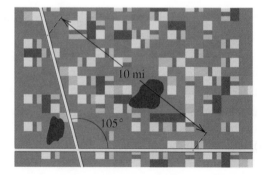

44 *Robotic design* Shown in the figure is a design for a robotic arm with two moving parts. The dimensions are chosen to emulate a human arm. The upper arm AC and lower arm CP rotate through angles θ_1 and θ_2, respectively, to hold an object at point $P(x, y)$.

(a) Show that $\angle ACP = 180° - (\theta_2 - \theta_1)$.

(b) Find $d(A, P)$, and then use part (a) and the law of cosines to show that

$$1 + \cos(\theta_2 - \theta_1) = \frac{x^2 + (y - 26)^2}{578}.$$

(c) If $x = 25$, $y = 4$, and $\theta_1 = 135°$, approximate θ_2.

Exercise 44

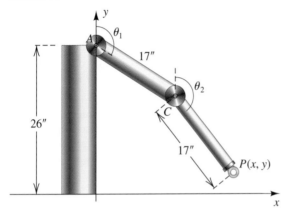

45 *Rescue efforts* A child is trapped 45 feet down an abandoned mine shaft that slants at an angle of 78° from the horizontal. A rescue tunnel is to be dug 50 feet from the shaft opening (see the figure).

(a) At what angle θ should the tunnel be dug?

(b) If the tunnel can be dug at a rate of 3 ft/hr, how many hours will it take to reach the child?

Exercise 45

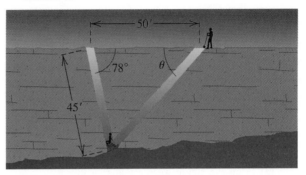

46 *Design for a jet fighter* Shown in the figure is a plan for the top of a wing of a jet fighter.

(a) Approximate angle ϕ.

(b) Approximate the area of quadrilateral $ABCD$.

(c) If the fuselage is 5.8 feet wide, approximate the wing span CC'.

Exercise 46

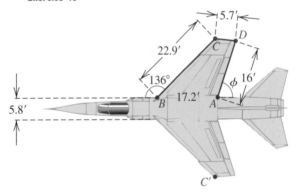

CHAPTER 4 DISCUSSION EXERCISES

1 *Mollweide's formula* The following equation, called Mollweide's formula, is sometimes used to check solutions to triangles because it involves all the angles and sides:

$$\frac{a + b}{c} = \frac{\cos \frac{1}{2}(\alpha - \beta)}{\sin \frac{1}{2}\gamma}$$

(a) Use the law of sines to show that

$$\frac{a + b}{c} = \frac{\sin \alpha + \sin \beta}{\sin \gamma}.$$

(b) Use a sum-to-product formula and a double-angle formula to verify Mollweide's formula.

2 Use the trigonometric form of a complex number to show that $z^{-n} = 1/z^n$, where n is a positive integer.

3 Discuss the algebraic and geometric similarities of the cube roots of any positive real number a.

4 Suppose that two vectors **v** and **w** have the same initial point, that the angle between them is θ, and that $\mathbf{v} \neq m\mathbf{w}$ (m is a real number).

(a) What is the geometric interpretation of $\mathbf{v} - \mathbf{w}$?

(b) How could you find $\|\mathbf{v} - \mathbf{w}\|$?

5 *A vector approach to the laws of sines and cosines*

(a) From the figure we see that $\mathbf{c} = \mathbf{b} + \mathbf{a}$. Use horizontal and vertical components to write **c** in terms of **i** and **j**.

Exercise 5

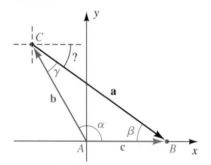

(b) Now find the magnitude of **c**, using the answer to part (a), and simplify to the point where you have proved the law of cosines.

(c) If **c** lies on the x-axis, then its **j**-component is zero. Use this fact to prove the law of sines.

6 *Euler's formula and other results* The following are some interesting and unexpected results involving complex numbers and topics that have been previously discussed.

(a) Leonhard Euler (1707–1783) gave us the following formula:

$$e^{i\theta} = \cos\theta + i\sin\theta$$

If we let $\theta = \pi$, we obtain $e^{i\pi} = -1$ or, equivalently,

$$e^{i\pi} + 1 = 0,$$

an equation relating five of the most important numbers in mathematics. Find $e^{2\pi i}$.

(b) We define the logarithm of a complex number $z \neq 0$ as follows:

$$\text{LN } z = \ln|z| + i(\theta + 2\pi n),$$

where ln is the natural logarithm function, θ is an argument of z, and n is an integer. The **principal value** of LN z is the value that corresponds to $n = 0$ and $-\pi < \theta \leq \pi$. Find the principal values of LN (-1) and LN i.

(c) We define the complex power w of a complex number $z \neq 0$ as follows:

$$z^w = e^{w \text{ LN } z}$$

We use principal values of LN z to find principal values of z^w. Find principal values of $\sqrt{i}$ and i^i.

EXPONENTIAL AND LOGARITHMIC FUNCTIONS

Exponential and logarithmic functions are transcendental functions, since they cannot be defined in terms of only addition, subtraction, multiplication, division, and rational powers of a variable x, as is the case for the algebraic functions considered in previous chapters. Such functions are of major importance in mathematics and have applications in almost every field of human endeavor. They are especially useful in the fields of chemistry, biology, physics, and engineering, where they help describe the manner in which quantities in nature grow or decay. As we shall see in this chapter, there is a close relationship between specific exponential and logarithmic functions—they are inverse functions of each other. (A portion of the material in this chapter overlaps with the brief survey given in Section 1.6.)

5.1 EXPONENTIAL FUNCTIONS

In previous chapters, we considered functions having terms of the form

$$\text{variable base}^{\text{constant power}},$$

such as x^2, $0.2x^{1.3}$, and $8x^{2/3}$. We now turn our attention to functions having terms of the form

$$\text{constant base}^{\text{variable power}},$$

such as 2^x, $(1.04)^{4x}$, and 3^{-x}. Let us begin by considering the function f defined by

$$f(x) = 2^x,$$

where x is restricted to *rational* numbers. (Recall that if $x = m/n$ for integers m and n with $n > 0$, then $2^x = 2^{m/n} = (\sqrt[n]{2})^m$.) Coordinates of several points on the graph of $y = 2^x$ are listed in the following table.

Figure I

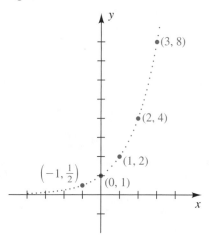

x	-10	-3	-2	-1	0	1	2	3	10
$y = 2^x$	$\frac{1}{1024}$	$\frac{1}{8}$	$\frac{1}{4}$	$\frac{1}{2}$	1	2	4	8	1024

Other values of y for x rational, such as $2^{1/3}$, $2^{-9/7}$, and $2^{5.143}$, can be approximated with a calculator. We can show algebraically that if x_1 and x_2 are rational numbers such that $x_1 < x_2$, then $2^{x_1} < 2^{x_2}$. Thus, f is an increasing function, and its graph rises. Plotting points leads to the sketch in Figure 1, where the small dots indicate that only the points with *rational* x-coordinates are on the graph. There is a *hole* in the graph whenever the x-coordinate of a point is irrational.

To extend the domain of f to all real numbers, it is necessary to define 2^x for every *irrational* exponent x. To illustrate, if we wish to define 2^{π}, we could use the nonterminating decimal representing $3.1415926 \ldots$ for π and consider the following *rational* powers of 2:

$$2^3, \quad 2^{3.1}, \quad 2^{3.14}, \quad 2^{3.141}, \quad 2^{3.1415}, \quad 2^{3.14159}, \quad \ldots$$

It can be shown, using calculus, that each successive power gets closer to a unique real number, denoted by 2^{π}. Thus,

$$2^x \to 2^{\pi} \quad \text{as} \quad x \to \pi, \quad \text{with } x \text{ rational.}$$

Figure 2

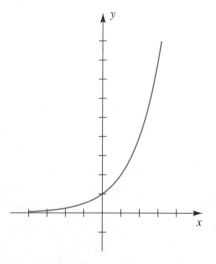

The same technique can be used for any other irrational power of 2. To sketch the graph of $y = 2^x$ with x *real,* we replace the holes in the graph in Figure 1 with points, and we obtain the graph in Figure 2. The function f defined by $f(x) = 2^x$ for every real number x is called the **exponential function** *with base 2.*

Let us next consider *any* base a, where a is a positive real number different from 1. As in the preceding discussion, to each real number x there corresponds exactly one positive number a^x such that the laws of exponents are true. Thus, as in the following chart, we may define a function f whose domain is $\mathbb{R}$ and range is the set of positive real numbers.

Terminology	Definition	Graph of f for $a > 1$	Graph of f for $0 < a < 1$
Exponential function f with base a	$f(x) = a^x$ for every x in $\mathbb{R}$, where $a > 0$ and $a \neq 1$		

The graphs in the chart show that if $a > 1$, then f is increasing on $\mathbb{R}$, and if $0 < a < 1$, then f is decreasing on $\mathbb{R}$. (These facts can be proved using calculus.) The graphs merely indicate the *general* appearance—the *exact* shape depends on the value of a. Note, however, that since $a^0 = 1$, the y-intercept is 1 for every a.

If $a > 1$, then as x *decreases* through negative values, the graph of f approaches the x-axis (see the third column in the chart). Thus, the x-axis is a *horizontal asymptote*. As x increases through positive values, the graph rises rapidly. This type of variation is characteristic of the **exponential law of growth,** and f is sometimes called a **growth function.**

If $0 < a < 1$, then as x *increases,* the graph of f approaches the x-axis asymptotically (see the last column in the chart). This type of variation is known as **exponential decay.**

When considering a^x we exclude the cases $a \leq 0$ and $a = 1$. Note that if $a < 0$, then a^x is not a real number for many values of x such as $\frac{1}{2}$, $\frac{3}{4}$, and $\frac{11}{6}$. If $a = 0$, then $a^0 = 0^0$ is undefined. Finally, if $a = 1$, then $a^x = 1$ for every x, and the graph of $y = a^x$ is a horizontal line.

The graph of an exponential function f is either increasing throughout its domain or decreasing throughout its domain. Thus, f is one-to-one by the theorem on page 68. Combining this result with the definition of a one-to-one function (see page 67) gives us parts (1) and (2) of the following theorem.

THEOREM: EXPONENTIAL FUNCTIONS ARE ONE-TO-ONE

The exponential function f given by

$$f(x) = a^x \quad \text{for} \quad 0 < a < 1 \quad \text{or} \quad a > 1$$

is one-to-one. Thus, the following equivalent conditions are satisfied for real numbers x_1 and x_2:

(1) If $x_1 \neq x_2$, then $a^{x_1} \neq a^{x_2}$.

(2) If $a^{x_1} = a^{x_2}$, then $x_1 = x_2$.

When using this theorem as a reason for a step in the solution to an example, we will state that *exponential functions are one-to-one.*

ILLUSTRATION

Exponential Functions Are One-to-One

➤ If $7^{3x} = 7^{2x+5}$, then $3x = 2x + 5$, or $x = 5$.

In the following example we solve a simple *exponential equation*—that is, an equation in which the variable appears in an exponent.

EXAMPLE 1   **Solving an exponential equation**

Solve the equation $3^{5x-8} = 9^{x+2}$.

Solution

$$
\begin{aligned}
3^{5x-8} &= 9^{x+2} & &\text{given} \\
3^{5x-8} &= (3^2)^{x+2} & &\text{express both sides with the same base} \\
3^{5x-8} &= 3^{2x+4} & &\text{law of exponents} \\
5x - 8 &= 2x + 4 & &\text{exponential functions are one-to-one} \\
3x &= 12 & &\text{subtract } 2x \text{ and add } 8 \\
x &= 4 & &\text{divide by } 3
\end{aligned}
$$

Note that the solution in Example 1 depended on the fact that the base 9 could be written as 3 to some power. We will consider only exponential equations of this type for now, but we will solve more general exponential equations later in the chapter.

In the next two examples we sketch the graphs of several different exponential functions.

Figure 3

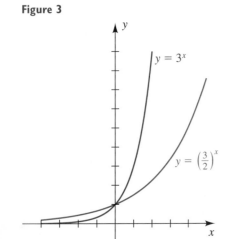

EXAMPLE 2   **Sketching graphs of exponential functions**

If $f(x) = \left(\frac{3}{2}\right)^x$ and $g(x) = 3^x$, sketch the graphs of f and g on the same coordinate plane.

Solution   Since $\frac{3}{2} > 1$ and $3 > 1$, each graph *rises* as x increases. The following table displays coordinates for several points on the graphs.

x	-2	-1	0	1	2	3	4
$y = \left(\frac{3}{2}\right)^x$	$\frac{4}{9} \approx 0.4$	$\frac{2}{3} \approx 0.7$	1	$\frac{3}{2}$	$\frac{9}{4} \approx 2.3$	$\frac{27}{8} \approx 3.4$	$\frac{81}{16} \approx 5.1$
$y = 3^x$	$\frac{1}{9} \approx 0.1$	$\frac{1}{3} \approx 0.3$	1	3	9	27	81

Plotting points and being familiar with the general graph of $y = a^x$ leads to the graphs in Figure 3.

Example 2 illustrates the fact that if $1 < a < b$, then $a^x < b^x$ for positive values of x and $b^x < a^x$ for negative values of x. In particular, since $\frac{3}{2} < 2 < 3$, the graph of $y = 2^x$ in Figure 2 lies between the graphs of f and g in Figure 3.

Figure 4

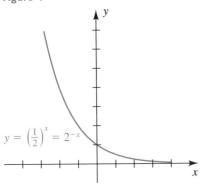

$y = \left(\frac{1}{2}\right)^x = 2^{-x}$

EXAMPLE 3   Sketching the graph of an exponential function

Sketch the graph of the equation $y = \left(\frac{1}{2}\right)^x$.

Solution   Since $0 < \frac{1}{2} < 1$, the graph *falls* as x increases. Coordinates of some points on the graph are listed in the following table.

x	-3	-2	-1	0	1	2	3
$y = \left(\frac{1}{2}\right)^x$	8	4	2	1	$\frac{1}{2}$	$\frac{1}{4}$	$\frac{1}{8}$

The graph is sketched in Figure 4. Since $\left(\frac{1}{2}\right)^x = (2^{-1})^x = 2^{-x}$, the graph is the same as the graph of the equation $y = 2^{-x}$. Note that the graph is a reflection through the y-axis of the graph of $y = 2^x$ in Figure 2.

Figure 5

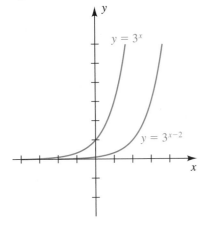

$y = 3^x$

$y = 3^{x-2}$

Equations of the form $y = a^u$, where u is some expression in x, occur in applications. The next two examples illustrate equations of this form.

EXAMPLE 4   Shifting graphs of exponential functions

Sketch the graph of the equation:

(a) $y = 3^{x-2}$ **(b)** $y = 3^x - 2$

Solution

(a) The graph of $y = 3^x$ was sketched in Figure 3 and is resketched in Figure 5. From the discussion of horizontal shifts in Section 1.4, we can obtain the graph of $y = 3^{x-2}$ by shifting the graph of $y = 3^x$ two units to the right, as shown in Figure 5.

The graph of $y = 3^{x-2}$ can also be obtained by plotting several points and using them as a guide to sketch an exponential-type curve.

(b) From the discussion of vertical shifts in Section 1.4, we can obtain the graph of $y = 3^x - 2$ by shifting the graph of $y = 3^x$ two units downward, as shown in Figure 6. Note that the y-intercept is -1 and the line $y = -2$ is a horizontal asymptote for the graph.

Figure 6

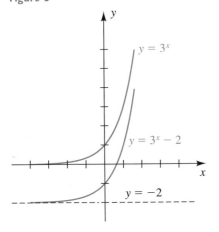

$y = 3^x$

$y = 3^x - 2$

$y = -2$

The bell-shaped graph of the function in the next example is similar to a *normal probability curve* used in statistical studies.

Figure 7

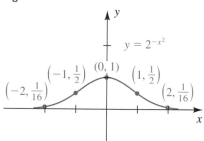

EXAMPLE 5   Sketching a bell-shaped graph

If $f(x) = 2^{-x^2}$, sketch the graph of f.

Solution   If we rewrite $f(x)$ as

$$f(x) = \frac{1}{2^{(x^2)}},$$

we see that as x increases through positive values, $f(x)$ decreases rapidly; hence the graph approaches the x-axis asymptotically. The maximum value of f is $f(0) = 1$. Since f is an even function, the graph is symmetric with respect to the y-axis. Some points on the graph are $(0, 1)$, $\left(1, \frac{1}{2}\right)$, and $\left(2, \frac{1}{16}\right)$. Plotting and using symmetry gives us the sketch in Figure 7.

APPLICATION

Bacterial Growth

Exponential functions may be used to describe the growth of certain populations. As an illustration, suppose it is observed experimentally that the number of bacteria in a culture doubles every day. If 1000 bacteria are present at the start, then we obtain the following table, where t is the time in days and $f(t)$ is the bacteria count at time t.

Figure 8

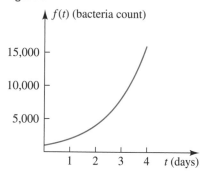

t (time in days)	0	1	2	3	4
$f(t)$ (bacteria count)	1000	2000	4000	8000	16,000

It appears that $f(t) = (1000)2^t$. With this formula we can predict the number of bacteria present at any time t. For example, at $t = 1.5 = \frac{3}{2}$,

$$f(t) = (1000)2^{3/2} \approx 2828.$$

The graph of f is sketched in Figure 8.

APPLICATION

Radioactive Decay

Certain physical quantities *decrease* exponentially. In such cases, if a is the base of the exponential function, then $0 < a < 1$. One of the most common examples of exponential decrease is the decay of a radioactive substance, or isotope. The **half-life** of an isotope is the time it takes for one-half the original amount in a given sample to decay. The half-life is the principal characteristic used to distinguish one radioactive substance from another. The polonium isotope ^{210}Po has a half-life of approximately 140 days; that is, given any amount, one-half of it will disintegrate in 140 days. If 20 milligrams of ^{210}Po is present initially, then the following table indicates the amount remaining after various intervals of time.

Figure 9

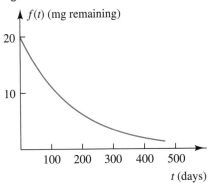

t (time in days)	0	140	280	420	560
$f(t)$ (mg remaining)	20	10	5	2.5	1.25

The sketch in Figure 9 illustrates the exponential nature of the disintegration.

Other radioactive substances have much longer half-lives. In particular, a by-product of nuclear reactors is the radioactive plutonium isotope ^{239}Pu, which has a half-life of approximately 24,000 years. It is for this reason that the disposal of radioactive waste is a major problem in modern society.

APPLICATION

Compound Interest

Compound interest provides a good illustration of exponential growth. If a sum of money P, the *principal*, is invested at a *simple* interest rate r, then the interest at the end of one interest period is the product Pr when r is expressed as a decimal. For example, if $P = \$1000$ and the interest rate is 9% per year, then $r = 0.09$, and the interest at the end of one year is $\$1000(0.09)$, or $\$90$.

If the interest is reinvested with the principal at the end of the interest period, then the new principal is

$$P + Pr \quad \text{or, equivalently,} \quad P(1 + r).$$

Note that to find the new principal we may multiply the original principal by $(1 + r)$. In the preceding example, the new principal is $\$1000(1.09)$, or $\$1090$.

After another interest period has elapsed, the new principal may be found by multiplying $P(1 + r)$ by $(1 + r)$. Thus, the principal after two interest periods is $P(1 + r)^2$. If we continue to reinvest, the principal after three periods is $P(1 + r)^3$; after four it is $P(1 + r)^4$; and, in general, the amount A accumulated after k interest periods is

$$A = P(1 + r)^k.$$

Interest accumulated by means of this formula is **compound interest.** Note that A is expressed in terms of an exponential function with base $1 + r$. The interest period may be measured in years, months, weeks, days, or any other suitable unit of time. When applying the formula for A, remember that *r is the interest rate per interest period expressed as a decimal.* For example, if the rate is stated as 6% *per year compounded monthly*, then the rate per month is $\frac{6}{12}\%$ or, equivalently, 0.5%. Thus, $r = 0.005$ and k is the number of months. If $\$100$ is invested at this rate, then the formula for A is

$$A = 100(1 + 0.005)^k = 100(1.005)^k.$$

In general, we have the following formula.

Compound Interest Formula

$$A = P\left(1 + \frac{r}{n}\right)^{nt},$$

where P = principal

r = annual interest rate expressed as a decimal

n = number of interest periods per year

t = number of years P is invested

A = amount after t years.

The next example illustrates a special case of the compound interest formula.

EXAMPLE 6   Using the compound interest formula

Suppose that $1000 is invested at an interest rate of 9% compounded monthly. Find the new amount of principal after 5 years, after 10 years, and after 15 years. Illustrate graphically the growth of the investment.

Solution   Applying the compound interest formula with $r = 0.09$, $n = 12$, and $P = \$1000$, we find that the amount after t years is

$$A = 1000\left(1 + \frac{0.09}{12}\right)^{12t} = 1000(1.0075)^{12t}.$$

Substituting $t = 5$, 10, and 15 and using a calculator, we obtain the following table.

Figure 10
Compound interest: $A = 1000(1.0075)^{12t}$

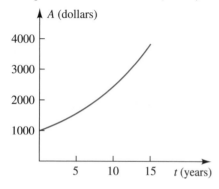

Number of years	Amount
5	$A = \$1000(1.0075)^{60} = \$1565.68$
10	$A = \$1000(1.0075)^{120} = \$2451.36$
15	$A = \$1000(1.0075)^{180} = \$3838.04$

The exponential nature of the increase is indicated by the fact that during the first five years, the growth in the investment is $565.68; during the second five-year period, the growth is $885.68; and during the last five-year period, it is $1386.68.

The sketch in Figure 10 illustrates the growth of $1000 invested over a period of 15 years.

We conclude this section with an example involving a graphing utility.

EXAMPLE 7   Estimating amounts of a
drug in the bloodstream

If an adult takes a 100-milligram tablet of a certain prescription drug orally, the rate R at which the drug enters the bloodstream t minutes later is predicted to be

$$R = 5(0.95)^t \text{ mg/min.}$$

It can be shown using calculus that the amount A of the drug in the bloodstream at time t can be approximated by

$$A = 97.4786[1 - (0.95)^t] \text{ mg.}$$

(a) Estimate how long it takes for 50 milligrams of the drug to enter the bloodstream.

(b) Estimate the number of milligrams of the drug in the bloodstream when the drug is entering at a rate of 3 mg/min.

Figure II
[0, 100] by [0, 100]

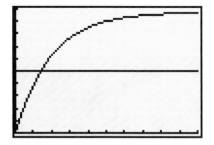

Figure 12
[0, 15] by [0, 10]

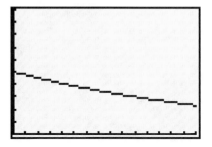

Solution

(a) We wish to determine t when A is equal to 50. Since the value of A cannot exceed 97.4786, we choose the viewing rectangle to be [0, 100] by [0, 100] with Xscl = Yscl = 10.

We next assign $97.4786[1 - (0.95)^x]$ to Y_1, assign 50 to Y_2, and graph Y_1 and Y_2, obtaining a display similar to that in Figure 11 (note that $x = t$). Using the zoom or intersect feature, we estimate that $A = 50$ mg when $x \approx 14$ min.

(b) We wish to determine t when R is equal to 3. Let us first assign $5(0.95)^x$ to Y_3. Since the maximum value of Y_3 is 5 (at $t = 0$), we use a viewing rectangle of dimensions [0, 15] by [0, 10] and obtain a display similar to that in Figure 12. Tracing Y_3 until $y = 3$ gives us $x \approx 9.96$. Thus, after almost 10 minutes, the drug will be entering the bloodstream at a rate of 3 mg/min. (Note that the initial rate, at $t = 0$, is 5 mg/min.) Finding the value of Y_1 at $x = 10$, we see that there is almost 39 milligrams of the drug in the bloodstream after 10 minutes.

5.1 EXERCISES

Exer. 1–8: Solve the equation.

1 $7^{x+6} = 7^{3x-4}$

2 $6^{7-x} = 6^{2x+1}$

3 $3^{2x+3} = 3^{(x^2)}$

4 $9^{(x^2)} = 3^{3x+2}$

5 $2^{-100x} = (0.5)^{x-4}$

6 $\left(\frac{1}{2}\right)^{6-x} = 2$

7 $4^{x-3} = 8^{4-x}$

8 $27^{x-1} = 9^{2x-3}$

9 Sketch the graph of f if $a = 2$.

(a) $f(x) = a^x$

(b) $f(x) = -a^x$

(c) $f(x) = 3a^x$

(d) $f(x) = a^{x+3}$

(e) $f(x) = a^x + 3$

(f) $f(x) = a^{x-3}$

(g) $f(x) = a^x - 3$

(h) $f(x) = a^{-x}$

(i) $f(x) = \left(\frac{1}{a}\right)^x$

(j) $f(x) = a^{3-x}$

10 Work Exercise 9 if $a = \frac{1}{2}$.

Exer. 11–20: Sketch the graph of f.

11 $f(x) = \left(\frac{2}{5}\right)^{-x}$

12 $f(x) = \left(\frac{2}{5}\right)^{x}$

13 $f(x) = -\left(\frac{1}{2}\right)^{x} + 4$

14 $f(x) = -3^{x} + 9$

15 $f(x) = 2^{|x|}$

16 $f(x) = 2^{-|x|}$

17 $f(x) = 3^{1-x^2}$

18 $f(x) = 2^{-(x+1)^2}$

19 $f(x) = 3^{x} + 3^{-x}$

20 $f(x) = 3^{x} - 3^{-x}$

21 Elk population One hundred elk, each 1 year old, are introduced into a game preserve. The number $N(t)$ alive after t years is predicted to be $N(t) = 100(0.9)^t$. Estimate the number alive after

(a) 1 year   (b) 5 years   (c) 10 years

22 Drug dosage A drug is eliminated from the body through urine. Suppose that for an initial dose of 10 milligrams, the amount $A(t)$ in the body t hours later is given by $A(t) = 10(0.8)^t$.

(a) Estimate the amount of the drug in the body 8 hours after the initial dose.

(b) What percentage of the drug still in the body is eliminated each hour?

23 Bacterial growth The number of bacteria in a certain culture increased from 600 to 1800 between 7:00 A.M. and 9:00 A.M. Assuming growth is exponential, the number $f(t)$ of bacteria t hours after 7:00 A.M. is given by $f(t) = 600(3)^{t/2}$.

(a) Estimate the number of bacteria in the culture at 8:00 A.M., 10:00 A.M., and 11:00 A.M.

(b) Sketch the graph of f for $0 \le t \le 4$.

24 Newton's law of cooling According to Newton's law of cooling, the rate at which an object cools is directly proportional to the difference in temperature between the object and the surrounding medium. The face of a household iron cools from $125°$ to $100°$ in 30 minutes in a room that remains at a constant temperature of $75°$. From calculus, the temperature $f(t)$ of the face after t hours of cooling is given by $f(t) = 50(2)^{-2t} + 75$.

(a) Assuming $t = 0$ corresponds to 1:00 P.M., approximate to the nearest tenth of a degree the temperature at 2:00 P.M., 3:30 P.M., and 4:00 P.M.

(b) Sketch the graph of f for $0 \le t \le 4$.

25 Radioactive decay The radioactive bismuth isotope ^{210}Bi has a half-life of 5 days. If there is 100 milligrams of ^{210}Bi present at $t = 0$, then the amount $f(t)$ remaining after t days is given by $f(t) = 100(2)^{-t/5}$.

(a) How much ^{210}Bi remains after 5 days? 10 days? 12.5 days?

(b) Sketch the graph of f for $0 \le t \le 30$.

26 Light penetration in an ocean An important problem in oceanography is to determine the amount of light that can penetrate to various ocean depths. The Beer-Lambert law asserts that the exponential function given by $I(x) = I_0 c^x$ is a model for this phenomenon (see the figure). For a certain location, $I(x) = 10(0.4)^x$ is the amount of light (in calories/cm^2/sec) reaching a depth of x meters.

(a) Find the amount of light at a depth of 2 meters.

(b) Sketch the graph of I for $0 \le x \le 5$.

Exercise 26

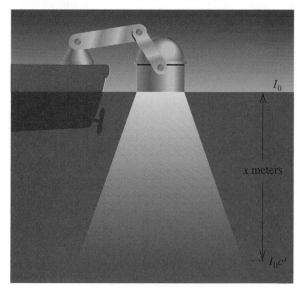

27 Decay of radium The half-life of radium is 1600 years. If the initial amount is q_0 milligrams, then the quantity $q(t)$ remaining after t years is given by $q(t) = q_0 2^{kt}$. Find k.

28 Dissolving salt in water If 10 grams of salt is added to a quantity of water, then the amount $q(t)$ that is undissolved after t minutes is given by $q(t) = 10\left(\frac{4}{5}\right)^t$. Sketch a graph that shows the value $q(t)$ at any time from $t = 0$ to $t = 10$.

29 Compound interest If $1000 is invested at a rate of 12% per year compounded monthly, find the principal after

(a) 1 month   (b) 6 months

(c) 1 year   (d) 20 years

30 Compound interest If a savings fund pays interest at a rate of 10% compounded semiannually, how much money invested now will amount to $5000 after 1 year?

31 Automobile trade-in value If a certain make of automobile is purchased for C dollars, its trade-in value $V(t)$ at the end of t years is given by $V(t) = 0.78C(0.85)^{t-1}$. If the original cost is $10,000, calculate, to the nearest dollar, the value after

(a) 1 year   (b) 4 years   (c) 7 years

32 Real estate appreciation If the value of real estate increases at a rate of 5% per year, after t years the value V of a house purchased for P dollars is $V = P(1.05)^t$. A graph for the value of a house purchased for $80,000 in 1986 is shown in the figure. Approximate the value of the house, to the nearest $1000, in the year 2010.

Exercise 32

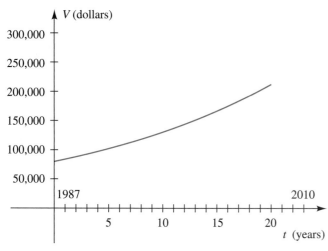

33 Manhattan Island The island of Manhattan was sold for $24 in 1626. How much would this amount have grown to by 2006 if it had been invested at 6% per year compounded quarterly?

34 Credit-card interest A certain department store requires its credit-card customers to pay interest on unpaid bills at the rate of 18% per year compounded monthly. If a customer buys a television set for $500 on credit and makes no payments for one year, how much is owed at the end of the year?

35 Depreciation The declining balance method is an accounting method in which the amount of depreciation taken each year is a fixed percentage of the present value of the item. If y is the value of the item in a given year, the depreciation taken is ay for some depreciation rate a with $0 < a < 1$, and the new value is $(1 - a)y$.

(a) If the initial value of the item is y_0, show that the value after n years of depreciation is $(1 - a)^n y_0$.

(b) At the end of T years, the item has a salvage value of s dollars. The taxpayer wishes to choose a depreciation rate such that the value of the item after T years will equal the salvage value (see the figure). Show that $a = 1 - \sqrt[T]{s/y_0}$.

Exercise 35

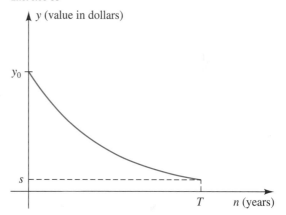

36 Language dating Glottochronology is a method of dating a language at a particular stage, based on the theory that over a long period of time linguistic changes take place at a fairly constant rate. Suppose that a language originally had N_0 basic words and that at time t, measured in millennia (1 millennium = 1000 years), the number $N(t)$ of basic words that remain in common use is given by $N(t) = N_0(0.805)^t$.

(a) Approximate the percentage of basic words lost every 100 years.

(b) If $N_0 = 200$, sketch the graph of N for $0 \le t \le 5$.

Exer. 37–40: Some lending institutions calculate the monthly payment M on a loan of L dollars at an interest rate r (expressed as a decimal) by using the formula

$$M = \frac{Lrk}{12(k - 1)},$$

where $k = [1 + (r/12)]^{12t}$ and t is the number of years that the loan is in effect.

37 Home mortgage

(a) Find the monthly payment on a 30-year $90,000 home mortgage if the interest rate is 12%.

(b) Find the total interest paid on the loan in part (a).

38 *Home mortgage*   Find the largest 25-year home mortgage that can be obtained at an interest rate of 10% if the monthly payment is to be $800.

39 *Car loan*   An automobile dealer offers customers no-down-payment 3-year loans at an interest rate of 15%. If a customer can afford to pay $220 per month, find the price of the most expensive car that can be purchased.

40 *Business loan*   The owner of a small business decides to finance a new computer by borrowing $3000 for 2 years at an interest rate of 12.5%.

  (a) Find the monthly payment.

  (b) Find the total interest paid on the loan.

c   **Exer. 41–42: Approximate the function at the value of *x* to four decimal places.**

41 (a) $f(x) = 13^{\sqrt{x+1.1}}$, $x = 3$

  (b) $g(x) = \left(\frac{5}{42}\right)^{-x}$, $x = 1.43$

  (c) $h(x) = (2^x + 2^{-x})^{2x}$, $x = 1.06$

42 (a) $f(x) = 2^{\sqrt[3]{1-x}}$, $x = 2.5$

  (b) $g(x) = \left(\frac{2}{25} + x\right)^{-3x}$, $x = 2.1$

  (c) $h(x) = \dfrac{3^{-x} + 5}{3^x - 16}$, $x = \sqrt{2}$

c   **Exer. 43–44: Sketch the graph of the equation. (a) Estimate *y* if *x* = 40. (b) Estimate *x* if *y* = 2.**

43 $y = (1.085)^x$ **44** $y = (1.0525)^x$

c   **Exer. 45–46: Use a graph to estimate the roots of the equation.**

45 $1.4x^2 - 2.2^x = 1$

46 $1.21^{3x} + 1.4^{-1.1x} - 2x = 0.5$

c   **Exer. 47–48: Graph *f* on the given interval. (a) Determine whether *f* is one-to-one. (b) Estimate the zeros of *f*.**

47 $f(x) = \dfrac{3.1^x - 2.5^{-x}}{2.7^x + 4.5^{-x}}$; $[-3, 3]$

48 $f(x) = \pi^{0.6x} - 1.3^{(x^{1.8})}$; $[-4, 4]$

  (*Hint:* Change $x^{1.8}$ to an equivalent form that is defined for $x < 0$.)

c   **Exer. 49–50: Graph *f* on the given interval. (a) Estimate where *f* is increasing or is decreasing. (b) Estimate the range of *f*.**

49 $f(x) = 0.7x^3 + 1.7^{(-1.8x)}$; $[-4, 1]$

50 $f(x) = \dfrac{3.1^{-x} - 4.1^x}{4.4^{-x} + 5.3^x}$; $[-3, 3]$

c   **51** *Trout population*   One thousand trout, each 1 year old, are introduced into a large pond. The number $N(t)$ still alive after t years is predicted to be given by the equation $N(t) = 1000(0.9)^t$. Use the graph of N to approximate when 500 trout will be alive.

c   **52** *Buying power*   An economist predicts that the buying power $B(t)$ of a dollar t years from now will be given by $B(t) = (0.95)^t$. Use the graph of B to approximate when the buying power will be half of what it is today.

c   **53** *Gompertz function*   The **Gompertz function,**

$$y = ka^{(b^x)} \quad \text{with } k > 0, 0 < a < 1, \text{ and } 0 < b < 1,$$

is sometimes used to describe the sales of a new product whose sales are initially large but then level off toward a maximum saturation level. Graph, on the same coordinate plane, the line $y = k$ and the Gompertz function with $k = 4$, $a = \frac{1}{8}$, and $b = \frac{1}{4}$. What is the significance of the constant k?

c   **54** *Logistic function*   The **logistic function,**

$$y = \frac{1}{k + ab^x} \quad \text{with } k > 0, a > 0, \text{ and } 0 < b < 1,$$

is sometimes used to describe the sales of a new product that experiences slower sales initially, followed by growth toward a maximum saturation level. Graph, on the same coordinate plane, the line $y = 1/k$ and the logistic function with $k = \frac{1}{4}$, $a = \frac{1}{8}$, and $b = \frac{5}{8}$. What is the significance of the value $1/k$?

c   **Exer. 55–56: If monthly payments *p* are deposited in a savings account paying an annual interest rate *r*, then the amount *A* in the account after *n* years is given by**

$$A = \frac{p\left(1 + \dfrac{r}{12}\right)\left[\left(1 + \dfrac{r}{12}\right)^{12n} - 1\right]}{\dfrac{r}{12}}.$$

Graph *A* for each value of *p* and *r*, and estimate *n* for *A* = $100,000.

55 $p = 100$, $r = 0.05$ **56** $p = 250$, $r = 0.09$

c   **57** *Government receipts*   Federal government receipts (in billions of dollars) for selected years are listed in the table (continued at the top of the next page).

Year	1910	1930	1950	1970
Receipts	0.7	4.6	39.4	192.8

Year	1980	1990	1995
Receipts	517.1	1031.3	1346.4

(a) Let $x = 0$ correspond to the year 1910. Plot the data, together with the functions f and g:

(1) $f(x) = 0.809(1.094)^x$

(2) $g(x) = 0.375x^2 - 18.4x + 88.1$

(b) Determine whether the exponential or quadratic function better models the data.

(c) Use your choice in part (b) to graphically estimate the year in which the federal government first collected $1 trillion.

c 58 *Epidemics*   In 1840, Britain experienced a bovine (cattle and oxen) epidemic called epizooty. The estimated number of new cases every 28 days is listed in the table. At the time, the *London Daily* made a dire prediction that the number of new cases would continue to increase indefinitely. William Farr correctly predicted when the number of new cases would peak. Of the two functions

$$f(t) = 653(1.028)^t$$

and $\qquad g(t) = 54{,}700e^{-(t-200)^2/7500}$

one models the newspaper's prediction and the other models Farr's prediction, where t is in days with $t = 0$ corresponding to August 12, 1840.

Date	New cases
Aug. 12	506
Sept. 9	1289
Oct. 7	3487
Nov. 4	9597
Dec. 2	18,817
Dec. 30	33,835
Jan. 27	47,191

(a) Graph each function, together with the data, in the viewing rectangle [0, 400] by [0, 60,000].

(b) Determine which function better models Farr's prediction.

(c) Determine the date on which the number of new cases peaked.

5.2 THE NATURAL EXPONENTIAL FUNCTION

The *compound interest formula* discussed in the preceding section is

$$A = P\left(1 + \frac{r}{n}\right)^{nt},$$

where P is the principal invested, r is the annual interest rate (expressed as a decimal), n is the number of interest periods per year, and t is the number of years that the principal is invested. The next example illustrates what happens if the rate and total time invested are fixed, but the *interest period* is varied.

EXAMPLE 1   Using the compound interest formula

Suppose $1000 is invested at a compound interest rate of 9%. Find the new amount of principal after one year if the interest is compounded quarterly, monthly, weekly, daily, hourly, and each minute.

Solution   If we let $P = \$1000$, $t = 1$, and $r = 0.09$ in the compound interest formula, then

$$A = \$1000\left(1 + \frac{0.09}{n}\right)^n$$

(continued)

for n interest periods per year. The values of n we wish to consider are listed in the following table, where we have assumed that there are 365 days in a year and hence $(365)(24) = 8760$ hours and $(8760)(60) = 525{,}600$ minutes. (In many business transactions an investment year is considered to be only 360 days.)

Interest period	Quarter	Month	Week	Day	Hour	Minute
n	4	12	52	365	8760	525,600

Using the compound interest formula (and a calculator), we obtain the amounts given in the following table.

Interest period	Amount after one year
Quarter	$\$1000\left(1 + \dfrac{0.09}{4}\right)^4 = \$1093.08$
Month	$\$1000\left(1 + \dfrac{0.09}{12}\right)^{12} = \$1093.81$
Week	$\$1000\left(1 + \dfrac{0.09}{52}\right)^{52} = \$1094.09$
Day	$\$1000\left(1 + \dfrac{0.09}{365}\right)^{365} = \$1094.16$
Hour	$\$1000\left(1 + \dfrac{0.09}{8760}\right)^{8760} = \$1094.17$
Minute	$\$1000\left(1 + \dfrac{0.09}{525{,}600}\right)^{525{,}600} = \$1094.17$

Note that, in the preceding example, after we reach an interest period of one hour, the number of interest periods per year has no effect on the final amount. If interest had been compounded each *second*, the result would still be $1094.17. (Some decimal places *beyond* the first two *do* change.) Thus, the amount approaches a fixed value as n increases. Interest is said to be **compounded continuously** if the number n of time periods per year increases without bound.

If we let $P = 1$, $r = 1$, and $t = 1$ in the compound interest formula, we obtain

$$A = \left(1 + \frac{1}{n}\right)^n.$$

The expression on the right-hand side of the equation is important in calculus. In Example 1 we considered a similar situation: as n increased, A ap-
proached a limiting value. The same phenomenon occurs for this formula,
as illustrated by the following table.

n	Approximation to $\left(1 + \dfrac{1}{n}\right)^n$
1	2.00000000
10	2.59374246
100	2.70481383
1000	2.71692393
10,000	2.71814593
100,000	2.71826824
1,000,000	2.71828047
10,000,000	2.71828169
100,000,000	2.71828181
1,000,000,000	2.71828183

In calculus it is shown that as n increases without bound, the value of the
expression $[1 + (1/n)]^n$ approaches a certain irrational number, denoted by
e. The number e arises in the investigation of many physical phenomena.
An approximation is $e \approx 2.71828$. Using the notation introduced in Section 1.4, we denote this fact as follows.

The Number e

If n is a positive integer, then
$$\left(1 + \frac{1}{n}\right)^n \to e \approx 2.71828 \quad \text{as} \quad n \to \infty.$$

In the following definition we use e as a base for an important exponential function.

**DEFINITION OF THE NATURAL
EXPONENTIAL FUNCTION**

The **natural exponential function** f is defined by
$$f(x) = e^x$$
for every real number x.

The natural exponential function is one of the most useful functions in
advanced mathematics and applications. Since $2 < e < 3$, the graph of

Figure 13

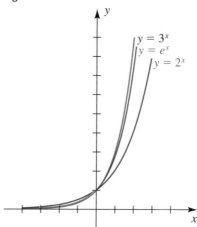

$y = e^x$ lies between the graphs of $y = 2^x$ and $y = 3^x$, as shown in Figure 13. Scientific and graphing calculators have an $\boxed{e^x}$ key for approximating values of the natural exponential function.

APPLICATION

Continuously Compounded Interest

The compound interest formula is

$$A = P\left(1 + \frac{r}{n}\right)^{nt}.$$

If we let $1/k = r/n$, then $k = n/r$, $n = kr$, and $nt = krt$, and we may rewrite the formula as

$$A = P\left(1 + \frac{1}{k}\right)^{krt} = P\left[\left(1 + \frac{1}{k}\right)^k\right]^{rt}.$$

For continuously compounded interest we let n (the number of interest periods per year) increase without bound, denoted by $n \to \infty$ or, equivalently, by $k \to \infty$. Using the fact that $[1 + (1/k)]^k \to e$ as $k \to \infty$, we see that

$$P\left[\left(1 + \frac{1}{k}\right)^k\right]^{rt} \to P[e]^{rt} = Pe^{rt} \quad \text{as} \quad k \to \infty.$$

This result gives us the following formula.

Continuously Compounded Interest Formula

$$A = Pe^{rt},$$

where P = principal

r = annual interest rate expressed as a decimal

t = number of years P is invested

A = amount after t years.

The next example illustrates a use of this formula.

EXAMPLE 2   **Using the continuously compounded interest formula**

Suppose \$20,000 is deposited in a money market account that pays interest at a rate of 8% per year compounded continuously. Determine the balance in the account after 5 years.

Solution   Applying the formula for continuously compounded interest with $P = 20,000$, $r = 0.08$, and $t = 5$, we have

$$A = Pe^{rt} = 20,000e^{0.08(5)} = 20,000e^{0.4}.$$

Using a calculator, we find that $A = \$29,836.49$.

The continuously compounded interest formula is just one specific case of the following law.

Law of Growth (or Decay) Formula

Let q_0 be the value of a quantity q at time $t = 0$ (that is, q_0 is the initial amount of q). If q changes instantaneously at a rate proportional to its current value, then

$$q = q_0 e^{rt},$$

where $r > 0$ is the rate of growth (or $r < 0$ is the rate of decay) of q.

EXAMPLE 3   **Predicting the population of a city**

The population of a city in 1970 was 153,800. Assuming that the population increases continuously at a rate of 5% per year, predict the population of the city in the year 2010.

Solution   We apply the growth formula $q = q_0 e^{rt}$ with initial population $q_0 = 153,800$, rate of growth $r = 0.05$, and time $t = 2010 - 1970 = 40$ years. Thus, a prediction for the population of the city in the year 2010 is

$$153,800e^{(0.05)(40)} = 153,800e^2 \approx 1,136,437.$$

The function f in the next example is important in advanced applications of mathematics.

EXAMPLE 4   **Sketching a graph involving two exponential functions**

Sketch the graph of f if

$$f(x) = \frac{e^x + e^{-x}}{2}.$$

Figure 14

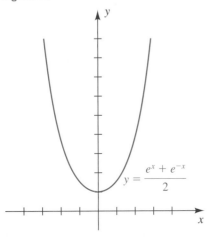

$$y = \frac{e^x + e^{-x}}{2}$$

Solution   Note that f is an even function, because

$$f(-x) = \frac{e^{-x} + e^{-(-x)}}{2} = \frac{e^{-x} + e^x}{2} = f(x).$$

Thus, the graph is symmetric with respect to the y-axis. Using a calculator, we obtain the following approximations of $f(x)$.

x	0	0.5	1.0	1.5	2.0
$f(x)$ (approx.)	1	1.13	1.54	2.35	3.76

Plotting points and using symmetry with respect to the y-axis gives us the sketch in Figure 14. The graph *appears* to be a parabola; however, this is not actually the case.

APPLICATION

Flexible Cables

Figure 15

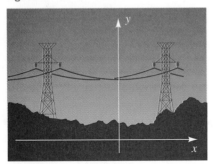

The function f of Example 4 occurs in applied mathematics and engineering, where it is called the **hyperbolic cosine function.** This function can be used to describe the shape of a uniform flexible cable or chain whose ends are supported from the same height, such as a telephone or power line (see Figure 15). If we introduce a coordinate system, as indicated in the figure, then it can be shown that an equation that corresponds to the shape of the cable is

$$y = \frac{a}{2}(e^{x/a} + e^{-x/a}),$$

where a is a real number. The graph is called a **catenary,** after the Latin word for *chain.* The function in Example 4 is the special case in which $a = 1$. See Discussion Exercise 1 at the end of this chapter for an application involving a catenary.

APPLICATION

Radiotherapy

Exponential functions play an important role in the field of *radiotherapy,* the treatment of tumors by radiation. The fraction of cells in a tumor that survive a treatment, called the *surviving fraction,* depends not only on the energy and nature of the radiation, but also on the depth, size, and characteristics of the tumor itself. The exposure to radiation may be thought of as a number of potentially damaging events, where at least one *hit* is required to kill a tumor cell. For instance, suppose that each cell has exactly one *target* that must be hit. If k denotes the average target size of a tumor cell and if x is the number of damaging events (the *dose*), then the surviving fraction $f(x)$ is given by

$$f(x) = e^{-kx}.$$

This is called the *one target–one hit surviving fraction.*

Suppose next that each cell has n targets and that each target must be hit once for the cell to die. In this case, the *n target–one hit surviving fraction* is given by

$$f(x) = 1 - (1 - e^{-kx})^n.$$

The graph of f may be analyzed to determine what effect increasing the dosage x will have on decreasing the surviving fraction of tumor cells. Note that $f(0) = 1$; that is, if there is no dose, then all cells survive. As an example, if $k = 1$ and $n = 2$, then

$$f(x) = 1 - (1 - e^{-x})^2$$
$$= 1 - (1 - 2e^{-x} + e^{-2x})$$
$$= 2e^{-x} - e^{-2x}.$$

A complete analysis of the graph of f requires calculus. The graph is sketched in Figure 16. The *shoulder* on the curve near the point $(0, 1)$ represents the threshold nature of the treatment—that is, a small dose results in very little tumor cell elimination. Note that for a large x, an increase in dosage has little effect on the surviving fraction. To determine the ideal dose to administer to a patient, specialists in radiation therapy must also take into account the number of healthy cells that are killed during a treatment.

Problems of the type illustrated in the next example occur in the study of calculus.

Figure 16

Surviving fraction of tumor cells after a radiation treatment

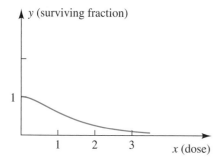

EXAMPLE 5   Finding zeros of a function involving exponentials

If $f(x) = x^2(-2e^{-2x}) + 2xe^{-2x}$, find the zeros of f.

Solution   We may factor $f(x)$ as follows:

$$f(x) = 2xe^{-2x} - 2x^2e^{-2x} \qquad \text{given}$$
$$= 2xe^{-2x}(1 - x) \qquad \text{factor out } 2xe^{-2x}$$

To find the zeros of f, we solve the equation $f(x) = 0$. Since $e^{-2x} > 0$ for every x, we see that $f(x) = 0$ if and only if $x = 0$ or $1 - x = 0$. Thus, the zeros of f are 0 and 1.

EXAMPLE 6   Sketching a Gompertz growth curve

In biology, the **Gompertz growth function** G, given by

$$G(t) = ke^{(-Ae^{-Bt})},$$

where k, A, and B are positive constants, is used to estimate the size of certain quantities at time t. The graph of G is called a **Gompertz growth curve.** The function is always positive and increasing, and as t increases without bound, $G(t)$ levels off and approaches the value k. Graph G on the interval $[0, 5]$ for $k = 1.1$, $A = 3.2$, and $B = 1.1$, and estimate the time t at which $G(t) = 1$.

Figure 17

[0, 5] by [0, 2]

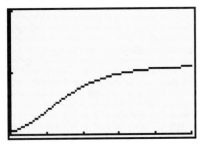

Solution   Some specific keystrokes for the TI-82/83 are given in Example 17 of Appendix I. We begin by assigning

$$1.1e^{(-3.2e^{-1.1t})}$$

to Y_1. Since we wish to graph G on the interval $[0, 5]$, we choose $Xmin = 0$ and $Xmax = 5$. Because $G(t)$ is always positive and does not exceed the value $k = 1.1$, we choose $Ymin = 0$ and $Ymax = 2$. Hence, the viewing rectangle dimensions are $[0, 5]$ by $[0, 2]$. Graphing G gives us a display similar to Figure 17. The endpoint values of the graph are approximately $(0, 0.045)$ and $(5, 1.086)$.

To determine the time when $y = G(t) = 1$, we use a zoom feature (or intersect feature, with $Y_2 = 1$) to obtain $x = t \approx 3.194$.

5.2 EXERCISES

Exer. 1–4: Use the graph of $y = e^x$ to help sketch the graph of f.

1   (a) $f(x) = e^{-x}$ (b) $f(x) = -e^x$

2   (a) $f(x) = e^{2x}$ (b) $f(x) = 2e^x$

3   (a) $f(x) = e^{x+4}$ (b) $f(x) = e^x + 4$

4   (a) $f(x) = e^{-2x}$ (b) $f(x) = -2e^x$

Exer. 5–6: If P dollars is deposited in a savings account that pays interest at a rate of $r\%$ per year compounded continuously, find the balance after t years.

5   $P = 1000,$ $r = 8\frac{1}{4},$ $t = 5$

6   $P = 100,$ $r = 12\frac{1}{2},$ $t = 10$

Exer. 7–8: How much money, invested at an interest rate of $r\%$ per year compounded continuously, will amount to A dollars after t years?

7   $A = 100,000,$ $r = 11,$ $t = 18$

8   $A = 15,000,$ $r = 9.5,$ $t = 4$

Exer. 9–10: An investment of P dollars increased to A dollars in t years. If interest was compounded continuously, find the interest rate.

9   $A = 13,464,$ $P = 1000,$ $t = 20$

10   $A = 890.20,$ $P = 400,$ $t = 16$

Exer. 11–12: Solve the equation.

11   $e^{(x^2)} = e^{7x-12}$ 12 $e^{3x} = e^{2x-1}$

Exer. 13–16: Find the zeros of f.

13   $f(x) = xe^x + e^x$

14   $f(x) = -x^2e^{-x} + 2xe^{-x}$

15   $f(x) = x^3(4e^{4x}) + 3x^2e^{4x}$

16   $f(x) = x^2(2e^{2x}) + 2xe^{2x} + e^{2x} + 2xe^{2x}$

Exer. 17–18: Simplify the expression.

17   $\dfrac{(e^x + e^{-x})(e^x + e^{-x}) - (e^x - e^{-x})(e^x - e^{-x})}{(e^x + e^{-x})^2}$

18   $\dfrac{(e^x - e^{-x})^2 - (e^x + e^{-x})^2}{(e^x + e^{-x})^2}$

19   **Crop growth**   An exponential function W such that $W(t) = W_0e^{kt}$ for $k > 0$ describes the first month of growth for crops such as maize, cotton, and soybeans. The function value $W(t)$ is the total weight in milligrams, W_0 is the weight on the day of emergence, and t is the time in days. If, for a species of soybean, $k = 0.2$ and $W_0 = 68$ mg, predict the weight at the end of 30 days.

20   **Crop growth**   Refer to Exercise 19. It is often difficult to measure the weight W_0 of a plant when it first emerges from the soil. If, for a species of cotton, $k = 0.21$ and the weight after 10 days is 575 milligrams, estimate W_0.

21   **U.S. population growth**   The 1980 population of the United States was approximately 227 million, and the population has been growing continuously at a rate of

0.7% per year. Predict the population $N(t)$ in the year 2010 if this growth trend continues.

22 Population growth in India The 1985 population estimate for India was 762 million, and the population has been growing continuously at a rate of about 2.2% per year. Assuming that this rapid growth rate continues, estimate the population $N(t)$ of India in the year 2010.

23 Longevity of halibut In fishery science, a cohort is the collection of fish that results from one annual reproduction. It is usually assumed that the number of fish $N(t)$ still alive after t years is given by an exponential function. For Pacific halibut, $N(t) = N_0 e^{-0.2t}$, where N_0 is the initial size of the cohort. Approximate the percentage of the original number still alive after 10 years.

24 Radioactive tracer The radioactive tracer ^{51}Cr can be used to locate the position of the placenta in a pregnant woman. Often the tracer must be ordered from a medical laboratory. If A_0 units (microcuries) are shipped, then because of radioactive decay, the number of units $A(t)$ present after t days is given by $A(t) = A_0 e^{-0.0249t}$.

(a) If 35 units are shipped and it takes 2 days for the tracer to arrive, approximately how many units will be available for the test?

(b) If 35 units are needed for the test, approximately how many units should be shipped?

25 Blue whale population growth In 1978, the population of blue whales in the southern hemisphere was thought to number 5000. Since whaling has been outlawed and an abundant food supply is available, the population $N(t)$ is expected to grow exponentially according to the formula $N(t) = 5000 e^{0.0036t}$, where t is in years and $t = 0$ corresponds to 1978. Predict the population in the year 2010.

26 Halibut growth The length (in centimeters) of many common commercial fish t years old can be approximated by a von Bertalanffy growth function of the form $f(t) = a(1 - be^{-kt})$, where a, b, and k are constants.

(a) For Pacific halibut, $a = 200$, $b = 0.956$, and $k = 0.18$. Estimate the length of a 10-year-old halibut.

(b) Use the graph of f to estimate the maximum attainable length of the Pacific halibut.

27 Atmospheric pressure Under certain conditions the atmospheric pressure p (in inches) at altitude h feet is given by $p = 29 e^{-0.000034h}$. What is the pressure at an altitude of 40,000 feet?

28 Polonium isotope decay If we start with c milligrams of the polonium isotope ^{210}Po, the amount remaining after t days may be approximated by $A = ce^{-0.00495t}$. If the initial amount is 50 milligrams, approximate, to the nearest hundredth, the amount remaining after
(a) 30 days   (b) 180 days   (c) 365 days

29 Growth of children The Jenss model is generally regarded as the most accurate formula for predicting the height of preschool children. If y is height (in centimeters) and x is age (in years), then
$$y = 79.041 + 6.39x - e^{3.261 - 0.993x}$$
for $\frac{1}{4} \le x \le 6$. From calculus, the rate of growth R (in cm/year) is given by $R = 6.39 + 0.993 e^{3.261 - 0.993x}$. Find the height and rate of growth of a typical 1-year-old child.

30 Particle velocity A very small spherical particle (on the order of 5 microns in diameter) is projected into still air with an initial velocity of v_0 m/sec, but its velocity decreases because of drag forces. Its velocity t seconds later is given by $v(t) = v_0 e^{-at}$ for some $a > 0$, and the distance $s(t)$ the particle travels is given by
$$s(t) = \frac{v_0}{a}(1 - e^{-at}).$$
The stopping distance is the total distance traveled by the particle.
(a) Find a formula that approximates the stopping distance in terms of v_0 and a.
(b) Use the formula in part (a) to estimate the stopping distance if $v_0 = 10$ m/sec and $a = 8 \times 10^5$.

31 Minimum wage In 1971 the minimum wage in the United States was $1.60 per hour. Assuming that the rate of inflation is 5% per year, find the equivalent minimum wage in the year 2010.

32 Land value In 1867 the United States purchased Alaska from Russia for $7,200,000. There is 586,400 square miles of land in Alaska. Assuming that the value of the land increases continuously at 3% per year and that land can be purchased at an equivalent price, determine the price of 1 acre in the year 2010. (One square mile is equivalent to 640 acres.)

Exer. 33–34: The *effective yield* (or effective annual interest rate) for an investment is the simple interest rate that would yield at the end of one year the same amount as is yielded by the compounded rate that is actually applied. Approximate, to the nearest 0.01%, the effective yield corresponding to an interest rate of r% per year compounded (a) quarterly and (b) continuously.

33 $r = 7$ **34** $r = 12$

[c] Exer. 35–36: Sketch the graph of the equation. (a) Estimate y if $x = 40$. (b) Estimate x if $y = 2$.

35 $y = e^{0.085x}$ 36 $y = e^{0.0525x}$

[c] Exer. 37–39: (a) Graph f using a graphing utility. (b) Sketch the graph of g by taking the reciprocals of y-coordinates in (a), *without* using a graphing utility.

37 $f(x) = \dfrac{e^x - e^{-x}}{2}$; $g(x) = \dfrac{2}{e^x - e^{-x}}$

38 $f(x) = \dfrac{e^x + e^{-x}}{2}$; $g(x) = \dfrac{2}{e^x + e^{-x}}$

39 $f(x) = \dfrac{e^x - e^{-x}}{e^x + e^{-x}}$; $g(x) = \dfrac{e^x + e^{-x}}{e^x - e^{-x}}$

40 **Probability density function** In statistics, the probability density function for the normal distribution is defined by

$$f(x) = \frac{1}{\sigma\sqrt{2\pi}} e^{-z^2/2} \quad \text{with} \quad z = \frac{x - \mu}{\sigma},$$

where μ and σ are real numbers (μ is the *mean* and σ^2 is the *variance* of the distribution). Sketch the graph of f for the case $\sigma = 1$ and $\mu = 0$.

[c] Exer. 41–42: Graph f and g on the same coordinate plane, and estimate the solutions of the equation $f(x) = g(x)$.

41 $f(x) = e^{0.5x} - e^{-0.4x}$; $g(x) = x^2 - 2$

42 $f(x) = 0.3e^x$; $g(x) = x^3 - x$

[c] Exer. 43–44: The functions f and g can be used to approximate e^x on the interval $[0, 1]$. Graph f, g, and $y = e^x$ on the same coordinate plane, and compare the accuracy of $f(x)$ and $g(x)$ as an approximation to e^x.

43 $f(x) = x + 1$; $g(x) = 1.72x + 1$

44 $f(x) = \frac{1}{2}x^2 + x + 1$; $g(x) = 0.84x^2 + 0.878x + 1$
    Note: Some specific keystrokes for the TI-82/83 are given in Example 18 of Appendix I.

[c] Exer. 45–46: Graph f, and estimate its zeros.

45 $f(x) = x^2e^x - xe^{(x^2)} + 0.1$

46 $f(x) = x^3e^x - x^2e^{2x} + 1$

[c] Exer. 47–48: Graph f on the interval $(0, 200]$. Find an approximate equation for the horizontal asymptote.

47 $f(x) = \left(1 + \dfrac{1}{x}\right)^x$ 48 $f(x) = \left(1 + \dfrac{2}{x}\right)^x$

[c] Exer. 49–50: Approximate the real root of the equation.

49 $e^{-x} = x$ 50 $e^{3x} = 5 - 2x$

[c] Exer. 51–52: Graph f, and determine where f is increasing or is decreasing.

51 $f(x) = xe^x$ 52 $f(x) = x^2e^{-2x}$

53 **Pollution from a smokestack** The concentration C (in units/m^3) of pollution near a ground-level point that is downwind from a smokestack source of height h is sometimes given by

$$C = \frac{Q}{\pi vab} e^{-y^2/(2a^2)} [e^{-(z-h)^2/(2b^2)} + e^{-(z+h)^2/(2b^2)}],$$

where Q is the source strength (in units/sec), v is the average wind velocity (in m/sec), z is the height (in meters) above the downwind point, y is the distance from the downwind point in the direction that is perpendicular to the wind (the cross-wind direction), and a and b are constants that depend on the downwind distance (see the figure).

(a) How does the concentration of pollution change at the ground-level, downwind position ($y = 0$ and $z = 0$) if the height of the smokestack is increased?

(b) How does the concentration of pollution change at ground level ($z = 0$) for a smokestack of fixed height h if a person moves in the cross-wind direction by increasing y?

Exercise 53

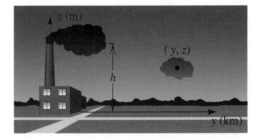

54 **Pollution concentration** Refer to Exercise 53. If the smokestack height is 100 meters and $b = 12$, use a graph to estimate the height z above the downwind point ($y = 0$) where the maximum pollution concentration occurs. (*Hint:* Let $h = 100$, $b = 12$, and graph the equation $C = e^{-(z-h)^2/(2b^2)} + e^{-(z+h)^2/(2b^2)}$.)

c 55 *Atmospheric density* The atmospheric density at altitude x is listed in the table.

Altitude (m)	0	2000	4000
Density (kg/m³)	1.225	1.007	0.819

Altitude (m)	6000	8000	10,000
Density (kg/m³)	0.660	0.526	0.414

(a) Find a function $f(x) = C_0 e^{kx}$ that approximates the density at altitude x, where C_0 and k are constants. Plot the data and f on the same coordinate axes.

(b) Use f to predict the density at 3000 and 9000 meters. Compare the predictions to the actual values of 0.909 and 0.467, respectively.

c 56 *Government spending* Federal government expenditures (in billions of dollars) for selected years are listed in the table.

Year	1910	1930	1950	1970
Expenditures	0.7	3.3	42.6	195.6

Year	1980	1990	1995
Expenditures	590.9	1252.7	1538.9

(a) Let $x = 0$ correspond to the year 1910. Find a function $A(x) = A_0 e^{kx}$ that approximates the data, where A_0 and k are constants. Plot the data and A on the same coordinate axes.

(b) Use A to predict graphically the year in which the federal government first spent \$1 trillion.

5.3 LOGARITHMIC FUNCTIONS

In Section 5.1 we observed that the exponential function given by $f(x) = a^x$ for $0 < a < 1$ or $a > 1$ is one-to-one. Hence, f has an inverse function f^{-1} (see Section 1.5). This inverse of the exponential function with base a is called the **logarithmic function with base a** and is denoted by **$\log_a$**. Its values are written $\log_a(x)$ or $\log_a x$, read "the logarithm of x with base a." Since, by the definition of an inverse function f^{-1},

$$y = f^{-1}(x) \qquad \text{if and only if} \qquad x = f(y),$$

the definition of $\log_a$ may be expressed as follows.

DEFINITION OF $\log_a$

Let a be a positive real number different from 1. The **logarithm of x with base a** is defined by

$$y = \log_a x \qquad \text{if and only if} \qquad x = a^y$$

for every $x > 0$ and every real number y.

Note that the two equations in the definition are equivalent. We call the first equation the **logarithmic form** and the second the **exponential form.**

You should strive to become an expert in changing each form into the other. The following diagram may help you achieve this goal.

Logarithmic form   **Exponential form**

$$\log_a x = y \qquad\qquad a^y = x$$

exponent

base

Observe that when forms are changed, *the bases of the logarithmic and exponential forms are the same.* The number y (that is, $\log_a x$) corresponds to the exponent in the exponential form. In words, $\log_a x$ is *the exponent to which the base a must be raised to obtain x.*

The following illustration contains examples of equivalent forms.

ILLUSTRATION

Equivalent Forms

     Logarithmic form   **Exponential form**

   $\log_5 u = 2$ $5^2 = u$

   $\log_b 8 = 3$ $b^3 = 8$

   $r = \log_p q$ $p^r = q$

   $w = \log_4 (2t + 3)$ $4^w = 2t + 3$

   $\log_3 x = 5 + 2z$ $3^{5+2z} = x$

The next example contains an application that involves changing from an exponential form to a logarithmic form.

EXAMPLE 1   **Changing exponential form to logarithmic form**

The number N of bacteria in a certain culture after t hours is given by $N = (1000)2^t$. Express t as a logarithmic function of N with base 2.

Solution $N = (1000)2^t$ given

$$\frac{N}{1000} = 2^t \qquad\qquad \text{isolate the exponential expression}$$

$$t = \log_2 \frac{N}{1000} \qquad \text{change to logarithmic form}$$

Some special cases of logarithms are given in the next example.

EXAMPLE 2   **Finding logarithms**

Find the number, if possible.

(a) $\log_{10} 100$ **(b)** $\log_2 \frac{1}{32}$ **(c)** $\log_9 3$ **(d)** $\log_7 1$ **(e)** $\log_3 (-2)$

Solution In each case we are given $\log_a x$ and must find the exponent y such that $a^y = x$. We obtain the following:

(a) $\log_{10} 100 = 2$ because $10^2 = 100$.

(b) $\log_2 \frac{1}{32} = -5$ because $2^{-5} = \frac{1}{32}$.

(c) $\log_9 3 = \frac{1}{2}$ because $9^{1/2} = 3$.

(d) $\log_7 1 = 0$ because $7^0 = 1$.

(e) $\log_3 (-2)$ is not possible because $3^y \neq -2$ for any real number y.

The following general properties follow from the interpretation of $\log_a x$ as an exponent.

Property of $\log_a x$	Reason	Illustration
(1) $\log_a 1 = 0$	$a^0 = 1$	$\log_3 1 = 0$
(2) $\log_a a = 1$	$a^1 = a$	$\log_{10} 10 = 1$
(3) $\log_a a^x = x$	$a^x = a^x$	$\log_2 8 = \log_2 2^3 = 3$
(4) $a^{\log_a x} = x$	as follows	$5^{\log_5 7} = 7$

The reason for property 4 follows directly from the definition of $\log_a$, since

$$\text{if} \quad y = \log_a x, \quad \text{then} \quad x = a^y, \quad \text{or} \quad x = a^{\log_a x}.$$

The logarithmic function with base a is the inverse of the exponential function with base a, so the graph of $y = \log_a x$ can be obtained by reflecting the graph of $y = a^x$ through the line $y = x$ (see Section 1.5). This procedure is illustrated in Figure 18 for the case $a > 1$. Note that the x-intercept of the graph is 1, the domain is the set of positive real numbers, the range is $\mathbb{R}$, and the y-axis is a vertical asymptote. Logarithms with base $0 < a < 1$ are seldom used, so we will not emphasize their graphs.

We see from Figure 18 that if $a > 1$, then $\log_a x$ is increasing on $(0, \infty)$ and hence is one-to-one by the theorem on page 68. Combining this result with parts (1) and (2) of the definition of one-to-one function on page 67 gives us the following theorem, which can also be proved if $0 < a < 1$.

Figure 18

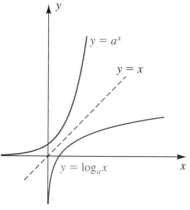

THEOREM: LOGARITHMIC FUNCTIONS ARE ONE-TO-ONE

The logarithmic function with base a is one-to-one. Thus, the following equivalent conditions are satisfied for positive real numbers x_1 and x_2:

(1) If $x_1 \neq x_2$, then $\log_a x_1 \neq \log_a x_2$.

(2) If $\log_a x_1 = \log_a x_2$, then $x_1 = x_2$.

When using this theorem as a reason for a step in the solution to an example, we will state that *logarithmic functions are one-to-one*.

In the following example we solve a simple *logarithmic equation*—that is, an equation involving a logarithm of an expression that contains a variable. Extraneous solutions may be introduced when logarithmic equations are solved. Hence, we must check solutions of logarithmic equations to make sure that we are taking logarithms of *only positive real numbers*; otherwise, a logarithmic function is not defined.

EXAMPLE 3   **Solving a logarithmic equation**

Solve the equation $\log_6 (4x - 5) = \log_6 (2x + 1)$.

Solution

$$\log_6 (4x - 5) = \log_6 (2x + 1) \quad \text{given}$$
$$4x - 5 = 2x + 1 \quad \text{logarithmic functions are one-to-one}$$
$$x = 3 \quad \text{solve for } x$$

✔ **Check $x = 3$**   LS: $\log_6 (4 \cdot 3 - 5) = \log_6 7$
RS: $\log_6 (2 \cdot 3 + 1) = \log_6 7$

Since $\log_6 7 = \log_6 7$ is a true statement, $x = 3$ is a solution.

In the next example we use the definition of logarithm to solve a logarithmic equation.

EXAMPLE 4   **Solving a logarithmic equation**

Solve the equation $\log_4 (5 + x) = 3$.

Solution $\log_4 (5 + x) = 3$ given
$5 + x = 4^3$ change to exponential form
$x = 59$ solve for x

✔ **Check $x = 59$**   LS: $\log_4 (5 + 59) = \log_4 64 = \log_4 4^3 = 3$
RS: 3

Since $3 = 3$ is a true statement, $x = 59$ is a solution.

We next sketch the graph of a specific logarithmic function.

EXAMPLE 5   **Sketching the graph of a logarithmic function**

Sketch the graph of f if $f(x) = \log_3 x$.

Solution   We will describe three methods for sketching the graph.

Method 1   Since the functions given by $\log_3 x$ and 3^x are inverses of each other, we proceed as we did for $y = \log_a x$ in Figure 18; that is, we first sketch the graph of $y = 3^x$ and then reflect it through the line $y = x$. This gives us the sketch in Figure 19. Note that points $(-1, 3^{-1})$, $(0, 1)$, and $(1, 3)$ on the graph of $y = 3^x$ reflect into the points $(3^{-1}, -1)$, $(1, 0)$, and $(3, 1)$ on the graph of $y = \log_3 x$.

Figure 19

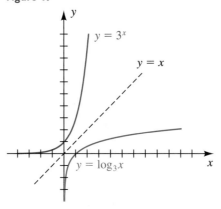

Method 2   We can find points on the graph of $y = \log_3 x$ by letting $x = 3^k$, where k is a real number, and then applying property 3 of logarithms on page 337, as follows:

$$y = \log_3 x = \log_3 3^k = k$$

Using this formula, we obtain the points on the graph listed in the following table.

$x = 3^k$	3^{-3}	3^{-2}	3^{-1}	3^0	3^1	3^2	3^3
$y = \log_3 x = k$	-3	-2	-1	0	1	2	3

This gives us the same points obtained using the first method.

Method 3   We can sketch the graph of $y = \log_3 x$ by sketching the graph of the equivalent exponential form $x = 3^y$.

As in the following examples, we often wish to sketch the graph of $f(x) = \log_a u$, where u is some expression involving x.

EXAMPLE 6   **Sketching the graph of a logarithmic function**

Sketch the graph of f if $f(x) = \log_3 |x|$ for $x \neq 0$.

Solution   The graph is symmetric with respect to the y-axis, since

$$f(-x) = \log_3 |-x| = \log_3 |x| = f(x).$$

If $x > 0$, then $|x| = x$ and the graph coincides with the graph of $y = \log_3 x$ sketched in Figure 19. Using symmetry, we reflect that part of the graph through the y-axis, obtaining the sketch in Figure 20.

Alternatively, we may think of this function as $g(x) = \log_3 x$ with $|x|$ substituted for x (refer to the discussion on page 63). Since all points on the graph of g have positive x-coordinates, we can obtain the graph of f by combining g with the reflection of g through the y-axis.

Figure 20

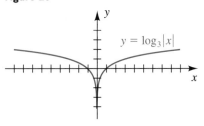

EXAMPLE 7   **Reflecting the graph of a logarithmic function**

Sketch the graph of f if $f(x) = \log_3 (-x)$.

Solution   The domain of f is the set of negative real numbers, since $\log_3 (-x)$ exists only if $-x > 0$ or, equivalently, $x < 0$. We can obtain the graph of f from the graph of $y = \log_3 x$ by replacing each point (x, y) in Figure 19 by $(-x, y)$. This is equivalent to reflecting the graph of $y = \log_3 x$ through the y-axis. The graph is sketched in Figure 21.

Another method is to change $y = \log_3 (-x)$ to the exponential form $3^y = -x$ and then sketch the graph of $x = -3^y$.

Figure 21

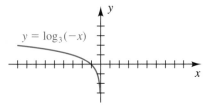

EXAMPLE 8   **Shifting graphs of logarithmic equations**

Sketch the graph of the equation:

(a) $y = \log_3 (x - 2)$ **(b)** $y = \log_3 x - 2$

Solution

(a) The graph of $y = \log_3 x$ was sketched in Figure 19 and is resketched in Figure 22. From the discussion of horizontal shifts in Section 1.4, we can obtain the graph of $y = \log_3 (x - 2)$ by shifting the graph of $y = \log_3 x$ two units to the right, as shown in Figure 22.

(b) From the discussion of vertical shifts in Section 1.4, the graph of $y = \log_3 x - 2$ can be obtained by shifting the graph of $y = \log_3 x$ two units downward, as shown in Figure 23. Note that the x-intercept is given by $\log_3 x = 2$, or $x = 3^2 = 9$.

Figure 22

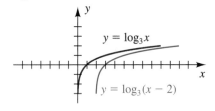

Figure 23

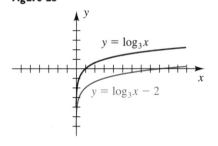

EXAMPLE 9 Reflecting the graph of a logarithmic function

Sketch the graph of f if $f(x) = \log_3 (2 - x)$.

Solution If we write

$$f(x) = \log_3 (2 - x) = \log_3 [-(x - 2)],$$

then, by applying the same technique used to obtain the graph of the equation $y = \log_3 (-x)$ in Example 7 (with x replaced by $x - 2$), we see that the graph of f is the reflection of the graph of $y = \log_3 (x - 2)$ through the vertical line $x = 2$. This gives us the sketch in Figure 24.

Another method is to change $y = \log_3 (2 - x)$ to the exponential form $3^y = 2 - x$ and then sketch the graph of $x = 2 - 3^y$.

Figure 24

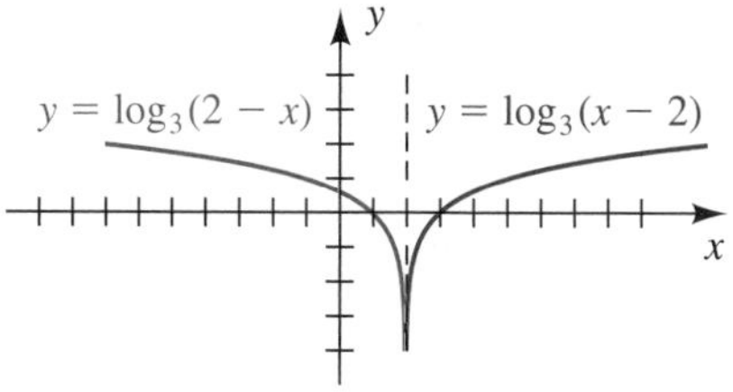

$y = \log_3 (2 - x)$ $y = \log_3 (x - 2)$

Before electronic calculators were invented, logarithms with base 10 were used for complicated numerical computations involving products, quotients, and powers of real numbers. Base 10 was used because it is well suited for numbers that are expressed in scientific form. Logarithms with base 10 are called **common logarithms.** The symbol **log x** is used as an abbreviation for $\log_{10} x$, just as $\sqrt{\ \ }$ is used as an abbreviation for $\sqrt[2]{\ \ }$.

DEFINITION OF COMMON LOGARITHM

$$\log x = \log_{10} x \qquad \text{for every} \qquad x > 0$$

Since inexpensive calculators are now available, there is no need for common logarithms as a tool for computational work. Base 10 does occur in applications, however, and hence many calculators have a $\boxed{\text{LOG}}$ key, which can be used to approximate common logarithms.

The natural exponential function is given by $f(x) = e^x$ (see Section 5.2). The logarithmic function with base e is called the **natural logarithmic function.** The symbol **ln x** (read "ell-en of x") is an abbreviation for $\log_e x$, and we refer to it as the **natural logarithm of x.** Thus, *the natural logarithmic function and the natural exponential function are inverse functions of each other.*

DEFINITION OF NATURAL LOGARITHM

$$\ln x = \log_e x \qquad \text{for every} \qquad x > 0$$

Many calculators have a key labeled $\boxed{\text{LN}}$, which can be used to approximate natural logarithms. The next illustration gives several examples of equivalent forms involving common and natural logarithms.

ILLUSTRATION

Equivalent Forms

Logarithmic form	Exponential form
$\log x = 2$	$10^2 = x$
$\log z = y + 3$	$10^{y+3} = z$
$\ln x = 2$	$e^2 = x$
$\ln z = y + 3$	$e^{y+3} = z$

To find x when given $\log x$ or $\ln x$, we may use the $\boxed{10^x}$ key or the $\boxed{e^x}$ key, respectively, on a calculator, as in the next example. If your calculator has an $\boxed{INV}$ key (for inverse), you may enter x and successively press $\boxed{INV}$ $\boxed{LOG}$ or $\boxed{INV}$ $\boxed{LN}$.

EXAMPLE 10   Solving a simple logarithmic equation

Find x if

(a) $\log x = 1.7959$ **(b)** $\ln x = 4.7$

Solution

(a) Changing $\log x = 1.7959$ to its equivalent exponential form gives us

$$x = 10^{1.7959}.$$

Evaluating the last expression to three-decimal-place accuracy yields

$$x \approx 62.503.$$

(b) Changing $\ln x = 4.7$ to its equivalent exponential form gives us

$$x = e^{4.7} \approx 109.95.$$

The following chart lists common and natural logarithmic forms for some of the properties discussed earlier.

Logarithms with base a	Common logarithms	Natural logarithms
$\log_a 1 = 0$	$\log 1 = 0$	$\ln 1 = 0$
$\log_a a = 1$	$\log 10 = 1$	$\ln e = 1$
$\log_a a^x = x$	$\log 10^x = x$	$\ln e^x = x$
$a^{\log_a x} = x$	$10^{\log x} = x$	$e^{\ln x} = x$

The next four examples illustrate applications of common and natural logarithms.

EXAMPLE 11   The Richter scale

On the Richter scale, the magnitude R of an earthquake of intensity I is given by

$$R = \log \frac{I}{I_0},$$

where I_0 is a certain minimum intensity.

(a) If the intensity of an earthquake is $1000 I_0$, find R.

(b) Express I in terms of R and I_0.

Solution

(a) $R = \log \dfrac{I}{I_0}$ given

$\quad\;\; = \log \dfrac{1000 I_0}{I_0}$ let $I = 1000 I_0$

$\quad\;\; = \log 1000$ cancel I_0

$\quad\;\; = \log 10^3$ $1000 = 10^3$

$\quad\;\; = 3$ $\log 10^x = x$ for every x

(b) $R = \log \dfrac{I}{I_0}$ given

$\quad \dfrac{I}{I_0} = 10^R$ change to exponential form

$\quad I = I_0 \cdot 10^R$ multiply by I_0

EXAMPLE 12   Newton's law of cooling

Newton's law of cooling states that the rate at which an object cools is directly proportional to the difference in temperature between the object and its surrounding medium. Newton's law can be used to show that under certain conditions the temperature T (in °C) of an object at time t (in hours) is given by $T = 75e^{-2t}$. Express t as a function of T.

Solution

$$T = 75e^{-2t} \qquad \text{given}$$

$$e^{-2t} = \frac{T}{75} \qquad \text{isolate the exponential expression}$$

$$-2t = \ln \frac{T}{75} \qquad \text{change to logarithmic form}$$

$$t = -\frac{1}{2} \ln \frac{T}{75} \qquad \text{divide by } -2$$

EXAMPLE 13   Approximating a doubling time

Assume that a population is growing continuously at a rate of 4% per year. Approximate the amount of time it takes for the population to double its size—that is, its **doubling time**.

Solution   Note that an initial population size is not given. Not knowing the initial size does not present a problem, however, since we wish only to determine the time needed to obtain a population size *relative* to the initial population size. Using the growth formula $q = q_0 e^{rt}$ with $r = 0.04$ gives us

$$2q_0 = q_0 e^{0.04t} \qquad \text{let } q = 2q_0$$

$$2 = e^{0.04t} \qquad \text{divide by } q_0 \ (q_0 \neq 0)$$

$$0.04t = \ln 2 \qquad \text{change to logarithmic form}$$

$$t = 25 \ln 2 \approx 17.3 \text{ yr.} \quad \text{multiply by } \frac{1}{0.04} = 25$$

The fact that q_0 did not have any effect on the answer indicates that the doubling time for a population of 1000 is the same as the doubling time for a population of 1,000,000 or any other reasonable initial population.

From the last example we may obtain a general formula for the doubling time of a population—namely,

$$rt = \ln 2 \qquad \text{or, equivalently,} \qquad t = \frac{\ln 2}{r}.$$

Since $\ln 2 \approx 0.69$, we see that the doubling time t for a growth of this type is approximately $0.69/r$. Because the numbers 70 and 72 are close to 69 but have more divisors, some resources refer to this doubling relationship as the **rule of 70** or the **rule of 72**. As an illustration of the rule of 72, if the growth rate of a population is 8%, then it takes about $72/8 = 9$ years for the population to double. More precisely, this value is

$$\frac{\ln 2}{8} \cdot 100 \approx 8.7 \text{ yr.}$$

EXAMPLE 14   Determining the half-life of a radioactive substance

A physicist finds that an unknown radioactive substance registers 2000 counts per minute on a Geiger counter. Ten days later the substance registers 1500 counts per minute. Using calculus, it can be shown that after t days the amount of radioactive material, and hence the number of counts per minute $N(t)$, is directly proportional to e^{ct} for some constant c. Determine the half-life of the substance.

Solution   Since $N(t)$ is directly proportional to e^{ct},

$$N(t) = ke^{ct},$$

where k is a constant. Letting $t = 0$ and using $N(0) = 2000$, we obtain

$$2000 = ke^{c \cdot 0} = k \cdot 1 = k.$$

Hence, the formula for $N(t)$ may be written

$$N(t) = 2000e^{ct}.$$

Since $N(10) = 1500$, we may determine c as follows:

$$1500 = 2000e^{c \cdot 10} \qquad \text{let } t = 10 \text{ in } N(t)$$

$$\tfrac{3}{4} = e^{10c} \qquad\qquad \text{isolate the exponential expression}$$

$$10c = \ln \tfrac{3}{4} \qquad\qquad \text{change to logarithmic form}$$

$$c = \tfrac{1}{10} \ln \tfrac{3}{4} \qquad\qquad \text{divide by 10}$$

Finally, since the half-life corresponds to the time t at which $N(t)$ is equal to 1000, we have the following:

$$1000 = 2000e^{ct} \qquad \text{let } N(t) = 1000$$

$$\tfrac{1}{2} = e^{ct} \qquad\qquad \text{isolate the exponential expression}$$

$$ct = \ln \tfrac{1}{2} \qquad\qquad \text{change to logarithmic form}$$

$$t = \frac{1}{c} \ln \frac{1}{2} \qquad\qquad \text{divide by } c$$

$$= \frac{1}{\tfrac{1}{10} \ln \tfrac{3}{4}} \ln \frac{1}{2} \qquad c = \tfrac{1}{10} \ln \tfrac{3}{4}$$

$$\approx 24 \text{ days} \qquad\qquad \text{approximate}$$

The following example is a good illustration of the power of a graphing utility, since it is impossible to find the exact solution using only algebraic methods.

EXAMPLE 15   Approximating a solution to an inequality

Graph $f(x) = \log (x + 1)$ and $g(x) = \ln (3 - x)$, and estimate the solution of the inequality $f(x) \geq g(x)$.

Figure 25
[−1, 3] by [−2, 2]

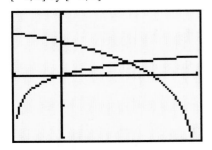

Solution   We begin by making the assignments

$$Y_1 = \log (x + 1) \qquad \text{and} \qquad Y_2 = \ln (3 - x).$$

Since the domain of f is $(-1, \infty)$ and the domain of g is $(-\infty, 3)$, we choose the viewing rectangle [−1, 3] by [−2, 2] and obtain the graph in Figure 25. Using a zoom or intersect feature, we find that the point of intersection is approximately (1.51, 0.40). Thus, the approximate solution of $f(x) \geq g(x)$ is the interval

$$1.51 < x < 3.$$

5.3 EXERCISES

Exer. 1–2: Change to logarithmic form.

1 (a) $4^3 = 64$ (b) $4^{-3} = \frac{1}{64}$ (c) $t^r = s$

(d) $3^x = 4 - t$ (e) $5^{7t} = \dfrac{a+b}{a}$ (f) $(0.7)^t = 5.3$

2 (a) $3^5 = 243$ (b) $3^{-4} = \frac{1}{81}$ (c) $c^p = d$

(d) $7^x = 100p$ (e) $3^{-2x} = \dfrac{P}{F}$ (f) $(0.9)^t = \frac{1}{2}$

Exer. 3–4: Change to exponential form.

3 (a) $\log_2 32 = 5$ (b) $\log_3 \frac{1}{243} = -5$

(c) $\log_t r = p$ (d) $\log_3 (x + 2) = 5$

(e) $\log_2 m = 3x + 4$ (f) $\log_b 512 = \frac{3}{2}$

4 (a) $\log_3 81 = 4$ (b) $\log_4 \frac{1}{256} = -4$

(c) $\log_v w = q$ (d) $\log_6 (2x - 1) = 3$

(e) $\log_4 p = 5 - x$ (f) $\log_a 343 = \frac{3}{4}$

Exer. 5–8: Solve for t using logarithms with base a.

5 $2a^{t/3} = 5$ 6 $3a^{4t} = 10$

7 $A = Ba^{Ct} + D$ 8 $L = Ma^{t/N} - P$

Exer. 9–10: Change to logarithmic form.

9 (a) $10^5 = 100,000$ (b) $10^{-3} = 0.001$

(c) $10^x = y + 1$ (d) $e^7 = p$

(e) $e^{2t} = 3 - x$

10 (a) $10^4 = 10,000$ (b) $10^{-2} = 0.01$

(c) $10^x = 38z$ (d) $e^4 = D$

(e) $e^{0.1t} = x + 2$

Exer. 11–12: Change to exponential form.

11 (a) $\log x = 50$ (b) $\log x = 20t$

(c) $\ln x = 0.1$ (d) $\ln w = 4 + 3x$

(e) $\ln (z - 2) = \frac{1}{6}$

12 (a) $\log x = -8$ (b) $\log x = y - 2$

(c) $\ln x = \frac{1}{2}$ (d) $\ln z = 7 + x$

(e) $\ln (t - 5) = 1.2$

Exer. 13–14: Find the number, if possible.

13 (a) $\log_5 1$ (b) $\log_3 3$ (c) $\log_4 (-2)$

(d) $\log_7 7^2$ (e) $3^{\log_3 8}$ (f) $\log_5 125$

(g) $\log_4 \frac{1}{16}$

14 (a) $\log_8 1$ (b) $\log_9 9$ (c) $\log_5 0$

(d) $\log_6 6^7$ (e) $5^{\log_5 4}$ (f) $\log_3 243$

(g) $\log_2 128$

Exer. 15–16: Find the number.

15 (a) $10^{\log 3}$ (b) $\log 10^5$ (c) $\log 100$

(d) $\log 0.0001$ (e) $e^{\ln 2}$ (f) $\ln e^{-3}$

(g) $e^{2 + \ln 3}$

16 (a) $10^{\log 7}$ (b) $\log 10^{-6}$ (c) $\log 100,000$

(d) $\log 0.001$ (e) $e^{\ln 8}$ (f) $\ln e^{2/3}$

(g) $e^{1 + \ln 5}$

Exer. 17–30: Solve the equation.

17 $\log_4 x = \log_4 (8 - x)$

18 $\log_3 (x + 4) = \log_3 (1 - x)$

19 $\log_5 (x - 2) = \log_5 (3x + 7)$

20 $\log_7 (x - 5) = \log_7 (6x)$

21 $\log x^2 = \log (-3x - 2)$ 22 $\ln x^2 = \ln (12 - x)$

23 $\log_3 (x - 4) = 2$ 24 $\log_2 (x - 5) = 4$

25 $\log_9 x = \frac{3}{2}$ 26 $\log_4 x = -\frac{3}{2}$

27 $\ln x^2 = -2$ 28 $\log x^2 = -4$

29 $e^{2 \ln x} = 9$ 30 $e^{-\ln x} = 0.2$

31 Sketch the graph of f if $a = 4$:

(a) $f(x) = \log_a x$ (b) $f(x) = -\log_a x$

(c) $f(x) = 2 \log_a x$ (d) $f(x) = \log_a (x + 2)$

(e) $f(x) = (\log_a x) + 2$ (f) $f(x) = \log_a (x - 2)$

(g) $f(x) = (\log_a x) - 2$ (h) $f(x) = \log_a |x|$

(i) $f(x) = \log_a (-x)$ (j) $f(x) = \log_a (3 - x)$

(k) $f(x) = |\log_a x|$

32 Work Exercise 31 if $a = 5$.

Exer. 33–36: Sketch the graph of f.

33 $f(x) = \log x$

34 $f(x) = \ln x$

35 $f(x) = \log_2 |x - 5|$

36 $f(x) = \log_3 |x + 1|$

Exer. 37–44: Shown in the figure is the graph of a function f. Express $f(x)$ in terms of logarithms with base 2.

37

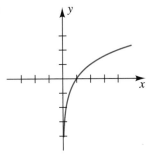

38

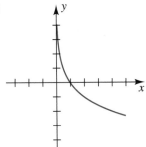

39

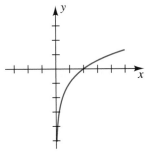

40

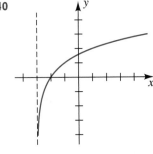

41

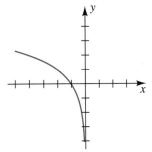

42

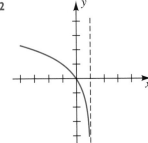

43

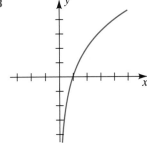

44

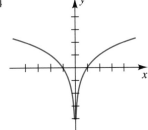

Exer. 45–46: Approximate x to three significant figures.

45 (a) $\log x = 3.6274$ (b) $\log x = 0.9469$
(c) $\log x = -1.6253$ (d) $\ln x = 2.3$
(e) $\ln x = 0.05$ (f) $\ln x = -1.6$

46 (a) $\log x = 1.8965$ **(b)** $\log x = 4.9680$

(c) $\log x = -2.2118$ **(d)** $\ln x = 3.7$

(e) $\ln x = 0.95$ **(f)** $\ln x = -5$

47 Radium decay If we start with q_0 milligrams of radium, the amount q remaining after t years is given by $q = q_0(2)^{-t/1600}$. Express t in terms of q and q_0.

48 Bismuth isotope decay The radioactive bismuth isotope ^{210}Bi disintegrates according to $Q = k(2)^{-t/5}$, where k is a constant and t is the time in days. Express t in terms of Q and k.

49 Electrical circuit A schematic of a simple electrical circuit consisting of a resistor and an inductor is shown in the figure. The current I at time t is given by the formula $I = 20e^{-Rt/L}$, where R is the resistance and L is the inductance. Solve this equation for t.

Exercise 49

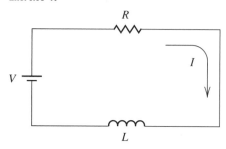

50 Electrical condenser An electrical condenser with initial charge Q_0 is allowed to discharge. After t seconds the charge Q is $Q = Q_0e^{kt}$, where k is a constant. Solve this equation for t.

51 Richter scale Use the Richter scale formula $R = \log (I/I_0)$ to find the magnitude of an earthquake that has an intensity

(a) 100 times that of I_0

(b) 10,000 times that of I_0

(c) 100,000 times that of I_0

52 Richter scale Refer to Exercise 51. The largest recorded magnitudes of earthquakes have been between 8 and 9 on the Richter scale. Find the corresponding intensities in terms of I_0.

53 Sound intensity The loudness of a sound, as experienced by the human ear, is based on its intensity level. A formula used for finding the intensity level α (in decibels) that corresponds to a sound intensity I is $\alpha = 10 \log (I/I_0)$, where I_0 is a special value of I agreed to be

the weakest sound that can be detected by the ear under certain conditions. Find α if

(a) I is 10 times as great as I_0

(b) I is 1000 times as great as I_0

(c) I is 10,000 times as great as I_0 (This is the intensity level of the average voice.)

54 Sound intensity Refer to Exercise 53. A sound intensity level of 140 decibels produces pain in the average human ear. Approximately how many times greater than I_0 must I be in order for α to reach this level?

55 U.S. population growth The population $N(t)$ (in millions) of the United States t years after 1980 may be approximated by the formula $N(t) = 227e^{0.007t}$. When will the population be twice what it was in 1980?

56 Population growth in India The population $N(t)$ (in millions) of India t years after 1985 may be approximated by the formula $N(t) = 762e^{0.022t}$. When will the population reach 1.5 billion?

57 Children's weight The Ehrenberg relation

$$\ln W = \ln 2.4 + (1.84)h$$

is an empirically based formula relating the height h (in meters) to the average weight W (in kilograms) for children 5 through 13 years old.

(a) Express W as a function of h that does not contain ln.

(b) Estimate the average weight of an 8-year-old child who is 1.5 meters tall.

58 Continuously compounded interest If interest is compounded continuously at the rate of 10% per year, approximate the number of years it will take an initial deposit of $6000 to grow to $25,000.

59 Air pressure The air pressure $p(h)$ (in lb/in^2) at an altitude of h feet above sea level may be approximated by the formula $p(h) = 14.7e^{-0.0000385h}$. At approximately what altitude h is the air pressure

(a) 10 lb/in^2?

(b) one-half its value at sea level?

60 Vapor pressure A liquid's vapor pressure P (in lb/in^2), a measure of its volatility, is related to its temperature T (in °F) by the Antoine equation

$$\log P = a + \frac{b}{c + T},$$

where a, b, and c are constants. Vapor pressure increases rapidly with an increase in temperature. Express P as a function of T.

61 **Elephant growth** The weight W (in kilograms) of a female African elephant at age t (in years) may be approximated by

$$W = 2600(1 - 0.51e^{-0.075t})^3.$$

(a) Approximate the weight at birth.

(b) Estimate the age of a female African elephant weighing 1800 kilograms by using (1) the accompanying graph and (2) the formula for W.

Exercise 61

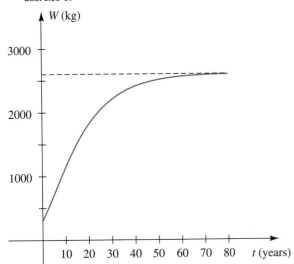

62 **Coal consumption** A country presently has coal reserves of 50 million tons. Last year 6.5 million tons of coal was consumed. Past years' data and population projections suggest that the rate of consumption R (in million tons/year) will increase according to the formula $R = 6.5e^{0.02t}$, and the total amount T (in million tons) of coal that will be used in t years is given by the formula $T = 325(e^{0.02t} - 1)$. If the country uses only its own resources, when will the coal reserves be depleted?

63 **Urban population density** An urban density model is a formula that relates the population density D (in thousands/mi^2) to the distance x (in miles) from the center of the city. The formula $D = ae^{-bx}$ for the central density a and coefficient of decay b has been found to be appropriate for many large U.S. cities. For the city of Atlanta in 1970, $a = 5.5$ and $b = 0.10$. At approximately what distance was the population density 2000 per square mile?

64 **Brightness of stars** Stars are classified into categories of brightness called magnitudes. The faintest stars, with

light flux L_0, are assigned a magnitude 6. Brighter stars of light flux L are assigned a magnitude m by means of the formula

$$m = 6 - (2.5) \log \frac{L}{L_0}.$$

(a) Find m if $L = 10^{0.4}L_0$.

(b) Solve the formula for L in terms of m and L_0.

65 **Radioactive iodine decay** Radioactive iodine ^{131}I is frequently used in tracer studies involving the thyroid gland. The substance decays according to the formula $A(t) = A_0 a^{-t}$, where A_0 is the initial dose and t is the time in days. Find a, assuming the half-life of ^{131}I is 8 days.

66 **Radioactive contamination** Radioactive strontium ^{90}Sr has been deposited in a large field by acid rain. If sufficient amounts make their way through the food chain to humans, bone cancer can result. It has been determined that the radioactivity level in the field is 2.5 times the safe level S. ^{90}Sr decays according to the formula

$$A(t) = A_0 e^{-0.0239t},$$

where A_0 is the amount currently in the field and t is the time in years. For how many years will the field be contaminated?

67 **Walking speed** In a survey of 15 cities ranging in population P from 300 to 3,000,000, it was found that the average walking speed S (in ft/sec) of a pedestrian could be approximated by $S = 0.05 + 0.86 \log P$.

(a) How does the population affect the average walking speed?

(b) For what population is the average walking speed 5 ft/sec?

68 **Computer chips** For manufacturers of computer chips, it is important to consider the fraction F of chips that will fail after t years of service. This fraction can sometimes be approximated by the formula $F = 1 - e^{-ct}$, where c is a positive constant.

(a) How does the value of c affect the reliability of a chip?

(b) If $c = 0.125$, after how many years will 35% of the chips have failed?

C **Exer. 69–70: Approximate the function at the value of x to four decimal places.**

69 (a) $f(x) = \ln(x + 1) + e^x$, $x = 2$

(b) $g(x) = \dfrac{(\log x)^2 - \log x}{4}$, $x = 3.97$

70 (a) $f(x) = \log(2x^2 + 1) - 10^{-x}$, $x = 1.95$

(b) $g(x) = \dfrac{x - 3.4}{\ln x + 4}$, $x = 0.55$

[c] **Exer. 71–72: Approximate the real root of the equation.**

71 $x \ln x = 1$ **72** $\ln x + x = 0$

[c] **Exer. 73–74: Graph f and g on the same coordinate plane, and estimate the solution of the inequality $f(x) \geq g(x)$.**

73 $f(x) = 2.2 \log(x + 2)$; $g(x) = \ln x$
Note: Some specific keystrokes for the TI-82/83 are given in Example 19 of Appendix I.

74 $f(x) = x \ln |x|$; $g(x) = 0.15e^x$

[c] **75** *Cholesterol level in women* Studies relating serum cholesterol level to coronary heart disease suggest that a

risk factor is the ratio x of the total amount C of cholesterol in the blood to the amount H of high-density lipoprotein cholesterol in the blood. For a female, the lifetime risk R of having a heart attack can be approximated by the formula

$$R = 2.07 \ln x - 2.04 \quad \text{provided} \quad 0 \leq R \leq 1.$$

For example, if $R = 0.65$, then there is a 65% chance that a woman will have a heart attack over an average lifetime.

(a) Calculate R for a female with $C = 242$ and $H = 78$.

(b) Graphically estimate x when the risk is 75%.

[c] **76** *Cholesterol level in men* Refer to Exercise 75. For a male, the risk can be approximated by the formula $R = 1.36 \ln x - 1.19$.

(a) Calculate R for a male with $C = 287$ and $H = 65$.

(b) Graphically estimate x when the risk is 75%.

5.4 PROPERTIES OF LOGARITHMS

In the preceding section we observed that $\log_a x$ can be interpreted as an exponent. Thus, it seems reasonable to expect that the laws of exponents can be used to obtain corresponding laws of logarithms. This is demonstrated in the proofs of the following laws, which are fundamental for all work with logarithms.

Laws of Logarithms

> If u and w denote positive real numbers, then
>
> **(1)** $\log_a (uw) = \log_a u + \log_a w$
>
> **(2)** $\log_a \left(\dfrac{u}{w}\right) = \log_a u - \log_a w$
>
> **(3)** $\log_a (u^c) = c \log_a u$ for every real number c

Proof For all three proofs, let

$$r = \log_a u \qquad \text{and} \qquad s = \log_a w.$$

The equivalent exponential forms are

$$u = a^r \qquad \text{and} \qquad w = a^s.$$

We now proceed as follows:

(1)	$uw = a^r a^s$	definition of u and w
	$uw = a^{r+s}$	law of exponents
	$\log_a (uw) = r + s$	change to logarithmic form
	$\log_a (uw) = \log_a u + \log_a w$	definition of r and s

(2) $$\frac{u}{w} = \frac{a^r}{a^s}$$ definition of u and w

$$\frac{u}{w} = a^{r-s}$$ law of exponents

$$\log_a \left(\frac{u}{w} \right) = r - s$$ change to logarithmic form

$$\log_a \left(\frac{u}{w} \right) = \log_a u - \log_a w$$ definition of r and s

(3) $$u^c = (a^r)^c$$ definition of u

$$u^c = a^{cr}$$ law of exponents

$$\log_a (u^c) = cr$$ change to logarithmic form

$$\log_a (u^c) = c \log_a u$$ definition of r ◢

The laws of logarithms for the special cases $a = 10$ (common logs) and $a = e$ (natural logs) are written as shown in the following chart.

Common logarithms	Natural logarithms
$\log (uw) = \log u + \log w$	$\ln (uw) = \ln u + \ln w$
$\log \left(\dfrac{u}{w} \right) = \log u - \log w$	$\ln \left(\dfrac{u}{w} \right) = \ln u - \ln w$
$\log (u^c) = c \log u$	$\ln (u^c) = c \ln u$

As indicated by the following warnings, there are no laws for expressing $\log_a (u + w)$ or $\log_a (u - w)$ in terms of simpler logarithms.

Warnings

$$\log_a (u + w) \neq \log_a u + \log_a w$$
$$\log_a (u - w) \neq \log_a u - \log_a w$$

The following examples illustrate uses of the laws of logarithms.

EXAMPLE 1   **Using laws of logarithms**

Express $\log_a \dfrac{x^3 \sqrt{y}}{z^2}$ in terms of logarithms of x, y, and z.

Solution   We write $\sqrt{y}$ as $y^{1/2}$ and use laws of logarithms:

$$\log_a \frac{x^3 \sqrt{y}}{z^2} = \log_a (x^3 y^{1/2}) - \log_a z^2 \qquad \text{law 2}$$

$$= \log_a x^3 + \log_a y^{1/2} - \log_a z^2 \qquad \text{law 1}$$

$$= 3 \log_a x + \tfrac{1}{2} \log_a y - 2 \log_a z \qquad \text{law 3}$$

(continued)

Note that if a term with a positive exponent (such as x^3) is in the numerator of the original expression, it will have a positive coefficient in the expanded form; and if it is in the denominator (such as z^2), it will have a negative coefficient in the expanded form.

EXAMPLE 2   Using laws of logarithms

Express as one logarithm:

$$\tfrac{1}{3} \log_a (x^2 - 1) - \log_a y - 4 \log_a z$$

Solution   We apply the laws of logarithms as follows:

$$
\begin{aligned}
&\tfrac{1}{3} \log_a (x^2 - 1) - \log_a y - 4 \log_a z \\
&= \log_a (x^2 - 1)^{1/3} - \log_a y - \log_a z^4 && \text{law 3} \\
&= \log_a \sqrt[3]{x^2 - 1} - (\log_a y + \log_a z^4) && \text{algebra} \\
&= \log_a \sqrt[3]{x^2 - 1} - \log_a (yz^4) && \text{law 1} \\
&= \log_a \frac{\sqrt[3]{x^2 - 1}}{yz^4} && \text{law 2}
\end{aligned}
$$

EXAMPLE 3   Solving a logarithmic equation

Solve the equation $\log_5 (2x + 3) = \log_5 11 + \log_5 3$.

Solution

$$
\begin{aligned}
\log_5 (2x + 3) &= \log_5 11 + \log_5 3 && \text{given} \\
\log_5 (2x + 3) &= \log_5 (11 \cdot 3) && \text{law 1 of logarithms} \\
2x + 3 &= 33 && \text{logarithmic functions are one-to-one} \\
x &= 15 && \text{solve for } x
\end{aligned}
$$

✔ **Check $x = 15$**   LS: $\log_5 (2 \cdot 15 + 3) = \log_5 33$
RS: $\log_5 11 + \log_5 3 = \log_5 (11 \cdot 3) = \log_5 33$

Since $\log_5 33 = \log_5 33$ is a true statement, $x = 15$ is a solution.

The laws of logarithms were proved for logarithms of *positive* real numbers u and w. If we apply these laws to equations in which u and w are expressions involving a variable, then extraneous solutions may occur. Answers should therefore be substituted for the variable in u and w to determine whether these expressions are defined.

EXAMPLE 4   Solving a logarithmic equation

Solve the equation $\log_2 x + \log_2 (x + 2) = 3$.

Solution

$$\log_2 x + \log_2 (x + 2) = 3 \qquad \text{given}$$

$$\log_2 [x(x + 2)] = 3 \qquad \text{law 1 of logarithms}$$

$$x(x + 2) = 2^3 \qquad \text{change to exponential form}$$

$$x^2 + 2x - 8 = 0 \qquad \text{multiply and set equal to 0}$$

$$(x - 2)(x + 4) = 0 \qquad \text{factor}$$

$$x - 2 = 0, \quad x + 4 = 0 \qquad \text{zero factor theorem}$$

$$x = 2, \qquad x = -4 \qquad \text{solve for } x$$

✔ **Check $x = 2$**   LS: $\log_2 2 + \log_2 (2 + 2) = 1 + \log_2 4$
$$= 1 + \log_2 2^2 = 1 + 2 = 3$$

RS: 3

Since $3 = 3$ is a true statement, $x = 2$ is a solution.

✔ **Check $x = -4$**   LS: $\log_2 (-4) + \log_2 (-4 + 2)$

Since logarithms of negative numbers are undefined, $x = -4$ is not a solution.

EXAMPLE 5   **Solving a logarithmic equation**

Solve the equation $\ln (x + 6) - \ln 10 = \ln (x - 1) - \ln 2$.

Solution

$$\ln (x + 6) - \ln (x - 1) = \ln 10 - \ln 2 \qquad \text{rearrange terms}$$

$$\ln \left(\frac{x + 6}{x - 1}\right) = \ln \frac{10}{2} \qquad \text{law 2 of logarithms}$$

$$\frac{x + 6}{x - 1} = 5 \qquad \text{ln is one-to-one}$$

$$x + 6 = 5x - 5 \qquad \text{multiply by } x - 1$$

$$x = \tfrac{11}{4} \qquad \text{solve for } x$$

✔ **Check**   Since both $\ln (x + 6)$ and $\ln (x - 1)$ are defined at $x = \tfrac{11}{4}$ (they are logarithms of positive real numbers) and since our algebraic steps are correct, it follows that $\tfrac{11}{4}$ is a solution of the given equation.

EXAMPLE 6   **Shifting the graph of a logarithmic equation**

Sketch the graph of $y = \log_3 (81x)$.

Figure 26

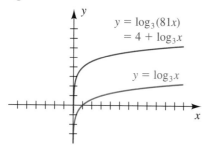

Solution   We may rewrite the equation as follows:

$$y = \log_3 (81x) \qquad \text{given}$$
$$= \log_3 81 + \log_3 x \quad \text{law 1 of logarithms}$$
$$= \log_3 3^4 + \log_3 x \quad 81 = 3^4$$
$$= 4 + \log_3 x \qquad \log_a a^x = x$$

Thus, we can obtain the graph of $y = \log_3 (81x)$ by vertically shifting the graph of $y = \log_3 x$ in Figure 19 upward four units. This gives us the sketch in Figure 26.

EXAMPLE 7   **Sketching graphs of logarithmic equations**

Sketch the graph of the equation:

(a) $y = \log_3 (x^2)$ **(b)** $y = 2 \log_3 x$

Solution

(a) Since $x^2 = |x|^2$, we may rewrite the given equation as

$$y = \log_3 |x|^2.$$

Using a law of logarithms, we have

$$y = 2 \log_3 |x|.$$

We can obtain the graph of $y = 2 \log_3 |x|$ by multiplying the y-coordinates of points on the graph of $y = \log_3 |x|$ in Figure 20 by 2. This gives us the graph in Figure 27(a).

Figure 27

(a)

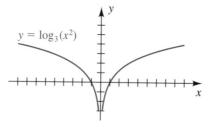

(b)

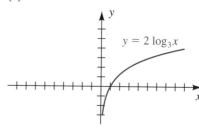

(b) If $y = 2 \log_3 x$, then x must be positive. Hence, the graph is identical to that part of the graph of $y = 2 \log_3 |x|$ in Figure 27(a) that lies to the right of the y-axis. This gives us Figure 27(b).

EXAMPLE 8 A relationship between selling price and demand

In the study of economics, the demand D for a product is often related to its selling price p by an equation of the form

$$\log_a D = \log_a c - k \log_a p,$$

where a, c, and k are positive constants.

(a) Solve the equation for D.

(b) How does increasing or decreasing the selling price affect the demand?

Solution

(a)
$$\log_a D = \log_a c - k \log_a p \qquad \text{given}$$

$$\log_a D = \log_a c - \log_a p^k \qquad \text{law 3 of logarithms}$$

$$\log_a D = \log_a \frac{c}{p^k} \qquad \text{law 2 of logarithms}$$

$$D = \frac{c}{p^k} \qquad \log_a \text{ is one-to-one}$$

(b) If the price p is increased, the denominator p^k in $D = c/p^k$ will also increase and hence the demand D for the product will decrease. If the price is decreased, then p^k will decrease and the demand D will increase.

5.4 EXERCISES

Exer. 1–8: Express in terms of logarithms of x, y, z, or w.

1 (a) $\log_4 (xz)$ (b) $\log_4 (y/x)$ (c) $\log_4 \sqrt[3]{z}$

2 (a) $\log_3 (xyz)$ (b) $\log_3 (xz/y)$ (c) $\log_3 \sqrt[5]{y}$

3 $\log_a \dfrac{x^3 w}{y^2 z^4}$

4 $\log_a \dfrac{y^5 w^2}{x^4 z^3}$

5 $\log \dfrac{\sqrt[3]{z}}{x\sqrt{y}}$

6 $\log \dfrac{\sqrt{y}}{x^4 \sqrt[3]{z}}$

7 $\ln \sqrt[4]{\dfrac{x^7}{y^5 z}}$

8 $\ln x \sqrt[3]{\dfrac{y^4}{z^5}}$

Exer. 9–16: Write the expression as one logarithm.

9 (a) $\log_3 x + \log_3 (5y)$ (b) $\log_3 (2z) - \log_3 x$

(c) $5 \log_3 y$

10 (a) $\log_4 (3z) + \log_4 x$ (b) $\log_4 x - \log_4 (7y)$

(c) $\frac{1}{3} \log_4 w$

11 $2 \log_a x + \frac{1}{3} \log_a (x - 2) - 5 \log_a (2x + 3)$

12 $5 \log_a x - \frac{1}{2} \log_a (3x - 4) - 3 \log_a (5x + 1)$

13 $\log (x^3 y^2) - 2 \log x\sqrt[3]{y} - 3 \log \left(\dfrac{x}{y}\right)$

14 $2 \log \dfrac{y^3}{x} - 3 \log y + \dfrac{1}{2} \log x^4 y^2$

15 $\ln y^3 + \frac{1}{3} \ln (x^3 y^6) - 5 \ln y$

16 $2 \ln x - 4 \ln (1/y) - 3 \ln (xy)$

Exer. 17–32: Solve the equation.

17 $\log_6 (2x - 3) = \log_6 12 - \log_6 3$

18 $\log_4 (3x + 2) = \log_4 5 + \log_4 3$

19 $2 \log_3 x = 3 \log_3 5$

20 $3 \log_2 x = 2 \log_2 3$

21 $\log x - \log (x + 1) = 3 \log 4$

22 $\log (x + 2) - \log x = 2 \log 4$

23 $\ln (-4 - x) + \ln 3 = \ln (2 - x)$

24 $\ln x + \ln (x + 6) = \frac{1}{2} \ln 9$

25 $\log_2 (x + 7) + \log_2 x = 3$

26 $\log_6 (x + 5) + \log_6 x = 2$

27 $\log_3 (x + 3) + \log_3 (x + 5) = 1$

28 $\log_3 (x - 2) + \log_3 (x - 4) = 2$

29 $\log (x + 3) = 1 - \log (x - 2)$

30 $\log (57x) = 2 + \log (x - 2)$

31 $\ln x = 1 - \ln (x + 2)$

32 $\ln x = 1 + \ln (x + 1)$

Exer. 33–44: Sketch the graph of f.

33 $f(x) = \log_3 (3x)$ 34 $f(x) = \log_4 (16x)$

35 $f(x) = 3 \log_3 x$ 36 $f(x) = \frac{1}{3} \log_3 x$

37 $f(x) = \log_3 (x^2)$ 38 $f(x) = \log_2 (x^2)$

39 $f(x) = \log_2 (x^3)$ 40 $f(x) = \log_3 (x^3)$

41 $f(x) = \log_2 \sqrt{x}$ 42 $f(x) = \log_2 \sqrt[3]{x}$

43 $f(x) = \log_3 \left(\dfrac{1}{x}\right)$ 44 $f(x) = \log_2 \left(\dfrac{1}{x}\right)$

Exer. 45–48: Shown in the figure is the graph of a function f. Express $f(x)$ as one logarithm with base 2.

45

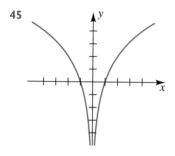

46

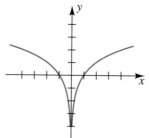

47

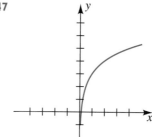

48

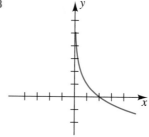

49 **Pareto's law**   Pareto's law for capitalist countries states that the relationship between annual income x and the number y of individuals whose income exceeds x is

$$\log y = \log b - k \log x,$$

where b and k are positive constants. Solve this equation for y.

50 **Price and demand**   If p denotes the selling price (in dollars) of a commodity and x is the corresponding demand (in number sold per day), then the relationship between p and x is sometimes given by $p = p_0 e^{-ax}$, where p_0 and a are positive constants. Express x as a function of p.

51 **Wind velocity**   If v denotes the wind velocity (in m/sec) at a height of z meters above the ground, then under certain conditions $v = c \ln (z/z_0)$, where c is a positive

constant and z_0 is the height at which the velocity is zero. Sketch the graph of this equation on a zv-plane for $c = 0.5$ and $z_0 = 0.1$ m.

52 *Eliminating pollution* If the pollution of Lake Erie were stopped suddenly, it has been estimated that the level y of pollutants would decrease according to the formula $y = y_0 e^{-0.3821t}$, where t is the time in years and y_0 is the pollutant level at which further pollution ceased. How many years would it take to clear 50% of the pollutants?

53 *Reaction to a stimulus* Let R denote the reaction of a subject to a stimulus of strength x. There are many possibilities for R and x. If the stimulus x is saltiness (in grams of salt per liter), R may be the subject's estimate of how salty the solution tasted, based on a scale from 0 to 10. One relationship between R and x is given by the Weber-Fechner formula, $R(x) = a \log (x/x_0)$, where a is a positive constant and x_0 is called the threshold stimulus.

(a) Find $R(x_0)$.

(b) Find a relationship between $R(x)$ and $R(2x)$.

54 *Electron energy* The energy $E(x)$ of an electron after passing through material of thickness x is given by $E(x) = E_0 e^{-x/x_0}$, where E_0 is the initial energy and x_0 is the radiation length.

(a) Express, in terms of E_0, the energy of an electron after it passes through material of thickness x_0.

(b) Express, in terms of x_0, the thickness at which the electron loses 99% of its initial energy.

55 *Ozone layer* One method of estimating the thickness of the ozone layer is to use the formula

$$\ln I_0 - \ln I = kx,$$

where I_0 is the intensity of a particular wavelength of light from the sun before it reaches the atmosphere, I is the intensity of the same wavelength after passing through a layer of ozone x centimeters thick, and k is the absorption constant of ozone for that wavelength. Suppose for a wavelength of 3176×10^{-8} cm with $k \approx 0.39$, I_0/I is measured as 1.12. Approximate the thickness of the ozone layer to the nearest 0.01 centimeter.

56 *Ozone layer* Refer to Exercise 55. Approximate the percentage decrease in the intensity of light with a wavelength of 3176×10^{-8} centimeter if the ozone layer is 0.24 centimeter thick.

[c] Exer. 57–58: Graph f and g on the same coordinate plane, and estimate the solution of the inequality $f(x) \geq g(x)$.

57 $f(x) = x^3 - 3.5x^2 + 3x$; $g(x) = \log 3x$

58 $f(x) = 3^{-0.5x}$; $g(x) = \log x$

[c] Exer. 59–60: Use a graph to estimate the roots of the equation on the given interval.

59 $e^{-x} - 2 \log (1 + x^2) + 0.5x = 0$;   [0, 8]

60 $0.3 \ln x + x^3 - 3.1x^2 + 1.3x + 0.8 = 0$;   (0, 3)

[c] Exer. 61–62: Graph f on the interval [0.2, 16]. (a) Estimate the intervals where f is increasing or is decreasing. (b) Estimate the maximum and minimum values of f on [0.2, 16].

61 $f(x) = 2 \log 2x - 1.5x + 0.1x^2$

62 $f(x) = 1.1^{3x} + x - 1.35^x - \log x + 5$

[c] Exer. 63–64: Solve the equation graphically.

63 $x \log x - \log x = 5$

64 $0.3e^x - \ln x = 4 \ln (x + 1)$

[c] Exer. 65–66: Bird calls decrease in intensity (loudness) as they travel through the atmosphere. The farther a bird is from an observer, the softer the sound. This decrease in intensity can be used to estimate the distance between an observer and a bird. A formula that can be used to measure this distance is

$$I = I_0 - 20 \log d - kd \quad \text{provided} \quad 0 \leq I \leq I_0,$$

where I_0 represents the intensity (in decibels) of the bird at a distance of one meter (I_0 is often known and usually depends only on the type of bird), I is the observed intensity at a distance d meters from the bird, and k is a positive constant that depends on the atmospheric conditions such as temperature and humidity. Given I_0, I, and k, graphically estimate the distance d between the bird and the observer.

65 $I_0 = 70$, $I = 20$, $k = 0.076$

66 $I_0 = 60$, $I = 15$, $k = 0.11$

5.5 EXPONENTIAL AND LOGARITHMIC EQUATIONS

In this section we shall consider various types of exponential and logarithmic equations and their applications. When solving an equation involving exponential expressions with constant bases and variables appearing in the exponent(s), we often *equate the logarithms of both sides* of the equation. When we do so, the variables in the exponent become multipliers, and the resulting equation is usually easier to solve. We will refer to this step as simply "take log of both sides."

EXAMPLE 1   **Solving an exponential equation**

Solve the equation $3^x = 21$.

Solution

$$3^x = 21 \qquad \text{given}$$
$$\log (3^x) = \log 21 \qquad \text{take log of both sides}$$
$$x \log 3 = \log 21 \qquad \text{law 3 of logarithms}$$
$$x = \frac{\log 21}{\log 3} \qquad \text{divide by log 3}$$

We could also have used natural logarithms to obtain

$$x = \frac{\ln 21}{\ln 3}.$$

Using a calculator gives us the approximate solution $x \approx 2.77$. A partial check is to note that since $3^2 = 9$ and $3^3 = 27$, the number x such that $3^x = 21$ must be between 2 and 3, somewhat closer to 3 than to 2.

We could also have solved the equation in Example 1 by changing the exponential form $3^x = 21$ to logarithmic form, as we did in Section 5.3, obtaining

$$x = \log_3 21.$$

This is, in fact, the solution of the equation; however, since calculators typically have keys only for log and ln, we cannot approximate $\log_3 21$ directly. The next theorem gives us a simple *change of base formula* for finding $\log_b u$ if $u > 0$ and b is *any* logarithmic base.

THEOREM: CHANGE OF BASE FORMULA

If $u > 0$ and if a and b are positive real numbers different from 1, then

$$\log_b u = \frac{\log_a u}{\log_a b}.$$

Proof   We begin with the equivalent equations

$$w = \log_b u \quad \text{and} \quad b^w = u$$

and proceed as follows:

$$b^w = u \qquad \text{given}$$
$$\log_a b^w = \log_a u \qquad \text{take } \log_a \text{ of both sides}$$
$$w \log_a b = \log_a u \qquad \text{law 3 of logarithms}$$
$$w = \frac{\log_a u}{\log_a b} \qquad \text{divide by } \log_a b$$

Since $w = \log_b u$, we obtain the formula.   ◢

The following special case of the change of base formula is obtained by letting $u = a$ and using the fact that $\log_a a = 1$:

$$\log_b a = \frac{1}{\log_a b}$$

The change of base formula is sometimes confused with law 2 of logarithms. The first of the following warnings could be remembered with the phrase "a quotient of logs is *not* the log of the quotient."

Warnings

$$\frac{\log_a u}{\log_a b} \neq \log_a \frac{u}{b}; \qquad \frac{\log_a u}{\log_a b} \neq \log_a (u - b)$$

The most frequently used special cases of the change of base formula are those for $a = 10$ (common logarithms) and $a = e$ (natural logarithms), as stated in the next box.

Special Change of Base Formulas

$$(1) \ \log_b u = \frac{\log_{10} u}{\log_{10} b} = \frac{\log u}{\log b} \qquad (2) \ \log_b u = \frac{\log_e u}{\log_e b} = \frac{\ln u}{\ln b}$$

Next, we will rework Example 1 using a change of base formula.

EXAMPLE 2   **Using a change of base formula**

Solve the equation $3^x = 21$.

Solution   We proceed as follows:

$$3^x = 21 \qquad \text{given}$$
$$x = \log_3 21 \qquad \text{change to logarithmic form}$$
$$= \frac{\log 21}{\log 3} \qquad \text{special change of base formula 1}$$

(continued)

Another method is to use special change of base formula 2, obtaining

$$x = \frac{\ln 21}{\ln 3}.$$

Logarithms with base 2 are used in computer science. The next example indicates how to approximate logarithms with base 2 using change of base formulas.

EXAMPLE 3   Approximating a logarithm with base 2

Approximate $\log_2 5$ using

(a) common logarithms   **(b)** natural logarithms

Solution   Using special change of base formulas 1 and 2, we obtain the following:

(a) $\log_2 5 = \dfrac{\log 5}{\log 2} \approx 2.322$

(b) $\log_2 5 = \dfrac{\ln 5}{\ln 2} \approx 2.322$

EXAMPLE 4   Solving an exponential equation

Solve the equation $5^{2x+1} = 6^{x-2}$.

Solution   We can use either common or natural logarithms. Using common logarithms gives us the following:

$5^{2x+1} = 6^{x-2}$	given
$\log (5^{2x+1}) = \log (6^{x-2})$	take log of both sides
$(2x + 1) \log 5 = (x - 2) \log 6$	law 3 of logarithms
$2x \log 5 + \log 5 = x \log 6 - 2 \log 6$	multiply
$2x \log 5 - x \log 6 = -\log 5 - 2 \log 6$	get all terms with x on one side
$x(\log 5^2 - \log 6) = -(\log 5 + \log 6^2)$	factor, and use law 3 of logarithms
$x = -\dfrac{\log (5 \cdot 36)}{\log \frac{25}{6}}$	solve for x, and use laws of logarithms

An approximation is $x \approx -3.64$.

EXAMPLE 5   Solving an exponential equation

Solve the equation $\dfrac{5^x - 5^{-x}}{2} = 3$.

Solution

$$\frac{5^x - 5^{-x}}{2} = 3 \qquad \text{given}$$

$$5^x - 5^{-x} = 6 \qquad \text{multiply by 2}$$

$$5^x - \frac{1}{5^x} = 6 \qquad \text{definition of negative exponent}$$

$$5^x(5^x) - \frac{1}{5^x}(5^x) = 6(5^x) \qquad \text{multiply by the lcd, } 5^x$$

$$5^{2x} - 6(5^x) - 1 = 0 \qquad \text{simplify and subtract } 6(5^x)$$

We recognize this form of the equation as a quadratic in 5^x and proceed as follows:

$$(5^x)^2 - 6(5^x) - 1 = 0 \qquad \text{law of exponents}$$

$$5^x = \frac{6 \pm \sqrt{36 + 4}}{2} \qquad \text{quadratic formula}$$

$$5^x = 3 \pm \sqrt{10} \qquad \text{simplify}$$

$$5^x = 3 + \sqrt{10} \qquad 5^x > 0, \text{ but } 3 - \sqrt{10} < 0$$

$$\log 5^x = \log(3 + \sqrt{10}) \qquad \text{take log of both sides}$$

$$x \log 5 = \log(3 + \sqrt{10}) \qquad \text{law 3 of logarithms}$$

$$x = \frac{\log(3 + \sqrt{10})}{\log 5} \qquad \text{divide by log 5}$$

An approximation is $x \approx 1.13$.

EXAMPLE 6   **Approximating light penetration in an ocean**

The Beer-Lambert law states that the amount of light I that penetrates to a depth of x meters in an ocean is given by $I = I_0 c^x$, where $0 < c < 1$ and I_0 is the amount of light at the surface.

(a) Solve for x in terms of common logarithms.

(b) If $c = \frac{1}{4}$, approximate the depth at which $I = 0.01I_0$. (This determines the zone where photosynthesis can take place.)

Solution

(a) $I = I_0 c^x \qquad \text{given}$

$$\frac{I}{I_0} = c^x \qquad \text{isolate the exponential expression}$$

$$x = \log_c \frac{I}{I_0} \qquad \text{change to logarithmic form}$$

$$= \frac{\log(I/I_0)}{\log c} \qquad \text{special change of base formula 1}$$

(continued)

(b) Letting $I = 0.01I_0$ and $c = \frac{1}{4}$ in the formula for x obtained in part (a), we have

$$x = \frac{\log\,(0.01I_0/I_0)}{\log\,\frac{1}{4}} = \frac{\log\,(0.01)}{\log\,1 - \log\,4} = \frac{\log\,10^{-2}}{0 - \log\,4} = \frac{-2}{-\log\,4} = \frac{2}{\log\,4}.$$

An approximation is $x \approx 3.32$ m.

EXAMPLE 7   **Comparing light intensities**

If a beam of light that has intensity I_0 is projected vertically downward into water, then its intensity $I(x)$ at a depth of x meters is $I(x) = I_0 e^{-1.4x}$ (see Figure 28). At what depth is the intensity one-half its value at the surface?

Figure 28

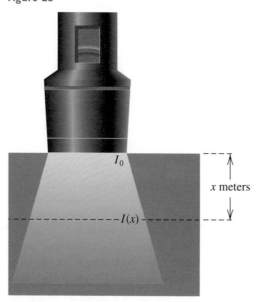

Solution   At the surface, $x = 0$, and the intensity is

$$I(0) = I_0 e^0 = I_0.$$

We wish to find the value of x such that $I(x) = \frac{1}{2}I_0$. This leads to the following:

$$
\begin{aligned}
I(x) &= \tfrac{1}{2}I_0 && \text{desired intensity} \\
I_0 e^{-1.4x} &= \tfrac{1}{2}I_0 && \text{formula for } I(x) \\
e^{-1.4x} &= \tfrac{1}{2} && \text{divide by } I_0 \ (I_0 \neq 0) \\
-1.4x &= \ln \tfrac{1}{2} && \text{change to logarithmic form} \\
x &= \frac{\ln \frac{1}{2}}{-1.4} && \text{divide by } -1.4
\end{aligned}
$$

An approximation is $x \approx 0.495$ m.

EXAMPLE 8   A logistic curve

A **logistic curve** is the graph of an equation of the form

$$y = \frac{k}{1 + be^{-cx}},$$

where k, b, and c are positive constants. Such curves are useful for describing a population y that grows rapidly initially, but whose growth rate decreases after x reaches a certain value. In a famous study of the growth of protozoa by Gause, a population of *Paramecium caudata* was found to be described by a logistic equation with $c = 1.1244$, $k = 105$, and x the time in days.

(a) Find b if the initial population was 3 protozoa.

(b) In the study, the maximum growth rate took place at $y = 52$. At what time x did this occur?

(c) Show that after a long period of time, the population described by any logistic curve approaches the constant k.

Solution

(a) Letting $c = 1.1244$ and $k = 105$ in the logistic equation, we obtain

$$y = \frac{105}{1 + be^{-1.1244x}}.$$

We now proceed as follows:

$$3 = \frac{105}{1 + be^0} = \frac{105}{1 + b} \qquad y = 3 \text{ when } x = 0$$

$$1 + b = 35 \qquad\qquad \text{multiply by } \frac{1 + b}{3}$$

$$b = 34 \qquad\qquad \text{solve for } b$$

(b) Using the fact that $b = 34$ leads to the following:

$$52 = \frac{105}{1 + 34e^{-1.1244x}} \qquad \text{let } y = 52 \text{ in part (a)}$$

$$1 + 34e^{-1.1244x} = \frac{105}{52} \qquad \text{multiply by } \frac{1 + 34e^{-1.1244x}}{52}$$

$$e^{-1.1244x} = \left(\frac{105}{52} - 1\right) \cdot \frac{1}{34} = \frac{53}{1768} \qquad \text{isolate } e^{-1.1244x}$$

$$-1.1244x = \ln \frac{53}{1768} \qquad \text{change to logarithmic form}$$

$$x = \frac{\ln \frac{53}{1768}}{-1.1244} \approx 3.12 \text{ days} \qquad \text{divide by } -1.1244$$

(c) As $x \to \infty$, $e^{-cx} \to 0$. Hence,

$$y = \frac{k}{1 + be^{-cx}} \to \frac{k}{1 + b \cdot 0} = k.$$

In the next example we graph the equation obtained in part (a) of the preceding example.

E X A M P L E 9   **Sketching the graph of a logistic curve**

Graph the logistic curve given by

$$y = \frac{105}{1 + 34e^{-1.1244x}},$$

and estimate the value of x for $y = 52$.

Solution   We begin by assigning

$$\frac{105}{1 + 34e^{-1.1244x}}$$

Figure 29
[0, 10] by [0, 105]

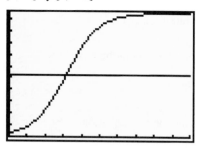

to Y_1 and 52 to Y_2. Since the time x is nonnegative, we choose Xmin = 0. We select Xmax = 10 in order to include the value of x found in part (b) of Example 8. By part (c), we know that the value of y cannot exceed 105. Thus, we choose Ymin = 0 and Ymax = 105 and obtain a display similar to Figure 29.

Using a zoom or intersect feature, we see that for $y = 52$, the value of x is approximately 3.12, which agrees with the approximation found in (b) of Example 8.

The following example shows how a change of base formula may be used to enable us to graph logarithmic functions with bases other than 10 and e on a graphing utility.

E X A M P L E 10   **Estimating points of intersection of logarithmic graphs**

Estimate the point of intersection of the graphs of

$$f(x) = \log_3 x \qquad \text{and} \qquad g(x) = \log_6 (x + 2).$$

Solution   Most graphing utilities are equipped to work with only common and natural logarithmic functions. Thus, we first use a change of base formula to rewrite f and g as

Figure 30
[−2, 4] by [−2, 2]

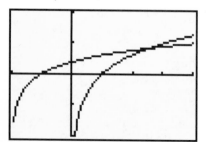

$$f(x) = \frac{\ln x}{\ln 3} \qquad \text{and} \qquad g(x) = \frac{\ln (x + 2)}{\ln 6}.$$

We next assign $(\ln x)/\ln 3$ and $(\ln (x + 2))/\ln 6$ to Y_1 and Y_2, respectively. After graphing Y_1 and Y_2 using a standard viewing rectangle, we see that there is a point of intersection in the first quadrant with $2 < x < 3$. Using a zoom or intersect feature, we find that the point of intersection is approximately (2.52, 0.84).

Figure 30 was obtained using viewing rectangle dimensions [−2, 4] by [−2, 2]. There are no other points of intersection, since f increases more rapidly than g for $x > 3$.

5.5 EXERCISES

Exer. 1–4: Find the exact solution and a two-decimal-place approximation for it by using (a) the method of Example 1 and (b) the method of Example 2.

1 $5^x = 8$

2 $4^x = 3$

3 $3^{4-x} = 5$

4 $\left(\frac{1}{3}\right)^x = 100$

Exer. 5–8: Evaluate using the change of base formula.

5 $\log_5 6$

6 $\log_2 20$

7 $\log_9 0.2$

8 $\log_6 \frac{1}{2}$

Exer. 9–10: Evaluate using the change of base formula (without a calculator).

9 $\dfrac{\log_5 16}{\log_5 4}$

10 $\dfrac{\log_7 243}{\log_7 3}$

Exer. 11–24: Find the exact solution, using common logarithms, and a two-decimal-place approximation of each solution, when appropriate.

11 $3^{x+4} = 2^{1-3x}$

12 $4^{2x+3} = 5^{x-2}$

13 $2^{2x-3} = 5^{x-2}$

14 $3^{2-3x} = 4^{2x+1}$

15 $2^{-x} = 8$

16 $2^{-x^2} = 5$

17 $\log x = 1 - \log (x - 3)$

18 $\log (5x + 1) = 2 + \log (2x - 3)$

19 $\log (x^2 + 4) - \log (x + 2) = 2 + \log (x - 2)$

20 $\log (x - 4) - \log (3x - 10) = \log (1/x)$

21 $5^x + 125(5^{-x}) = 30$

22 $3(3^x) + 9(3^{-x}) = 28$

23 $4^x - 3(4^{-x}) = 8$

24 $2^x - 6(2^{-x}) = 6$

Exer. 25–30: Solve the equation without using a calculator.

25 $\log (x^2) = (\log x)^2$

26 $\log \sqrt{x} = \sqrt{\log x}$

27 $\log (\log x) = 2$

28 $\log \sqrt{x^3 - 9} = 2$

29 $x^{\sqrt{\log x}} = 10^8$

30 $\log (x^3) = (\log x)^3$

Exer. 31–34: Use common logarithms to solve for x in terms of y.

31 $y = \dfrac{10^x + 10^{-x}}{2}$

32 $y = \dfrac{10^x - 10^{-x}}{2}$

33 $y = \dfrac{10^x - 10^{-x}}{10^x + 10^{-x}}$

34 $y = \dfrac{10^x + 10^{-x}}{10^x - 10^{-x}}$

Exer. 35–38: Use natural logarithms to solve for x in terms of y.

35 $y = \dfrac{e^x - e^{-x}}{2}$

36 $y = \dfrac{e^x + e^{-x}}{2}$

37 $y = \dfrac{e^x + e^{-x}}{e^x - e^{-x}}$

38 $y = \dfrac{e^x - e^{-x}}{e^x + e^{-x}}$

Exer. 39–40: Sketch the graph of f, and use the change of base formula to approximate the y-intercept.

39 $f(x) = \log_2 (x + 3)$

40 $f(x) = \log_3 (x + 5)$

Exer. 41–42: Sketch the graph of f, and use the change of base formula to approximate the x-intercept.

41 $f(x) = 4^x - 3$

42 $f(x) = 3^x - 6$

Exer. 43–46: Chemists use a number denoted by pH to describe quantitatively the acidity or basicity of solutions. By definition, pH $= -\log [H^+]$, where $[H^+]$ is the hydrogen ion concentration in moles per liter.

43 Approximate the pH of each substance.

(a) vinegar: $[H^+] \approx 6.3 \times 10^{-3}$

(b) carrots: $[H^+] \approx 1.0 \times 10^{-5}$

(c) sea water: $[H^+] \approx 5.0 \times 10^{-9}$

44 Approximate the hydrogen ion concentration $[H^+]$ of each substance.

(a) apples: pH ≈ 3.0

(b) beer: pH ≈ 4.2

(c) milk: pH ≈ 6.6

45 A solution is considered basic if $[H^+] < 10^{-7}$ or acidic if $[H^+] > 10^{-7}$. Find the corresponding inequalities involving pH.

46 Many solutions have a pH between 1 and 14. Find the corresponding range of $[H^+]$.

47 *Compound interest* Use the compound interest formula to determine how long it will take for a sum of money to double if it is invested at a rate of 6% per year compounded monthly.

48 *Compound interest* Solve the compound interest formula

$$A = P\left(1 + \frac{r}{n}\right)^{nt}$$

for t by using natural logarithms.

49 Photic zone Refer to Example 6. The most important zone in the sea from the viewpoint of marine biology is the photic zone, in which photosynthesis takes place. The photic zone ends at the depth where about 1% of the surface light penetrates. In very clear waters in the Caribbean, 50% of the light at the surface reaches a depth of about 13 meters. Estimate the depth of the photic zone.

50 Photic zone In contrast to the situation described in the previous exercise, in parts of New York harbor, 50% of the surface light does not reach a depth of 10 centimeters. Estimate the depth of the photic zone.

51 Drug absorption If a 100-milligram tablet of an asthma drug is taken orally and if none of the drug is present in the body when the tablet is first taken, the total amount A in the bloodstream after t minutes is predicted to be

$$A = 100[1 - (0.9)^t] \quad \text{for} \quad 0 \le t \le 10.$$

(a) Sketch the graph of the equation.

(b) Determine the number of minutes needed for 50 milligrams of the drug to have entered the bloodstream.

52 Drug dosage A drug is eliminated from the body through urine. Suppose that for a dose of 10 milligrams, the amount $A(t)$ remaining in the body t hours later is given by $A(t) = 10(0.8)^t$ and that in order for the drug to be effective, at least 2 milligrams must be in the body.

(a) Determine when 2 milligrams is left in the body.

(b) What is the half-life of the drug?

53 Genetic mutation The basic source of genetic diversity is mutation, or changes in the chemical structure of genes. If a gene mutates at a constant rate m and if other evolutionary forces are negligible, then the frequency F of the original gene after t generations is given by $F = F_0(1 - m)^t$, where F_0 is the frequency at $t = 0$.

(a) Solve the equation for t using common logarithms.

(b) If $m = 5 \times 10^{-5}$, after how many generations does $F = \frac{1}{2}F_0$?

C 54 Employee productivity Certain learning processes may be illustrated by the graph of an equation of the form $f(x) = a + b(1 - e^{-cx})$, where a, b, and c are positive constants. Suppose a manufacturer estimates that a new employee can produce five items the first day on the job. As the employee becomes more proficient, the daily production increases until a certain maximum production is reached. Suppose that on the nth day on

the job, the number $f(n)$ of items produced is approximated by

$$f(n) = 3 + 20(1 - e^{-0.1n}).$$

(a) Estimate the number of items produced on the fifth day, the ninth day, the twenty-fourth day, and the thirtieth day.

(b) Sketch the graph of f from $n = 0$ to $n = 30$. (Graphs of this type are called *learning curves* and are used frequently in education and psychology.)

(c) What happens as n increases without bound?

55 Height of trees The growth in height of trees is frequently described by a logistic equation. Suppose the height h (in feet) of a tree at age t (in years) is

$$h = \frac{120}{1 + 200e^{-0.2t}},$$

as illustrated by the graph in the figure.

(a) What is the height of the tree at age 10?

(b) At what age is the height 50 feet?

Exercise 55

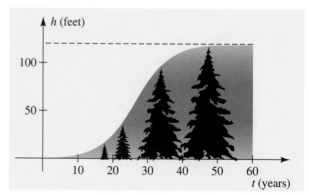

56 Employee productivity Manufacturers sometimes use empirically based formulas to predict the time required to produce the nth item on an assembly line for an integer n. If $T(n)$ denotes the time required to assemble the nth item and T_1 denotes the time required for the first, or prototype, item, then typically $T(n) = T_1 n^{-k}$ for some positive constant k.

(a) For many airplanes, the time required to assemble the second airplane, $T(2)$, is equal to $(0.80)T_1$. Find the value of k.

(b) Express, in terms of T_1, the time required to assemble the fourth airplane.

(c) Express, in terms of $T(n)$, the time $T(2n)$ required to assemble the $(2n)$th airplane.

57 *Vertical wind shear* Refer to Exercises 65–66 in Appendix VI. If v_0 is the wind speed at height h_0 and if v_1 is the wind speed at height h_1, then the vertical wind shear can be described by the equation

$$\frac{v_0}{v_1} = \left(\frac{h_0}{h_1}\right)^P,$$

where P is a constant. During a one-year period in Montreal, the maximum vertical wind shear occurred when the winds at the 200-foot level were 25 mi/hr while the winds at the 35-foot level were 6 mi/hr. Find P for these conditions.

58 *Vertical wind shear* Refer to Exercise 57. The average vertical wind shear is given by the equation

$$s = \frac{v_1 - v_0}{h_1 - h_0}.$$

Suppose that the velocity of the wind increases with increasing altitude and that all values for wind speeds taken at the 35-foot and 200-foot altitudes are greater than 1 mi/hr. Does increasing the value of P produce larger or smaller values of s?

Exer. 59–60: An economist suspects that the following data points lie on the graph of $y = c2^{kx}$, where c and k are constants. If the data points have three-decimal-place accuracy, is this suspicion correct?

59 (0, 4), (1, 3.249), (2, 2.639), (3, 2.144)

60 (0, −0.3), (0.5, −0.345), (1, −0.397), (1.5, −0.551), (2, −0.727)

Exer. 61–62: It is suspected that the following data points lie on the graph of $y = c \log(kx + 10)$, where c and k are constants. If the data points have three-decimal-place accuracy, is this suspicion correct?

61 (0, 1.5), (1, 1.619), (2, 1.720), (3, 1.997)

62 (0, 0.7), (1, 0.782), (2, 0.847), (3, 0.900), (4, 0.945)

Exer. 63–64: Approximate the function at the value of x to four decimal places.

63 $h(x) = \log_4 x - 2 \log_8 1.2x;$ $x = 5.3$

64 $h(x) = 3 \log_3 (2x - 1) + 7 \log_2 (x + 0.2);$ $x = 52.6$

Exer. 65–66: Use a graph to estimate the roots of the equation on the given interval.

65 $x - \ln(0.3x) - 3 \log_3 x = 0;$ (0, 9)

66 $2 \log 2x - \log_3 x^2 = 0;$ (0, 3)

Exer. 67–68: Graph f and g on the same coordinate plane, and estimate the solution of the equation $f(x) = g(x)$.

67 $f(x) = x;$ $g(x) = 3 \log_2 x$

68 $f(x) = x;$ $g(x) = -x^2 - \log_5 x$

Exer. 69–70: Graph f and g on the same coordinate plane, and estimate the solution of the inequality $f(x) > g(x)$.

69 $f(x) = 3^{-x} - 4^{0.2x};$ $g(x) = \ln(1.2) - x$

70 $f(x) = 3 \log_4 x - \log x;$ $g(x) = e^x - 0.25x^4$

71 *Human memory* A group of elementary students were taught long division over a one-week period. Afterward, they were given a test. The average score was 85. Each week thereafter, they were given an equivalent test, without any review. Let $n(t)$ represent the average score after $t \geq 0$ weeks. Graph each $n(t)$, and determine which function best models the situation.

(1) $n(t) = 85e^{t/3}$
(2) $n(t) = 70 + 10 \ln(t + 1)$
(3) $n(t) = 86 - e^t$
(4) $n(t) = 85 - 15 \ln(t + 1)$

72 *Cooling* A jar of boiling water at 212°F is set on a table in a room with a temperature of 72°F. If $T(t)$ represents the temperature of the water after t hours, graph $T(t)$ and determine which function best models the situation.

(1) $T(t) = 212 - 50t$
(2) $T(t) = 140e^{-t} + 72$
(3) $T(t) = 212e^{-t}$
(4) $T(t) = 72 + 10 \ln(140t + 1)$

73 *Ozone layer* Measurements made by the total ozone mapping spectrometer aboard NASA's Nimbus-7 weather satellite showed that ozone levels in the stratosphere are decreasing at a higher rate than anticipated. In April 1993, ozone levels in most of the northern hemisphere measured 11% below levels for April 1992. According to some experts, the cause of this rapid depletion may be the volcanic eruption of Mount Pinatubo in the Philippines in June 1991, rather than the thinning of the ozone layer.

(a) Assuming that the ozone level in April 1992 is the normal level and that the 1993 rate of decrease continues, use a table to numerically predict the year in which the ozone layer will be 50% of its normal level.

(b) Determine the year in part (a) using algebraic methods.

[c] 74 *Skin cancer* Ultraviolet-B radiation from the sun is the primary cause of nonmelanotic skin cancer. Much of this radiation is filtered out by ozone in the atmosphere. In Toronto, the amount of ultraviolet-B radiation increased by 7% from the summer of 1988 to the summer of 1989. If this increase in ultraviolet light is permanent, it may cause the likelihood of contracting skin cancer over a lifetime to increase by the same percentage.

(a) If this rate of increase continues, graphically approximate how many years it will take for the likelihood of contracting skin cancer in Toronto to double.

(b) Determine the year in part (a) using algebraic methods.

CHAPTER 5 REVIEW EXERCISES

Exer. 1–16: Sketch the graph of f.

1 $f(x) = 3^{x+2}$

2 $f(x) = \left(\frac{3}{5}\right)^x$

3 $f(x) = \left(\frac{3}{2}\right)^{-x}$

4 $f(x) = 3^{-2x}$

5 $f(x) = 3^{-x^2}$

6 $f(x) = 1 - 3^{-x}$

7 $f(x) = e^{x/2}$

8 $f(x) = \frac{1}{2}e^x$

9 $f(x) = e^{x-2}$

10 $f(x) = e^{2-x}$

11 $f(x) = \log_6 x$

12 $f(x) = \log_6 (36x)$

13 $f(x) = \log_4 (x^2)$

14 $f(x) = \log_4 \sqrt[3]{x}$

15 $f(x) = \log_2 (x + 4)$

16 $f(x) = \log_2 (4 - x)$

Exer. 17–18: Evaluate without using a calculator.

17 (a) $\log_2 \frac{1}{16}$ (b) $\log_\pi 1$ (c) $\ln e$
   (d) $6^{\log_6 4}$ (e) $\log 1{,}000{,}000$ (f) $10^{3 \log 2}$
   (g) $\log_4 2$

18 (a) $\log_5 \sqrt[3]{5}$ (b) $\log_5 1$ (c) $\log 10$
   (d) $e^{\ln 5}$ (e) $\log \log 10^{10}$ (f) $e^{2 \ln 5}$
   (g) $\log_{27} 3$

Exer. 19–36: Solve the equation without using a calculator.

19 $2^{3x-1} = \frac{1}{2}$

20 $\log \sqrt{x} = \log (x - 6)$

21 $\log_8 (x - 5) = \frac{2}{3}$

22 $\log_4 (x + 1) = 2 + \log_4 (3x - 2)$

23 $2 \ln (x + 3) - \ln (x + 1) = 3 \ln 2$

24 $\log \sqrt[4]{x + 1} = \frac{1}{2}$

25 $2^{5-x} = 6$

26 $3^{(x^2)} = 7$

27 $2^{5x+3} = 3^{2x+1}$

28 $\log_3 (3x) = \log_3 x + \log_3 (4 - x)$

29 $\log_4 x = \sqrt[3]{\log_4 x}$

30 $e^{x+\ln 4} = 3e^x$

31 $10^{2 \log x} = 5$

32 $e^{\ln (x+1)} = 3$

33 $x^2(-2xe^{-x^2}) + 2xe^{-x^2} = 0$

34 $e^x + 2 = 8e^{-x}$

35 (a) $\log x^2 = \log (6 - x)$ (b) $2 \log x = \log (6 - x)$

36 (a) $\ln (e^x)^2 = 16$ (b) $\ln e^{(x^2)} = 16$

37 Express $\log x^4 \sqrt[3]{y^2/z}$ in terms of logarithms of x, y, and z.

38 Express $\log (x^2/y^3) + 4 \log y - 6 \log \sqrt{xy}$ as one logarithm.

Exer. 39–40: Use common logarithms to solve the equation for x in terms of y.

39 $y = \dfrac{1}{10^x + 10^{-x}}$

40 $y = \dfrac{1}{10^x - 10^{-x}}$

Exer. 41–42: Approximate x to three significant figures.

41 (a) $x = \ln 6.6$ (b) $\log x = 1.8938$
   (c) $\ln x = -0.75$

42 (a) $x = \log 8.4$ (b) $\log x = -2.4260$
   (c) $\ln x = 1.8$

Exer. 43–44: (a) Find the domain and range of the function. (b) Find the inverse of the function and its domain and range.

43 $y = \log_2 (x + 1)$

44 $y = 2^{3-x} - 2$

45 *Bacteria growth* The number of bacteria in a certain culture at time t (in hours) is given by $Q(t) = 2(3^t)$, where $Q(t)$ is measured in thousands.

(a) What is the number of bacteria at $t = 0$?

(b) Find the number after 10 minutes, 30 minutes, and 1 hour.

46 *Compound interest*   If $1000 is invested at a rate of 12% compounded quarterly, what is the principal after one year?

47 *Radioactive iodine decay*   Radioactive iodine ^{131}I, which is frequently used in tracer studies involving the thyroid gland, decays according to $N = N_0(0.5)^{t/8}$, where N_0 is the initial dose and t is the time in days.

(a) Sketch the graph of the equation if $N_0 = 64$.

(b) Find the half-life of ^{131}I.

48 *Trout population*   A pond is stocked with 1000 trout. Three months later, it is estimated that 600 remain. Find a formula of the form $N = N_0 a^{ct}$ that can be used to estimate the number of trout remaining after t months.

49 *Continuously compounded interest*   Ten thousand dollars is invested in a savings fund in which interest is compounded continuously at the rate of 11% per year.

(a) When will the account contain $35,000?

(b) How long does it take for money to double in the account?

50 *Electrical current*   The current $I(t)$ in a certain electrical circuit at time t is given by $I(t) = I_0 e^{-Rt/L}$, where R is the resistance, L is the inductance, and I_0 is the initial current at $t = 0$. Find the value of t, in terms of L and R, for which $I(t)$ is 1% of I_0.

51 *Sound intensity*   The sound intensity level formula is $\alpha = 10 \log (I/I_0)$.

(a) Solve for I in terms of α and I_0.

(b) Show that a one-decibel rise in the intensity level α corresponds to a 26% increase in the intensity I.

52 *Fish growth*   The length L of a fish is related to its age by means of the von Bertalanffy growth formula

$$L = a(1 - be^{-kt}),$$

where a, b, and k are positive constants that depend on the type of fish. Solve this equation for t to obtain a formula that can be used to estimate the age of a fish from a length measurement.

53 *Earthquake area in the West*   In the western United States, the area A (in mi^2) affected by an earthquake is related to the magnitude R of the quake by the formula

$$R = 2.3 \log (A + 3000) - 5.1.$$

Solve for A in terms of R.

54 *Earthquake area in the East*   Refer to Exercise 53. For the eastern United States, the area-magnitude formula has the form

$$R = 2.3 \log (A + 34,000) - 7.5.$$

If A_1 is the area affected by an earthquake of magnitude R in the West and A_2 is the area affected by a similar quake in the East, find a formula for A_1/A_2 in terms of R.

55 *Earthquake area in the Central states*   Refer to Exercise 53. For the Rocky Mountain and Central states, the area-magnitude formula has the form

$$R = 2.3 \log (A + 14,000) - 6.6.$$

If an earthquake has magnitude 4 on the Richter scale, estimate the area A of the region that will feel the quake.

56 *Atmospheric pressure*   Under certain conditions, the atmospheric pressure p at altitude h is given by the formula $p = 29e^{-0.000034h}$. Express h as a function of p.

57 *Rocket velocity*   A rocket of mass m_1 is filled with fuel of initial mass m_2. If frictional forces are disregarded, the total mass m of the rocket at time t after ignition is related to its upward velocity v by $v = -a \ln m + b$, where a and b are constants. At ignition time $t = 0$, $v = 0$ and $m = m_1 + m_2$. At burnout, $m = m_1$. Use this information to find a formula, in terms of one logarithm, for the velocity of the rocket at burnout.

58 *Earthquake frequency*   Let n be the average number of earthquakes per year that have magnitudes between R and $R + 1$ on the Richter scale. A formula that approximates the relationship between n and R is

$$\log n = 7.7 - (0.9)R.$$

(a) Solve the equation for n in terms of R.

(b) Find n if $R = 4, 5$, and 6.

59 *Earthquake energy*   The energy E (in ergs) released during an earthquake of magnitude R may be approximated by using the formula

$$\log E = 11.4 + (1.5)R.$$

(a) Solve for E in terms of R.

(b) Find the energy released during the famous Alaskan quake of 1964, which measured 8.4 on the Richter scale.

60 *Radioactive decay*   A certain radioactive substance decays according to the formula $q(t) = q_0 e^{-0.0063t}$, where q_0 is the initial amount of the substance and t is the time in days. Approximate the half-life of the substance.

61 *Children's growth* The Count Model is a formula that can be used to predict the height of preschool children. If h is height (in centimeters) and t is age (in years), then

$$h = 70.228 + 5.104t + 9.222 \ln t$$

for $\frac{1}{4} \leq t \leq 6$. From calculus, the rate of growth R (in cm/year) is given by $R = 5.104 + (9.222/t)$. Predict the height and rate of growth of a typical 2-year-old.

62 *Electrical circuit* The current I in a certain electrical circuit at time t is given by

$$I = \frac{V}{R}(1 - e^{-Rt/L}),$$

where V is the electromotive force, R is the resistance, and L is the inductance. Solve the equation for t.

63 *Carbon 14 dating* The technique of carbon 14 (^{14}C) dating is used to determine the age of archaeological and geological specimens. The formula $T = -8310 \ln x$ is sometimes used to predict the age T (in years) of a bone fossil, where x is the percentage (expressed as a decimal) of ^{14}C still present in the fossil.

  (a) Estimate the age of a bone fossil that contains 4% of the ^{14}C found in an equal amount of carbon in present-day bone.

  (b) Approximate the percentage of ^{14}C present in a fossil that is 10,000 years old.

64 *Population of Kenya* Based on present birth and death rates, the population of Kenya is expected to increase according to the formula $N = 20.2e^{0.041t}$, with N in millions and $t = 0$ corresponding to 1985. How many years will it take for the population to double?

65 *Language history* Refer to Exercise 36 of Section 5.1. If a language originally had N_0 basic words of which $N(t)$ are still in use, then $N(t) = N_0(0.805)^t$, where time t is measured in millennia. After how many years are one-half the basic words still in use?

CHAPTER 5 DISCUSSION EXERCISES

c **1 *Catenary*** Refer to the catenary discussion on page 330 and Figure 15.

  (a) Describe the graph of the displayed equation for increasing values of a.

  (b) Find an equation of the cable in the figure such that the lowest point on the cable is 30 feet off the ground and the difference between the highest point on the cable (where it is connected to the tower) and the lowest point is less than 2 feet, provided that the towers are 40 feet apart.

2 Refer to Exercise 62 of Section 5.3. Discuss how to solve this exercise *without* the use of the formula for the total amount T. Proceed with your solution, and compare your answer to the answer arrived at using the formula for T.

c **3** Shown in the figure is a graph of $f(x) = (\ln x)/x$ for $x > 0$. The maximum value of $f(x)$ occurs at $x = e$.

  (a) The integers 2 and 4 have the unusual property that $2^4 = 4^2$. Show that if $x^y = y^x$ for positive real numbers x and y, then $(\ln x)/x = (\ln y)/y$.

  (b) Use the graph of f (a table is helpful) to find another pair of real numbers x and y (to two decimal places) such that $x^y \approx y^x$.

  (c) Explain why many pairs of real numbers satisfy the equation $x^y = y^x$.

Exercise 3

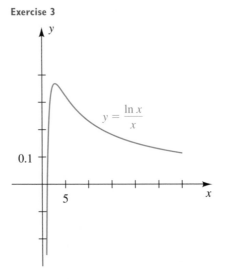

c **4 (a)** Compare the results of Exercise 43 in Section 5.1 and Exercise 35 in Section 5.2. Explain the difference between the two functions.

(b) Now suppose you are investing money at 8.5% compounded monthly. How would a graph of this growth compare with the other two graphs?

(c) Using the function described in part (b), mentally estimate the answers to parts (a) and (b) of Exercise 35 in Section 5.2, and explain why you believe they are correct before actually calculating them.

5 Since $y = \log_3 (x^2)$ is equivalent to $y = 2 \log_3 x$ by law 3 of logarithms, why aren't the graphs in Examples 7(a) and 7(b) of Section 5.4 the same?

c 6 *Unpaid balance on a mortgage* When lending institutions loan money, they expect to receive a return equivalent to the amount given by the compound interest formula. The borrower accumulates money "against" the original amount by making a monthly payment M that accumulates according to

$$\frac{12M[(1 + r/12)^{12t} - 1]}{r},$$

where r is the annual rate of interest and t is the number of years of the mortgage.

(a) Create a formula for the unpaid balance U of a loan.

(b) Graph the unpaid balance for the home mortgage loan in Exercise 37(a) of Section 5.1.

(c) What is the unpaid balance after 10 years? Estimate the number of years it will take to pay off one-half of the loan.

(d) Discuss the conditions your graph must satisfy to be correct.

(e) Discuss the validity of your results obtained from the graph.

c 7 Discuss *how many times* the graphs of $y = 0.01(1.001)^x$ and $y = x^3 - 99x^2 - 100x$ intersect. Approximate the points of intersection. In general, compare the growth of polynomial functions and exponential functions.

c 8 Discuss *how many times* the graphs of
$$y = x \quad \text{and} \quad y = (\ln x)^4$$
intersect. Approximate the points of intersection. What can you conclude about the growth of $y = x$ and $y = (\ln x)^n$, where n is a positive integer, as x increases without bound?

c 9 *Salary increases* Suppose you just started a job at $40,000 per year. In 5 years, you are scheduled to be making $60,000 per year. Determine the annual exponential rate of increase that describes this situation. Assume that the same exponential rate of increase will continue for 40 years. Using the rule of 70 (page 344), mentally estimate your annual salary in 40 years, and compare the estimate to an actual computation.

c 10 *Energy release* Consider these three events:

(1) On May 18, 1980, the volcanic eruption of Mount St. Helens in Washington released about 1.7×10^{18} joules of energy.

(2) When a 1-megaton nuclear bomb detonates, it releases about 4×10^{15} joules of energy.

(3) The 1989 San Francisco earthquake registered 7.1 on the Richter scale.

(a) Make some comparisons (i.e., how many of one event is equivalent to another) in terms of energy released. (*Hint:* Refer to Exercise 59 in the Chapter 5 Review Exercises.) *Note:* The atomic bombs dropped in World War II were 1-kiloton bombs (1000 1-kiloton bombs = 1 1-megaton bomb).

(b) What reading on the Richter scale would be equivalent to the Mount St. Helens eruption? Has there ever been a reading that high?

c 11 *Dow Jones average* The Dow Jones industrial average is an index of 30 of America's biggest corporations and is the most common measure of stock performance in the United States. The following table contains some 1000-point milestone dates for the Dow.

Dow Jones average	Day first reached	Number of days from previous milestone
1003.16	11/14/72	—
2002.25	1/8/87	5168
3004.46	4/17/91	1560
4003.33	2/23/95	1408
5023.55	11/21/95	271
6010.00	10/14/96	328
7022.44	2/13/97	122
8038.88	7/16/97	153
9033.23	4/6/98	264

Find an exponential model for these data, and use it to predict when the Dow will reach 15,000. Find the average yearly rate of return according to the Dow. Discuss some of the practical considerations pertaining to these calculations.

TOPICS FROM ANALYTIC GEOMETRY

Plane geometry includes the study of figures—such as lines, circles, and triangles—that lie in a plane. Theorems are proved by reasoning deductively from certain postulates. In analytic geometry, plane geometric figures are investigated by introducing coordinate systems and then using equations and formulas. If the study of analytic geometry were to be summarized by means of one statement, perhaps the following would be appropriate: Given an equation, find its graph, and conversely, given a graph, find its equation. In this chapter we shall apply coordinate methods to several basic plane figures.

6.1   PARABOLAS

The *conic sections,* also called *conics,* can be obtained by intersecting a double-napped right circular cone with a plane. By varying the position of the plane, we obtain a *circle,* an *ellipse,* a *parabola,* or a *hyperbola,* as illustrated in Figure 1.

Figure 1

(a) Circle   **(b)** Ellipse   **(c)** Parabola   **(d)** Hyperbola

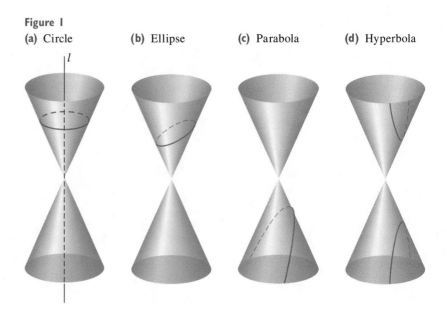

Degenerate conics are obtained if the plane intersects the cone in only one point or along either one or two lines that lie on the cone. Conic sections were studied extensively by the ancient Greeks, who discovered properties that enable us to state their definitions in terms of points and lines, as we do in our discussion.

From our work in Section 1.2 we may conclude that if $a \neq 0$, the graph of $y = ax^2 + bx + c$ is a *parabola* with a vertical axis. We shall next state a general definition of a parabola and derive equations for parabolas that have either a vertical axis or a horizontal axis.

DEFINITION OF A PARABOLA

A **parabola** is the set of all points in a plane equidistant from a fixed point F (the **focus**) and a fixed line l (the **directrix**) that lie in the plane.

Figure 2

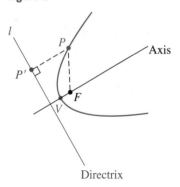

Figure 3

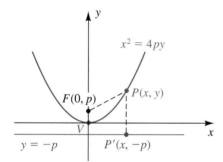

We shall assume that F is not on l, for this would result in a line. If P is a point in the plane and P' is the point on l determined by a line through P that is perpendicular to l (see Figure 2), then, by the preceding definition, P is on the parabola if and only if the distances $d(P, F)$ and $d(P, P')$ are equal. The **axis** of the parabola is the line through F that is perpendicular to the directrix. The **vertex** of the parabola is the point V on the axis half-way from F to l. The vertex is the point on the parabola that is closest to the directrix.

To obtain a simple equation for a parabola, place the y-axis along the axis of the parabola, with the origin at the vertex V, as shown in Figure 3. In this case, the focus F has coordinates $(0, p)$ for some real number $p \neq 0$, and the equation of the directrix is $y = -p$. (The figure shows the case $p > 0$.) By the distance formula, a point $P(x, y)$ is on the parabola if and only if $d(P, F) = d(P, P')$—that is, if

$$\sqrt{(x - 0)^2 + (y - p)^2} = \sqrt{(x - x)^2 + (y + p)^2}.$$

We square both sides and simplify:

$$x^2 + (y - p)^2 = (y + p)^2$$
$$x^2 + y^2 - 2py + p^2 = y^2 + 2py + p^2$$
$$x^2 = 4py$$

An equivalent equation for the parabola is

$$y = \frac{1}{4p}x^2.$$

We have shown that the coordinates of every point (x, y) on the parabola satisfy $x^2 = 4py$. Conversely, if (x, y) is a solution of $x^2 = 4py$, then by reversing the previous steps we see that the point (x, y) is on the parabola.

If $p > 0$, the parabola opens upward, as in Figure 3. If $p < 0$, the parabola opens downward. The graph is symmetric with respect to the y-axis, since substitution of $-x$ for x does not change the equation $x^2 = 4py$.

If we interchange the roles of x and y, we obtain

$$y^2 = 4px \qquad \text{or, equivalently,} \qquad x = \frac{1}{4p}y^2.$$

This is an equation of a parabola with vertex at the origin, focus $F(p, 0)$, and opening right if $p > 0$ or left if $p < 0$. The equation of the directrix is $x = -p$.

For convenience we often refer to "the parabola $x^2 = 4py$" (or $y^2 = 4px$) instead of "the parabola with equation $x^2 = 4py$" (or $y^2 = 4px$).

The next chart summarizes our discussion.

Parabolas with Vertex $V(0, 0)$

Equation, focus, directrix	Graph for $p > 0$	Graph for $p < 0$
$x^2 = 4py$ or $y = \dfrac{1}{4p}x^2$ Focus: $F(0, p)$ Directrix: $y = -p$		
$y^2 = 4px$ or $x = \dfrac{1}{4p}y^2$ Focus: $F(p, 0)$ Directrix: $x = -p$		

We see from the chart that for any nonzero real number a, the graph of the **standard equation** $y = ax^2$ or $x = ay^2$ is a parabola with vertex $V(0, 0)$. Moreover, $a = 1/(4p)$ or, equivalently, $p = 1/(4a)$, where $|p|$ is the distance between the focus F and vertex V. To find the directrix l, recall that l is also a distance $|p|$ from V.

EXAMPLE 1   **Finding the focus and directrix of a parabola**

Find the focus and directrix of the parabola $y = -\frac{1}{6}x^2$, and sketch its graph.

Solution   The equation has the form $y = ax^2$, with $a = -\frac{1}{6}$. As in the preceding chart, $a = 1/(4p)$, and hence

$$p = \frac{1}{4a} = \frac{1}{4\left(-\frac{1}{6}\right)} = \frac{1}{-\frac{4}{6}} = -\frac{3}{2}.$$

Figure 4

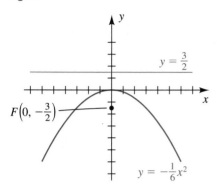

Thus, the parabola opens downward and has focus $F\left(0, -\frac{3}{2}\right)$, as illustrated in Figure 4. The directrix is the horizontal line $y = \frac{3}{2}$, which is a distance $\frac{3}{2}$ above V, as shown in the figure.

EXAMPLE 2   Finding an equation of a parabola satisfying prescribed conditions

(a) Find an equation of a parabola that has vertex at the origin, opens right, and passes through the point $P(7, -3)$.

(b) Find the focus of the parabola.

Solution

Figure 5

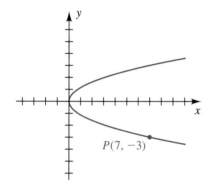

(a) The parabola is sketched in Figure 5. An equation of a parabola with vertex at the origin that opens right has the form $x = ay^2$ for some number a. If $P(7, -3)$ is on the graph, then we can substitute 7 for x and -3 for y to find a:

$$7 = a(-3)^2, \quad \text{or} \quad a = \tfrac{7}{9}$$

Hence, an equation for the parabola is $x = \frac{7}{9}y^2$.

(b) The focus is a distance p to the right of the vertex. Since $a = \frac{7}{9}$, we have

$$p = \frac{1}{4a} = \frac{1}{4\left(\frac{7}{9}\right)} = \frac{9}{28}.$$

Thus, the focus has coordinates $\left(\frac{9}{28}, 0\right)$.

If we take a standard equation of a parabola (of the form $x^2 = 4py$) and replace x with $x - h$ and y with $y - k$, then

$$x^2 = 4py \quad \text{becomes} \quad (x - h)^2 = 4p(y - k). \qquad (*)$$

From our discussion of translations in Section 1.4, we recognize that the graph of the second equation can be obtained from the graph of the first equation by shifting it h units to the right and k units up—thereby moving the vertex from $(0, 0)$ to (h, k). Squaring the left side of $(*)$ and simplifying leads to an equation of the form $y = ax^2 + bx + c$, where a, b, and c are real numbers.

Similarly, if we begin with $(y - k)^2 = 4p(x - h)$, it may be written in the form $x = ay^2 + by + c$. In the following chart $V(h, k)$ has been placed in the first quadrant, but the information given in the leftmost column holds true regardless of the position of V.

Parabolas with Vertex $V(h, k)$

Equation, focus, directrix	Graph for $p > 0$	Graph for $p < 0$
$(x - h)^2 = 4p(y - k)$ or $y = ax^2 + bx + c$, where $p = \dfrac{1}{4a}$ Focus: $F(h, k + p)$ Directrix: $y = k - p$		
$(y - k)^2 = 4p(x - h)$ or $x = ay^2 + by + c$, where $p = \dfrac{1}{4a}$ Focus: $F(h + p, k)$ Directrix: $x = h - p$		

EXAMPLE 3   **Sketching a parabola with a horizontal axis**

Discuss and sketch the graph of $2x = y^2 + 8y + 22$.

Solution   The equation can be written in the form shown in the second row of the preceding chart, $x = ay^2 + by + c$, so we see from the chart that the graph is a parabola with a horizontal axis. We first write the given equation as

$$y^2 + 8y + \underline{\quad} = 2x - 22 + \underline{\quad}$$

and then complete the square by adding $\left[\frac{1}{2}(8)\right]^2 = 16$ to both sides:

$$y^2 + 8y + 16 = 2x - 6$$
$$(y + 4)^2 = 2(x - 3)$$

Figure 6

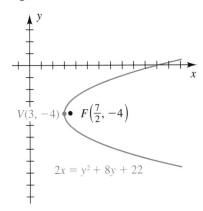

$2x = y^2 + 8y + 22$

Referring to the last chart, we see that $h = 3$, $k = -4$, and $4p = 2$ or, equivalently, $p = \frac{1}{2}$. This gives us the following:

The vertex $V(h, k)$ is $V(3, -4)$.

The focus is $F(h + p, k) = F\left(3 + \frac{1}{2}, -4\right)$, or $F\left(\frac{7}{2}, -4\right)$.

The directrix is $x = h - p = 3 - \frac{1}{2}$, or $x = \frac{5}{2}$.

The parabola is sketched in Figure 6.

EXAMPLE 4 Finding an equation of a parabola given its vertex and directrix

A parabola has vertex $V(-4, 2)$ and directrix $y = 5$. Express the equation of the parabola in the form $y = ax^2 + bx + c$.

Solution The vertex and directrix are shown in Figure 7. The dashes indicate a possible position for the parabola. The last chart shows that an equation of the parabola is

$$(x - h)^2 = 4p(y - k),$$

with $h = -4$ and $k = 2$ and with p equal to *negative* 3, since V is 3 units *below* the directrix. This gives us

$$(x + 4)^2 = -12(y - 2).$$

The last equation can be expressed in the form $y = ax^2 + bx + c$, as follows:

$$x^2 + 8x + 16 = -12y + 24$$
$$12y = -x^2 - 8x + 8$$
$$y = -\tfrac{1}{12}x^2 - \tfrac{2}{3}x + \tfrac{2}{3}$$

Figure 7

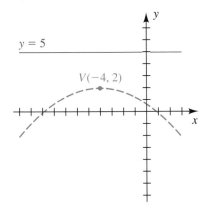

$y = 5$

$V(-4, 2)$

Figure 8

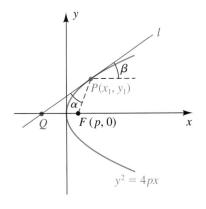

$y^2 = 4px$

An important property is associated with a tangent line to a parabola. (A *tangent line* to a parabola is a line that has exactly one point in common with the parabola but does not cut through the parabola.) Suppose l is the tangent line at a point $P(x_1, y_1)$ on the graph of $y^2 = 4px$, and let F be the focus. As in Figure 8, let α denote the angle between l and the line segment FP, and let β denote the angle between l and the indicated horizontal half-line with endpoint P. It can be shown that $\alpha = \beta$. This *reflective property* has many applications. For example, the shape of the mirror in a searchlight is obtained by revolving a parabola about its axis. The resulting three-dimensional surface is said to be *generated* by the parabola and is called a **paraboloid.** The **focus** of the paraboloid is the same as the focus of the generating parabola. If a light source is placed at F, then, by a law of physics (the angle of reflection equals the angle of incidence), a beam of

Figure 9
(a) Searchlight mirror

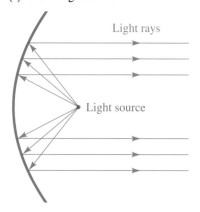

Light rays

Light source

(b) Telescope mirror

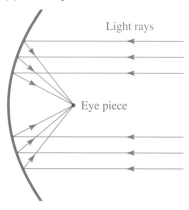

Light rays

Eye piece

light will be reflected along a line parallel to the axis (see Figure 9(a)). The same principle is used in the construction of mirrors for telescopes or solar ovens—a beam of light coming toward the parabolic mirror and parallel to the axis will be reflected into the focus (see Figure 9(b)). Antennas for radar systems, radio telescopes, and field microphones used at football games also make use of this property.

EXAMPLE 5 Locating the focus of a satellite TV antenna

The interior of a satellite TV antenna is a dish having the shape of a (finite) paraboloid that has diameter 12 feet and is 2 feet deep, as shown in Figure 10. Find the distance from the center of the dish to the focus.

Figure 10

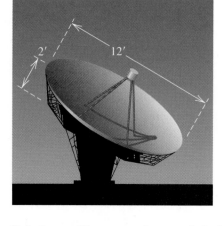

Figure 11

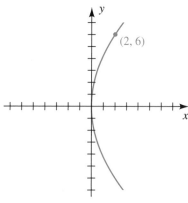

Solution The generating parabola is sketched on an xy-plane in Figure 11, where we have taken the vertex of the parabola at the origin and its axis along the x-axis. An equation of the parabola is $y^2 = 4px$, where p is the required distance from the center of the dish to the focus. Since the point $(2, 6)$ is on the parabola, we obtain

$$6^2 = 4p \cdot 2, \quad \text{or} \quad p = \tfrac{36}{8} = 4.5 \text{ ft.}$$

In the next example we use a graphing utility to sketch the graph of a parabola with a horizontal axis.

EXAMPLE 6 Graphing half-parabolas

Graph $x = y^2 + 2y - 4$.

Solution Some specific keystrokes for the TI-82/83 are given in Example 16 of Appendix I. The graph is a parabola with a horizontal axis.

Since y is not a function of x, we will solve the equation for y and obtain two equations (much as we did with circles in Example 11 of Section 1.2). We begin by solving the equivalent equation

$$y^2 + 2y - 4 - x = 0$$

for y in terms of x by using the quadratic formula, with $a = 1$, $b = 2$, and $c = -4 - x$:

$$y = \frac{-2 \pm \sqrt{2^2 - 4(1)(-4 - x)}}{2(1)} \qquad \text{quadratic formula}$$

$$= \frac{-2 \pm \sqrt{20 + 4x}}{2} \qquad \text{simplify}$$

$$= -1 \pm \sqrt{x + 5} \qquad \text{factor out } \sqrt{4}; \text{ simplify}$$

The last equation, $y = -1 \pm \sqrt{x + 5}$, represents the top half of the parabola ($y = -1 + \sqrt{x + 5}$) and the bottom half ($y = -1 - \sqrt{x + 5}$). Note that $y = -1$ is the axis of the parabola.

Next, we make the assignments

$$Y_1 = \sqrt{x + 5}, \qquad Y_2 = -1 + Y_1, \qquad \text{and} \qquad Y_3 = -1 - Y_1.$$

We select the functions Y_2 and Y_3 to be graphed with a standard viewing rectangle, and we obtain a display similar to Figure 12.

If your graphing calculator has an inverse graphing feature, you can make the assignment $Y_1 = x^2 + 2x - 4$ (the given equation, with x in place of y) and simply graph the inverse of Y_1 to obtain the graph in Figure 12.

Figure 12
[−15, 15] by [−10, 10]

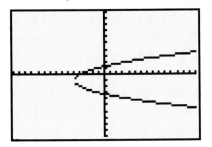

6.1 EXERCISES

Exer. 1–12: Find the vertex, focus, and directrix of the parabola. Sketch its graph, showing the focus and the directrix.

1 $8y = x^2$

2 $20x = y^2$

3 $2y^2 = -3x$

4 $x^2 = -3y$

5 $(x + 2)^2 = -8(y - 1)$

6 $(x - 3)^2 = \frac{1}{2}(y + 1)$

7 $(y - 2)^2 = \frac{1}{4}(x - 3)$

8 $(y + 1)^2 = -12(x + 2)$

9 $y = x^2 - 4x + 2$

10 $y^2 + 14y + 4x + 45 = 0$

11 $x^2 + 20y = 10$

12 $y^2 - 4y - 2x - 4 = 0$

Exer. 13–16: Find an equation for the parabola shown in the figure.

13

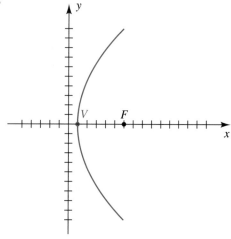

14

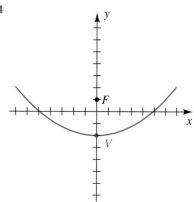

15

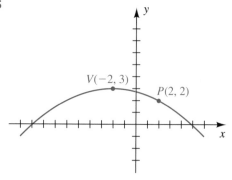

16

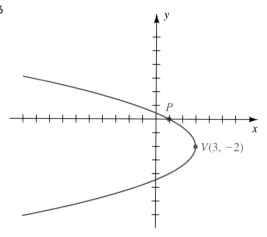

Exer. 17–28: Find an equation of the parabola that satisfies the given conditions.

17   Focus $F(2, 0)$,   directrix $x = -2$

18   Focus $F(0, -4)$,   directrix $y = 4$

19   Focus $F(6, 4)$,   directrix $y = -2$

20   Focus $F(-3, -2)$,   directrix $y = 1$

21   Vertex $V(3, -5)$,   directrix $x = 2$

22   Vertex $V(-2, 3)$,   directrix $y = 5$

23   Vertex $V(-1, 0)$,   focus $F(-4, 0)$

24   Vertex $V(1, -2)$,   focus $F(1, 0)$

25   Vertex at the origin, symmetric to the y-axis, and passing through the point $(2, -3)$

26   Vertex at the origin, symmetric to the y-axis, and passing through the point $(6, 3)$

27   Vertex $V(-3, 5)$, axis parallel to the x-axis, and passing through the point $(5, 9)$

28   Vertex $V(3, -2)$, axis parallel to the x-axis, and y-intercept 1

Exer. 29–32: Find an equation for the set of points in an xy-plane that are equidistant from the point P and the line l.

29   $P(0, 5)$; l: $y = -3$ 30 $P(7, 0)$; l: $x = 1$

31   $P(-6, 3)$; l: $x = -2$ 32 $P(5, -2)$; l: $y = 4$

Exer. 33–36: Find an equation for the indicated half of the parabola.

33 Lower half of $(y + 1)^2 = x + 3$

34 Upper half of $(y - 2)^2 = x - 4$

35 Right half of $(x + 1)^2 = y - 4$

36 Left half of $(x + 3)^2 = y + 2$

Exer. 37–38: Find an equation for the parabola that has a vertical axis and passes through the given points.

37 $P(2, 5), \quad Q(-2, -3), \quad R(1, 6)$

38 $P(3, -1), \quad Q(1, -7), \quad R(-2, 14)$

Exer. 39–40: Find an equation for the parabola that has a horizontal axis and passes through the given points.

39 $P(-1, 1), \quad Q(11, -2), \quad R(5, -1)$

40 $P(2, 1), \quad Q(6, 2), \quad R(12, -1)$

41 *Telescope mirror* A mirror for a reflecting telescope has the shape of a (finite) paraboloid of diameter 8 inches and depth 1 inch. How far from the center of the mirror will the incoming light collect?

Exercise 41

42 *Antenna dish* A satellite antenna dish has the shape of a paraboloid that is 10 feet across at the open end and is 3 feet deep. At what distance from the center of the dish should the receiver be placed to receive the greatest intensity of sound waves?

43 *Searchlight reflector* A searchlight reflector has the shape of a paraboloid, with the light source at the focus. If the reflector is 3 feet across at the opening and 1 foot deep, where is the focus?

44 *Flashlight mirror* A flashlight mirror has the shape of a paraboloid of diameter 4 inches and depth $\frac{3}{4}$ inch, as shown in the figure. Where should the bulb be placed so that the emitted light rays are parallel to the axis of the paraboloid?

Exercise 44

45 *Receiving dish* A sound receiving dish used at outdoor sporting events is constructed in the shape of a paraboloid, with its focus 5 inches from the vertex. Determine the width of the dish if the depth is to be 2 feet.

46 *Receiving dish* Work Exercise 45 if the receiver is 9 inches from the vertex.

47 *Parabolic reflector*

(a) The focal length of the (finite) paraboloid in the figure is the distance p between its vertex and focus. Express p in terms of r and h.

(b) A reflector is to be constructed with a focal length of 10 feet and a depth of 5 feet. Find the radius of the reflector.

Exercise 47

48 *Confocal parabolas* The parabola $y^2 = 4p(x + p)$ has its focus at the origin and axis along the x-axis. By assigning different values to p, we obtain a family of confocal parabolas, as shown in the figure. Such families occur in the study of electricity and magnetism. Show

that there are exactly two parabolas in the family that pass through a given point $P(x_1, y_1)$ if $y_1 \neq 0$.

Exercise 48

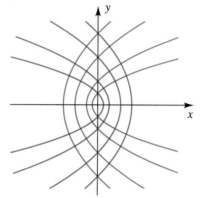

49 Jodrell Bank radio telescope A radio telescope has the shape of a paraboloid of revolution, with focal length p and diameter of base $2a$. From calculus, the surface area S available for collecting radio waves is

$$S = \frac{8\pi p^2}{3}\left[\left(1 + \frac{a^2}{4p^2}\right)^{3/2} - 1\right].$$

One of the largest radio telescopes, located in Jodrell Bank, Cheshire, England, has diameter 250 feet and focal length 50 feet. Approximate S to the nearest square foot.

50 Satellite path A satellite will travel in a parabolic path near a planet if its velocity v in meters per second satisfies the equation $v = \sqrt{2k/r}$, where r is the distance in meters between the satellite and the center of the planet and k is a positive constant. The planet will be located at the focus of the parabola, and the satellite will

pass by the planet once. Suppose a satellite is designed to follow a parabolic path and travel within 58,000 miles of Mars, as shown in the figure.

(a) Determine an equation of the form $x = ay^2$ that describes its flight path.

(b) For Mars, $k = 4.28 \times 10^{13}$. Approximate the maximum velocity of the satellite.

(c) Find the velocity of the satellite when its y-coordinate is 100,000 miles.

Exercise 50

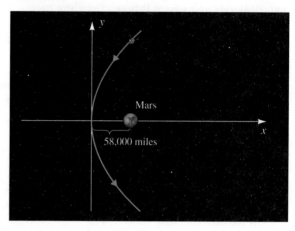

Mars

58,000 miles

c **Exer. 51–52: Graph the equation.**

51 $x = -y^2 + 2y + 5$ 52 $x = 2y^2 + 3y - 7$

c **Exer. 53–54: Graph the parabolas on the same coordinate plane, and estimate the points of intersection.**

53 $y = x^2 - 2.1x - 1$; $x = y^2 + 1$

54 $y = -2.1x^2 + 0.1x + 1.2$; $x = 0.6y^2 + 1.7y - 1.1$

6.2 ELLIPSES

An ellipse may be defined as follows. (*Foci* is the plural of *focus*.)

DEFINITION OF AN ELLIPSE

An **ellipse** is the set of all points in a plane, the sum of whose distances from two fixed points (the **foci**) in the plane is a positive constant.

Figure 13

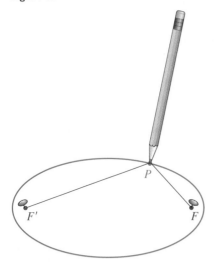

Figure 13

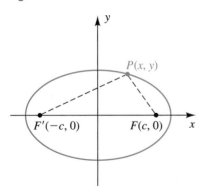

Figure 14

We can construct an ellipse on paper as follows: Insert two pushpins in the paper at any points F and F', and fasten the ends of a piece of string to the pins. After looping the string around a pencil and drawing it tight, as at point P in Figure 13, move the pencil, keeping the string tight. The sum of the distances $d(P, F)$ and $d(P, F')$ is the length of the string and hence is constant; thus, the pencil will trace out an ellipse with foci at F and F'. The midpoint of the segment $F'F$ is called the **center** of the ellipse. By changing the positions of F and F' while keeping the length of the string fixed, we can vary the shape of the ellipse considerably. If F and F' are far apart so that $d(F, F')$ is almost the same as the length of the string, the ellipse is flat. If $d(F, F')$ is close to zero, the ellipse is almost circular. If $F = F'$, we obtain a circle with center F.

To obtain a simple equation for an ellipse, choose the x-axis as the line through the two foci F and F', with the center of the ellipse at the origin. If F has coordinates $(c, 0)$ with $c > 0$, then, as in Figure 14, F' has coordinates $(-c, 0)$. Hence, the distance between F and F' is $2c$. The constant sum of the distances of P from F and F' will be denoted by $2a$. To obtain points that are not on the x-axis, we must have $2a > 2c$—that is, $a > c$. By definition, $P(x, y)$ is on the ellipse if and only if the following equivalent equations are true:

$$d(P, F) + d(P, F') = 2a$$

$$\sqrt{(x - c)^2 + (y - 0)^2} + \sqrt{(x + c)^2 + (y - 0)^2} = 2a$$

$$\sqrt{(x - c)^2 + y^2} = 2a - \sqrt{(x + c)^2 + y^2}$$

Squaring both sides of the last equation gives us

$$x^2 - 2cx + c^2 + y^2 = 4a^2 - 4a\sqrt{(x + c)^2 + y^2} + x^2 + 2cx + c^2 + y^2,$$

or

$$a\sqrt{(x + c)^2 + y^2} = a^2 + cx.$$

Squaring both sides again yields

$$a^2(x^2 + 2cx + c^2 + y^2) = a^4 + 2a^2cx + c^2x^2,$$

or

$$x^2(a^2 - c^2) + a^2y^2 = a^2(a^2 - c^2).$$

Dividing both sides by $a^2(a^2 - c^2)$, we obtain

$$\frac{x^2}{a^2} + \frac{y^2}{a^2 - c^2} = 1.$$

Recalling that $a > c$ and therefore $a^2 - c^2 > 0$, we let

$$b = \sqrt{a^2 - c^2}, \quad \text{or} \quad b^2 = a^2 - c^2.$$

This substitution gives us the equation

$$\frac{x^2}{a^2} + \frac{y^2}{b^2} = 1.$$

Since $c > 0$ and $b^2 = a^2 - c^2$, it follows that $a^2 > b^2$ and hence $a > b$.

We have shown that the coordinates of every point (x, y) on the ellipse in Figure 15 satisfy the equation $(x^2/a^2) + (y^2/b^2) = 1$. Conversely, if (x, y) is a solution of this equation, then by reversing the preceding steps we see that the point (x, y) is on the ellipse.

Figure 15

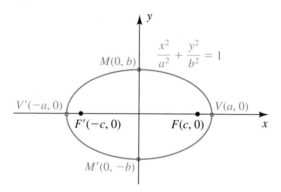

We may find the x-intercepts of the ellipse by letting $y = 0$ in the equation. Doing so gives us $x^2/a^2 = 1$, or $x^2 = a^2$. Consequently, the x-intercepts are a and $-a$. The corresponding points $V(a, 0)$ and $V'(-a, 0)$ on the graph are called the **vertices** of the ellipse (see Figure 15). The line segment $V'V$ is called the **major axis.** Similarly, letting $x = 0$ in the equation, we obtain $y^2/b^2 = 1$, or $y^2 = b^2$. Hence, the y-intercepts are b and $-b$. The segment between $M'(0, -b)$ and $M(0, b)$ is called the **minor axis** of the ellipse. The major axis is always longer than the minor axis, since $a > b$.

Applying tests for symmetry, we see that the ellipse is symmetric with respect to the x-axis, the y-axis, and the origin.

Similarly, if we take the foci on the y-axis, we obtain the equation

Figure 16

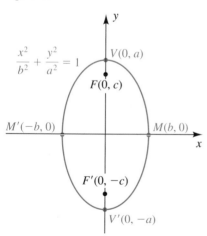

$$\frac{x^2}{b^2} + \frac{y^2}{a^2} = 1.$$

In this case, the vertices of the ellipse are $(0, \pm a)$ and the endpoints of the minor axis are $(\pm b, 0)$, as shown in Figure 16.

The preceding discussion may be summarized as follows.

Standard Equations of an Ellipse with Center at the Origin

The graph of

$$\frac{x^2}{a^2} + \frac{y^2}{b^2} = 1 \qquad \text{or} \qquad \frac{x^2}{b^2} + \frac{y^2}{a^2} = 1,$$

where $a > b > 0$, is an ellipse with center at the origin. The length of the major axis is $2a$, and the length of the minor axis is $2b$. The foci are a distance c from the origin, where $c^2 = a^2 - b^2$.

We have shown that an equation of an ellipse with center at the origin and foci on a coordinate axis can always be written in the form

$$\frac{x^2}{p} + \frac{y^2}{q} = 1, \qquad \text{or} \qquad qx^2 + py^2 = pq,$$

with p and q positive and $p \neq q$. If $p > q$, the major axis is on the x-axis; if $p < q$, the major axis is on the y-axis. It is unnecessary to memorize these facts, because in any given problem the major axis can be determined by examining the x- and y-intercepts.

EXAMPLE 1 Sketching an ellipse with center at the origin

Sketch the graph of $2x^2 + 9y^2 = 18$, and find the foci.

Solution The graph is an ellipse with center at the origin and foci on a coordinate axis. To find the x-intercepts, we let $y = 0$, obtaining

$$2x^2 = 18, \qquad \text{or} \qquad x = \pm 3.$$

To find the y-intercepts, we let $x = 0$, obtaining

$$9y^2 = 18, \qquad \text{or} \qquad y = \pm\sqrt{2}.$$

We are now able to sketch the ellipse (Figure 17). Since $\sqrt{2} \approx 1.4 < 3$, the major axis is on the x-axis.

To find the foci, we let $a = 3$ and $b = \sqrt{2}$ and calculate

$$c^2 = a^2 - b^2 = 3^2 - (\sqrt{2})^2 = 7.$$

Thus, $c = \sqrt{7}$, and the foci are $(\pm\sqrt{7}, 0)$.

Figure 17

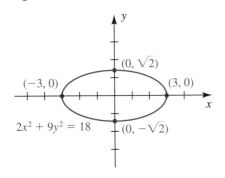

EXAMPLE 2 Sketching an ellipse with center at the origin

Sketch the graph of $9x^2 + 4y^2 = 25$, and find the foci.

Solution As in Example 1, the graph is an ellipse with center at the origin and foci on a coordinate axis. To find the x-intercepts, we let $y = 0$, obtaining

$$9x^2 = 25, \qquad \text{or} \qquad x = \pm\tfrac{5}{3}.$$

(continued)

To find the y-intercepts, we let $x = 0$, obtaining

$$4y^2 = 25, \quad \text{or} \quad y = \pm\tfrac{5}{2}.$$

These calculations give us the sketch in Figure 18. Since $\frac{5}{3} < \frac{5}{2}$, the major axis is on the y-axis.

Figure 18

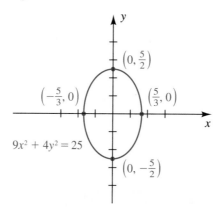

To find the foci, we let $a = \frac{5}{2}$ and $b = \frac{5}{3}$ and calculate

$$c^2 = a^2 - b^2 = \left(\tfrac{5}{2}\right)^2 - \left(\tfrac{5}{3}\right)^2 = \tfrac{125}{36}.$$

Thus, $c = \sqrt{125/36} = 5\sqrt{5}/6 \approx 1.86$, and the foci are $(0, \pm 5\sqrt{5}/6)$.

E X A M P L E   3   **Finding an equation of an ellipse given its vertices and foci**

Find an equation of the ellipse with vertices $(\pm 4, 0)$ and foci $(\pm 2, 0)$.

Solution   Since the foci are on the x-axis and are equidistant from the origin, the major axis is on the x-axis and the ellipse has center $(0, 0)$. Thus, a general equation of an ellipse is

$$\frac{x^2}{a^2} + \frac{y^2}{b^2} = 1.$$

Since the vertices are $(\pm 4, 0)$, we conclude that $a = 4$. Since the foci are $(\pm 2, 0)$, we have $c = 2$. Hence,

$$b^2 = a^2 - c^2 = 4^2 - 2^2 = 12,$$

and an equation of the ellipse is

$$\frac{x^2}{16} + \frac{y^2}{12} = 1.$$

In certain applications it is necessary to work with only one-half of an ellipse. The next example indicates how to find equations in such cases.

EXAMPLE 4   Finding equations for half-ellipses

Find equations for the upper half, lower half, left half, and right half of the ellipse $9x^2 + 4y^2 = 25$.

Solution   The graph of the entire ellipse was sketched in Figure 18. To find equations for the upper and lower halves, we solve for y in terms of x, as follows:

$$9x^2 + 4y^2 = 25 \qquad\qquad\text{given}$$

$$y^2 = \frac{25 - 9x^2}{4} \qquad\qquad\text{solve for } y^2$$

$$y = \pm\sqrt{\frac{25 - 9x^2}{4}} = \pm\frac{1}{2}\sqrt{25 - 9x^2} \quad\text{take the square root}$$

Since $\sqrt{25 - 9x^2} \geq 0$, it follows that equations for the upper and lower halves are $y = \frac{1}{2}\sqrt{25 - 9x^2}$ and $y = -\frac{1}{2}\sqrt{25 - 9x^2}$, respectively.

To find equations for the left and right halves, we use a procedure similar to that above and solve for x in terms of y, obtaining

$$x = \pm\sqrt{\frac{25 - 4y^2}{9}} = \pm\frac{1}{3}\sqrt{25 - 4y^2}.$$

The left half of the ellipse has the equation $x = -\frac{1}{3}\sqrt{25 - 4y^2}$, and the right half is given by $x = \frac{1}{3}\sqrt{25 - 4y^2}$.

If we take a standard equation of an ellipse $(x^2/a^2 + y^2/b^2 = 1)$ and replace x with $x - h$ and y with $y - k$, then

$$\frac{x^2}{a^2} + \frac{y^2}{b^2} = 1 \quad\text{becomes}\quad \frac{(x - h)^2}{a^2} + \frac{(y - k)^2}{b^2} = 1. \qquad (*)$$

The graph of $(*)$ is an ellipse with center (h, k). Squaring terms in $(*)$ and simplifying gives us an equation of the form

$$Ax^2 + Cy^2 + Dx + Ey + F = 0,$$

where the coefficients are real numbers and both A and C are positive. Conversely, if we start with such an equation, then by completing squares we can obtain a form that helps give us the center of the ellipse and the lengths of the major and minor axes. This technique is illustrated in the next example.

EXAMPLE 5   Sketching an ellipse with center (h, k)

Discuss and sketch the graph of the equation

$$16x^2 + 9y^2 + 64x - 18y - 71 = 0.$$

Solution   We begin by grouping the terms containing x and those containing y:

$$(16x^2 + 64x) + (9y^2 - 18y) = 71$$

Next, we factor out the coefficients of x^2 and y^2 as follows:

$$16(x^2 + 4x + \underline{}) + 9(y^2 - 2y + \underline{}) = 71$$

We now complete the squares for the expressions within parentheses:

$$16(x^2 + 4x + 4) + 9(y^2 - 2y + 1) = 71 + \underline{16 \cdot 4} + \underline{9 \cdot 1}$$

By adding 4 to the expression within the first parentheses we have added 64 to the left side of the equation, and hence we must compensate by adding 64 to the right side. Similarly, by adding 1 to the expression within the second parentheses we have added 9 to the left side, and consequently we must also add 9 to the right side. The last equation may be written

$$16(x + 2)^2 + 9(y - 1)^2 = 144.$$

Dividing by 144 to obtain 1 on the right side gives us

$$\frac{(x + 2)^2}{9} + \frac{(y - 1)^2}{16} = 1.$$

The graph of the last equation is an ellipse with center $C(-2, 1)$ and major axis on the vertical line $x = -2$ (since $9 < 16$). Using $a = 4$ and $b = 3$ gives us the ellipse in Figure 19.

Figure 19

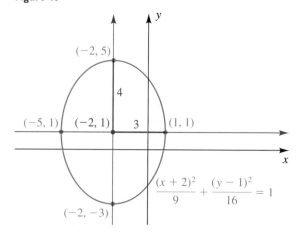

To find the foci, we first calculate

$$c^2 = a^2 - b^2 = 4^2 - 3^2 = 7.$$

The distance from the center of the ellipse to the foci is $c = \sqrt{7}$. Since the center is $(-2, 1)$, the foci are $(-2, 1 \pm \sqrt{7})$.

Graphing calculators and computer programs are sometimes unable to plot the graphs of an equation of the form

$$Ax^2 + Cy^2 + Dx + Ey + F = 0,$$

such as that considered in the last example. In these cases we must first solve the equation for y in terms of x and then plot the two resulting functions, as illustrated in the next example.

EXAMPLE 6 **Graphing half-ellipses**

Sketch the graph of $3x^2 + 4y^2 + 12x - 8y + 9 = 0$.

Solution The equation may be regarded as a quadratic equation in y of the form $ay^2 + by^2 + c = 0$ by rearranging terms as follows:

$$4y^2 - 8y + (3x^2 + 12x + 9) = 0$$

Applying the quadratic formula to the previous equation, with $a = 4$, $b = -8$, and $c = 3x^2 + 12x + 9$, gives us

$$y = \frac{-(-8) \pm \sqrt{(-8)^2 - 4(4)(3x^2 + 12x + 9)}}{2(4)}$$

$$= \frac{8 \pm \sqrt{64 - 16(3x^2 + 12x + 9)}}{8}$$

$$= 1 \pm \tfrac{1}{8}\sqrt{64 - 16(3x^2 + 12x + 9)}.$$

Figure 20
$[-6, 6]$ by $[-4, 4]$

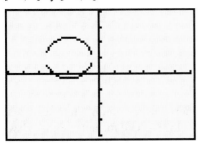

Note that we did not completely simplify the radicand, since we will be using a graphing utility.

As in Example 6 of the previous section, we now make the assignments

$$Y_1 = \tfrac{1}{8}\sqrt{64 - 16(3x^2 + 12x + 9)}, \quad Y_2 = 1 + Y_1, \quad \text{and} \quad Y_3 = 1 - Y_1.$$

Finally, we select Y_2 and Y_3 to be graphed and obtain a display similar to Figure 20.

Ellipses can be very flat or almost circular. To obtain information about the *roundness* of an ellipse, we sometimes use the term *eccentricity*, which is defined as follows, with a, b, and c having the same meanings as before.

**DEFINITION OF
ECCENTRICITY**

The **eccentricity** e of an ellipse is

$$e = \frac{\text{distance from center to focus}}{\text{distance from center to vertex}} = \frac{c}{a} = \frac{\sqrt{a^2 - b^2}}{a}.$$

Consider the ellipse $(x^2/a^2) + (y^2/b^2) = 1$, and suppose that the length $2a$ of the major axis is fixed and the length $2b$ of the minor axis is variable (note that $0 < b < a$). Since b^2 is positive, $a^2 - b^2 < a^2$ and hence $\sqrt{a^2 - b^2} < a$. Dividing both sides of the last inequality by a gives us $\sqrt{a^2 - b^2}/a < 1$, or $0 < e < 1$. If b is close to 0 (c is close to a), then $\sqrt{a^2 - b^2} \approx a$, $e \approx 1$, and the ellipse is very flat. This case is illustrated in Figure 21(a), with $a = 2$, $b = 0.3$, and $e \approx 0.99$. If b is close to a (c is close to 0), then $\sqrt{a^2 - b^2} \approx 0$, $e \approx 0$, and the ellipse is almost circular. This case is illustrated in Figure 21(b), with $a = 2$, $b = 1.9999$, and $e \approx 0.01$.

Figure 21
(a) Eccentricity almost 1   (b) Eccentricity almost 0

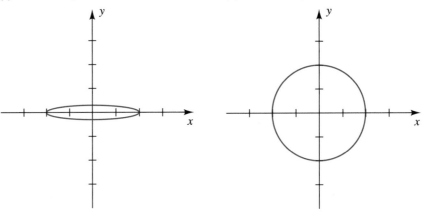

After many years of analyzing an enormous amount of empirical data, the German astronomer Johannes Kepler (1571–1630) formulated three laws that describe the motion of planets about the sun. Kepler's first law states that the orbit of each planet in the solar system is an ellipse with the sun at one focus. Most of these orbits are almost circular, so their corresponding eccentricities are close to 0. To illustrate, for Earth, $e \approx 0.017$; for Mars, $e \approx 0.093$; and for Uranus, $e \approx 0.046$. The orbits of Mercury and Pluto are less circular, with eccentricities of 0.206 and 0.249, respectively.

Many comets have elliptical orbits with the sun at a focus. In this case the eccentricity e is close to 1, and the ellipse is very flat. In the next example we use the **astronomical unit** (AU)—that is, the average distance from the earth to the sun—to specify large distances (1 AU $\approx$ 93,000,000 mi).

EXAMPLE 7   Approximating a distance in an elliptical path

Halley's comet has an elliptical orbit with eccentricity $e = 0.967$. The closest that Halley's comet comes to the sun is 0.587 AU. Approximate the maximum distance of the comet from the sun, to the nearest 0.1 AU.

Solution   Figure 22 illustrates the orbit of the comet, where c is the distance from the center of the ellipse to a focus (the sun) and $2a$ is the length of the major axis.

Figure 22

Halley's comet

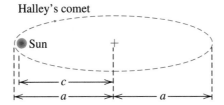

Since $a - c$ is the minimum distance between the sun and the comet, we have (in AU)

$$a - c = 0.587, \quad \text{or} \quad a = c + 0.587.$$

Since $e = c/a = 0.967$, we obtain the following:

$$
\begin{aligned}
c &= 0.967a && \text{multiply by } a \\
&= 0.967(c + 0.587) && \text{substitute for } a \\
&\approx 0.967c + 0.568 && \text{multiply} \\
c - 0.967c &\approx 0.568 && \text{subtract } 0.967c \\
c(1 - 0.967) &\approx 0.568 && \text{factor out } c \\
c &\approx \frac{0.568}{0.033} \approx 17.2 && \text{solve for } c
\end{aligned}
$$

Since $a = c + 0.587$, we obtain

$$a \approx 17.2 + 0.587 \approx 17.8,$$

and the maximum distance between the sun and the comet is

$$a + c \approx 17.8 + 17.2 = 35.0 \text{ AU}.$$

Figure 23

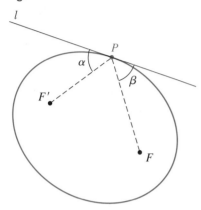

Figure 24

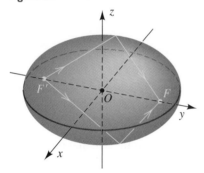

An ellipse has a *reflective property* analogous to that of the parabola discussed at the end of the previous section. To illustrate, let l denote the tangent line at a point P on an ellipse with foci F and F', as shown in Figure 23. If α is the acute angle between $F'P$ and l and if β is the acute angle between FP and l, it can be shown that $\alpha = \beta$. Thus, if a ray of light or sound emanates from one focus, it is reflected to the other focus. This property is used in the design of certain types of optical equipment.

If the ellipse with center O and foci F' and F on the x-axis is revolved about the x-axis, as illustrated in Figure 24, we obtain a three-dimensional surface called an **ellipsoid**. The upper half or lower half is a **hemi-ellipsoid**, as is the right half or left half. Sound waves or other impulses that are emitted from the focus F' will be reflected off the ellipsoid into the focus F. This property is used in the design of *whispering galleries*—structures with ellipsoidal ceilings, in which a person who whispers at one focus can be heard at the other focus. Examples of whispering galleries may be found in the Rotunda of the Capitol Building in Washington, D.C., and in the Mormon Tabernacle in Salt Lake City.

The reflective property of ellipsoids (and hemi-ellipsoids) is used in modern medicine in a device called a *lithotripter,* which disintegrates kidney stones by means of high-energy underwater shock waves. After taking extremely accurate measurements, the operator positions the patient so that the stone is at a focus. Ultra–high frequency shock waves are then produced at the other focus, and reflected waves break up the kidney stone. Recovery time with this technique is usually 3–4 days, instead of the 2–3 weeks with conventional surgery. Moreover, the mortality rate is less than 0.01%, as compared to 2–3% for traditional surgery (see Exercises 49–50).

6.2 EXERCISES

Exer. 1–14: Find the vertices and foci of the ellipse. Sketch its graph, showing the foci.

1 $\dfrac{x^2}{9} + \dfrac{y^2}{4} = 1$

2 $\dfrac{x^2}{25} + \dfrac{y^2}{16} = 1$

3 $\dfrac{x^2}{15} + \dfrac{y^2}{16} = 1$

4 $\dfrac{x^2}{45} + \dfrac{y^2}{49} = 1$

5 $4x^2 + y^2 = 16$

6 $y^2 + 9x^2 = 9$

7 $4x^2 + 25y^2 = 1$

8 $10y^2 + x^2 = 5$

9 $\dfrac{(x - 3)^2}{16} + \dfrac{(y + 4)^2}{9} = 1$

10 $\dfrac{(x + 2)^2}{25} + \dfrac{(y - 3)^2}{4} = 1$

11 $4x^2 + 9y^2 - 32x - 36y + 64 = 0$

12 $x^2 + 2y^2 + 2x - 20y + 43 = 0$

13 $25x^2 + 4y^2 - 250x - 16y + 541 = 0$

14 $4x^2 + y^2 = 2y$

Exer. 15–18: Find an equation for the ellipse shown in the figure.

15

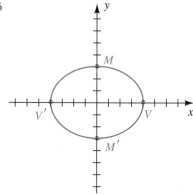

16

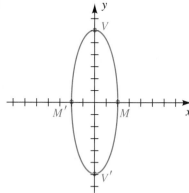

17

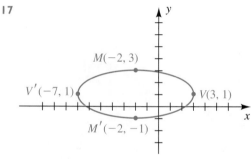

18

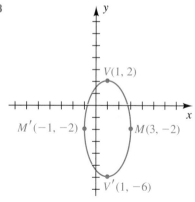

Exer. 19–30: Find an equation for the ellipse that has its center at the origin and satisfies the given conditions.

19 Vertices $V(\pm 8, 0)$, foci $F(\pm 5, 0)$

20 Vertices $V(0, \pm 7)$, foci $F(0, \pm 2)$

21 Vertices $V(0, \pm 5)$, minor axis of length 3

22 Foci $F(\pm 3, 0)$, minor axis of length 2

23 Vertices $V(0, \pm 6)$, passing through $(3, 2)$

24 Passing through $(2, 3)$ and $(6, 1)$

25 Eccentricity $\frac{3}{4}$, vertices $V(0, \pm 4)$

26 Eccentricity $\frac{1}{2}$, vertices on the x-axis, passing through $(1, 3)$

27 x-intercepts ± 2, y-intercepts $\pm \frac{1}{3}$

28 x-intercepts $\pm \frac{1}{2}$, y-intercepts ± 4

29 Horizontal major axis of length 8, minor axis of length 5

30 Vertical major axis of length 7, minor axis of length 6

Exer. 31–32: Find the points of intersection of the graphs of the equations. Sketch both graphs on the same coordinate plane, and show the points of intersection.

31 $\begin{cases} x^2 + 4y^2 = 20 \\ x + 2y = 6 \end{cases}$ 32 $\begin{cases} x^2 + 4y^2 = 36 \\ x^2 + y^2 = 12 \end{cases}$

Exer. 33–36: Find an equation for the set of points in an xy-plane such that the sum of the distances from F and F' is k.

33 $F(3, 0)$, $F'(-3, 0)$; $k = 10$

34 $F(12, 0)$, $F'(-12, 0)$; $k = 26$

35 $F(0, 15)$, $F'(0, -15)$; $k = 34$

36 $F(0, 8)$, $F'(0, -8)$; $k = 20$

Exer. 37–44: Determine whether the graph of the equation is the upper, lower, left, or right half of an ellipse, and find an equation for the ellipse.

37 $y = 11\sqrt{1 - \dfrac{x^2}{49}}$

38 $y = -6\sqrt{1 - \dfrac{x^2}{25}}$

39 $x = -\frac{1}{3}\sqrt{9 - y^2}$

40 $x = \frac{4}{5}\sqrt{25 - y^2}$

41 $x = 1 + 2\sqrt{1 - \dfrac{(y + 2)^2}{9}}$

42 $x = -2 - 5\sqrt{1 - \dfrac{(y - 1)^2}{16}}$

43 $y = 2 - 7\sqrt{1 - \dfrac{(x + 1)^2}{9}}$

44 $y = -1 + \sqrt{1 - \dfrac{(x - 3)^2}{16}}$

45 *Dimensions of an arch* An arch of a bridge is semielliptical, with major axis horizontal. The base of the arch is 30 feet across, and the highest part of the arch is 10 feet above the horizontal roadway, as shown in the figure. Find the height of the arch 6 feet from the center of the base.

Exercise 45

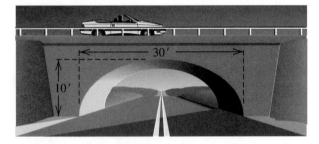

46 *Designing a bridge* A bridge is to be constructed across a river that is 200 feet wide. The arch of the bridge is to be semielliptical and must be constructed so that a ship less than 50 feet wide and 30 feet high can pass safely through the arch, as shown in the figure.

(a) Find an equation for the arch.

(b) Approximate the height of the arch in the middle of the bridge.

Exercise 46

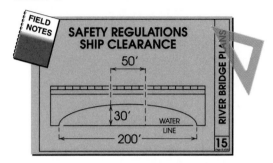

47 *Earth's orbit* Assume that the length of the major axis of the earth's orbit is 186,000,000 miles and that the eccentricity is 0.017. Approximate, to the nearest 1000 miles, the maximum and minimum distances between the earth and the sun.

48 *Mercury's orbit* The planet Mercury travels in an elliptical orbit that has eccentricity 0.206 and major axis of length 0.774 AU. Find the maximum and minimum distances between Mercury and the sun.

49 *Elliptical reflector* The basic shape of an elliptical reflector is a hemi-ellipsoid of height h and diameter k, as shown in the figure. Waves emitted from focus F will reflect off the surface into focus F'.

(a) Express the distances $d(V, F)$ and $d(V, F')$ in terms of h and k.

(b) An elliptical reflector of height 17 centimeters is to be constructed so that waves emitted from F are reflected to a point F' that is 32 centimeters from V. Find the diameter of the reflector and the location of F.

Exercise 49

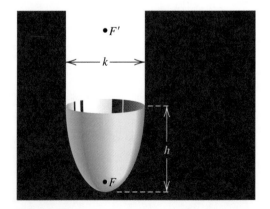

50 *Lithotripter operation*   A lithotripter of height 15 centimeters and diameter 18 centimeters is to be constructed (see the figure). High-energy underwater shock waves will be emitted from the focus F that is closest to the vertex V.

(a) Find the distance from V to F.

(b) How far from V (in the vertical direction) should a kidney stone be located?

Exercise 50

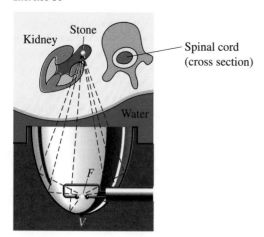

51 *Whispering gallery*   The ceiling of a whispering gallery has the shape of the hemi-ellipsoid shown in Figure 24, with the highest point of the ceiling 15 feet above the elliptical floor and the vertices of the floor 50 feet apart. If two people are standing at the foci F' and F, how far from the vertices are their feet?

52 An ellipse has a vertex at the origin and foci $F_1(p, 0)$ and $F_2(p + 2c, 0)$, as shown in the figure. If the focus at F_1 is fixed and (x, y) is on the ellipse, show that y^2

approaches $4px$ as $c \to \infty$. (Thus, as $c \to \infty$, the ellipse takes on the shape of a parabola.)

Exercise 52

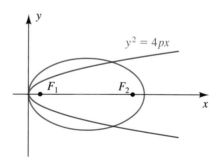

c **Exer. 53–54: The planets move around the sun in elliptical orbits. Given the semimajor axis a in millions of kilometers and eccentricity e, graph the orbit for the planet. Center the major axis on the x-axis, and plot the location of the sun at one focus.**

53 *Earth's path* $a = 149.6$, $e = 0.093$

54 *Pluto's path* $a = 5913$, $e = 0.249$

c **Exer. 55–58: Graph the ellipses on the same coordinate plane, and estimate their points of intersection.**

55 $\dfrac{x^2}{2.9} + \dfrac{y^2}{2.1} = 1;$ $\dfrac{x^2}{4.3} + \dfrac{(y - 2.1)^2}{4.9} = 1$

56 $\dfrac{x^2}{3.9} + \dfrac{y^2}{2.4} = 1;$ $\dfrac{(x + 1.9)^2}{4.1} + \dfrac{y^2}{2.5} = 1$

57 $\dfrac{(x + 0.1)^2}{1.7} + \dfrac{y^2}{0.9} = 1;$ $\dfrac{x^2}{0.9} + \dfrac{(y - 0.25)^2}{1.8} = 1$

58 $\dfrac{x^2}{3.1} + \dfrac{(y - 0.2)^2}{2.8} = 1;$ $\dfrac{(x + 0.23)^2}{1.8} + \dfrac{y^2}{4.2} = 1$

6.3   HYPERBOLAS

The definition of a hyperbola is similar to that of an ellipse. The only change is that instead of using the *sum* of distances from two fixed points, we use the *difference*.

DEFINITION OF A HYPERBOLA

A **hyperbola** is the set of all points in a plane, the difference of whose distances from two fixed points (the **foci**) in the plane is a positive constant.

Figure 25

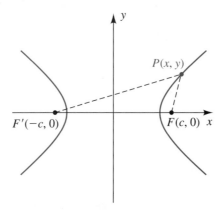

To find a simple equation for a hyperbola, we choose a coordinate system with foci at $F(c, 0)$ and $F'(-c, 0)$ and denote the (constant) distance by $2a$. The midpoint of the segment $F'F$ (the origin) is called the **center** of the hyperbola. Referring to Figure 25, we see that a point $P(x, y)$ is on the hyperbola if and only if either one of the following is true:

(1) $d(P, F') - d(P, F) = 2a$ or   **(2)** $d(P, F) - d(P, F') = 2a$

If P is not on the x-axis, then from Figure 25 we see that

$$d(P, F) < d(F', F) + d(P, F'),$$

because the length of one side of a triangle is always less than the sum of the lengths of the other two sides. Similarly,

$$d(P, F') < d(F', F) + d(P, F).$$

Equivalent forms for the previous two inequalities are

$$d(P, F) - d(P, F') < d(F', F)   \text{and}   d(P, F') - d(P, F) < d(F', F).$$

Since the differences on the left sides of these inequalities both equal $2a$ and since $d(F', F) = 2c$, the last two inequalities imply that $2a < 2c$, or $a < c$. (Recall that for ellipses we had $a > c$.)

Next, equations (1) and (2) may be replaced by the single equation

$$\left| d(P, F) - d(P, F') \right| = 2a.$$

Using the distance formula to find $d(P, F)$ and $d(P, F')$, we obtain an equation of the hyperbola:

$$\left| \sqrt{(x - c)^2 + (y - 0)^2} - \sqrt{(x + c)^2 + (y - 0)^2} \right| = 2a$$

Employing the type of simplification procedure that we used to derive an equation for an ellipse, we can rewrite the preceding equation as

$$\frac{x^2}{a^2} - \frac{y^2}{c^2 - a^2} = 1.$$

Finally, if we let

$$b^2 = c^2 - a^2 \quad \text{with} \quad b > 0$$

in the preceding equation, we obtain

$$\frac{x^2}{a^2} - \frac{y^2}{b^2} = 1.$$

We have shown that the coordinates of every point (x, y) on the hyperbola in Figure 25 satisfy the equation $(x^2/a^2) - (y^2/b^2) = 1$. Conversely, if (x, y) is a solution of this equation, then by reversing steps we see that the point (x, y) is on the hyperbola.

Applying tests for symmetry, we see that the hyperbola is symmetric with respect to both axes and the origin. We may find the x-intercepts of the hyperbola by letting $y = 0$ in the equation. Doing so gives us $x^2/a^2 = 1$, or $x^2 = a^2$, and consequently the x-intercepts are a and $-a$. The corresponding points $V(a, 0)$ and $V'(-a, 0)$ on the graph are called the **vertices** of the hyperbola (see Figure 26). The line segment $V'V$ is called the **transverse axis**. The graph has no y-intercept, since the equation $-y^2/b^2 = 1$ has the *complex* solutions $y = \pm bi$. The points $W(0, b)$ and $W'(0, -b)$ are endpoints of the **conjugate axis** $W'W$. The points W and W' are not on the hyperbola; however, as we shall see, they are useful for describing the graph.

Solving the equation $(x^2/a^2) - (y^2/b^2) = 1$ for y gives us

$$y = \pm \frac{b}{a} \sqrt{x^2 - a^2}.$$

If $x^2 - a^2 < 0$ or, equivalently, $-a < x < a$, then there are no points (x, y) on the graph. There *are* points $P(x, y)$ on the graph if $x \geq a$ or $x \leq -a$.

It can be shown that *the lines $y = \pm(b/a)x$ are asymptotes for the hyperbola*. These asymptotes serve as excellent guides for sketching the graph. A convenient way to sketch the asymptotes is to first plot the vertices $V(a, 0)$, $V'(-a, 0)$ and the points $W(0, b)$, $W'(0, -b)$ (see Figure 26). If vertical and horizontal lines are drawn through these endpoints of the transverse and conjugate axes, respectively, then the diagonals of the resulting **auxiliary rectangle** have slopes b/a and $-b/a$. Hence, by extending these diagonals we obtain the asymptotes $y = \pm(b/a)x$. The hyperbola is then sketched as in Figure 26, using the asymptotes as guides. The two parts that make up the hyperbola are called the **right branch** and the **left branch** of the hyperbola.

Figure 26

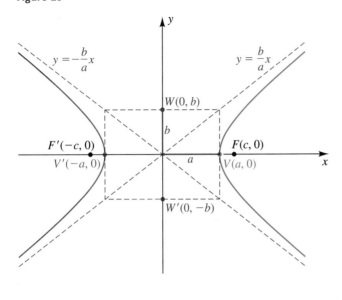

Figure 27

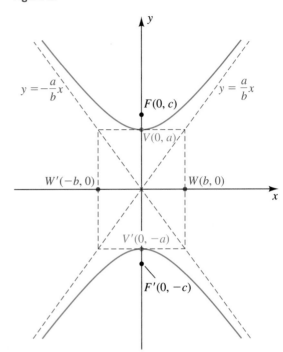

Similarly, if we take the foci on the y-axis, we obtain the equation

$$\frac{y^2}{a^2} - \frac{x^2}{b^2} = 1.$$

In this case, the vertices of the hyperbola are $(0, \pm a)$ and the endpoints of the conjugate axis are $(\pm b, 0)$, as shown in Figure 27. The asymptotes are $y = \pm(a/b)x$ (*not* $y = \pm(b/a)x$, as in the previous case), and we now refer to the **upper branch** and **lower branch** of the hyperbola.

The preceding discussion may be summarized as follows.

Standard Equations of a Hyperbola with Center at the Origin

The graph of

$$\frac{x^2}{a^2} - \frac{y^2}{b^2} = 1 \quad \text{or} \quad \frac{y^2}{a^2} - \frac{x^2}{b^2} = 1$$

is a hyperbola with center at the origin. The length of the transverse axis is $2a$, and the length of the conjugate axis is $2b$. The foci are a distance c from the origin, where $c^2 = a^2 + b^2$.

We have shown that an equation of a hyperbola with center at the origin and foci on a coordinate axis can always be written in the form

$$\frac{x^2}{p} + \frac{y^2}{q} = 1, \quad \text{or} \quad qx^2 + py^2 = pq,$$

where p and q have opposite signs. The vertices are on the x-axis if p is positive or on the y-axis if q is positive.

 EXAMPLE 1 Sketching a hyperbola with center at the origin

Sketch the graph of $9x^2 - 4y^2 = 36$. Find the foci and equations of the asymptotes.

Solution From the remarks preceding this example, the graph is a hyperbola with center at the origin. To express the given equation in a standard form, we divide both sides by 36 and simplify, obtaining

$$\frac{x^2}{4} - \frac{y^2}{9} = 1.$$

Comparing $(x^2/4) - (y^2/9) = 1$ to $(x^2/a^2) - (y^2/b^2) = 1$, we see that $a^2 = 4$ and $b^2 = 9$; that is, $a = 2$ and $b = 3$. The hyperbola has its vertices on the x-axis, since there are x-intercepts and no y-intercepts. The vertices $(\pm 2, 0)$ and the endpoints $(0, \pm 3)$ of the conjugate axis determine the auxiliary rectangle whose diagonals (extended) give us the asymptotes. The graph of the equation is sketched in Figure 28.

To find the foci, we calculate

$$c^2 = a^2 + b^2 = 4 + 9 = 13.$$

Thus, $c = \sqrt{13}$, and the foci are $(\pm\sqrt{13}, 0)$.

The equations of the asymptotes, $y = \pm\frac{3}{2}x$, can be found by referring to the graph or to the equations $y = \pm(b/a)x$.

Figure 28

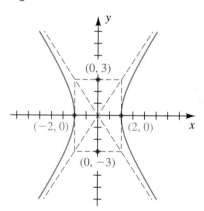

The preceding example indicates that for hyperbolas it is not always true that $a > b$, as is the case for ellipses. In fact, we may have $a < b$, $a > b$, or $a = b$.

EXAMPLE 2 Sketching a hyperbola with center at the origin

Sketch the graph of $4y^2 - 2x^2 = 1$. Find the foci and equations of the asymptotes.

Solution To express the given equation in a standard form, we write

$$\frac{y^2}{\frac{1}{4}} - \frac{x^2}{\frac{1}{2}} = 1.$$

(continued)

Figure 29

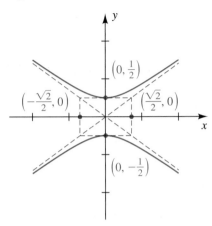

Thus,

$$a^2 = \tfrac{1}{4}, \qquad b^2 = \tfrac{1}{2}, \qquad \text{and} \qquad c^2 = a^2 + b^2 = \tfrac{3}{4},$$

and consequently

$$a = \frac{1}{2}, \qquad b = \frac{1}{\sqrt{2}} = \frac{\sqrt{2}}{2}, \qquad \text{and} \qquad c = \frac{\sqrt{3}}{2}.$$

The hyperbola has its vertices on the y-axis, since there are y-intercepts and no x-intercepts. The vertices are $\left(0, \pm\tfrac{1}{2}\right)$, the endpoints of the conjugate axes are $(\pm\sqrt{2}/2, 0)$, and the foci are $(0, \pm\sqrt{3}/2)$. The graph is sketched in Figure 29.

To find the equations of the asymptotes, we refer to the figure or use $y = \pm(a/b)x$, obtaining $y = \pm(\sqrt{2}/2)x$.

EXAMPLE 3   **Finding an equation of a hyperbola satisfying prescribed conditions**

A hyperbola has vertices $(\pm 3, 0)$ and passes through the point $P(5, 2)$. Find its equation, foci, and asymptotes.

Figure 30

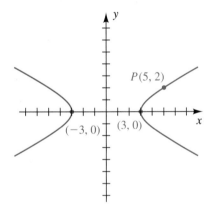

Solution   We begin by sketching a hyperbola with vertices $(\pm 3, 0)$ that passes through the point $P(5, 2)$, as in Figure 30.

An equation of the hyperbola has the form

$$\frac{x^2}{3^2} - \frac{y^2}{b^2} = 1.$$

Since $P(5, 2)$ is on the hyperbola, the x- and y-coordinates satisfy this equation; that is,

$$\frac{5^2}{3^2} - \frac{2^2}{b^2} = 1.$$

Solving for b^2 gives us $b^2 = \tfrac{9}{4}$, and hence an equation for the hyperbola is

$$\frac{x^2}{9} - \frac{y^2}{\tfrac{9}{4}} = 1$$

or, equivalently,

$$x^2 - 4y^2 = 9.$$

To find the foci, we first calculate

$$c^2 = a^2 + b^2 = 9 + \tfrac{9}{4} = \tfrac{45}{4}.$$

Hence, $c = \sqrt{\tfrac{45}{4}} = \tfrac{3}{2}\sqrt{5} \approx 3.35$, and the foci are $\left(\pm\tfrac{3}{2}\sqrt{5}, 0\right)$.

The general equations of the asymptotes are $y = \pm(b/a)x$. Substituting $a = 3$ and $b = \tfrac{3}{2}$ gives us $y = \pm\tfrac{1}{2}x$.

The next example indicates how to find equations for certain parts of a hyperbola.

EXAMPLE 4   Finding equations of portions of a hyperbola

The hyperbola $9x^2 - 4y^2 = 36$ was discussed in Example 1. Solve the equation as indicated, and describe the resulting graph.

(a) For x in terms of y **(b)** For y in terms of x

Solution

(a) We solve for x in terms of y as follows:

$$9x^2 - 4y^2 = 36 \qquad \text{given}$$
$$x^2 = \frac{36 + 4y^2}{9} \qquad \text{solve for } x^2$$
$$x = \pm\tfrac{2}{3}\sqrt{9 + y^2} \qquad \text{factor out 4, and take the square root}$$

Figure 31

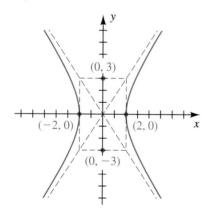

The graph of the equation $x = \tfrac{2}{3}\sqrt{9 + y^2}$ is the right branch of the hyperbola sketched in Example 1 (and repeated in Figure 31), and the graph of $x = -\tfrac{2}{3}\sqrt{9 + y^2}$ is the left branch.

(b) We solve for y in terms of x as follows:

$$9x^2 - 4y^2 = 36 \qquad \text{given}$$
$$y^2 = \frac{9x^2 - 36}{4} \qquad \text{solve for } y^2$$
$$y = \pm\tfrac{3}{2}\sqrt{x^2 - 4} \qquad \text{factor out 9, and take the square root}$$

The graph of $y = \tfrac{3}{2}\sqrt{x^2 - 4}$ is the upper half of the right and left branches, and the graph of $y = -\tfrac{3}{2}\sqrt{x^2 - 4}$ is the lower half of these branches.

As was the case for ellipses, we may use translations to help sketch hyperbolas that have centers at some point $(h, k) \neq (0, 0)$. The following example illustrates this technique.

EXAMPLE 5   Sketching a hyperbola with center (h, k)

Discuss and sketch the graph of the equation

$$9x^2 - 4y^2 - 54x - 16y + 29 = 0.$$

Solution   We arrange our work using a procedure similar to that used for ellipses in Example 5 of the previous section:

$$(9x^2 - 54x) + (-4y^2 - 16y) = -29 \qquad \text{group terms}$$

$$9(x^2 - 6x + \underline{}) - 4(y^2 + 4y + \underline{}) = -29 \qquad \text{factor out 9 and } -4$$

$$9(x^2 - 6x + 9) - 4(y^2 + 4y + 4) = -29 + \underline{9 \cdot 9} - \underline{4 \cdot 4}$$

$$\text{complete the squares}$$

$$9(x - 3)^2 - 4(y + 2)^2 = 36 \qquad \text{factor, and simplify}$$

$$\frac{(x - 3)^2}{4} - \frac{(y + 2)^2}{9} = 1 \qquad \text{divide by 36}$$

The last equation indicates that the hyperbola has center $C(3, -2)$ with vertices and foci on the horizontal line $y = -2$, because the term containing x is positive. We also know that

$$a^2 = 4, \qquad b^2 = 9, \qquad \text{and} \qquad c^2 = a^2 + b^2 = 13.$$

Hence,

$$a = 2, \qquad b = 3, \qquad \text{and} \qquad c = \sqrt{13}.$$

As illustrated in Figure 32, the vertices are $(3 \pm 2, -2)$—that is, $(5, -2)$ and $(1, -2)$. The endpoints of the conjugate axis are $(3, -2 \pm 3)$—that is, $(3, 1)$ and $(3, -5)$. The foci are $(3 \pm \sqrt{13}, -2)$, and equations of the asymptotes are

$$y + 2 = \pm\tfrac{3}{2}(x - 3).$$

Figure 32

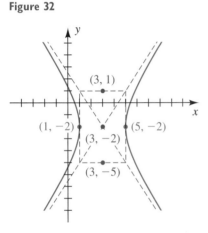

The results of Sections 6.1 through 6.3 indicate that the graph of every equation of the form

$$Ax^2 + Cy^2 + Dx + Ey + F = 0$$

is a conic, except for certain degenerate cases in which a point, one or two lines, or no graph is obtained. Although we have considered only special examples, our methods can be applied to any such equation. If A and C are equal and not 0, then the graph, when it exists, is a circle or, in exceptional cases, a point. If A and C are unequal but have the same sign, an equation is obtained whose graph, when it exists, is an ellipse (or a point). If A and C have opposite signs, an equation of a hyperbola is obtained or possibly, in the degenerate case, two intersecting straight lines. If either A or C (but not both) is 0, the graph is a parabola or, in certain cases, a pair of parallel lines.

We shall conclude this section with an application involving hyperbolas.

E X A M P L E   6   **Locating a ship**

Coast Guard station A is 200 miles directly east of another station B. A ship is sailing on a line parallel to and 50 miles north of the line through A and B. Radio signals are sent out from A and B at the rate of 980 ft/μsec (microsecond). If, at 1:00 P.M., the signal from B reaches the ship 400 microseconds after the signal from A, locate the position of the ship at that time.

Figure 33

(a)

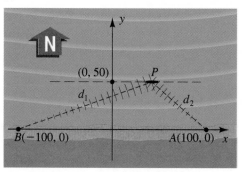

(b)

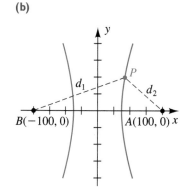

Solution   Let us introduce a coordinate system, as shown in Figure 33(a), with the stations at points A and B on the x-axis and the ship at P on the line $y = 50$. Since at 1:00 P.M. it takes 400 microseconds longer for the signal to arrive from B than from A, the difference $d_1 - d_2$ in the indicated distances at that time is

$$d_1 - d_2 = (980)(400) = 392{,}000 \text{ ft.}$$

Dividing by 5280 (ft/mi) gives us

$$d_1 - d_2 = \frac{392{,}000}{5280} = 74.\overline{24} \text{ mi.}$$

At 1:00 P.M., point P is on the right branch of a hyperbola whose equation is $(x^2/a^2) - (y^2/b^2) = 1$ (see Figure 33(b)), consisting of all points whose difference in distances from the foci B and A is $d_1 - d_2$. In our derivation of the equation $(x^2/a^2) - (y^2/b^2) = 1$, we let $d_1 - d_2 = 2a$; it follows that in the present situation

$$a = \frac{74.\overline{24}}{2} = 37.\overline{12} \quad \text{and} \quad a^2 \approx 1378.$$

Since the distance c from the origin to either focus is 100,

$$b^2 = c^2 - a^2 \approx 10{,}000 - 1378, \quad \text{or} \quad b^2 \approx 8622.$$

Hence, an (approximate) equation for the hyperbola that has foci A and B and passes through P is

$$\frac{x^2}{1378} - \frac{y^2}{8622} = 1.$$

(continued)

If we let $y = 50$ (the y-coordinate of P), we obtain

$$\frac{x^2}{1378} - \frac{2500}{8622} = 1.$$

Solving for x gives us $x \approx 42.16$. Rounding off to the nearest mile, we find that the coordinates of P are approximately $(42, 50)$.

An extension of the method used in Example 6 is the basis for the navigational system LORAN (for Long Range Navigation). This system involves two pairs of radio transmitters, such as those located at T, T' and S, S' in Figure 34. Suppose that signals sent out by the transmitters at T and T' reach a radio receiver in a ship located at some point P. The difference in the times of arrival of the signals can be used to determine the difference in the distances of P from T and T'. Thus, P lies on one branch of a hyperbola with foci at T and T'. Repeating this process for the other pair of transmitters, we see that P also lies on one branch of a hyperbola with foci at S and S'. The intersection of these two branches determines the position of P.

Figure 34

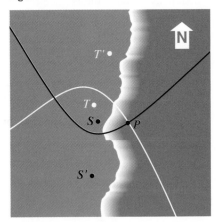

Figure 35

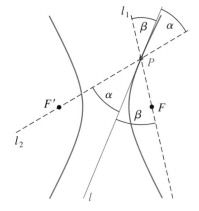

A hyperbola has a *reflective property* analogous to that of the ellipse discussed in the previous section. To illustrate, let l denote the tangent line at a point P on a hyperbola with foci F and F', as shown in Figure 35. If α is the acute angle between $F'P$ and l and if β is the acute angle between FP and l, it can be shown that $\alpha = \beta$. If a ray of light is directed along the line l_1 toward F, it will be reflected back at P along the line l_2 toward F'. This property is used in the design of telescopes of the Cassegrain type (see Exercise 52).

6.3 EXERCISES

Exer. 1–16: Find the vertices, the foci, and the equations of the asymptotes of the hyperbola. Sketch its graph, showing the asymptotes and the foci.

1 $\dfrac{x^2}{9} - \dfrac{y^2}{4} = 1$

2 $\dfrac{y^2}{49} - \dfrac{x^2}{16} = 1$

3 $\dfrac{y^2}{9} - \dfrac{x^2}{4} = 1$

4 $\dfrac{x^2}{49} - \dfrac{y^2}{16} = 1$

5 $x^2 - \dfrac{y^2}{24} = 1$

6 $y^2 - \dfrac{x^2}{15} = 1$

7 $y^2 - 4x^2 = 16$

8 $x^2 - 2y^2 = 8$

9 $16x^2 - 36y^2 = 1$

10 $y^2 - 16x^2 = 1$

11 $\dfrac{(y + 2)^2}{9} - \dfrac{(x + 2)^2}{4} = 1$

12 $\dfrac{(x - 3)^2}{25} - \dfrac{(y - 1)^2}{4} = 1$

13 $144x^2 - 25y^2 + 864x - 100y - 2404 = 0$

14 $y^2 - 4x^2 - 12y - 16x + 16 = 0$

15 $4y^2 - x^2 + 40y - 4x + 60 = 0$

16 $25x^2 - 9y^2 + 100x - 54y + 10 = 0$

Exer. 17–20: Find an equation for the hyperbola shown in the figure.

17

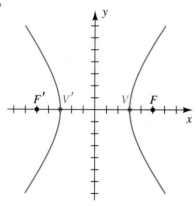

18

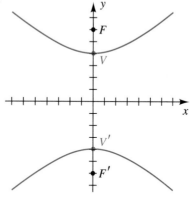

19

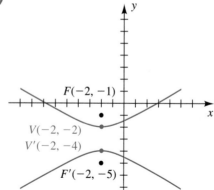

20

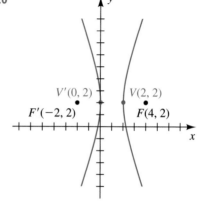

Exer. 21–32: Find an equation for the hyperbola that has its center at the origin and satisfies the given conditions.

21 Foci $F(0, \pm4)$,   vertices $V(0, \pm1)$

22 Foci $F(\pm8, 0)$,   vertices $V(\pm5, 0)$

23 Foci $F(\pm5, 0)$,   vertices $V(\pm3, 0)$

24 Foci $F(0, \pm3)$,   vertices $V(0, \pm2)$

25 Foci $F(0, \pm5)$,   conjugate axis of length 4

26 Vertices $V(\pm4, 0)$,   passing through $(8, 2)$

27 Vertices $V(\pm3, 0)$,   asymptotes $y = \pm2x$

28 Foci $F(0, \pm10)$,   asymptotes $y = \pm\frac{1}{3}x$

29 x-intercepts ±5,   asymptotes $y = \pm2x$

30 y-intercepts ±2,   asymptotes $y = \pm\frac{1}{4}x$

31 Vertical transverse axis of length 10, conjugate axis of length 14

32 Horizontal transverse axis of length 6, conjugate axis of length 2

Exer. 33–34: Find the points of intersection of the graphs of the equations. Sketch both graphs on the same coordinate plane, and show the points of intersection.

33 $\begin{cases} y^2 - 4x^2 = 16 \\ y - x = 4 \end{cases}$ 34 $\begin{cases} x^2 - y^2 = 4 \\ y^2 - 3x = 0 \end{cases}$

Exer. 35–38: Find an equation for the set of points in an xy-plane such that the difference of the distances from F and F' is k.

35 $F(13, 0)$, $F'(-13, 0)$; $k = 24$

36 $F(5, 0)$, $F'(-5, 0)$; $k = 8$

37 $F(0, 10)$, $F'(0, -10)$; $k = 16$

38 $F(0, 17)$, $F'(0, -17)$; $k = 30$

Exer. 39–46: Describe the part of a hyperbola given by the equation.

39 $x = \frac{5}{4}\sqrt{y^2 + 16}$ 40 $x = -\frac{5}{4}\sqrt{y^2 + 16}$

41 $y = \frac{3}{7}\sqrt{x^2 + 49}$ 42 $y = -\frac{3}{7}\sqrt{x^2 + 49}$

43 $y = -\frac{9}{4}\sqrt{x^2 - 16}$ 44 $y = \frac{9}{4}\sqrt{x^2 - 16}$

45 $x = -\frac{2}{3}\sqrt{y^2 - 36}$ 46 $x = \frac{2}{3}\sqrt{y^2 - 36}$

47 The graphs of the equations

$$\frac{x^2}{a^2} - \frac{y^2}{b^2} = 1 \quad \text{and} \quad \frac{x^2}{a^2} - \frac{y^2}{b^2} = -1$$

are called *conjugate hyperbolas*. Sketch the graphs of both equations on the same coordinate plane, with $a = 5$ and $b = 3$, and describe the relationship between the two graphs.

48 Find an equation of the hyperbola with foci $(h \pm c, k)$ and vertices $(h \pm a, k)$, where

$$0 < a < c \quad \text{and} \quad c^2 = a^2 + b^2.$$

49 *Alpha particles* In 1911, the physicist Ernest Rutherford (1871–1937) discovered that if alpha particles are shot toward the nucleus of an atom, they are eventually repulsed away from the nucleus along hyperbolic paths. The figure illustrates the path of a particle that starts toward the origin along the line $y = \frac{1}{2}x$ and comes within 3 units of the nucleus. Find an equation of the path.

Exercise 49

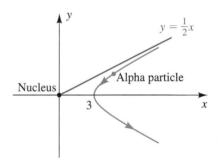

50 *Airplane maneuver* An airplane is flying along the hyperbolic path illustrated in the figure. If an equation of the path is $2y^2 - x^2 = 8$, determine how close the airplane comes to a town located at $(3, 0)$. (*Hint:* Let S denote the square of the distance from a point (x, y) on the path to $(3, 0)$, and find the minimum value of S.)

Exercise 50

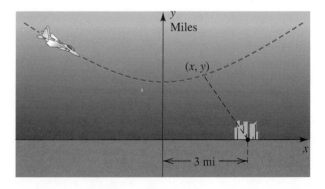

51 *Locating a ship* A ship is traveling a course that is 100 miles from, and parallel to, a straight shoreline. The ship sends out a distress signal that is received by two Coast Guard stations A and B, located 200 miles apart, as shown in the figure. By measuring the difference in signal reception times, it is determined that the ship is 160 miles closer to B than to A. Where is the ship?

Exercise 51

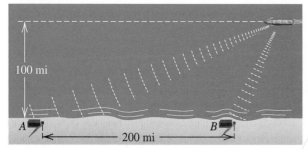

52 *Design of a telescope* The Cassegrain telescope design (dating back to 1672) makes use of the reflective properties of both the parabola and the hyperbola. Shown in the figure is a (split) parabolic mirror, with focus at F_1 and axis along the line l, and a second hyperbolic mirror, with one focus also at F_1 and transverse axis along l. Where do incoming light waves parallel to the common axis finally collect?

Exercise 52

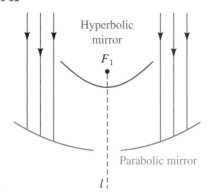

c Exer. 53–54: Graph the hyperbolas on the same coordinate plane, and estimate their first-quadrant point of intersection.

53 $\dfrac{(y - 0.1)^2}{1.6} - \dfrac{(x + 0.2)^2}{0.5} = 1;$

$\dfrac{(y - 0.5)^2}{2.7} - \dfrac{(x - 0.1)^2}{5.3} = 1$

54 $\dfrac{(x - 0.1)^2}{0.12} - \dfrac{y^2}{0.1} = 1; \dfrac{x^2}{0.9} - \dfrac{(y - 0.3)^2}{2.1} = 1$

c Exer. 55–56: Graph the hyperbolas on the same coordinate plane, and determine the number of points of intersection.

55 $\dfrac{(x - 0.3)^2}{1.3} - \dfrac{y^2}{2.7} = 1; \dfrac{y^2}{2.8} - \dfrac{(x - 0.2)^2}{1.2} = 1$

56 $\dfrac{(x + 0.2)^2}{1.75} - \dfrac{(y - 0.5)^2}{1.6} = 1;$

$\dfrac{(x - 0.6)^2}{2.2} - \dfrac{(y + 0.4)^2}{2.35} = 1$

57 *Comet's path* Comets can travel in elliptical, parabolic, or hyperbolic paths around the sun. If a comet travels in a parabolic or hyperbolic path, it will pass by the sun once and never return. Suppose that a comet's coordinates in miles can be described by the equation

$$\frac{x^2}{26 \times 10^{14}} - \frac{y^2}{18 \times 10^{14}} = 1 \quad \text{for} \quad x > 0,$$

where the sun is located at a focus, as shown in the figure.

(a) Approximate the coordinates of the sun.

(b) For the comet to maintain a hyperbolic trajectory, the minimum velocity v of the comet, in meters per second, must satisfy $v > \sqrt{2k/r}$, where r is the distance between the comet and the center of the sun in meters and $k = 1.325 \times 10^{20}$ is a constant. Determine v when r is minimum.

Exercise 57

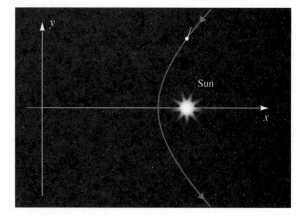

6.4 PLANE CURVES AND PARAMETRIC EQUATIONS

If f is a function, the graph of the equation $y = f(x)$ is often called a *plane curve*. However, this definition is restrictive, because it excludes many useful graphs. The following definition is more general.

DEFINITION OF PLANE CURVE

A **plane curve** is a set C of ordered pairs $(f(t), g(t))$, where f and g are functions defined on an interval I.

For simplicity, we often refer to a plane curve as a **curve**. The **graph** of C in the preceding definition consists of all points $P(t) = (f(t), g(t))$ in an xy-plane, for t in I. We shall use the term *curve* interchangeably with *graph of a curve*. We sometimes regard the point $P(t)$ as tracing the curve C as t varies through the interval I.

Figure 36

(a) Curve

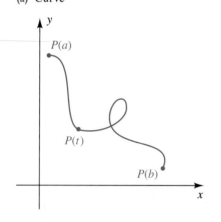

(b) Closed curve

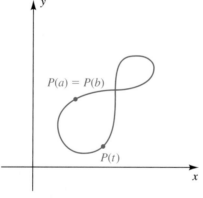

(c) Simple closed curve

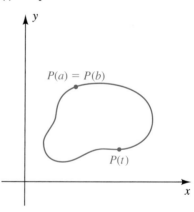

The graphs of several curves are sketched in Figure 36, where I is a closed interval $[a, b]$—that is, $a \le t \le b$. In (a) of the figure, $P(a) \ne P(b)$, and $P(a)$ and $P(b)$ are called the **endpoints** of C. The curve in (a) intersects itself; that is, two different values of t produce the same point. If $P(a) = P(b)$, as in Figure 36(b), then C is a **closed curve.** If $P(a) = P(b)$ and C does not intersect itself at any other point, as in Figure 36(c), then C is a **simple closed curve.**

A convenient way to represent curves is given in the next definition.

DEFINITION OF PARAMETRIC
EQUATIONS

Let C be the curve consisting of all ordered pairs $(f(t), g(t))$, where f and g are defined on an interval I. The equations

$$x = f(t), \quad y = g(t),$$

for t in I, are **parametric equations** for C with **parameter** t.

The curve C in this definition is referred to as a **parametrized curve,** and the parametric equations are a **parametrization** for C. We often use the notation

$$x = f(t), \quad y = g(t); \quad t \text{ in } I$$

to indicate the domain I of f and g. Sometimes it may be possible to eliminate the parameter and obtain a familiar equation in x and y for C. In simple cases we can sketch a graph of a parametrized curve by plotting points and connecting them in order of increasing t, as illustrated in the next example.

EXAMPLE 1   Sketching the graph of a parametrized curve

Sketch the graph of the curve C that has the parametrization

$$x = 2t, \quad y = t^2 - 1; \quad -1 \le t \le 2.$$

Solution   We use the parametric equations to tabulate coordinates of points $P(x, y)$ on C, as follows.

t	-1	$-\frac{1}{2}$	0	$\frac{1}{2}$	1	$\frac{3}{2}$	2
x	-2	-1	0	1	2	3	4
y	0	$-\frac{3}{4}$	-1	$-\frac{3}{4}$	0	$\frac{5}{4}$	3

Figure 37
$x = 2t, y = t^2 - 1; -1 \le t \le 2$

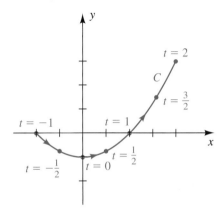

Plotting points leads to the sketch in Figure 37. The arrowheads on the graph indicate the direction in which $P(x, y)$ traces the curve as t *increases* from -1 to 2.

We may obtain a more familiar description of the graph by eliminating the parameter. Solving the first parametric equation for t, we obtain $t = \frac{1}{2}x$. Substituting this expression for t in the second equation gives us

$$y = \left(\tfrac{1}{2}x\right)^2 - 1.$$

The graph of this equation in x and y is a parabola symmetric with respect to the y-axis with vertex $(0, -1)$. However, since $x = 2t$ and $-1 \le t \le 2$, we see that $-2 \le x \le 4$ for points (x, y) on C, and hence C is that part of the parabola between the points $(-2, 0)$ and $(4, 3)$ shown in Figure 37.

As indicated by the arrowheads in Figure 37, the point $P(x, y)$ traces the curve C from *left to right* as t increases. The parametric equations

$$x = -2t, \quad y = t^2 - 1; \quad -2 \le t \le 1$$

give us the same graph; however, as t increases, $P(x, y)$ traces the curve from *right to left*. For other parametrizations, the point $P(x, y)$ may oscillate back and forth as t increases.

The **orientation** of a parametrized curve C is the direction determined by *increasing* values of the parameter. We often indicate an orientation by placing arrowheads on C, as in Figure 37. If $P(x, y)$ moves back and forth as t increases, we may place arrows *alongside* of C.

As we have observed, a curve may have different orientations, depending on the parametrization. To illustrate, the curve C in Example 1 is given parametrically by any of the following:

$$x = 2t, \quad y = t^2 - 1; \quad -1 \le t \le 2$$
$$x = t, \quad y = \tfrac{1}{4}t^2 - 1; \quad -2 \le t \le 4$$
$$x = t^3, \quad y = \tfrac{1}{4}t^6 - 1; \quad \sqrt[3]{-2} \le t \le \sqrt[3]{4}$$

The next example demonstrates that it is sometimes useful to eliminate the parameter *before* plotting points.

EXAMPLE 2 Describing the motion of a point

A point moves in a plane such that its position $P(x, y)$ at time t is given by

$$x = a \cos t, \quad y = a \sin t; \quad t \text{ in } \mathbb{R},$$

where $a > 0$. Describe the motion of the point.

Solution When x and y contain trigonometric functions of t, we can often eliminate the parameter t by isolating the trigonometric functions, squaring both sides of the equations, and then using one of the Pythagorean identities, as follows:

$$x = a \cos t, \quad y = a \sin t \qquad \text{given}$$

$$\frac{x}{a} = \cos t, \quad \frac{y}{a} = \sin t \qquad \text{isolate } \cos t \text{ and } \sin t$$

$$\frac{x^2}{a^2} = \cos^2 t, \quad \frac{y^2}{a^2} = \sin^2 t \qquad \text{square both sides}$$

$$\frac{x^2}{a^2} + \frac{y^2}{a^2} = 1 \qquad \cos^2 t + \sin^2 t = 1$$

$$x^2 + y^2 = a^2 \qquad \text{multiply by } a^2$$

This shows that the point $P(x, y)$ moves on the circle C of radius a with center at the origin (see Figure 38). The point is at $A(a, 0)$ when $t = 0$, at $(0, a)$ when $t = \pi/2$, at $(-a, 0)$ when $t = \pi$, at $(0, -a)$ when $t = 3\pi/2$, and back

Figure 38
$x = a \cos t, y = a \sin t; t \text{ in } \mathbb{R}$

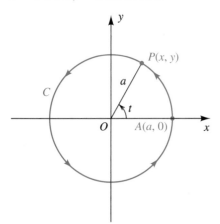

at $A(a, 0)$ when $t = 2\pi$. Thus, P moves around C in a counterclockwise direction, making one revolution every 2π units of time. The orientation of C is indicated by the arrowheads in the figure.

Note that in this example we may interpret t geometrically as the radian measure of the angle generated by the line segment OP.

E X A M P L E   3   Sketching the graph of a parametrized curve

Sketch the graph of the curve C that has the parametrization

$$x = -2 + t^2, \quad y = 1 + 2t^2; \quad t \text{ in } \mathbb{R},$$

and indicate the orientation.

Solution   To eliminate the parameter, we use the first equation to obtain $t^2 = x + 2$ and then substitute for t^2 in the second equation. Thus,

$$y = 1 + 2(x + 2).$$

The graph of the last equation is the line of slope 2 through the point $(-2, 1)$, as indicated by the dashes in Figure 39(a). However, since $t^2 \geq 0$, we see from the parametric equations for C that

$$x = -2 + t^2 \geq -2 \quad \text{and} \quad y = 1 + 2t^2 \geq 1.$$

Thus, the graph of C is that part of the line to the right of $(-2, 1)$ (the point corresponding to $t = 0$), as shown in Figure 39(b). The orientation is indicated by the arrows alongside of C. As t increases in the interval $(-\infty, 0]$, $P(x, y)$ moves down the curve toward the point $(-2, 1)$. As t increases in $[0, \infty)$, $P(x, y)$ moves up the curve away from $(-2, 1)$.

Figure 39

(a)   (b)

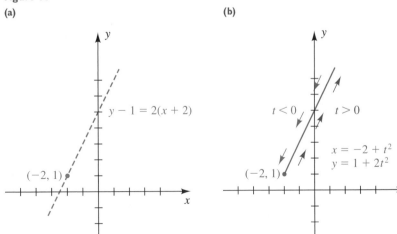

If a curve C is described by an equation $y = f(x)$ for some function f, then an easy way to obtain parametric equations for C is to let

$$x = t, \quad y = f(t),$$

where t is in the domain of f. For example, if $y = x^3$, then parametric equations are

$$x = t, \quad y = t^3; \quad t \text{ in } \mathbb{R}.$$

We can use many different substitutions for x, provided that as t varies through some interval, x takes on every value in the domain of f. Thus, the graph of $y = x^3$ is also given by

$$x = t^{1/3}, \quad y = t; \quad t \text{ in } \mathbb{R}.$$

Note, however, that the parametric equations

$$x = \sin t, \quad y = \sin^3 t; \quad t \text{ in } \mathbb{R}$$

give only that part of the graph of $y = x^3$ between the points $(-1, -1)$ and $(1, 1)$.

EXAMPLE 4   Finding parametric equations for a line

Find three parametrizations for the line of slope m through the point (x_1, y_1).

Solution   By the point-slope form, an equation for the line is

$$y - y_1 = m(x - x_1). \tag{$*$}$$

If we let $x = t$, then $y - y_1 = m(t - x_1)$ and we obtain the parametrization

$$x = t, \quad y = y_1 + m(t - x_1); \quad t \text{ in } \mathbb{R}.$$

We obtain another parametrization for the line if we let $x - x_1 = t$ in $(*)$. In this case $y - y_1 = mt$, and we have

$$x = x_1 + t, \quad y = y_1 + mt; \quad t \text{ in } \mathbb{R}.$$

As a third illustration, if we let $x - x_1 = \tan t$ in $(*)$, then

$$x = x_1 + \tan t, \quad y = y_1 + m \tan t; \quad -\frac{\pi}{2} < t < \frac{\pi}{2}.$$

There are many other parametrizations for the line.

Parametric equations of the form

$$x = a \sin \omega_1 t, \quad y = b \cos \omega_2 t; \quad t \geq 0,$$

where a, b, ω_1, and ω_2 are constants, occur in electrical theory. The variables x and y usually represent voltages or currents at time t. The resulting curve is often difficult to sketch; however, using an oscilloscope and imposing voltages or currents on the input terminals, we can represent the graph, a **Lissajous figure,** on the screen of the oscilloscope. Graphing utilities are very helpful in obtaining these complicated graphs.

EXAMPLE 5   Graphing a Lissajous figure

Sketch the graph of the Lissajous figure that has the parametrization

$$x = \sin 2t, \quad y = \cos t; \quad 0 \le t \le 2\pi.$$

Determine the values of t that correspond to the curve in each quadrant.

Solution   Some specific keystrokes for the TI-82/83 are given in Example 27 of Appendix I. We first need to set our graphing utility in a parametric mode. Next we make the assignments

$$X_{1T} = \sin 2t \qquad \text{and} \qquad Y_{1T} = \cos t,$$

where the subscript 1T on X and Y indicates that X_{1T} and Y_{1T} represent the first *pair* of parametric equations.

When graphing parametric equations, we need to assign minimum (Tmin) and maximum (Tmax) values to the parameter t, in addition to viewing rectangle dimensions. We also need to select an increment, or step value (Tstep), for t. A typical value for Tstep is $\pi/30 \approx 0.105$. If a smaller value of Tstep is chosen, the accuracy of the sketch is increased, but so is the amount of time needed to sketch the graph.

For this example, we use Tmin = 0, Tmax = 2π, and Tstep = 0.1. Since x and y are between -1 and 1, we will assign -1 to Ymin and 1 to Ymax. To maintain our $3:2$ screen proportion, we select -1.5 for Xmin and 1.5 for Xmax, and then we graph X_{1T} and Y_{1T} to obtain the Lissajous figure in Figure 40.

Referring to the parametric equations, we see that as t increases from 0 to $\pi/2$, the point $P(x, y)$ starts at $(0, 1)$ and traces the part of the curve in quadrant I (in a generally clockwise direction). As t increases from $\pi/2$ to π, $P(x, y)$ traces the part in quadrant III (in a counterclockwise direction). For $\pi < t < 3\pi/2$, we obtain the part in quadrant IV; and $3\pi/2 < t < 2\pi$ gives us the part in quadrant II.

Figure 40
$[-1.5, 1.5]$ by $[-1, 1]$

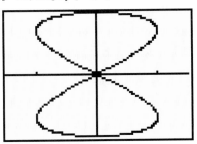

EXAMPLE 6   Finding parametric equations for a cycloid

The curve traced by a fixed point P on the circumference of a circle as the circle rolls along a line in a plane is called a **cycloid.** Find parametric equations for a cycloid.

Solution Suppose the circle has radius a and that it rolls along (and above) the x-axis in the positive direction. If one position of P is the origin, then Figure 41 displays part of the curve and a possible position of the circle. The v-shaped part of the curve at $x = 2\pi a$ is called a **cusp.**

Figure 41

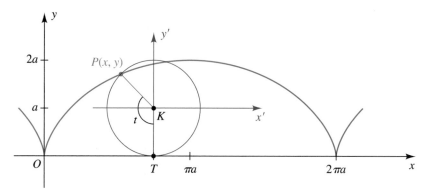

Let K denote the center of the circle and T the point of tangency with the x-axis. We introduce, as a parameter t, the radian measure of angle TKP. The distance the circle has rolled is $d(O, T) = at$ (formula for the length of a circular arc). Consequently, the coordinates of K are $(x, y) = (at, a)$. If we consider an $x'y'$-coordinate system with origin at $K(at, a)$ and if $P(x', y')$ denotes the point P relative to this system, then, by adding x' and y' to the x- and y-coordinates of K, we obtain

Figure 42

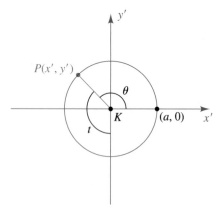

$$x = at + x', \qquad y = a + y'.$$

If, as in Figure 42, θ denotes an angle in standard position on the $x'y'$-plane, then $\theta + t = 3\pi/2$ or, equivalently, $\theta = (3\pi/2) - t$. Hence,

$$x' = a \cos \theta = a \cos\left(\frac{3\pi}{2} - t\right) = -a \sin t$$

$$y' = a \sin \theta = a \sin\left(\frac{3\pi}{2} - t\right) = -a \cos t,$$

and substitution in $x = at + x'$, $y = a + y'$ gives us parametric equations for the cycloid:

$$x = a(t - \sin t), \quad y = a(1 - \cos t); \quad t \text{ in } \mathbb{R}$$

If $a < 0$, then the graph of $x = a(t - \sin t)$, $y = a(1 - \cos t)$ is the inverted cycloid that results if the circle of Example 6 rolls *below* the x-axis. This curve has a number of important physical properties. To illustrate, suppose a thin wire passes through two fixed points A and B, as shown in

Figure 43

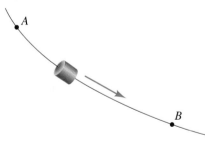

Figure 43, and that the shape of the wire can be changed by bending it in any manner. Suppose further that a bead is allowed to slide along the wire and the only force acting on the bead is gravity. We now ask which of all the possible paths will allow the bead to slide from A to B in the least amount of time. It is natural to believe that the desired path is the straight line segment from A to B; however, this is not the correct answer. The path that requires the least time coincides with the graph of an inverted cycloid with A at the origin. Because the velocity of the bead increases more rapidly along the cycloid than along the line through A and B, the bead reaches B more rapidly, even though the distance is greater.

There is another interesting property of this **curve of least descent.** Suppose that A is the origin and B is the point with x-coordinate $\pi|a|$ — that is, the lowest point on the cycloid in the first arc to the right of A. If the bead is released at *any* point between A and B, it can be shown that the *time* required for it to reach B is always the *same*.

Variations of the cycloid occur in applications. For example, if a motorcycle wheel rolls along a straight road, then the curve traced by a fixed point on one of the spokes is a cycloidlike curve. In this case the curve does not have cusps, nor does it intersect the road (the x-axis) as does the graph of a cycloid. If the wheel of a train rolls along a railroad track, then the curve traced by a fixed point on the circumference of the wheel (which extends below the track) contains loops at regular intervals. Other cycloids are defined in Exercises 39 and 40.

6.4 EXERCISES

Exer. 1–22: Find an equation in x and y whose graph contains the points on the curve C. Sketch the graph of C, and indicate the orientation.

1 $x = t - 2$,	$y = 2t + 3$;	$0 \le t \le 5$
2 $x = 1 - 2t$,	$y = 1 + t$;	$-1 \le t \le 4$
3 $x = t^2 + 1$,	$y = t^2 - 1$;	$-2 \le t \le 2$
4 $x = t^3 + 1$,	$y = t^3 - 1$;	$-2 \le t \le 2$
5 $x = 4t^2 - 5$,	$y = 2t + 3$;	t in $\mathbb{R}$
6 $x = t^3$,	$y = t^2$;	t in $\mathbb{R}$
7 $x = e^t$,	$y = e^{-2t}$;	t in $\mathbb{R}$
8 $x = \sqrt{t}$,	$y = 3t + 4$;	$t \ge 0$
9 $x = 2 \sin t$,	$y = 3 \cos t$;	$0 \le t \le 2\pi$
10 $x = \cos t - 2$,	$y = \sin t + 3$;	$0 \le t \le 2\pi$

11 $x = \sec t$,	$y = \tan t$;	$-\pi/2 < t < \pi/2$
12 $x = \cos 2t$,	$y = \sin t$;	$-\pi \le t \le \pi$
13 $x = t^2$,	$y = 2 \ln t$;	$t > 0$
14 $x = \cos^3 t$,	$y = \sin^3 t$;	$0 \le t \le 2\pi$
15 $x = \sin t$,	$y = \csc t$;	$0 < t \le \pi/2$
16 $x = e^t$,	$y = e^{-t}$;	t in $\mathbb{R}$
17 $x = t$,	$y = \sqrt{t^2 - 1}$;	$|t| \ge 1$
18 $x = -2\sqrt{1 - t^2}$,	$y = t$;	$|t| \le 1$
19 $x = t$,	$y = \sqrt{t^2 - 2t + 1}$;	$0 \le t \le 4$
20 $x = 2t$,	$y = 8t^3$;	$-1 \le t \le 1$
21 $x = (t + 1)^3$,	$y = (t + 2)^2$;	$0 \le t \le 2$
22 $x = \tan t$,	$y = 1$;	$-\pi/2 < t < \pi/2$

Exer. 23–24: Curves C_1, C_2, C_3, and C_4 are given parametrically, for t in $\mathbb{R}$. Sketch their graphs, and indicate orientations.

23 C_1: $x = t^2$, $y = t$
     C_2: $x = t^4$, $y = t^2$
     C_3: $x = \sin^2 t$, $y = \sin t$
     C_4: $x = e^{2t}$, $y = -e^t$

24 C_1: $x = t$, $y = 1 - t$
     C_2: $x = 1 - t^2$, $y = t^2$
     C_3: $x = \cos^2 t$, $y = \sin^2 t$
     C_4: $x = \ln t - t$, $y = 1 + t - \ln t$; $t > 0$

Exer. 25–26: The parametric equations specify the position of a moving point $P(x, y)$ at time t. Sketch the graph, and indicate the motion of P as t increases.

25 (a) $x = \cos t$, $y = \sin t$; $0 \leq t \leq \pi$

     (b) $x = \sin t$, $y = \cos t$; $0 \leq t \leq \pi$

     (c) $x = t$, $y = \sqrt{1 - t^2}$; $-1 \leq t \leq 1$

26 (a) $x = t^2$, $y = 1 - t^2$; $0 \leq t \leq 1$

     (b) $x = 1 - \ln t$, $y = \ln t$; $1 \leq t \leq e$

     (c) $x = \cos^2 t$, $y = \sin^2 t$; $0 \leq t \leq 2\pi$

27 Show that

$$x = a \cos t + h, \quad y = b \sin t + k; \quad 0 \leq t \leq 2\pi$$

are parametric equations of an ellipse with center (h, k) and axes of lengths $2a$ and $2b$.

28 Show that

$$x = a \sec t + h, \quad y = b \tan t + k;$$
$$-\pi/2 < t < 3\pi/2 \text{ and } t \neq \pi/2$$

are parametric equations of a hyperbola with center (h, k), transverse axis of length $2a$, and conjugate axis of length $2b$. Determine the values of t for each branch.

Exer. 29–30: (a) Find three parametrizations that give the same graph as the given equation. (b) Find three parametrizations that give only a portion of the graph of the given equation.

29 $y = x^2$

30 $y = \ln x$

31 Refer to Example 5.

     (a) Describe the Lissajous figure given by $f(t) = a \sin \omega t$ and $g(t) = b \cos \omega t$ for $t \geq 0$ and $a \neq b$.

     (b) Suppose $f(t) = a \sin \omega_1 t$ and $g(t) = b \sin \omega_2 t$, where ω_1 and ω_2 are positive rational numbers, and write ω_2/ω_1 as m/n for positive integers m and n. Show

that if $p = 2\pi n/\omega_1$, then both $f(t + p) = f(t)$ and $g(t + p) = g(t)$. Conclude that the curve retraces itself every p units of time.

32 Shown in the figure is the Lissajous figure given by

$$x = 2 \sin 3t, \quad y = 3 \sin 1.5t; \quad t \geq 0.$$

Find the period of the figure—that is, the length of the smallest t-interval that traces the curve.

Exercise 32

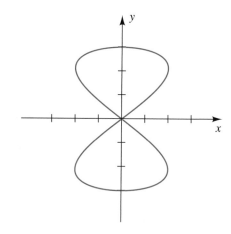

c **Exer. 33–36:** Lissajous figures are used in the study of electrical circuits to determine the phase difference ϕ between a known voltage $V_1(t) = A \sin (\omega t)$ and an unknown voltage $V_2(t) = B \sin (\omega t + \phi)$ having the same frequency. The voltages are graphed parametrically as $x = V_1(t)$ and $y = V_2(t)$. If ϕ is acute, then

$$\phi = \sin^{-1} \frac{y_{\text{int}}}{y_{\text{max}}},$$

where y_{int} is the nonnegative y-intercept and y_{max} is the maximum y-value on the curve.

(a) Graph the parametric curve $x = V_1(t)$ and $y = V_2(t)$ for the specified range of t.

(b) Use the graph to approximate ϕ in degrees.

33 $V_1(t) = 3 \sin (240\pi t)$, $V_2(t) = 4 \sin (240\pi t)$;
$$0 \leq t \leq 0.01$$

34 $V_1(t) = 6 \sin (120\pi t)$, $V_2(t) = 5 \cos (120\pi t)$;
$$0 \leq t \leq 0.02$$

35 $V_1(t) = 80 \sin (60\pi t)$, $V_2(t) = 70 \cos (60\pi t - \pi/3)$;
$$0 \leq t \leq 0.035$$

36 $V_1(t) = 163 \sin (120\pi t)$, $V_2(t) = 163 \sin (120\pi t + \pi/4)$;
$$0 \leq t \leq 0.02$$

c Exer. 37–38: Graph the Lissajous figure in the viewing rectangle [−1, 1] by [−1, 1] for the specified range of t.

37 $x(t) = \sin(6\pi t)$, $y(t) = \cos(5\pi t)$; $0 \le t \le 2$

38 $x(t) = \sin(4t)$, $y(t) = \sin(3t + \pi/6)$; $0 \le t \le 6.5$

39 A circle C of radius b rolls on the outside of the circle $x^2 + y^2 = a^2$, and $b < a$. Let P be a fixed point on C, and let the initial position of P be $A(a, 0)$, as shown in the figure. If the parameter t is the angle from the positive x-axis to the line segment from O to the center of C, show that parametric equations for the curve traced by P (an *epicycloid*) are

$$x = (a + b) \cos t - b \cos \left(\frac{a + b}{b} t \right),$$

$$y = (a + b) \sin t - b \sin \left(\frac{a + b}{b} t \right);$$

$$0 \le t \le 2\pi.$$

Exercise 39

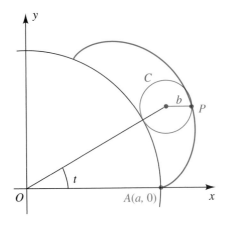

40 If the circle C of Exercise 39 rolls on the inside of the second circle (see the figure), then the curve traced by P is a *hypocycloid*.

(a) Show that parametric equations for this curve are

$$x = (a - b) \cos t + b \cos \left(\frac{a - b}{b} t \right),$$

$$y = (a - b) \sin t - b \sin \left(\frac{a - b}{b} t \right);$$

$$0 \le t \le 2\pi.$$

(b) If $b = \frac{1}{4}a$, show that $x = a \cos^3 t$, $y = a \sin^3 t$ and sketch the graph.

Exercise 40

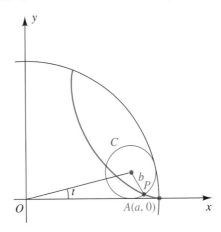

41 If $b = \frac{1}{3}a$ in Exercise 39, find parametric equations for the epicycloid and sketch the graph.

42 The radius of circle B is one-third that of circle A. How many revolutions will circle B make as it rolls around circle A until it reaches its starting point? (*Hint:* Use Exercise 41.)

c Exer. 43–46: Graph the curve.

43 $x = 3 \sin^5 t$, $y = 3 \cos^5 t$; $0 \le t \le 2\pi$

44 $x = 8 \cos t - 2 \cos 4t$,
$y = 8 \sin t - 2 \sin 4t$; $0 \le t \le 2\pi$

45 $x = 3t - 2 \sin t$, $y = 3 - 2 \cos t$; $-8 \le t \le 8$

46 $x = 2t - 3 \sin t$, $y = 2 - 3 \cos t$; $-8 \le t \le 8$

c Exer. 47–50: Graph the given curves on the same coordinate plane, and describe the shape of the resulting figure.

47 C_1: $x = 2 \sin 3t$, $y = 3 \cos 2t$; $-\pi/2 \le t \le \pi/2$
C_2: $x = \frac{1}{4} \cos t + \frac{3}{4}$, $y = \frac{1}{4} \sin t + \frac{3}{2}$; $0 \le t \le 2\pi$
C_3: $x = \frac{1}{4} \cos t - \frac{3}{4}$, $y = \frac{1}{4} \sin t + \frac{3}{2}$; $0 \le t \le 2\pi$
C_4: $x = \frac{3}{4} \cos t$, $y = \frac{1}{4} \sin t$; $0 \le t \le 2\pi$
C_5: $x = \frac{1}{4} \cos t$, $y = \frac{1}{8} \sin t + \frac{3}{4}$; $\pi \le t \le 2\pi$

48 C_1: $x = \frac{3}{2} \cos t + 1$, $y = \sin t - 1$; $-\pi/2 \le t \le \pi/2$
C_2: $x = \frac{3}{2} \cos t + 1$, $y = \sin t + 1$; $-\pi/2 \le t \le \pi/2$
C_3: $x = 1$, $y = 2 \tan t$; $-\pi/4 \le t \le \pi/4$

49 C_1: $x = \tan t$, $y = 3 \tan t$; $0 \le t \le \pi/4$
C_2: $x = 1 + \tan t$, $y = 3 - 3 \tan t$; $0 \le t \le \pi/4$
C_3: $x = \frac{1}{2} + \tan t$, $y = \frac{3}{2}$; $0 \le t \le \pi/4$

50 C_1: $x = 1 + \cos t$, $y = 1 + \sin t$; $\pi/3 \le t \le 2\pi$
C_2: $x = 1 + \tan t$, $y = 1$; $0 \le t \le \pi/4$

6.5 POLAR COORDINATES

In a rectangular coordinate system, the ordered pair (a, b) denotes the point whose directed distances from the x- and y-axes are b and a, respectively. Another method for representing points is to use *polar coordinates*. We begin with a fixed point O (the **origin,** or **pole**) and a directed half-line (the **polar axis**) with endpoint O. Next we consider any point P in the plane different from O. If, as illustrated in Figure 44, $r = d(O, P)$ and θ denotes the measure of any angle determined by the polar axis and OP, then r and θ are **polar coordinates** of P and the symbols (r, θ) or $P(r, \theta)$ are used to denote P. As usual, θ is considered positive if the angle is generated by a counterclockwise rotation of the polar axis and negative if the rotation is clockwise. Either radian or degree measure may be used for θ.

The polar coordinates of a point are not unique. For example, $(3, \pi/4)$, $(3, 9\pi/4)$, and $(3, -7\pi/4)$ all represent the same point (see Figure 45). We shall also allow r to be negative. In this case, instead of measuring $|r|$ units along the terminal side of the angle θ, we measure along the half-line with endpoint O that has direction *opposite* that of the terminal side. The points corresponding to the pairs $(-3, 5\pi/4)$ and $(-3, -3\pi/4)$ are also plotted in Figure 45.

Figure 44

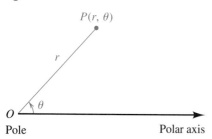

Figure 45

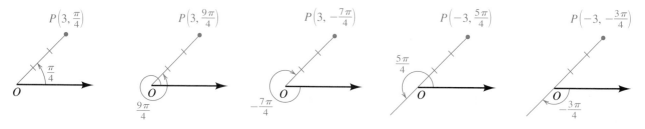

We agree that the pole O has polar coordinates $(0, \theta)$ for *any* θ. An assignment of ordered pairs of the form (r, θ) to points in a plane is a **polar coordinate system,** and the plane is an **$r\theta$-plane.**

A **polar equation** is an equation in r and θ. A **solution** of a polar equation is an ordered pair (a, b) that leads to equality if a is substituted for r and b for θ. The **graph** of a polar equation is the set of all points (in an $r\theta$-plane) that correspond to the solutions.

The simplest polar equations are $r = a$ and $\theta = a$, where a is a nonzero real number. Since the solutions of the polar equation $r = a$ are of the form (a, θ) for *any* angle θ, it follows that the graph is a circle of radius $|a|$ with center at the pole. A graph for $a > 0$ is sketched in Figure 46. The same graph is obtained for $r = -a$.

Figure 46

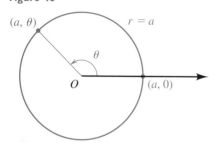

Figure 47

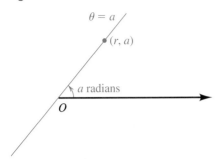

The solutions of the polar equation $\theta = a$ are of the form (r, a) for *any* real number r. Since the coordinate a (the angle) is constant, the graph of $\theta = a$ is a line through the origin, as illustrated in Figure 47 for an acute angle a.

In the following examples we obtain the graphs of polar equations by plotting points. As you proceed through this section, you should try to recognize forms of polar equations so that you will be able to sketch their graphs by plotting few, if any, points.

> **EXAMPLE 1**   Sketching the graph of a polar equation

Sketch the graph of the polar equation $r = 4 \sin \theta$.

Solution   The following table displays some solutions of the equation. We have included a third row in the table that contains one-decimal-place approximations to r.

θ	0	$\dfrac{\pi}{6}$	$\dfrac{\pi}{4}$	$\dfrac{\pi}{3}$	$\dfrac{\pi}{2}$	$\dfrac{2\pi}{3}$	$\dfrac{3\pi}{4}$	$\dfrac{5\pi}{6}$	π
r	0	2	$2\sqrt{2}$	$2\sqrt{3}$	4	$2\sqrt{3}$	$2\sqrt{2}$	2	0
r (approx.)	0	2	2.8	3.4	4	3.4	2.8	2	0

Figure 48

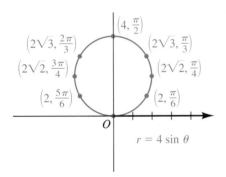

$r = 4 \sin \theta$

The points in an $r\theta$-plane that correspond to the pairs in the table appear to lie on a circle of radius 2, and we draw the graph accordingly (see Figure 48). As an aid to plotting points, we have extended the polar axis in the negative direction and introduced a vertical line through the pole (this line is the graph of the equation $\theta = \pi/2$).

The proof that the graph of $r = 4 \sin \theta$ is a circle is given in Example 8. Additional points obtained by letting θ vary from π to 2π lie on the same circle. For example, the solution $(-2, 7\pi/6)$ gives us the same point as $(2, \pi/6)$; the point corresponding to $(-2\sqrt{2}, 5\pi/4)$ is the same as that obtained from $(2\sqrt{2}, \pi/4)$; and so on. If we let θ increase through all real numbers, we obtain the same points again and again because of the periodicity of the sine function.

> **EXAMPLE 2**   Sketching the graph of a polar equation

Sketch the graph of the polar equation $r = 2 + 2 \cos \theta$.

Solution   Since the cosine function decreases from 1 to -1 as θ varies from 0 to π, it follows that r decreases from 4 to 0 in this θ-interval. The following table exhibits some solutions of $r = 2 + 2 \cos \theta$, together with one-decimal-place approximations to r.

(continued)

θ	0	$\dfrac{\pi}{6}$	$\dfrac{\pi}{4}$	$\dfrac{\pi}{3}$	$\dfrac{\pi}{2}$	$\dfrac{2\pi}{3}$	$\dfrac{3\pi}{4}$	$\dfrac{5\pi}{6}$	π
r	4	$2+\sqrt{3}$	$2+\sqrt{2}$	3	2	1	$2-\sqrt{2}$	$2-\sqrt{3}$	0
r (approx.)	4	3.7	3.4	3	2	1	0.6	0.3	0

Figure 49

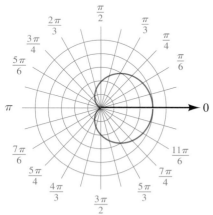

Plotting points in an $r\theta$-plane leads to the upper half of the graph sketched in Figure 49. (We have used polar coordinate graph paper, which displays lines through O at various angles and concentric circles with centers at the pole.)

If θ increases from π to 2π, then $\cos\theta$ increases from -1 to 1, and consequently r increases from 0 to 4. Plotting points for $\pi \le \theta \le 2\pi$ gives us the lower half of the graph.

The same graph may be obtained by taking other intervals of length 2π for θ.

The heart-shaped graph in Example 2 is a **cardioid.** In general, the graph of any of the polar equations in Figure 50, with $a \ne 0$, is a cardioid.

Figure 50

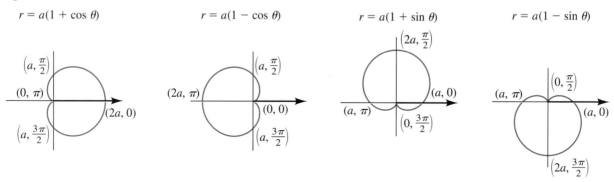

$r = a(1 + \cos\theta)$ $r = a(1 - \cos\theta)$ $r = a(1 + \sin\theta)$ $r = a(1 - \sin\theta)$

If a and b are not zero, then the graphs of the following polar equations are **limaçons:**

$$r = a + b\cos\theta \qquad r = a + b\sin\theta$$

Note that the special limaçons in which $|a| = |b|$ are cardioids.

Using the θ-interval $[0, 2\pi]$ is usually sufficient to graph polar equations. For equations with more complex graphs, it is often helpful to graph by using subintervals of $[0, 2\pi]$ that are determined by the θ-values that make $r = 0$—that is, the **pole values.** We will demonstrate this technique in the next example.

E X A M P L E 3   Sketching the graph of a polar equation

Sketch the graph of the polar equation $r = 2 + 4 \cos \theta$.

Solution   We first find the pole values by solving the equation $r = 0$:

$$2 + 4 \cos \theta = 0$$
$$\cos \theta = -\tfrac{1}{2}$$
$$\theta = \frac{2\pi}{3}, \frac{4\pi}{3}$$

We next construct a table of θ-values from 0 to 2π, using subintervals determined by the quadrantal angles and the pole values. The row numbers on the left-hand side correspond to the numbers in Figure 51.

Figure 51

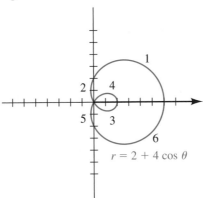

$r = 2 + 4 \cos \theta$

	θ	$\cos \theta$	$4 \cos \theta$	$r = 2 + 4 \cos \theta$
(1)	$0 \to \pi/2$	$1 \to 0$	$4 \to 0$	$6 \to 2$
(2)	$\pi/2 \to 2\pi/3$	$0 \to -1/2$	$0 \to -2$	$2 \to 0$
(3)	$2\pi/3 \to \pi$	$-1/2 \to -1$	$-2 \to -4$	$0 \to -2$
(4)	$\pi \to 4\pi/3$	$-1 \to -1/2$	$-4 \to -2$	$-2 \to 0$
(5)	$4\pi/3 \to 3\pi/2$	$-1/2 \to 0$	$-2 \to 0$	$0 \to 2$
(6)	$3\pi/2 \to 2\pi$	$0 \to 1$	$0 \to 4$	$2 \to 6$

You should verify the table entries with the figure, especially for rows 3 and 4 (in which the value of r is negative). The graph is called a limaçon with an inner loop.

The following chart summarizes the four categories of limaçons according to the ratio of a and b in the listed general equations.

Limaçons $a \pm b \cos \theta$, $a \pm b \sin \theta$ $(a > 0, b > 0)$

Name	Limaçon with an inner loop	Cardioid	Limaçon with a dimple	Convex limaçon
Condition	$\dfrac{a}{b} < 1$	$\dfrac{a}{b} = 1$	$1 < \dfrac{a}{b} < 2$	$\dfrac{a}{b} \geq 2$
Specific graph				
Specific equation	$r = 2 + 4 \cos \theta$	$r = 4 + 4 \cos \theta$	$r = 6 + 4 \cos \theta$	$r = 8 + 4 \cos \theta$

EXAMPLE 4   **Sketching the graph of a polar equation**

Sketch the graph of the polar equation $r = a \sin 2\theta$ for $a > 0$.

Solution   The following table contains θ-intervals and the corresponding values of r. The row numbers on the left-hand side correspond to the numbers in Figure 52.

Figure 52

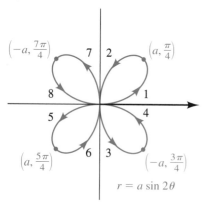

$r = a \sin 2\theta$

	θ	2θ	$\sin 2\theta$	$r = a \sin 2\theta$
(1)	$0 \rightarrow \pi/4$	$0 \rightarrow \pi/2$	$0 \rightarrow 1$	$0 \rightarrow a$
(2)	$\pi/4 \rightarrow \pi/2$	$\pi/2 \rightarrow \pi$	$1 \rightarrow 0$	$a \rightarrow 0$
(3)	$\pi/2 \rightarrow 3\pi/4$	$\pi \rightarrow 3\pi/2$	$0 \rightarrow -1$	$0 \rightarrow -a$
(4)	$3\pi/4 \rightarrow \pi$	$3\pi/2 \rightarrow 2\pi$	$-1 \rightarrow 0$	$-a \rightarrow 0$
(5)	$\pi \rightarrow 5\pi/4$	$2\pi \rightarrow 5\pi/2$	$0 \rightarrow 1$	$0 \rightarrow a$
(6)	$5\pi/4 \rightarrow 3\pi/2$	$5\pi/2 \rightarrow 3\pi$	$1 \rightarrow 0$	$a \rightarrow 0$
(7)	$3\pi/2 \rightarrow 7\pi/4$	$3\pi \rightarrow 7\pi/2$	$0 \rightarrow -1$	$0 \rightarrow -a$
(8)	$7\pi/4 \rightarrow 2\pi$	$7\pi/2 \rightarrow 4\pi$	$-1 \rightarrow 0$	$-a \rightarrow 0$

You should verify the table entries with the figure, especially for rows 3, 4, 7, and 8 (in which the value of r is negative).

The graph in Example 4 is a **four-leafed rose.** In general, a polar equation of the form

$$r = a \sin n\theta \qquad \text{or} \qquad r = a \cos n\theta$$

for any positive integer n greater than 1 and any nonzero real number a has a graph that consists of a number of loops through the origin. If n is even, there are $2n$ loops, and if n is odd, there are n loops.

The graph of the polar equation $r = a\theta$ for any nonzero real number a is a **spiral of Archimedes.** The case $a = 1$ is considered in the next example.

Figure 53

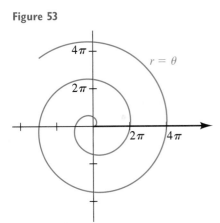

EXAMPLE 5   Sketching the graph of a spiral of Archimedes

Sketch the graph of the polar equation $r = \theta$ for $\theta \geq 0$.

Solution   The graph consists of all points that have polar coordinates of the form (c, c) for every real number $c \geq 0$. Thus, the graph contains the points $(0, 0)$, $(\pi/2, \pi/2)$, (π, π), and so on. As θ increases, r increases at the same rate, and the spiral winds around the origin in a counterclockwise direction, intersecting the polar axis at $0, 2\pi, 4\pi, \ldots$, as illustrated in Figure 53.

If θ is allowed to be negative, then as θ decreases through negative values, the resulting spiral winds around the origin and is the symmetric image, with respect to the vertical axis, of the curve sketched in Figure 53.

Let us next superimpose an xy-plane on an $r\theta$-plane so that the positive x-axis coincides with the polar axis. Any point P in the plane may then be assigned rectangular coordinates (x, y) or polar coordinates (r, θ). If $r > 0$, we have a situation similar to that illustrated in Figure 54(a); if $r < 0$, we have that shown in (b) of the figure. In Figure 54(b), for later purposes, we

Figure 54
(a) $r > 0$ (b) $r < 0$

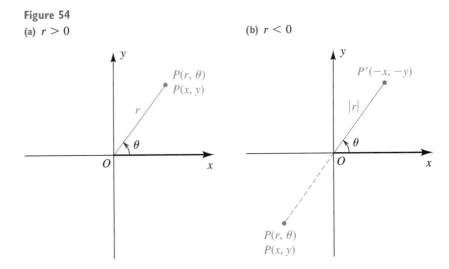

have also plotted the point P', having polar coordinates $(|r|, \theta)$ and rectangular coordinates $(-x, -y)$.

The following result specifies relationships between (x, y) and (r, θ), where it is assumed that the positive x-axis coincides with the polar axis.

Relationships Between Rectangular and Polar Coordinates

> The rectangular coordinates (x, y) and polar coordinates (r, θ) of a point P are related as follows:
>
> **(1)** $x = r \cos \theta, \quad y = r \sin \theta$
>
> **(2)** $r^2 = x^2 + y^2, \quad \tan \theta = \dfrac{y}{x}$ if $x \neq 0$

Proof   Although we have pictured θ as an acute angle in Figure 54, the discussion that follows is valid for all angles. If $r > 0$, as in Figure 54(a), then $\cos \theta = x/r$ and $\sin \theta = y/r$, and hence

$$x = r \cos \theta, \qquad y = r \sin \theta.$$

If $r < 0$, then $|r| = -r$, and from Figure 54(b) we see that

$$\cos \theta = \frac{-x}{|r|} = \frac{-x}{-r} = \frac{x}{r}, \qquad \sin \theta = \frac{-y}{|r|} = \frac{-y}{-r} = \frac{y}{r}.$$

Multiplication by r gives us relationship 1, and therefore these formulas hold if r is either positive or negative. If $r = 0$, then the point is the pole and we again see that the formulas in (1) are true.

The formulas in relationship 2 follow readily from Figure 54.   ◢

We may use the preceding result to change from one system of coordinates to the other.

Figure 55

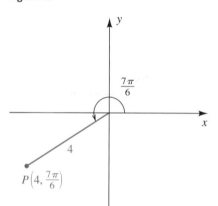

$P\left(4, \frac{7\pi}{6}\right)$

EXAMPLE 6   Changing polar coordinates to rectangular coordinates

If $(r, \theta) = (4, 7\pi/6)$ are polar coordinates of a point P, find the rectangular coordinates of P.

Solution   The point P is plotted in Figure 55. Substituting $r = 4$ and $\theta = 7\pi/6$ in relationship 1 of the preceding result, we obtain the following:

$$x = r \cos \theta = 4 \cos (7\pi/6) = 4(-\sqrt{3}/2) = -2\sqrt{3}$$
$$y = r \sin \theta = 4 \sin (7\pi/6) = 4(-1/2) = -2$$

Hence, the rectangular coordinates of P are $(x, y) = (-2\sqrt{3}, -2)$.

EXAMPLE 7   Changing rectangular coordinates
to polar coordinates

If $(x, y) = (-1, \sqrt{3})$ are rectangular coordinates of a point P, find three different pairs of polar coordinates (r, θ) for P.

Figure 56

(a)

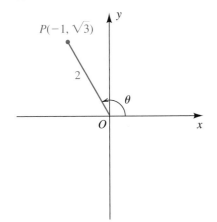

(b)

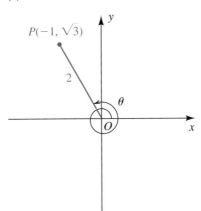

(c)

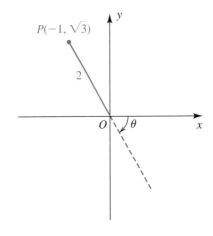

Solution   Three possibilities for θ are illustrated in Figure 56(a)–(c). Using $x = -1$ and $y = \sqrt{3}$ in relationship 2 of the previous result, we obtain

$$r^2 = x^2 + y^2 = (-1)^2 + (\sqrt{3})^2 = 4,$$

and since r is positive in Figure 56(a), $r = 2$. Using

$$\tan \theta = \frac{y}{x} = \frac{\sqrt{3}}{-1} = -\sqrt{3},$$

we see that the reference angle for θ is $\theta_R = \pi/3$, and hence

$$\theta = \pi - \frac{\pi}{3} = \frac{2\pi}{3}.$$

Thus, $(2, 2\pi/3)$ is one pair of polar coordinates for P.

Referring to Figure 56(b) and the values obtained for (a), we obtain

$$r = 2 \quad \text{and} \quad \theta = \frac{2\pi}{3} + 2\pi = \frac{8\pi}{3}.$$

Hence, $(2, 8\pi/3)$ is another pair of polar coordinates for P.

In Figure 56(c), $\theta = -\pi/3$. In this case we use $r = -2$ to obtain $(-2, -\pi/3)$ as a third pair of polar coordinates for P.

We may use the relationships between rectangular and polar coordinates to transform a polar equation to an equation in x and y, and vice versa. This procedure is illustrated in the next three examples.

Figure 57

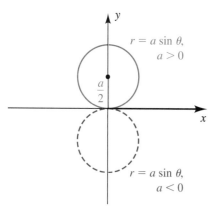

EXAMPLE 8   Changing a polar equation
to an equation in x and y

Find an equation in x and y that has the same graph as the polar equation $r = a \sin \theta$, with $a \neq 0$. Sketch the graph.

Solution   A formula that relates $\sin \theta$ and y is given by $y = r \sin \theta$. To introduce the expression $r \sin \theta$ into the equation $r = a \sin \theta$, we multiply both sides by r, obtaining

$$r^2 = ar \sin \theta.$$

Next, if we substitute $x^2 + y^2$ for r^2 and $r \sin \theta$ for y, the last equation becomes

$$x^2 + y^2 = ay,$$

or $$x^2 + y^2 - ay = 0.$$

Completing the square in y gives us

$$x^2 + y^2 - ay + \left(\frac{a}{2}\right)^2 = \left(\frac{a}{2}\right)^2,$$

or $$x^2 + \left(y - \frac{a}{2}\right)^2 = \left(\frac{a}{2}\right)^2.$$

In the xy-plane, the graph of the last equation is a circle with center $(0, a/2)$ and radius $|a|/2$, as illustrated in Figure 57 for the case $a > 0$ (the solid circle) and $a < 0$ (the dashed circle).

Figure 58

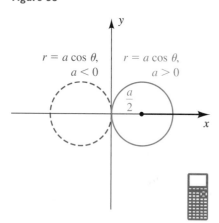

Using the same method as in the preceding example, we can show that the graph of $r = a \cos \theta$, with $a \neq 0$, is a circle of radius $a/2$ of the type illustrated in Figure 58.

EXAMPLE 9   Changing an equation in x and y
to a polar equation

(a) Find a polar equation for the hyperbola $x^2 - y^2 = 16$.
(b) Use the polar equation in part (a) and a graphing calculator to sketch the hyperbola.

Solution

(a) Using the formulas $x = r \cos \theta$ and $y = r \sin \theta$, we obtain the following polar equations:

$$(r \cos \theta)^2 - (r \sin \theta)^2 = 16 \qquad \text{substitute for } x \text{ and } y$$
$$r^2 \cos^2 \theta - r^2 \sin^2 \theta = 16 \qquad \text{square the terms}$$
$$r^2(\cos^2 \theta - \sin^2 \theta) = 16 \qquad \text{factor out } r^2$$
$$r^2 \cos 2\theta = 16 \qquad \text{double-angle formula}$$
$$r^2 = \frac{16}{\cos 2\theta} \qquad \text{divide by } \cos 2\theta$$

The division by $\cos 2\theta$ is allowable because $\cos 2\theta \neq 0$. (Note that if $\cos 2\theta = 0$, then $r^2 \cos 2\theta \neq 16$.) We may also write the polar equation as $r^2 = 16 \sec 2\theta$.

(b) Solving $r^2 = 16/\cos 2\theta$ for r gives us

$$r = \frac{\pm 4}{\sqrt{\cos 2\theta}}.$$

We assign $4/\sqrt{\cos 2\theta}$ to r_1 and $-r_1$ to r_2. Graphing r_1 and r_2 on the θ-interval $[0, \pi]$ leads to Figure 59. Using an increment of 0.05 for θ allows us to observe the orientation of the graph of the hyperbola.

Figure 59
$[-9, 9]$ by $[-6, 6]$

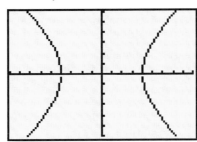

E X A M P L E   10   Finding a polar equation of a line

Find a polar equation of an arbitrary line.

Solution   Every line in an xy-coordinate plane is the graph of a linear equation that can be written in the form $ax + by = c$. Using the formulas $x = r \cos \theta$ and $y = r \sin \theta$ gives us the following equivalent polar equations:

$$ar \cos \theta + br \sin \theta = c \qquad \text{substitute for } x \text{ and } y$$
$$r(a \cos \theta + b \sin \theta) = c \qquad \text{factor out } r$$

If $a \cos \theta + b \sin \theta \neq 0$, the last equation may be written as follows:

$$r = \frac{c}{a \cos \theta + b \sin \theta}$$

If we superimpose an xy-plane on an $r\theta$-plane, then the graph of a polar equation may be symmetric with respect to the x-axis (the polar axis), the y-axis (the line $\theta = \pi/2$), or the origin (the pole). Some typical symmetries are illustrated in Figure 60. The next result summarizes these symmetries.

Figure 60   Symmetries of graphs of polar equations

(a) Polar axis   **(b)** Line $\theta = \pi/2$ **(c)** Pole

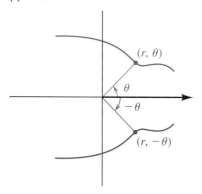 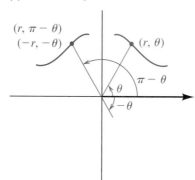 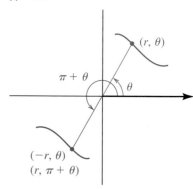

Tests for Symmetry

> **(1)** The graph of $r = f(\theta)$ is symmetric with respect to the polar axis if substitution of $-\theta$ for θ leads to an equivalent equation.
>
> **(2)** The graph of $r = f(\theta)$ is symmetric with respect to the vertical line $\theta = \pi/2$ if substitution of either (a) $\pi - \theta$ for θ or (b) $-r$ for r and $-\theta$ for θ leads to an equivalent equation.
>
> **(3)** The graph of $r = f(\theta)$ is symmetric with respect to the pole if substitution of either (a) $\pi + \theta$ for θ or (b) $-r$ for r leads to an equivalent equation.

To illustrate, since $\cos(-\theta) = \cos\theta$, the graph of the polar equation $r = 2 + 4\cos\theta$ in Example 3 is symmetric with respect to the polar axis, by test 1. Since $\sin(\pi - \theta) = \sin\theta$, the graph in Example 1 is symmetric with respect to the line $\theta = \pi/2$, by test 2. The graph of the four-leafed rose in Example 4 is symmetric with respect to the polar axis, the line $\theta = \pi/2$, and the pole. Other tests for symmetry may be stated; however, those we have listed are among the easiest to apply.

Unlike the graph of an equation in x and y, the graph of a polar equation $r = f(\theta)$ can be symmetric with respect to the polar axis, the line $\theta = \pi/2$, or the pole *without* satisfying one of the preceding tests for symmetry. This is true because of the many different ways of specifying a point in polar coordinates.

Another difference between rectangular and polar coordinate systems is that the points of intersection of two graphs cannot always be found by

Figure 61

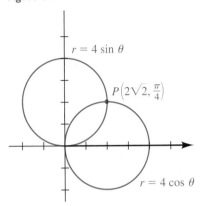

$r = 4 \sin \theta$

$P\left(2\sqrt{2}, \frac{\pi}{4}\right)$

$r = 4 \cos \theta$

solving the polar equations simultaneously. To illustrate, from Example 1, the graph of $r = 4 \sin \theta$ is a circle of diameter 4 with center at $(2, \pi/2)$ (see Figure 61). Similarly, the graph of $r = 4 \cos \theta$ is a circle of diameter 4 with center at $(2, 0)$ on the polar axis. Referring to Figure 61, we see that the coordinates of the point of intersection $P(2\sqrt{2}, \pi/4)$ in quadrant I satisfy both equations; however, the origin O, which is on each circle, *cannot* be found by solving the equations simultaneously. Thus, in searching for points of intersection of polar graphs, it is sometimes necessary to refer to the graphs themselves, *in addition* to solving the two equations simultaneously. An alternative method is to use different (equivalent) equations for the graphs.

6.5 EXERCISES

Exer. 1–34: Sketch the graph of the polar equation.

1 $r = 5$

2 $r = -2$

3 $\theta = -\pi/6$

4 $\theta = \pi/4$

5 $r = 3 \cos \theta$

6 $r = -2 \sin \theta$

7 $r = 4 \cos \theta + 2 \sin \theta$

8 $r = 6 \cos \theta - 2 \sin \theta$

9 $r = 4(1 - \sin \theta)$

10 $r = 3(1 + \cos \theta)$

11 $r = -6(1 + \cos \theta)$

12 $r = 2(1 + \sin \theta)$

13 $r = 2 + 4 \sin \theta$

14 $r = 1 + 2 \cos \theta$

15 $r = \sqrt{3} - 2 \sin \theta$

16 $r = 2\sqrt{3} - 4 \cos \theta$

17 $r = 2 - \cos \theta$

18 $r = 5 + 3 \sin \theta$

19 $r = 4 \csc \theta$

20 $r = -3 \sec \theta$

21 $r = 8 \cos 3\theta$

22 $r = 2 \sin 4\theta$

23 $r = 3 \sin 2\theta$

24 $r = 8 \cos 5\theta$

25 $r^2 = 4 \cos 2\theta$ (lemniscate)

26 $r^2 = -16 \sin 2\theta$

27 $r = 2^\theta, \theta \geq 0$ (spiral)

28 $r = e^{2\theta}, \theta \geq 0$ (logarithmic spiral)

29 $r = 2\theta, \theta \geq 0$

30 $r\theta = 1, \theta > 0$ (spiral)

31 $r = 6 \sin^2 (\theta/2)$

32 $r = -4 \cos^2 (\theta/2)$

33 $r = 2 + 2 \sec \theta$ (conchoid)

34 $r = 1 - \csc \theta$

Exer. 35–40: Change the polar coordinates to rectangular coordinates.

35 (a) $(3, \pi/4)$ (b) $(-1, 2\pi/3)$

36 (a) $(5, 5\pi/6)$ (b) $(-6, 7\pi/3)$

37 (a) $(8, -2\pi/3)$ (b) $(-3, 5\pi/3)$

38 (a) $(4, -\pi/4)$ (b) $(-2, 7\pi/6)$

39 $\left(6, \arctan \frac{3}{4}\right)$ 40 $\left(10, \arccos \left(-\frac{1}{3}\right)\right)$

Exer. 41–44: Change the rectangular coordinates to polar coordinates with $r > 0$ and $0 \leq \theta \leq 2\pi$.

41 (a) $(-1, 1)$ (b) $(-2\sqrt{3}, -2)$

42 (a) $(3\sqrt{3}, 3)$ (b) $(2, -2)$

43 (a) $(7, -7\sqrt{3})$ (b) $(5, 5)$

44 (a) $(-2\sqrt{2}, -2\sqrt{2})$ (b) $(-4, 4\sqrt{3})$

45 Which polar coordinates represent the same point as $(3, \pi/3)$?

(a) $(3, 7\pi/3)$ (b) $(3, -\pi/3)$ (c) $(-3, 4\pi/3)$

(d) $(3, -2\pi/3)$ (e) $(-3, -2\pi/3)$ (f) $(-3, -\pi/3)$

46 Which polar coordinates represent the same point as $(4, -\pi/2)$?

(a) $(4, 5\pi/2)$ (b) $(4, 7\pi/2)$ (c) $(-4, -\pi/2)$

(d) $(4, -5\pi/2)$ (e) $(-4, -3\pi/2)$ (f) $(-4, \pi/2)$

Exer. 47–56: Find a polar equation that has the same graph as the equation in x and y.

47 $x = -3$ **48** $y = 2$

49 $x^2 + y^2 = 16$ **50** $x^2 = 8y$

51 $2y = -x$ **52** $y = 6x$

53 $y^2 - x^2 = 4$ **54** $xy = 8$

55 $(x - 1)^2 + y^2 = 1$

56 $(x + 2)^2 + (y - 3)^2 = 13$

Exer. 57–74: Find an equation in x and y that has the same graph as the polar equation. Use it to help sketch the graph in an $r\theta$-plane.

57 $r \cos \theta = 5$ **58** $r \sin \theta = -2$

59 $r - 6 \sin \theta = 0$ **60** $r = 2$

61 $\theta = \pi/4$ **62** $r = 4 \sec \theta$

63 $r^2(4 \sin^2 \theta - 9 \cos^2 \theta) = 36$

64 $r^2(\cos^2 \theta + 4 \sin^2 \theta) = 16$

65 $r^2 \cos 2\theta = 1$ **66** $r^2 \sin 2\theta = 4$

67 $r(\sin \theta - 2 \cos \theta) = 6$

68 $r(3 \cos \theta - 4 \sin \theta) = 12$

69 $r(\sin \theta + r \cos^2 \theta) = 1$ **70** $r(r \sin^2 \theta - \cos \theta) = 3$

71 $r = 8 \sin \theta - 2 \cos \theta$ **72** $r = 2 \cos \theta - 4 \sin \theta$

73 $r = \tan \theta$ **74** $r = 6 \cot \theta$

75 If $P_1(r_1, \theta_1)$ and $P_2(r_2, \theta_2)$ are points in an $r\theta$-plane, use the law of cosines to prove that

$$[d(P_1, P_2)]^2 = r_1^2 + r_2^2 - 2r_1r_2 \cos (\theta_2 - \theta_1).$$

76 Prove that the graph of each polar equation is a circle, and find its center and radius.

(a) $r = a \sin \theta, a \neq 0$

(b) $r = b \cos \theta, b \neq 0$

(c) $r = a \sin \theta + b \cos \theta, a \neq 0$ and $b \neq 0$

Exer. 77–78: Refer to Exercise 81 in Section 2.7. Suppose that a radio station has two broadcasting towers located along a north-south line and that the towers are separated by a distance of $\frac{1}{2}\lambda$, where λ is the wavelength of the station's broadcasting signal. Then the intensity I of the signal in the direction θ can be expressed by the given equation, where I_0 is the maximum intensity of the signal.

(a) Plot I using polar coordinates with $I_0 = 5$ for $\theta \in [0, 2\pi]$.

(b) Determine the directions in which the radio signal has maximum and minimum intensity.

77 $I = \frac{1}{2}I_0[1 + \cos (\pi \sin \theta)]$

78 $I = \frac{1}{2}I_0[1 + \cos (\pi \sin 2\theta)]$

C **Exer. 79–80:** Graph the polar equation for the indicated values of θ, and use the graph to determine symmetries.

79 $r = 2 \sin^2 \theta \tan^2 \theta;$ $-\pi/3 \leq \theta \leq \pi/3$

(*Note:* Some specific keystrokes for the TI-82/83 are given in Example 28 of Appendix I.)

80 $r = \dfrac{4}{1 + \sin^2 \theta};$ $0 \leq \theta \leq 2\pi$

C **Exer. 81–82:** Graph the polar equations on the same coordinate plane, and estimate the points of intersection of the graphs.

81 $r = 8 \cos 3\theta,$ $r = 4 - 2.5 \cos \theta$

82 $r = 2 \sin^2 \theta,$ $r = \frac{3}{4}(\theta + \cos^2 \theta)$

6.6 POLAR EQUATIONS OF CONICS

The following theorem combines the definitions of parabola, ellipse, and hyperbola into a unified description of the conic sections. The constant e in the statement of the theorem is the **eccentricity** of the conic. The point F is a **focus** of the conic, and the line l is a **directrix**. Possible positions of F and l are illustrated in Figure 62.

Figure 62

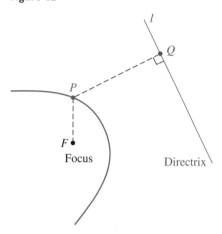

THEOREM ON CONICS

Let F be a fixed point and l a fixed line in a plane. The set of all points P in the plane, such that the ratio $d(P, F)/d(P, Q)$ is a positive constant e with $d(P, Q)$ the distance from P to l, is a conic section. The conic is a parabola if $e = 1$, an ellipse if $0 < e < 1$, and a hyperbola if $e > 1$.

Figure 63

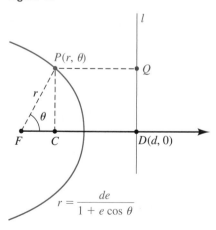

$$r = \frac{de}{1 + e\cos\theta}$$

Proof   If $e = 1$, then $d(P, F) = d(P, Q)$, and, by definition, the resulting conic is a parabola with focus F and directrix l.

Suppose next that $0 < e < 1$. It is convenient to introduce a polar coordinate system in the plane with F as the pole and l perpendicular to the polar axis at the point $D(d, 0)$, with $d > 0$, as illustrated in Figure 63. If $P(r, \theta)$ is a point in the plane such that $d(P, F)/d(P, Q) = e < 1$, then P lies to the left of l. Let C be the projection of P on the polar axis. Since

$$d(P, F) = r \qquad \text{and} \qquad d(P, Q) = d - r\cos\theta,$$

it follows that P satisfies the condition in the theorem if and only if the following are true:

$$\frac{r}{d - r\cos\theta} = e \quad ,$$

$$r = de - er\cos\theta$$

$$r(1 + e\cos\theta) = de$$

$$r = \frac{de}{1 + e\cos\theta}$$

The same equations are obtained if $e = 1$; however, there is no point (r, θ) on the graph if $1 + \cos\theta = 0$.

An equation in x and y corresponding to $r = de - er \cos \theta$ is

$$\sqrt{x^2 + y^2} = de - ex.$$

Squaring both sides and rearranging terms leads to

$$(1 - e^2)x^2 + 2de^2x + y^2 = d^2e^2.$$

Completing the square and simplifying, we obtain

$$\left(x + \frac{de^2}{1 - e^2}\right)^2 + \frac{y^2}{1 - e^2} = \frac{d^2e^2}{(1 - e^2)^2}.$$

Finally, dividing both sides by $d^2e^2/(1 - e^2)^2$ gives us an equation of the form

$$\frac{(x - h)^2}{a^2} + \frac{y^2}{b^2} = 1,$$

with $h = -de^2/(1 - e^2)$. Consequently, the graph is an ellipse with center at the point $(h, 0)$ on the x-axis and with

$$a^2 = \frac{d^2e^2}{(1 - e^2)^2} \qquad \text{and} \qquad b^2 = \frac{d^2e^2}{1 - e^2}.$$

Since

$$c^2 = a^2 - b^2 = \frac{d^2e^4}{(1 - e^2)^2},$$

we obtain $c = de^2/(1 - e^2)$, and hence $|h| = c$. This proves that F is a focus of the ellipse. It also follows that $e = c/a$. A similar proof may be given for the case $e > 1$.   ◀

We also can show that every conic that is not degenerate may be described by means of the statement in the theorem on conics. This gives us a formulation of conic sections that is equivalent to the one used previously. Since the theorem includes all three types of conics, it is sometimes regarded as a definition for the conic sections.

If we had chosen the focus F to the *right* of the directrix, as illustrated in Figure 64 (with $d > 0$), then the equation $r = de/(1 - e \cos \theta)$ would have resulted. (Note the minus sign in place of the plus sign.) Other sign changes occur if d is allowed to be negative.

If we had taken l *parallel* to the polar axis through one of the points $(d, \pi/2)$ or $(d, 3\pi/2)$, as illustrated in Figure 65, then the corresponding equations would have contained $\sin \theta$ instead of $\cos \theta$.

Figure 64

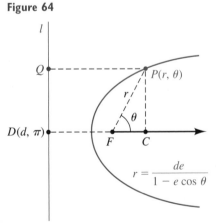

$$r = \frac{de}{1 - e \cos \theta}$$

Figure 65

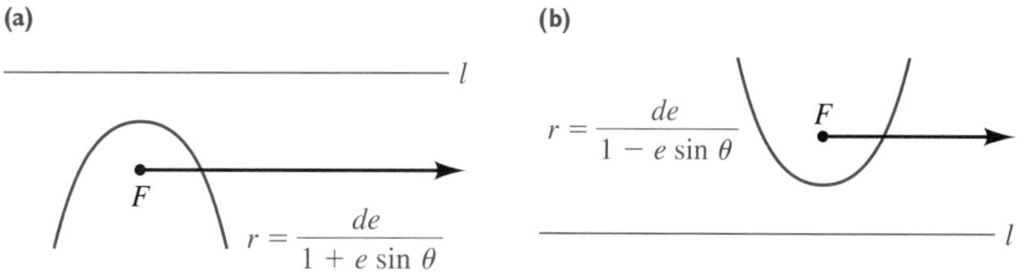

(a)

$$r = \frac{de}{1 + e \sin \theta}$$

(b)

$$r = \frac{de}{1 - e \sin \theta}$$

The following theorem summarizes our discussion.

THEOREM ON POLAR EQUATIONS OF CONICS

A polar equation that has one of the four forms

$$r = \frac{de}{1 \pm e \cos \theta} \quad \text{or} \quad r = \frac{de}{1 \pm e \sin \theta}$$

is a conic section. The conic is a parabola if $e = 1$, an ellipse if $0 < e < 1$, or a hyperbola if $e > 1$.

EXAMPLE 1   Sketching the graph of a polar equation of an ellipse

Describe and sketch the graph of the polar equation

$$r = \frac{10}{3 + 2 \cos \theta}.$$

Solution   We first divide the numerator and denominator of the fraction by 3 to obtain the constant term 1 in the denominator:

$$r = \frac{\frac{10}{3}}{1 + \frac{2}{3} \cos \theta}$$

This equation has one of the forms in the preceding theorem, with $e = \frac{2}{3}$. Thus, the graph is an ellipse with focus F at the pole and major axis along the polar axis. We find the endpoints of the major axis by letting $\theta = 0$ and $\theta = \pi$. This gives us the points $V(2, 0)$ and $V'(10, \pi)$. Hence,

$$2a = d(V', V) = 12, \quad \text{or} \quad a = 6.$$

The center of the ellipse is the midpoint $(4, \pi)$ of the segment $V'V$. Using the fact that $e = c/a$, we obtain

$$c = ae = 6\left(\tfrac{2}{3}\right) = 4.$$

Hence, $b^2 = a^2 - c^2 = 6^2 - 4^2 = 36 - 16 = 20.$

Thus, $b = \sqrt{20}$. The graph is sketched in Figure 66. For reference, we have superimposed an xy-coordinate system on the polar system.

Figure 66

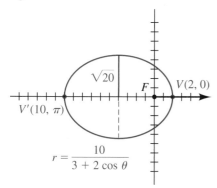

$$r = \frac{10}{3 + 2 \cos \theta}$$

EXAMPLE 2   Sketching the graph of a polar equation of a hyperbola

Describe and sketch the graph of the polar equation

$$r = \frac{10}{2 + 3 \sin \theta}.$$

Solution   To express the equation in the proper form, we divide the numerator and denominator of the fraction by 2:

$$r = \frac{5}{1 + \frac{3}{2} \sin \theta}$$

Thus, $e = \frac{3}{2}$, and, by the theorem on polar equations of conics, the graph is a hyperbola with a focus at the pole. The expression $\sin \theta$ tells us that the transverse axis of the hyperbola is perpendicular to the polar axis. To find the vertices, we let $\theta = \pi/2$ and $\theta = 3\pi/2$ in the given equation. This gives us the points $V(2, \pi/2)$ and $V'(-10, 3\pi/2)$. Hence,

$$2a = d(V, V') = 8, \quad \text{or} \quad a = 4.$$

The points $(5, 0)$ and $(5, \pi)$ on the graph can be used to sketch the lower branch of the hyperbola. The upper branch is obtained by symmetry, as illustrated in Figure 67. If we desire more accuracy or additional information, we calculate

$$c = ae = 4\left(\tfrac{3}{2}\right) = 6$$

and

$$b^2 = c^2 - a^2 = 6^2 - 4^2 = 36 - 16 = 20.$$

Asymptotes may then be constructed in the usual way.

Figure 67

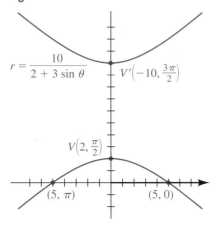

$$r = \frac{10}{2 + 3 \sin \theta}$$

$V'\left(-10, \frac{3\pi}{2}\right)$

$V\left(2, \frac{\pi}{2}\right)$

$(5, \pi)$ $(5, 0)$

EXAMPLE 3   Sketching the graph of a polar equation of a parabola

Sketch the graph of the polar equation

$$r = \frac{15}{4 - 4 \cos \theta}.$$

Solution   To obtain the proper form, we divide the numerator and denominator by 4:

$$r = \frac{\frac{15}{4}}{1 - \cos \theta}$$

Figure 68

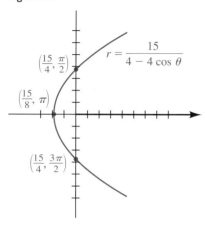

Consequently, $e = 1$, and, by the theorem on polar equations of conics, the graph is a parabola with focus at the pole. We may obtain a sketch by plotting the points that correspond to the quadrantal angles indicated in the following table.

θ	0	$\dfrac{\pi}{2}$	π	$\dfrac{3\pi}{2}$
r	undefined	$\frac{15}{4}$	$\frac{15}{8}$	$\frac{15}{4}$

Note that there is no point on the graph corresponding to $\theta = 0$, since the denominator $1 - \cos\theta$ is 0 for that value. Plotting the three points and using the fact that the graph is a parabola with focus at the pole gives us the sketch in Figure 68.

If we desire only a rough sketch of a conic, then the technique employed in Example 3 is recommended. To use this method, we plot (if possible) points corresponding to $\theta = 0$, $\pi/2$, π, and $3\pi/2$. These points, together with the type of conic (obtained from the value of the eccentricity e), readily lead to the sketch.

E X A M P L E 4   **Expressing a polar equation of a conic in terms of x and y**

Find an equation in x and y that has the same graph as the polar equation

$$r = \frac{15}{4 - 4\cos\theta}.$$

Solution

$$r(4 - 4\cos\theta) = 15 \qquad \text{multiply by the lcd}$$
$$4r - 4r\cos\theta = 15 \qquad \text{distribute}$$
$$4(\pm\sqrt{x^2 + y^2}) - 4x = 15 \qquad \text{substitute for } r \text{ and } r\cos\theta$$
$$4(\pm\sqrt{x^2 + y^2}) = 15 + 4x \qquad \text{isolate the radical term}$$
$$16(x^2 + y^2) = 225 + 120x + 16x^2 \qquad \text{square both sides}$$
$$16y^2 = 225 + 120x \qquad \text{simplify}$$

We may write the last equation as $x = \frac{16}{120}y^2 - \frac{225}{120}$ or, simplified, $x = \frac{2}{15}y^2 - \frac{15}{8}$. We recognize this equation as that of a parabola with vertex $V\left(-\frac{15}{8}, 0\right)$ and opening to the right. Its graph on an xy-coordinate system would be the same as the graph in Figure 68.

| EXAMPLE 5 | Finding a polar equation of a conic satisfying prescribed conditions |

Find a polar equation of the conic with a focus at the pole, eccentricity $e = \frac{1}{2}$, and directrix $r = -3 \sec \theta$.

Solution The equation $r = -3 \sec \theta$ of the directrix may be written $r \cos \theta = -3$, which is equivalent to $x = -3$ in a rectangular coordinate system. This gives us the situation illustrated in Figure 64, with $d = 3$. Hence, a polar equation has the form

$$r = \frac{de}{1 - e \cos \theta}.$$

We now substitute $d = 3$ and $e = \frac{1}{2}$:

$$r = \frac{3\left(\frac{1}{2}\right)}{1 - \frac{1}{2} \cos \theta} \qquad \text{or, equivalently,} \qquad r = \frac{3}{2 - \cos \theta}$$

6.6 EXERCISES

Exer. 1–12: Find the eccentricity, and classify the conic. Sketch the graph, and label the vertices.

1 $r = \dfrac{12}{6 + 2 \sin \theta}$ 2 $r = \dfrac{12}{6 - 2 \sin \theta}$

3 $r = \dfrac{12}{2 - 6 \cos \theta}$ 4 $r = \dfrac{12}{2 + 6 \cos \theta}$

5 $r = \dfrac{3}{2 + 2 \cos \theta}$ 6 $r = \dfrac{3}{2 - 2 \sin \theta}$

7 $r = \dfrac{4}{\cos \theta - 2}$ 8 $r = \dfrac{4 \sec \theta}{2 \sec \theta - 1}$

9 $r = \dfrac{6 \csc \theta}{2 \csc \theta + 3}$ 10 $r = \dfrac{8 \csc \theta}{2 \csc \theta - 5}$

11 $r = \dfrac{4 \csc \theta}{1 + \csc \theta}$

12 $r = \csc \theta (\csc \theta - \cot \theta)$

Exer. 13–24: Find equations in x and y for the polar equations in Exercises 1–12.

Exer. 25–32: Find a polar equation of the conic with focus at the pole that has the given eccentricity and equation of directrix.

25 $e = \frac{1}{3}$, $r = 2 \sec \theta$ 26 $e = 1$, $r \cos \theta = 5$

27 $e = \frac{4}{3}$, $r \cos \theta = -3$ 28 $e = 3$, $r = -4 \sec \theta$

29 $e = 1$, $r \sin \theta = -2$ 30 $e = 4$, $r = -3 \csc \theta$

31 $e = \frac{2}{5}$, $r = 4 \csc \theta$ 32 $e = \frac{3}{4}$, $r \sin \theta = 5$

Exer. 33–34: Find a polar equation of the parabola with focus at the pole and the given vertex.

33 $V\left(4, \dfrac{\pi}{2}\right)$ 34 $V(5, 0)$

Exer. 35–36: An ellipse has a focus at the pole with the given center C and vertex V. Find (a) the eccentricity and (b) a polar equation for the ellipse.

35 $C\left(3, \dfrac{\pi}{2}\right)$, $V\left(1, \dfrac{3\pi}{2}\right)$ 36 $C(2, \pi)$, $V(1, 0)$

37 Kepler's first law Kepler's first law asserts that planets travel in elliptical orbits with the sun at one focus. To find an equation of an orbit, place the pole O at the center of the sun and the polar axis along the major axis of the ellipse (see the figure).

(a) Show that an equation of the orbit is

$$r = \frac{(1 - e^2)a}{1 - e \cos \theta},$$

where e is the eccentricity and $2a$ is the length of the major axis.

(b) The perihelion distance r_{per} and aphelion distance r_{aph} are defined as the minimum and maximum distances, respectively, of a planet from the sun. Show that $r_{per} = a(1 - e)$ and $r_{aph} = a(1 + e)$.

Exercise 37

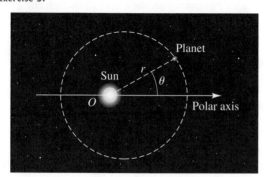

38 Kepler's first law Refer to Exercise 37. The planet Pluto travels in an elliptical orbit of eccentricity 0.249. If the perihelion distance is 29.62 AU, find a polar equation for the orbit and estimate the aphelion distance.

C **Exer. 39–42: Polar equations of conics can be used to describe the motion of comets. These paths can be graphed using the polar equation**

$$r = \frac{r_{per}(1 + e)}{1 - e \cos \theta},$$

where e is the eccentricity of the conic and r_{per} is the perihelion distance measured in AU.

(a) **For each comet, determine whether its trajectory is elliptical, parabolic, or hyperbolic.**

(b) **The orbit of Saturn has $r_{per} = 9.006$ and $e = 0.056$. Graph both the motion of the comet and the orbit of Saturn in the specified veiwing rectangle.**

39 Halley's Comet $r_{per} = 0.5871$, $e = 0.9673$,
$[-36, 36]$ by $[-24, 24]$

40 Encke's Comet $r_{per} = 0.3317$, $e = 0.8499$,
$[-18, 18]$ by $[-12, 12]$

41 Comet 1959 III $r_{per} = 1.251$, $e = 1.003$,
$[-18, 18]$ by $[-12, 12]$

42 Comet 1973.99 $r_{per} = 0.142$, $e = 1.000$,
$[-18, 18]$ by $[-12, 12]$

CHAPTER 6 REVIEW EXERCISES

Exer. 1–16: Find the vertices and foci of the conic, and sketch its graph.

1 $y^2 = 64x$

2 $y = 8x^2 + 32x + 33$

3 $9y^2 = 144 - 16x^2$

4 $9y^2 = 144 + 16x^2$

5 $x^2 - y^2 - 4 = 0$

6 $25x^2 + 36y^2 = 1$

7 $25y = 100 - x^2$

8 $3x^2 + 4y^2 - 18x + 8y + 19 = 0$

9 $x^2 - 9y^2 + 8x + 90y - 210 = 0$

10 $x = 2y^2 + 8y + 3$

11 $4x^2 + 9y^2 + 24x - 36y + 36 = 0$

12 $4x^2 - y^2 - 40x - 8y + 88 = 0$

13 $y^2 - 8x + 8y + 32 = 0$

14 $4x^2 + y^2 - 24x + 4y + 36 = 0$

15 $x^2 - 9y^2 + 8x + 7 = 0$

16 $y^2 - 2x^2 + 6y + 8x - 3 = 0$

Exer. 17–18: Find the standard equation of a parabola with a vertical axis that satisfies the given conditions.

17 x-intercepts -10 and -4, y-intercept 80

18 x-intercepts -11 and 3, passing through $(2, 39)$

Exer. 19–28: Find an equation for the conic that satisfies the given conditions.

19 Hyperbola, with vertices $V(0, \pm 7)$ and endpoints of conjugate axis $(\pm 3, 0)$

20 Parabola, with focus $F(-4, 0)$ and directrix $x = 4$

21 Parabola, with focus $F(0, -10)$ and directrix $y = 10$

22 Parabola, with vertex at the origin, symmetric to the x-axis, and passing through the point $(5, -1)$

23 Ellipse, with vertices $V(0, \pm 10)$ and foci $F(0, \pm 5)$

24 Hyperbola, with foci $F(\pm 10, 0)$ and vertices $V(\pm 5, 0)$

25 Hyperbola, with vertices $V(0, \pm 6)$ and asymptotes $y = \pm 9x$

26 Ellipse, with foci $F(\pm 2, 0)$ and passing through the point $(2, \sqrt{2})$

27 Ellipse, with eccentricity $\frac{2}{3}$ and endpoints of minor axis $(\pm 5, 0)$

28 Ellipse, with eccentricity $\frac{3}{4}$ and foci $F(\pm 12, 0)$

29 (a) Determine A so that the point $(2, -3)$ is on the conic $Ax^2 + 2y^2 = 4$.

(b) Is the conic an ellipse or a hyperbola?

30 If a square with sides parallel to the coordinate axes is inscribed in the ellipse $(x^2/a^2) + (y^2/b^2) = 1$, express the area A of the square in terms of a and b.

31 Find the standard equation of the circle that has center at the focus of the parabola $y = \frac{1}{8}x^2$ and passes through the origin.

32 *Focal length and angular velocity* A cylindrical container, partially filled with mercury, is rotated about its axis so that the angular speed of each cross section is ω radians/second. From physics, the function f, whose graph generates the inside surface of the mercury (see the figure), is given by

$$f(x) = \tfrac{1}{64}\omega^2 x^2 + k,$$

where k is a constant. Determine the angular speed ω that will result in a focal length of 2 feet.

Exercise 32

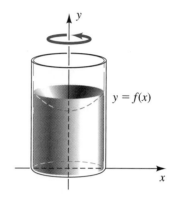

$y = f(x)$

Exer. 33–37: Find an equation in x and y whose graph contains the points on the curve C. Sketch the graph of C, and indicate the orientation.

33 $x = 3 + 4t,$ $\quad y = t - 1;$ $\quad -2 \le t \le 2$

34 $x = \sqrt{-t},$ $\quad y = t^2 - 4;$ $\quad t \le 0$

35 $x = \cos^2 t - 2,$ $\quad y = \sin t + 1;$ $\quad 0 \le t \le 2\pi$

36 $x = \sqrt{t},$ $\quad y = 2^{-t};$ $\quad t \ge 0$

37 $x = \dfrac{1}{t} + 1,$ $\quad y = \dfrac{2}{t} - t;$ $\quad 0 < t \le 4$

38 Curves C_1, C_2, C_3, and C_4 are given parametrically for t in $\mathbb{R}$. Sketch their graphs, and discuss their similarities and differences.

$C_1: x = t,$ $\qquad\qquad y = \sqrt{16 - t^2}$
$C_2: x = -\sqrt{16 - t},$ $\quad y = -\sqrt{t}$
$C_3: x = 4\cos t,$ $\qquad y = 4\sin t$
$C_4: x = e^t,$ $\qquad\qquad y = -\sqrt{16 - e^{2t}}$

39 Change $(5, 7\pi/4)$ to rectangular coordinates.

40 Change $(2\sqrt{3}, -2)$ to polar coordinates with $r > 0$ and $0 \le \theta < 2\pi$.

Exer. 41–52: Sketch the graph of the polar equation.

41 $r = -4\sin\theta$ $\qquad\qquad$ 42 $r = 8\sec\theta$

43 $r = 3\sin 5\theta$ $\qquad\qquad$ 44 $r = 6 - 3\cos\theta$

45 $r = 3 - 3\sin\theta$ $\qquad\qquad$ 46 $r = 2 + 4\cos\theta$

47 $r^2 = 9\sin 2\theta$ $\qquad\qquad$ 48 $2r = \theta$

49 $r = \dfrac{8}{1 - 3\sin\theta}$ $\qquad\qquad$ 50 $r = 6 - r\cos\theta$

51 $r = \dfrac{6}{3 + 2\cos\theta}$

52 $r = \dfrac{-6\csc\theta}{1 - 2\csc\theta}$

Exer. 53–56: Find a polar equation that has the same graph as the equation in x and y.

53 $y^2 = 4x$

54 $x^2 + y^2 - 3x + 4y = 0$

55 $2x - 3y = 8$

56 $x^2 + y^2 = 2xy$

Exer. 57–62: Find an equation in x and y that has the same graph as the polar equation.

57 $r^2 = \tan\theta$

58 $r = 2\cos\theta + 3\sin\theta$

59 $r^2 = 4\sin 2\theta$

60 $\theta = \sqrt{3}$

61 $r = 5\sec\theta + 3r\sec\theta$

62 $r^2 \sin\theta = 6\csc\theta + r\cot\theta$

CHAPTER 6   DISCUSSION   EXERCISES

1 On a parabola, the line segment through the focus, perpendicular to the axis, and intercepted by the parabola is called the *focal chord* or *latus rectum*. The length of the focal chord is called the *focal width*. Find a formula for the focal width w in terms of the focal length $|p|$.

2 On the graph of a hyperbola with center at the origin O, draw a circle with center at the origin and radius $r = d(O, F)$, where F denotes a focus of the hyperbola. What relationship do you observe?

3 A point $P(x, y)$ is on an ellipse if and only if

$$d(P, F) + d(P, F') = 2a.$$

If $b^2 = a^2 - c^2$, derive the general equation of an ellipse—that is,

$$\frac{x^2}{a^2} + \frac{y^2}{b^2} = 1.$$

4 A point $P(x, y)$ is on a hyperbola if and only if

$$\left| d(P, F) - d(P, F') \right| = 2a.$$

If $c^2 = a^2 + b^2$, derive the general equation of a hyperbola—that is,

$$\frac{x^2}{a^2} - \frac{y^2}{b^2} = 1.$$

5 A point $P(x, y)$ is the same distance from $(4, 0)$ as it is from the circle $x^2 + y^2 = 4$, as illustrated in the figure.

Show that the collection of all such points forms a branch of a hyperbola, and sketch its graph.

Exercise 5

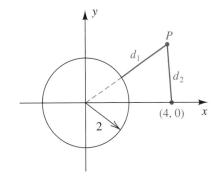

6 Sketch the graphs of $r = f(\theta) = 2 + 4\cos\theta$, $r = f(\theta - \alpha)$, and $r = f(\theta + \alpha)$ for $\alpha = \pi/4$. Try as many values of α as necessary to generalize results concerning the graphs of $r = f(\theta - \alpha)$ and $r = f(\theta + \alpha)$, where $\alpha > 0$.

7 *Generalized roses* Examine the graph of $r = \sin n\theta$ for odd values of n and even values of n. Derive an expression for the *leaf angle* (the number of degrees between consecutive pole values). What other generalizations do you observe? How do the graphs change if sin is replaced by cos?

USING A TI-82/83 GRAPHING CALCULATOR

This appendix provides instructions for performing some basic operations on a TI-82 or TI-83 graphing calculator, as well as for using some advanced features. Most of the examples in this appendix are taken from examples or exercises in the text, but the solutions presented are not the same as the solutions that appear in the text. The solutions here highlight TI-82/83 features and illustrate the specific keystrokes, whereas the solutions in the text are of a more general nature. For example, the reasons for selecting a particular viewing rectangle may be discussed more thoroughly in the text than in this appendix.

Working through this appendix is an excellent way to familiarize yourself with the TI-82/83 and how a graphing calculator is used in this text—thereby reducing the time needed for in-class instruction in using the graphing calculator. The *Guidebook* that comes with the calculator has a section entitled "Getting Started." Consulting that section first will help you become familiar with the keyboard and with calculating arithmetic expressions.

At first we will provide every keystroke needed to obtain the screens shown in the figures (the screens are computer-generated TI-82 facsimiles; the TI-83 screens are similar). As we progress, we will assume that displaying *all* keystroke sequences is not necessary. When different keystrokes are needed to perform an operation on the TI-83, the difference is noted in blue in the margin near the first occurrence and the affected keystrokes are printed in blue throughout. Some of the examples may use functions or concepts that are not used in this book or in your course; however, by working through them you will gain insight into the many capabilities of the graphing calculator.

The first example involves creating a table of values for an equation; it is similar to Exercises 90 and 91 in Section 1.1.

E X A M P L E 1   Creating a table of values

Of equations (1)–(4), choose the one that best describes the following table of data.

x	1	2	3	4	5
y	1	8	17	28	41

(1) $y = 7x - 6$

(2) $y = 27\sqrt{x} - 26$

(3) $y = x^2 + 4x - 4$

(4) $y = x^3 + x^2 - 3x + 2$

Solution   After pressing the [ON] key, assign the right sides of equations (1)–(4) to Y_1–Y_4, respectively, using the following keystroke sequence. (Note that the 7 in the Y_1 assignment represents a data entry, as opposed to menu choice [7].)

Make Y assignments

Y_1: [Y=] [CLEAR] 7 [X,T,θ] [1] [−] 6 [ENTER]

Y_2: [CLEAR] 27 [2nd] [√][2] [X,T,θ] [3] [−] 26 [ENTER]

Y_3: [CLEAR] [X,T,θ] [x^2] [+] 4 [X,T,θ] [−] 4 [ENTER]

Y_4: [CLEAR] [X,T,θ] [MATH] 3 [+] [X,T,θ] [x^2] [−] 3
[X,T,θ] [+] 2 [ENTER]

[1]**TI-83 Note:** The [X,T,θ] key is labeled [X,T,θ,n].

[2]**TI-83 Note:** The [√] key supplies an opening parenthesis, [(], for the radicand; you need to supply the closing parenthesis, [)]. Other keys such as [LOG] and [LN] operate in a similar manner.

[3]**TI-83 Note:** Press [)].

[4]**TI-83 Note:** TblStart is used in place of TblMin.

Figure 1(a) shows the current screen. To create a table, press [2nd] [TblSet]. Now assign 1 to TblMin[4] (the value of x that the table will begin with) by pressing 1 [ENTER]. Next, assign 1 to ΔTbl (the difference between consecutive values), as shown in Figure 1(b).

Figure 1

(a) Assignments

(b) TblSet values

(c) Values of x, Y_1, and Y_2

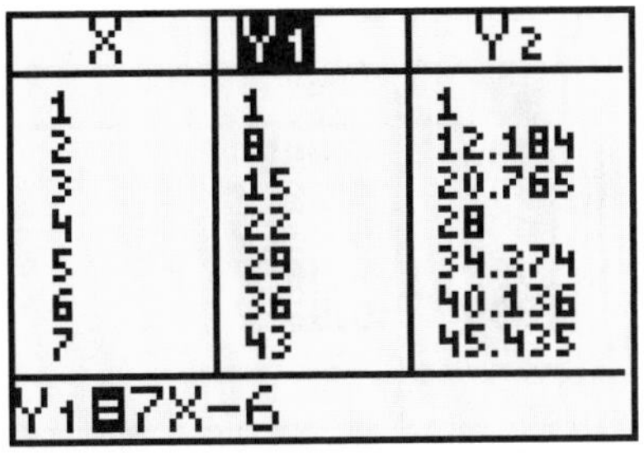

(d) Values of x, Y_3, and Y_4

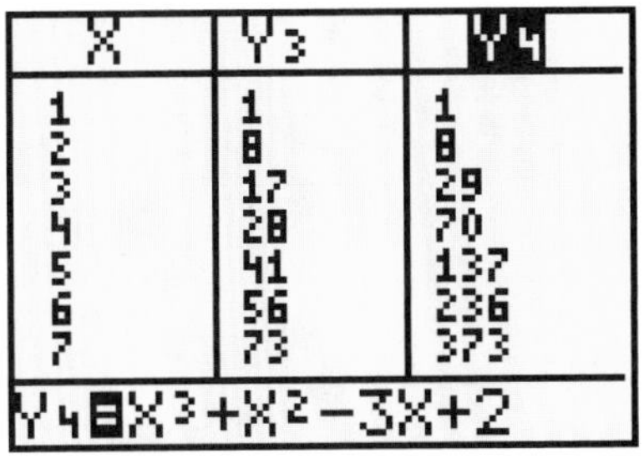

X	Y₁	Y₂
1	1	1
2	8	12.184
3	15	20.765
4	22	28
5	29	34.374
6	36	40.136
7	43	45.435

Y₁**⊟**7X-6

X	Y₃	Y₄
1	1	1
2	8	8
3	17	29
4	28	70
5	41	137
6	56	236
7	73	373

Y₄**⊟**X³+X²-3X+2

To see the values of x, Y_1, and Y_2, press $\boxed{\text{2nd}}$ $\boxed{\text{TABLE}}$ (see Figure 1(c)). Press the right arrow key $\boxed{\triangleright}$ and the up arrow key $\boxed{\triangle}$ to highlight the Y_1 column. Press $\boxed{\triangleright}$ three times to see the values of x, Y_3, and Y_4, as shown in Figure 1(d). The values in the Y_3 column are an exact match with the values given in the table, so equation (3) best describes the data.

The next example (Exercise 89(a) in Section 1.2) illustrates how to plot points and set viewing rectangle parameters.

EXAMPLE 2   **Plotting points**

The table lists the numbers of basic and pay cable television subscribers from 1990 to 1994.

Year	Subscribers
1990	54,871,330
1991	55,786,390
1992	57,211,600
1993	58,834,440
1994	59,332,200

Plot the data in the viewing rectangle [1988, 1996] by [54×10^6, 61×10^6].

Solution   Begin by clearing Lists 1 and 2 and then entering the years in List 1 and the numbers of subscribers in List 2. Clear Y_1–Y_4 as in Example 1.

Clear lists $\boxed{\text{STAT}}$ $\boxed{4}$ $\boxed{\text{2nd}}$ $\boxed{\text{L1}}$ $\boxed{,}$ $\boxed{\text{2nd}}$ $\boxed{\text{L2}}$ $\boxed{\text{ENTER}}$

(continued)

Figure 2

(a) Data in lists

(b) Viewing rectangle values

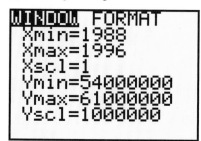

(c) Plot choices

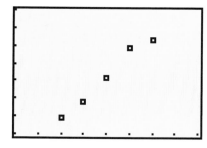

(d) Graph of data

[1988, 1996] by [54×10^6, 61×10^6]

Enter data

STAT 1 1990 ▽ (or use ENTER) 1991 ▽ 1992 ▽ 1993 ▽
1994 ▽ △ (5 times) ▷ 54871330 ▽ 55786390 ▽
57211600 ▽ 58834440 ▽ 59332200 ▽ (see Figure 2(a))

Select VR (viewing rectangle)

WINDOW ▽ [5] 1988 ▽ 1996 ▽ 1 ▽ 54 2nd EE 6 ▽ 61
2nd EE 6 ▽ 1 2nd EE 6 ENTER (see Figure 2(b))

[5]**TI-83 Note:** It is not necessary to press ▽ after pressing WINDOW .

(The terms Xmin, Xmax, etc., were discussed on page 27.)

Next, turn on Plot 1 and use the default choices for Type, Xlist, Ylist, and Mark.

Mark plot choices

2nd STAT PLOT 1 ENTER (see Figure 2(c))

Lastly, press GRAPH to obtain Figure 2(d). Note that pressing TRACE and using ◁ and ▷ displays the *x*- and *y*-values of the data points at the bottom of the screen. The P1 in the upper right-hand corner[6] stands for Plot 1.

[6]**TI-83 Note:** P1 is in the upper left-hand corner.

Before continuing, turn off Plot 1 by entering

2nd STAT PLOT 1 ▷ ENTER .

E X A M P L E 3   Sketching two simple graphs

Sketch the graphs of $y = x$ and $y = x^2$ on the same screen.

Solution   First make the assignments

$$Y_1 = x \quad \text{and} \quad Y_2 = x^2,$$

using the following keystroke sequence (clear Y_3 and Y_4):

Make Y assignments

Y_1: | Y= | | CLEAR | | X,T,θ | | ENTER |
Y_2: | CLEAR | | X,T,θ | | x^2 | | ENTER |

Figure 3(a) shows the current screen.

Figure 3

(a) Assignments

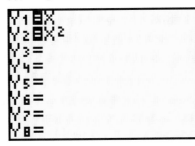

(b) Standard viewing rectangle

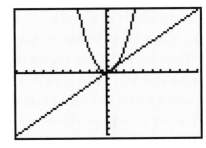

$[-10, 10]$ by $[-10, 10]$

(c) "Square" standard viewing rectangle

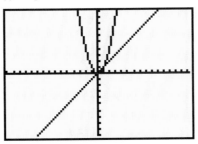

$[-15, 15]$ by $[-10, 10]$

(d) Viewing rectangle information

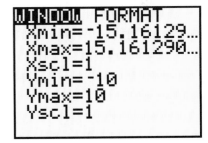

Next, choose a standard viewing rectangle by pressing | ZOOM | | 6 |; the resulting screen is shown in Figure 3(b). Note that the graph of $y = x$ does not appear to make a 45° angle from the positive x-axis (as it should). You can remedy this situation by pressing | ZOOM | | 5 |, which gives a screen that provides a true proportion; this screen is shown in Figure 3(c). (We will list the rounded dimensions of $[-15, 15]$ by $[-10, 10]$ rather than $[-15.2, 15.2]$ by $[-10, 10]$.) By pressing | WINDOW |, you can display the current viewing rectangle dimensions, as shown in Figure 3(d).

(continued)

Through the *value feature* you can easily find the value of Y_1 or Y_2 at $x = 3$ by entering 2nd CALC 1 3 ENTER △. By alternately pressing △ and ▽, you can see the values 3 for Y_1 and 9 for Y_2. Note that the upper right-hand corner of the screen displays the number 1 or 2, depending on whether the cursor is on Y_1 or Y_2, respectively.

We often need to determine the coordinates of a point to some specified accuracy. The next example illustrates one technique for doing so.

EXAMPLE 4   **Practicing the trace and zoom features**

Familiarize yourself with the trace and zoom features by finding the points of intersection of the graphs in Example 3.

Solution   Press GRAPH to get the last graphs drawn on the screen. Now press TRACE. Note the blinking cursor on your screen and its x- and y-coordinates at the bottom of the screen. Press ◁ until the cursor is close to (or on) (0, 0). The current screen is shown in Figure 4(a). Now press ZOOM 2 ENTER to *zoom in* on the position of the cursor (see Figure 4(b)). The actual viewing rectangle dimensions of Figure 4(b) depend on the original position of the cursor and the values of the zoom factors (see choice 4 under the MEMORY submenu of the ZOOM menu).

Figure 4

(a) After pressing TRACE   (b) After a zoom-in

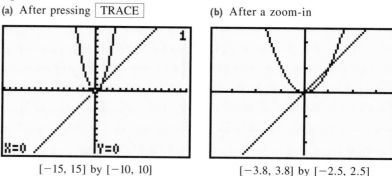

[−15, 15] by [−10, 10]   [−3.8, 3.8] by [−2.5, 2.5]

Having seen that (0, 0) is a likely candidate for a point of intersection, try to "zero in" on the other point of intersection. Press TRACE again, press ◁ or ▷ as many times as necessary to get to about $x = 0.5$, and then press ▽ and △, noting that the cursor jumps between the graphs of Y_1 and Y_2. Now press ◁ or ▷ to get closer to the point of intersection. You can get as close to the point as desired by repeating the process of alternately using the key sequence ZOOM 2 ENTER and pressing TRACE with ◁ or ▷. This process is referred to in the text as *using the zoom feature*. The other point of intersection is actually (1, 1).

We will now examine specific keystrokes and screens for some of the examples in the text. The next example is Example 10 in Section 1.2.

EXAMPLE 5 **Estimating points of intersection of graphs**

Estimate the points of intersection of the graphs of $y = 2x - 1$ and $y = x^2 - 3$.

Solution

Make Y assignments

Y_1: | Y= | | CLEAR | 2 | X,T,θ | | $-$ | 1 | $\triangledown$ |

Y_2: | CLEAR | | X,T,θ | | x^2 | | $-$ | 3 (see Figure 5(a))

Select VR

| ZOOM | | 6 | | ZOOM | | 5 | (see Figure 5(b))

Now try another approach to zooming in on a region—using a *box*.

Select box corners

| ZOOM | | 1 | | $\triangle$ (8 times) | | $\triangleright$ (5 times) | | ENTER | | $\triangleright$ (8 times) |
| $\triangle$ (8 times) | (see Figure 5(c))
| ENTER | (see Figure 5(d))

Figure 5

(a) Assignments

(b) Standard viewing rectangle

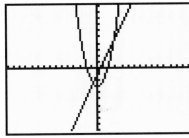

$[-15, 15]$ by $[-10, 10]$

(c) Selecting box corners

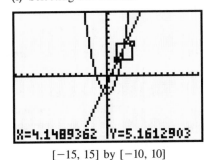

$[-15, 15]$ by $[-10, 10]$

(d) Enhanced view

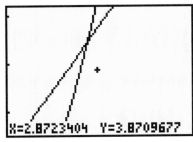

$[1.6, 4.2]$ by $[2.6, 5.2]$

You may now trace and zoom as shown in Example 4 or by using the box method. The approximate coordinates for the point of intersection in the first quadrant are $(2.7, 4.5)$. To determine coordinates for the point of intersection in the third quadrant, press | ZOOM | | 6 | and then proceed as before to obtain the approximate coordinates $(-0.7, -2.5)$.

In the following example we will rework Example 5 using a different feature on the TI-82/83.

E X A M P L E 6   **Estimating points of intersection of graphs using the intersect feature**

Estimate the points of intersection of the graphs of $y = 2x - 1$ and $y = x^2 - 3$.

Solution   Begin by making the same Y assignments and selecting a square viewing rectangle, as in Example 5. Now press [2nd] [CALC] [5] to obtain the intersect option. In replying to the "First curve?" inquiry, just press [ENTER] to indicate that graph 1 is the first curve. Similarly, press [ENTER] again to indicate that graph 2 is the second curve.

The "Guess?" prompt now appears on the screen and requires you to move the cursor close to a point of intersection. Press [▷] eight times to get close to the point of intersection in the first quadrant. Now press [ENTER] to obtain the approximate coordinates,

$$X = 2.7320508 \quad \text{and} \quad Y = 4.4641016.$$

In the next example we will rework Example 5 one more time, using yet another feature on the TI-82/83.

E X A M P L E 7   **Estimating points of intersection of graphs using the solve feature**

Estimate the points of intersection of the graphs of $y = 2x - 1$ and $y = x^2 - 3$.

Solution   This time use the solve feature on the TI-82/83. You need to enter one of the following expressions on the home screen:

$$\text{Solve } (expression, \ variable, \ guess)$$

or   Solve $(expression, \ variable, \ guess, \ \{lower, \ upper\})$

Use $Y_1 - Y_2$ for *expression*, since *expression* is assumed to be equal to 0. The *variable* is x, and a reasonable *guess* for x for the point of intersection in the first quadrant is 3. *Lower* and *upper* are optional (2 and 4 would be appropriate values if you were going to enter values).

Enter parameters

[2nd] [QUIT] [CLEAR] [MATH] [0] [7] [2nd] [Y-VARS] [8] [1] [1] [−]
[2nd] [Y-VARS] [1] [2] [,] [X,T,θ] [,] [3] [)] [ENTER]

After completing this computation, the TI-82/83 has the value of the x-coordinate of the solution (approximately 2.732) stored in ANS. To easily find the corresponding value of the y-coordinate of the solution, store ANS in X and query Y_1 (or Y_2).

[7]**TI-83 Note:** Use the key sequence [2nd] [CATALOG] [S] [▽] (24 times) [ENTER] for [MATH] [0].
[8]**TI-83 Note:** Use the key sequence [VARS] [▷] for [2nd] [Y-VARS] (here and in the next line).

Find y-coordinate | STO▷ | | X,T,θ | | ENTER | | 2nd | | Y-VARS | | 1 | | 1 | | ENTER |

You can see that the *y*-coordinate of the solution is approximately 4.464.

The next example (Example 11 in Section 1.2) demonstrates one method of graphing circles on the TI-82/83.

E X A M P L E 8   **Estimating points of intersection of graphs**

Estimate the points of intersection of the circles

$$x^2 + y^2 = 25 \qquad \text{and} \qquad x^2 + y^2 - 4y = 12.$$

Solution   Solving the equation $x^2 + y^2 = 25$ for *y* in terms of *x* gives us $y = \pm\sqrt{25 - x^2}$. Assign $\sqrt{25 - x^2}$ to Y_1 and $-Y_1$ to Y_2.

Make Y assignments Y_1: | Y= | | CLEAR | | 2nd | | √ | (25 | − | | X,T,θ | | x² |) ▽

                         Y_2: | CLEAR | | (−) | | 2nd | | Y-VARS | | 1 | | 1 | ▽

Note the difference between the subtraction key | − | and the negation key | (−) |. The subtraction key is used for operations such as $3 - 4$, whereas the negation key is used to enter an expression such as -2. Solving $x^2 + y^2 - 4y = 12$ for *y* in terms of *x* gives $y = 2 \pm \sqrt{16 - x^2}$. Assign $\sqrt{16 - x^2}$ to Y_3, $2 + Y_3$ to Y_4, and $2 - Y_3$ to Y_5.

Make Y assignments Y_3: | CLEAR | | 2nd | | √ | (16 | − | | X,T,θ | | x² |) ▽

                         Y_4: | CLEAR | 2 | + | | 2nd | | Y-VARS | | 1 | | 3 | ▽

                         Y_5: | CLEAR | 2 | − | | 2nd | | Y-VARS | | 1 | | 3 |

Turn off Y_3 | △ | | △ | | △ | | ◁ | | ENTER |

Note that the = sign by Y_3 is not shaded (see Figure 6(a)).

Graph Y_1, Y_2, Y_4, *and* Y_5 | ZOOM | | 6 | (oval shape) | ZOOM | | 5 | (circle shape) (see Figure 6(b))

Figure 6

(a) Assignments

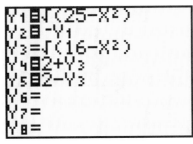

(b) Graph of Y_1, Y_2, Y_4, and Y_5

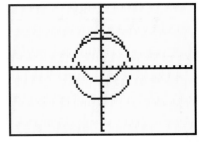

$[-15, 15]$ by $[-10, 10]$

(continued)

Now trace and zoom in on the point of intersection in the first quadrant; the point's coordinates are approximately (3.8, 3.25). Since both circles are symmetric with respect to the y-axis, the other point of intersection is approximately $(-3.8, 3.25)$. Alternatively, you could use the intersect feature (with Y_1 as the first curve and Y_4 as the second curve) or the command Solve $(Y_1 - Y_4, X, 4)$ to find the approximate x-coordinate, 3.79967.

We proceed to parts (a) and (b) of Example 11 in Section 1.3.

EXAMPLE 9   **Analyzing the graph of a function**

Let $f(x) = x^{2/3} - 3$. Sketch the graph of f, and find $f(-2)$.

[9]**TI-83 Note:** The entire graph is produced by the TI-83.
[10]**TI-83 Note:** Press $\boxed{)}$.

Solution   The assignment of $X^\wedge(2/3) - 3$ to Y_1 produces only the portion of the graph with $x \geq 0$.[9] Three ways to represent the function so as to obtain the entire graph of f are $\sqrt[3]{X^2} - 3$, $(X^\wedge(1/3))^2 - 3$, and $(X^\wedge 2)^\wedge(1/3) - 3$. The first representation will be used here to demonstrate use of the special cube root key. As a general rule, however, it is best to represent $x^{m/n}$ as $(X^\wedge(1/n))^\wedge m$. Be sure to clear (or turn off) Y_2–Y_5 after entering Y_1.

Make Y assignments Y_1: $\boxed{Y=}$ $\boxed{CLEAR}$ $\boxed{MATH}$ $\boxed{4}$ $\boxed{X,T,\theta}$ $\boxed{x^2}$[10] $\boxed{-}$ 3 (see Figure 7(a))

Select square VR $\boxed{ZOOM}$ $\boxed{6}$ $\boxed{ZOOM}$ $\boxed{5}$ (see Figure 7(b))

Find $f(-2)$ by storing -2 in memory location X and then querying the value of Y_1. Alternatively, you could use the value feature with $x = -2$ to find $f(-2)$.

Store -2 in X $\boxed{2nd}$ $\boxed{QUIT}$ $\boxed{CLEAR}$ -2 $\boxed{STO\triangleright}$ $\boxed{X,T,\theta}$ $\boxed{ENTER}$

Query Y_1 $\boxed{2nd}$ $\boxed{Y\text{-VARS}}$ $\boxed{1}$ $\boxed{1}$ $\boxed{ENTER}$

Figure 7

(a) Assignment

(b) Graph of f

(c) Finding the function value

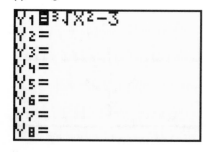

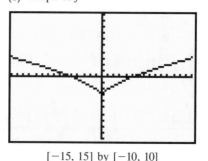

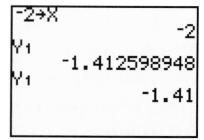

$[-15, 15]$ by $[-10, 10]$

Figure 7(c) shows the current screen and indicates that $f(-2) \approx -1.41$. To obtain exactly two digits after the decimal point, press the following key sequence:

Change precision　　| MODE | | ▽ | | ▷ | | ▷ | | ▷ | | ENTER | | 2nd | | QUIT |

You can now recall the last operation by pressing | 2nd | | ENTRY |; then press | ENTER | to obtain the value -1.41 (also shown in Figure 7(c)). Change the precision back to "Float" before proceeding by pressing | MODE | | ▽ | | ENTER |.

The next example shows how to examine a function of a function—that is, a **composite function.**

E X A M P L E 10　　**Graphically analyzing a composite function**

Let $f(x) = x^2 - 16$ and $g(x) = \sqrt{x}$. Sketch $y = (f \circ g)(x)$, and use the graph to find the domain of $f \circ g$. Find $f(g(3))$.

Solution　　Begin by making the assignment $Y_1 = \sqrt{x}$ and the assignment $Y_2 = (Y_1)^2 - 16$.

Make Y assignments　　Y_1: | Y= | | CLEAR | | 2nd | | √ | | X,T,θ |[11] | ENTER |

Y_2: | 2nd | | Y-VARS | | 1 | | 1 | | x^2 | | − | 16 | ENTER | (see Figure 8(a))

[11]**TI-83 Note:** Press |) |.

Since you want only the graph of Y_2, turn off Y_1 before graphing.

Turn off Y_1　　| △ | | △ | | ◁ | | ENTER | (done in Figure 8(a))

After pressing | ZOOM | | 5 | to obtain Figure 8(b), change to a different viewing rectangle to see a more complete picture of the graph of Y_2.

Change VR　　| WINDOW | | ▽ | −10 | ▽ | 50 | ▽ | 5 | ▽ | −20 | ▽ | 20 | ▽ | 5

(see Figure 8(c))

Figure 8

(a) Assignments

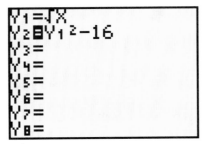

(b) Standard viewing rectangle

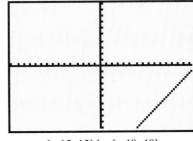

$[-15, 15]$ by $[-10, 10]$

(continued)

Figure 8 (continued)
(c) Changing window values

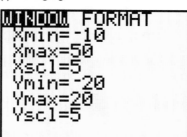

(d) Graph of $y = (f \circ g)(x)$

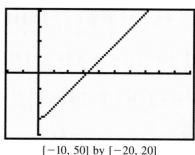

$[-10, 50]$ by $[-20, 20]$

Now press ⬚ GRAPH ⬚ to obtain Figure 8(d). The graph is a half-line with endpoint $(0, -16)$, indicating that the domain of Y_2 is all $x \geq 0$. Note that the viewing rectangle selected gives a 60×40 graph and maintains a true proportion, since the line appears to be at a $45°$ angle from the horizontal (a slope of 1). It is fairly easy to see that the x-intercept is between 15 and 20 (it is actually 16), as shown in Figure 8(d).

To find $f(g(3))$, use the value feature with $x = 3$. You could also store 3 in X and then query Y_2, obtaining -13.

Find $f(g(3))$ ⬚ 2nd ⬚ ⬚ CALC ⬚ ⬚ 1 ⬚ ⬚ 3 ⬚ ⬚ ENTER ⬚

The next example (using the function in Exercise 27 in Section 1.5) shows how to graph the inverse of a function using the TI-82/83 feature "DrawInv."

E X A M P L E   11   **Graphing the inverse of a function using the draw inverse feature**

Sketch the graph of the inverse of $f(x) = \sqrt{3 - x}$.

Solution   Begin by assigning f to Y_1. Be sure to clear the other Y values.

Make Y assignments Y_1: ⬚ Y= ⬚ ⬚ CLEAR ⬚ ⬚ 2nd ⬚ ⬚ √ ⬚ ⬚ (⬚ ⬚ 3 ⬚ ⬚ − ⬚ ⬚ X,T,θ ⬚ ⬚) ⬚ ⬚ ENTER ⬚

Graphing Y_1 with a standard viewing rectangle gives Figure 9(a). It appears that the domain of f is $x \leq 3$ and the range of f is $y \geq 0$. To graph the inverse of f without actually finding f^{-1}, use the inverse feature as follows:

Draw inverse   ⬚ 2nd ⬚ ⬚ DRAW ⬚ ⬚ 8 ⬚ ⬚ 2nd ⬚ ⬚ Y-VARS ⬚ ⬚ 1 ⬚ ⬚ 1 ⬚ ⬚ ENTER ⬚ (see Figure 9(b))

Figure 9

(a) Graph of f

(b) Graph of f and f^{-1}

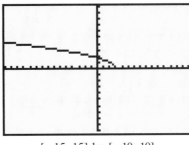

$[-15, 15]$ by $[-10, 10]$

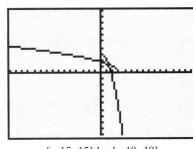

$[-15, 15]$ by $[-10, 10]$

The figure shows that the graph of f^{-1} is the reflection of the graph of f through the line $y = x$, the domain of f^{-1} is $x \geq 0$, and the range of f^{-1} is $y \leq 3$.

[12]**TI-83 Note:** The zero feature is analogous to the root feature.

The next example shows how to find the zeros of a function by using a TI-82 root feature.[12]

E X A M P L E 12   **Estimating zeros of a function using the root feature**

Estimate the zeros of $f(x) = x^3 + 0.2x^2 - 2.6x + 1.1$.

Solution   Begin by assigning f to Y_1.

Make Y assignment

Y_1: | Y= | | CLEAR | | X,T,θ | | MATH | | 3 | | + | .2 | X,T,θ | | x^2 |
| − | 2.6 | X,T,θ | | + | 1.1 | ENTER |

Graphing Y_1 using | ZOOM | | 6 | shows that all three zeros of f are between $x = -3$ and $x = 3$. Assign new window parameters of Xmin $= -3$, Xmax $= 3$, Xscl $= 1$, Ymin $= -2$, Ymax $= 2$, and Yscl $= 1$, and then press | GRAPH | to obtain Figure 10(a).

Figure 10

(a) Graph of Y_1

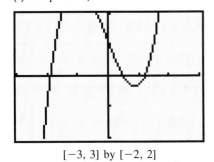

$[-3, 3]$ by $[-2, 2]$

(b) Lower bound estimate

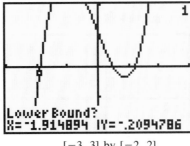

Lower Bound?
X=-1.914894 Y=-.2094786

$[-3, 3]$ by $[-2, 2]$

(c) Estimate of leftmost zero

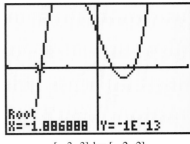

Root
X=-1.886888 Y=-1E-13

$[-3, 3]$ by $[-2, 2]$

(continued)

TI-83 Note: Left and right bound are analogous to lower and upper bound.

To activate the root feature, press $\boxed{\text{2nd}}$ $\boxed{\text{CALC}}$ $\boxed{2}$. To obtain a value for "Lower Bound?," [13] press $\boxed{\triangleleft}$ until the blinking cursor appears below the x-axis near $x = -2$ (see Figure 10(b)). Press $\boxed{\text{ENTER}}$; this assigns the value of X shown on the screen as the lower bound estimate. Now press $\boxed{\triangleright}$ until the cursor is above the x-axis. Again press $\boxed{\text{ENTER}}$ to assign the current value of X as the upper bound estimate. Continuing, press $\boxed{\triangleleft}$ so that the cursor is close to the x-axis (no specific placement is necessary) and then press $\boxed{\text{ENTER}}$—the x-value on the screen is the "Guess?" requested by the calculator.

[14]**TI-83 Note:** The value of y is 0.

Figure 10(c) shows that the leftmost zero is $x \approx -1.886888$ (note that the value of y is extremely small, but not zero[14]). Repeating the root procedure for the other two zeros yields the approximate values 0.48507595 and 1.2018125.

The next example involves a polynomial that has both small and large roots, making the selection of a viewing rectangle rather difficult. Since you should be familiar by now with assigning simple expressions to Y values and selecting various viewing rectangles, we omit some of the keystrokes.

E X A M P L E   13   **Exploring the graph of a polynomial**

Find the zeros of $f(x) = x^3 - 1000x^2 - x + 1000$.

Solution   Assign $x^3 - 1000x^2 - x + 1000$ to Y_1 and use a standard viewing rectangle to obtain Figure 11(a).

By using synthetic division, it can be determined that the roots are ± 1 and 1000. Some experimentation suggests selection of the viewing rectangle $[-200, 1100]$ by $[-20, 10]$ with Xscl = 100 and Yscl = 1. The graph in Figure 11(b) clearly shows the root at $x = 1000$ but obscures the roots at $x = \pm 1$.

Figure II

(a) Small view of f

(b) Larger view of f

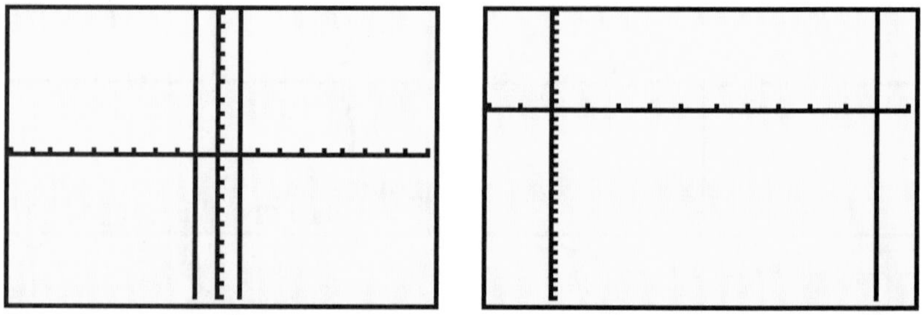

$[-10, 10]$ by $[-10, 10]$

$[-200, 1100]$ by $[-20, 10]$

(c) Assignments

(d) Compressed graph of f

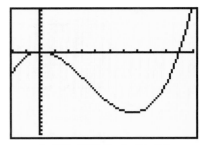

[-200, 1100] by [-20, 10]

The next step is to try to find a multiplier that will compress the graph of Y_1 so that the turning points (low and high points) are visible. Again, some trial and error reveals that 10^{-7} is an acceptable multiplier of Y_1. The keystroke sequence for Y_2 is as follows:

Make Y *assignment* Y_2:   Y= ▽ ▽ 10 ^ (-7) 2nd Y-VARS 1 1

Turn off Y_1 △ △ ◁ ENTER (see Figure 11(c))

Now graph Y_2 to obtain Figure 11(d). Experiment by changing the exponent on the multiplying factor, keeping in mind that Figure 11(d) is the graph of f with a vertical compression factor of 10^7 (ten million!).

The next example shows how to find the low and high points of a graph by using the TI-82/83 features "fMin" and "fMax."

EXAMPLE 14   **Finding low and high points using the fMin and fMax features**

Find the low and high points of $y = x - x^3$ on the interval [-1, 1].

Solution   Begin by assigning $x - x^3$ to Y_1 and setting Xmin $= -1$, Xmax $= 1$, Xscl $= 0.1$, Ymin $= -1$, Ymax $= 1$, and Yscl $= 0.1$. Graphing at this time gives Figure 12(a). The figure shows a low point to the left of the y-axis (between $x = -1$ and $x = 0$) and a high point to the right of the y-axis (between $x = 0$ and $x = 1$). To find the low point, enter fMin(Y_1, X, -1, 0) on the home screen. The specific keystrokes are as follows:

Using fMin   2nd QUIT CLEAR MATH 6 2nd Y-VARS 1 1 ,
X,T,θ , -1 , 0) ENTER

The result that is returned is the x-value of the low point—in this case, $x \approx -0.577$. If you store this value in X and then query Y_1, you find that the associated y-value is approximately -0.385 (see Figure 12(b)). You can find the coordinates of the high point in a similar fashion, using

(continued)

Figure 12

(a) Graph of Y_1

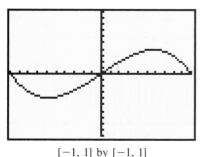

[−1, 1] by [−1, 1]

(b) Using fMin to find a low point

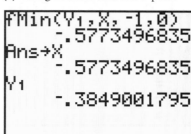

(c) Using fMax to find a high point

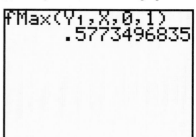

fMax(Y_1, X, 0, 1) (choice 7 on the MATH menu) as shown in Figure 12(c). Since f is an odd function, the approximate coordinates (0.577, 0.385) of the high point are simply the negatives of those of the low point.

The next example illustrates a method for graphing rational functions and finding equations of the vertical asymptotes.

E X A M P L E   15   Sketching the graph of a rational function

Sketch the graph of

$$f(x) = \frac{x^2 - x}{9x^3 - 9x^2 - 22x + 8},$$

and find equations of the vertical asymptotes.

Solution   Begin by making the assignments

$$Y_1 = x^2 - x, \qquad Y_2 = 9x^3 - 9x^2 - 22x + 8, \qquad \text{and} \qquad Y_3 = Y_1/Y_2.$$

Now turn off Y_1 and Y_2 (see Figure 13(a)), and graph Y_3 using ZOOM 6 . This graph gives virtually no indication of the true shape of f, so change the viewing rectangle to [−2, 3] by [−1, 1] and regraph Y_3 to obtain Figure 13(b). Note the near-vertical lines on the screen. These lines result from the calculator's trying to draw a line from one pixel on the screen to another. To eliminate the near-vertical lines, change to *dot mode,* as follows.

Select Dot   MODE ▽ ▽ ▽ ▽ ▷ ENTER

Press GRAPH to redraw Y_3. Figure 13(c) gives a reasonable view of the graph of f and indicates that there are three vertical asymptotes. Since there is no point to zoom in on in order to find an equation of the vertical asymptotes, abandon the graph of Y_3 and examine the graph of Y_2 (the denominator), looking for its zeros. Graphing only Y_2 (turn off Y_3) with the same viewing rectangle, but using *connected mode,* gives Figure 13(d).

Figure 13

(a) Assignments

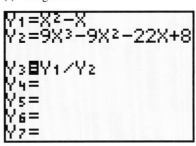

(b) Connected-mode view of f

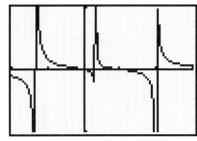

$[-2, 3]$ by $[-1, 1]$

(c) Dot-mode view of f

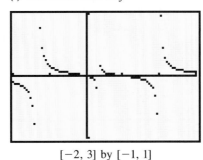

$[-2, 3]$ by $[-1, 1]$

(d) Graph of the denominator

$[-2, 3]$ by $[-1, 1]$

The zeros in Figure 13(d) have the approximate values $-1.33, 0.33$, and 2.00. From algebra, the rational root theorem indicates that the zeros may be $-\frac{4}{3}, \frac{1}{3}$, and 2. By storing each of these values in X and querying Y_2 (obtaining 0 each time), you can make reasonably sure that these are the zeros of Y_2. Since none of these zeros is a zero of the numerator, you have the following equations of the vertical asymptotes for f:

$$x = -\tfrac{4}{3}, \qquad x = \tfrac{1}{3}, \qquad \text{and} \qquad x = 2$$

In the next example (Example 6 in Section 6.1) we demonstrate how to use the TI-82/83 to sketch the graph of a parabola that has a horizontal axis.

EXAMPLE 16 Graphing half-parabolas

Sketch the graph of $x = y^2 + 2y - 4$.

Solution Begin by solving the equivalent equation

$$y^2 + 2y - 4 - x = 0$$

for y in terms of x by using the quadratic formula with $a = 1$, $b = 2$, and $c = -4 - x$, obtaining $y = -1 \pm \sqrt{x + 5}$. The last equation represents the top and bottom halves of the parabola in a function form.

(continued)

Next, make the assignments

$$Y_1 = \sqrt{x + 5}, \qquad Y_2 = -1 + Y_1, \qquad \text{and} \qquad Y_3 = -1 - Y_1.$$

Make Y assignments

Y_1:
Y= | CLEAR | 2nd | √ | (| X,T,θ | + | 5 |) | ENTER

Y_2:
CLEAR | −1 | + | 2nd | Y-VARS | 1 | 1 | ENTER

Y_3:
CLEAR | −1 | − | 2nd | Y-VARS | 1 | 1 | (see Figure 14(a))

Turn off Y_1, choose a standard viewing rectangle, and graph Y_2 and Y_3 to obtain Figure 14(b).

Figure 14

(a) Assignments

(b) Graph of the parabola

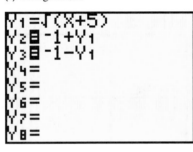

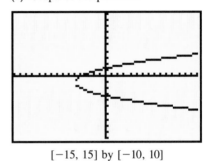

$[-15, 15]$ by $[-10, 10]$

The TI-82/83 has an interesting process for handling any statements involving the relational operators ($=, \neq, >, \geq, <, \leq$), which are found by pressing 2nd TEST . To illustrate this process, we will examine the statements

$$5 > 3, \qquad 3 = 2, \qquad \text{and} \qquad 2 \leq 2.$$

Figure 15

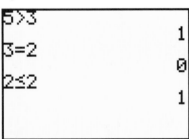

These statements and their results are shown in Figure 15. The result of 1 in the second line of Figure 15 indicates that the statement $5 > 3$ is true, whereas the 0 in the fourth line indicates that the statement $3 = 2$ is false. We can also compare function values—say, $Y_1 > Y_2$—and then graph the result. This procedure (referred to as the *Boolean method*) is demonstrated in the next example.

Example 17 (Example 6 in Section 5.2) involves the graph of a function in which the natural exponential function plays a key role.

EXAMPLE 17   Sketching a Gompertz growth curve

Sketch the graph of the Gompertz growth function G, given by

$$G(t) = ke^{(-Ae^{-Bt})},$$

on the interval $[0, 5]$ for $k = 1.1$, $A = 3.2$, and $B = 1.1$, and estimate the time t (to three decimal places) at which $G(t) = 1$.

Figure 16

(a) Assignment

(b) Graph of G

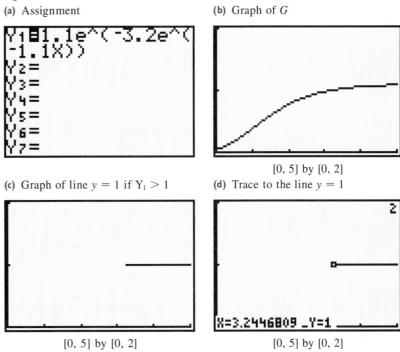

[0, 5] by [0, 2]

(c) Graph of line $y = 1$ if $Y_1 > 1$

(d) Trace to the line $y = 1$

[0, 5] by [0, 2]

[0, 5] by [0, 2]

Solution   Begin by making the assignment (using x for t)

$$Y_1 = 1.1e^{(-3.2e^{-1.1t})}.$$

Make Y assignment Y_1: $\boxed{Y=}$ $\boxed{\text{CLEAR}}$ 1.1 $\boxed{\text{2nd}}$ $\boxed{e^x}$ $\boxed{(}$ -3.2 $\boxed{\text{2nd}}$ $\boxed{e^x}$
$\boxed{(}$ -1.1 $\boxed{\text{X,T,}\theta}$ $\boxed{)}$ $\boxed{)}$ $\boxed{\text{ENTER}}$ (see Figure 16(a))

From the discussion on page 331 you know that the maximum value of G is 1.1, so choose the viewing rectangle [0, 5] by [0, 2]. Graphing G gives Figure 16(b). The endpoint values of the graph are approximately (0, 0.045) and (5, 1.086).

To determine the time when $y = G(t) = 1$, use the Boolean method. First change to dot mode and then assign $Y_1 > 1$ to Y_2.

Make Y assignment Y_2: $\boxed{Y=}$ $\boxed{\triangledown}$ $\boxed{\triangledown}$ $\boxed{\text{2nd}}$ $\boxed{\text{Y-VARS}}$ $\boxed{1}$ $\boxed{1}$ $\boxed{\text{2nd}}$ $\boxed{\text{TEST}}$ $\boxed{3}$ 1

Thus, when $Y_1 \leq 1$, the graph of Y_2 will be the horizontal line $y = 0$, and when $Y_1 > 1$, the graph of Y_2 will be the horizontal line $y = 1$. Turning off Y_1 and graphing Y_2 gives Figure 16(c). Press $\boxed{\text{TRACE}}$ and note that the displayed y-coordinate is 0. Press $\boxed{\triangleright}$ successively until the y-coordinate becomes 1 (see Figure 16(d)). Now press $\boxed{\text{ZOOM}}$ $\boxed{2}$ $\boxed{\text{ENTER}}$ and repeat the trace and zoom procedure until you obtain the desired three-decimal-place accuracy for $x = t \approx 3.194$.

(continued)

The Boolean method is presented merely as an alternative method of solution. It may be most useful in situations in which it is difficult to determine how many times one graph intersects another.

In the next example (Exercise 44 in Section 5.2), we will use the table feature to compare some function values.

E X A M P L E   18   **Comparing function values using the table feature**

Compare the accuracy of

$$f(x) = \tfrac{1}{2}x^2 + x + 1 \qquad \text{and} \qquad g(x) = 0.84x^2 + 0.878x + 1$$

as an approximation to e^x on the interval [0, 1].

Solution   Begin by assigning $f(x)$ to Y_1, $g(x)$ to Y_2, and e^x to Y_3. Set the window parameters as Xmin = 0, Xmax = 1, Xscl = 0.1, Ymin = 1, Ymax = 3, and Yscl = 0.1. Now graph all three functions, as shown in Figure 17(a). All three graphs nearly coincide on [0, 0.5], but Y_2 is a much better approximation to e^x on the interval [0.5, 1] than Y_1.

To determine the better approximation on the interval [0, 0.5], make use of a table.

Table setup   [2nd] [TblSet] [0] [▽] .05

Turn on Auto for both Indpnt and Depend. Turn off Y_2 and press [2nd] [TABLE] to obtain Figure 17(b). The table shows that Y_1 closely approximates Y_3. Now turn off Y_1, turn on Y_2, and press [2nd] [TABLE] to obtain Figure 17(c).

Although the values of Y_2 are close to Y_3, they are not as close as those of Y_1 if x is near 0. Hence, Y_1 is the better approximation to e^x for x near 0, but Y_2 is better for x near 1.

Figure 17

(a) Graph of Y_1, Y_2, and Y_3

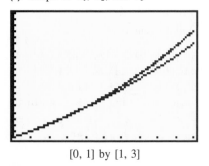

[0, 1] by [1, 3]

(b) Table values for Y_1 and Y_3

X	Y1	Y3
0	1	1
.05	1.0513	1.0513
.1	1.105	1.1052
.15	1.1613	1.1618
.2	1.22	1.2214
.25	1.2813	1.284
.3	1.345	1.3499

X=0

(c) Table values for Y_2 and Y_3

X	Y2	Y3
0	1	1
.05	1.046	1.0513
.1	1.0962	1.1052
.15	1.1506	1.1618
.2	1.2092	1.2214
.25	1.272	1.284
.3	1.339	1.3499

Y2■.84X²+.878X+1

The following example (Exercise 73 in Section 5.3) is a good illustration of the power of a graphing utility, since it is impossible to find the exact solution using only algebraic methods.

E X A M P L E   19   **Approximating a solution to an inequality**

Graph $f(x) = 2.2 \log (x + 2)$ and $g(x) = \ln x$, and estimate the solution of the inequality $f(x) \geq g(x)$.

Solution   Begin by making the assignments

$$Y_1 = 2.2 \log (x + 2) \qquad \text{and} \qquad Y_2 = \ln x.$$

Make Y assignments Y_1: | Y= | CLEAR | 2.2 | LOG | (| X,T,θ | + | 2 |) | ENTER |

[15]**TI-83 Note:** Press |) | . Y_2: | CLEAR | LN | X,T,θ |[15] | ENTER | (see Figure 18(a))

Since the domain of f is $(-2, \infty)$ and the domain of g is $(0, \infty)$, select Xmin $= -2$. Choosing the viewing rectangle $[-2, 16]$ by $[-4, 8]$ with Xscl $=$ Yscl $= 1$ and graphing Y_1 and Y_2 in connected mode gives Figure 18(b).

It is difficult to see whether the graph of f intersects the graph of g in Figure 18(b), so apply the Boolean method to help find any intersection points. Turn off Y_1 and Y_2 and assign $Y_1 \geq Y_2$ to Y_3.

Make Y assignment Y_3: | 2nd | Y-VARS | 1 | 1 | 2nd | TEST | 4 | 2nd | Y-VARS | 1 | 2 |

Now graph Y_3 in dot mode to obtain Figure 18(c). It is now easy to determine the interval on which $f(x) \geq g(x)$; the interval corresponds to the line segment shown in Figure 18(c). By tracing to and then zooming in on the right endpoint of the line segment of $y = 1$ in Figure 18(c), you can determine this interval to be $(0, 14.90)$.

As a variation of the assignment for Y_3, you could assign $(Y_1 \geq Y_2) + 1$ to Y_3 and obtain the graph in Figure 18(d). The advantage of using this assignment is that the graph of Y_3 is always visible, since it never coincides with the x-axis.

Figure 18

(a) Assignments

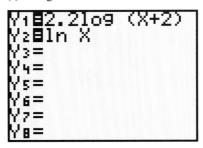

(b) Graph of f and g

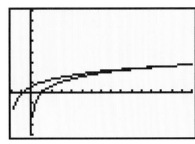

$[-2, 16]$ by $[-4, 8]$

(continued)

Figure 18　(continued)
(c) Graph of $Y_3 = Y_1 \geq Y_2$

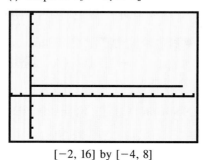

$[-2, 16]$ by $[-4, 8]$

(d) Graph of $Y_3 = (Y_1 \geq Y_2) + 1$

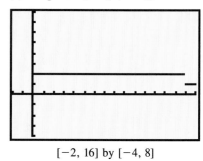

$[-2, 16]$ by $[-4, 8]$

Another (more common) approach to solving an inequality of the form $f(x) \geq g(x)$ (as in Example 19) is to examine the equivalent inequality,

$$f(x) - g(x) \geq 0.$$

We can then assign $f(x) - g(x)$ to Y_1 and determine all values of x that correspond to nonnegative values of Y_1 by using the root feature.

The next example (Example 9 in Section 1.4) shows how we can utilize the Boolean method in graphing a piecewise-defined function.

EXAMPLE 20　Sketching the graph of a piecewise-defined function

Sketch the graph of the function f if

$$f(x) = \begin{cases} 2x + 3 & \text{if } x < 0 \\ x^2 & \text{if } 0 \leq x < 2 \\ 1 & \text{if } x \geq 2 \end{cases}$$

Solution　Begin by making the assignment

$$Y_1 = \underbrace{(2x + 3)(x < 0)}_{\text{first piece}} + \underbrace{x^2(0 \leq x)(x < 2)}_{\text{second piece}} + \underbrace{1(x \geq 2)}_{\text{third piece}}.$$

Make Y assignment　Y_1:

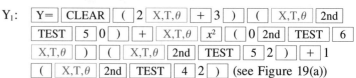

(see Figure 19(a))

As the variable x takes on values from Xmin to Xmax, the inequality $x < 0$ in the first piece will have a value of either 1 (if $x < 0$) or 0 (if $x \geq 0$). This value is multiplied by the value of $2x + 3$ and assigned to Y_1. In the second piece, note that *both* $0 \leq x$ and $x < 2$ must be true for the value of x^2 to be assigned to Y_1. The general idea is that each piece is "on" only when x takes the associated domain values.

Figure 19

(a) Assignment

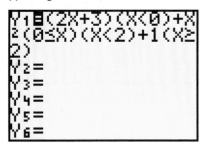

(b) Graph of f in connected mode

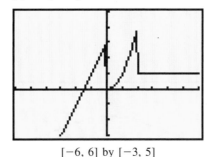

$[-6, 6]$ by $[-3, 5]$

(c) Graph of f in dot mode

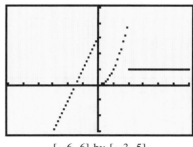

$[-6, 6]$ by $[-3, 5]$

Graphing the function using the viewing rectangle dimensions $[-6, 6]$ by $[-3, 5]$ exhibits the most important features of the graph (see Figure 19(b)). Since Figure 19(b) was drawn in connected mode, the calculator included lines between the endpoints of the pieces. To eliminate these lines, redraw the graph in dot mode, as shown in Figure 19(c). Note that the graphing calculator makes no distinction between including and excluding an endpoint (some software packages do).

An alternative method for representing the function f is to assign each piece to a Y-value, as follows:

$$Y_1 = (2x + 3)(x < 0), \quad Y_2 = x^2(0 \le x)(x < 2), \quad Y_3 = 1(x \ge 2)$$

Pressing ⬚ GRAPH ⬚ now will result in the calculator's graphing three screens, which is rather slow. An improvement in speed results from graphing $Y_4 = Y_1 + Y_2 + Y_3$ to obtain the graph of f (be sure to turn off Y_1, Y_2, and Y_3).

Yet another method for representing the function f is to assign each piece to a Y-value using division, as follows:

$$Y_1 = (2x + 3)/(x < 0), \qquad Y_2 = x^2/((0 \le x)(x < 2)), \qquad Y_3 = 1/(x \ge 2)$$

Graphing the three Y-values gives the graph of f once more. The advantage of this method when you are using the connected mode is apparent —try it!

When you are finding values of the trigonometric functions on the TI-82, there is one entry format that yields an unexpected result. Evaluating the expression tan 57°20′ in degree mode gives the incorrect value 1.73. To correctly evaluate tan 57°20′, enter the expression tan 57′20′ (no ° symbol) to obtain the approximation 1.56. Both the ° symbol and the ′ symbol are found under the ANGLE menu. On the TI-83, evaluating tan (57°20′) yields the correct value.

The result discussed in the next example (Example 5 in Section 2.3) plays an important role in advanced mathematics.

EXAMPLE 21   Sketching the graph of $f(x) = (\sin x)/x$

If $f(x) = (\sin x)/x$, sketch the graph of f on $[-\pi, \pi]$, and investigate the behavior of $f(x)$ as $x \to 0^-$ and as $x \to 0^+$.

Figure 20

(a) Assignment

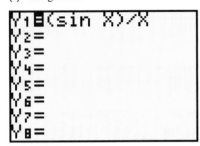

(b) Graph of f

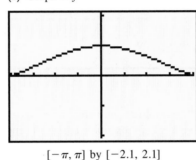

$[-\pi, \pi]$ by $[-2.1, 2.1]$

(c) Zoom-in at the point $(0, 1)$

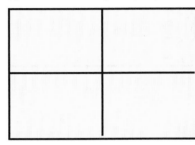

$[-0.001, 0.001]$ by $[0.999, 1.001]$

Solution   Note that f is undefined at $x = 0$, because substitution of 0 for x yields the meaningless expression $0/0$. Begin by assigning f to Y_1.

Make Y assignment Y_1: [Y=] [CLEAR] [SIN] [X,T,θ] [¹⁶] [÷] [X,T,θ] (see Figure 20(a))

[16]**TI-83 Note:** Press [ )].

The screen has a 3:2 (horizontal:vertical) proportion, so use the viewing rectangle $[-\pi, \pi]$ by $[-2.1, 2.1]$ $\left(\text{since } \frac{2}{3}\pi \approx 2.1\right)$. Selecting radian mode and connected mode and graphing Y_1 gives Figure 20(b). To get a better view of the graph of f when x is close to 0, choose the viewing rectangle $[-0.001, 0.001]$ by $[0.999, 1.001]$ to obtain Figure 20(c). Using tracing and zoom features suggests that

$$\text{as} \quad x \to 0^-, \quad f(x) \to 1 \quad \text{and as} \quad x \to 0^+, \quad f(x) \to 1.$$

There is a hole in the graph at the point $(0, 1)$; however, most graphing utilities are not capable of showing this fact.

This graphical technique does not *prove* that $f(x) \to 1$ as $x \to 0$, but it does make it appear highly probable. A rigorous proof, based on the definition of $\sin x$ and geometric considerations, can be found in calculus texts.

The next example (Example 7 in Section 2.7) discusses *simultaneous* plotting of functions—an alternative method for graphing several functions (as opposed to *sequential* plotting).

EXAMPLE 22   Sketching the graph of a sum of
two trigonometric functions

Sketch the graphs of $y = \cos x$, $y = \sin x$, and $y = \cos x + \sin x$ on the same coordinate plane for $0 \le x \le 3\pi$.

Solution Make the following assignments:

$$Y_1 = \cos x, \qquad Y_2 = \sin x, \qquad \text{and} \qquad Y_3 = Y_1 + Y_2$$

Make Y assignments

Y_1: | Y= | CLEAR | COS | X,T,θ |[17] | ENTER |

[17]**TI-83 Note:** Press |) |.

Y_2: | CLEAR | SIN | X,T,θ |[17] | ENTER |

Y_3: | CLEAR | 2nd | Y-VARS | 1 | 1 | + | 2nd | Y-VARS | 1 | 2 |

(see Figure 21(a))

Figure 21

(a) Assignments

(b) Graph of Y_1, Y_2, and Y_3

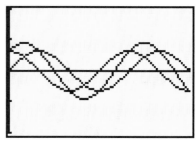

$[0, 3\pi]$ by $[-\pi, \pi]$

(c) Enhanced graph of Y_1, Y_2, and Y_3

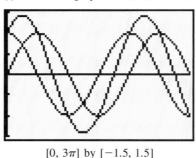

$[0, 3\pi]$ by $[-1.5, 1.5]$

(d) Graphing in "Simul" mode

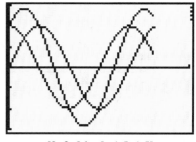

$[0, 3\pi]$ by $[-1.5, 1.5]$

Since you want a 3:2 (horizontal:vertical) screen proportion, choose the viewing rectangle $[0, 3\pi]$ by $[-\pi, \pi]$, obtaining Figure 21(b). The clarity of the graph can be enhanced by changing the viewing rectangle to $[0, 3\pi]$ by $[-1.5, 1.5]$, as in Figure 21(c).

Figure 21(c) was obtained using sequential plotting—that is, the graphs of Y_1, Y_2, and Y_3 were plotted in their respective order. To change from the sequential mode to a simultaneous mode in which the three graphs are plotted at the same time, use the following key sequence:

Select "Simul"

| MODE | ▽ (5 times) | ▷ | ENTER |

Now press | GRAPH |, and you will see Y_1, Y_2, and Y_3 plotted at the same time—that is, in the simultaneous mode (see Figure 21(d)). Select sequential mode before proceeding.

A topic that you may study in calculus is the relationship between a curve and its associated *circle of curvature,* or *osculating circle.* For now, think of the circle of curvature of a curve as the circle that best coincides with the curve at a point that is on the curve.

E X A M P L E   23   **Finding points of intersection of two nearly identical graphs**

For the graph of $y = \sec x$ and the point $P(0, 1)$, the equation of the circle of curvature is $x^2 + (y - 2)^2 = 1$. Determine the portion of the interval $(-\pi/2, \pi/2)$ on which the graph of $y = \sec x$ is below the graph of the lower half of its circle of curvature.

Solution   Solving the equation $x^2 + (y - 2)^2 = 1$ for y gives $y = 2 \pm \sqrt{1 - x^2}$, where the plus sign corresponds to the upper half of the circle and the minus sign corresponds to the lower half of the circle. Hence, make the assignments

$$Y_1 = 1/\cos x \qquad \text{and} \qquad Y_2 = 2 - \sqrt{1 - x^2}.$$

Make Y assignments

Y_1: $\boxed{Y=}$ $\boxed{\text{CLEAR}}$ $\boxed{1}$ $\boxed{\div}$ $\boxed{\text{COS}}$ $\boxed{X,T,\theta}$ ^{18} $\boxed{\text{ENTER}}$

Y_2: $\boxed{\text{CLEAR}}$ $\boxed{2}$ $\boxed{-}$ $\boxed{\text{2nd}}$ $\boxed{\sqrt{\ }}$ $\boxed{(}$ $\boxed{1}$ $\boxed{-}$ $\boxed{X,T,\theta}$ $\boxed{x^2}$ $\boxed{)}$ $\boxed{\text{ENTER}}$

$\boxed{\text{CLEAR}}$ (see Figure 22(a))

^{18}**TI-83 Note:** Press $\boxed{)}$.

Figure 22
(a) Assignments

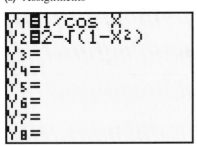

(b) Graph of Y_1 and Y_2

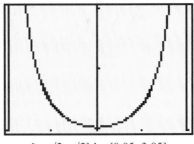

$[-\pi/2, \pi/2]$ by $[0.95, 3.05]$

(c) Graph of $Y_3 = (Y_1 < Y_2) + 1$

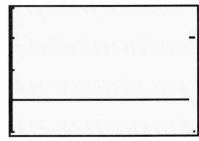

$[0, 1]$ by $[0.5, 2.5]$

To maintain a 3:2 proportion on the screen, select the viewing rectangle $[-\pi/2, \pi/2]$ by $[0.95, 3.05]$ $\left(\text{since } \frac{2}{3}(\pi/2) \approx 1.05 \text{ and the } y\text{-coordi-}\right.$ nate of the center of the circle of curvature is 2$\left.\right)$. Pressing $\boxed{\text{GRAPH}}$ gives Figure 22(b), in which the two graphs nearly coincide.

Using a trace and zoom routine near P yields two nearly identical and nearly horizontal lines. Once again, use the Boolean method to help determine the points of intersection of Y_1 and Y_2, assigning $(Y_1 < Y_2) + 1$ to Y_3.

Make Y assignment

Y_3:

Since the domain of Y_2 is $[-1, 1]$, the graphs of Y_1 and Y_2 are symmetric with respect to the y-axis, and you desire to see only the y-values 1 and 2, choose the viewing rectangle $[0, 1]$ by $[0.5, 2.5]$. Switching to dot mode, turning off Y_1 and Y_2, and graphing only Y_3 gives Figure 22(c).

Figure 22(c) indicates that the graph of $y = \sec x$ is *above* the graph of the semicircle on the interval $(0, 0.977)$ and *below* it on $(0.997, 1]$. Thus, the graph of $y = \sec x$ is lower than the graph of $y = 2 - \sqrt{1 - x^2}$ on $[-1, -0.977) \cup (0.997, 1]$.

The next example (Example 11 in Section 3.6) shows how a graphing utility can be extremely helpful in determining whether an equation involving inverse trigonometric functions is an identity and, if the equation is not an identity, in finding solutions of the equation.

E X A M P L E 24   **Investigating an equation**

Determine whether the equation

$$\arctan x = \frac{\arcsin x}{\arccos x}$$

is an identity. If it is not an identity, then approximate the values of x for which the equation is true—that is, solve the equation.

Solution   Begin by making the assignments

$$Y_1 = \tan^{-1} x \qquad \text{and} \qquad Y_2 = \sin^{-1} x/\cos^{-1} x$$

Make Y assignments Y_1: | Y= | | CLEAR | | 2nd | | TAN⁻¹ | | X,T,θ |[19] | ENTER |

Y_2: | CLEAR | | 2nd | | SIN⁻¹ | | X,T,θ |[19] | ÷ | | 2nd | | COS⁻¹ |

[19]**TI-83 Note:** Press |) | . | X,T,θ |[19] | ENTER | | CLEAR | (see Figure 23(a))

Figure 23

(a) Assignments

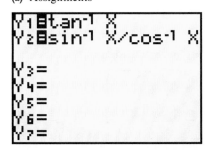

(b) Select window values

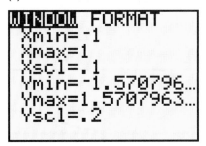

(c) Graph of Y_1 and Y_2

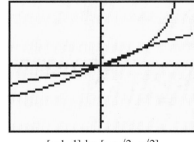

$[-1, 1]$ by $[-\pi/2, \pi/2]$

Since the domain of $\sin^{-1}$ and $\cos^{-1}$ is $[-1, 1]$ and the range of $\tan^{-1}$ is $(-\pi/2, \pi/2)$, choose the viewing rectangle dimensions $[-1, 1]$ by $[-\pi/2, \pi/2]$ with Xscl $= 0.1$ and Yscl $= 0.2$ (see Figure 23(b)).

(continued)

Graphing Y_1 and Y_2 in the connected mode gives Figure 23(c). Since the graphs representing Y_1 and Y_2 are not the same, you know that *the given equation is not an identity.* However, because the graphs intersect at least twice, you know that the equation has at least two solutions. A little exploration shows that there are exactly two points of intersection.

It appears that $x = 0$ is a solution, and a quick check in the given equation verifies that this is true: arctan $0 = 0$ and (arcsin 0)/arccos $0 = 0/(\pi/2) = 0$. To estimate the point of intersection in the first quadrant, use the intersect feature to determine that the point has the appropriate coordinates (0.450, 0.423). Hence

$$x = 0 \qquad \text{and} \qquad x \approx 0.450$$

are the values of x for which the given equation is true.

The next example demonstrates the use of the shade feature in solving a two-dimensional inequality.

E X A M P L E 25   **Solving an inequality using the shade feature**
Graph the inequality $64y^3 - x^3 \le e^{1-2x}$.

Solution   First solve the given inequality for y, and then assign that expression to Y_1.

$$64y^3 - x^3 \le e^{1-2x} \qquad\qquad \text{given}$$

$$y^3 \le \frac{e^{1-2x} + x^3}{64} \qquad\qquad \text{add } x^3 \text{, and divide by 64}$$

$$y \le \tfrac{1}{4}(e^{1-2x} + x^3)^{1/3} \qquad \text{take the cube root}$$

Make Y assignment Y_1: $\boxed{Y=}$ $\boxed{\text{CLEAR}}$ $\boxed{.25}$ $\boxed{(}$ $\boxed{\text{2nd}}$ $\boxed{e^x}$ $\boxed{(}$ $\boxed{1}$ $\boxed{-}$ $\boxed{2}$ $\boxed{X,T,\theta}$ $\boxed{)}$ $\boxed{+}$
$\boxed{X,T,\theta}$ $\boxed{\text{MATH}}$ $\boxed{3}$ $\boxed{)}$ $\boxed{\wedge}$ $\boxed{(}$ $\boxed{1}$ $\boxed{\div}$ $\boxed{3}$ $\boxed{)}$ $\boxed{\text{ENTER}}$
(see Figure 24(a))

Figure 24

(a) Assignment

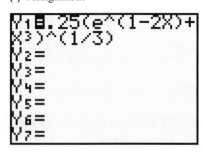

(b) Graph of Y_1

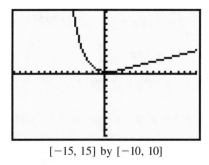

$[-15, 15]$ by $[-10, 10]$

(c) Graph with Shade

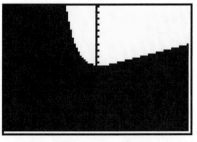

$[-15, 15]$ by $[-10, 10]$

A graph of Y_1 with viewing rectangle $[-15, 15]$ by $[-10, 10]$ is shown in Figure 24(b). Since you want to shade the portion of the plane below the graph, use the following shade command,[20] entered on the home screen:

$$\text{Shade } (Y\text{min}, Y_1, 1, X\text{min}, X\text{max})$$

[20]**TI-83 Note:** Use the command Shade $(Y\text{min}, Y_1)$.

Shade instructions

| 2nd | QUIT | CLEAR | 2nd | DRAW | 7 | VARS | 1 | 4 | , |

| 2nd | Y-VARS | 1 | 1 | [21] | , | 1 | , | VARS | 1 | 1 | , | VARS | 1 | 2 |

[21]**TI-83 Note:** Omit the rest of the keystrokes on this line.

|) | ENTER | (see Figure 24(c))

The five shade parameters can be interpreted as follows: lower boundary, upper boundary, resolution of shading (with 1 being the most dense and 9 being the least dense), left boundary, and right boundary.

The solution of the inequality includes the graph of Y_1 and the shaded region.

We now examine how to *solve a system of equations* on the TI-82/83.

E X A M P L E 26   **Solving a system of equations**

Solve the system of equations

$$\begin{cases} 3.1x + 6.7y - 8.7z = 1.5 \\ 4.1x - 5.1y + 0.2z = 2.1 \\ 0.6x + 1.1y - 7.4z = 3.9 \end{cases}$$

Solution   Make the assignments

$$[A] = \begin{bmatrix} 3.1 & 6.7 & -8.7 \\ 4.1 & -5.1 & 0.2 \\ 0.6 & 1.1 & -7.4 \end{bmatrix} \quad \text{and} \quad [B] = \begin{bmatrix} 1.5 \\ 2.1 \\ 3.9 \end{bmatrix}.$$

Make matrix assignments

[A]: | MATRX | ◁ | 1 | 3 | ENTER | 3 | ENTER | 3.1 | ENTER | 6.7 |
| ENTER | −8.7 | ENTER | 4.1 | ENTER | −5.1 | ENTER | .2 |
| ENTER | .6 | ENTER | 1.1 | ENTER | −7.4 | ENTER |
(see Figure 25(a))

[B]: | MATRX | ◁ | 2 | 3 | ENTER | 1 | ENTER | 1.5 | ENTER | 2.1 |
| ENTER | 3.9 | ENTER | (see Figure 25(b))

Solve the system by finding the product $[A]^{-1}[B]$.

Find the inverse of a matrix

| 2nd | QUIT | CLEAR | MATRX | 1 | x^{-1} | × |
| MATRX | 2 | ENTER |

(continued)

Figure 25

(a) Matrix [A]

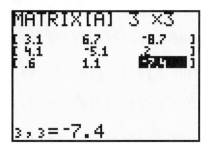

(b) Matrix [B]

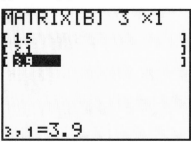

(c) [A]⁻¹ × [B]

Figure 25(c) indicates that the solution of the system is

$$x \approx -0.1081, \qquad y \approx -0.5227, \qquad \text{and} \qquad z \approx -0.6135.$$

Also shown in Figure 25(c) is the display for finding the determinant of [A].

Find the determinant of a matrix

MATRX ▷ 1 MATRX 1 ENTER

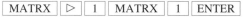

The next example (Example 5 in Section 6.4) demonstrates how to graph parametric equations on the TI-82/83.

E X A M P L E 27   **Graphing a set of parametric equations**

Sketch the graph of the Lissajous figure that has the parametrization

$$x = \sin 2t, \quad y = \cos t; \quad 0 \le t \le 2\pi.$$

Solution   You first need to set the calculator in a parametric mode. The "Par" mode is found under the MODE menu next to "Func."

Select "Par"

MODE ▽ (3 times) ▷ ENTER (see Figure 26(a))

Figure 26

(a) Changing to "Par" mode

(b) Parametric assignments

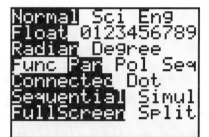

(c) Parametric window values

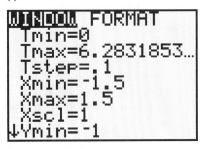

(d) Lissajous figure

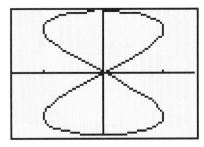

$[-1.5, 1.5]$ by $[-1, 1]$

Next make the assignments

$$X_{1T} = \sin 2t \qquad \text{and} \qquad Y_{1T} = \cos t,$$

where the subscript "1T" on X and Y indicates that X_{1T} and Y_{1T} represent the first *pair* of parametric equations.

Make parametric assignments

[22]**TI-83 Note:** Press |) |.

X_{1T}: | Y= | | CLEAR | | SIN | 2 | X,T,θ | [22] | ENTER |

Y_{1T}: | CLEAR | | COS | | X,T,θ | [22] | ENTER | (see Figure 26(b))

When graphing parametric equations, you need to assign minimum (Tmin) and maximum (Tmax) values to the parameter t, in addition to viewing rectangle dimensions. You also need to select an increment, or step value (Tstep), for t. A typical value for Tstep is $\pi/30 \approx 0.105$. If a smaller value of Tstep is chosen, the accuracy of the sketch is increased, but so is the amount of time needed to sketch the graph.

For this example, use Tmin $= 0$, Tmax $= 2\pi$, and Tstep $= 0.1$. Since x and y are between -1 and 1, assign -1 to Ymin and 1 to Ymax. To maintain the 3:2 screen proportion, select -1.5 for Xmin and 1.5 for Xmax. The specific keystrokes are as follows:

Select window values

| WINDOW | ▽ 0 ▽ 2 | 2nd | ∧ | ▽ .1 ▽ -1.5 ▽ 1.5
▽ 1 ▽ -1 ▽ 1 ▽ 1 (see Figure 26(c))

Pressing | GRAPH | gives the Lissajous figure in Figure 26(d). Pressing | TRACE | reveals the current values of X, Y, and T. Experiment with the | ◁ | and | ▷ | cursor keys to familiarize yourself with the orientation of the curve.

Using parametric equations on a graphing calculator can greatly simplify the representation of equations that are not in function form. For example, to graph the parabola in Example 16, you make the assignments

$$X_{1T} = t^2 + 2t - 4 \qquad \text{and} \qquad Y_{1T} = t$$

in "Par" mode. Next, since the axis of symmetry is $y = -1$ and since $y = t$, set Tmin and Ymin to -5, Tmax and Ymax to 3, Xmin to -6, Xmax to 6, and Tstep to 0.1.

Graphing gives a figure similar to Figure 14(b). To graph selected portions of the parabola, vary the values of Tmin and Tmax. Try a variation of this graph using Tmin $= -4$ and Tmax $= 2$.

The next example (Exercise 79 in Section 6.5) demonstrates how to graph a polar equation on the TI-82/83.

E X A M P L E 28　　**Graphing a polar equation**

Graph $r = 2 \sin^2 \theta \tan^2 \theta$ for $-\pi/3 \le \theta \le \pi/3$.

Solution　　First select "Pol" under the MODE menu and then assign the function to r1 under the "Y=" menu.

Make polar assignment　　r1: | Y= | CLEAR | 2 | (| SIN | X,T,θ |) |[23] x^2
　　　　　　　　　　　　　　| (| TAN | X,T,θ |) |[23] x^2 (see Figure 27(a))

[23]**TI-83 Note:** Press |) |.

Next assign $-\pi/3$ to θmin and $\pi/3$ to θmax.

Figure 27

(a) Polar assignment

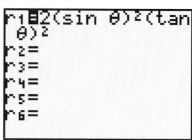

(b) Polar window values

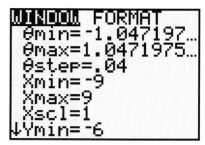

(c) Graph of $r = 2 \sin^2 \theta \tan^2 \theta$

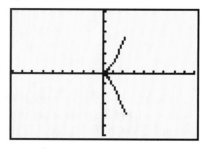

$[-9, 9]$ by $[-6, 6]$

(d) Graph of $r = 6 \sin 2\theta$

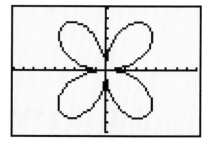

$[-9, 9]$ by $[-6, 6]$

Select window values
| WINDOW | $\triangledown$ | (−) | 2nd | ^ | ÷ | 3 | $\triangledown$ | 2nd | ^ | ÷ | 3 |

$\triangledown$ .04 $\triangledown$ −9 $\triangledown$ 9 $\triangledown$ 1 $\triangledown$ −6 $\triangledown$ 6 $\triangledown$ 1 (see Figure 27(b))

Press GRAPH to obtain Figure 27(c). Pressing TRACE displays the values for X, Y, and θ (same as T).

You may want to practice with another figure, such as a four-leafed rose. If so, go back and assign 6 sin 2θ to r1, 0 to θmin, and 2π to θmax (see Figure 27(d)).

The TI-82 and TI-83 have many features beyond the ones we have discussed, such as the ability to split the screen and store and recall graphs, but the purpose of this appendix is not to illustrate every feature of the calculator. Rather, we hope that you have become familiar with the most common features needed for studying the material in this text.

COMMON GRAPHS AND THEIR EQUATIONS

(Graphs of conics appear on the back endpaper of this text.)

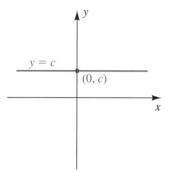

Horizontal line; constant function

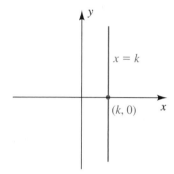

Vertical line

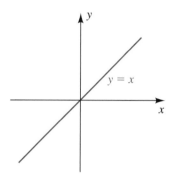

Identity function

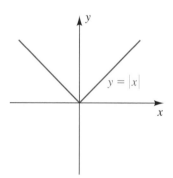

Absolute value function

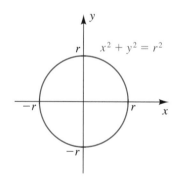

Circle with center (0, 0) and radius r

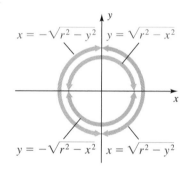

Semicircles

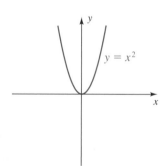

Parabola with vertical axis; squaring function

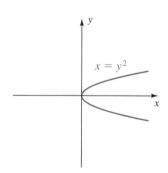

Parabola with horizontal axis

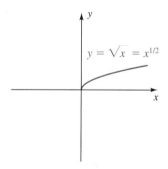

Square root function

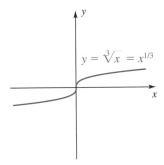

Cube root function

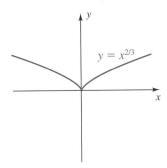

A graph with a cusp at the origin

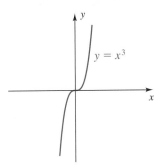

Cubing function

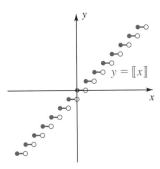

Greatest integer function

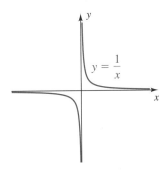

Reciprocal function

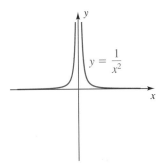

A rational function

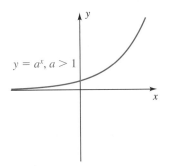

Exponential growth function
(includes natural exponential
function)

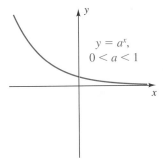

Exponential decay function

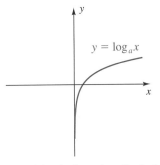

Logarithmic function (includes
common and natural logarithmic
functions)

A SUMMARY OF GRAPH TRANSFORMATIONS

The graph of $y = f(x)$ is shown in black in each figure. The domain of f is $[-1, 3]$ and the range of f is $[-4, 3]$.

$$y = g(x) = f(x) + 3$$

The graph of f is shifted vertically upward 3 units.
Domain of g: $[-1, 3]$ Range of g: $[-1, 6]$

$$y = h(x) = f(x) - 4$$

The graph of f is shifted vertically downward 4 units.
Domain of h: $[-1, 3]$ Range of h: $[-8, -1]$

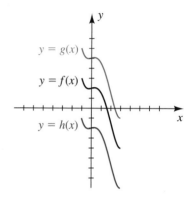

$$y = g(x) = f(x - 3)$$

The graph of f is shifted horizontally to the right 3 units.
Domain of g: $[2, 6]$ Range of g: $[-4, 3]$

$$y = h(x) = f(x + 6)$$

The graph of f is shifted horizontally to the left 6 units.
Domain of h: $[-7, -3]$ Range of h: $[-4, 3]$

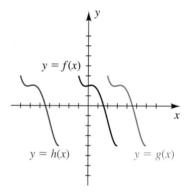

$$y = g(x) = 2f(x) \quad [2 > 1]$$

The graph of f is stretched vertically by a factor of 2.
Domain of g: $[-1, 3]$ Range of g: $[-8, 6]$

$$y = h(x) = \tfrac{1}{2}f(x) \quad \left[\tfrac{1}{2} < 1\right]$$

The graph of f is compressed vertically by a factor of 2.
Domain of h: $[-1, 3]$ Range of h: $\left[-2, \tfrac{3}{2}\right]$

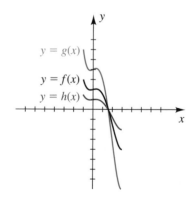

$$y = g(x) = f(2x) \quad [2 > 1]$$

The graph of f is compressed horizontally by a factor of 2.
Domain of g: $\left[-\frac{1}{2}, \frac{3}{2}\right]$ Range of g: $[-4, 3]$

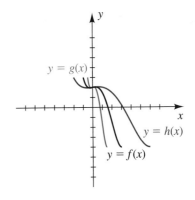

$$y = h(x) = f\left(\tfrac{1}{2}x\right) \quad \left[\tfrac{1}{2} < 1\right]$$

The graph of f is stretched horizontally by a factor of 2.
Domain of h: $[-2, 6]$ Range of h: $[-4, 3]$

$$y = g(x) = -f(x)$$

The graph of f is reflected through the x-axis.
Domain of g: $[-1, 3]$ Range of g: $[-3, 4]$

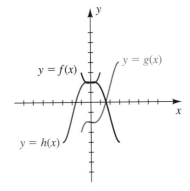

$$y = h(x) = f(-x)$$

The graph of f is reflected through the y-axis.
Domain of h: $[-3, 1]$ Range of h: $[-4, 3]$

$$y = g(x) = |f(x)|$$

Reflect points on f with negative y-values through the x-axis.
Domain of g: $[-1, 3]$ Range of g: $[0, 4]$

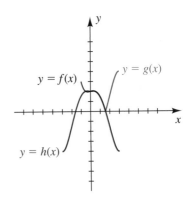

$$y = h(x) = f(|x|)$$

Reflect points on f with positive x-values through the y-axis.
Domain of h: $[-3, 3]$ Range of h: $[-4, 3]$ at most.
                         In this case, the range is a
                         subset of $[-4, 3]$.

GRAPHS OF TRIGONOMETRIC FUNCTIONS AND THEIR INVERSES

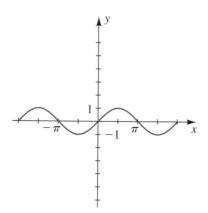

$y = \sin x$

Domain: $\mathbb{R}$

Range: $[-1, 1]$

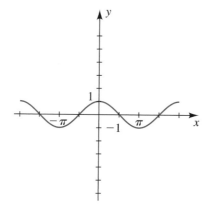

$y = \cos x$

Domain: $\mathbb{R}$

Range: $[-1, 1]$

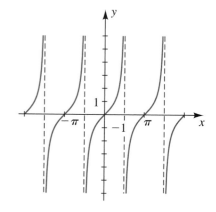

$y = \tan x$

Domain: $x \neq \dfrac{\pi}{2} + \pi n$

Range: $\mathbb{R}$

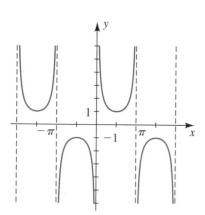

$y = \csc x$

Domain: $x \neq \pi n$

Range: $(-\infty, -1] \cup [1, \infty)$

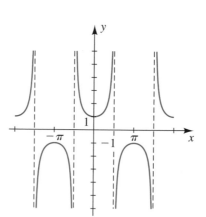

$y = \sec x$

Domain: $x \neq \dfrac{\pi}{2} + \pi n$

Range: $(-\infty, -1] \cup [1, \infty)$

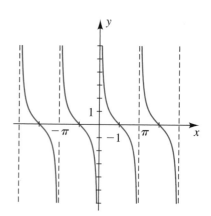

$y = \cot x$

Domain: $x \neq \pi n$

Range: $\mathbb{R}$

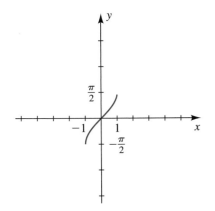

$y = \sin^{-1} x$
Domain: $[-1, 1]$

Range: $\left[-\dfrac{\pi}{2}, \dfrac{\pi}{2} \right]$

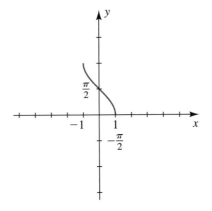

$y = \cos^{-1} x$
Domain: $[-1, 1]$

Range: $[0, \pi]$

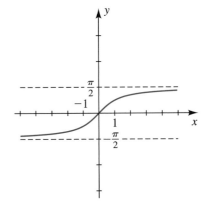

$y = \tan^{-1} x$
Domain: $\mathbb{R}$

Range: $\left(-\dfrac{\pi}{2}, \dfrac{\pi}{2} \right)$

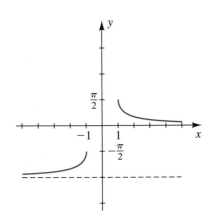

$y = \csc^{-1} x$
Domain: $(-\infty, -1] \cup [1, \infty)$

Range: $\left(-\pi, -\dfrac{\pi}{2} \right] \cup \left(0, \dfrac{\pi}{2} \right]$

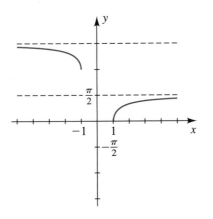

$y = \sec^{-1} x$
Domain: $(-\infty, -1] \cup [1, \infty)$

Range: $\left[0, \dfrac{\pi}{2} \right) \cup \left[\pi, \dfrac{3\pi}{2} \right)$

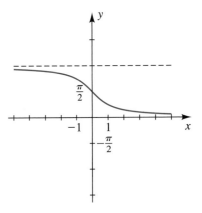

$y = \cot^{-1} x$
Domain: $\mathbb{R}$

Range: $(0, \pi)$

COMPLEX NUMBERS

Complex numbers are needed to find solutions of equations that cannot be solved using only the set $\mathbb{R}$ of real numbers. The following chart illustrates several simple quadratic equations and the types of numbers required for solutions.

Equation	Solutions	Type of numbers required
$x^2 = 9$	$3, -3$	Integers
$x^2 = \frac{9}{4}$	$\frac{3}{2}, -\frac{3}{2}$	Rational numbers
$x^2 = 5$	$\sqrt{5}, -\sqrt{5}$	Irrational numbers
$x^2 = -9$	?	Complex numbers

The solutions of the first three equations in the chart are in $\mathbb{R}$; however, since squares of real numbers are never negative, $\mathbb{R}$ does not contain the solutions of $x^2 = -9$. To solve this equation, we need the **complex number system** $\mathbb{C}$, which contains both $\mathbb{R}$ and numbers whose squares are negative.

We begin by introducing the **imaginary unit,** denoted by i, which has the following properties.

Properties of i

$$i = \sqrt{-1}, \qquad i^2 = -1$$

Because its square is negative, the letter i does not represent a real number. It is a new mathematical entity that will enable us to obtain $\mathbb{C}$. Since i, together with $\mathbb{R}$, is to be contained in $\mathbb{C}$, we must consider products of the form bi for a real number b and also expressions of the form $a + bi$ for real numbers a and b. The next chart provides definitions we shall use.

Terminology	Definition	Example(s)
Complex number	$a + bi$, where a and b are real numbers and $i^2 = -1$	$3, 2 + i, 2i$
Imaginary number	$a + bi$ with $b \neq 0$	$3 + 2i, -4i$
Pure imaginary number	bi with $b \neq 0$	$-4i, \sqrt{3}\,i, i$
Equality	$a + bi = c + di$ if and only if $a = c$ and $b = d$	$x + yi = 3 + 4i$ iff $x = 3$ and $y = 4$
Sum	$(a + bi) + (c + di) = (a + c) + (b + d)i$	see Example 1(a)
Product	$(a + bi)(c + di) = (ac - bd) + (ad + bc)i$	see Example 1(b)

Note that the pure imaginary numbers are a subset of the imaginary numbers and the imaginary numbers are a subset of the complex numbers. We use the phrase *nonreal complex number* interchangeably with *imaginary number.*

It is not necessary to memorize the definitions of addition and multiplication of complex numbers given in the preceding chart. Instead, *we may treat all symbols as having properties of real numbers, with exactly one exception: We replace i^2 by -1.* Thus, for the product $(a + bi)(c + di)$ we simply use the distributive laws and the fact that

$$(bi)(di) = bdi^2 = bd(-1) = -bd.$$

EXAMPLE 1   Addition and multiplication of complex numbers

Express in the form $a + bi$, where a and b are real numbers:

(a) $(3 + 4i) + (2 + 5i)$ **(b)** $(3 + 4i)(2 + 5i)$

Solution

(a) $(3 + 4i) + (2 + 5i) = (3 + 2) + (4 + 5)i = 5 + 9i$

(b) $(3 + 4i)(2 + 5i) = (3 + 4i)2 + (3 + 4i)(5i)$

$$= 6 + 8i + 15i + 20i^2$$
$$= 6 + 23i + 20(-1)$$
$$= -14 + 23i$$

The set $\mathbb{R}$ of real numbers may be identified with the set of complex numbers of the form $a + 0i$. It is also convenient to denote the complex number $0 + bi$ by bi. Thus,

$$(a + 0i) + (0 + bi) = (a + 0) + (0 + b)i = a + bi.$$

Hence, we may regard $a + bi$ as the sum of two complex numbers a and bi (that is, $a + 0i$ and $0 + bi$). For the complex number $a + bi$, we call a the **real part** and b the **imaginary part.**

We can now solve an equation such as $x^2 = -9$. Specifically, since

$$(3i)(3i) = 3^2 i^2 = 9(-1) = -9,$$

we see that one solution is $3i$ and another is $-3i$.

In the next chart we define the difference of complex numbers and multiplication of a complex number by a real number.

Terminology	Definition
Difference	$(a + bi) - (c + di) = (a - c) + (b - d)i$
Multiplication by a real number k	$k(a + bi) = ka + (kb)i$

If we are asked to write an expression in the form $a + bi$, we shall also accept the form $a - di$, since $a - di = a + (-d)i$.

EXAMPLE 2   Operations with complex numbers

Express in the form $a + bi$, where a and b are real numbers:

(a) $4(2 + 5i) - (3 - 4i)$ **(b)** $(4 - 3i)(2 + i)$

(c) $i(3 - 2i)^2$ **(d)** i^{51}

Solution

(a) $4(2 + 5i) - (3 - 4i) = 8 + 20i - 3 + 4i = 5 + 24i$

(b) $(4 - 3i)(2 + i) = 8 - 6i + 4i - 3i^2 = 11 - 2i$

(c) $i(3 - 2i)^2 = i(9 - 12i + 4i^2) = i(5 - 12i) = 5i - 12i^2 = 12 + 5i$

(d) Taking successive powers of i, we obtain

$$i^1 = i, \quad i^2 = -1, \quad i^3 = -i, \quad i^4 = 1,$$

and then the cycle starts over:

$$i^5 = i, \quad i^6 = i^2 = -1, \quad \text{and so on}$$

In particular,

$$i^{51} = i^{48} i^3 = (i^4)^{12} i^3 = (1)^{12} i^3 = (1)(-i) = -i.$$

The following concept has important uses in working with complex numbers.

DEFINITION OF THE CONJUGATE OF A COMPLEX NUMBER

If $z = a + bi$ is a complex number, then its **conjugate**, denoted by $\bar{z}$, is $a - bi$.

Since $a - bi = a + (-bi)$, it follows that the conjugate of $a - bi$ is

$$a - (-bi) = a + bi.$$

Therefore, $a + bi$ and $a - bi$ are *conjugates of each other.* Some properties of conjugates are given in Exercises 55–60.

ILLUSTRATION

Conjugates

Complex number	Conjugate
$5 + 7i$	$5 - 7i$
$5 - 7i$	$5 + 7i$
$4i$	$-4i$
3	3

The following two properties are consequences of the definitions of the sum and the product of complex numbers.

Properties of conjugates	Illustration
$(a + bi) + (a - bi) = 2a$	$(4 + 3i) + (4 - 3i) = 4 + 4 = 2 \cdot 4$
$(a + bi)(a - bi) = a^2 + b^2$	$(4 + 3i)(4 - 3i) = 4^2 - (3i)^2 = 4^2 - 3^2 i^2 = 4^2 + 3^2$

Note that *the sum and the product of a complex number and its conjugate are real numbers.* Conjugates are useful for finding the **multiplicative inverse** of $a + bi$, $1/(a + bi)$, or for simplifying the quotient of two complex numbers. As illustrated in the next example, we may think of these types of simplifications as merely *rationalizing the denominator,* since we are multiplying the quotient by the conjugate of the denominator divided by itself.

EXAMPLE 3   Quotients of complex numbers

Express in the form $a + bi$, where a and b are real numbers:

(a) $\dfrac{1}{9 + 2i}$ (b) $\dfrac{7 - i}{3 - 5i}$

Solution

(a) $\dfrac{1}{9 + 2i} = \dfrac{1}{9 + 2i} \cdot \dfrac{9 - 2i}{9 - 2i} = \dfrac{9 - 2i}{81 + 4} = \dfrac{9}{85} - \dfrac{2}{85}i$

(b) $\dfrac{7 - i}{3 - 5i} = \dfrac{7 - i}{3 - 5i} \cdot \dfrac{3 + 5i}{3 + 5i} = \dfrac{21 + 35i - 3i - 5i^2}{9 + 25}$

$\qquad\qquad = \dfrac{26 + 32i}{34} = \dfrac{13}{17} + \dfrac{16}{17}i$

If p is a positive real number, then the equation $x^2 = -p$ has solutions in $\mathbb{C}$. One solution is $\sqrt{p}\, i$, since

$$(\sqrt{p}\, i)^2 = (\sqrt{p}\,)^2 i^2 = p(-1) = -p.$$

Similarly, $-\sqrt{p}\, i$ is also a solution.

The definition of $\sqrt{-r}$ in the next chart is motivated by $(\sqrt{r}\, i)^2 = -r$ for $r > 0$. When using this definition, take care *not* to write $\sqrt{ri}$ when $\sqrt{r}\, i$ is intended.

Terminology	Definition	Illustrations
Principal square root $\sqrt{-r}$ for $r > 0$	$\sqrt{-r} = \sqrt{r}\, i$	$\sqrt{-9} = \sqrt{9}\, i = 3i$ $\sqrt{-5} = \sqrt{5}\, i$ $\sqrt{-1} = \sqrt{1}\, i = i$

The radical sign must be used with caution when the radicand is negative. For example, the formula $\sqrt{a}\, \sqrt{b} = \sqrt{ab}$, which holds for positive real numbers, is *not* true when a and b are both negative, as shown below:

$$\sqrt{-3}\, \sqrt{-3} = (\sqrt{3}\, i)(\sqrt{3}\, i) = (\sqrt{3}\,)^2 i^2 = 3(-1) = -3$$

But $$\sqrt{(-3)(-3)} = \sqrt{9} = 3.$$

Hence, $$\sqrt{-3}\, \sqrt{-3} \neq \sqrt{(-3)(-3)}.$$

If only *one* of a or b is negative, then $\sqrt{a}\, \sqrt{b} = \sqrt{ab}$. In general, we shall not apply laws of radicals if radicands are negative. Instead, we shall change the form of radicals before performing any operations, as illustrated in the next example.

EXAMPLE 4   Working with square roots of negative numbers

Express in the form $a + bi$, where a and b are real numbers:

$$(5 - \sqrt{-9})(-1 + \sqrt{-4})$$

Solution   First we use the definition $\sqrt{-r} = \sqrt{r}\, i$, and then we simplify:

$$(5 - \sqrt{-9})(-1 + \sqrt{-4}) = (5 - \sqrt{9}\, i)(-1 + \sqrt{4}\, i)$$
$$= (5 - 3i)(-1 + 2i)$$
$$= -5 + 10i + 3i - 6i^2$$
$$= -5 + 13i + 6 = 1 + 13i$$

Recall that if the discriminant $b^2 - 4ac$ of the quadratic equation $ax^2 + bx + c = 0$ is negative, then there are no real roots of the equation.

In fact, the solutions of the equation are two *imaginary* numbers. Moreover, the solutions are conjugates of each other, as shown in the next example.

EXAMPLE 5 **A quadratic equation with complex solutions**

Solve the equation $5x^2 + 2x + 1 = 0$.

Solution Applying the quadratic formula with $a = 5$, $b = 2$, and $c = 1$, we see that

$$x = \frac{-2 \pm \sqrt{2^2 - 4(5)(1)}}{2(5)} \qquad x = \frac{-b \pm \sqrt{b^2 - 4ac}}{2a}$$

$$= \frac{-2 \pm \sqrt{-16}}{10} = \frac{-2 \pm 4i}{10} = \frac{-1 \pm 2i}{5} = -\frac{1}{5} \pm \frac{2}{5}i.$$

Thus, the solutions of the equation are $-\frac{1}{5} + \frac{2}{5}i$ and $-\frac{1}{5} - \frac{2}{5}i$.

EXAMPLE 6 **An equation with complex solutions**

Solve the equation $x^3 - 1 = 0$.

Solution Using the difference of two cubes factoring formula (see inside front cover), we write $x^3 - 1 = 0$ as

$$(x - 1)(x^2 + x + 1) = 0.$$

Setting each factor equal to zero and solving the resulting equations, we obtain the solutions

$$1, \quad \frac{-1 \pm \sqrt{1 - 4}}{2} = \frac{-1 \pm \sqrt{3}i}{2}$$

or, equivalently,

$$1, \quad -\frac{1}{2} + \frac{\sqrt{3}}{2}i, \quad -\frac{1}{2} - \frac{\sqrt{3}}{2}i.$$

Since the number 1 is called the **unit real number** and the given equation may be written as $x^3 = 1$, we call these three solutions the **cube roots of unity.**

Note that $x^2 + 1$ is irreducible over the *real* numbers. However, if we factor over the *complex* numbers, then $x^2 + 1$ may be factored as follows:

$$x^2 + 1 = (x + i)(x - i)$$

APPENDIX V EXERCISES

Exer. 1–34: Write the expression in the form $a + bi$, where a and b are real numbers.

1 $(5 - 2i) + (-3 + 6i)$ 2 $(-5 + 7i) + (4 + 9i)$

3 $(7 - 6i) - (-11 - 3i)$ 4 $(-3 + 8i) - (2 + 3i)$

5 $(3 + 5i)(2 - 7i)$ 6 $(-2 + 6i)(8 - i)$

7 $(1 - 3i)(2 + 5i)$ 8 $(8 + 2i)(7 - 3i)$

9 $(5 - 2i)^2$ 10 $(6 + 7i)^2$

11 $i(3 + 4i)^2$ 12 $i(2 - 7i)^2$

13 $(3 + 4i)(3 - 4i)$ 14 $(4 + 9i)(4 - 9i)$

15 i^{43} 16 i^{92}

17 i^{73} 18 i^{66}

19 $\dfrac{3}{2 + 4i}$ 20 $\dfrac{5}{2 - 7i}$

21 $\dfrac{1 - 7i}{6 - 2i}$ 22 $\dfrac{2 + 9i}{-3 - i}$

23 $\dfrac{-4 + 6i}{2 + 7i}$ 24 $\dfrac{-3 - 2i}{5 + 2i}$

25 $\dfrac{4 - 2i}{-5i}$ 26 $\dfrac{-2 + 6i}{3i}$

27 $(2 + 5i)^3$ 28 $(3 - 2i)^3$

29 $(2 - \sqrt{-4})(3 - \sqrt{-16})$

30 $(-3 + \sqrt{-25})(8 - \sqrt{-36})$

31 $\dfrac{4 + \sqrt{-81}}{7 - \sqrt{-64}}$ 32 $\dfrac{5 - \sqrt{-121}}{1 + \sqrt{-25}}$

33 $\dfrac{\sqrt{-36}\,\sqrt{-49}}{\sqrt{-16}}$ 34 $\dfrac{\sqrt{-25}}{\sqrt{-16}\,\sqrt{-81}}$

Exer. 35–38: Find the values of x and y, where x and y are real numbers.

35 $8 + (3x + y)i = 2x - 4i$

36 $(x - y) + 3i = 7 + yi$

37 $(3x + 2y) - y^3 i = 9 - 27i$

38 $x^3 + (2x - y)i = -8 - 3i$

Exer. 39–54: Find the solutions of the equation.

39 $x^2 - 6x + 13 = 0$ 40 $x^2 - 2x + 26 = 0$

41 $x^2 + 4x + 13 = 0$ 42 $x^2 + 8x + 17 = 0$

43 $x^2 - 5x + 20 = 0$ 44 $x^2 + 3x + 6 = 0$

45 $4x^2 + x + 3 = 0$ 46 $-3x^2 + x - 5 = 0$

47 $x^3 + 125 = 0$ 48 $x^3 - 27 = 0$

49 $x^4 = 256$ 50 $x^4 = 81$

51 $4x^4 + 25x^2 + 36 = 0$

52 $27x^4 + 21x^2 + 4 = 0$

53 $x^3 + 3x^2 + 4x = 0$

54 $8x^3 - 12x^2 + 2x - 3 = 0$

Exer. 55–60: Verify the property.

55 $\overline{z + w} = \overline{z} + \overline{w}$

56 $\overline{z - w} = \overline{z} - \overline{w}$

57 $\overline{z \cdot w} = \overline{z} \cdot \overline{w}$

58 $\overline{z/w} = \overline{z}/\overline{w}$

59 $\overline{z} = z$ if and only if z is real.

60 $\overline{z^2} = (\overline{z})^2$

LINES

One of the basic concepts in geometry is that of a *line*. In this appendix we will restrict our discussion to lines that lie in a coordinate plane. This will allow us to use algebraic methods to study their properties. Two of our principal objectives may be stated as follows:

(1) Given a line *l* in a coordinate plane, find an equation whose graph corresponds to *l*.

(2) Given an equation of a line *l* in a coordinate plane, sketch the graph of the equation.

The following concept is fundamental to the study of lines.

DEFINITION OF SLOPE OF A LINE

Let *l* be a line that is not parallel to the *y*-axis, and let $P_1(x_1, y_1)$ and $P_2(x_2, y_2)$ be distinct points on *l*. The **slope *m*** of *l* is

$$m = \frac{y_2 - y_1}{x_2 - x_1}.$$

If *l* is parallel to the *y*-axis, then the slope of *l* is not defined.

Typical points P_1 and P_2 on a line *l* are shown in Figure 1. The numerator $y_2 - y_1$ in the formula for *m* is the vertical change in direction from P_1 to P_2 and may be positive, negative, or zero. The denominator $x_2 - x_1$ is the horizontal change from P_1 to P_2, and it may be positive or negative, but never zero, because *l* is not parallel to the *y*-axis if a slope exists. In Figure 1(a) the slope is positive, and we say that the line *rises*. In Figure 1(b) the slope is negative, and the line *falls*.

489

Figure I

(a) Positive slope (line rises)

(b) Negative slope (line falls)

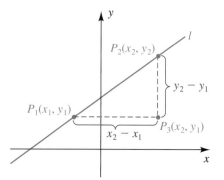

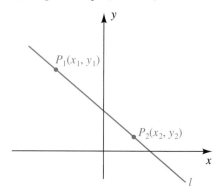

In finding the slope of a line it is immaterial which point we label as P_1 and which as P_2, since

$$\frac{y_2 - y_1}{x_2 - x_1} = \frac{y_2 - y_1}{x_2 - x_1} \cdot \frac{(-1)}{(-1)} = \frac{y_1 - y_2}{x_1 - x_2}.$$

If the points are labeled so that $x_1 < x_2$, as in Figure 1, then $x_2 - x_1 > 0$, and hence the slope is positive, negative, or zero, depending on whether $y_2 > y_1$, $y_2 < y_1$, or $y_2 = y_1$, respectively.

The definition of slope is independent of the two points that are chosen on l. If other points $P_1'(x_1', y_1')$ and $P_2'(x_2', y_2')$ are used, then, as in Figure 2, the triangle with vertices P_1', P_2', and $P_3'(x_2', y_1')$ is similar to the triangle with vertices P_1, P_2, and $P_3(x_2, y_1)$. Since the ratios of corresponding sides of similar triangles are equal,

$$\frac{y_2 - y_1}{x_2 - x_1} = \frac{y_2' - y_1'}{x_2' - x_1'}.$$

Figure 2

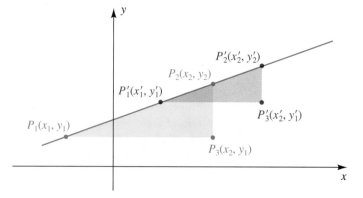

E X A M P L E 1 Finding slopes

Sketch the line through each pair of points, and find its slope:

(a) $A(-1, 4)$ and $B(3, 2)$ **(b)** $A(2, 5)$ and $B(-2, -1)$

(c) $A(4, 3)$ and $B(-2, 3)$ **(d)** $A(4, -1)$ and $B(4, 4)$

Solution The lines are sketched in Figure 3. We use the definition of slope to find the slope of each line.

Figure 3

(a) $m = -\frac{1}{2}$ **(b)** $m = \frac{3}{2}$ **(c)** $m = 0$ **(d)** m undefined

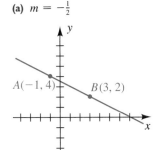 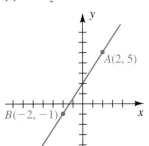 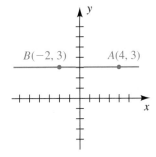 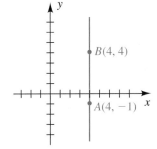

(a) $m = \dfrac{2 - 4}{3 - (-1)} = \dfrac{-2}{4} = -\dfrac{1}{2}$

(b) $m = \dfrac{5 - (-1)}{2 - (-2)} = \dfrac{6}{4} = \dfrac{3}{2}$

(c) $m = \dfrac{3 - 3}{-2 - 4} = \dfrac{0}{-6} = 0$

(d) The slope is undefined because the line is parallel to the y-axis. Note that if the formula for m is used, the denominator is zero.

E X A M P L E 2 Sketching a line with a given slope

Sketch a line through $P(2, 1)$ that has

(a) slope $\frac{5}{3}$ **(b)** slope $-\frac{5}{3}$

Solution If the slope of a line is a/b and b is positive, then for every change of b units in the horizontal direction, the line rises or falls $|a|$ units, depending on whether a is positive or negative, respectively.

(a) If $P(2, 1)$ is on the line and $m = \frac{5}{3}$, we can obtain another point on the line by starting at P and moving 3 units to the right and 5 units upward. This gives us the point $Q(5, 6)$, and the line is determined as in Figure 4(a).

(b) If $P(2, 1)$ is on the line and $m = -\frac{5}{3}$, we move 3 units to the right and 5 units downward, obtaining the line through $Q(5, -4)$, as in Figure 4(b).

(continued)

Figure 4

(a) $m = \frac{5}{3}$

(b) $m = -\frac{5}{3}$

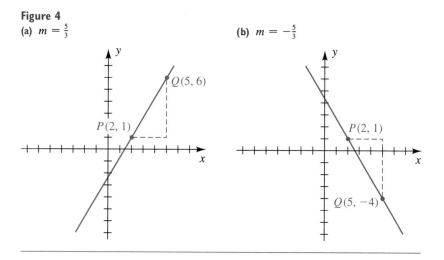

The diagram in Figure 5 indicates the slopes of several lines through the origin. The line that lies on the x-axis has slope $m = 0$. If this line is rotated about O in the counterclockwise direction (as indicated by the blue arrow), the slope is positive and increases, reaching the value 1 when the line bisects the first quadrant and continuing to increase as the line gets closer to the y-axis. If we rotate the line of slope $m = 0$ in the *clockwise* direction (as indicated by the red arrow), the slope is negative, reaching the value -1 when the line bisects the second quadrant and becoming large and negative as the line gets closer to the y-axis.

Figure 5

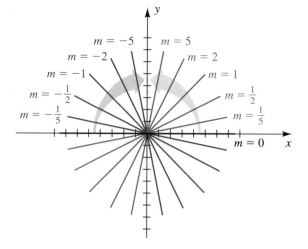

Lines that are horizontal or vertical have simple equations, as indicated in the following chart.

Terminology	Definition	Graph	Equation
Horizontal line	A line parallel to the x-axis		$y = b$
Vertical line	A line parallel to the y-axis		$x = a$

A common error is to regard the graph of $y = b$ as consisting of only the one point $(0, b)$. If we express the equation in the form $0 \cdot x + y = b$, we see that the value of x is immaterial; thus, the graph of $y = b$ consists of the points (x, b) for *every* x and hence is a horizontal line. Similarly, the graph of $x = a$ is the vertical line consisting of all points (a, y), where y is a real number.

Figure 6

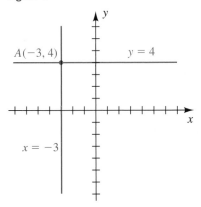

E X A M P L E 3   **Finding equations of horizontal and vertical lines**

Find an equation of the line through $A(-3, 4)$ that is parallel to

(a) the x-axis   **(b)** the y-axis

Solution   The two lines are sketched in Figure 6. As indicated in the preceding chart, the equations are $y = 4$ for part (a) and $x = -3$ for part (b).

Let us next find an equation of a line l through a point $P_1(x_1, y_1)$ with slope m. If $P(x, y)$ is any point with $x \neq x_1$ (see Figure 7), then P is on l if and only if the slope of the line through P_1 and P is m—that is, if

Figure 7

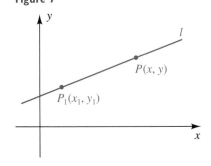

$$\frac{y - y_1}{x - x_1} = m.$$

This equation may be written in the form

$$y - y_1 = m(x - x_1).$$

Note that (x_1, y_1) is a solution of the last equation, and hence the points on l are precisely the points that correspond to the solutions. This equation for l is referred to as the **point-slope form.**

Point-Slope Form for the Equation of a Line

> An equation for the line through the point (x_1, y_1) with slope m is
>
> $$y - y_1 = m(x - x_1).$$

The point-slope form is only one possibility for an equation of a line. There are many equivalent equations. We sometimes simplify the equation obtained using the point-slope form to either

$$ax + by = c \quad \text{or} \quad ax + by + d = 0,$$

where a, b, and c are integers with no common factor, $a > 0$, and $d = -c$.

EXAMPLE 4   **Finding an equation of a line through two points**

Find an equation of the line through $A(1, 7)$ and $B(-3, 2)$.

Solution   The line is sketched in Figure 8. The formula for the slope m gives us

$$m = \frac{7 - 2}{1 - (-3)} = \frac{5}{4}.$$

We may use the coordinates of either A or B for (x_1, y_1) in the point-slope form. Using $A(1, 7)$ gives us

$$
\begin{aligned}
y - 7 &= \tfrac{5}{4}(x - 1) &&\text{point-slope form} \\
4(y - 7) &= 5(x - 1) &&\text{multiply by 4} \\
4y - 28 &= 5x - 5 &&\text{multiply factors} \\
-5x + 4y &= 23 &&\text{subtract } 5x \text{ and add 28} \\
5x - 4y &= -23 &&\text{multiply by } -1
\end{aligned}
$$

The last equation is one of the desired forms for an equation of a line. Another is $5x - 4y + 23 = 0$.

Figure 8

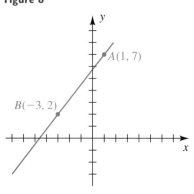

The point-slope form for the equation of a line may be rewritten as $y = mx - mx_1 + y_1$, which is of the form

$$y = mx + b$$

Figure 9

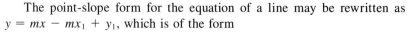

with $b = -mx_1 + y_1$. The real number b is the y-intercept of the graph, as indicated in Figure 9. Since the equation $y = mx + b$ displays the slope m and y-intercept b of l, it is called the **slope-intercept form** for the equation of a line. Conversely, if we start with $y = mx + b$, we may write

$$y - b = m(x - 0).$$

Comparing this equation with the point-slope form, we see that the graph is a line with slope m and passing through the point $(0, b)$. We have proved the following result.

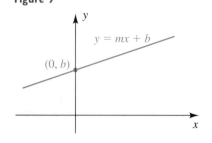

Slope-Intercept Form for the Equation of a Line

> The graph of $y = mx + b$ is a line having slope m and y-intercept b.

EXAMPLE 5 Expressing an equation in slope-intercept form

Express the equation $2x - 5y = 8$ in slope-intercept form.

Solution Our goal is to solve the given equation for y to obtain the form $y = mx + b$. Doing so gives us the following equivalent equations:

$$2x - 5y = 8, \quad -5y = -2x + 8, \quad y = \tfrac{2}{5}x + \left(-\tfrac{8}{5}\right)$$

The last equation is the slope-intercept form $y = mx + b$ with slope $m = \tfrac{2}{5}$ and y-intercept $b = -\tfrac{8}{5}$.

It follows from the point-slope form that every line is a graph of an equation

$$ax + by = c,$$

where a, b, and c are real numbers and a and b are not both zero. We call such an equation a **linear equation** in x and y. Let us show, conversely, that the graph of $ax + by = c$, with a and b not both zero, is always a line. If $b \neq 0$, we may solve for y, obtaining

$$y = \left(-\frac{a}{b}\right)x + \frac{c}{b},$$

which, by the slope-intercept form, is an equation of a line with slope $-a/b$ and y-intercept c/b. If $b = 0$ but $a \neq 0$, we may solve for x, obtaining $x = c/a$, which is the equation of a vertical line with x-intercept c/a. This discussion establishes the following result.

General Form for the Equation of a Line

> The graph of a linear equation $ax + by = c$ is a line, and conversely, every line is the graph of a linear equation.

For simplicity, we use the terminology *the line* $ax + by = c$ rather than *the line with equation* $ax + by = c$.

EXAMPLE 6 Sketching the graph of a linear equation

Sketch the graph of $2x - 5y = 8$.

Solution We know from the preceding discussion that the graph is a line, so it is sufficient to find two points on the graph. Let us find the x- and

(continued)

Figure 10

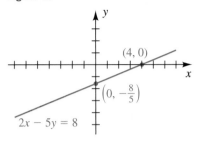

$2x - 5y = 8$

y-intercepts by substituting $y = 0$ and $x = 0$, respectively, in the given equation, $2x - 5y = 8$.

$$x\text{-}intercept: \quad \text{If } y = 0, \text{ then } 2x = 8, \text{ or } x = 4.$$

$$y\text{-}intercept: \quad \text{If } x = 0, \text{ then } -5y = 8, \text{ or } y = -\tfrac{8}{5}.$$

Plotting the intercepts $(4, 0)$ and $\left(0, -\tfrac{8}{5}\right)$ and drawing a line through them gives us the graph in Figure 10.

The following theorem specifies the relationship between **parallel lines** (lines in a plane that do not intersect) and slope.

THEOREM ON SLOPES OF PARALLEL LINES

> Two nonvertical lines are parallel if and only if they have the same slope.

Figure 11

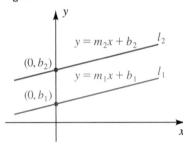

$y = m_2x + b_2$ l_2

$(0, b_2)$

$y = m_1x + b_1$ l_1

$(0, b_1)$

Proof   Let l_1 and l_2 be distinct lines of slopes m_1 and m_2, respectively. If the y-intercepts are b_1 and b_2 (see Figure 11), then, by the slope-intercept form, the lines have equations

$$y = m_1x + b_1 \quad \text{and} \quad y = m_2x + b_2.$$

The lines intersect at some point (x, y) if and only if the values of y are equal for some x—that is, if

$$m_1x + b_1 = m_2x + b_2,$$

or

$$(m_1 - m_2)x = b_2 - b_1.$$

The last equation can be solved for x if and only if $m_1 - m_2 \neq 0$. We have shown that the lines l_1 and l_2 intersect if and only if $m_1 \neq m_2$. Hence, they do *not* intersect (are parallel) if and only if $m_1 = m_2$.   ◢

EXAMPLE 7   Finding an equation of a line parallel to a given line

Find an equation of the line through $P(5, -7)$ that is parallel to the line $6x + 3y = 4$.

Solution   We first express $6x + 3y = 4$ in slope-intercept form, obtaining $y = -2x + \tfrac{4}{3}$. Thus, the slope of the given line is -2. Since parallel lines have the same slope, the required line also has slope -2. Using the point $P(5, -7)$ in the point-slope form gives us

$$y - (-7) = -2(x - 5),$$

which can be written as

$$y = -2x + 3.$$

Figure 12

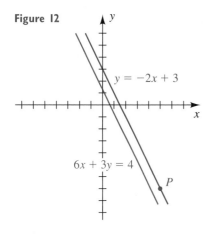

$y = -2x + 3$

$6x + 3y = 4$

P

The last equation is in slope-intercept form and shows that the parallel line we have found has y-intercept 3. This line and the given line are sketched in Figure 12.

As an alternative solution, we might use the fact that lines of the form $6x + 3y = k$ have the same slope as the given line and hence are parallel to it. Substituting $x = 5$ and $y = -7$ into the equation $6x + 3y = k$ gives us $6(5) + 3(-7) = k$ or, equivalently, $k = 9$. The equation $6x + 3y = 9$ is equivalent to $y = -2x + 3$.

The next theorem gives us information about **perpendicular lines** (lines that intersect at a right angle).

THEOREM ON SLOPES OF PERPENDICULAR LINES

Two lines with slope m_1 and m_2 are perpendicular if and only if

$$m_1 m_2 = -1.$$

Figure 13

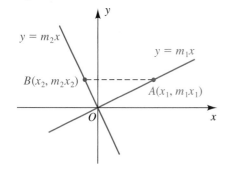

$y = m_2 x$

$B(x_2, m_2 x_2)$

$y = m_1 x$

$A(x_1, m_1 x_1)$

O

Proof   For simplicity, let us consider the special case of two lines that intersect at the origin O, as illustrated in Figure 13. Equations of these lines are $y = m_1 x$ and $y = m_2 x$. If, as in the figure, we choose points $A(x_1, m_1 x_1)$ and $B(x_2, m_2 x_2)$ different from O on the lines, then the lines are perpendicular if and only if angle AOB is a right angle. Applying the Pythagorean theorem, we know that angle AOB is a right angle if and only if

$$[d(A, B)]^2 = [d(O, B)]^2 + [d(O, A)]^2$$

or, by the distance formula,

$$(x_2 - x_1)^2 + (m_2 x_2 - m_1 x_1)^2 = x_2^2 + (m_2 x_2)^2 + x_1^2 + (m_1 x_1)^2.$$

Squaring terms, simplifying, and factoring gives us

$$-2m_1 m_2 x_1 x_2 - 2x_1 x_2 = 0$$
$$-2x_1 x_2 (m_1 m_2 + 1) = 0.$$

Since both x_1 and x_2 are not zero, we may divide both sides by $-2x_1 x_2$, obtaining $m_1 m_2 + 1 = 0$. Thus, the lines are perpendicular if and only if $m_1 m_2 = -1$.

The same type of proof may be given if the lines intersect at *any* point (a, b).   ◢

A convenient way to remember the conditions on slopes of perpendicular lines is to note that m_1 and m_2 must be *negative reciprocals* of each other—that is, $m_1 = -1/m_2$ and $m_2 = -1/m_1$.

Figure 14

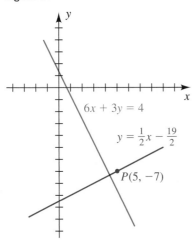

$6x + 3y = 4$

$y = \frac{1}{2}x - \frac{19}{2}$

$P(5, -7)$

EXAMPLE 8   Finding an equation of a line perpendicular to a given line

Find the slope-intercept form for the line through $P(5, -7)$ that is perpendicular to the line $6x + 3y = 4$.

Solution   We considered the line $6x + 3y = 4$ in Example 7 and found that its slope is -2. Hence, the slope of the required line is the negative reciprocal $-[1/(-2)]$, or $\frac{1}{2}$. Using $P(5, -7)$ gives us the following:

$$y - (-7) = \tfrac{1}{2}(x - 5) \quad \text{point-slope form}$$
$$y + 7 = \tfrac{1}{2}x - \tfrac{5}{2} \quad \text{simplify}$$
$$y = \tfrac{1}{2}x - \tfrac{19}{2} \quad \text{put in slope-intercept form}$$

The last equation is in slope-intercept form and shows that the perpendicular line has y-intercept $-\frac{19}{2}$. This line and the given line are sketched in Figure 14.

EXAMPLE 9   Finding an equation of a perpendicular bisector

Given $A(-3, 1)$ and $B(5, 4)$, find the general form of the perpendicular bisector l of the line segment AB.

Figure 15

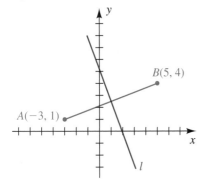

$B(5, 4)$

$A(-3, 1)$

l

Solution   The line segment AB and its perpendicular bisector l are shown in Figure 15. We calculate the following, where M is the midpoint of AB:

Coordinates of M: $\left(\dfrac{-3 + 5}{2}, \dfrac{1 + 4}{2}\right) = \left(1, \dfrac{5}{2}\right)$ midpoint formula

Slope of AB: $\dfrac{4 - 1}{5 - (-3)} = \dfrac{3}{8}$ slope formula

Slope of l: $-\dfrac{1}{\frac{3}{8}} = -\dfrac{8}{3}$ negative reciprocal of $\frac{3}{8}$

Using the point $M\left(1, \frac{5}{2}\right)$ and slope $-\frac{8}{3}$ gives us the following equivalent equations for l:

$$y - \tfrac{5}{2} = -\tfrac{8}{3}(x - 1) \quad \text{point-slope form}$$
$$6y - 15 = -16(x - 1) \quad \text{multiply by the lcd, 6}$$
$$6y - 15 = -16x + 16 \quad \text{multiply}$$
$$16x + 6y = 31 \quad \text{put in general form}$$

Two variables x and y are **linearly related** if $y = ax + b$, where a and b are real numbers and $a \neq 0$. Linear relationships between variables occur frequently in applied problems. The following example gives one illustration.

EXAMPLE 10 Relating air temperature to altitude

The relationship between the air temperature T (in °F) and the altitude h (in feet above sea level) is approximately linear for $0 \le h \le 20{,}000$. If the temperature at sea level is 60°, an increase of 5000 feet in altitude lowers the air temperature about 18°.

(a) Express T in terms of h, and sketch the graph on an hT-coordinate system.

(b) Approximate the air temperature at an altitude of 15,000 feet.

(c) Approximate the altitude at which the temperature is 0°.

Solution

(a) If T is linearly related to h, then

$$T = ah + b$$

for some constants a and b (a represents the slope and b the T-intercept). Since $T = 60°$ when $h = 0$ ft (sea level), the T-intercept is 60, and the temperature T for $0 \le h \le 20{,}000$ is given by

$$T = ah + 60.$$

From the given data, we note that when the altitude $h = 5000$ ft, the temperature $T = 60° - 18° = 42°$. Hence, we may find a as follows:

$$42 = a(5000) + 60 \qquad \text{let } T = 42 \text{ and } h = 5000$$

$$a = \frac{42 - 60}{5000} = -\frac{9}{2500} \qquad \text{solve for } a$$

Substituting for a in $T = ah + 60$ gives us the following formula for T:

$$T = -\tfrac{9}{2500}h + 60$$

The graph is sketched in Figure 16, with different scales on the axes.

(b) Using the last formula for T obtained in part (a), we find that the temperature (in °F) when $h = 15{,}000$ is

$$T = -\tfrac{9}{2500}(15{,}000) + 60 = -54 + 60 = 6.$$

(c) To find the altitude h that corresponds to $T = 0°$, we proceed as follows:

$$T = -\tfrac{9}{2500}h + 60 \qquad \text{from part (a)}$$

$$0 = -\tfrac{9}{2500}h + 60 \qquad \text{let } T = 0$$

$$\tfrac{9}{2500}h = 60 \qquad \text{add } \tfrac{9}{2500}h$$

$$h = 60 \cdot \tfrac{2500}{9} \qquad \text{multiply by } \tfrac{2500}{9}$$

$$h = \frac{50{,}000}{3} \approx 16{,}667 \text{ ft} \qquad \text{simplify and approximate}$$

Figure 16

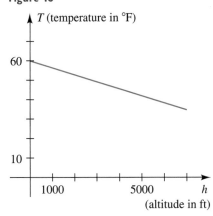

In some applications, numbers that occur in equations of lines are irrational or involve approximate data. In such cases it may be sufficient to estimate the coordinates of the point using a graphing utility, as in the next example.

EXAMPLE 11   Estimating the point of intersection of two lines

Suppose the following lines were obtained using approximate data:

$$1.018x + 0.230y = 0.447$$
$$1.847x + 4.538y = 1.414$$

Use a graphing utility to estimate the coordinates of the point of intersection to two decimal places.

Figure 17
$[-15, 15]$ by $[-10, 10]$

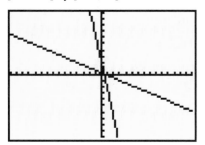

Solution   Solving each equation for y, we obtain

$$y = (0.447 - 1.018x)/0.230,$$
$$y = (1.414 - 1.847x)/4.538.$$

We next assign the preceding expressions for y to Y_1 and Y_2, respectively. A display of the graph (similar to Figure 17) shows that there is a point of intersection near the origin. Using the zoom or intersect feature yields the approximate solution $(0.41, 0.15)$.

Problems similar to this one are often encountered in statistical analysis when the *method of least squares* is employed to find a *linear regression line*.

APPENDIX VI   EXERCISES

Exer. 1–6: Sketch the line through A and B, and find its slope m.

1 $A(-3, 2)$, $B(5, -4)$ **2** $A(4, -1)$, $B(-6, -3)$

3 $A(2, 5)$, $B(-7, 5)$ **4** $A(5, -1)$, $B(5, 6)$

5 $A(-3, 2)$, $B(-3, 5)$ **6** $A(4, -2)$, $B(-3, -2)$

Exer. 7–10: Use slopes to show that the points are vertices of the specified polygon.

7 $A(-3, 1)$, $B(5, 3)$, $C(3, 0)$, $D(-5, -2)$;   parallelogram

8 $A(2, 3)$, $B(5, -1)$, $C(0, -6)$, $D(-6, 2)$;   trapezoid

9 $A(6, 15)$, $B(11, 12)$, $C(-1, -8)$, $D(-6, -5)$;   rectangle

10 $A(1, 4)$, $B(6, -4)$, $C(-15, -6)$;   right triangle

11 If three consecutive vertices of a parallelogram are $A(-1, -3)$, $B(4, 2)$, and $C(-7, 5)$, find the fourth vertex.

12 Let $A(x_1, y_1)$, $B(x_2, y_2)$, $C(x_3, y_3)$, and $D(x_4, y_4)$ denote the vertices of an arbitrary quadrilateral. Show that the line segments joining midpoints of adjacent sides form a parallelogram.

Exer. 13–14: Sketch the graph of $y = mx$ for the given values of m.

13 $m = 3, -2, \frac{2}{3}, -\frac{1}{4}$ **14** $m = 5, -3, \frac{1}{2}, -\frac{1}{3}$

Exer. 15–16: Sketch the graph of the line through P for each value of m.

15 $P(3, 1)$; $m = \frac{1}{2}, -1, -\frac{1}{5}$

16 $P(-2, 4)$; $m = 1, -2, -\frac{1}{2}$

Exer. 17–18: Sketch the graphs of the lines on the same coordinate plane.

17 $y = x + 3$, $y = x + 1$, $y = -x + 1$

18 $y = -2x - 1$, $y = -2x + 3$, $y = \frac{1}{2}x + 3$

Exer. 19–30: Find a general form of an equation of the line through the point A that satisfies the given condition.

19 $A(5, -2)$

   (a) parallel to the y-axis

   (b) perpendicular to the y-axis

20 $A(-4, 2)$

   (a) parallel to the x-axis

   (b) perpendicular to the x-axis

21 $A(5, -3)$;   slope -4 22 $A(-1, 4)$;   slope $\frac{2}{3}$

23 $A(4, 0)$;   slope -3 24 $A(0, -2)$;   slope 5

25 $A(4, -5)$;   through $B(-3, 6)$

26 $A(-1, 6)$; x-intercept 5

27 $A(2, -4)$;   parallel to the line $5x - 2y = 4$

28 $A(-3, 5)$;   parallel to the line $x + 3y = 1$

29 $A(7, -3)$;   perpendicular to the line $2x - 5y = 8$

30 $A(4, 5)$;   perpendicular to the line $3x + 2y = 7$

Exer. 31–34: Find the slope-intercept form of the line that satisfies the given conditions.

31 x-intercept 4, y-intercept -3

32 x-intercept -5, y-intercept -1

33 Through $A(5, 2)$ and $B(-1, 4)$

34 Through $A(-2, 1)$ and $B(3, 7)$

Exer. 35–36: Find a general form of an equation for the perpendicular bisector of the segment AB.

35 $A(3, -1)$, $B(-2, 6)$ 36 $A(4, 2)$, $B(-2, 10)$

Exer. 37–38: Find an equation for the line that bisects the given quadrants.

37 II and IV   38 I and III

Exer. 39–42: Use the slope-intercept form to find the slope and y-intercept of the given line, and sketch its graph.

39 $2x = 15 - 3y$ 40 $7x = -4y - 8$

41 $4x - 3y = 9$ 42 $x - 5y = -15$

Exer. 43–44: Find an equation of the line shown in the figure.

43 (a)

(b)
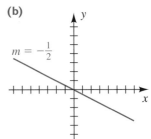

(c)
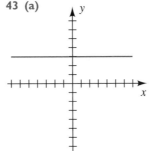

(d)
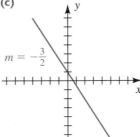

44 (a)
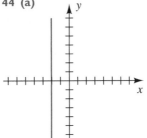

(b)
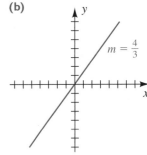

(c)
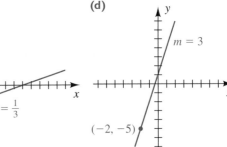

(d)
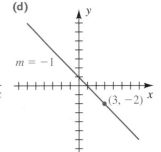

Exer. 45–46: If a line *l* has nonzero *x*- and *y*-intercepts *a* and *b*, respectively, then its *intercept form* is

$$\frac{x}{a} + \frac{y}{b} = 1.$$

Find the intercept form for the given line.

45 $4x - 2y = 6$ 46 $x - 3y = -2$

47 Find an equation of the circle that has center $C(3, -2)$ and is tangent to the line $y = 5$.

48 Find an equation of the line that is tangent to the circle $x^2 + y^2 = 25$ at the point $P(3, 4)$.

49 *Fetal growth*   The growth of a fetus more than 12 weeks old can be approximated by the formula $L = 1.53t - 6.7$, where L is the length (in centimeters) and t is the age (in weeks). Prenatal length can be determined by ultrasound. Approximate the age of a fetus whose length is 28 centimeters.

50 *Estimating salinity*   Salinity of the ocean refers to the amount of dissolved material found in a sample of seawater. Salinity S can be estimated from the amount C of chlorine in seawater using $S = 0.03 + 1.805C$, where S and C are measured by weight in parts per thousand. Approximate C if S is 0.35.

51 *Weight of a humpback whale*   The expected weight W (in tons) of a humpback whale can be approximated from its length L (in feet) by using $W = 1.70L - 42.8$ for $30 \le L \le 50$.

   (a) Estimate the weight of a 40-foot humpback whale.

   (b) If the error in estimating the length could be as large as 2 feet, what is the corresponding error for the weight estimate?

52 *Growth of a blue whale*   Newborn blue whales are approximately 24 feet long and weigh 3 tons. Young whales are nursed for 7 months, and by the time of weaning they often are 53 feet long and weigh 23 tons. Let L and W denote the length (in feet) and the weight (in tons), respectively, of a whale that is t months of age.

   (a) If L and t are linearly related, express L in terms of t.

   (b) What is the daily increase in the length of a young whale? (Use 1 month = 30 days.)

   (c) If W and t are linearly related, express W in terms of t.

   (d) What is the daily increase in the weight of a young whale?

53 *Baseball stats*   Suppose a major league baseball player has hit 5 home runs in the first 14 games, and he keeps up this pace throughout the 162-game season.

   (a) Express the number y of home runs in terms of the number x of games played.

   (b) How many home runs will the player hit for the season?

54 *Cheese production*   A cheese manufacturer produces 18,000 pounds of cheese from January 1 through March 24. Suppose that this rate of production continues for the remainder of the year.

   (a) Express the number y of pounds of cheese produced in terms of the number x of the day in a 365-day year.

   (b) Predict, to the nearest pound, the number of pounds produced for the year.

55 *Childhood weight*   A baby weighs 10 pounds at birth, and three years later the child's weight is 30 pounds. Assume that childhood weight W (in pounds) is linearly related to age t (in years).

   (a) Express W in terms of t.

   (b) What is W on the child's sixth birthday?

   (c) At what age will the child weigh 70 pounds?

   (d) Sketch, on a tW-plane, a graph that shows the relationship between W and t for $0 \le t \le 12$.

56 *Loan repayment*   A college student receives an interest-free loan of $8250 from a relative. The student will repay $125 per month until the loan is paid off.

   (a) Express the amount P (in dollars) remaining to be paid in terms of time t (in months).

   (b) After how many months will the student owe $5000?

   (c) Sketch, on a tP-plane, a graph that shows the relationship between P and t for the duration of the loan.

57 *Vaporizing water*   The amount of heat H (in joules) required to convert one gram of water into vapor is linearly related to the temperature T (in °C) of the atmosphere. At 10°C this conversion requires 2480 joules, and each increase in temperature of 15°C lowers the amount of heat needed by 40 joules. Express H in terms of T.

58 *Aerobic power*   In exercise physiology, aerobic power P is defined in terms of maximum oxygen intake. For altitudes up to 1800 meters, aerobic power is optimal— that is, 100%. Beyond 1800 meters, P decreases linearly from the maximum of 100% to a value near 40% at 5000 meters.

   (a) Express aerobic power P in terms of altitude h (in meters) for $1800 \le h \le 5000$.

   (b) Estimate aerobic power in Mexico City (altitude: 2400 meters), the site of the 1968 Summer Olympic Games.

59 *Urban heat island* The urban heat island phenomenon has been observed in Tokyo. The average temperature was 13.5°C in 1915, and since then has risen 0.032°C per year.

(a) Assuming that temperature T (in °C) is linearly related to time t (in years) and that $t = 0$ corresponds to 1915, express T in terms of t.

(b) Predict the average temperature in the year 2000.

60 *Rising ground temperature* In 1870 the average ground temperature in Paris was 11.8°C. Since then it has risen at a nearly constant rate, reaching 13.5°C in 1969.

(a) Express the temperature T (in °C) in terms of time t (in years), where $t = 0$ corresponds to the year 1870 and $0 \le t \le 99$.

(b) During what year was the average ground temperature 12.5°C?

61 *Business expenses* The owner of an ice cream franchise must pay the parent company $1000 per month plus 5% of the monthly revenue R. Operating cost of the franchise includes a fixed cost of $2600 per month for items such as utilities and labor. The cost of ice cream and supplies is 50% of the revenue.

(a) Express the owner's monthly expense E in terms of R.

(b) Express the monthly profit P in terms of R.

(c) Determine the monthly revenue needed to break even.

62 *Drug dosage* Pharmacological products must specify recommended dosages for adults and children. Two formulas for modification of adult dosage levels for young children are

$$\text{Cowling's rule:} \quad y = \tfrac{1}{24}(t + 1)a$$

and   Friend's rule: $y = \tfrac{2}{25}ta,$

where a denotes adult dose (in milligrams) and t denotes the age of the child (in years).

(a) If $a = 100$, graph the two linear equations on the same coordinate plane for $0 \le t \le 12$.

(b) For what age do the two formulas specify the same dosage?

63 *Video game* In the video game shown in the figure, an airplane flies from left to right along the path given by $y = 1 + (1/x)$ and shoots bullets in the tangent direction at creatures placed along the x-axis at $x = 1, 2, 3, 4$.

From calculus, the slope of the tangent line to the path at $P(1, 2)$ is $m = -1$ and at $Q(\tfrac{3}{2}, \tfrac{5}{3})$ is $m = -\tfrac{4}{9}$. Determine whether a creature will be hit if bullets are shot when the airplane is at

(a) P (b) Q

Exercise 63

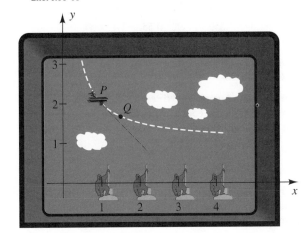

64 *Temperature scales* The relationship between the temperature reading F on the Fahrenheit scale and the temperature reading C on the Celsius scale is given by $C = \tfrac{5}{9}(F - 32)$.

(a) Find the temperature at which the reading is the same on both scales.

(b) When is the Fahrenheit reading twice the Celsius reading?

65 *Vertical wind shear* Vertical wind shear occurs when wind speed varies at different heights above the ground. Wind shear is of great importance to pilots during take-offs and landings. If the wind speed is v_1 at height h_1 and v_2 at height h_2, then the average wind shear s is given by the slope formula

$$s = \frac{v_2 - v_1}{h_2 - h_1}.$$

If the wind speed at ground level is 22 mi/hr and s has been determined to be 0.07, find the wind speed 185 feet above the ground.

66 *Vertical wind shear* In the study of vertical wind shear, the formula

$$\frac{v_1}{v_2} = \left(\frac{h_1}{h_2}\right)^P$$

is sometimes used, where P is a variable that depends on the terrain and structures near ground level. In Montreal, average daytime values for P with north winds over 29 mi/hr were determined to be 0.13. If a 32 mi/hr north wind is measured 20 feet above the ground, approximate the average wind shear (see Exercise 65) between 20 feet and 200 feet.

Exer. 67–68: The given points were found using empirical methods. Determine whether they lie on the same line $y = ax + b$, and if so, find the values of a and b.

67 $A(-1.3, -1.3598)$, $B(-0.55, -1.11905)$, $C(1.2, -0.5573)$, $D(3.25, 0.10075)$

68 $A(-0.22, 1.6968)$, $B(-0.12, 1.6528)$, $C(1.3, 1.028)$, $D(1.45, 0.862)$

Exer. 69–70: Graph the lines on the same coordinate plane, and find the coordinates of the points of intersection (the coordinates are integers).

69 $x - 3y = -58$; $3x - y = -70$

70 $x + 10y = 123$; $2x - y = -6$

Exer. 71–72: Graph the lines on the same coordinate plane, and estimate the coordinates of the points of intersection. Identify the polygon determined by the lines.

71 $2x - y = -1$; $x + 2y = -2$; $3x + y = 11$

72 $10x - 42y = -7.14$; $8.4x + 2y = -3.8$; $0.5x - 2.1y = 2.73$; $16.8x + 4y = 14$

Exer. 73–74: Determine a line in the form $y = ax + b$ that approximately models the data in each table. Plot the line together with the data on the same coordinate axes. *Note:* For exercises requiring an approximate model, answers may vary depending on the data points selected.

73

x	y
-7	-25
-5.8	-21
-5	-18.5
-4	-15.4
0.6	-0.58
1.8	3.26
3	7.1
4.6	12.2

74

x	y
0.4	2.88
1.2	2.45
2.2	1.88
3.6	1.12
4.4	0.68
6.2	-0.30

75 *Super Bowl TV costs* The following table gives the cost (in thousands of dollars) for a 30-second television advertisement during the Super Bowl for various years.

Year	Cost
1982	325
1983	400
1985	510
1986	550
1987	600

(a) Plot the data on the xy-plane.

(b) Determine a line in the form $y = ax + b$, where x is the year and y is the cost that models the data. Plot the line together with the data on the same coordinate axes. Answers may vary.

(c) Use the line from part (b) to predict the cost of a 30-second commercial in 1984 and 1995. Compare your answers to the actual values of $450,000 and $1,000,000, respectively.

76 *Record times in the mile* The world record times (in seconds) for the mile run are listed in the table.

Year	Time
1954	238.0
1957	237.2
1958	234.5
1962	234.4
1964	234.1
1965	233.6
1966	231.3
1967	231.1
1975	229.4
1979	229.1
1980	228.8
1981	227.3

(a) Plot the data.

(b) Find a line of the form $T = aY + b$ that approximates these data, where T is the time and Y is the year. Graph this line together with the data on the same coordinate axes.

(c) Use the line to predict the record time in 1985, and compare it with the actual record of 226.3 seconds.

(d) Interpret the slope of this line.

ANSWERS TO SELECTED EXERCISES

A *Student's Solutions Manual* to accompany this textbook is available from your college bookstore. The guide contains detailed solutions to approximately one-third of the exercises, as well as strategies for solving other exercises in the text.

CHAPTER 1

EXERCISES 1.1

1 (a) Negative   (b) Positive   (c) Negative
  (d) Positive
3 (a) $<$ (b) $>$ (c) $=$
5 (a) $>$ (b) $>$ (c) $>$
7 (a) $x < 0$ (b) $y \geq 0$ (c) $q \leq \pi$
  (d) $2 < d < 4$ (e) $t \geq 5$ (f) $-z \leq 3$
  (g) $\dfrac{p}{q} \leq 7$ (h) $\dfrac{1}{w} \geq 9$ (i) $|x| > 7$
9 (a) 5 (b) 3 (c) 11
11 (a) -15 (b) -3 (c) 11
13 (a) $4 - \pi$ (b) $4 - \pi$ (c) $1.5 - \sqrt{2}$
15 (a) 4 (b) 12 (c) 12 (d) 8
17 (a) 10 (b) 9 (c) 9 (d) 19
19 $|7 - x| < 5$ 21 $|-3 - x| \geq 8$
23 $|x - 4| \leq 3$ 25 $-x - 3$ 27 $2 - x$
29 $b - a$ 31 $x^2 + 4$ 33 $\dfrac{26}{7}$ 35 $0, 2, -\dfrac{3}{4}$
37 $\dfrac{4}{3}$ 39 $-\dfrac{3}{2}, \dfrac{4}{3}$ 41 $-\dfrac{6}{5}, \dfrac{2}{3}$ 43 $\pm\dfrac{3}{5}$
45 $3 \pm \sqrt{17}$ 47 $-2 \pm \sqrt{2}$ 49 $\dfrac{3}{4} \pm \dfrac{1}{4}\sqrt{41}$
51 $\pm 3, \pm 4$
53 $(-\infty, -2)$ 55 $[4, \infty)$

57 $(-2, 4]$ 59 $[3, 7]$

61 $(0, \pi)$

63 $-5 < x \leq 8$ 65 $-4 \leq x \leq -1$ 67 $x \geq 4$
69 $x < -5$ 71 $0 \leq x \leq 2\pi$ 73 $(12, \infty)$
75 $[9, 19)$ 77 $\left(-\dfrac{2}{3}, \infty\right)$ 79 $\left(\dfrac{4}{3}, \infty\right)$

81 All real numbers except 1
83 Construct a right triangle with sides of lengths $\sqrt{2}$ and 1. The hypotenuse will have length $\sqrt{3}$. Next, construct a right triangle with sides of lengths $\sqrt{3}$ and $\sqrt{2}$. The hypotenuse will have length $\sqrt{5}$.

85

Height	Weight
64	137
65	141
66	145
67	148
68	152
69	156
70	160
71	164
72	168
73	172
74	176
75	180
76	184
77	188
78	192
79	196

87 $6\dfrac{2}{3}$ yr
89 (a) 1525.7; 1454.7
  (b) As people age, they require fewer calories. Coefficients of w and h are positive because large people require more calories.
91 (4)   93 (a) (2)   (b) 860 min

EXERCISES 1.2

1 (a) The line parallel to the y-axis that intersects the x-axis at $(-2, 0)$
  (b) The line parallel to the x-axis that intersects the y-axis at $(0, 3)$
  (c) All points to the right of and on the y-axis
  (d) All points in quadrants I and III
  (e) All points below the x-axis
  (f) All points on the y-axis

3 (a) $\sqrt{29}$ **(b)** $\left(5, -\dfrac{1}{2}\right)$

5 $d(A, C)^2 = d(A, B)^2 + d(B, C)^2$; area $= 28$

7 $d(A, B) = d(B, C) = d(C, D) = d(D, A)$ and
$d(A, C)^2 = d(A, B)^2 + d(B, C)^2$

9 $(13, -28)$ **11** $d(A, C) = d(B, C) = \sqrt{145}$

13 $(0, 3 + \sqrt{11}), (0, 3 - \sqrt{11})$

15 $a < \dfrac{2}{5}$ or $a > 4$

17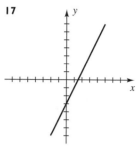

19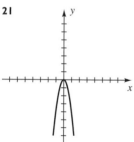

21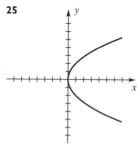

23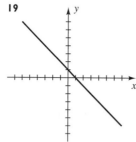

25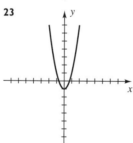

27

29

31

33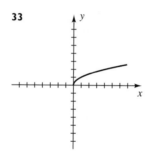

35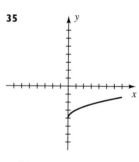

37 (a) 21, 23   **(b)** 25, 27   **(c)** 29

39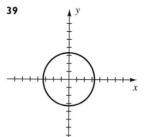

41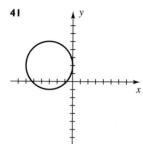

43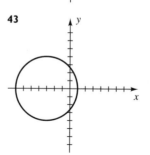

45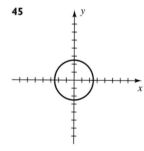

47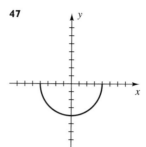

49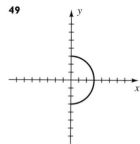

51 $(x - 2)^2 + (y + 3)^2 = 25$ **53** $\left(x - \dfrac{1}{4}\right)^2 + y^2 = 5$

55 $(x + 4)^2 + (y - 6)^2 = 41$

57 $(x + 3)^2 + (y - 6)^2 = 9$

59 $(x + 4)^2 + (y - 4)^2 = 16$

61 $(x - 1)^2 + (y - 2)^2 = 34$ **63** $C(2, -3); r = 7$

65 $C(0, -2); r = 11$ **67** $C(3, -1); r = \dfrac{1}{2}\sqrt{70}$

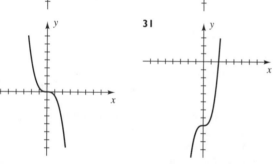

69 $C(-2, 1)$; $r = 0$ (a point)

71 Not a circle, since r^2 cannot equal -2

73 $y = \sqrt{36 - x^2}$; $y = -\sqrt{36 - x^2}$;
$x = \sqrt{36 - y^2}$; $x = -\sqrt{36 - y^2}$

75 $y = -1 + \sqrt{49 - (x - 2)^2}$;
$y = -1 - \sqrt{49 - (x - 2)^2}$;
$x = 2 + \sqrt{49 - (y + 1)^2}$; $x = 2 - \sqrt{49 - (y + 1)^2}$

77 (a) Inside (b) On (c) Outside

79 (a) 2 (b) $3 \pm \sqrt{5}$

81 $(x + 2)^2 + (y - 3)^2 = 25$ **83** $\sqrt{5}$

85 $[-15, -3) \cup (2, 15]$ **87** $(-1, 0) \cup (0, 1)$

89 (a)

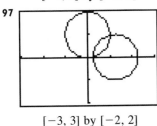

[1988, 1996] by [54×10^6, 61×10^6]

(b) The number is increasing.

91 (2)

93

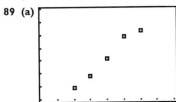

[−6, 6] by [−4, 4]

$-1.2, 0.5, 1.6$

95

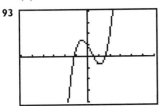

[−3, 3] by [−2, 2]

$(0.6, 0.8), (-0.6, -0.8)$

97

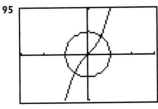

[−3, 3] by [−2, 2]

$(0.999, 0.968),$
$(0.251, 0.032)$

99

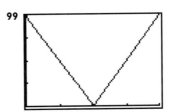

[0, 4] by [0, 4]

101 (a) 1126 ft (b) $-42°C$

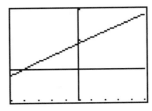

[−50, 50] by [900, 1200]

EXERCISES 1.3

1 $-6, -4, -24$ **3** $-12, -22, -36$

5 (a) $5a - 2$ (b) $-5a - 2$ (c) $-5a + 2$
(d) $5a + 5h - 2$ (e) $5a + 5h - 4$ (f) 5

7 (a) $a^2 - a + 3$ (b) $a^2 + a + 3$
(c) $-a^2 + a - 3$ (d) $a^2 + 2ah + h^2 - a - h + 3$
(e) $a^2 + h^2 - a - h + 6$ (f) $2a + h - 1$

9 $2x + h$ **11** $\dfrac{1}{\sqrt{x - 3} + \sqrt{a - 3}}$

13 (a) $\dfrac{4}{a^2}$ (b) $\dfrac{1}{4a^2}$ (c) $4a$ (d) $2a$

15 (a) $\dfrac{2a}{a^2 + 1}$ (b) $\dfrac{a^2 + 1}{2a}$ (c) $\dfrac{2\sqrt{a}}{a + 1}$ (d) $\dfrac{\sqrt{2a^3 + 2a}}{a^2 + 1}$

17 (a) $[-3, 4]$ (b) $[-2, 2]$ (c) 0 (d) $-1, \dfrac{1}{2}, 2$

(e) $\left(-1, \dfrac{1}{2}\right) \cup (2, 4]$

19 $\left[-\dfrac{7}{2}, \infty\right)$ **21** $[-3, 3]$

23 All real numbers except $-2, 0,$ and 2

25 $\left[\dfrac{3}{2}, 4\right) \cup (4, \infty)$ **27** $(2, \infty)$ **29** $[-2, 2]$

31 (a)

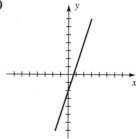

(b) $D = (-\infty, \infty)$,
$R = (-\infty, \infty)$
(c) Increasing on
$(-\infty, \infty)$

41 $f(x) = \dfrac{1}{6}x + \dfrac{3}{2}$ **43** Yes   **45** No   **47** Yes

49 No   **51** No   **53** $V = 4x(15 - x)(10 - x)$

55 (a) $y = \dfrac{500}{x}$ **(b)** $C = 300x + \dfrac{100{,}000}{x} - 600$

57 $S(h) = 6h - 50$

59 (a) $y = 2.5t + 33$

33 (a)

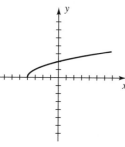

(b) $D = (-\infty, \infty)$,
$R = (-\infty, 4]$
(c) Increasing on
$(-\infty, 0]$,
decreasing on
$[0, \infty)$

(b)

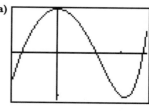

The yearly increase
in height

35 (a)

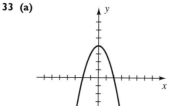

(b) $D = [-4, \infty)$,
$R = [0, \infty)$
(c) Increasing on
$[-4, \infty)$

(c) 58 in.

61 $d = 2\sqrt{t^2 + 2500}$

63 (a) $y = \sqrt{h^2 + 2hr}$ **(b)** 1280.6 mi

65 $d = \sqrt{90{,}400 + x^2}$ **67** 20 lb

69 $\dfrac{50}{9}$ ohms   **71** d is multiplied by 9.

37 (a)

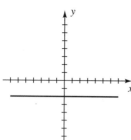

(b) $D = (-\infty, \infty)$,
$R = \{-2\}$
(c) Constant on
$(-\infty, \infty)$

73 (a)

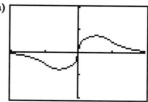

$[-2, 2]$ by $[-2, 2]$

(b) $[-0.75, 0.75]$
(c) Decreasing on
$[-2, -0.55]$
and on $[0.55, 2]$,
increasing on
$[-0.55, 0.55]$

39 (a)

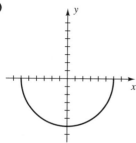

(b) $D = [-6, 6]$,
$R = [-6, 0]$
(c) Decreasing on
$[-6, 0]$,
increasing on
$[0, 6]$

75 (a)

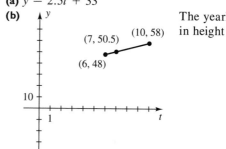

$[-0.7, 1.4]$ by $[-1.1, 1]$

(b) $[-1.03, 1]$
(c) Increasing on
$[-0.7, 0]$ and
on $[1.06, 1.4]$,
decreasing on
$[0, 1.06]$

77 (a) 8   **(b)** ± 8 **(c)** No real solutions   **(d)** 625
(e) No real solutions

79 (a) $f(x) = \dfrac{2857}{3}x - \dfrac{5,636,795}{3}$

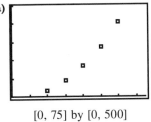

[1984, 2005] by [10,000, 30,000]

(b) Average annual increase in the price paid for a new car

(c) 1999

81 (a) 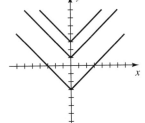   (b) No

[0, 75] by [0, 500]

EXERCISES 1.4

1 Odd   **3** Even   **5** Neither   **7** Even   **9** Odd

11 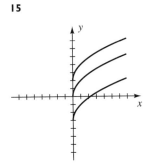   **13**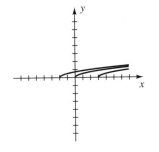

15 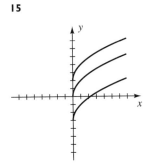   **17**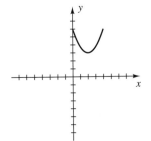

19 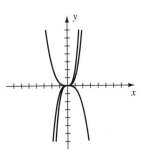   **21**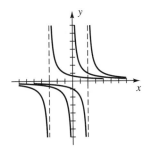

23 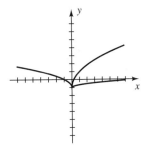   **25**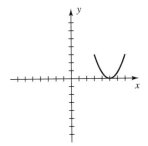

27 $(7, -3)$

29 (a) 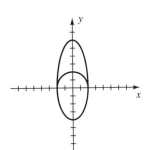   (b)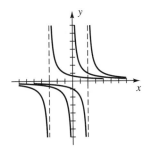

(c) 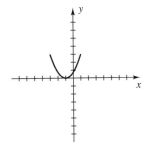   (d)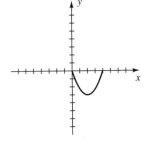

(e)

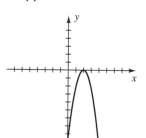

(f)

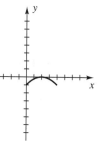

33 (a) $y = f(x + 4)$ **(b)** $y = f(x) + 1$
    (c) $y = f(-x)$

35

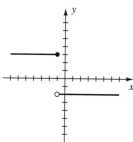

37

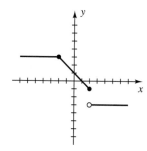

(g)

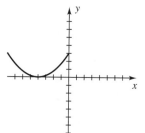

(h)

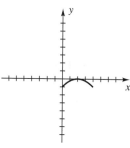

39

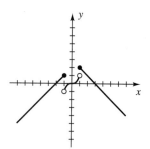

(i)

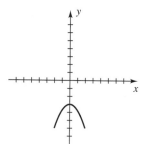

(j)

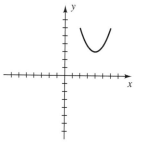

41 (a)

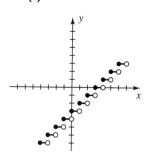

(b)

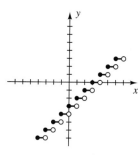

(k)

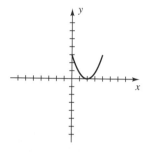

(l)

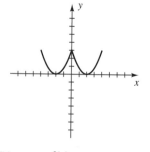

(c)

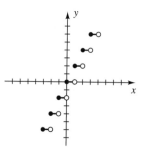

(d)

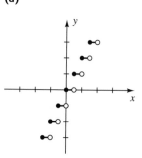

31 (a) $y = f(x + 9) + 1$ **(b)** $y = -f(x)$
    (c) $y = -f(x + 7) - 1$

(e)

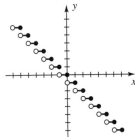

43 If $x > 0$, two different points on the graph have x-coordinate x.

45

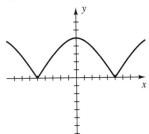

47

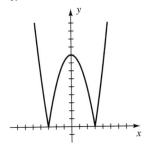

49

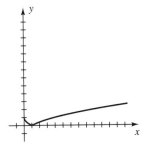

51 (a) $D = [-2, 6]$, $R = [-16, 8]$
(b) $D = [-4, 12]$, $R = [-4, 8]$
(c) $D = [1, 9]$, $R = [-3, 9]$
(d) $D = [-4, 4]$, $R = [-7, 5]$
(e) $D = [-6, 2]$, $R = [-4, 8]$
(f) $D = [-2, 6]$, $R = [-8, 4]$
(g) $D = [-6, 6]$, $R = [-4, 8]$
(h) $D = [-2, 6]$, $R = [0, 8]$

53 $T(x) = \begin{cases} 0.15x & \text{if } x \leq 20{,}000 \\ 0.20x - 1000 & \text{if } x > 20{,}000 \end{cases}$

55 $\begin{cases} 1.20x & \text{if } 0 \leq x \leq 10{,}000 \\ 1.50x - 3000 & \text{if } 10{,}000 < x \leq 15{,}000 \\ 1.80x - 7500 & \text{if } x > 15{,}000 \end{cases}$

57 $(-3.12, 22.00)$

59 $(-\infty, -3) \cup (-3, 1.87) \cup (4.13, \infty)$

61

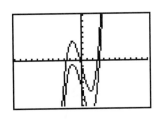

$[-12, 12]$ by $[-8, 8]$

63

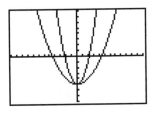

$[-12, 12]$ by $[-8, 8]$

65

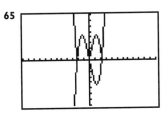

$[-12, 12]$ by $[-8, 8]$

67 (a) 194.80, 234.80

(b) $C_1(x) = \begin{cases} 119.80 & \text{if } 0 \leq x \leq 200 \\ 119.80 + 0.25(x - 200) & \text{if } x > 200 \end{cases}$

$C_2(x) = 159.8 + 0.15x$ for $x \geq 0$

(c)

x	Y_1	Y_2
100	119.8	174.8
200	119.8	189.8
300	144.8	204.8
400	169.8	219.8
500	194.8	234.8
600	219.8	249.8
700	244.8	264.8
800	269.8	279.8
900	294.8	294.8
1000	319.8	309.8
1100	344.8	324.8
1200	369.8	339.8

(d) I if $x \in [0, 900)$, II if $x > 900$

EXERCISES 1.5

I Yes   **3** No   **5** Yes   **7** No   **9** No
II Yes

Exer. 13–16: Show that $f(g(x)) = x = g(f(x))$.

13

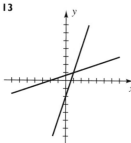

15

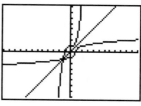

17 $f^{-1}(x) = \dfrac{x-5}{3}$ **19** $f^{-1}(x) = \dfrac{2x+1}{3x}$

21 $f^{-1}(x) = \dfrac{5x+2}{2x-3}$ **23** $f^{-1}(x) = -\sqrt{\dfrac{2-x}{3}}$

25 $f^{-1}(x) = \sqrt[3]{\dfrac{x+5}{2}}$ **27** $f^{-1}(x) = 3 - x^2,\, x \geq 0$

29 $f^{-1}(x) = (x-1)^3$ **31** $f^{-1}(x) = x$

33 $f^{-1}(x) = -\sqrt{x+4}$

35 (a)

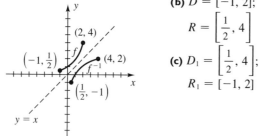

(b) $D = [-1, 2]$;
$R = \left[\dfrac{1}{2}, 4\right]$

(c) $D_1 = \left[\dfrac{1}{2}, 4\right]$;
$R_1 = [-1, 2]$

37 (a)

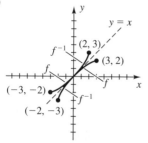

(b) $D = [-3, 3]$;
$R = [-2, 2]$
(c) $D_1 = [-2, 2]$;
$R_1 = [-3, 3]$

39 (a) Since f is one-to-one, an inverse exists;
$$f^{-1}(x) = \dfrac{x-b}{a}$$
(b) No; not one-to-one

41 (c) The graph of f is symmetric about the line $y = x$.
Thus, $f(x) = f^{-1}(x)$.

43 Yes

45

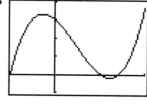

$[-1, 2]$ by $[-1, 4]$

(a) $[-0.27, 1.22]$
(b) $D = [-0.20, 3.31]$;
$R = [-0.27, 1.22]$

47

$[-12, 12]$ by $[-8, 8]$

$f^{-1}(x) = x^3 + 1$

49 (a) 805 ft³/min
(b) $V^{-1}(x) = \dfrac{1}{35}x$. Given an air circulation of x cubic
feet per minute, $V^{-1}(x)$ computes the maximum
number of people that should be in the restaurant
at one time.
(c) 67

EXERCISES 1.6

I (a)

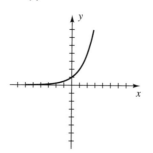

(b)

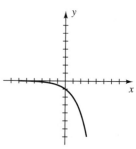

(c)

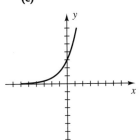

(d)

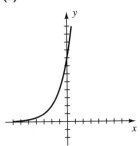

3

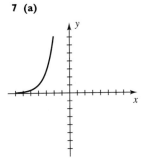

5

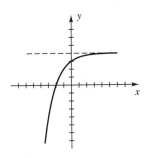

(e)

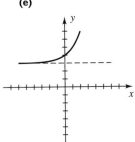

(f)

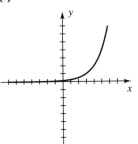

7 (a)

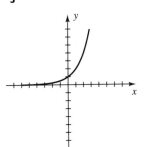

(b)

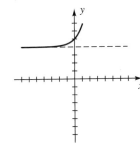

9 (a) $\log_4 64 = 3$ **(b)** $\log_4 \dfrac{1}{64} = -3$ **(c)** $\log_t s = r$

11 (a) $2^5 = 32$ **(b)** $3^{-5} = \dfrac{1}{243}$ **(c)** $t^p = r$

13 (a) $\log 100{,}000 = 5$ **(b)** $\log 0.001 = -3$
(c) $\log (y + 1) = x$

15 (a) $10^{50} = x$ **(b)** $10^{20t} = x$ **(c)** $e^{0.1} = x$

17 (a) 0   **(b)** 1   **(c)** Not possible   **(d)** 2   **(e)** 8
(f) 3   **(g)** -2

19 (a) 3   **(b)** 5   **(c)** 2   **(d)** -4 **(e)** 2
(f) -3 **(g)** $3e^2$

(g)

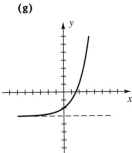

(h)

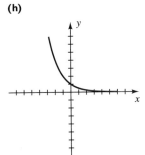

(i)

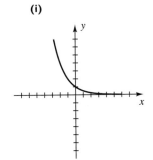

(j)

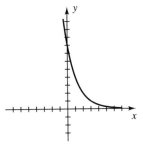

21 (a)

(b)

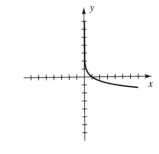

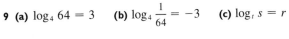

(c)

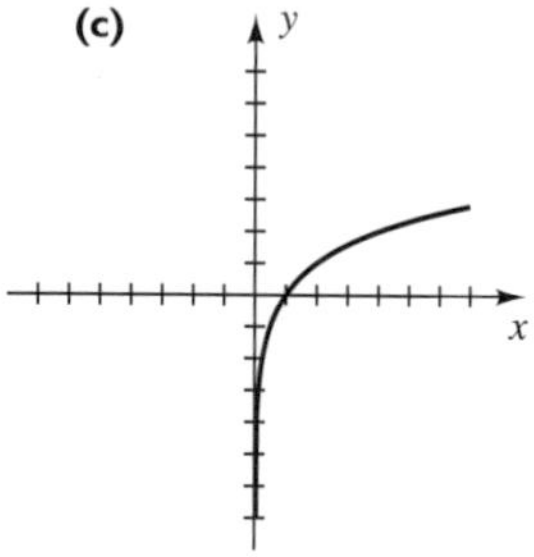

(d)

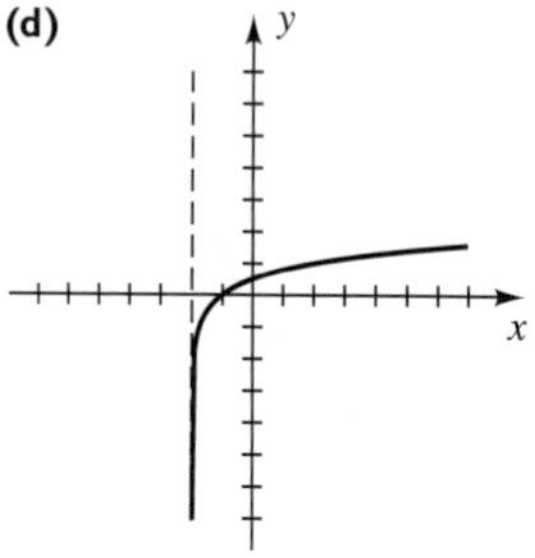

(e)

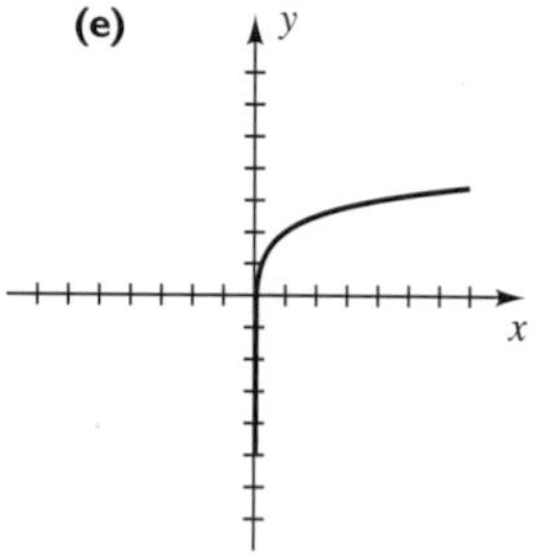

(f)

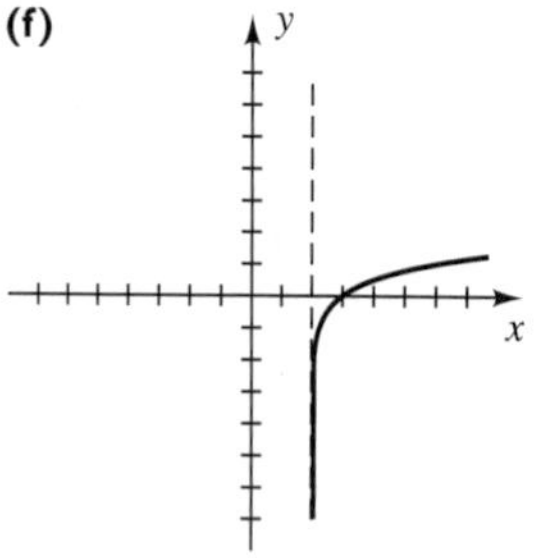

(g)

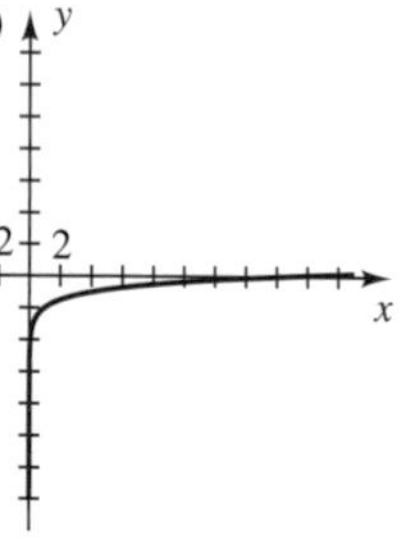

(h)

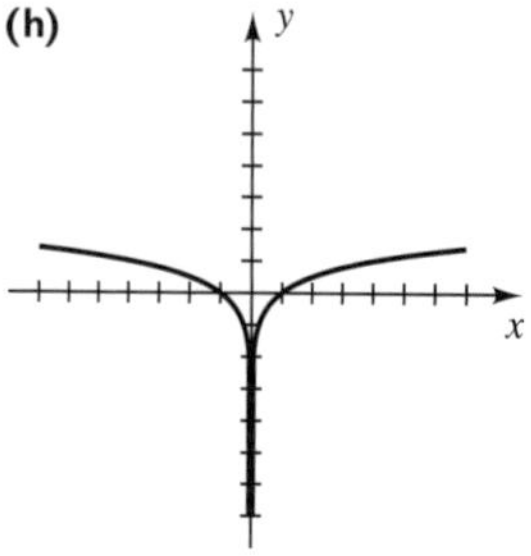

(i)

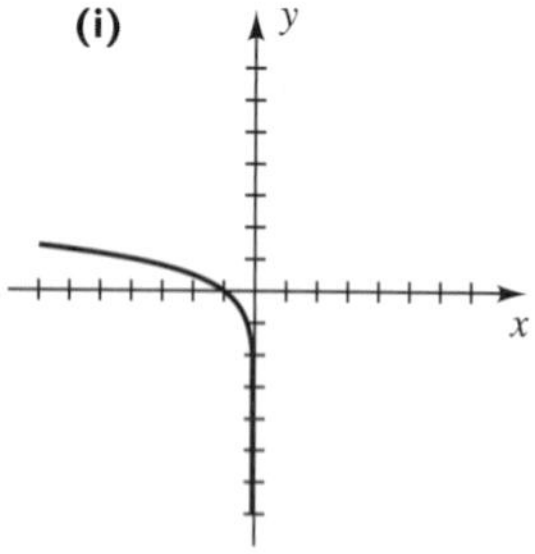

(j)

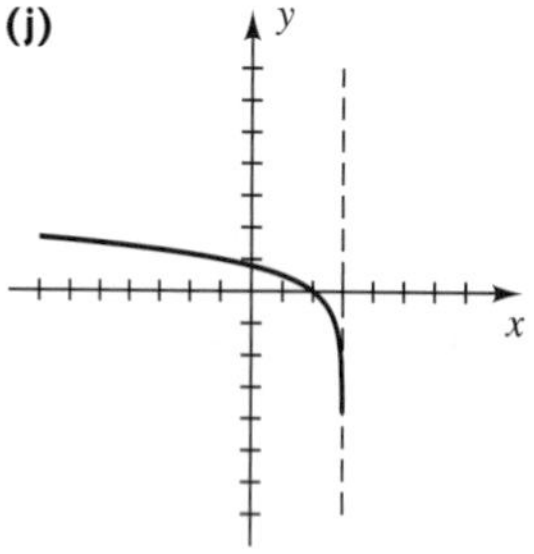

(k)

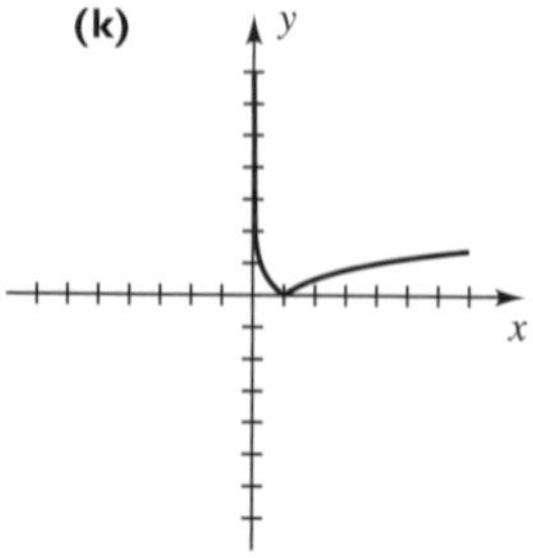

23

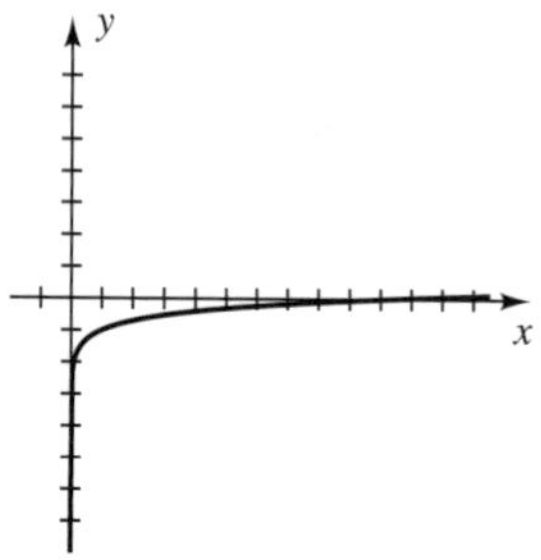

25

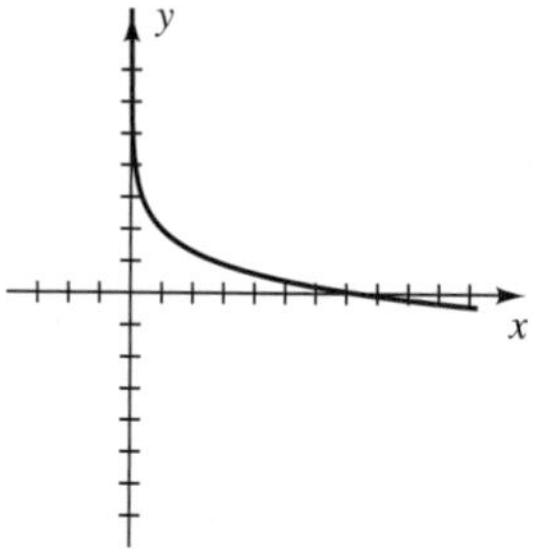

27 **(a)** $\log_4 x + \log_4 z$ **(b)** $\log_4 y - \log_4 x$

(c) $\frac{1}{3} \log_4 z$

29 $3 \log_a x + \log_a w - 2 \log_a y - 4 \log_a z$

31 1.1133 **33** −0.7325 **35** 2

37 y-intercept $= \log_2 3 \approx 1.5850$

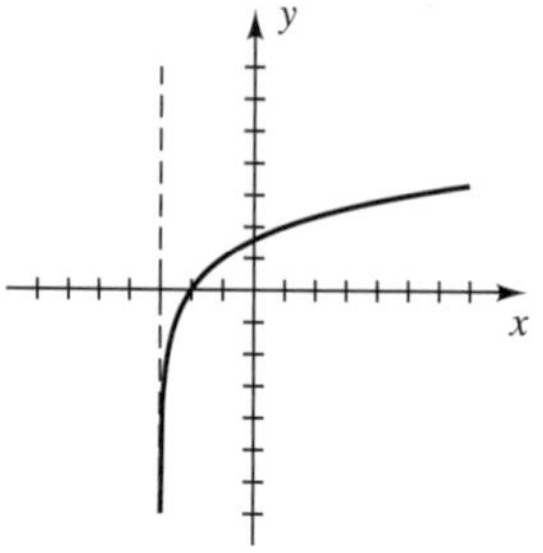

39 x-intercept $= \log_4 3 \approx 0.7925$

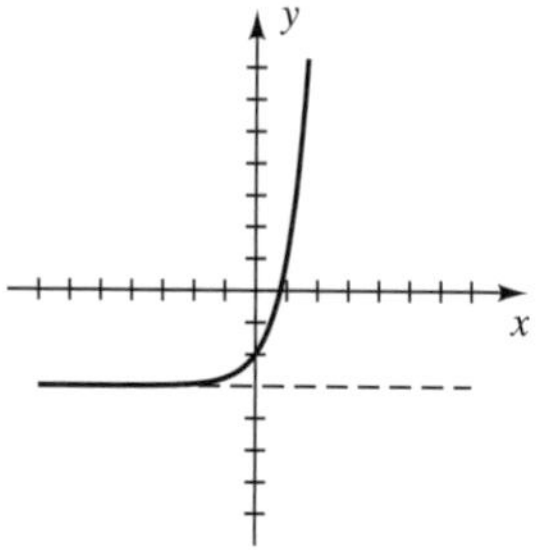

CHAPTER 1 REVIEW EXERCISES

1 **(a)** $<$ **(b)** $>$ **(c)** $>$

2 **(a)** $x < 0$ **(b)** $\frac{1}{3} < a < \frac{1}{2}$ **(c)** $|x| \le 4$

3 **(a)** 7 **(b)** −1 **(c)** $\frac{1}{6}$ **4** **(a)** 5 **(b)** 5 **(c)** 7

5 $\frac{3}{7}$ **6** $-\frac{26}{3}$ **7** $-4, \frac{3}{2}$ **8** $\frac{1}{2} \pm \frac{1}{2}\sqrt{21}$

9 $\left(\frac{2}{3}, \infty\right)$ **10** $\left(-\frac{11}{4}, \frac{9}{4}\right)$

11 **(a)** $(-\infty, 3)$

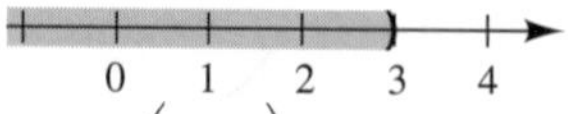

(b) $[-3, 3]$

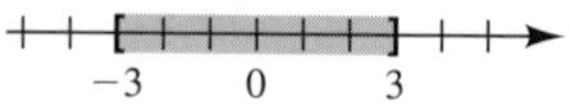

(c) $\left(0, \frac{\pi}{2}\right)$

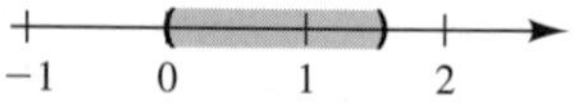

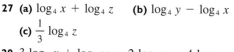

12 (a) $x \geq -5$ **(b)** $-2 < x \leq 2$ **(c)** $-\dfrac{\pi}{2} \leq x < \dfrac{3\pi}{2}$

13 The points in quadrants II and IV

14 (a) $\sqrt{265}$ **(b)** $\left(-\dfrac{13}{2}, 1\right)$

15 x-intercept: -5
    y-intercept: none

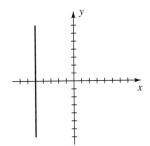

16 x-intercept: none
    y-intercept: 3.5

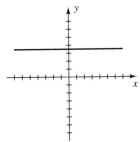

21 x-intercept: 1
    y-intercept: 1

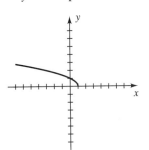

22 x-intercept: 1
    y-intercept: -1

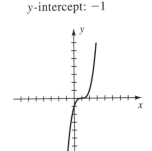

17 x-intercept: 1.6
    y-intercept: 4

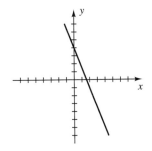

18 x-intercept: 4
    y-intercept: $-\dfrac{4}{3}$

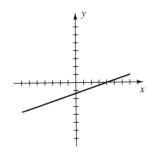

23 x-intercepts: ± 4
    y-intercepts: ± 4

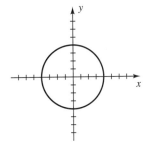

24 x-intercept: none
    y-intercept: 8

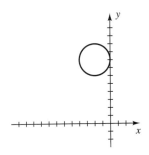

25 $(x - 7)^2 + (y + 4)^2 = 149$
26 $(x - 3)^2 + (y + 2)^2 = 169$
27 $C(0, 6)$; $r = \sqrt{5}$
28 $C(-3, 2)$; $r = \dfrac{1}{2}\sqrt{13}$

29 (a) $\dfrac{1}{2}$ **(b)** $-\dfrac{1}{\sqrt{2}}$ **(c)** 0 **(d)** $-\dfrac{x}{\sqrt{3 - x}}$

   (e) $-\dfrac{x}{\sqrt{x + 3}}$ **(f)** $\dfrac{x^2}{\sqrt{x^2 + 3}}$ **(g)** $\dfrac{x^2}{x + 3}$

30 (a) $D = \left[\dfrac{4}{3}, \infty\right)$; $R = [0, \infty)$

   (b) $D =$ All real numbers except -3; $R = (0, \infty)$

19 x-intercept: 0
    y-intercept: 0

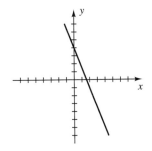

20 x-intercept: 0
    y-intercept: 0

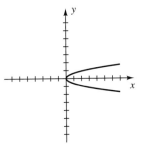

31 $-2a - h + 1$ **32 (a)** Yes **(b)** No

33 (a)

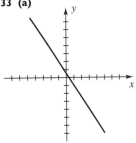

   (b) $D = \mathbb{R}$; $R = \mathbb{R}$
   (c) Decreasing on $(-\infty, \infty)$

34 (a)

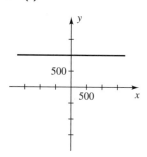

(b) $D = \mathbb{R}$;
$R = \{1000\}$

(c) Constant on
$(-\infty, \infty)$

(c)

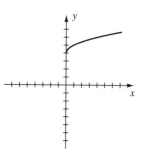

(d)

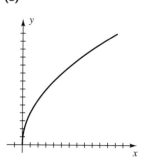

35 (a)

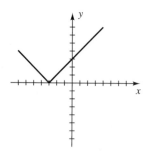

(b) $D = \mathbb{R}$;
$R = [0, \infty)$

(c) Decreasing on
$(-\infty, -3]$,
increasing on
$[-3, \infty)$

(e)

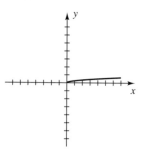

(f)

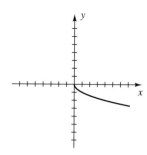

36 (a)

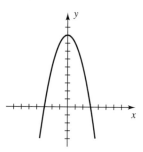

(b) $D = \mathbb{R}$;
$R = (-\infty, 9]$

(c) Increasing on
$(-\infty, 0]$,
decreasing on
$[0, \infty)$

39 (a)

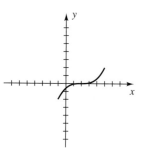

(b)

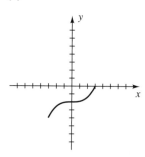

37 (a) Odd   **(b)** Neither   **(c)** Even

38 (a)

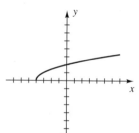

(b)

(c)

(d)

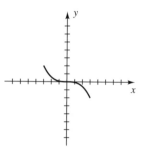
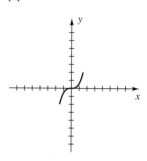

(e)

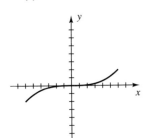

(f)

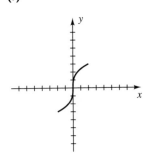

45

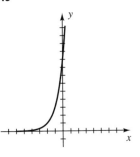

46

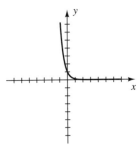

(g)

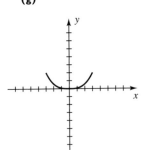

(h)

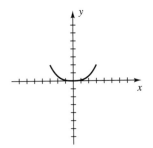

47

48

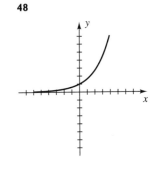

49

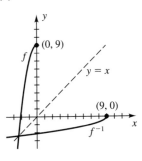

40 $(x + 2)^2 + (y - 1)^2 = 25$

41 (a) $f^{-1}(x) = \dfrac{10 - x}{15}$ **42 (a)** $f^{-1}(x) = -\sqrt{\dfrac{9 - x}{2}}$

(b)

(b)

50 (a) -4 **(b)** 0 **(c)** 1 **(d)** 4 **(e)** 6 **(f)** 8

(g) $\dfrac{1}{2}$

51 (a) $\dfrac{1}{3}$ **(b)** 0 **(c)** 1 **(d)** 5 **(e)** 1 **(f)** 25

(g) $\dfrac{1}{3}$

52 $4 \log x + \dfrac{2}{3} \log y - \dfrac{1}{3} \log z$

53 (a) $D = (-1, \infty), R = \mathbb{R}$

(b) $y = 2^x - 1, D = \mathbb{R}, R = (-1, \infty)$

54 (a) $D = \mathbb{R}, R = (-2, \infty)$

(b) $y = 3 - \log_2 (x + 2), D = (-2, \infty), R = \mathbb{R}$

43 (a) 2 **(b)** 4 **(c)** 2 **(d)** 2 **(e)** $x > 2$

44 (a) 5 **(b)** 7 **(c)** 4

(d) Not enough information is given.

55 Between 36.1 ft and 60.1 ft
56 (a) 273 ft   **(b)** 2008
57 (a) $V = 6000t + 89,000$ **(b)** $2\dfrac{1}{3}$

58 (a) $F = \dfrac{9}{5}C + 32$ **(b)** $1.8°F$

59 (a) $40.96°F$ **(b)** 6909 ft
60 (a) 3405 ft   **(b)** $80.4°F$
61 $37°F$

62 (a) $y = \dfrac{bh}{a - b}$ **(b)** $V = \dfrac{1}{3}\pi h(a^2 + ab + b^2)$

(c) $\dfrac{200}{7\pi} \approx 9.1$ ft

63 Increases 250%   **64** $A = \dfrac{\sqrt{3}}{4}s^2$ **65** (3)

CHAPTER 1 DISCUSSION EXERCISES

1 1 gallon ≈ 0.13368 ft^3; 586.85 ft^2
2 Either $a = 0$ or $b = 0$

3 (a) $\dfrac{ac + bd}{a^2 + b^2} + \dfrac{ad - bc}{a^2 + b^2}i$ **(b)** Yes

(c) a and b cannot both be 0
4 (a) 11,006 ft

(b) $h = \dfrac{1}{6}(2497D - 497G - 64,000)$

5 $R(x_3, y_3) =$

$$\left(\left(1 - \dfrac{m}{n}\right)x_1 + \dfrac{m}{n}x_2, \left(1 - \dfrac{m}{n}\right)y_1 + \dfrac{m}{n}y_2\right)$$

7 $2ax + ah + b$

8 (a) $g(x) = -\dfrac{1}{2}x + 3$ **(b)** $g(x) = -\dfrac{1}{2}x - 3$

(c) $g(x) = -\dfrac{1}{2}x + 7$ **(d)** $g(x) = -\dfrac{1}{2}x$

(e) $g(x) = 2x + 6$
9 $f(x) = 40 - 20[\![-x/15]\!]$

10 $x = \dfrac{0.4996 + \sqrt{(-0.4996)^2 - 4(0.0833)(3.5491 - D)}}{2(0.0833)}$

11 (b) $f(x) = \begin{cases} 0.132(x - 1)^2 + 0.7 & \text{if } 1 \le x \le 6 \\ -0.517x + 7.102 & \text{if } 6 < x \le 12 \end{cases}$

(c)

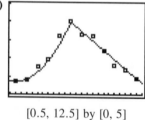

[0.5, 12.5] by [0, 5]

12

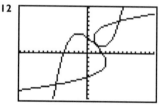

$[-15, 15]$ by $[-10, 10]$

CHAPTER 2

EXERCISES 2.1

Exer. 1–4: The answers are not unique.

1 (a) $480°, 840°, -240°, -600°$
(b) $495°, 855°, -225°, -585°$
(c) $330°, 690°, -390°, -750°$
3 (a) $260°, 980°, -100°, -460°$

(b) $\dfrac{17\pi}{6}, \dfrac{29\pi}{6}, -\dfrac{7\pi}{6}, \dfrac{19\pi}{6}$

(c) $\dfrac{7\pi}{4}, \dfrac{15\pi}{4}, -\dfrac{9\pi}{4}, -\dfrac{17\pi}{4}$

5 (a) $84°42'26''$ **(b)** $57.5°$
7 (a) $131°8'23''$ **(b)** $43.58°$

9 (a) $\dfrac{5\pi}{6}$ **(b)** $-\dfrac{\pi}{3}$ **(c)** $\dfrac{5\pi}{4}$

11 (a) $\dfrac{5\pi}{2}$ **(b)** $\dfrac{2\pi}{5}$ **(c)** $\dfrac{5\pi}{9}$

13 (a) $120°$ **(b)** $330°$ **(c)** $135°$
15 (a) $-630°$ **(b)** $1260°$ **(c)** $20°$
17 $114°35'30''$ **19** $286°28'44''$ **21** $37.6833°$
23 $115.4408°$ **25** $63°10'8''$ **27** $310°37'17''$
29 2.5 cm
31 (a) $2\pi \approx 6.28$ cm   **(b)** $8\pi \approx 25.13$ cm^2

33 (a) $1.75; \dfrac{315}{\pi} \approx 100.27°$ **(b)** 14 cm^2

35 (a) $\dfrac{20\pi}{9} \approx 6.98$ m   **(b)** $\dfrac{80\pi}{9} \approx 27.93$ m^2

37 In miles: **(a)** 4189   **(b)** 3142   **(c)** 2094
(d) 698   **(e)** 70

39 $\dfrac{1}{8}$ radian $\approx 7°10'$ **41** 7.29×10^{-5}

43 (a) 80π **(b)** $\dfrac{100\pi}{3} \approx 104.72$

45 (a) $\dfrac{200\pi}{3}, 90\pi$ **(b)** $\dfrac{100\pi}{3}, \dfrac{105\pi}{4}$

47 (a) $\dfrac{21\pi}{8} \approx 8.25$ ft   **(b)** $\dfrac{2}{3}d$

49 Large   **51** 192.08

EXERCISES 2.2

Note: Answers are in the order *sin, cos, tan, cot, sec, csc* for any exercises that require the values of the six trigonometric functions.

1 $\dfrac{8}{17}, -\dfrac{15}{17}, -\dfrac{8}{15}, -\dfrac{15}{8}, -\dfrac{17}{15}, \dfrac{17}{8}$

3 $-\dfrac{7}{25}, \dfrac{24}{25}, -\dfrac{7}{24}, -\dfrac{24}{7}, \dfrac{25}{24}, -\dfrac{25}{7}$

5 (a) $\left(-\dfrac{3}{5}, -\dfrac{4}{5}\right)$ **(b)** $\left(-\dfrac{3}{5}, -\dfrac{4}{5}\right)$ **(c)** $\left(\dfrac{3}{5}, -\dfrac{4}{5}\right)$

   (d) $\left(-\dfrac{3}{5}, \dfrac{4}{5}\right)$

7 (a) $\left(\dfrac{12}{13}, \dfrac{5}{13}\right)$ **(b)** $\left(\dfrac{12}{13}, \dfrac{5}{13}\right)$ **(c)** $\left(-\dfrac{12}{13}, \dfrac{5}{13}\right)$

   (d) $\left(\dfrac{12}{13}, -\dfrac{5}{13}\right)$

Note: U denotes *undefined.*

9 (a) $(1, 0)$; 0, 1, 0, U, 1, U
   (b) $(-1, 0)$; 0, -1, 0, U, -1, U
11 (a) $(0, -1)$; -1, 0, U, 0, U, -1
   (b) $(0, 1)$; 1, 0, U, 0, U, 1

13 (a) $\left(\dfrac{\sqrt{2}}{2}, \dfrac{\sqrt{2}}{2}\right)$; $\dfrac{\sqrt{2}}{2}, \dfrac{\sqrt{2}}{2}$, 1, 1, $\sqrt{2}, \sqrt{2}$

   (b) $\left(-\dfrac{\sqrt{2}}{2}, \dfrac{\sqrt{2}}{2}\right)$; $\dfrac{\sqrt{2}}{2}, -\dfrac{\sqrt{2}}{2}$, -1, -1, $-\sqrt{2}, \sqrt{2}$

15 (a) $\left(-\dfrac{\sqrt{2}}{2}, -\dfrac{\sqrt{2}}{2}\right)$; $-\dfrac{\sqrt{2}}{2}, -\dfrac{\sqrt{2}}{2}$, 1, 1, $-\sqrt{2}, -\sqrt{2}$

   (b) $\left(\dfrac{\sqrt{2}}{2}, -\dfrac{\sqrt{2}}{2}\right)$; $-\dfrac{\sqrt{2}}{2}, \dfrac{\sqrt{2}}{2}$, -1, -1, $\sqrt{2}, -\sqrt{2}$

17 (a) IV   **(b)** III   **(c)** II   **(d)** III

19 $\cot t = \dfrac{\sqrt{1 - \sin^2 t}}{\sin t}$ **21** $\sec t = \dfrac{1}{\sqrt{1 - \sin^2 t}}$

23 $\sin t = \dfrac{\sqrt{\sec^2 t - 1}}{\sec t}$ **25** $\dfrac{3}{5}, -\dfrac{4}{5}, -\dfrac{3}{4}, -\dfrac{4}{3}, -\dfrac{5}{4}, \dfrac{5}{3}$

27 $-\dfrac{5}{13}, \dfrac{12}{13}, -\dfrac{5}{12}, -\dfrac{12}{5}, \dfrac{13}{12}, -\dfrac{13}{5}$

29 $-\dfrac{\sqrt{8}}{3}, -\dfrac{1}{3}, \sqrt{8}, \dfrac{1}{\sqrt{8}}, -3, -\dfrac{3}{\sqrt{8}}$

31 $\dfrac{\sqrt{15}}{4}, -\dfrac{1}{4}, -\sqrt{15}, -\dfrac{1}{\sqrt{15}}, -4, \dfrac{4}{\sqrt{15}}$

33 (a) -1 **(b)** -4
35 (a) 5   **(b)** 5
37 $1 - \sin t \cos t$ **39** $\sin t$

Exer. 41–62: Typical verifications are given.

41 $\cos t \sec t = \cos t \, (1/\cos t) = 1$
43 $\sin t \sec t = \sin t \, (1/\cos t) = \sin t/\cos t = \tan t$
45 $\dfrac{\csc t}{\sec t} = \dfrac{1/\sin t}{1/\cos t} = \dfrac{\cos t}{\sin t} = \cot t$
47 $(1 + \cos 2t)(1 - \cos 2t) = 1 - \cos^2 2t = \sin^2 2t$
49 $\cos^2 t \, (\sec^2 t - 1) = \cos^2 t \, (\tan^2 t) = \cos^2 t \cdot \dfrac{\sin^2 t}{\cos^2 t}$

$\qquad = \sin^2 t$

51 $\dfrac{\sin (t/2)}{\csc (t/2)} + \dfrac{\cos (t/2)}{\sec (t/2)} = \dfrac{\sin (t/2)}{1/\sin (t/2)} + \dfrac{\cos (t/2)}{1/\cos (t/2)}$

$\qquad = \sin^2 (t/2) + \cos^2 (t/2) = 1$

53 $(1 + \sin t)(1 - \sin t) = 1 - \sin^2 t = \cos^2 t = \dfrac{1}{\sec^2 t}$

55 $\sec t - \cos t = \dfrac{1}{\cos t} - \cos t = \dfrac{1 - \cos^2 t}{\cos t} = \dfrac{\sin^2 t}{\cos t}$

$\qquad = \dfrac{\sin t}{\cos t} \cdot \sin t = \tan t \sin t$

57 $(\cot t + \csc t)(\tan t - \sin t)$
$\quad = \cot t \tan t - \cot t \sin t + \csc t \tan t - \csc t \sin t$
$\quad = \dfrac{1}{\tan t} \tan t - \dfrac{\cos t}{\sin t} \sin t + \dfrac{1}{\sin t}\dfrac{\sin t}{\cos t} - \dfrac{1}{\sin t} \sin t$
$\quad = 1 - \cos t + \dfrac{1}{\cos t} - 1 = -\cos t + \sec t$
$\qquad\qquad = \sec t - \cos t$

59 $\sec^2 3t \csc^2 3t = (1 + \tan^2 3t)(1 + \cot^2 3t)$
$\qquad = 1 + \tan^2 3t + \cot^2 3t + 1$
$\qquad = \sec^2 3t + \csc^2 3t$

61 $\log \csc t = \log \left(\dfrac{1}{\sin t}\right) = \log 1 - \log \sin t$
$\qquad = 0 - \log \sin t = -\log \sin t$

63 $-\tan t$ **65** $\sec t$ **67** $-\sin \dfrac{t}{2}$

69 (a) -0.8 **(b)** -0.9 **(c)** 0.5, 2.6
71 (a) -0.7 **(b)** 0.4   **(c)** 2.2, 4.1
73 (a)

Time	T	H	Time	T	H
12 A.M.	60	60	12 P.M.	60	60
3 A.M.	52	74	3 P.M.	68	46
6 A.M.	48	80	6 P.M.	72	40
9 A.M.	52	74	9 P.M.	68	46

(b) Max: 72°F at 6:00 P.M., 80% at 6:00 A.M.;
min: 48°F at 6:00 A.M., 40% at 6:00 P.M.

EXERCISES 2.3

I (a) -1 **(b)** $-\dfrac{\sqrt{2}}{2}$ **(c)** -1

3 (a) 1 **(b)** -1 **(c)** 1

Exer. 5–10: Typical verifications are given.

5 $\sin(-t)\sec(-t) = (-\sin t)\sec t$
$= (-\sin t)(1/\cos t)$
$= -\tan t$

7 $\dfrac{\cot(-t)}{\csc(-t)} = \dfrac{-\cot t}{-\csc t} = \dfrac{\cos t/\sin t}{1/\sin t} = \cos t$

9 $\dfrac{1}{\cos(-t)} - \tan(-t)\sin(-t)$

$$= \frac{1}{\cos t} - (-\tan t)(-\sin t)$$

$$= \frac{1}{\cos t} - \frac{\sin t}{\cos t}\sin t$$

$$= \frac{1-\sin^2 t}{\cos t} = \frac{\cos^2 t}{\cos t} = \cos t$$

II (a) 0 **(b)** -1 **13 (a)** $\dfrac{\sqrt{2}}{2}$ **(b)** -1

15 (a) 1 **(b)** $-\infty$ **17 (a)** -1 **(b)** ∞

19 (a) ∞ **(b)** $\sqrt{2}$ **21 (a)** $-\infty$ **(b)** 1

23 $\dfrac{3\pi}{2}, \dfrac{7\pi}{2}$ **25** $\dfrac{\pi}{4}, \dfrac{3\pi}{4}, \dfrac{9\pi}{4}, \dfrac{11\pi}{4}$ **27** $0, 2\pi, 4\pi$

29 $\dfrac{3\pi}{4}, \dfrac{5\pi}{4}, \dfrac{11\pi}{4}, \dfrac{13\pi}{4}$ **31** $\dfrac{\pi}{4}, \dfrac{5\pi}{4}$

33 (a) $-\dfrac{7\pi}{4}, -\dfrac{5\pi}{4}, \dfrac{\pi}{4}, \dfrac{3\pi}{4}$

(b) $-\dfrac{7\pi}{4} < t < -\dfrac{5\pi}{4}, \dfrac{\pi}{4} < t < \dfrac{3\pi}{4}$

(c) $-2\pi \le t < -\dfrac{7\pi}{4}, -\dfrac{5\pi}{4} < t < \dfrac{\pi}{4}$, and

$\dfrac{3\pi}{4} < t \le 2\pi$

35 (a) $-\dfrac{5\pi}{4}, -\dfrac{3\pi}{4}, \dfrac{3\pi}{4}, \dfrac{5\pi}{4}$

(b) $-2\pi \le t < -\dfrac{5\pi}{4}, -\dfrac{3\pi}{4} < t < \dfrac{3\pi}{4}$, and

$\dfrac{5\pi}{4} < t \le 2\pi$

(c) $-\dfrac{5\pi}{4} < t < -\dfrac{3\pi}{4}$ and $\dfrac{3\pi}{4} < t < \dfrac{5\pi}{4}$

37
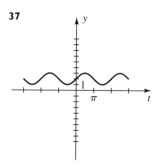

39
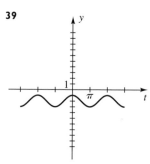

41
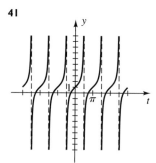

43
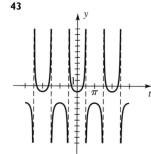

45 (a) $\left[-2\pi, -\dfrac{3\pi}{2}\right), \left(-\dfrac{3\pi}{2}, -\pi\right], \left[0, \dfrac{\pi}{2}\right), \left(\dfrac{\pi}{2}, \pi\right]$

(b) $\left[-\pi, -\dfrac{\pi}{2}\right), \left(-\dfrac{\pi}{2}, 0\right], \left[\pi, \dfrac{3\pi}{2}\right), \left(\dfrac{3\pi}{2}, 2\pi\right]$

47 (a) The tangent function increases on *all* intervals
on which it is defined. Between -2π and 2π,

these intervals are $\left[-2\pi, -\dfrac{3\pi}{2}\right), \left(-\dfrac{3\pi}{2}, -\dfrac{\pi}{2}\right),$

$\left(-\dfrac{\pi}{2}, \dfrac{\pi}{2}\right), \left(\dfrac{\pi}{2}, \dfrac{3\pi}{2}\right),$ and $\left(\dfrac{3\pi}{2}, 2\pi\right]$.

(b) The tangent function is *never* decreasing on any
interval for which it is defined.

51 $\pm 0.72, \pm 1.62, \pm 2.61,$
± 2.98

53 $(\pm 2.03, 1.82);$
$(\pm 4.91, -4.81)$

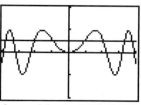

$[-\pi, \pi]$ by $[-2.09, 2.09]$

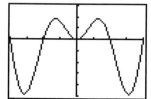

$[-2\pi, 2\pi]$ by $[-5.19, 3.19]$

55 0 **57** 1 **59** 1

EXERCISES 2.4

1 $-\dfrac{3}{5}, \dfrac{4}{5}, -\dfrac{3}{4}, -\dfrac{4}{3}, \dfrac{5}{4}, -\dfrac{5}{3}$

3 $-\dfrac{5}{\sqrt{29}}, -\dfrac{2}{\sqrt{29}}, \dfrac{5}{2}, \dfrac{2}{5}, -\dfrac{\sqrt{29}}{2}, -\dfrac{\sqrt{29}}{5}$

5 $\dfrac{4}{\sqrt{17}}, -\dfrac{1}{\sqrt{17}}, -4, -\dfrac{1}{4}, -\sqrt{17}, \dfrac{\sqrt{17}}{4}$

7 $\dfrac{4}{5}, \dfrac{3}{5}, \dfrac{4}{3}, \dfrac{3}{4}, \dfrac{5}{3}, \dfrac{5}{4}$

9 $-\dfrac{7}{\sqrt{53}}, -\dfrac{2}{\sqrt{53}}, \dfrac{7}{2}, \dfrac{2}{7}, -\dfrac{\sqrt{53}}{2}, -\dfrac{\sqrt{53}}{7}$

Note: U denotes *undefined.*

11 (a) 1, 0, U, 0, U, 1   **(b)** 0, 1, 0, U, 1, U
   (c) −1, 0, U, 0, U, −1   **(d)** 0, −1, 0, U, −1, U

13 $\dfrac{3}{5}, \dfrac{4}{5}, \dfrac{3}{4}, \dfrac{4}{3}, \dfrac{5}{4}, \dfrac{5}{3}$ **15** $\dfrac{5}{13}, \dfrac{12}{13}, \dfrac{5}{12}, \dfrac{12}{5}, \dfrac{13}{12}, \dfrac{13}{5}$

17 $\dfrac{\sqrt{11}}{6}, \dfrac{5}{6}, \dfrac{\sqrt{11}}{5}, \dfrac{5}{\sqrt{11}}, \dfrac{6}{5}, \dfrac{6}{\sqrt{11}}$

19 $\dfrac{4}{5}, \dfrac{3}{5}, \dfrac{4}{3}, \dfrac{3}{4}, \dfrac{5}{3}, \dfrac{5}{4}$

21 $\dfrac{2}{5}, \dfrac{\sqrt{21}}{5}, \dfrac{2}{\sqrt{21}}, \dfrac{\sqrt{21}}{2}, \dfrac{5}{\sqrt{21}}, \dfrac{5}{2}$

23 $\dfrac{a}{\sqrt{a^2+b^2}}, \dfrac{b}{\sqrt{a^2+b^2}}, \dfrac{a}{b}, \dfrac{b}{a}, \dfrac{\sqrt{a^2+b^2}}{b}, \dfrac{\sqrt{a^2+b^2}}{a}$

25 $\dfrac{b}{c}, \dfrac{\sqrt{c^2-b^2}}{c}, \dfrac{b}{\sqrt{c^2-b^2}}, \dfrac{\sqrt{c^2-b^2}}{b}, \dfrac{c}{\sqrt{c^2-b^2}}, \dfrac{c}{b}$

27 $x = 8; y = 4\sqrt{3}$ **29** $x = 7\sqrt{2}; y = 7$
31 $x = 4\sqrt{3}; y = 4$ **33** $200\sqrt{3} \approx 346.4$ ft
35 192 ft   **37** 1.02 m

EXERCISES 2.5

1 (a) 60°   **(b)** 20°   **(c)** 22°   **(d)** 60°

3 (a) $\dfrac{\pi}{4}$ **(b)** $\dfrac{\pi}{3}$ **(c)** $\dfrac{\pi}{6}$ **(d)** $\dfrac{\pi}{4}$

5 (a) $\pi - 3 \approx 8.1°$ **(b)** $\pi - 2 \approx 65.4°$
   (c) $2\pi - 5.5 \approx 44.9°$ **(d)** $32\pi - 100 \approx 30.4°$

7 (a) $\dfrac{\sqrt{3}}{2}$ **(b)** $\dfrac{\sqrt{2}}{2}$ **9 (a)** $-\dfrac{\sqrt{3}}{2}$ **(b)** $\dfrac{1}{2}$

11 (a) $-\dfrac{\sqrt{3}}{3}$ **(b)** $-\sqrt{3}$ **13 (a)** $-\dfrac{\sqrt{3}}{3}$ **(b)** $\sqrt{3}$

15 (a) -2 **(b)** $\dfrac{2}{\sqrt{3}}$ **17 (a)** $-\dfrac{2}{\sqrt{3}}$ **(b)** 2

19 (a) 0.958   **(b)** 0.778   **21 (a)** 0.387   **(b)** 0.472
23 (a) 2.650   **(b)** 3.179   **25 (a)** 30.46°   **(b)** 30°27′
27 (a) 74.88°   **(b)** 74°53′
29 (a) 24.94°   **(b)** 24°57′
31 (a) 76.38°   **(b)** 76°23′
33 (a) 0.9899   **(b)** −0.1097   **(c)** −0.1425
   (d) 0.7907   **(e)** −11.2493   **(f)** 1.3677
35 (a) 214.3°, 325.7°   **(b)** 41.5°, 318.5°
   (c) 70.3°, 250.3°   **(d)** 133.8°, 313.8°
   (e) 153.6°, 206.4°   **(f)** 42.3°, 137.7°
37 (a) 0.43, 2.71   **(b)** 1.69, 4.59   **(c)** 1.87, 5.01
   (d) 0.36, 3.50   **(e)** 0.96, 5.32   **(f)** 3.35, 6.07
39 0.28 cm
41 (a) The maximum occurs when the sun is rising in the east.
   (b) $\dfrac{\sqrt{2}}{4} \approx 35\%$
43 $(9, 9\sqrt{3})$

EXERCISES 2.6

1 (a) 4, 2π **(b)** 1, $\dfrac{\pi}{2}$

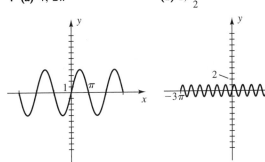

(c) $\dfrac{1}{4}$, 2π **(d)** 1, 8π

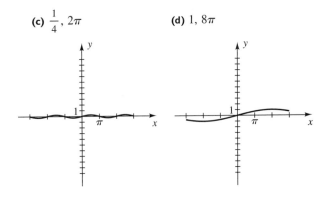

(e) $2, 8\pi$

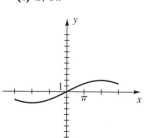

(f) $\dfrac{1}{2}, \dfrac{\pi}{2}$

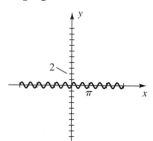

(e) $2, 6\pi$

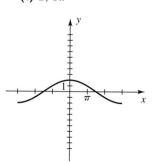

(f) $\dfrac{1}{2}, \dfrac{2\pi}{3}$

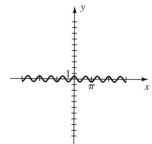

(g) $4, 2\pi$

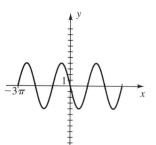

(h) $1, \dfrac{\pi}{2}$

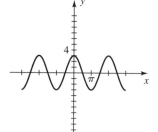

(g) $3, 2\pi$

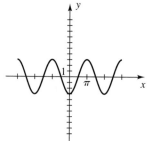

(h) $1, \dfrac{2\pi}{3}$

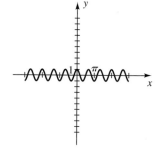

3 (a) $3, 2\pi$

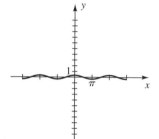

(b) $1, \dfrac{2\pi}{3}$

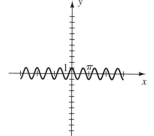

5 $1, 2\pi, \dfrac{\pi}{2}$

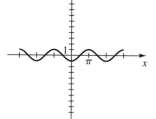

7 $3, 2\pi, -\dfrac{\pi}{6}$

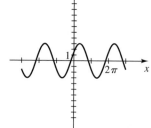

(c) $\dfrac{1}{3}, 2\pi$

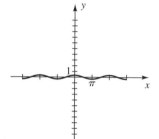

(d) $1, 6\pi$

9 $1, 2\pi, -\dfrac{\pi}{2}$

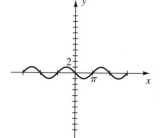

11 $4, 2\pi, \dfrac{\pi}{4}$

13 $1, \pi, \dfrac{\pi}{2}$

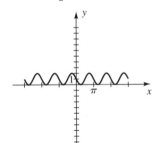

15 $1, \dfrac{2\pi}{3}, -\dfrac{\pi}{3}$

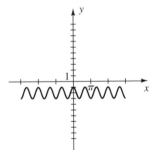

29 $3, 4\pi, \dfrac{\pi}{2}$

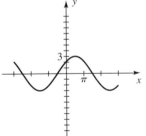

31 $5, 6\pi, -\dfrac{\pi}{2}$

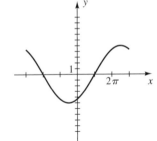

17 $2, \dfrac{2\pi}{3}, \dfrac{\pi}{3}$

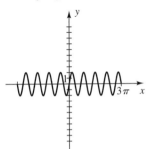

19 $1, 4\pi, \dfrac{2\pi}{3}$

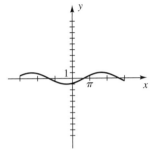

33 $3, 2, -4$

35 $\sqrt{2}, 4, \dfrac{1}{2}$

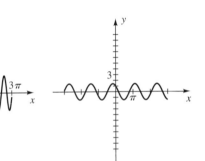

21 $6, 2, 0$

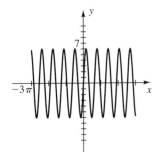

23 $2, 4, 0$

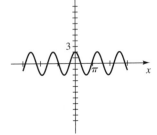

37 $2, \pi, \dfrac{\pi}{2}$

39 $5, \pi, -\pi$

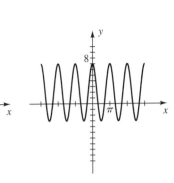

25 $\dfrac{1}{2}, 1, 0$

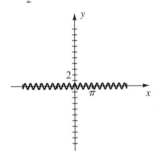

27 $5, \dfrac{2\pi}{3}, \dfrac{\pi}{6}$

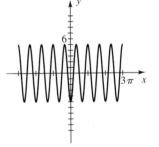

41 (a) $4, 2\pi, -\pi$ (b) $y = 4 \sin (x + \pi)$

43 (a) $2, 4, -3$ (b) $y = 2 \sin \left(\dfrac{\pi}{2} x + \dfrac{3\pi}{2} \right)$

45 4π **47** $a = 8, b = 4\pi$

49

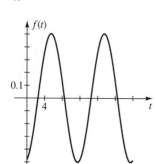

51

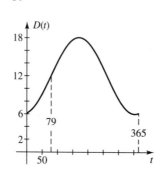

(b)

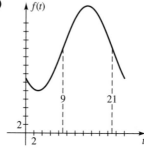

53 The temperature is 20°F at 9:00 A.M. ($t = 0$). It increases to a high of 35°F at 3:00 P.M. ($t = 6$) and then decreases to 20°F at 9:00 P.M. ($t = 12$). It continues to decrease to a low of 5°F at 3:00 A.M. ($t = 18$). It then rises to 20°F at 9:00 A.M. ($t = 24$).

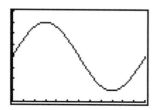

[0, 24] by [0, 40]

55 (a) $f(t) = 10 \sin\left[\dfrac{\pi}{12}(t - 10)\right] + 0$, with $a = 10$,

$b = \dfrac{\pi}{12}$, $c = -\dfrac{5\pi}{6}$, $d = 0$

(b)

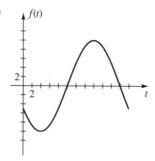

57 (a) $f(t) = 10 \sin\left[\dfrac{\pi}{12}(t - 9)\right] + 20$, with $a = 10$,

$b = \dfrac{\pi}{12}$, $c = -\dfrac{3\pi}{4}$, $d = 20$

59 (a)

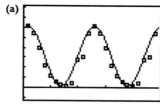

[0.5, 24.5] by [−1, 8]

(b) $P(t) = 2.95 \sin\left(\dfrac{\pi}{6}t + \dfrac{\pi}{3}\right) + 3.15$

61 (a)

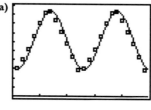

[0.5, 24.5] by [0, 20]

(b) $D(t) = 6.42 \sin\left(\dfrac{\pi}{6}t - \dfrac{2\pi}{3}\right) + 12.3$

63 As $x \to 0^-$ or as $x \to 0^+$, y oscillates between -1 and 1 and does not approach a unique value.

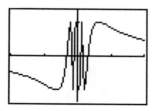

[−2, 2] by [−1.33, 1.33]

65 As $x \to 0^-$ or as $x \to 0^+$, y appears to approach 2.

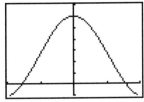

$[-2, 2]$ by $[-0.33, 2.33]$

67 $y = 4$

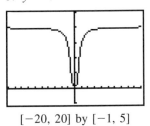

$[-20, 20]$ by $[-1, 5]$

69 $[-\pi, -1.63] \cup$
    $[-0.45, 0.61] \cup$
    $[1.49, 2.42]$

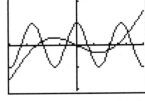

$[-\pi, \pi]$ by $[-2.09, 2.09]$

EXERCISES 2.7

1 π

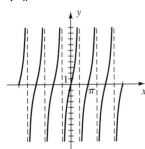

3 π

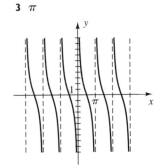

5 2π

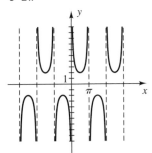

7 2π

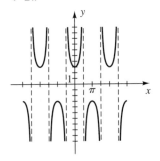

9 π

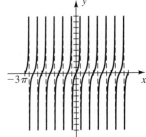

11 $\dfrac{\pi}{2}$

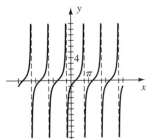

13 4π

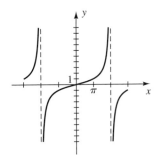

15 $\dfrac{\pi}{2}$

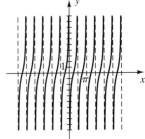

17 2π

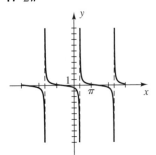

19 π

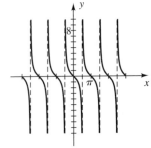

21 $\dfrac{\pi}{2}$

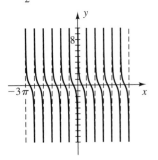

23 3π

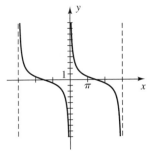

25 $\dfrac{\pi}{2}$

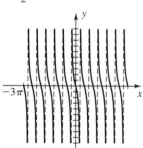

27 2π

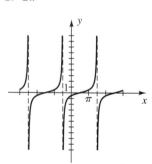

41 π

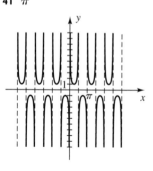

43 6π

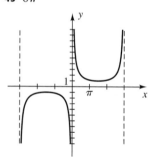

29 2π

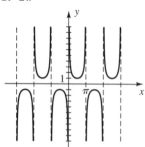

31 π

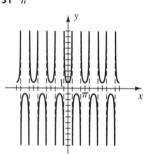

45 π

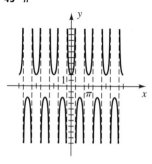

47 4π

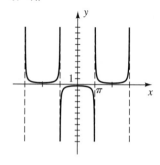

33 6π

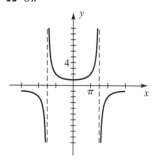

35 π

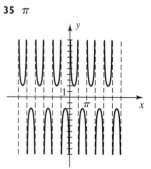

49 2

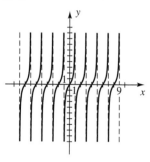

51 1

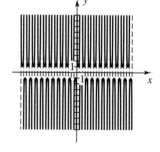

53 $y = -\cot\left(x + \dfrac{\pi}{2}\right)$

37 4π

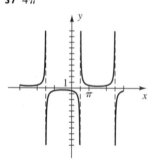

39 2π

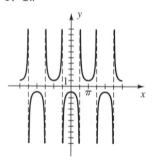

55

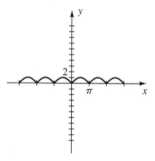

57

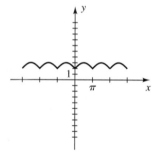

59

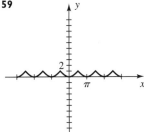

61

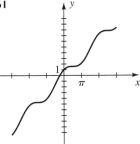

63

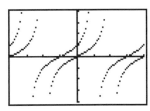

65

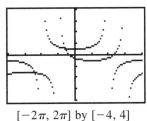

67

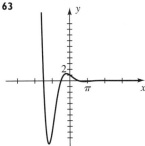

$[-2\pi, 2\pi]$ by $[-4, 4]$

69

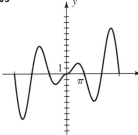

$[-2\pi, 2\pi]$ by $[-4, 4]$

71

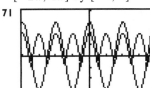

$[-2\pi, 2\pi]$ by $[-4, 4]$

73 $e^{-x/4}$

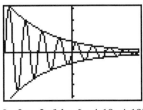

$[-2\pi, 2\pi]$ by $[-4.19, 4.19]$

75 $(-2.76, 3.09)$;
$(1.23, -3.68)$

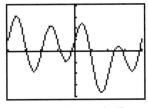

$[-\pi, \pi]$ by $[-4, 4]$

77 $[-0.70, 0.12]$

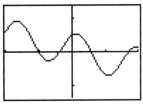

$[-2, 2]$ by $[-1.33, 1.33]$

79 $[-\pi, -1.31] \cup$
$[0.11, 0.95] \cup$
$[2.39, \pi]$

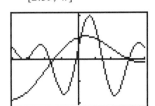

$[-\pi, \pi]$ by $[-2.09, 2.09]$

81 (a) I_0 **(b)** $0.044I_0$ **(c)** $0.603I_0$

83 (a) $A_0e^{-\alpha z}$ **(b)** $\dfrac{\alpha}{k}z_0$ **(c)** $\dfrac{\ln 2}{\alpha}$

EXERCISES 2.8

1 $\beta = 60°$, $a = \dfrac{20}{3}\sqrt{3}$, $c = \dfrac{40}{3}\sqrt{3}$

3 $\alpha = 45°$, $a = b = 15\sqrt{2}$

5 $\alpha = \beta = 45°$, $c = 5\sqrt{2}$

7 $\alpha = 60°$, $\beta = 30°$, $a = 15$

9 $\beta = 53°$, $a \approx 18$, $c \approx 30$

11 $\alpha = 18°9'$, $a \approx 78.7$, $c \approx 252.6$

13 $\alpha \approx 29°$, $\beta \approx 61°$, $c \approx 51$

15 $\alpha \approx 69°$, $\beta \approx 21°$, $a \approx 5.4$ **17** $b = c \cos \alpha$

19 $a = b \cot \beta$ **21** $c = a \csc \alpha$

23 $b = \sqrt{c^2 - a^2}$

25 $250\sqrt{3} + 4 \approx 437$ ft **27** 28,800 ft **29** 160 m

31 9659 ft **33 (a)** 58 ft **(b)** 27 ft **35** 51°20′

37 16.3° **39** 2063 ft **41** 1,459,379 ft²

43 21.8° **45** 20.2 m **47** 29.7 km **49** 3944 mi

51 126 mi/hr

53 (a) 45%

(b) Each satellite has a signal range of more than 120°.

55 $h = d \sin \alpha + c$ **57** $h = \dfrac{d}{\cot \alpha - \cot \beta}$

59 $h = d(\tan \beta - \tan \alpha)$

61 N70°E; N40°W; S15°W; S25°E

63 (a) 55 mi **(b)** S63°E **65** 324 mi

67 Amplitude, 10 cm; period, $\dfrac{1}{3}$ sec; frequency, 3 osc/sec. The point is at the origin at $t = 0$. It moves upward with decreasing speed, reaching the point with coordinate 10 at $t = \dfrac{1}{12}$. It then reverses direction and moves downward, gaining speed until it reaches the origin at $t = \dfrac{1}{6}$. It continues downward with decreasing speed, reaching the point with coordinate -10 at $t = \dfrac{1}{4}$. It then reverses direction and moves upward with increasing speed, returning to the origin at $t = \dfrac{1}{3}$.

69 Amplitude, 4 cm; period, $\dfrac{4}{3}$ sec; frequency, $\dfrac{3}{4}$ osc/sec. The motion is similar to that in Exercise 67; however, the point starts 4 units above the origin and moves downward, reaching the origin at $t = \dfrac{1}{3}$ and the point with coordinate -4 at $t = \dfrac{2}{3}$. It then reverses direction and moves upward, reaching the origin at $t = 1$ and its initial point at $t = \dfrac{4}{3}$.

71 $d = 5 \cos \dfrac{2\pi}{3} t$

73 (a) $y = 25 \cos \dfrac{\pi}{15} t$

(b) 324,000 ft

CHAPTER 2 REVIEW EXERCISES

1 $\dfrac{11\pi}{6}, \dfrac{9\pi}{4}, -\dfrac{5\pi}{6}, \dfrac{4\pi}{3}, \dfrac{\pi}{5}$

2 $810°, -120°, 315°, 900°, 36°$

3 (a) 0.1   (b) 0.2 m^2

4 (a) $\dfrac{35\pi}{12}$ cm   (b) $\dfrac{175\pi}{16}$ cm^2

5 $(-1, 0)$; $(0, -1)$; $(0, 1)$; $\left(-\dfrac{\sqrt{2}}{2}, -\dfrac{\sqrt{2}}{2}\right)$; $(1, 0)$; $\left(\dfrac{\sqrt{3}}{2}, \dfrac{1}{2}\right)$

6 $\left(\dfrac{3}{5}, \dfrac{4}{5}\right)$; $\left(\dfrac{3}{5}, \dfrac{4}{5}\right)$; $\left(-\dfrac{3}{5}, \dfrac{4}{5}\right)$; $\left(-\dfrac{3}{5}, \dfrac{4}{5}\right)$

7 (a) $-\dfrac{4}{5}, \dfrac{3}{5}, -\dfrac{4}{3}, -\dfrac{3}{4}, \dfrac{5}{3}, -\dfrac{5}{4}$

(b) $\dfrac{2}{\sqrt{13}}, -\dfrac{3}{\sqrt{13}}, -\dfrac{2}{3}, -\dfrac{3}{2}, -\dfrac{\sqrt{13}}{3}, \dfrac{\sqrt{13}}{2}$

8 (a) II   (b) III   (c) IV

9 $\tan t = \sqrt{\sec^2 t - 1}$ **10** $\cot t = \sqrt{\csc^2 t - 1}$

Exer. 11–20: Typical verifications are given.

11 $\sin t \,(\csc t - \sin t) = \sin t \csc t - \sin^2 t$
$= 1 - \sin^2 t = \cos^2 t$

12 $\cos t \,(\tan t + \cot t) = \cos t \cdot \dfrac{\sin t}{\cos t} + \cos t \cdot \dfrac{\cos t}{\sin t}$
$= \sin t + \dfrac{\cos^2 t}{\sin t} = \dfrac{\sin^2 t + \cos^2 t}{\sin t}$
$= \dfrac{1}{\sin t} = \csc t$

13 $(\cos^2 t - 1)(\tan^2 t + 1) = (\cos^2 t - 1)(\sec^2 t)$
$= \cos^2 t \sec^2 t - \sec^2 t$
$= 1 - \sec^2 t$

14 $\dfrac{\sec t - \cos t}{\tan t} = \dfrac{\dfrac{1}{\cos t} - \cos t}{\dfrac{\sin t}{\cos t}} = \dfrac{\dfrac{1 - \cos^2 t}{\cos t}}{\dfrac{\sin t}{\cos t}} = \dfrac{\dfrac{\sin^2 t}{\cos t}}{\dfrac{\sin t}{\cos t}}$
$= \dfrac{\dfrac{\sin t}{\cos t}}{\dfrac{1}{\cos t}} = \dfrac{\tan t}{\sec t}$

15 $\dfrac{1 + \tan^2 t}{\tan^2 t} = \dfrac{1}{\tan^2 t} + \dfrac{\tan^2 t}{\tan^2 t} = \cot^2 t + 1 = \csc^2 t$

16 $\dfrac{\sec t + \csc t}{\sec t - \csc t} = \dfrac{\dfrac{1}{\cos t} + \dfrac{1}{\sin t}}{\dfrac{1}{\cos t} - \dfrac{1}{\sin t}} = \dfrac{\dfrac{\sin t + \cos t}{\cos t \sin t}}{\dfrac{\sin t - \cos t}{\cos t \sin t}}$
$= \dfrac{\sin t + \cos t}{\sin t - \cos t}$

17 $\dfrac{\cot t - 1}{1 - \tan t} = \dfrac{\dfrac{\cos t}{\sin t} - 1}{1 - \dfrac{\sin t}{\cos t}} = \dfrac{\dfrac{\cos t - \sin t}{\sin t}}{\dfrac{\cos t - \sin t}{\cos t}}$
$= \dfrac{(\cos t - \sin t)\cos t}{(\cos t - \sin t)\sin t} = \dfrac{\cos t}{\sin t} = \cot t$

18 $\dfrac{1 + \sec t}{\tan t + \sin t} = \dfrac{1 + \dfrac{1}{\cos t}}{\dfrac{\sin t}{\cos t} + \sin t \cos t} = \dfrac{\dfrac{\cos t + 1}{\cos t}}{\dfrac{\sin t \,(1 + \cos t)}{\cos t}}$
$= \dfrac{1}{\sin t} = \csc t$

19 $\dfrac{\tan (-t) + \cot (-t)}{\tan t} = \dfrac{-\tan t - \cot t}{\tan t} = -\dfrac{\tan t}{\tan t} - \dfrac{\cot t}{\tan t}$
$= -1 - \cot^2 t = -(1 + \cot^2 t)$
$= -\csc^2 t$

20 $-\dfrac{1}{\csc(-t)} - \dfrac{\cot(-t)}{\sec(-t)} = -\dfrac{1}{-\csc t} - \dfrac{-\cot t}{\sec t}$

$$= \sin t + \dfrac{\cos t/\sin t}{1/\cos t}$$

$$= \sin t + \dfrac{\cos^2 t}{\sin t} = \dfrac{\sin^2 t + \cos^2 t}{\sin t}$$

$$= \dfrac{1}{\sin t} = \csc t$$

21 (a) $-\dfrac{4}{5}, \dfrac{3}{5}, -\dfrac{4}{3}, -\dfrac{3}{4}, \dfrac{5}{3}, -\dfrac{5}{4}$

(b) $\dfrac{2}{\sqrt{13}}, -\dfrac{3}{\sqrt{13}}, -\dfrac{2}{3}, -\dfrac{3}{2}, -\dfrac{\sqrt{13}}{3}, \dfrac{\sqrt{13}}{2}$

(c) $-1, 0, U, 0, U, -1$

22 $\dfrac{\sqrt{33}}{7}, \dfrac{4}{7}, \dfrac{\sqrt{33}}{4}, \dfrac{4}{\sqrt{33}}, \dfrac{7}{4}, \dfrac{7}{\sqrt{33}}$

23 $x = 6\sqrt{3};\ y = 3\sqrt{3}$　　**24** $x = \dfrac{7}{2}\sqrt{2};\ y = \dfrac{7}{2}\sqrt{2}$

25 (a) $\dfrac{\pi}{4}, \dfrac{\pi}{6}, \dfrac{\pi}{8}$　　**(b)** $65°, 43°, 8°$

26 (a) $1, 0, U, 0, U, 1$

(b) $\dfrac{\sqrt{2}}{2}, -\dfrac{\sqrt{2}}{2}, -1, -1, -\sqrt{2}, \sqrt{2}$

(c) $0, 1, 0, U, 1, U$

(d) $-\dfrac{1}{2}, \dfrac{\sqrt{3}}{2}, -\dfrac{\sqrt{3}}{3}, -\sqrt{3}, \dfrac{2}{\sqrt{3}}, -2$

27 (a) $-\dfrac{\sqrt{2}}{2}$　**(b)** $-\dfrac{\sqrt{3}}{3}$　**(c)** $-\dfrac{1}{2}$　**(d)** -2

(e) -1　**(f)** $-\dfrac{2}{\sqrt{3}}$

28 $310.5°$

29 $5, 2\pi$　　　　　　　**30** $\dfrac{2}{3}, 2\pi$

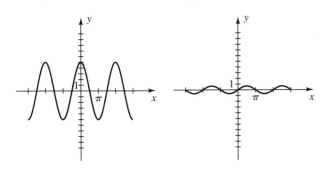

31 $\dfrac{1}{3}, \dfrac{2\pi}{3}$　　　　　**32** $\dfrac{1}{2}, 6\pi$

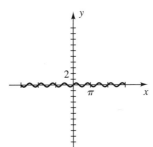

33 $3, 4\pi$　　　　　　　**34** $4, \pi$

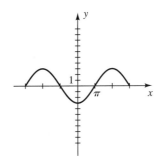

35 $2, 2$　　　　　　　　**36** $4,4$

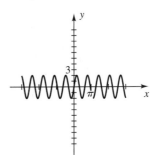

37 (a) $1.43, 2$　**(b)** $y = 1.43 \sin \pi x$

38 (a) $3.27, 3\pi$　**(b)** $y = -3.27 \sin \dfrac{2}{3}x$

39 (a) $3, \dfrac{4\pi}{3}$　**(b)** $y = -3 \cos \dfrac{3}{2}x$

40 (a) $2, \dfrac{\pi}{2}$　**(b)** $y = 2 \cos \dfrac{\pi}{2}x$

41

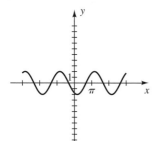

42

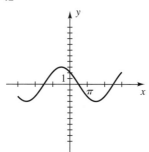

49

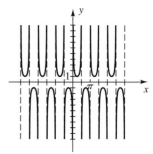

50

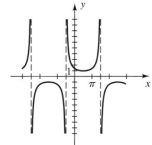

43

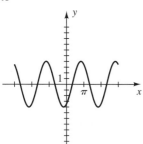

44

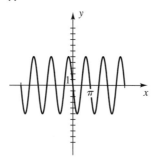

51

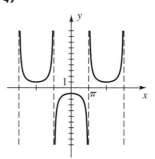

52

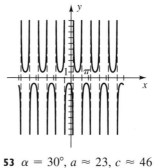

45

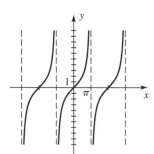

46

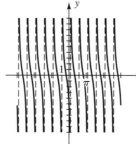

53 $\alpha = 30°$, $a \approx 23$, $c \approx 46$
54 $\beta = 35°20'$, $a \approx 310$, $c \approx 380$
55 $\alpha \approx 68°$, $\beta \approx 22°$, $c \approx 67$
56 $\alpha \approx 13°$, $\beta \approx 77°$, $b = 40$
57 (a) $\dfrac{109\pi}{6}$ **(b)** 440.2
58 1048 ft
59 0.093 mi/sec   **60** 52°
61 Approximately 67,900,000 mi
62 $\dfrac{6\pi}{5}$ radians $= 216°$ **63** 250 ft
64 (a) 231.0 ft   **(b)** 434.5
65 (b) 2 mi
66 (a) $T = h + d(\cos \alpha \tan \theta - \sin \alpha)$ **(b)** 22.54 ft
67 (a) $\dfrac{25}{3}\sqrt{3} \approx 14.43$ ft-candles   **(b)** 37.47°
68 (b) 4.69
69 (a) 74.05 in.   **(b)** 24.75 in.
70 (a) $S = 4a^2 \sin \theta$ **(b)** $V = \dfrac{4}{3}a^3 \sin^2 \theta \cos \theta$
71 (a) $h = R \sec \dfrac{s}{R} - R$ **(b)** $h \approx 1650$ ft

47

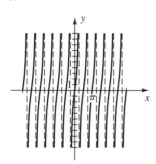

48

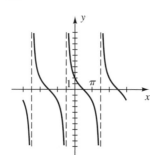

72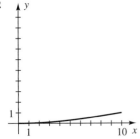

73 $y = 98.6 + (0.3) \sin\left(\dfrac{\pi}{12}t - \dfrac{11\pi}{12}\right)$

74 (a) 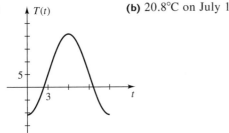   **(b)** 20.8°C on July 1

75 (a) 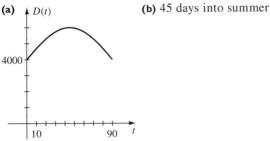   **(b)** 45 days into summer

76 (a) The cork is in simple harmonic motion.
    (b) $1 \le t \le 2$

CHAPTER 2 DISCUSSION EXERCISES

3 None
5 The values of y_1, y_2, and y_3 are very close to each other near $x = 0$.
6 (a) $x \approx -0.4161$, $y \approx 0.9093$
    (b) $x \approx -0.8838$, $y \approx -0.4678$
7 (a) $x \approx 1.8415$, $y \approx -0.5403$
    (b) $x \approx -1.2624$, $y \approx 0.9650$
8 (a) $\dfrac{500\pi}{3}$ rad/sec   **(b)** $D(t) = 5 \cos\left(\dfrac{500\pi}{3}t\right) + 18$
    (c) 10 revolutions

CHAPTER 3

EXERCISES 3.1

Exer. 1–50: Typical verifications are given for Exercises 1, 5, 9, ..., 49.

1 $\csc\theta - \sin\theta = \dfrac{1}{\sin\theta} - \sin\theta = \dfrac{1 - \sin^2\theta}{\sin\theta} = \dfrac{\cos^2\theta}{\sin\theta}$
$$= \dfrac{\cos\theta}{\sin\theta}\cos\theta = \cot\theta\cos\theta$$

5 $\dfrac{\csc^2\theta}{1 + \tan^2\theta} = \dfrac{\csc^2\theta}{\sec^2\theta} = \dfrac{1/\sin^2\theta}{1/\cos^2\theta} = \dfrac{\cos^2\theta}{\sin^2\theta}$
$$= \left(\dfrac{\cos\theta}{\sin\theta}\right)^2 = \cot^2\theta$$

9 $\dfrac{1}{1 - \cos\gamma} + \dfrac{1}{1 + \cos\gamma} = \dfrac{1 + \cos\gamma + 1 - \cos\gamma}{1 - \cos^2\gamma} = \dfrac{2}{\sin^2\gamma}$
$$= 2\csc^2\gamma$$

13 $\csc^4 t - \cot^4 t = (\csc^2 t + \cot^2 t)(\csc^2 t - \cot^2 t)$
$$= (\csc^2 t + \cot^2 t)(1)$$
$$= \csc^2 t + \cot^2 t$$

17 $\dfrac{\tan^2 x}{\sec x + 1} = \dfrac{\sec^2 x - 1}{\sec x + 1} = \dfrac{(\sec x + 1)(\sec x - 1)}{\sec x + 1}$
$$= \sec x - 1 = \dfrac{1}{\cos x} - 1 = \dfrac{1 - \cos x}{\cos x}$$

21 $\sin^4 r - \cos^4 r = (\sin^2 r - \cos^2 r)(\sin^2 r + \cos^2 r)$
$$= (\sin^2 r - \cos^2 r)(1)$$
$$= \sin^2 r - \cos^2 r$$

25 $(\sec t + \tan t)^2 = \left(\dfrac{1}{\cos t} + \dfrac{\sin t}{\cos t}\right)^2 = \left(\dfrac{1 + \sin t}{\cos t}\right)^2$
$$= \dfrac{(1 + \sin t)^2}{\cos^2 t} = \dfrac{(1 + \sin t)^2}{1 - \sin^2 t}$$
$$= \dfrac{(1 + \sin t)^2}{(1 + \sin t)(1 - \sin t)} = \dfrac{1 + \sin t}{1 - \sin t}$$

29 $\dfrac{1 + \csc\beta}{\cot\beta + \cos\beta} = \dfrac{1 + \dfrac{1}{\sin\beta}}{\dfrac{\cos\beta}{\sin\beta} + \cos\beta} = \dfrac{\dfrac{\sin\beta + 1}{\sin\beta}}{\dfrac{\cos\beta + \cos\beta\sin\beta}{\sin\beta}}$
$$= \dfrac{\sin\beta + 1}{\cos\beta(1 + \sin\beta)} = \dfrac{1}{\cos\beta} = \sec\beta$$

33 $\displaystyle RS = \frac{\tan \alpha + \tan \beta}{1 - \tan \alpha \tan \beta} = \frac{\dfrac{\sin \alpha}{\cos \alpha} + \dfrac{\sin \beta}{\cos \beta}}{1 - \dfrac{\sin \alpha}{\cos \alpha} \cdot \dfrac{\sin \beta}{\cos \beta}}$

$\displaystyle = \frac{\dfrac{\sin \alpha \cos \beta + \cos \alpha \sin \beta}{\cos \alpha \cos \beta}}{\dfrac{\cos \alpha \cos \beta - \sin \alpha \sin \beta}{\cos \alpha \cos \beta}}$

$\displaystyle = \frac{\sin \alpha \cos \beta + \cos \alpha \sin \beta}{\cos \alpha \cos \beta - \sin \alpha \sin \beta}$

$= LS$

37 $\displaystyle \frac{1}{\tan \beta + \cot \beta} = \frac{1}{\dfrac{\sin \beta}{\cos \beta} + \dfrac{\cos \beta}{\sin \beta}} = \frac{1}{\dfrac{\sin^2 \beta + \cos^2 \beta}{\cos \beta \sin \beta}}$

$= \sin \beta \cos \beta$

41 $RS = \sec^4 \phi - 4 \tan^2 \phi = (\sec^2 \phi)^2 - 4 \tan^2 \phi$
$= (1 + \tan^2 \phi)^2 - 4 \tan^2 \phi$
$= 1 + 2 \tan^2 \phi + \tan^4 \phi - 4 \tan^2 \phi$
$= 1 - 2 \tan^2 \phi + \tan^4 \phi$
$= (1 - \tan^2 \phi)^2 = LS$

45 $\log 10^{\tan t} = \log_{10} 10^{\tan t} = \tan t$, since $\log_a a^x = x$.

49 $\ln |\sec \theta + \tan \theta| = \ln \left| \dfrac{(\sec \theta + \tan \theta)(\sec \theta - \tan \theta)}{\sec \theta - \tan \theta} \right|$

$= \ln \left| \dfrac{\sec^2 \theta - \tan^2 \theta}{\sec \theta - \tan \theta} \right|$

$= \ln \left| \dfrac{1}{\sec \theta - \tan \theta} \right|$

$= \ln |1| - \ln |\sec \theta - \tan \theta|$

$= -\ln |\sec \theta - \tan \theta|$

Exer. 51–62: A typical value of t or θ and the resulting nonequality are given.

51 $\pi, -1 \neq 1$ **53** $\dfrac{3\pi}{2}, 1 \neq -1$ **55** $\dfrac{\pi}{4}, 2 \neq 1$

57 $\pi, -1 \neq 1$ **59** $\dfrac{\pi}{4}, \cos \sqrt{2} \neq 1$ **61** $\pi, -5 \neq 0$

63 $a^3 \cos^3 \theta$ **65** $a \tan \theta \sin \theta$ **67** $a \sec \theta$

69 $\dfrac{1}{a^2} \cos^2 \theta$ **71** $a \tan \theta$ **73** $a^4 \sec^3 \theta \tan \theta$

75 The graph of f appears to be that of $y = g(x) = -1$.

$\displaystyle \frac{\sin^2 x - \sin^4 x}{(1 - \sec^2 x)\cos^4 x} = \frac{\sin^2 x(1 - \sin^2 x)}{-\tan^2 x \cos^4 x}$

$\displaystyle = \frac{\sin^2 x \cos^2 x}{-(\sin^2 x/\cos^2 x)\cos^4 x}$

$\displaystyle = \frac{\sin^2 x \cos^2 x}{-\sin^2 x \cos^2 x}$

$= -1$

77 The graph of f appears to be that of $y = g(x) = \cos x$.

$\sec x(\sin x \cos x + \cos^2 x) - \sin x$
$\quad = \sec x \cos x(\sin x + \cos x) - \sin x$
$\quad = (\sin x + \cos x) - \sin x = \cos x$

EXERCISES 3.2

Exer. 1–36: n denotes any integer.

1 $\dfrac{5\pi}{4} + 2\pi n, \dfrac{7\pi}{4} + 2\pi n$ **3** $\dfrac{\pi}{3} + \pi n$

5 $\dfrac{\pi}{3} + 2\pi n, \dfrac{5\pi}{3} + 2\pi n$

7 No solution, since $\dfrac{\pi}{2} > 1$.

9 All θ except $\theta = \dfrac{\pi}{2} + \pi n$

11 $\dfrac{\pi}{12} + \pi n, \dfrac{11\pi}{12} + \pi n$ **13** $\dfrac{\pi}{2} + 3\pi n$

15 $-\dfrac{\pi}{12} + 2\pi n, \dfrac{7\pi}{12} + 2\pi n$

17 $\dfrac{\pi}{4} + \pi n, \dfrac{7\pi}{12} + \pi n$ **19** $\dfrac{2\pi}{3} + 2\pi n, \dfrac{4\pi}{3} + 2\pi n$

21 $\dfrac{\pi}{4} + \dfrac{\pi}{2} n$ **23** $2\pi n, \dfrac{3\pi}{2} + 2\pi n$

25 $\dfrac{\pi}{3} + \pi n, \dfrac{2\pi}{3} + \pi n$ **27** $\dfrac{4\pi}{3} + 2\pi n, \dfrac{5\pi}{3} + 2\pi n$

29 $\dfrac{\pi}{6} + \pi n, \dfrac{5\pi}{6} + \pi n$ **31** $\dfrac{7\pi}{6} + 2\pi n, \dfrac{11\pi}{6} + 2\pi n$

33 $\dfrac{\pi}{12} + \pi n, \dfrac{5\pi}{12} + \pi n$ **35** $e^{(\pi/2) + \pi n}$

37 $\dfrac{3\pi}{8}, \dfrac{7\pi}{8}, \dfrac{11\pi}{8}, \dfrac{15\pi}{8}$ **39** $\dfrac{\pi}{3}, \dfrac{2\pi}{3}, \dfrac{4\pi}{3}, \dfrac{5\pi}{3}$

41 $\dfrac{\pi}{6}, \dfrac{5\pi}{6}, \dfrac{3\pi}{2}$ **43** $0, \pi, \dfrac{\pi}{4}, \dfrac{3\pi}{4}, \dfrac{5\pi}{4}, \dfrac{7\pi}{4}$

45 $\dfrac{\pi}{2}, \dfrac{3\pi}{2}, \dfrac{2\pi}{3}, \dfrac{4\pi}{3}$ **47** No solution   **49** $\dfrac{11\pi}{6}, \dfrac{\pi}{2}$

51 $0, \dfrac{\pi}{2}$ **53** $\dfrac{\pi}{4}, \dfrac{5\pi}{4}$

55 All α in $[0, 2\pi)$ except $0, \dfrac{\pi}{2}, \pi$, and $\dfrac{3\pi}{2}$

57 $\dfrac{\pi}{2}, \dfrac{3\pi}{2}, \dfrac{7\pi}{6}, \dfrac{11\pi}{6}$ **59** $\dfrac{3\pi}{4}, \dfrac{7\pi}{4}$

61 $15°30', 164°30'$ **63** $135°, 315°, 116°30', 296°30'$

65 $41°50', 138°10', 194°30', 345°30'$ **67** 10

69 (a)

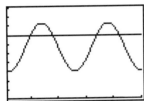

[1, 25] by [0, 100]
(b) July: 83°F; Oct.: 56.5°F   **(c)** May through Sept.
71 $t \approx 3.50$ and $t \approx 8.50$ **73 (a)** 3.29   **(b)** 4

75 (a)

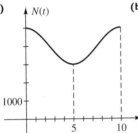

(b) $0 \leq t < \dfrac{5}{3}$ and

$\dfrac{25}{3} < t \leq 10$

77 $A\left(-\dfrac{4\pi}{3}, -\dfrac{2\pi}{3} + \dfrac{1}{2}\sqrt{3}\right), B\left(-\dfrac{2\pi}{3}, -\dfrac{\pi}{3} - \dfrac{1}{2}\sqrt{3}\right),$

$C\left(\dfrac{2\pi}{3}, \dfrac{\pi}{3} + \dfrac{1}{2}\sqrt{3}\right), D\left(\dfrac{4\pi}{3}, \dfrac{2\pi}{3} - \dfrac{1}{2}\sqrt{3}\right)$

79 $\dfrac{7}{360}$ **81** $[0, 1.27] \cup [5.02, 2\pi]$

83 $(0.39, 1.96) \cup (2.36, 3.53) \cup (5.11, 5.50)$

85

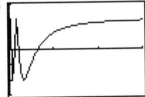

[0, 3] by [−1.5, 1.5]
(a) 0.6366   **(b)** Approaches $y = 1$
(c) An infinite number of zeros
87 5.400   **89** 3.619
91 −1.48, 1.08   **93** ±1.00

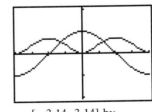

[−3.14, 3.14] by   [−3.14, 3.14] by
[−2.09, 2.09]   [−2.09, 2.09]

95 ±0.64, ±2.42

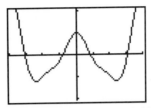

[−3.14, 3.14] by [−2.09, 2.09]

97 (a) 37.6°   **(b)** 52.5°

EXERCISES 3.3

1 (a) $\cos 43°23'$ **(b)** $\sin 16°48'$ **(c)** $\cot \dfrac{\pi}{3}$

   (d) $\csc 72.72°$

3 (a) $\sin \dfrac{3\pi}{20}$ **(b)** $\cos\left(\dfrac{2\pi - 1}{4}\right)$ **(c)** $\cot\left(\dfrac{\pi - 2}{2}\right)$

   (d) $\sec\left(\dfrac{\pi}{2} - 0.53\right)$

5 (a) $\dfrac{\sqrt{2} + \sqrt{3}}{2}$ **(b)** $\dfrac{\sqrt{6} - \sqrt{2}}{4}$

7 (a) $\sqrt{3} + 1$ **(b)** $-2 - \sqrt{3}$

9 (a) $\dfrac{\sqrt{2} - 1}{2}$ **(b)** $\dfrac{\sqrt{6} + \sqrt{2}}{4}$

11 $\cos 25°$ **13** $\sin(-5°)$ **15** $\sin(-5)$

17 (a) $\dfrac{77}{85}$ **(b)** $\dfrac{36}{85}$ **(c)** I

19 (a) $-\dfrac{24}{25}$ **(b)** $-\dfrac{24}{7}$ **(c)** IV

21 (a) $\dfrac{3\sqrt{21} - 8}{25} \approx 0.23$ **(b)** $\dfrac{4\sqrt{21} + 6}{25} \approx 0.97$ **(c)** I

23 $\sin(\theta + \pi) = \sin\theta\cos\pi + \cos\theta\sin\pi$
$= \sin\theta(-1) + \cos\theta(0) = -\sin\theta$

25 $\sin\left(x - \dfrac{5\pi}{2}\right) = \sin x \cos\dfrac{5\pi}{2} - \cos x \sin\dfrac{5\pi}{2}$

$= -\cos x$

27 $\cos(\theta - \pi) = \cos\theta\cos\pi + \sin\theta\sin\pi = -\cos\theta$

29 $\cos\left(x + \dfrac{3\pi}{2}\right) = \cos x \cos\dfrac{3\pi}{2} - \sin x \sin\dfrac{3\pi}{2}$

$= \sin x$

31 $\tan\left(x - \dfrac{\pi}{2}\right) = \dfrac{\sin\left(x - \dfrac{\pi}{2}\right)}{\cos\left(x - \dfrac{\pi}{2}\right)}$

$= \dfrac{\sin x \cos \dfrac{\pi}{2} - \cos x \sin \dfrac{\pi}{2}}{\cos x \cos \dfrac{\pi}{2} + \sin x \sin \dfrac{\pi}{2}}$

$= \dfrac{-\cos x}{\sin x} = -\cot x$

33 $\tan\left(\theta + \dfrac{\pi}{2}\right) = \cot\left[\dfrac{\pi}{2} - \left(\theta + \dfrac{\pi}{2}\right)\right]$

$= \cot(-\theta) = -\cot \theta$

35 $\sin\left(\theta + \dfrac{\pi}{4}\right) = \sin \theta \cos \dfrac{\pi}{4} + \cos \theta \sin \dfrac{\pi}{4}$

$= \dfrac{\sqrt{2}}{2} \sin \theta + \dfrac{\sqrt{2}}{2} \cos \theta$

$= \dfrac{\sqrt{2}}{2}(\sin \theta + \cos \theta)$

37 $\tan\left(u + \dfrac{\pi}{4}\right) = \dfrac{\tan u + \tan \dfrac{\pi}{4}}{1 - \tan u \tan \dfrac{\pi}{4}} = \dfrac{1 + \tan u}{1 - \tan u}$

39 $\cos(u + v) + \cos(u - v)$

$= (\cos u \cos v - \sin u \sin v) +$
$\quad (\cos u \cos v + \sin u \sin v)$
$= 2 \cos u \cos v$

41 $\sin(u + v) \cdot \sin(u - v)$

$= (\sin u \cos v + \cos u \sin v) \cdot$
$\quad (\sin u \cos v - \cos u \sin v)$
$= \sin^2 u \cos^2 v - \cos^2 u \sin^2 v$
$= \sin^2 u (1 - \sin^2 v) - (1 - \sin^2 u) \sin^2 v$
$= \sin^2 u - \sin^2 u \sin^2 v - \sin^2 v + \sin^2 u \sin^2 v$
$= \sin^2 u - \sin^2 v$

43 $\dfrac{1}{\cot \alpha - \cot \beta} = \dfrac{1}{\dfrac{\cos \alpha}{\sin \alpha} - \dfrac{\cos \beta}{\sin \beta}}$

$= \dfrac{1}{\dfrac{\cos \alpha \sin \beta - \cos \beta \sin \alpha}{\sin \alpha \sin \beta}}$

$= \dfrac{\sin \alpha \sin \beta}{\sin(\beta - \alpha)}$

45 $\sin(u + v + w)$

$= \sin[(u + v) + w]$
$= \sin(u + v) \cos w + \cos(u + v) \sin w$
$= (\sin u \cos v + \cos u \sin v) \cos w +$
$\quad (\cos u \cos v - \sin u \sin v) \sin w$
$= \sin u \cos v \cos w + \cos u \sin v \cos w +$
$\quad \cos u \cos v \sin w - \sin u \sin v \sin w$

47 $\cot(u + v) = \dfrac{\cos(u + v)}{\sin(u + v)}$

$= \dfrac{(\cos u \cos v - \sin u \sin v)(1/\sin u \sin v)}{(\sin u \cos v + \cos u \sin v)(1/\sin u \sin v)}$

$= \dfrac{\cot u \cot v - 1}{\cot v + \cot u}$

49 $\sin(u - v) = \sin[u + (-v)]$
$= \sin u \cos(-v) + \cos u \sin(-v)$
$= \sin u \cos v - \cos u \sin v$

51 $\dfrac{f(x + h) - f(x)}{h} = \dfrac{\cos(x + h) - \cos x}{h}$

$= \dfrac{\cos x \cos h - \sin x \sin h - \cos x}{h}$

$= \dfrac{\cos x \cos h - \cos x}{h} - \dfrac{\sin x \sin h}{h}$

$= \cos x \left(\dfrac{\cos h - 1}{h}\right) - \sin x \left(\dfrac{\sin h}{h}\right)$

53 $0, \dfrac{\pi}{3}, \dfrac{2\pi}{3}$ **55** $\dfrac{\pi}{6}, \dfrac{\pi}{2}, \dfrac{5\pi}{6}$

57 $\dfrac{\pi}{12}, \dfrac{5\pi}{12}; \dfrac{3\pi}{4}$ is extraneous

59 (a) $f(x) = 2 \cos\left(2x - \dfrac{\pi}{6}\right)$ (b) $2, \pi, \dfrac{\pi}{12}$

(c)

61 (a) $f(x) = 2\sqrt{2} \cos\left(3x + \dfrac{\pi}{4}\right)$ (b) $2\sqrt{2}, \dfrac{2\pi}{3}, -\dfrac{\pi}{12}$

(c)

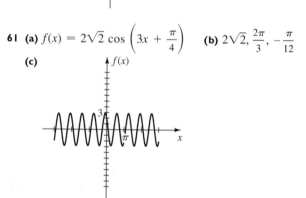

63 $y = 10\sqrt{41} \cos\left(60\pi t - \tan^{-1} \dfrac{5}{4}\right)$

$\approx 10\sqrt{41} \cos(60\pi t - 0.8961)$

65 (a) $y = \sqrt{13}\cos(t - C)$ with $\tan C = \dfrac{3}{2}$; $\sqrt{13}$, 2π

(b) $t = C + \dfrac{\pi}{2} + \pi n \approx 2.55 + \pi n$ for every non-negative integer n

67 (a) $p(t) = A \sin \omega t + B \sin(\omega t + \tau)$
$= A \sin \omega t + B(\sin \omega t \cos \tau + \cos \omega t \sin \tau)$
$= (B \sin \tau) \cos \omega t + (A + B \cos \tau) \sin \omega t$
$= a \cos \omega t + b \sin \omega t$
with $a = B \sin \tau$ and $b = A + B \cos \tau$

(b) $C^2 = (B \sin \tau)^2 + (A + B \cos \tau)^2$
$= B^2 \sin^2 \tau + A^2 + 2AB \cos \tau + B^2 \cos^2 \tau$
$= A^2 + B^2(\sin^2 \tau + \cos^2 \tau) + 2AB \cos \tau$
$= A^2 + B^2 + 2AB \cos \tau$

69 (a) $C^2 = A^2 + B^2 + 2AB \cos \tau \le A^2 + B^2 + 2AB$,
since $\cos \tau \le 1$ and $A > 0$, $B > 0$. Thus,
$C^2 \le (A + B)^2$, and hence $C \le A + B$.

(b) $0, 2\pi$ **(c)** $\cos \tau > -B/(2A)$

71 $(-2.97, -2.69)$, $(-1.00, -0.37)$, $(0.17, 0.46)$, $(2.14, 2.77)$

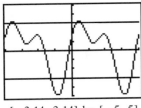

$[-3.14, 3.14]$ by $[-5, 5]$

EXERCISES 3.4

1 $\dfrac{24}{25}$, $-\dfrac{7}{25}$, $-\dfrac{24}{7}$ **3** $-\dfrac{4}{9}\sqrt{2}$, $-\dfrac{7}{9}$, $\dfrac{4}{7}\sqrt{2}$

5 $\dfrac{1}{10}\sqrt{10}$, $\dfrac{3}{10}\sqrt{10}$, $\dfrac{1}{3}$

7 $-\dfrac{1}{2}\sqrt{2 + \sqrt{2}}$, $\dfrac{1}{2}\sqrt{2 - \sqrt{2}}$, $-\sqrt{2} - 1$

9 (a) $\dfrac{1}{2}\sqrt{2 - \sqrt{2}}$ **(b)** $\dfrac{1}{2}\sqrt{2 - \sqrt{3}}$ **(c)** $\sqrt{2} + 1$

11 $\sin 10\theta = \sin(2 \cdot 5\theta) = 2 \sin 5\theta \cos 5\theta$

13 $4 \sin \dfrac{x}{2} \cos \dfrac{x}{2} = 2 \cdot 2 \sin \dfrac{x}{2} \cos \dfrac{x}{2} = 2 \sin\left(2 \cdot \dfrac{x}{2}\right)$
$= 2 \sin x$

15 $(\sin t + \cos t)^2 = \sin^2 t + 2 \sin t \cos t + \cos^2 t$
$= 1 + \sin 2t$

17 $\sin 3u = \sin(2u + u)$
$= \sin 2u \cos u + \cos 2u \sin u$
$= (2 \sin u \cos u) \cos u + (1 - 2 \sin^2 u) \sin u$
$= 2 \sin u \cos^2 u + \sin u - 2 \sin^3 u$
$= 2 \sin u(1 - \sin^2 u) + \sin u - 2 \sin^3 u$
$= 2 \sin u - 2 \sin^3 u + \sin u - 2 \sin^3 u$
$= 3 \sin u - 4 \sin^3 u = \sin u(3 - 4 \sin^2 u)$

19 $\cos 4\theta = \cos(2 \cdot 2\theta) = 2 \cos^2 2\theta - 1$
$= 2(2 \cos^2 \theta - 1)^2 - 1$
$= 2(4 \cos^4 \theta - 4 \cos^2 \theta + 1) - 1$
$= 8 \cos^4 \theta - 8 \cos^2 \theta + 1$

21 $\sin^4 t = (\sin^2 t)^2 = \left(\dfrac{1 - \cos 2t}{2}\right)^2$
$= \dfrac{1}{4}(1 - 2 \cos 2t + \cos^2 2t)$
$= \dfrac{1}{4} - \dfrac{1}{2}\cos 2t + \dfrac{1}{4}\left(\dfrac{1 + \cos 4t}{2}\right)$
$= \dfrac{1}{4} - \dfrac{1}{2}\cos 2t + \dfrac{1}{8} + \dfrac{1}{8}\cos 4t$
$= \dfrac{3}{8} - \dfrac{1}{2}\cos 2t + \dfrac{1}{8}\cos 4t$

23 $\sec 2\theta = \dfrac{1}{\cos 2\theta} = \dfrac{1}{2 \cos^2 \theta - 1} = \dfrac{1}{2\left(\dfrac{1}{\sec^2 \theta}\right) - 1}$
$= \dfrac{1}{\dfrac{2 - \sec^2 \theta}{\sec^2 \theta}} = \dfrac{\sec^2 \theta}{2 - \sec^2 \theta}$

25 $2 \sin^2 2t + \cos 4t = 2 \sin^2 2t + \cos(2 \cdot 2t)$
$= 2 \sin^2 2t + (1 - 2 \sin^2 2t) = 1$

27 $\tan 3u = \tan(2u + u) = \dfrac{\tan 2u + \tan u}{1 - \tan 2u \tan u}$
$= \dfrac{\dfrac{2 \tan u}{1 - \tan^2 u} + \tan u}{1 - \dfrac{2 \tan u}{1 - \tan^2 u} \cdot \tan u}$
$= \dfrac{\dfrac{2 \tan u + \tan u - \tan^3 u}{1 - \tan^2 u}}{\dfrac{1 - \tan^2 u - 2 \tan^2 u}{1 - \tan^2 u}}$
$= \dfrac{3 \tan u - \tan^3 u}{1 - 3 \tan^2 u} = \dfrac{\tan u(3 - \tan^2 u)}{1 - 3 \tan^2 u}$

29 $\dfrac{3}{8} + \dfrac{1}{2}\cos \theta + \dfrac{1}{8}\cos 2\theta$

31 $\dfrac{3}{8} - \dfrac{1}{2}\cos 4x + \dfrac{1}{8}\cos 8x$ **33** $0, \pi, \dfrac{2\pi}{3}, \dfrac{4\pi}{3}$

35 $\dfrac{\pi}{3}, \dfrac{5\pi}{3}, \pi$ **37** $0, \pi$ **39** $0, \dfrac{\pi}{3}, \dfrac{5\pi}{3}$

43 (a) $1.20, 5.09$

(b) $P\left(\dfrac{2\pi}{3}, -1.5\right)$, $Q(\pi, -1)$, $R\left(\dfrac{4\pi}{3}, -1.5\right)$

45 (a) $-\dfrac{3\pi}{2}, -\dfrac{\pi}{2}, \dfrac{\pi}{2}, \dfrac{3\pi}{2}$

(b) $0, \pm\pi, \pm2\pi, \pm\dfrac{\pi}{4}, \pm\dfrac{3\pi}{4}, \pm\dfrac{5\pi}{4}, \pm\dfrac{7\pi}{4}$

47 (b) Yes, point B is 25 miles from A.

49 (a) $V = \dfrac{5}{2}\sin\theta$ **(b)** 53.13°   **51 (b)** 12.43 mm

53 The graph of f appears to be that of
$y = g(x) = \tan x$.
$$\frac{\sin 2x + \sin x}{\cos 2x + \cos x + 1} = \frac{2\sin x \cos x + \sin x}{(2\cos^2 x - 1) + \cos x + 1}$$
$$= \frac{\sin x(2\cos x + 1)}{\cos x(2\cos x + 1)} = \frac{\sin x}{\cos x} = \tan x$$

55 $-3.55, 5.22$

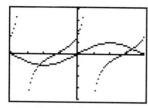

$[-2\pi, 2\pi]$ by $[-4, 4]$

57 $-2.03, -0.72,$ **59** -2.59
$0.58, 2.62$

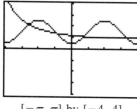

$[-\pi, \pi]$ by $[-4, 4]$

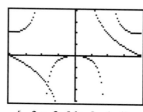

$[-2\pi, 2\pi]$ by $[-4, 4]$

EXERCISES 3.5

1 $\dfrac{1}{2}\cos 4t - \dfrac{1}{2}\cos 10t$ **3** $\dfrac{1}{2}\cos 2u + \dfrac{1}{2}\cos 10u$

5 $\sin 12\theta + \sin 6\theta$ **7** $\dfrac{3}{2}\sin 3x + \dfrac{3}{2}\sin x$

9 $2\sin 4\theta \cos 2\theta$ **11** $-2\sin 4x \sin x$

13 $-2\cos 5t \sin 2t$ **15** $2\cos\dfrac{3}{2}x \cos\dfrac{1}{2}x$

17 $\dfrac{\sin 4t + \sin 6t}{\cos 4t - \cos 6t} = \dfrac{2\sin 5t \cos t}{2\sin 5t \sin t} = \cot t$

19 $\dfrac{\sin u + \sin v}{\cos u + \cos v} = \dfrac{2\sin\dfrac{1}{2}(u+v)\cos\dfrac{1}{2}(u-v)}{2\cos\dfrac{1}{2}(u+v)\cos\dfrac{1}{2}(u-v)}$
$$= \tan\dfrac{1}{2}(u+v)$$

21 $\dfrac{\sin u - \sin v}{\sin u + \sin v} = \dfrac{2\cos\dfrac{1}{2}(u+v)\sin\dfrac{1}{2}(u-v)}{2\sin\dfrac{1}{2}(u+v)\cos\dfrac{1}{2}(u-v)}$
$$= \cot\dfrac{1}{2}(u+v)\tan\dfrac{1}{2}(u-v)$$
$$= \dfrac{\tan\dfrac{1}{2}(u-v)}{\tan\dfrac{1}{2}(u+v)}$$

23 $4\cos x \cos 2x \sin 3x = 2\cos 2x (2\sin 3x \cos x)$
$= 2\cos 2x (\sin 4x + \sin 2x)$
$= (2\cos 2x \sin 4x) + (2\cos 2x \sin 2x)$
$= [\sin 6x - \sin(-2x)] + (\sin 4x - \sin 0)$
$= \sin 2x + \sin 4x + \sin 6x$

25 $\dfrac{1}{2}\sin[(a+b)x] + \dfrac{1}{2}\sin[(a-b)x]$ **27** $\dfrac{\pi}{4}n$

29 $\dfrac{\pi}{2}n$ **31** $\dfrac{\pi}{2} + \pi n, \dfrac{\pi}{12} + \dfrac{\pi}{2}n, \dfrac{5\pi}{12} + \dfrac{\pi}{2}n$

33 $\dfrac{\pi}{7} + \dfrac{2\pi}{7}n, \dfrac{2\pi}{3}n$ **35** $\dfrac{\pi}{4}, \dfrac{3\pi}{4}, \dfrac{5\pi}{4}, \dfrac{7\pi}{4}, \dfrac{\pi}{2}, \dfrac{3\pi}{2}$

37 $0, \pm\pi, \pm2\pi, \pm\dfrac{\pi}{4}, \pm\dfrac{3\pi}{4}, \pm\dfrac{5\pi}{4}, \pm\dfrac{7\pi}{4}$

39 $f(x) = \dfrac{1}{2}\sin\dfrac{\pi n}{l}(x+kt) + \dfrac{1}{2}\sin\dfrac{\pi n}{l}(x-kt)$

41 (a) $0, \pm1.05, \pm1.57, \pm2.09, \pm3.14$
(b) $0, \pm\dfrac{\pi}{3}, \pm\dfrac{\pi}{2}, \pm\dfrac{2\pi}{3}, \pm\pi$

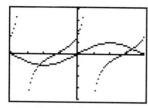

$[-3.14, 3.14]$ by $[-2.09, 2.09]$

43 The graph of f appears to be that of
$y = g(x) = \tan 2x$.

$$\frac{\sin x + \sin 2x + \sin 3x}{\cos x + \cos 2x + \cos 3x} = \frac{\sin 2x + (\sin 3x + \sin x)}{\cos 2x + (\cos 3x + \cos x)}$$

$$= \frac{\sin 2x + 2 \sin 2x \cos x}{\cos 2x + 2 \cos 2x \cos x}$$

$$= \frac{\sin 2x(1 + 2 \cos x)}{\cos 2x(1 + 2 \cos x)}$$

$$= \frac{\sin 2x}{\cos 2x} = \tan 2x$$

EXERCISES 3.6

1 (a) $-\dfrac{\pi}{4}$ (b) $\dfrac{2\pi}{3}$ (c) $-\dfrac{\pi}{3}$

3 (a) $\dfrac{\pi}{3}$ (b) $\dfrac{\pi}{4}$ (c) $\dfrac{\pi}{6}$

5 (a) Not defined   (b) Not defined   (c) $\dfrac{\pi}{4}$

7 (a) $-\dfrac{3}{10}$ (b) $\dfrac{1}{2}$ (c) 14

9 (a) $\dfrac{\pi}{3}$ (b) $\dfrac{5\pi}{6}$ (c) $-\dfrac{\pi}{6}$

11 (a) $-\dfrac{\pi}{4}$ (b) $\dfrac{3\pi}{4}$ (c) $-\dfrac{\pi}{4}$

13 (a) $\dfrac{\sqrt{3}}{2}$ (b) $\dfrac{\sqrt{2}}{2}$ (c) Not defined

15 (a) $\dfrac{\sqrt{5}}{2}$ (b) $\dfrac{\sqrt{34}}{5}$ (c) $\dfrac{4}{\sqrt{15}}$

17 (a) $\dfrac{\sqrt{3}}{2}$ (b) 0 (c) $-\dfrac{77}{36}$

19 (a) $-\dfrac{24}{25}$ (b) $-\dfrac{161}{289}$ (c) $\dfrac{24}{7}$

21 (a) $-\dfrac{1}{10}\sqrt{2}$ (b) $\dfrac{4}{17}\sqrt{17}$ (c) $\dfrac{1}{2}$

23 $\dfrac{x}{\sqrt{x^2 + 1}}$ **25** $\dfrac{\sqrt{x^2 + 4}}{2}$ **27** $2x\sqrt{1 - x^2}$

29 $\sqrt{\dfrac{1 + x}{2}}$ **31** (a) $-\dfrac{\pi}{2}$ (b) 0 (c) $\dfrac{\pi}{2}$

33

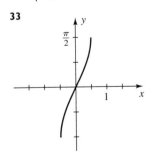

35

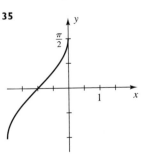

37

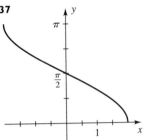

39

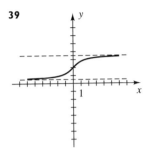

41

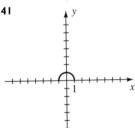

43 (a) $2 \le x \le 4$

(b) $-\dfrac{\pi}{4} \le y \le \dfrac{\pi}{4}$

(c) $x = \sin 2y + 3$

45 (a) $-\dfrac{3}{2} \le x \le \dfrac{3}{2}$ (b) $0 \le y \le 4\pi$

(c) $x = \dfrac{3}{2} \cos \dfrac{1}{4} y$

47 $x = \sin^{-1}(-y - 3)$ **49** $x = \cos^{-1}\left[\dfrac{1}{2}(15 - y)\right]$

51 $x = x_R$ or $x = \pi - x_R$, where $x_R = \sin^{-1}\left(\dfrac{3}{4} \sin y\right)$

53 $\cos^{-1}(-1 + \sqrt{2}) \approx 1.1437$,
$2\pi - \cos^{-1}(-1 + \sqrt{2}) \approx 5.1395$

55 $\tan^{-1}\dfrac{1}{4}(-9 + \sqrt{57}) \approx -0.3478$,

$\tan^{-1}\dfrac{1}{4}(-9 - \sqrt{57}) \approx -1.3337$

57 $\cos^{-1}\dfrac{1}{5}\sqrt{15} \approx 0.6847$, $\cos^{-1}\left(-\dfrac{1}{5}\sqrt{15}\right) \approx 2.4569$,

$\cos^{-1}\dfrac{1}{3}\sqrt{3} \approx 0.9553$, $\cos^{-1}\left(-\dfrac{1}{3}\sqrt{3}\right) \approx 2.1863$

59 $\sin^{-1}\left(\pm\dfrac{1}{6}\sqrt{30}\right) \approx \pm1.1503$

61 $\cos^{-1}\left(-\dfrac{3}{5}\right) \approx 2.2143$, $\cos^{-1}\dfrac{1}{3} \approx 1.2310$,

$2\pi - \cos^{-1}\left(-\dfrac{3}{5}\right) \approx 4.0689$, $2\pi - \cos^{-1}\dfrac{1}{3} \approx 5.0522$

63 $\cos^{-1}\dfrac{2}{3} \approx 0.8411$, $2\pi - \cos^{-1}\dfrac{2}{3} \approx 5.4421$,

$\dfrac{\pi}{3} \approx 1.0472$, $\dfrac{5\pi}{3} \approx 5.2360$

65 (a) 1.65 m   **(b)** 0.92 m   **(c)** 0.43 m   **67** 3.07°

69 (a) $\alpha = \theta - \sin^{-1}\dfrac{d}{k}$ **(b)** 40°

71 Let $\alpha = \sin^{-1} x$ and $\beta = \tan^{-1}\dfrac{x}{\sqrt{1+x^2}}$ with

$-\dfrac{\pi}{2} < \alpha < \dfrac{\pi}{2}$ and $-\dfrac{\pi}{2} < \beta < \dfrac{\pi}{2}$. Thus, $\sin\alpha = x$

and $\sin\beta = x$. Since the sine function is one-to-one

on $\left(-\dfrac{\pi}{2}, \dfrac{\pi}{2}\right)$, we have $\alpha = \beta$.

73 Let $\alpha = \arcsin(-x)$ and $\beta = \arcsin x$ with

$-\dfrac{\pi}{2} \le \alpha \le \dfrac{\pi}{2}$ and $-\dfrac{\pi}{2} \le \beta \le \dfrac{\pi}{2}$. Thus,

$\sin\alpha = -x$ and $\sin\beta = x$. Consequently,

$\sin\alpha = -\sin\beta = \sin(-\beta)$. Since the sine function

is one-to-one on $\left[-\dfrac{\pi}{2}, \dfrac{\pi}{2}\right]$, we have $\alpha = -\beta$.

75 Let $\alpha = \arctan x$ and $\beta = \arctan(1/x)$. Since

$x > 0$, we have $0 < \alpha < \dfrac{\pi}{2}$ and $0 < \beta < \dfrac{\pi}{2}$, and

hence $0 < \alpha + \beta < \pi$. Thus,

$\tan(\alpha+\beta) = \dfrac{\tan\alpha + \tan\beta}{1 - \tan\alpha\tan\beta} = \dfrac{x + (1/x)}{1 - x\cdot(1/x)} =$

$\dfrac{x + (1/x)}{0}$. Since the denominator is 0, $\tan(\alpha+\beta)$ is

undefined and hence $\alpha + \beta = \dfrac{\pi}{2}$.

77 Domain: [0, 2];   **79** 0.29

range: $\left[-\dfrac{\pi}{2}, \pi\right]$

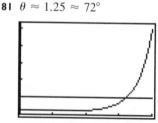

[−3, 6] by [−2, 4]

[−3, 3] by [−2, 2]

81 $\theta \approx 1.25 \approx 72°$

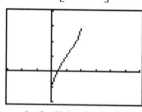

[0, 1.57] by [0, 1.05]

83 $\tan^{-1} 1 = 45°$ **85** $\tan^{-1}\dfrac{1}{2} \approx 26.6°$

CHAPTER 3 REVIEW EXERCISES

1 $(\cot^2 x + 1)(1 - \cos^2 x) = (\csc^2 x)(\sin^2 x) = 1$

2 $\cos\theta + \sin\theta\tan\theta = \cos\theta + \sin\theta\cdot\dfrac{\sin\theta}{\cos\theta}$

$= \dfrac{\cos^2\theta + \sin^2\theta}{\cos\theta} = \dfrac{1}{\cos\theta} = \sec\theta$

3 $\dfrac{(\sec^2\theta - 1)\cot\theta}{\tan\theta\sin\theta + \cos\theta} = \dfrac{(\tan^2\theta)\cot\theta}{\dfrac{\sin\theta}{\cos\theta}\cdot\sin\theta + \cos\theta}$

$= \dfrac{\tan\theta}{\dfrac{\sin^2\theta + \cos^2\theta}{\cos\theta}} = \dfrac{\sin\theta/\cos\theta}{1/\cos\theta}$

$= \sin\theta$

4 $(\tan x + \cot x)^2 = \left(\dfrac{\sin x}{\cos x} + \dfrac{\cos x}{\sin x}\right)^2$

$= \left(\dfrac{\sin^2 x + \cos^2 x}{\cos x \sin x}\right)^2$

$= \dfrac{1}{\cos^2 x \sin^2 x} = \sec^2 x \csc^2 x$

5 $\dfrac{1}{1 + \sin t}\cdot\dfrac{1 - \sin t}{1 - \sin t} = \dfrac{1 - \sin t}{1 - \sin^2 t} = \dfrac{1 - \sin t}{\cos^2 t}$

$= \dfrac{1 - \sin t}{\cos t}\cdot\dfrac{1}{\cos t}$

$= \left(\dfrac{1}{\cos t} - \dfrac{\sin t}{\cos t}\right)\cdot\sec t$

$= (\sec t - \tan t)\sec t$

6 $\dfrac{\sin(\alpha - \beta)}{\cos(\alpha + \beta)} = \dfrac{(\sin\alpha\cos\beta - \cos\alpha\sin\beta)/\cos\alpha\cos\beta}{(\cos\alpha\cos\beta - \sin\alpha\sin\beta)/\cos\alpha\cos\beta}$

$= \dfrac{\tan\alpha - \tan\beta}{1 - \tan\alpha\tan\beta}$

7 $\tan 2u = \dfrac{2\tan u}{1 - \tan^2 u} = \dfrac{2\cdot\dfrac{1}{\cot u}}{1 - \dfrac{1}{\cot^2 u}} = \dfrac{\dfrac{2}{\cot u}}{\dfrac{\cot^2 u - 1}{\cot^2 u}}$

$= \dfrac{2\cot u}{\cot^2 u - 1} = \dfrac{2\cot u}{(\csc^2 u - 1) - 1} = \dfrac{2\cot u}{\csc^2 u - 2}$

8 $\cos^2\dfrac{v}{2} = \dfrac{1 + \cos v}{2} = \dfrac{1 + \dfrac{1}{\sec v}}{2} = \dfrac{\dfrac{\sec v + 1}{\sec v}}{2}$

$= \dfrac{1 + \sec v}{2\sec v}$

9 $\dfrac{\tan^3\phi - \cot^3\phi}{\tan^2\phi + \csc^2\phi}$

$= \dfrac{(\tan\phi - \cot\phi)[(\tan^2\phi + \tan\phi\cot\phi + \cot^2\phi)]}{[\tan^2\phi + (1 + \cot^2\phi)]}$

$= \tan\phi - \cot\phi$

10 $\text{LS} = \dfrac{\sin u + \sin v}{\csc u + \csc v} = \dfrac{\sin u + \sin v}{\dfrac{1}{\sin u} + \dfrac{1}{\sin v}} = \dfrac{\sin u + \sin v}{\dfrac{\sin v + \sin u}{\sin u \sin v}}$

$$= \sin u \sin v$$

$\text{RS} = \dfrac{1 - \sin u \sin v}{-1 + \csc u \csc v} = \dfrac{1 - \sin u \sin v}{-1 + \dfrac{1}{\sin u \sin v}}$

$$= \dfrac{1 - \sin u \sin v}{\dfrac{1 - \sin u \sin v}{\sin u \sin v}}$$

$$= \sin u \sin v$$

Since the LS and RS equal the same expression and the steps are reversible, the identity is verified.

11 $\left(\dfrac{\sin^2 x}{\tan^4 x}\right)^3 \left(\dfrac{\csc^3 x}{\cot^6 x}\right)^2 = \left(\dfrac{\sin^6 x}{\tan^{12} x}\right)\left(\dfrac{\csc^6 x}{\cot^{12} x}\right) = \dfrac{(\sin x \csc x)^6}{(\tan x \cot x)^{12}}$

$$= \dfrac{(1)^6}{(1)^{12}} = 1$$

12 $\dfrac{\cos \gamma}{1 - \tan \gamma} + \dfrac{\sin \gamma}{1 - \cot \gamma} = \dfrac{\cos \gamma}{\dfrac{\cos \gamma - \sin \gamma}{\cos \gamma}} + \dfrac{\sin \gamma}{\dfrac{\sin \gamma - \cos \gamma}{\sin \gamma}}$

$$= \dfrac{\cos^2 \gamma}{\cos \gamma - \sin \gamma} + \dfrac{\sin^2 \gamma}{\sin \gamma - \cos \gamma}$$

$$= \dfrac{\cos^2 \gamma - \sin^2 \gamma}{\cos \gamma - \sin \gamma}$$

$$= \dfrac{(\cos \gamma + \sin \gamma)(\cos \gamma - \sin \gamma)}{\cos \gamma - \sin \gamma}$$

$$= \cos \gamma + \sin \gamma$$

13 $\dfrac{\cos (-t)}{\sec (-t) + \tan (-t)} = \dfrac{\cos t}{\sec t - \tan t} = \dfrac{\cos t}{\dfrac{1}{\cos t} - \dfrac{\sin t}{\cos t}}$

$$= \dfrac{\cos t}{\dfrac{1 - \sin t}{\cos t}} = \dfrac{\cos^2 t}{1 - \sin t} = \dfrac{1 - \sin^2 t}{1 - \sin t}$$

$$= \dfrac{(1 - \sin t)(1 + \sin t)}{1 - \sin t} = 1 + \sin t$$

14 $\dfrac{\cot (-t) + \csc (-t)}{\sin (-t)} = \dfrac{-\cot t - \csc t}{-\sin t} = \dfrac{\dfrac{\cos t}{\sin t} + \dfrac{1}{\sin t}}{\sin t}$

$$= \dfrac{\cos t + 1}{\sin^2 t} = \dfrac{\cos t + 1}{1 - \cos^2 t}$$

$$= \dfrac{\cos t + 1}{(1 - \cos t)(1 + \cos t)} = \dfrac{1}{1 - \cos t}$$

15 $\sqrt{\dfrac{1 - \cos t}{1 + \cos t}} = \sqrt{\dfrac{(1 - \cos t)}{(1 + \cos t)} \cdot \dfrac{(1 - \cos t)}{(1 - \cos t)}}$

$$= \sqrt{\dfrac{(1 - \cos t)^2}{1 - \cos^2 t}}$$

$$= \sqrt{\dfrac{(1 - \cos t)^2}{\sin^2 t}} = \dfrac{|1 - \cos t|}{|\sin t|} = \dfrac{1 - \cos t}{|\sin t|},$$

since $(1 - \cos t) \geq 0$.

16 $\sqrt{\dfrac{1 - \sin \theta}{1 + \sin \theta}} = \sqrt{\dfrac{(1 - \sin \theta)}{(1 + \sin \theta)} \cdot \dfrac{(1 + \sin \theta)}{(1 + \sin \theta)}}$

$$= \sqrt{\dfrac{1 - \sin^2 \theta}{(1 + \sin \theta)^2}}$$

$$= \sqrt{\dfrac{\cos^2 \theta}{(1 + \sin \theta)^2}} = \dfrac{|\cos \theta|}{|1 + \sin \theta|}$$

$$= \dfrac{|\cos \theta|}{1 + \sin \theta},$$

since $(1 + \sin \theta) \geq 0$.

17 $\cos\left(x - \dfrac{5\pi}{2}\right) = \cos x \cos \dfrac{5\pi}{2} + \sin x \sin \dfrac{5\pi}{2} = \sin x$

18 $\tan\left(x + \dfrac{3\pi}{4}\right) = \dfrac{\tan x + \tan \dfrac{3\pi}{4}}{1 - \tan x \tan \dfrac{3\pi}{4}} = \dfrac{\tan x - 1}{1 + \tan x}$

19 $\dfrac{1}{4} \sin 4\beta = \dfrac{1}{4} \sin (2 \cdot 2\beta) = \dfrac{1}{4}(2 \sin 2\beta \cos 2\beta)$

$$= \dfrac{1}{2}(2 \sin \beta \cos \beta)(\cos^2 \beta - \sin^2 \beta)$$

$$= \sin \beta \cos^3 \beta - \cos \beta \sin^3 \beta$$

20 $\tan \dfrac{1}{2} \theta = \dfrac{1 - \cos \theta}{\sin \theta} = \dfrac{1}{\sin \theta} - \dfrac{\cos \theta}{\sin \theta} = \csc \theta - \cot \theta$

21 $\sin 8\theta = 2 \sin 4\theta \cos 4\theta$

$\qquad = 2(2 \sin 2\theta \cos 2\theta)(1 - 2 \sin^2 2\theta)$

$\qquad = 8 \sin \theta \cos \theta(1 - 2 \sin^2 \theta)[1 - 2(2 \sin \theta \cos \theta)^2]$

$\qquad = 8 \sin \theta \cos \theta(1 - 2 \sin^2 \theta)(1 - 8 \sin^2 \theta \cos^2 \theta)$

22 Let $\alpha = \arctan x$ and $\beta = \arctan \dfrac{2x}{1 - x^2}$. Because

$-1 < x < 1$, $-\dfrac{\pi}{4} < \alpha < \dfrac{\pi}{4}$. Thus, $\tan \alpha = x$ and

$\tan \beta = \dfrac{2x}{1 - x^2} = \dfrac{2 \tan \alpha}{1 - \tan^2 \alpha} = \tan 2\alpha$. Since the

tangent function is one-to-one on $\left(-\dfrac{\pi}{2}, \dfrac{\pi}{2}\right)$, we

have $\beta = 2\alpha$ or, equivalently, $\alpha = \dfrac{1}{2}\beta$.

23 $\dfrac{\pi}{2}, \dfrac{3\pi}{2}, \dfrac{\pi}{4}, \dfrac{7\pi}{4}, \dfrac{3\pi}{4}, \dfrac{5\pi}{4}$　　**24** $\dfrac{7\pi}{6}, \dfrac{11\pi}{6}$　　**25** $0, \pi$

26 $\dfrac{\pi}{4}, \dfrac{3\pi}{4}, \dfrac{5\pi}{4}, \dfrac{7\pi}{4}$ **27** $0, \pi, \dfrac{2\pi}{3}, \dfrac{4\pi}{3}$

28 $\dfrac{\pi}{2}, \dfrac{3\pi}{2}, \dfrac{\pi}{4}, \dfrac{5\pi}{4}, \dfrac{3\pi}{4}, \dfrac{7\pi}{4}$ **29** $\dfrac{7\pi}{6}, \dfrac{11\pi}{6}, \dfrac{\pi}{2}$

30 $\dfrac{2\pi}{3}, \dfrac{4\pi}{3}, \pi$ **31** $\dfrac{\pi}{6}, \dfrac{5\pi}{6}, \dfrac{\pi}{3}, \dfrac{5\pi}{3}$

32 All x in $[0, 2\pi)$ except $\dfrac{\pi}{4}, \dfrac{3\pi}{4}, \dfrac{5\pi}{4}, \dfrac{7\pi}{4}$

33 $\dfrac{\pi}{3}, \dfrac{5\pi}{3}$ **34** $0, \dfrac{\pi}{3}, \dfrac{2\pi}{3}, \pi, \dfrac{4\pi}{3}, \dfrac{5\pi}{3}$

35 $\dfrac{3}{4}, \dfrac{7}{4}, \dfrac{11}{4}, \dfrac{15}{4}, \dfrac{19}{4}, \dfrac{23}{4}$ **36** $0, \pi, \dfrac{\pi}{3}, \dfrac{5\pi}{3}$

37 $\dfrac{\pi}{3}, \dfrac{5\pi}{3}$ **38** $\dfrac{\pi}{6}, \dfrac{5\pi}{6}, \dfrac{7\pi}{6}, \dfrac{11\pi}{6}$

39 $0, \dfrac{\pi}{8}, \dfrac{3\pi}{8}, \dfrac{5\pi}{8}, \dfrac{7\pi}{8}, \pi, \dfrac{9\pi}{8}, \dfrac{11\pi}{8}, \dfrac{13\pi}{8}, \dfrac{15\pi}{8}$

40 $\dfrac{\pi}{5}, \dfrac{3\pi}{5}, \pi, \dfrac{7\pi}{5}, \dfrac{9\pi}{5}$ **41** $\dfrac{\sqrt{6} - \sqrt{2}}{4}$

42 $-2 - \sqrt{3}$ **43** $\dfrac{\sqrt{2} - \sqrt{6}}{4}$ **44** $\dfrac{2}{\sqrt{2 - \sqrt{2}}}$

45 $\dfrac{84}{85}$ **46** $-\dfrac{13}{85}$ **47** $-\dfrac{84}{13}$ **48** $-\dfrac{36}{77}$

49 $\dfrac{36}{85}$ **50** $-\dfrac{36}{85}$ **51** $\dfrac{240}{289}$ **52** $-\dfrac{161}{289}$

53 $\dfrac{24}{7}$ **54** $\dfrac{1}{10}\sqrt{10}$ **55** $\dfrac{1}{3}$ **56** $\dfrac{5}{34}\sqrt{34}$

57 (a) $\dfrac{1}{2}\cos 3t - \dfrac{1}{2}\cos 11t$

     (b) $\dfrac{1}{2}\cos \dfrac{1}{12}u + \dfrac{1}{2}\cos \dfrac{5}{12}u$

     (c) $3 \sin 8x - 3 \sin 2x$ **(d)** $2 \sin 10\theta - 2 \sin 4\theta$

58 (a) $2 \sin 5u \cos 3u$ **(b)** $2 \sin \dfrac{11}{2}\theta \sin \dfrac{5}{2}\theta$

     (c) $2 \cos \dfrac{9}{40}t \sin \dfrac{1}{40}t$ **(d)** $6 \cos 4x \cos 2x$

59 $\dfrac{\pi}{6}$ **60** $\dfrac{\pi}{4}$ **61** $\dfrac{\pi}{3}$ **62** π **63** $-\dfrac{\pi}{4}$

64 $\dfrac{3\pi}{4}$ **65** $\dfrac{1}{2}$ **66** 2 **67** Not defined   **68** $\dfrac{\pi}{2}$

69 $\dfrac{240}{289}$ **70** $-\dfrac{7}{25}$

71

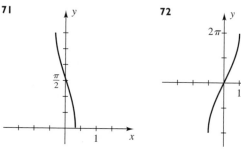

72

73

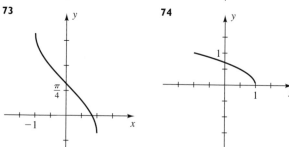

74

75 $\cos (\alpha + \beta + \gamma) = \cos [(\alpha + \beta) + \gamma]$
$= \cos (\alpha + \beta) \cos \gamma - \sin (\alpha + \beta) \sin \gamma$
$= (\cos \alpha \cos \beta - \sin \alpha \sin \beta) \cos \gamma -$
$(\sin \alpha \cos \beta + \cos \alpha \sin \beta) \sin \gamma$
$= \cos \alpha \cos \beta \cos \gamma - \sin \alpha \sin \beta \cos \gamma -$
$\sin \alpha \cos \beta \sin \gamma - \cos \alpha \sin \beta \sin \gamma$

76 (b) $t = 0, \pm\dfrac{\pi}{4b}$ **(c)** $\dfrac{2}{3}\sqrt{2}A$

77 $\pm\dfrac{\pi}{4}, \pm\dfrac{3\pi}{4}, \pm\dfrac{5\pi}{4}, \pm\dfrac{7\pi}{4}, \pm\dfrac{\pi}{3}, \pm\dfrac{5\pi}{3}$

78 (a) $x = 2d \tan \dfrac{1}{2}\theta$ **(b)** $d \le 1000$ ft

79 (a) $d = r\left(\sec \dfrac{1}{2}\theta - 1\right)$ **(b)** $43°$

80 (a) $78.7°$ **(b)** $61.4°$

CHAPTER 3 DISCUSSION EXERCISES

1 *Hint:* Factor $\sin^3 x - \cos^3 x$ as the difference of cubes.

2 $\sqrt{a^2 - x^2}$
$= \begin{cases} a \cos \theta & \text{if } 0 \le \theta \le \pi/2 \text{ or } 3\pi/2 \le \theta < 2\pi \\ -a \cos \theta & \text{if } \pi/2 < \theta < 3\pi/2 \end{cases}$

3 45; approximately 6.164

4 The difference quotient for the sine function appears to be the cosine function.

5 *Hint:* Write the equation in the form $\dfrac{\pi}{4} + \alpha = 4\theta$, and take the tangent of both sides.

6 (a) The **inverse sawtooth function,** denoted by saw^{-1}, is defined by $y = \text{saw}^{-1} x$ iff $x = \text{saw } y$ for $-1 \le x \le 1$ and $-2 \le y \le 2$.
(b) $0.85; -0.4$
(c) $\text{saw}(\text{saw}^{-1} x) = x$ if $-1 \le x \le 1$; $\text{saw}^{-1}(\text{saw } y) = y$ if $-2 \le y \le 2$
(d)

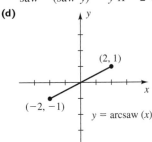

$y = \text{arcsaw}(x)$

CHAPTER 4

EXERCISES 4.1

1 $\beta = 62°$, $b \approx 14.1$, $c \approx 15.6$
3 $\gamma = 100°10'$, $b \approx 55.1$, $c \approx 68.7$
5 $\beta = 76°30'$, $a \approx 13.6$, $c \approx 17.8$
7 No triangle exists.
9 $\alpha \approx 77°30'$, $\beta \approx 49°10'$, $b \approx 108$; $\alpha \approx 102°30'$, $\beta \approx 24°10'$, $b \approx 59$
11 $\alpha \approx 82.54°$, $\beta \approx 49.72°$, $b \approx 100.85$; $\alpha \approx 97.46°$, $\beta \approx 34.80°$, $b \approx 75.45$
13 $\beta \approx 53°40'$, $\gamma \approx 61°10'$, $c \approx 20.6$
15 $\alpha \approx 25.993°$, $\gamma \approx 32.383°$, $a \approx 0.146$ **17** 219 yd
19 (a) 1.6 mi   **(b)** 0.6 mi   **21** 2.7 mi   **23** 628 m
25 3.7 mi from A and 5.4 mi from B **27** 350 ft
29 (a) 18.7   **(b)** 814   **31** (3949.9, 2994.2)

EXERCISES 4.2

1 $a \approx 26$, $\beta \approx 41°$, $\gamma \approx 79°$
3 $b \approx 180$, $\alpha \approx 25°$, $\gamma \approx 5°$
5 $c \approx 2.75$, $\alpha \approx 21°10'$, $\beta \approx 43°40'$
7 $\alpha \approx 29°$, $\beta \approx 47°$, $\gamma \approx 104°$
9 $\alpha \approx 12°30'$, $\beta \approx 136°30'$, $\gamma \approx 31°00'$ **11** 196 ft
13 24 mi   **15** 39 mi   **17** 2.3 mi   **19** N55°31'E
21 63.7 ft from first and third base; 66.8 ft from second base
23 37,039 ft $\approx$ 7 mi

25 *Hint:* Use the formula $\sin\dfrac{\theta}{2} = \sqrt{\dfrac{1 - \cos\theta}{2}}$.

27 (a) $72°, 108°, 36°$ **(b)** 0.62   **(c)** 0.59, 0.36

Exer. 29–36: The answer is in square units.

29 260   **31** 11.21   **33** 13.1   **35** 517.0
37 1.62 acres   **39** 123.4 ft^2

EXERCISES 4.3

1 5   **3** $\sqrt{85}$ **5** 8   **7** 1   **9** 0

Note: Point P is the point corresponding to the geometric representation.

11 $P(4, 2)$ **13** $P(3, -5)$ **15** $P(-3, 6)$
17 $P(-6, 4)$ **19** $P(0, 2)$

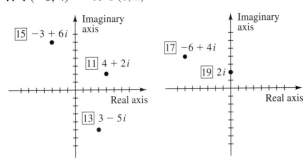

21 $\sqrt{2} \text{ cis } \dfrac{7\pi}{4}$ **23** $8 \text{ cis } \dfrac{5\pi}{6}$ **25** $4 \text{ cis } \dfrac{\pi}{6}$
27 $4\sqrt{2} \text{ cis } \dfrac{5\pi}{4}$ **29** $20 \text{ cis } \dfrac{3\pi}{2}$ **31** $12 \text{ cis } 0$
33 $7 \text{ cis } \pi$ **35** $6 \text{ cis } \dfrac{\pi}{2}$ **37** $10 \text{ cis } \dfrac{4\pi}{3}$
39 $\sqrt{5} \text{ cis } \left(\tan^{-1}\dfrac{1}{2}\right)$
41 $\sqrt{10} \text{ cis } \left[\tan^{-1}\left(-\dfrac{1}{3}\right) + \pi\right]$
43 $\sqrt{34} \text{ cis } \left(\tan^{-1}\dfrac{3}{5} + \pi\right)$
45 $5 \text{ cis } \left[\tan^{-1}\left(-\dfrac{3}{4}\right) + 2\pi\right]$
47 $2\sqrt{2} + 2\sqrt{2}i$ **49** $-3 + 3\sqrt{3}i$ **51** -5
53 $5 + 3i$ **55** $2 - i$ **57** $-2, i$
59 $10\sqrt{3} - 10i$, $-\dfrac{2}{5}\sqrt{3} + \dfrac{2}{5}i$ **61** $40, \dfrac{5}{2}$
63 $8 - 4i$, $\dfrac{8}{5} + \dfrac{4}{5}i$ **67** $17.21 + 24.57i$
69 $11.01 + 9.24i$ **71** $\sqrt{365}$ ohms   **73** 70.43 volts

EXERCISES 4.4

1 $-972 - 972i$ **3** $-32i$ **5** -8

7 $-\dfrac{1}{2}\sqrt{2} - \dfrac{1}{2}\sqrt{2}\,i$ **9** $-\dfrac{1}{2} - \dfrac{1}{2}\sqrt{3}\,i$

11 $-64\sqrt{3} - 64i$ **13** $\pm\left(\dfrac{1}{2}\sqrt{6} + \dfrac{1}{2}\sqrt{2}\,i\right)$

15 $\pm\left(\dfrac{\sqrt[4]{2}}{2} + \dfrac{\sqrt[4]{18}}{2}i\right),\ \pm\left(\dfrac{\sqrt[4]{18}}{2} - \dfrac{\sqrt[4]{2}}{2}i\right)$

17 $3i,\ \pm\dfrac{3}{2}\sqrt{3} - \dfrac{3}{2}i$

19 $\pm1,\ \dfrac{1}{2} \pm \dfrac{1}{2}\sqrt{3}\,i,$ **21** $\sqrt[10]{2}$ cis θ with $\theta = 9°,$
$-\dfrac{1}{2} \pm \dfrac{1}{2}\sqrt{3}\,i$ $81°, 153°, 225°, 297°$

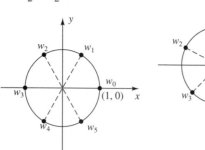

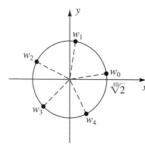

23 $\pm2, \pm2i$ **25** $\pm2i, \sqrt{3} \pm i, -\sqrt{3} \pm i$
27 $2i, \pm\sqrt{3} - i$
29 3 cis θ with $\theta = 0°, 72°, 144°, 216°, 288°$

EXERCISES 4.5

1 $\langle 3, 1\rangle, \langle 1, -7\rangle, \langle 13, 8\rangle, \langle 3, -32\rangle$
3 $\langle -15, 6\rangle, \langle 1, -2\rangle, \langle -68, 28\rangle, \langle 12, -12\rangle$
5 $4\mathbf{i} - 3\mathbf{j}, -2\mathbf{i} + 7\mathbf{j}, 19\mathbf{i} - 17\mathbf{j}, -11\mathbf{i} + 33\mathbf{j}$

7 Terminal points are   **9** Terminal points are
    $(3, 2), (-1, 5), (2, 7),$ $(-4, 6), (-2, 3),$
    $(6, 4), (3, -15).$ $(-6, 9), (-8, 12),$
                        $(6, -9).$

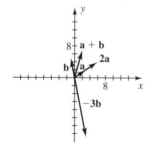

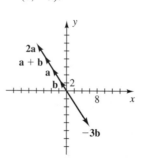

11 $-\mathbf{b}$ **13** $\mathbf{f}$ **15** $-\dfrac{1}{2}\mathbf{e}$

17 $\mathbf{a} + (\mathbf{b} + \mathbf{c}) = \langle a_1, a_2\rangle + (\langle b_1, b_2\rangle + \langle c_1, c_2\rangle)$
$= \langle a_1, a_2\rangle + \langle b_1 + c_1, b_2 + c_2\rangle$
$= \langle a_1 + b_1 + c_1, a_2 + b_2 + c_2\rangle$
$= \langle a_1 + b_1, a_2 + b_2\rangle + \langle c_1, c_2\rangle$
$= (\langle a_1, a_2\rangle + \langle b_1, b_2\rangle) + \langle c_1, c_2\rangle$
$= (\mathbf{a} + \mathbf{b}) + \mathbf{c}$

19 $\mathbf{a} + (-\mathbf{a}) = \langle a_1, a_2\rangle + (-\langle a_1, a_2\rangle)$
$= \langle a_1, a_2\rangle + \langle -a_1, -a_2\rangle$
$= \langle a_1 - a_1, a_2 - a_2\rangle$
$= \langle 0, 0\rangle = \mathbf{0}$

21 $(mn)\mathbf{a} = (mn)\langle a_1, a_2\rangle$
$= \langle (mn)a_1, (mn)a_2\rangle$
$= \langle mna_1, mna_2\rangle$
$= m\langle na_1, na_2\rangle$ or $n\langle ma_1, ma_2\rangle$
$= m(n\langle a_1, a_2\rangle)$ or $n(m\langle a_1, a_2\rangle)$
$= m(n\mathbf{a})$ or $n(m\mathbf{a})$

23 $0\mathbf{a} = 0\langle a_1, a_2\rangle = \langle 0a_1, 0a_2\rangle = \langle 0, 0\rangle = \mathbf{0}.$
Also, $m\mathbf{0} = m\langle 0, 0\rangle = \langle m0, m0\rangle = \langle 0, 0\rangle = \mathbf{0}.$

25 $-(\mathbf{a} + \mathbf{b}) = -(\langle a_1, a_2\rangle + \langle b_1, b_2\rangle)$
$= -(\langle a_1 + b_1, a_2 + b_2\rangle)$
$= \langle -(a_1 + b_1), -(a_2 + b_2)\rangle$
$= \langle -a_1 - b_1, -a_2 - b_2\rangle$
$= \langle -a_1, -a_2\rangle + \langle -b_1, -b_2\rangle$
$= -\mathbf{a} + (-\mathbf{b}) = -\mathbf{a} - \mathbf{b}$

27 $\|2\mathbf{v}\| = \|2\langle a, b\rangle\| = \|\langle 2a, 2b\rangle\| = \sqrt{(2a)^2 + (2b)^2}$
$= \sqrt{4a^2 + 4b^2} = 2\sqrt{a^2 + b^2} = 2\|\langle a, b\rangle\|$
$= 2\|\mathbf{v}\|$

29 $3\sqrt{2}; \dfrac{7\pi}{4}$ **31** $5, \pi$ **33** $\sqrt{41};\ \tan^{-1}\left(-\dfrac{5}{4}\right) + \pi$

35 $18; \dfrac{3\pi}{2}$ **37** 102 lb   **39** 7.2 kg
41 89 kg; S66°W   **43** 5.8 lb; $129°$
45 $40.96; 28.68$ **47** $-6.18; 19.02$
49 (a) $\mathbf{F} = \langle 7, 2\rangle$ (b) $\mathbf{G} = -\mathbf{F} = \langle -7, -2\rangle$
51 (a) $\mathbf{F} \approx \langle -5.86, 1.13\rangle$
    (b) $\mathbf{G} = -\mathbf{F} \approx \langle 5.86, -1.13\rangle$
53 $\sin^{-1}(0.4) \approx 23.6°$ **55** $56°; 232$ mi/hr
57 420 mi/hr; $244°$ **59** N22°W
61 $\mathbf{v}_1 \approx 4.1\mathbf{i} - 7.10\mathbf{j}; \mathbf{v}_2 \approx 0.98\mathbf{i} - 3.67\mathbf{j}$
63 (a) $(24.51, 20.57)$ (b) $(-24.57, 18.10)$
65 28.2 lb/person

EXERCISES 4.6

1 (a) 24 (b) $\cos^{-1}\left(\dfrac{24}{\sqrt{29}\sqrt{45}}\right) \approx 48°22'$

3 (a) -14 (b) $\cos^{-1}\left(\dfrac{-14}{\sqrt{17}\sqrt{13}}\right) \approx 160°21'$

5 (a) 45   **(b)** $\cos^{-1}\left(\dfrac{45}{\sqrt{81}\sqrt{41}}\right) \approx 38°40'$

7 (a) $-\dfrac{149}{5}$ **(b)** $\cos^{-1}\left(\dfrac{-149/5}{\sqrt{149}\sqrt{149/25}}\right) = 180°$

9 $\langle 4, -1\rangle \cdot \langle 2, 8\rangle = 0$ **11** $(-4\mathbf{j})\cdot(-7\mathbf{i}) = 0$

13 Opposite   **15** Same   **17** $\dfrac{6}{5}$ **19** $\pm\dfrac{3}{8}$

21 (a) -23 **(b)** -23 **23** -51

25 $17/\sqrt{26} \approx 3.33$ **27** 2.2   **29** 7

31 28   **33** 12

35 $\mathbf{a}\cdot\mathbf{a} = \langle a_1, a_2\rangle \cdot \langle a_1, a_2\rangle = a_1^2 + a_2^2$
$$= (\sqrt{a_1^2 + a_2^2})^2 = \|\mathbf{a}\|^2$$

37 $(m\mathbf{a})\cdot\mathbf{b} = (m\langle a_1, a_2\rangle)\cdot\langle b_1, b_2\rangle$
$$= \langle ma_1, ma_2\rangle \cdot \langle b_1, b_2\rangle$$
$$= ma_1 b_1 + ma_2 b_2$$
$$= m(a_1 b_1 + a_2 b_2) = m(\mathbf{a}\cdot\mathbf{b})$$

39 $\mathbf{0}\cdot\mathbf{a} = \langle 0, 0\rangle \cdot \langle a_1, a_2\rangle = 0(a_1) + 0(a_2)$
$$= 0 + 0 = 0$$

41 $1000\sqrt{3} \approx 1732$ ft-lb

43 (a) $\mathbf{v} = (93 \times 10^6)\mathbf{i} + (0.432 \times 10^6)\mathbf{j}$;
$\mathbf{w} = (93 \times 10^6)\mathbf{i} - (0.432 \times 10^6)\mathbf{j}$
(b) 0.53°

45 $\left\langle \dfrac{4}{5}, \dfrac{3}{5}\right\rangle$ **47** 2.6   **49** 24.33

51 $16\sqrt{3} \approx 27.7$ horsepower

CHAPTER 4 REVIEW EXERCISES

1 $a = \sqrt{43}, \beta = \cos^{-1}\left(\dfrac{4}{43}\sqrt{43}\right), \gamma = \cos^{-1}\left(\dfrac{5}{86}\sqrt{43}\right)$

2 $\alpha = 60°, \beta = 90°, b = 4; \alpha = 120°, \beta = 30°, b = 2$

3 $\gamma = 75°, a = 50\sqrt{6}, c = 50(1 + \sqrt{3})$

4 $\alpha = \cos^{-1}\left(\dfrac{7}{8}\right), \beta = \cos^{-1}\left(\dfrac{11}{16}\right), \gamma = \cos^{-1}\left(-\dfrac{1}{4}\right)$

5 $\alpha = 38°, a \approx 8.0, c \approx 13$

6 $\gamma \approx 19°10', \beta \approx 137°20', b \approx 258$

7 $\alpha \approx 24°, \gamma \approx 41°, b \approx 10.1$

8 $\alpha \approx 42°, \beta \approx 87°, \gamma \approx 51°$ **9** 290   **10** 10.9

11 $10\sqrt{2}$ cis $\dfrac{3\pi}{4}$ **12** 4 cis $\dfrac{5\pi}{3}$ **13** 17 cis π

14 12 cis $\dfrac{3\pi}{2}$ **15** 10 cis $\dfrac{7\pi}{6}$

16 $\sqrt{41}$ cis $\left(\tan^{-1}\dfrac{5}{4}\right)$ **17** $10\sqrt{3} - 10i$

18 $12 + 5i$ **19** $-12 - 12\sqrt{3}\,i, -\dfrac{3}{2}$

20 $-4\sqrt{2}\,i, -2\sqrt{2}$ **21** $-512i$ **22** i

23 $-972 + 972i$ **24** $-2^{19} - 2^{19}\sqrt{3}\,i$

25 $-3, \dfrac{3}{2} \pm \dfrac{3}{2}\sqrt{3}\,i$

26 (a) 2^{24} **(b)** $\sqrt[3]{2}$ cis θ with $\theta = 100°, 220°, 340°$

27 2 cis θ with $\theta = 0°, 72°, 144°, 216°, 288°$

28 Terminal points are
$(-2, -3), (-6, 13),$
$(-8, 10), (-1, 4).$

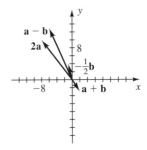

29 (a) $12\mathbf{i} + 19\mathbf{j}$ **(b)** $-8\mathbf{i} + 13\mathbf{j}$ **(c)** $\sqrt{40} \approx 6.32$
(d) $\sqrt{29} - \sqrt{17} \approx 1.26$

30 Circle with center (a_1, a_2) and radius c

31 The vectors $\mathbf{a}, \mathbf{b},$ and $\mathbf{a} - \mathbf{b}$ form a triangle with the vector $\mathbf{a} - \mathbf{b}$ opposite angle θ. The conclusion is a direct application of the law of cosines with sides $\|\mathbf{a}\|, \|\mathbf{b}\|,$ and $\|\mathbf{a} - \mathbf{b}\|$.

32 $\langle 14\cos 40°, -14\sin 40°\rangle$ **33** 109; S78°E

34 183°; 70 mi/hr

35 (a) 10   **(b)** $\cos^{-1}\left(\dfrac{10}{\sqrt{13}\sqrt{17}}\right) \approx 47°44'$ **(c)** $\dfrac{10}{\sqrt{13}}$

36 (a) 80   **(b)** $\cos^{-1}\left(\dfrac{40}{\sqrt{40}\sqrt{50}}\right) \approx 26°34'$ **(c)** $\sqrt{40}$

37 56   **38** 47.6°   **39** 53,000,000 mi

40 (a) 449 ft   **(b)** 434 ft

41 (a) 33 mi, 41 mi   **(b)** 30 mi   **42** 204

43 1 hour and 16 minutes   **44 (c)** 158°

45 (a) 47°   **(b)** 20

46 (a) 72°   **(b)** 181.6 ft^2 **(c)** 37.6 ft

CHAPTER 4 DISCUSSION EXERCISES

4 (b) *Hint:* Law of cosines

5 (a) $(\|\mathbf{b}\|\cos\alpha + \|\mathbf{a}\|\cos\beta)\mathbf{i} +$
$(\|\mathbf{b}\|\sin\alpha - \|\mathbf{a}\|\sin\beta)\mathbf{j}$

6 (a) 1   **(b)** $\pi i; \dfrac{\pi}{2}i$ **(c)** $\dfrac{\sqrt{2}}{2} + \dfrac{\sqrt{2}}{2}i; e^{-\pi/2} \approx 0.2079$

CHAPTER 5

EXERCISES 5.1

1 5   **3** −1, 3   **5** $-\dfrac{4}{99}$ **7** $\dfrac{18}{5}$

9 (a)

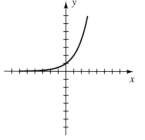

(b)

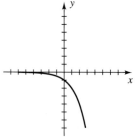

(c)

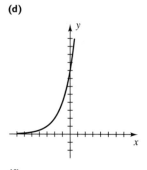

(d)

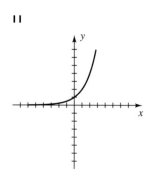

(e)

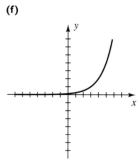

(f)

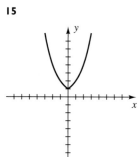

(g)

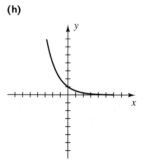

(h)

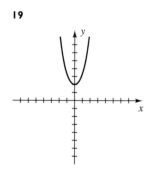

(i)

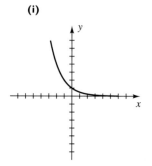

(j)

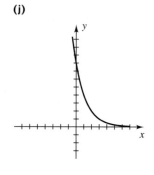

11

13

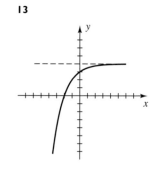

15

17

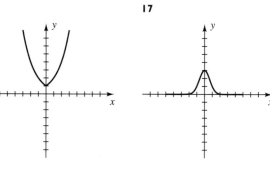

19

21 (a) 90   **(b)** 59   **(c)** 35
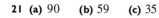

23 **(a)** 1039; 3118; 5400
(b)

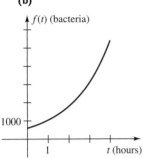

25 **(a)** 50 mg; 25 mg;
$$\frac{25}{2}\sqrt{2} \approx 17.7 \text{ mg}$$

(b)

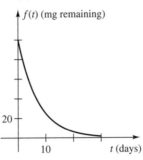

27 $-\dfrac{1}{1600}$

29 **(a)** $1010.00   **(b)** $1061.52   **(c)** $1126.83
(d) $10,892.55

31 **(a)** $7800   **(b)** $4790   **(c)** $2942

33 $161,657,351,965.80

35 **(a)** Examine the pattern formed by the value y in the year n.
(b) Solve $s = (1 - a)^T y_0$ for a.

37 **(a)** $925.75   **(b)** $243,270

39 $6346.40

41 **(a)** 180.1206   **(b)** 20.9758   **(c)** 7.3639
(a) 26.13   **(b)** 8.50

43

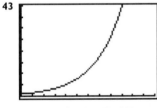

[0, 60] by [0, 40]

45

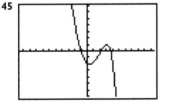

$-1.02, 2.14, 3.62$

[−10.5, 10.5] by [−7, 7]

47

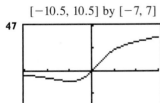

(a) Not one-to-one
(b) 0

[−3, 3] by [−2, 2]

49

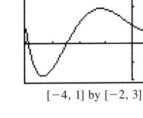

[−4, 1] by [−2, 3]

(a) Increasing: $[-3.37, -1.19] \cup [0.52, 1]$;
decreasing: $[-4, -3.37] \cup [-1.19, 0.52]$
(b) $[-1.79, 1.94]$

51 6.58 yr

53

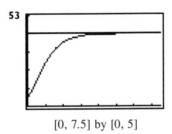

[0, 7.5] by [0, 5]

The maximum number
of sales approaches k.

55

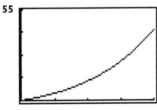

[0, 40] by [0, 200,000]

After approximately
32.8 yr

57 **(a)**

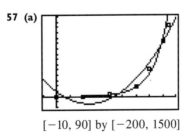

[−10, 90] by [−200, 1500]

(b) Exponential function f **(c)** 1989

EXERCISES 5.2

I (a) 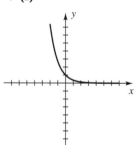   **(b)**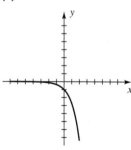

3 (a) 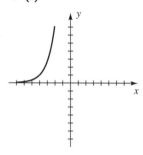   **(b)**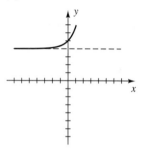

5 $1510.59   **7** $13,806.92   **9** 13%   **11** 3, 4

13 −1   **15** $-\dfrac{3}{4}, 0$ **17** $\dfrac{4}{(e^x + e^{-x})^2}$ **19** 27.43 g

21 280.0 million   **23** 13.5%   **25** 5610

27 7.44 in.   **29** 75.77 cm; 15.98 cm/yr

31 $11.25 per hr   **33 (a)** 7.19%   **(b)** 7.25%

(a) 29.96   **(b)** 8.15

35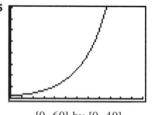

[0, 60] by [0, 40]

37 (a)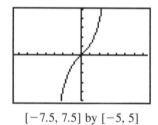

[−7.5, 7.5] by [−5, 5]

(b)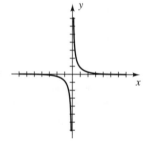

39 (a)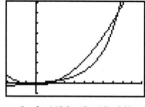

[−4.5, 4.5] by [−3, 3]

(b)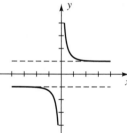

41 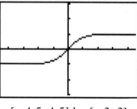   −1.04, 2.11

[−3, 11] by [−10, 80]

43  $f(x)$ is closer to e^x if $x \approx 0$; $g(x)$ is closer to e^x if $x \approx 1$.

[0, 4.5] by [0, 3]

45 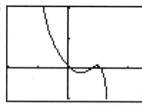   0.11, 0.79, 1.13

[−2, 2.5] by [−1, 2]

47 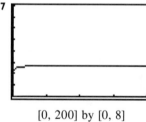 $y \approx 2.71 \approx e$

[0, 200] by [0, 8]

49 0.567

51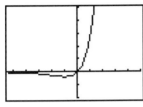

$[-5.5, 5]$ by $[-2, 5]$

Increasing on $[-1, \infty)$; decreasing on $(-\infty, -1]$

53 (a) As h increases, C decreases.
(b) As y increases, C decreases.
55 (a) $f(x) = 1.225e^{-0.0001085x}$

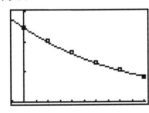

$[-1000, 10,100]$ by $[0, 1.5]$
(b) 0.885, 0.461

EXERCISES 5.3

1 (a) $\log_4 64 = 3$ **(b)** $\log_4 \dfrac{1}{64} = -3$ **(c)** $\log_t s = r$

(d) $\log_3 (4 - t) = x$ **(e)** $\log_5 \dfrac{a + b}{a} = 7t$

(f) $\log_{0.7} (5.3) = t$

3 (a) $2^5 = 32$ **(b)** $3^{-5} = \dfrac{1}{243}$ **(c)** $t^p = r$

(d) $3^5 = (x + 2)$ **(e)** $2^{3x+4} = m$ **(f)** $b^{3/2} = 512$

5 $t = 3 \log_a \dfrac{5}{2}$ **7** $t = \dfrac{1}{C} \log_a \left(\dfrac{A - D}{B} \right)$

9 (a) $\log 100{,}000 = 5$ **(b)** $\log 0.001 = -3$
(c) $\log (y + 1) = x$ **(d)** $\ln p = 7$
(e) $\ln (3 - x) = 2t$

11 (a) $10^{50} = x$ **(b)** $10^{20t} = x$ **(c)** $e^{0.1} = x$
(d) $e^{4+3x} = w$ **(e)** $e^{1/6} = z - 2$

13 (a) 0 **(b)** 1 **(c)** Not possible **(d)** 2 **(e)** 8
(f) 3 **(g)** -2

15 (a) 3 **(b)** 5 **(c)** 2 **(d)** -4 **(e)** 2
(f) -3 **(g)** $3e^2$

17 4 **19** No solution **21** $-1, -2$ **23** 13

25 27 **27** $\pm \dfrac{1}{e}$ **29** 3

31 (a)

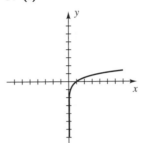

(b)

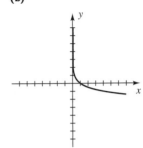

(c)

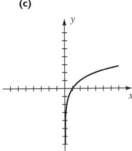

(d)

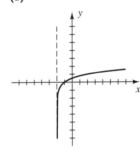

(e)

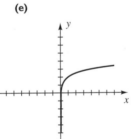

(f)

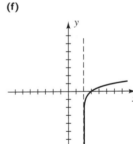

(g)

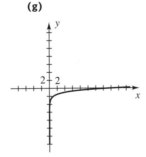

(h)

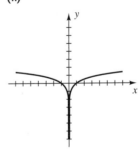

(i)

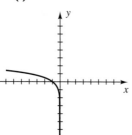

(j)

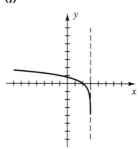

(k)

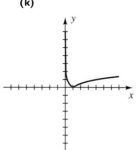

33

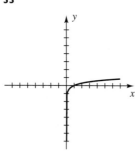

35

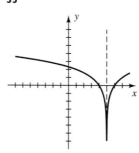

37 $f(x) = \log_2 x$ **39** $f(x) = \log_2 x - 1$
41 $f(x) = \log_2 (-x)$ **43** $f(x) = 2 \log_2 x$
45 (a) 4240   **(b)** 8.85   **(c)** 0.0237   **(d)** 9.97
    (e) 1.05   **(f)** 0.202
47 $t = -1600 \log_2 \left(\dfrac{q}{q_0}\right)$ **49** $t = -\dfrac{L}{R} \ln \left(\dfrac{I}{20}\right)$
51 (a) 2   **(b)** 4   **(c)** 5
53 (a) 10   **(b)** 30   **(c)** 40   **55** In the year 2079
57 (a) $W = 2.4e^{1.84h}$ **(b)** 37.92 kg
59 (a) 10,007 ft   **(b)** 18,004 ft
61 (a) 305.9 kg   **(b)** (1) 20 yr (2) 19.8 yr
63 10.1 mi   **65** $2^{1/8} \approx 1.09$

67 (a) Pedestrians have faster average walking speeds
    in large cities.
    (b) 570,000
69 (a) 8.4877   **(b)** −0.0601
71 1.763   **73** (0, 14.90)

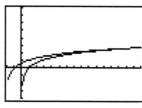

$[-2, 16]$ by $[-4, 8]$
75 (a) 30%   **(b)** 3.85

EXERCISES 5.4

1 (a) $\log_4 x + \log_4 z$ **(b)** $\log_4 y - \log_4 x$
    (c) $\dfrac{1}{3} \log_4 z$
3 $3 \log_a x + \log_a w - 2 \log_a y - 4 \log_a z$
5 $\dfrac{1}{3} \log z - \log x - \dfrac{1}{2} \log y$
7 $\dfrac{7}{4} \ln x - \dfrac{5}{4} \ln y - \dfrac{1}{4} \ln z$
9 (a) $\log_3 (5xy)$ **(b)** $\log_3 \dfrac{2z}{x}$ **(c)** $\log_3 y^5$
11 $\log_a \dfrac{x^2 \sqrt[3]{x-2}}{(2x+3)^5}$ **13** $\log \dfrac{y^{13/3}}{x^2}$ **15** $\ln x$ **17** $\dfrac{7}{2}$
19 $5\sqrt{5}$ **21** No solution   **23** −7   **25** 1
27 −2   **29** $\dfrac{-1 + \sqrt{65}}{2}$ **31** $-1 + \sqrt{1 + e}$
33   **35**

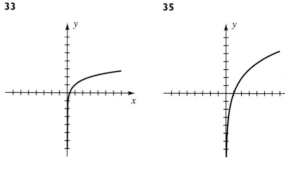

37

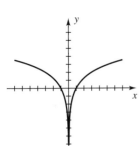

39

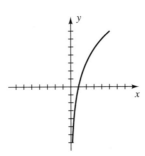

59

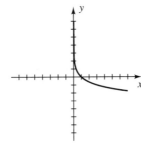

1.41, 6.59

$[0, 8]$ by $[-1.67, 3.67]$

61

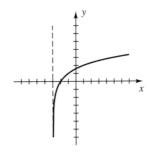

(a) Increasing on
$[0.2, 0.63]$ and
$[6.87, 16]$;
decreasing on
$[0.63, 6.87]$

(b) 4.61; -3.31

$[0.2, 16]$ by $[-4.77, 5.77]$

63 6.94   **65** 115 m

41

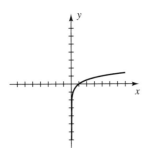

43

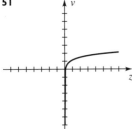

45 $f(x) = \log_2 x^2$ **47** $f(x) = \log_2 (8x)$ **49** $y = \dfrac{b}{x^k}$

51

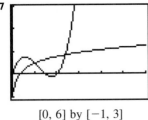

53 (a) 0   (b) $R(2x) = R(x) + a \log 2$ **55** 0.29 cm

57

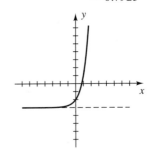

$(0, 1.01] \cup [2.4, \infty)$

$[0, 6]$ by $[-1, 3]$

EXERCISES 5.5

1 $\dfrac{\log 8}{\log 5} \approx 1.29$ **3** $4 - \dfrac{\log 5}{\log 3} \approx 2.54$ **5** 1.1133

7 -0.7325 **9** 2   **11** $\dfrac{\log (2/81)}{\log 24} \approx -1.16$

13 $\dfrac{\log (8/25)}{\log (4/5)} \approx 5.11$ **15** -3 **17** 5

19 $\dfrac{2}{3} \sqrt{\dfrac{101}{11}} \approx 2.02$ **21** 1, 2

23 $\dfrac{\log (4 + \sqrt{19})}{\log 4} \approx 1.53$ **25** 1 or 100   **27** 10^{100}

29 10,000   **31** $x = \log (y \pm \sqrt{y^2 - 1})$

33 $x = \dfrac{1}{2} \log \left(\dfrac{1 + y}{1 - y}\right)$ **35** $x = \ln (y + \sqrt{y^2 + 1})$

37 $x = \dfrac{1}{2} \ln \left(\dfrac{y + 1}{y - 1}\right)$

39 y-intercept $= \log_2 3$ **41** x-intercept $= \log_4 3$
     ≈ 1.5850 ≈ 0.7925

43 (a) 2.2   **(b)** 5   **(c)** 8.3

45 Basic if pH > 7, acidic if pH < 7

47 11.58 yr $\approx$ 11 yr 7 mo   **49** 86.4 m

51 (a)   **(b)** 6.58 min

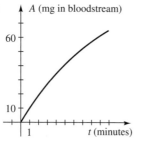

53 (a) $t = \dfrac{\log (F/F_0)}{\log (1 - m)}$ **(b)** After 13,863 generations

55 (a) 4.28 ft   **(b)** 24.8 yr   **57** $\dfrac{\ln (25/6)}{\ln (200/35)} \approx 0.82$

59 The suspicion is correct.

61 The suspicion is incorrect.   **63** -0.5764

65 None

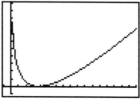

$[-1, 17]$ by $[-1, 11]$

67 1.37, 9.94

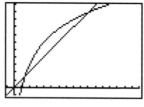

$[-1, 17]$ by $[-1, 11]$

69

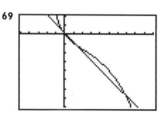

$(-\infty, -0.32) \cup$
$(1.52, 6.84)$

$[-5, 10]$ by $[-8, 2]$

71 (4)   **73 (a)** 1998   **(b)** 1998

CHAPTER 5 REVIEW EXERCISES

1

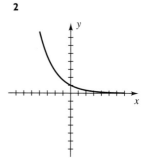

2

3

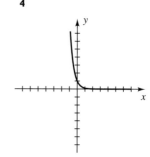

4

5

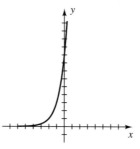

6

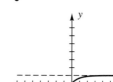

7

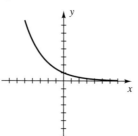

8

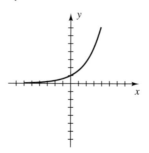

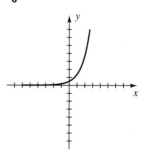

9

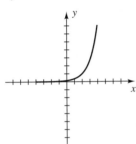

10

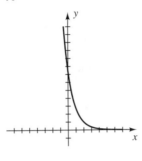

11

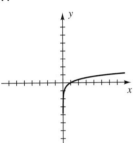

12

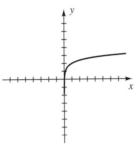

13

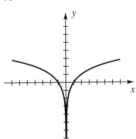

14

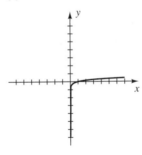

15

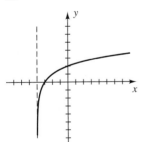

16

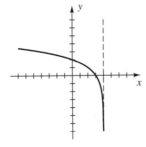

17 (a) -4 **(b)** 0 **(c)** 1 **(d)** 4 **(e)** 6 **(f)** 8

(g) $\dfrac{1}{2}$

18 (a) $\dfrac{1}{3}$ **(b)** 0 **(c)** 1 **(d)** 5 **(e)** 1 **(f)** 25

(g) $\dfrac{1}{3}$

19 0 **20** 9 **21** 9 **22** $\dfrac{33}{47}$ **23** 1 **24** 99

25 $5 - \dfrac{\log 6}{\log 2}$ **26** $\pm\sqrt{\dfrac{\log 7}{\log 3}}$ **27** $\dfrac{\log (3/8)}{\log (32/9)}$ **28** 1

29 $\dfrac{1}{4}, 1, 4$ **30** No solution   **31** $\sqrt{5}$ **32** 2

33 $0, \pm1$ **34** $\ln 2$ **35 (a)** $-3, 2$ **(b)** 2

36 (a) 8 **(b)** ±4 **37** $4 \log x + \dfrac{2}{3} \log y - \dfrac{1}{3} \log z$

38 $-\log (xy^2)$ **39** $x = \log \left(\dfrac{1 \pm \sqrt{1 - 4y^2}}{2y} \right)$

40 If $y < 0$, then $x = \log \left(\dfrac{1 - \sqrt{1 + 4y^2}}{2y} \right)$.

If $y > 0$, then $x = \log \left(\dfrac{1 + \sqrt{1 + 4y^2}}{2y} \right)$.

41 (a) 1.89 **(b)** 78.3 **(c)** 0.472

42 (a) 0.924 **(b)** 0.00375 **(c)** 6.05

43 (a) $D = (-1, \infty), R = \mathbb{R}$

(b) $y = 2^x - 1, D = \mathbb{R}, R = (-1, \infty)$

44 (a) $D = \mathbb{R}, R = (-2, \infty)$

(b) $y = 3 - \log_2 (x + 2), D = (-2, \infty), R = \mathbb{R}$

45 (a) 2000

(b) $2000(3^{1/6}) \approx 2401; 2000(3^{1/2}) \approx 3464; 6000$

46 $\$1125.51$

47 (a)   **(b)** 8 days

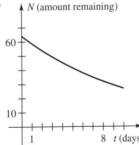

48 $N = 1000\left(\dfrac{3}{5}\right)^{t/3}$

49 (a) After 11.39 yr   **(b)** 6.30 yr

50 $t = (\ln 100) \dfrac{L}{R} \approx 4.6 \dfrac{L}{R}$

51 (a) $I = I_0 10^{\alpha/10}$

(b) Examine $I(\alpha + 1)$, where $I(\alpha)$ is the intensity corresponding to α decibels.

52 $t = -\dfrac{1}{k} \ln\left(\dfrac{a - L}{ab}\right)$ **53** $A = 10^{(R+5.1)/2.3} - 3000$

54 $\dfrac{A_1}{A_2} = \dfrac{10^{(R+5.1)/2.3} - 3000}{10^{(R+7.5)/2.3} - 34{,}000}$ **55** $26{,}615.9 \text{ mi}^2$

56 $h = \dfrac{\ln(29/p)}{0.000034}$ **57** $v = a \ln\left(\dfrac{m_1 + m_2}{m_1}\right)$

58 (a) $n = 10^{7.7-0.9R}$ **(b)** $12{,}589$; 1585; 200

59 (a) $E = 10^{11.4+1.5R}$ **(b)** 10^{24} ergs   **60** 110 days

61 86.8 cm; 9.715 cm/yr   **62** $t = -\dfrac{L}{R} \ln\left(\dfrac{V - RI}{V}\right)$

63 (a) $26{,}749$ yr   **(b)** 30% **64** 16.91 yr

65 3196 yr

7 $(-0.9999011, 0.00999001)$, $(-0.0001, 0.01)$, $(100, 0.01105111)$, and $(36{,}102.844, 4.6928 \times 10^{13})$. Exponential function values (with base > 1) are greater than polynomial function values (with leading term positive) for very large values of x.

8 (x, x) with $x \approx 0.44239443$, 4.1770774, and $5{,}503.6647$. The y-values for $y = x$ eventually will be larger than the y-values for $y = (\ln x)^n$.

9 8.447177%; $\$1{,}025{,}156.25$

10 (a) 3.5 earthquakes $= 1$ bomb, 425 bombs $= 1$ eruption
  (b) 9.22; no

11 October 8, 2007; about 8.65%

CHAPTER 5 DISCUSSION EXERCISES

1 (a) Graph flattens

  (b) $y = \dfrac{101}{2}(e^{x/101} + e^{-x/101}) - 71$

2 7.16 yr

3 (a) *Hint:* Take the natural logarithm of both sides first.

  (b) 2.50 and 2.97

  (c) Note that $f(e) = \dfrac{1}{e}$. Any horizontal line $y = k$, with $0 < k < \dfrac{1}{e}$, will intersect the graph at points $\left(x_1, \dfrac{\ln x_1}{x_1}\right)$ and $\left(x_2, \dfrac{\ln x_2}{x_2}\right)$, where $1 < x_1 < e$ and $x_2 > e$.

4 (a) The difference is in the compounding.

  (b) Closer to the second function

  (c) 29 and 8.2; 29.61 and 8.18

5 *Hint:* Check the restrictions for the logarithm laws.

6 (a) $U = P\left(1 + \dfrac{r}{12}\right)^{12t} - \dfrac{12M[(1 + r/12)^{12t} - 1]}{r}$

  (b)

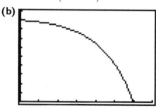

$[0, 35]$ by $[0, 100{,}000]$

  (c) $\$84{,}076.50$; 24.425 yr

CHAPTER 6

EXERCISES 6.1

1 $V(0, 0)$; $F(0, 2)$; $y = -2$ **3** $V(0, 0)$; $F\left(-\dfrac{3}{8}, 0\right)$; $x = \dfrac{3}{8}$

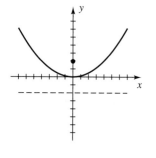

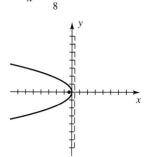

5 $V(-2, 1)$; $F(-2, -1)$; $y = 3$ **7** $V(3, 2)$; $F\left(\dfrac{49}{16}, 2\right)$; $x = \dfrac{47}{16}$

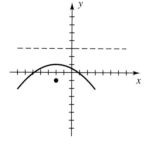

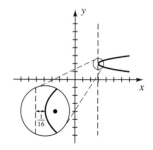

9 $V(2, -2)$; $F\left(2, -\dfrac{7}{4}\right)$;

$y = -\dfrac{9}{4}$

11 $V\left(0, \dfrac{1}{2}\right)$; $F\left(0, -\dfrac{9}{2}\right)$;

$y = \dfrac{11}{2}$

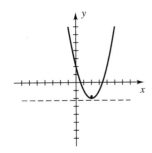

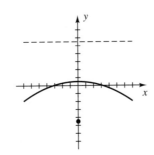

EXERCISES 6.2

1 $V(\pm 3, 0)$; $F(\pm\sqrt{5}, 0)$

3 $V(0, \pm 4)$; $F(0, \pm 1)$

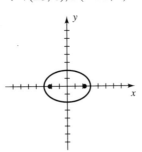

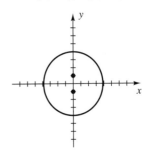

13 $y^2 = 20(x - 1)$ **15** $(x + 2)^2 = -16(y - 3)$

17 $y^2 = 8x$ **19** $(x - 6)^2 = 12(y - 1)$

21 $(y + 5)^2 = 4(x - 3)$ **23** $y^2 = -12(x + 1)$

25 $3x^2 = -4y$ **27** $(y - 5)^2 = 2(x + 3)$

29 $x^2 = 16(y - 1)$ **31** $(y - 3)^2 = -8(x + 4)$

33 $y = -\sqrt{x + 3} - 1$ **35** $x = \sqrt{y - 4} - 1$

37 $y = -x^2 + 2x + 5$ **39** $x = y^2 - 3y + 1$

41 4 in.   **43** $\dfrac{9}{16}$ ft from the center of the paraboloid

45 $2\sqrt{480} \approx 43.82$ in.

47 (a) $p = \dfrac{r^2}{4h}$ **(b)** $10\sqrt{2}$ ft   **49** 64,968 ft^2

51

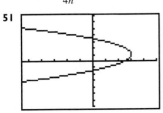

$[-11, 10]$ by $[-7, 7]$

53 $(2.08, -1.04)$, $(2.92, 1.38)$

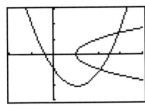

$[-2, 4]$ by $[-3, 3]$

5 $V(0, \pm 4)$; $F(0, \pm 2\sqrt{3})$

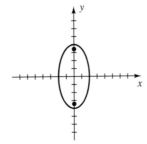

7 $V\left(\pm\dfrac{1}{2}, 0\right)$;

$F\left(\pm\dfrac{1}{10}\sqrt{21}, 0\right)$

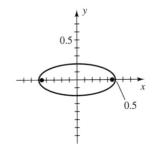

9 $V(3 \pm 4, -4)$;

$F(3 \pm \sqrt{7}, -4)$

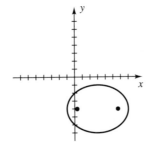

11 $V(4 \pm 3, 2)$;

$F(4 \pm \sqrt{5}, 2)$

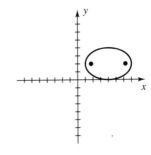

13 $V(5, 2 \pm 5)$;
$F(5, 2 \pm \sqrt{21})$

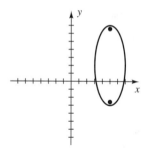

15 $\dfrac{x^2}{4} + \dfrac{y^2}{36} = 1$ **17** $\dfrac{(x + 2)^2}{25} + \dfrac{(y - 1)^2}{4} = 1$

19 $\dfrac{x^2}{64} + \dfrac{y^2}{39} = 1$ **21** $\dfrac{4x^2}{9} + \dfrac{y^2}{25} = 1$

23 $\dfrac{8x^2}{81} + \dfrac{y^2}{36} = 1$ **25** $\dfrac{x^2}{7} + \dfrac{y^2}{16} = 1$

27 $\dfrac{x^2}{4} + 9y^2 = 1$ **29** $\dfrac{x^2}{16} + \dfrac{4y^2}{25} = 1$

31 $(2, 2)$, $(4, 1)$

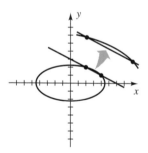

33 $\dfrac{x^2}{25} + \dfrac{y^2}{16} = 1$ **35** $\dfrac{x^2}{64} + \dfrac{y^2}{289} = 1$

37 Upper half of $\dfrac{x^2}{49} + \dfrac{y^2}{121} = 1$

39 Left half of $x^2 + \dfrac{y^2}{9} = 1$

41 Right half of $\dfrac{(x - 1)^2}{4} + \dfrac{(y + 2)^2}{9} = 1$

43 Lower half of $\dfrac{(x + 1)^2}{9} + \dfrac{(y - 2)^2}{49} = 1$

45 $\sqrt{84} \approx 9.2$ ft   **47** 94,581,000; 91,419,000

49 (a) $d = h - \sqrt{h^2 - \dfrac{1}{4}k^2}$; $d' = h + \sqrt{h^2 - \dfrac{1}{4}k^2}$

(b) 16 cm; 2 cm from V

51 5 ft

53

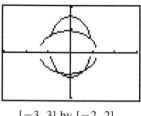

$[-300, 300]$ by $[-200, 200]$

55 $(\pm 1.540, 0.618)$ **57** $(-0.88, 0.76)$,
$\qquad\qquad\qquad\qquad\qquad (-0.48, -0.91)$,
$\qquad\qquad\qquad\qquad\qquad (0.58, -0.81)$,
$\qquad\qquad\qquad\qquad\qquad (0.92, 0.59)$

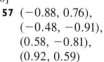

$[-6, 6]$ by $[-2, 6]$

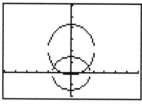

$[-3, 3]$ by $[-2, 2]$

EXERCISES 6.3

1 $V(\pm 3, 0)$; $F(\pm \sqrt{13}, 0)$;   **3** $V(0, \pm 3)$; $F(0, \pm \sqrt{13})$;
$\qquad y = \pm \dfrac{2}{3}x$ $\qquad\qquad y = \pm \dfrac{3}{2}x$

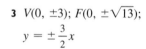

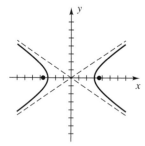

5 $V(\pm 1, 0)$; $F(\pm 5, 0)$;   **7** $V(0, \pm 4)$; $F(0, \pm 2\sqrt{5})$;
$\qquad y = \pm \sqrt{24}x$ $\qquad\qquad y = \pm 2x$

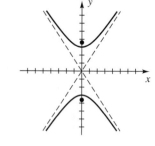

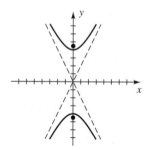

9 $V\left(\pm\dfrac{1}{4}, 0\right)$;

$F\left(\pm\dfrac{1}{12}\sqrt{13}, 0\right)$;

$y = \pm\dfrac{2}{3}x$

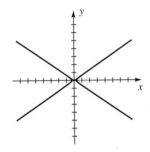

11 $V(-2, -2 \pm 3)$;
$F(-2, -2 \pm \sqrt{13})$;

$(y + 2) = \pm\dfrac{3}{2}(x + 2)$

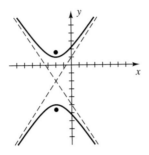

13 $V(-3 \pm 5, -2)$;
$F(-3 \pm 13, -2)$;

$(y + 2) = \pm\dfrac{12}{5}(x + 3)$

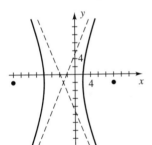

15 $V(-2, -5 \pm 3)$;
$F(-2, -5 \pm 3\sqrt{5})$;

$(y + 5) = \pm\dfrac{1}{2}(x + 2)$

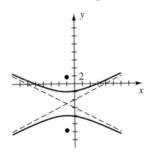

17 $\dfrac{x^2}{9} - \dfrac{y^2}{16} = 1$ **19** $(y + 3)^2 - \dfrac{(x + 2)^2}{3} = 1$

21 $y^2 - \dfrac{x^2}{15} = 1$ **23** $\dfrac{x^2}{9} - \dfrac{y^2}{16} = 1$ **25** $\dfrac{y^2}{21} - \dfrac{x^2}{4} = 1$

27 $\dfrac{x^2}{9} - \dfrac{y^2}{36} = 1$ **29** $\dfrac{x^2}{25} - \dfrac{y^2}{100} = 1$

31 $\dfrac{y^2}{25} - \dfrac{x^2}{49} = 1$ **33** $(0, 4), \left(\dfrac{8}{3}, \dfrac{20}{3}\right)$

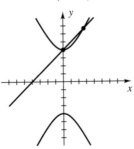

35 $\dfrac{x^2}{144} - \dfrac{y^2}{25} = 1$ **37** $\dfrac{y^2}{64} - \dfrac{x^2}{36} = 1$

39 Right branch of $\dfrac{x^2}{25} - \dfrac{y^2}{16} = 1$

41 Upper branch of $\dfrac{y^2}{9} - \dfrac{x^2}{49} = 1$

43 Lower halves of the branches of $\dfrac{x^2}{16} - \dfrac{y^2}{81} = 1$

45 Left halves of the branches of $\dfrac{y^2}{36} - \dfrac{x^2}{16} = 1$

47 The graphs have the
same asymptotes.

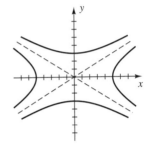

49 $x = \sqrt{9 + 4y^2}$

51 If a coordinate system similar to that in Example 6
is introduced, then the ship's coordinates are

$\left(\dfrac{80}{3}\sqrt{34}, 100\right) \approx (155.5, 100)$.

53 $(0.741, 2.206)$ **55** None

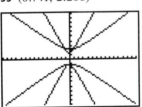   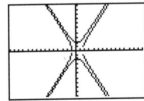

[-15, 15] by [-10, 10]   [-15, 15] by [-10, 10]
57 (a) $(6.63 \times 10^7, 0)$ **(b)** $v > 103{,}600$ m/sec

EXERCISES 6.4

1 $y = 2x + 7$

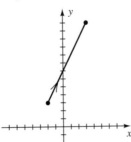

3 $y = x - 2$

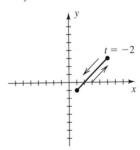

5 $(y - 3)^2 = x + 5$

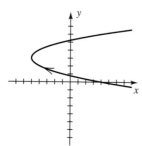

7 $y = 1/x^2$

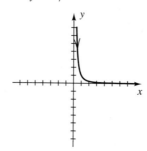

21 $y = (x^{1/3} + 1)^2$

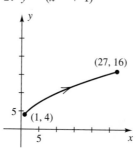

(27, 16)

(1, 4)

9 $\dfrac{x^2}{4} + \dfrac{y^2}{9} = 1$

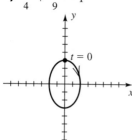

$t = 0$

11 $x^2 - y^2 = 1$

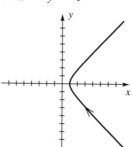

23 C_1

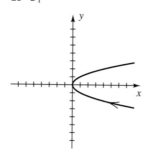

C_2

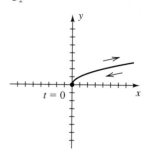

$t = 0$

13 $y = \ln x$

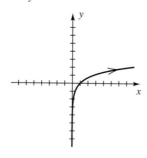

15 $y = 1/x$

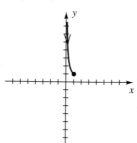

C_3

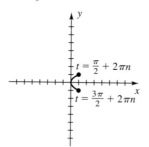

$t = \dfrac{\pi}{2} + 2\pi n$

$t = \dfrac{3\pi}{2} + 2\pi n$

C_4

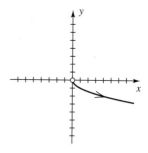

17 $y = \sqrt{x^2 - 1}$

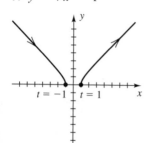

$t = -1 \quad t = 1$

19 $y = |x - 1|$

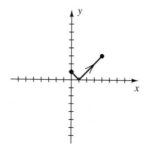

25 (a)

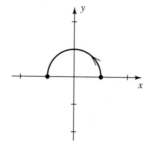

(b)

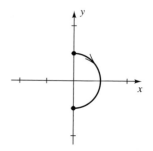

(c)

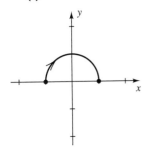

29 Answers are not unique.

(a) (1) $x = t,$ $y = t^2;$ $t \in \mathbb{R}$

(2) $x = \tan t,$ $y = \tan^2 t;$ $-\dfrac{\pi}{2} < t < \dfrac{\pi}{2}$

(3) $x = t^3,$ $y = t^6;$ $t \in \mathbb{R}$

(b) (1) $x = e^t,$ $y = e^{2t};$ $t \in \mathbb{R}$ (only gives $x > 0$)

(2) $x = \sin t,$ $y = \sin^2 t;$ $t \in \mathbb{R}$ (only gives $-1 \le x \le 1$)

(3) $x = \tan^{-1} t, y = (\tan^{-1} t)^2; t \in \mathbb{R}$ $\left(\text{only gives} -\dfrac{\pi}{2} < x < \dfrac{\pi}{2}\right)$

31 (a) The figure is an ellipse with center $(0, 0)$ and axes of length $2a$ and $2b$.

33 (a)

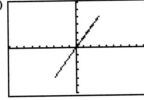

$[-9, 9]$ by $[-6, 6]$

(b) $0°$

35 (a)

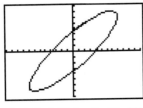

$[-120, 120]$ by $[-80, 80]$

(b) $30°$

37

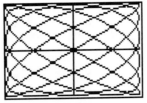

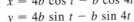

$[-1, 1]$ by $[-1, 1]$

41 $x = 4b \cos t - b \cos 4t,$
$y = 4b \sin t - b \sin 4t$

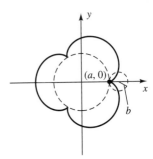

43

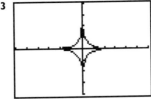

$[-6, 6]$ by $[-4, 4]$

45

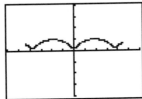

$[-30, 30]$ by $[-20, 20]$

47 A mask with a mouth, nose, and eyes

49 The letter A

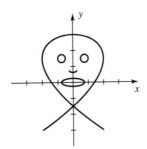

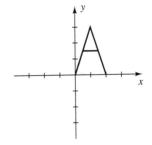

EXERCISES 6.5

1
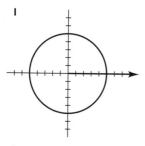

3
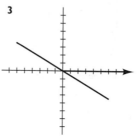

21
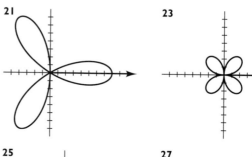

23
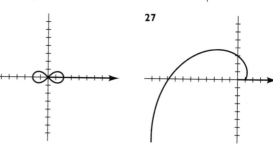

5
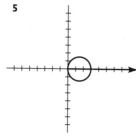

7
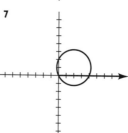

25
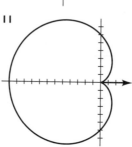

27
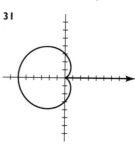

9
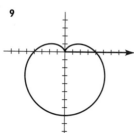

11
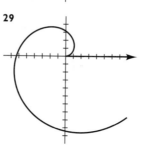

29
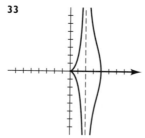

31

13
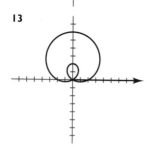

15
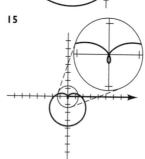

33

17
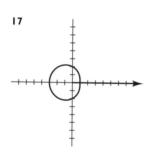

19
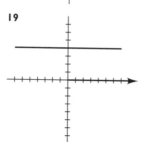

35 (a) $\left(\frac{3}{2}\sqrt{2}, \frac{3}{2}\sqrt{2}\right)$ **(b)** $\left(\frac{1}{2}, -\frac{1}{2}\sqrt{3}\right)$

37 (a) $(-4, -4\sqrt{3})$ **(b)** $\left(-\frac{3}{2}, \frac{3}{2}\sqrt{3}\right)$

39 $\left(\frac{24}{5}, \frac{18}{5}\right)$ **41 (a)** $\left(\sqrt{2}, \frac{3\pi}{4}\right)$ **(b)** $\left(4, \frac{7\pi}{6}\right)$

43 (a) $\left(14, \frac{5\pi}{3}\right)$ **(b)** $\left(5\sqrt{2}, \frac{\pi}{4}\right)$ **45** (a), (c), (e)

47 $r = -3 \sec \theta$ **49** $r = 4$ **51** $\theta = \tan^{-1}\left(-\frac{1}{2}\right)$

53 $r^2 = -4 \sec 2\theta$ **55** $r = 2 \cos \theta$

57 $x = 5$

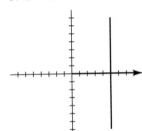

59 $x^2 + (y - 3)^2 = 9$

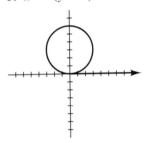

73 $y^2 = \dfrac{x^4}{1 - x^2}$

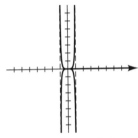

75 Let $P_1(r_1, \theta_1)$ and $P_2(r_2, \theta_2)$ be points in an $r\theta$-plane. Let $a = r_1$, $b = r_2$, $c = d(P_1, P_2)$, and $\gamma = \theta_2 - \theta_1$. Substituting into the law of cosines, $c^2 = a^2 + b^2 - 2ab \cos \gamma$, gives us the formula.

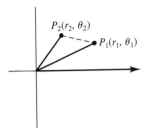

61 $y = x$

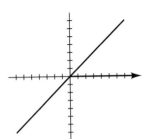

63 $\dfrac{y^2}{9} - \dfrac{x^2}{4} = 1$

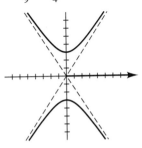

65 $x^2 - y^2 = 1$

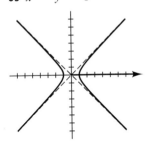

67 $y - 2x = 6$

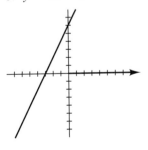

77 (a)

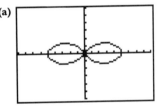

$[-9, 9]$ by $[-6, 6]$

(b) Max: east-west direction; Min: north-south direction

79 Symmetric with respect to the polar axis

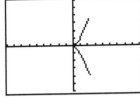

$[-9, 9]$ by $[-6, 6]$

69 $y = -x^2 + 1$

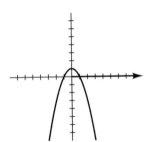

71 $(x + 1)^2 + (y - 4)^2 = 17$

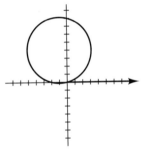

81 The approximate polar coordinates are $(1.75, \pm 0.45)$, $(4.49, \pm 1.77)$, and $(5.76, \pm 2.35)$.

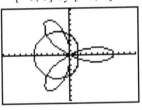

$[-12, 12]$ by $[-9, 9]$

EXERCISES 6.6

1 $\dfrac{1}{3}$, ellipse

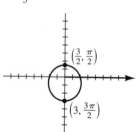

3 3, hyperbola

5 1, parabola

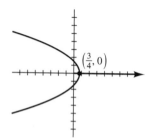

7 $\dfrac{1}{2}$, ellipse

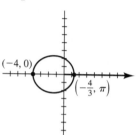

9 $\dfrac{3}{2}$, hyperbola

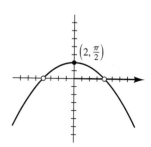

11 1, parabola

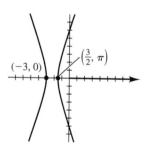

13 $9x^2 + 8y^2 + 12y - 36 = 0$

15 $8x^2 - y^2 + 36x + 36 = 0$

17 $4y^2 + 12x - 9 = 0$ **19** $3x^2 + 4y^2 + 8x - 16 = 0$

21 $4x^2 - 5y^2 + 36y - 36 = 0; x \neq \pm 3$

23 $x^2 + 8y - 16 = 0; x \neq \pm 4$ **25** $r = \dfrac{2}{3 + \cos\theta}$

27 $r = \dfrac{12}{3 - 4\cos\theta}$ **29** $r = \dfrac{2}{1 - \sin\theta}$

31 $r = \dfrac{8}{5 + 2\sin\theta}$ **33** $r = \dfrac{8}{1 + \sin\theta}$

35 (a) $\dfrac{3}{4}$ (b) $r = \dfrac{7}{4 - 3\sin\theta}$

39 (a) Elliptical
(b)

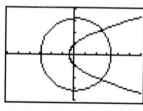

$[-36, 36]$ by $[-24, 24]$

41 (a) Hyperbolic
(b)

$[-18, 18]$ by $[-12, 12]$

CHAPTER 6 REVIEW EXERCISES

1 $V(0, 0); F(16, 0)$

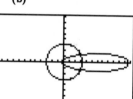

2 $V(-2, 1); F\left(-2, \dfrac{33}{32}\right)$

3 $V(0, \pm 4); F(0, \pm\sqrt{7})$

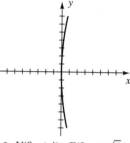

4 $V(0, \pm 4); F(0, \pm 5)$

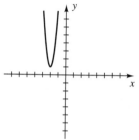

5 $V(\pm 2, 0); F(\pm 2\sqrt{2}, 0)$

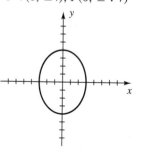

6 $V\left(\pm\dfrac{1}{5}, 0\right);$

$F\left(\pm\dfrac{1}{30}\sqrt{11}, 0\right)$

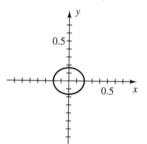

7 $V(0, 4)$; $F\left(0, -\dfrac{9}{4}\right)$

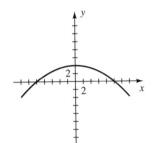

8 $V(3 \pm 2, -1)$;
$F(3 \pm 1, -1)$

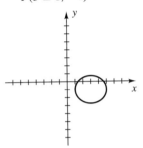

15 $V(-4 \pm 3, 0)$;
$F(-4 \pm \sqrt{10}, 0)$

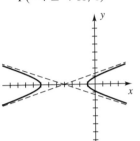

16 $V(2, -3 \pm 2)$;
$F(2, -3 \pm \sqrt{6})$

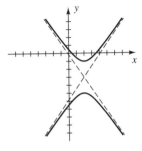

9 $V(-4 \pm 1, 5)$;
$F\left(-4 \pm \dfrac{1}{3}\sqrt{10}, 5\right)$

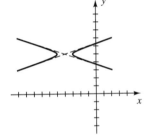

10 $V(-5, -2)$;
$F\left(-\dfrac{39}{8}, -2\right)$

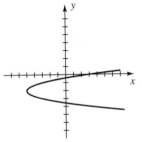

17 $y = 2(x + 7)^2 - 18$ **18** $y = -3(x + 4)^2 + 147$

19 $\dfrac{y^2}{49} - \dfrac{x^2}{9} = 1$ **20** $y^2 = -16x$ **21** $x^2 = -40y$

22 $x = 5y^2$ **23** $\dfrac{x^2}{75} + \dfrac{y^2}{100} = 1$ **24** $\dfrac{x^2}{25} - \dfrac{y^2}{75} = 1$

25 $\dfrac{y^2}{36} - \dfrac{x^2}{\frac{4}{9}} = 1$ **26** $\dfrac{x^2}{8} + \dfrac{y^2}{4} = 1$ **27** $\dfrac{x^2}{25} + \dfrac{y^2}{45} = 1$

28 $\dfrac{x^2}{256} + \dfrac{y^2}{112} = 1$ **29 (a)** $-\dfrac{7}{2}$ **(b)** Hyperbola

30 $A = \dfrac{4a^2b^2}{a^2 + b^2}$ **31** $x^2 + (y - 2)^2 = 4$

32 $2\sqrt{2}$ rad/sec ≈ 0.45 rev/sec

33 $x = 4y + 7$ **34** $y = x^4 - 4$

11 $V(-3 \pm 3, 2)$;
$F(-3 \pm \sqrt{5}, 2)$

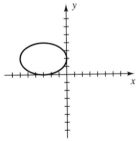

12 $V(5, -4 \pm 2)$;
$F(5, -4 \pm \sqrt{5})$

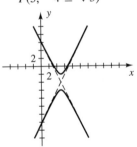

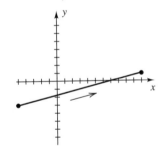

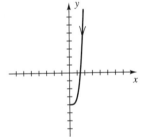

13 $V(2, -4)$; $F(4, -4)$

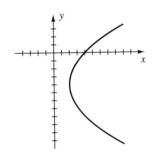

14 $V(3, -2 \pm 2)$;
$F(3, -2 \pm \sqrt{3})$

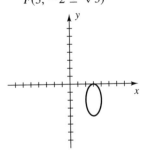

35 $(y - 1)^2 = -(x + 1)$

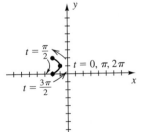

36 $y = 2^{-x^2}$

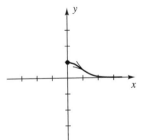

37 $y = \dfrac{2x^2 - 4x + 1}{x - 1}$

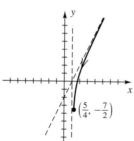

$\left(\dfrac{5}{4}, -\dfrac{7}{2}\right)$

43

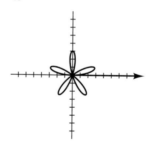

44

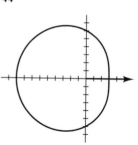

38 C_1

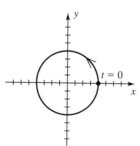

C_2

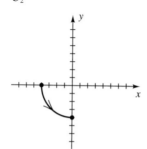

45

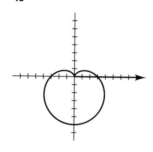

46

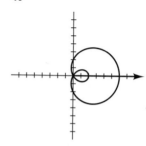

C_3

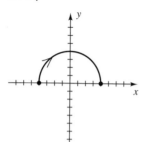

$t = 0$

C_4

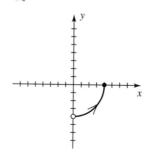

47

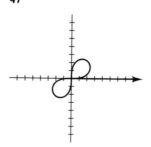

48

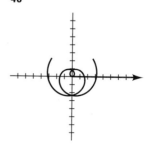

39 $\left(\dfrac{5}{2}\sqrt{2}, -\dfrac{5}{2}\sqrt{2}\right)$ **40** $\left(4, \dfrac{11\pi}{6}\right)$

41

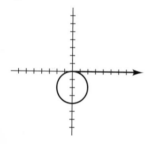

42

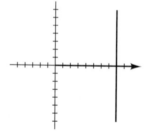

49

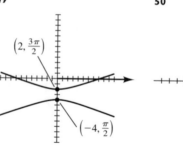

$\left(2, \dfrac{3\pi}{2}\right)$

$\left(-4, \dfrac{\pi}{2}\right)$

50

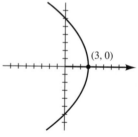

$(3, 0)$

51 $\frac{2}{3}$, ellipse

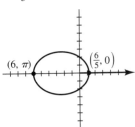

52 $\frac{1}{2}$, ellipse

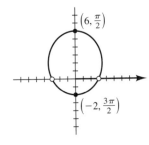

53 $r = 4 \cot \theta \csc \theta$ **54** $r = 3 \cos \theta - 4 \sin \theta$

55 $r(2 \cos \theta - 3 \sin \theta) = 8$ **56** $\theta = \frac{\pi}{4}$

57 $x^3 + xy^2 = y$ **58** $x^2 + y^2 = 2x + 3y$

59 $(x^2 + y^2)^2 = 8xy$ **60** $y = (\tan \sqrt{3})\, x$

61 $8x^2 + 9y^2 + 10x - 25 = 0$ **62** $y^2 = 6 + x$

CHAPTER 6 DISCUSSION EXERCISES

1 $w = 4|p|$

2 The circle goes through both foci and all four vertices of the auxiliary rectangle.

5 $\dfrac{(x - 2)^2}{1} - \dfrac{y^2}{3} = 1,\ x \geq 3,$

or

$x = 2 + \sqrt{1 + \dfrac{y^2}{3}}$

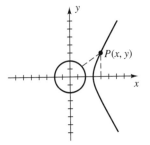

6 The graph of $r = f(\theta - \alpha)$ is the graph of $r = f(\theta)$ rotated counterclockwise through an angle α, whereas the graph of $r = f(\theta + \alpha)$ is rotated clockwise.

7 $(180/n)°$

APPENDIX V EXERCISES

1 $2 + 4i$ **3** $18 - 3i$ **5** $41 - 11i$ **7** $17 - i$

9 $21 - 20i$ **11** $-24 - 7i$ **13** 25 **15** $-i$

17 i **19** $\dfrac{3}{10} - \dfrac{3}{5}i$ **21** $\dfrac{1}{2} - i$ **23** $\dfrac{34}{53} + \dfrac{40}{53}i$

25 $\dfrac{2}{5} + \dfrac{4}{5}i$ **27** $-142 - 65i$ **29** $-2 - 14i$

31 $-\dfrac{44}{113} + \dfrac{95}{113}i$ **33** $\dfrac{21}{2}i$ **35** $x = 4,\ y = -16$

37 $x = 1,\ y = 3$ **39** $3 \pm 2i$ **41** $-2 \pm 3i$

43 $\dfrac{5}{2} \pm \dfrac{1}{2}\sqrt{55}\, i$ **45** $-\dfrac{1}{8} \pm \dfrac{1}{8}\sqrt{47}\, i$

47 $-5, \dfrac{5}{2} \pm \dfrac{5}{2}\sqrt{3}\, i$ **49** $\pm 4,\ \pm 4i$ **51** $\pm 2i,\ \pm \dfrac{3}{2}i$

53 $0, -\dfrac{3}{2} \pm \dfrac{1}{2}\sqrt{7}\, i$

55 If $w = c + di$, then $\overline{z + w} = \overline{(a + bi) + (c + di)}$
$= \overline{(a + c) + (b + d)i} = (a + c) - (b + d)i$
$= (a - bi) + (c - di) = \bar{z} + \bar{w}.$

57 $\overline{z \cdot w} = \overline{(a + bi) \cdot (c + di)}$
$= \overline{(ac - bd) + (ad + bc)i}$
$= (ac - bd) - (ad + bc)i$
$= ac - adi - bd - bci$
$= a(c - di) - bi(c - di)$
$= (a - bi) \cdot (c - di) = \bar{z} \cdot \bar{w}$

59 If $\bar{z} = z$, then $a - bi = a + bi$ and hence $-bi = bi$, or $2bi = 0$. Thus, $b = 0$ and $z = a$ is real. Conversely, if z is real, then $b = 0$ and hence $\bar{z} = \overline{a + 0i} = a - 0i = a + 0i = z.$

APPENDIX VI EXERCISES

1 $m = -\dfrac{3}{4}$ **3** $m = 0$

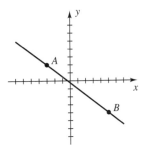

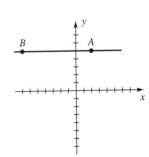

5 m is undefined

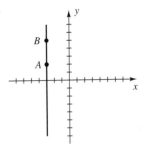

7 The slopes of opposite sides are equal.
9 The slopes of opposite sides are equal, and the slopes of two adjacent sides are negative reciprocals.
11 $(-12, 0)$

13

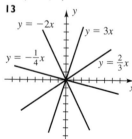

15

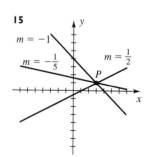

17

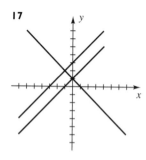

19 (a) $x = 5$ **(b)** $y = -2$ **21** $4x + y = 17$
23 $3x + y = 12$ **25** $11x + 7y = 9$
27 $5x - 2y = 18$ **29** $5x + 2y = 29$
31 $y = \dfrac{3}{4}x - 3$ **33** $y = -\dfrac{1}{3}x + \dfrac{11}{3}$
35 $5x - 7y = -15$ **37** $y = -x$
39 $m = -\dfrac{2}{3}, b = 5$ **41** $m = \dfrac{4}{3}, b = -3$

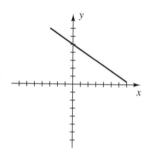

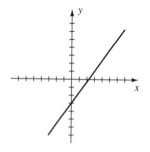

43 (a) $y = 3$ **(b)** $y = -\dfrac{1}{2}x$ **(c)** $y = -\dfrac{3}{2}x + 1$
(d) $y + 2 = -(x - 3)$
45 $\dfrac{x}{3/2} + \dfrac{y}{-3} = 1$ **47** $(x - 3)^2 + (y + 2)^2 = 49$

49 Approximately 23 weeks
51 (a) 25.2 tons   **(b)** As large as 3.4 tons
53 (a) $y = \dfrac{5}{14}x$ **(b)** 58
55 (a) $W = \dfrac{20}{3}t + 10$
(b) 50 lb
(c) 9 yr
(d)

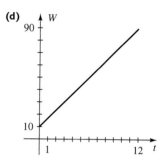

57 $H = -\dfrac{8}{3}T + \dfrac{7520}{3}$
59 (a) $T = 0.032t + 13.5$ **(b)** 16.22°C
61 (a) $E = 0.55R + 3600$ **(b)** $P = 0.45R - 3600$
(c) $8000
63 (a) Yes: the creature at $x = 3$ **(b)** No
65 34.95 mi/hr   **67** $a = 0.321; b = -0.9425$
69 $(-19, 13)$

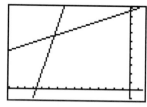

$[-30, 3]$ by $[-2, 20]$

71 $(-0.8, -0.6), (4.8, -3.4), (2, 5)$; right isosceles triangle

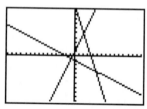

$[-15, 15]$ by $[-10, 10]$

73 $y = 3.2x - 2.6$

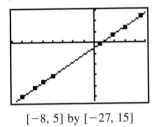

[−8, 5] by [−27, 15]

75 (b) $y = 55x - 108,685$

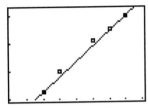

[1980, 1988] by [300, 625]
(c) $435,000; $1,040,000

Index of Applications

INDEX

PARABOLA

$$x^2 = 4py$$

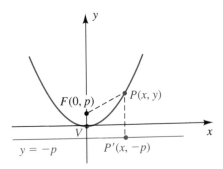

ELLIPSE

$$\frac{x^2}{a^2} + \frac{y^2}{b^2} = 1 \quad \text{with} \quad a^2 = b^2 + c^2$$

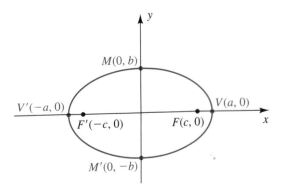

HYPERBOLA

$$\frac{x^2}{a^2} - \frac{y^2}{b^2} = 1 \quad \text{with} \quad c^2 = a^2 + b^2$$

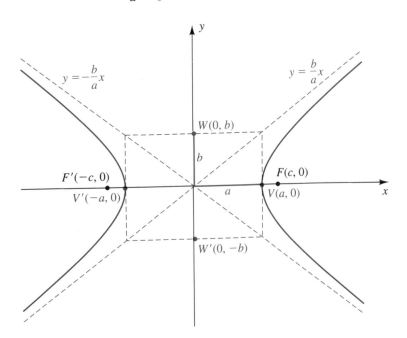

TRIGONOMETRY

OMETRIC FUNCTIONS

OF ACUTE ANGLES

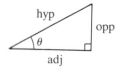

$$\sin \theta = \frac{\text{opp}}{\text{hyp}} \qquad \csc \theta = \frac{\text{hyp}}{\text{opp}}$$

$$\cos \theta = \frac{\text{adj}}{\text{hyp}} \qquad \sec \theta = \frac{\text{hyp}}{\text{adj}}$$

$$\tan \theta = \frac{\text{opp}}{\text{adj}} \qquad \cot \theta = \frac{\text{adj}}{\text{opp}}$$

OF ARBITRARY ANGLES

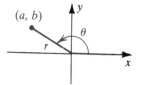

$$\sin \theta = \frac{b}{r} \qquad \csc \theta = \frac{r}{b}$$

$$\cos \theta = \frac{a}{r} \qquad \sec \theta = \frac{r}{a}$$

$$\tan \theta = \frac{b}{a} \qquad \cot \theta = \frac{a}{b}$$

OF REAL NUMBERS

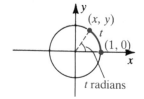

$$\sin t = y \qquad \csc t = \frac{1}{y}$$

$$\cos t = x \qquad \sec t = \frac{1}{x}$$

$$\tan t = \frac{y}{x} \qquad \cot t = \frac{x}{y}$$

SPECIAL RIGHT TRIANGLES

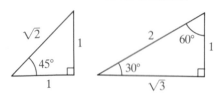

OBLIQUE TRIANGLE

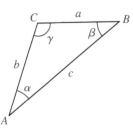

AREA

$$\mathcal{A} = \frac{1}{2}bc \sin \alpha$$

$$\mathcal{A} = \frac{1}{2}ac \sin \beta$$

$$\mathcal{A} = \frac{1}{2}ab \sin \gamma$$

$$\mathcal{A} = \sqrt{s(s-a)(s-b)(s-c)},$$

where $s = \frac{1}{2}(a + b + c)$ (*Heron's Formula*)

LAW OF COSINES

$$a^2 = b^2 + c^2 - 2bc \cos \alpha$$
$$b^2 = a^2 + c^2 - 2ac \cos \beta$$
$$c^2 = a^2 + b^2 - 2ab \cos \gamma$$

LAW OF SINES

$$\frac{\sin \alpha}{a} = \frac{\sin \beta}{b} = \frac{\sin \gamma}{c}$$

SPECIAL VALUES OF TRIGONOMETRIC FUNCTIONS

θ (degrees)	θ (radians)	$\sin \theta$	$\cos \theta$	$\tan \theta$	$\cot \theta$	$\sec \theta$	$\csc \theta$
0°	0	0	1	0	—	1	—
30°	$\frac{\pi}{6}$	$\frac{1}{2}$	$\frac{\sqrt{3}}{2}$	$\frac{\sqrt{3}}{3}$	$\sqrt{3}$	$\frac{2\sqrt{3}}{3}$	2
45°	$\frac{\pi}{4}$	$\frac{\sqrt{2}}{2}$	$\frac{\sqrt{2}}{2}$	1	1	$\sqrt{2}$	$\sqrt{2}$
60°	$\frac{\pi}{3}$	$\frac{\sqrt{3}}{2}$	$\frac{1}{2}$	$\sqrt{3}$	$\frac{\sqrt{3}}{3}$	2	$\frac{2\sqrt{3}}{3}$
90°	$\frac{\pi}{2}$	1	0	—	0	—	1

GREEK ALPHABET

Letter	Name	Letter	Name
A α	alpha	N ν	nu
B β	beta	Ξ ξ	xi
Γ γ	gamma	O o	omicron
Δ δ	delta	Π π	pi
E ϵ	epsilon	P ρ	rho
Z ζ	zeta	Σ σ	sigma
H η	eta	T τ	tau
Θ θ	theta	Υ υ	upsilon
I ι	iota	Φ ϕ (φ)	phi
K κ	kappa	X χ	chi
Λ λ	lambda	Ψ ψ	psi
M μ	mu	Ω ω	omega